EXPERIMENTAL TECHNIQUES IN HIGH-ENERGY NUCLEAR AND PARTICLE PHYSICS

Second Edition

EXPERIMENTAL TECHNIQUES IN HIGH-ENERGY NUCLEAR AND PARTICLE PHYSICS

Second Edition

Edited by

Thomas Ferbel

University of Rochester, New York

World Scientific

Singapore • New Jersey • London • Hong Kong

Published by

World Scientific Publishing Co. Pte. Ltd.
P O Box 128, Farrer Road, Singapore 912805
USA office: Suite 1B, 1060 Main Street, River Edge, NJ 07661
UK office: 57 Shelton Street, Covent Garden, London WC2H 9HE

British Library Cataloguing-in-Publication Data
A catalogue record for this book is available from the British Library.

First published in 1987 by Addison-Wesley Publishing Company, Inc.

Published in 1991 by World Scientific Publishing Co. Pte. Ltd.
Reprinted in 1999.

EXPERIMENTAL TECHNIQUES IN HIGH-ENERGY NUCLEAR AND PARTICLE PHYSICS

ISBN 981-02-0867-7
ISBN 981-02-0868-5 (pbk)

Printed in Singapore.

CONTENTS

PREFACE TO SECOND EDITION

With the depletion of the original 1000 copies of "Experimental Techniques in High Energy Physics" published by Addison-Wesley in 1987, the publishers did not think that the demand for the book was sufficiently great to warrant a re-issue. I consequently turned to Dr. K. K. Phua, whose World Scientific Publishing Company is known to consider other than purely financial reasons for determining what they publish. We rapidly agreed that the excellent contents of "Experimental Techniques" deserved another lease on life.

The new edition has several changes that I should note. Firstly, Willis is now at Columbia University and Kleinknecht has moved to Mainz (and his book "Detectors for Particle Radiation" is now available in English from Cambridge University Press). Secondly, another book on detectors, R. Fernow's "Introduction to Experimental Particle Physics", has been published (Cambridge University Press). Finally, I felt it was important to provide an article on calorimetry that conveyed the sense of some of the latest developments in the field, as pioneered by R. Wigmans, J. Brau, H. Brückmann, and their colleagues, and I have chosen a paper by Dr. Wigmans that appeared in *Nucl. Inst. and Meth.* **A265**, 273 (1988) to represent these new directions. (Wigmans has recently moved to Texas Tech, Lubbock.) It has been a pleasure to work with World Scientific on this new edition, and I thank Dr. K. K. Phua and Helene D'Cruz for their cooperation. I also thank Chris Fabjan for his advice, and Richard Wigmans and Elsevier Science Publishers for allowing me to include the NIM article in this volume.

T. Ferbel
Rochester, New York
June, 1991

PREFACE

When I was asked by Richard Mixter of Benjamin/Cummings to write a book on techniques of experimental particle physics, or to suggest someone who might be interested in such a project, it became clear after several minutes of conversation that it would be exceedingly hard to write a reasonably comprehensive treatise on such a broad subject. On the other hand, many excellent articles came to mind that could be compiled into one or two volumes, which could, at least partly, fill the clear need. The present book is an attempt to provide graduate students and practicing experimenters with a compact source on some of the ingenious ideas and techniques developed for modern experiments in particle physics.

Among the problems encountered in assembling such a series of articles into one volume are: which articles to choose, and how to avoid excessive repetition? It should be obvious that a compendium of reviews and sundry manuscripts cannot possibly have the intellectual coherence of excellent books such as that by P. Rice-Evans on Spark, Streamer, Proportional and Drift Chambers (Richelieu Press-1974), or by K. Kleinknecht on Particle Detectors (Teubner-Verlag-1985; unfortunately, at this time, this is available only in German). In the present compilation, I have concentrated primarily on articles on calorimetry and on charged-particle tracking, and I hope that I have succeeded in minimizing overlap and in maximizing the quality of content.

I have picked Kleinknecht's review "Particle Detectors" [Physics Reports *84*, 85 (1982)] to lead off the volume. It is a fine general introduction to the spectrometer systems and various specific tools used in nuclear and particle physics. This is followed by Sauli's unpublished classic lectures on proportional and drift chambers [CERN Report 77-09 (1977)], which have been corrected by Sauli for minor typographical errors in the original manuscript. The recently published review by Charpak and Sauli on high resolution electronic detectors [Ann. Rev. Nuc. Sci. *34*, 285 (1984)] provides the latest information on a variety of high resolution devices that have applicability in high energy physics and other disciplines; besides treating gaseous chambers, the authors have a very useful section on solid state detectors. The next article is a thorough treatment of calorimetry by Fabjan, based on his lectures at St. Croix [*Techniques and Concepts of High Energy Physics-III*, Plenum Corp, T. Ferbel, ed (1985)]. A companion piece to Fabjan's article is U. Amaldi's excellent review of sources of fluctuations in calorimetry [Phys. Scripta *23*, 409 (1981)]. This is followed by Allison's and Wright's paper on the physics and applications of energy loss, Cherenkov and transition radiation (*Formulae and Methods in Experimental Data Evaluation*, Vol. 2, El-42, R. K. Bock, et al, eds. (1984), EPS/CERN). Next are two articles on photo-sensitive detection schemes; one, on ring imaging counters by Seguinot, Ypsilantis, et al [Nuc. Inst. and Methods *200*, 219 (1982)] and the other, a brief semi-historical review of recent developments in this area by D. Anderson [IEEE Trans. Nuc. Sci *32*, 495 (1985)]. This is followed by two articles concerning liquid-argon calorimetry; one is the classic by Willis and Radeka [Nuc. Inst. and Methods *120*, 221 (1974)], and the other is the manuscript by Doke on properties of noble liquids that is published in a not too accessible journal [Portuglal Phys. *12*, 9 (1981)]. To round out this volume, I have included Radeka's paper on signal, noise and resolution in detectors [IEEE Trans. Nuc. Sci. *21*, 51 (1974)] and James' review of Monte Carlo methods [Rep. Prog. Phys. *43*, 1145 (1980)].

Initially, I had hoped to include several other outstanding papers in this compilation. Iwata's exhaustive masterpiece on calorimetry (Nagoya Report DPNU-B-80), Eichinger's and Regler's review of track fitting (CERN Report 81-06), and something recent on time projection chambers (e.g., from M. Shapiro's thesis, LBL Report 18820), were among the possibilities. Pressures of time and the realities of the publishing world have, however, precluded the preparation of a much longer tome. There is, consequently, now a clear need for a companion piece to this book, and I hope that some enterprising individual will make the effort to collect such a volume.

I wish to thank the authors and original publishers of the articles that appear in this book for their cooperation in this venture, and I am particularly grateful to Chris Fabjan, George Fanourakis and Joey Huston for advice on the contents.

T. Ferbel
Rochester, New York
September, 1985

Reprinted, with permission, from *Physics Reports*, Vol. 84, pp. 85-161 (1982).

PARTICLE DETECTORS

K. KLEINKNECHT

Institut für Physik der Universität Dortmund, Dortmund, Germany

1. Introduction

Progress in experimental particle physics has always been closely linked to improvements in accelerator and detector technology. The search for small or point-like constituents of matter required the study of scattering and annihilation processes at ever larger center-of-mass energies. This was achieved either by large fixed-target accelerators, like the 400 GeV Proton Synchrotrons at CERN and Fermilab and the 30 GeV electron LINAC at SLAC, or by storage rings both for protons (CERN ISR) and electron–positron pairs (SPEAR and PEP at SLAC, DORIS and PETRA at DESY, CESR at Cornell). The highest c.m. energy available now is 540 GeV provided by the antiproton–proton collider at CERN. Progress in accelerator technology includes the invention of stochastic and electron cooling of antiproton beams and the development of superconducting pulsed dipole magnets for the 800 GeV Tevatron at Fermilab.

Experiments are based on the ability of the researcher to detect particles produced by these accelerators or storage rings. The detecting equipment has undergone three major developments during the past ten years: the *size* of experiments has been increasing; for fixed-target experiments, this is a natural consequence of the larger momenta of particles involved and the correspondingly larger lever arms for magnetic analysis required. Also, larger target masses were needed for reactions with small cross-section, as in neutrino physics. For storage ring detectors, the large size is dictated by the necessity to cover most of the 4π solid angle around the interaction point. The second development concerns the *speed* of data acquisition: while pulsed devices like bubble or spark chambers were limited to 1–10 recorded events per accelerator pulse, the invention of proportional and drift chambers has increased this rate by a factor of 100 enabling experiments with 10^8 recorded events. In parallel with this goes the third evolution: the increase in *complexity* of the detectors. The number of independent analog informations from a large experiment can reach 10^4, and after digitization this yields up to 10^5 bits of information for one event from such a detector. Experiments of this type have become possible because the *reliability* of the equipment has increased considerably during this time and because fast on-line computers enable permanent *control* and monitoring of the detector.

In spite of the complexity of large experiments, the basic principles of detectors are simple. In this article, I go through these principles and some of the newer developments, following the list of physical quantities measured by the detector: position, time, mass, energy and momentum.

2. Position measurement

2.1. *Physical processes for detection*

The physical processes which enable us to detect particles are different for neutral and charged particles. Photons can interact by photoelectric or Compton effect or by pair creation, where the latter process dominates at energies above 100 MeV. The resulting electrons and positrons can be detected by their electromagnetic interaction. Neutrons of high energy will produce a shower of hadrons when colliding with detector material, thus enabling the detection of charged secondaries. Neutrinos interact by weak interaction conserving lepton number, producing hadrons and a charged or neutral lepton.

In contrast to these neutral particles, charged particles can be detected directly by their electromagnetic interaction with the atomic electrons of the detector material. In these collisions, the energy loss of a heavy charged particle with mass $m > m_e$ by ionization is given by the Bethe–Bloch-formula [BE 30, BE 32, BE 33, ST 71]:

$$-\mathrm{d}E/\mathrm{d}x = (4\pi r_e^2 m_e c^2 N_0 Z z^2/A\beta^2)[\ln(2m_e c^2\beta^2/((1-\beta^2)I)) - \beta^2] ,$$

where x is the thickness of material traversed in g cm^{-2}, N_0 is Avogrado's number, Z and A are atomic and mass numbers of the material, ze and $v = \beta c$ the charge and velocity of the moving particle, m_e the electron mass, $r_e = 2.8$ fm the classical electron radius and I an effective atomic ionization potential ranging from 13.5 eV in hydrogen to 1 keV in lead. The dependence of this energy loss on particle velocity is characterized by the $1/\beta^2$ variation at low energies, by a minimum at $\beta\gamma = p/mc \approx 4$ and finally at high energies by the relativistic rise by a factor which, for gases, is around 1.5. Fig. 1 shows this behaviour as measured [LE 78a] in an argon–methane mixture. The minimum value of the energy loss around $\beta\gamma = 4$ is 1.5 MeV/(g cm^{-2}) for iron and 1.8 MeV/(g cm^{-2}) for carbon.

The energy loss by ionization is distributed statistically around the mean loss described by the Bethe–Bloch-formula, the distribution being asymmetric with a tail [LA 44] at high losses due to δ ray production and distant collisions.

The calculation of the energy loss distribution was first done by Landau [LA 44] and Sternheimer [ST 52] assuming that the Rutherford term in the cross-section is the only source of the fluctua-

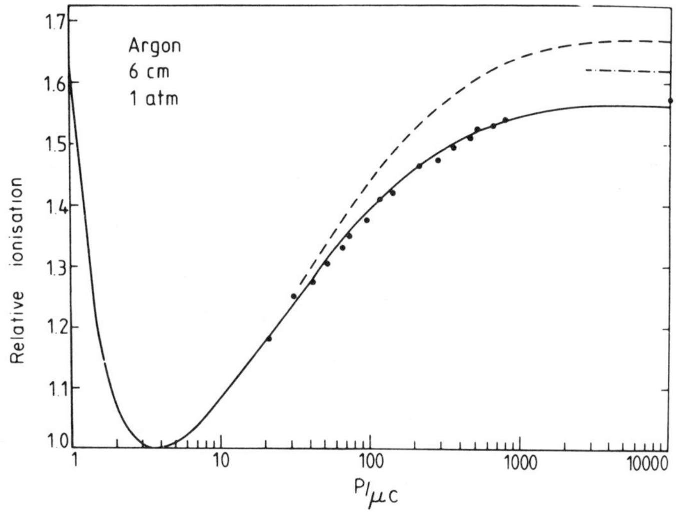

Fig. 1. Ionization energy loss in argon at atmospheric pressure relative to value at $\beta\gamma = p/cm \approx 4$. Points from [LE 78a] in argon/5% CH_4; dashed line: calculation of [ST 52]; dash–dot line: [ER 77]; solid line: photo absorption ionization model [CO 75, CO 76, AL 80].

tions and that its behaviour in the region of binding energies is described by a mean ionization potential. Fig. 2 shows that this model reproduces poorly the measured [HA 73] energy loss distribution for thin (1.5 cm) gas layers, but that more refined calculations including the shell structure of the atoms (Photo Absorption Ionization Model, PAI [CO 75, CO 76, AL 80]) give a satisfactory description of the data. The relativistic rise of the energy loss as measured in fig. 1 is also reproduced well by these models, while former calculations gave a rise which was too large by 10–15%.

The average energy needed for creating an electron–ion pair is fairly similar in different gases, viz. 40 eV/pair in helium and 26 eV/pair in argon, while it is much smaller in solids, e.g. 3 eV/pair for Si, such that for solids the number of pairs is larger and the statistical fluctuations in energy loss are smaller. However, the technical problems with the production of large volumes of purified semiconductors have limited the use of Si- and Ge-counters to low-energy high resolution γ spectroscopy. For detection of charged particles in large area detectors we are left with ionization in gases, mainly noble gases, and in liquid or solid scintillators converting the ionization energy loss into visible light.

2.2. Proportional chambers

Proportional tubes had been used since a long time. A cylindrical tube (radius r_a) on negative potential and a central wire (radius r_i) on positive potential create an electric field of the form

$$E(r) = V_0/(r \ln(r_a/r_i)) ,$$

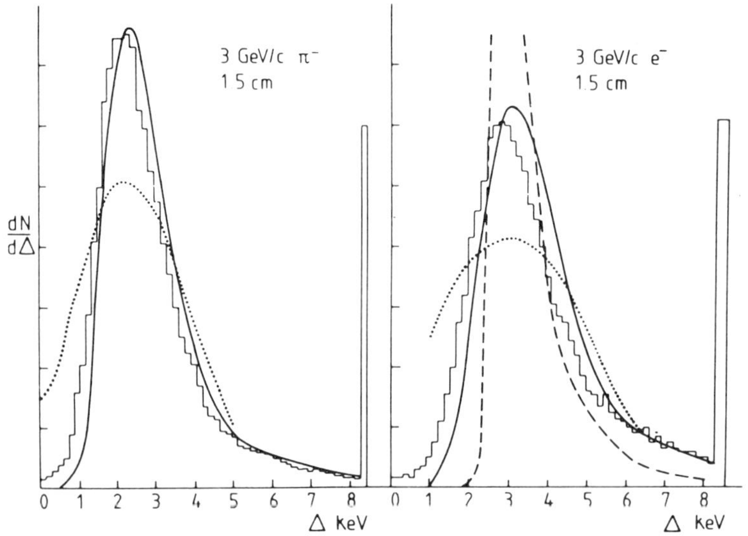

Fig. 2. Ionization energy loss distribution for pions and electrons in 1.5 cm of argon: 5% CH_4 at atmospheric pressure. Data of [HA 73]. Dashed curve of [LA 44, MA 69], dotted curve of [LA 44, BL 50], solid curve of [AL 80].

which reaches 10^4-10^5 V/cm near the anode wire. An electron liberated in an ionization process gains the kinetic energy ΔT between two collisions at radial distances r_1 and r_2

$$\Delta T = e \int_{r_1}^{r_2} E(r)\, \mathrm{d}r \,,$$

and if ΔT exceeds the ionization energy of the gas atoms, a secondary ionization can take place. A chain of such processes leads to an avalanche of secondary electrons and ions. The number of secondary electrons per primary electron (gas amplification α) reaches 10^4-10^6 in the proportional region, where α is independent of the number of primary electrons.

The field configuration in a proportional chamber with many anode wires in a plane between two cathode planes is shown in fig. 3. The discovery of Charpak et al. [CH 68] was that these separate anode wires act as independent detectors. The capacitive coupling of negative pulses from one wire to the next is negligible compared to the positive pulse on the neighbour wires induced by the moving avalanche. The main contribution does not come from the fast-moving electrons but from the much (1000 times) slower ions in the drop-shaped avalanche (fig. 4). The time structure of the negative pulse induced on the anode wire by the avalanches has been clarified subsequently [FI 75]. It consists of several pulses induced by different avalanches created by different primary ionization electrons drifting one after another into the high field region near the anode wire. These pulses have typically a rise-time of 0.1 ns from the electron part of the avalanche

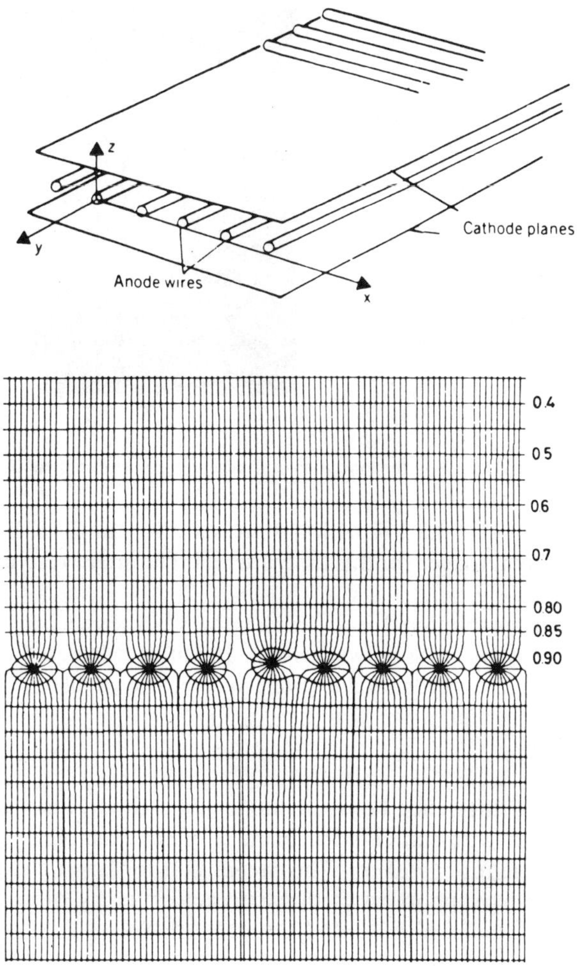

Fig. 3. Field configuration in a proportional chamber; field and equipotential lines are drawn [CH 70b].

and a decay time of 30 ns. Fig. 5 shows an oscilloscope picture with a time resolution sufficient to resolve these separate pulses, which in normal applications are integrated into one pulse by slower amplifiers.

As a practical example for operating proportional chamber systems, one of the first spectrometers with large chambers [SC 71] used a wire distance of 2 mm, gold-plated Tungsten wires of 20 μm, a gap between signal wires and cathode plane of 6 mm and an argon–isobutane gas mixture. The amplifiers [CU 71], based on MECL 1035 chips, had a threshold of 200 μV on 2 kΩ and an effective resolving time of 30 ns. The detection efficiency for this system, shown in fig. 6, allows

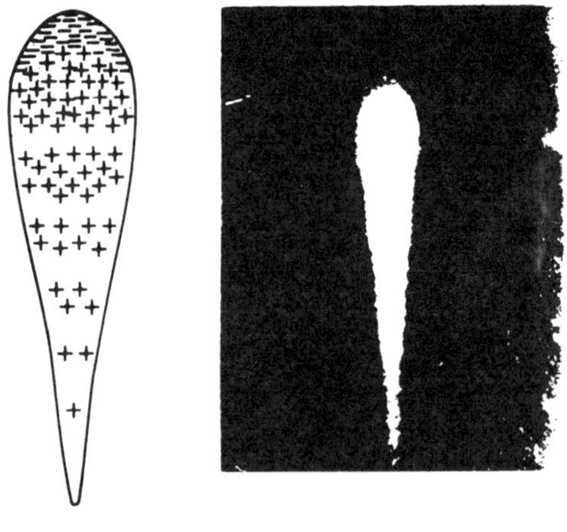

Fig. 4. Shape of the ionization avalanche in a high electric field [LO 61].

operating at full efficiency at 4.3 kV with a 40 ns sensitive time, i.e. the gate for the wire signals was opened by an external trigger for this time interval. The space resolution for this wire distance is $\sigma \sim 0.7$ mm.

One of the problems encountered in large chambers is the mechanical instability of the signal wires due to the electrostatic force between wires. It can be calculated [TR 69] that the system is stable if the wire tension T exceeds a value given by the wire geometry

$$T > (V\,l/2\,\pi a)^2\, 4\pi\epsilon_0 \ ,$$

where V is the potential difference between anode and cathode, l is the length of the signal wire, and a is the gap between wire and cathode plane. For the example above, with a wire tension of 50 p, the wires are stable for $l \leq 60$ cm. This means that for larger chambers the wires have to be supported every 60 cm by support wires threaded across, or by other methods. An inefficiency of 10% in a region of 5 mm width around the support wire is a consequence of some of these schemes [KL 70].

An enormous increase in spatial resolution of proportional chambers can be achieved by using the information from pulses induced on the cathode plane [RA 74, CH 78a]. For this purpose, at least one of the cathode planes is made of strips perpendicular to the direction of anode wires (see fig. 7). The pulses induced by the avalanche on the individual strips vary with the distance of the strip from the avalanche, and the center of gravity of the integrated pulse heights is a measure of the avalanche position. Fig. 8 demonstrates the precision which can be obtained with this center-of-gravity method. A soft X-ray produces ionization at three positions separated by 200 μm. The center of gravity y of each avalanche and the integrated charge c are measured, and the distribution in y shows three peaks with a variance of 35 μm. Most of this resolution error comes from the range of the original photoelectron. This impressive accuracy, however, is achieved with a detector where both mechanical construction and electronic pulse-height processing is costly.

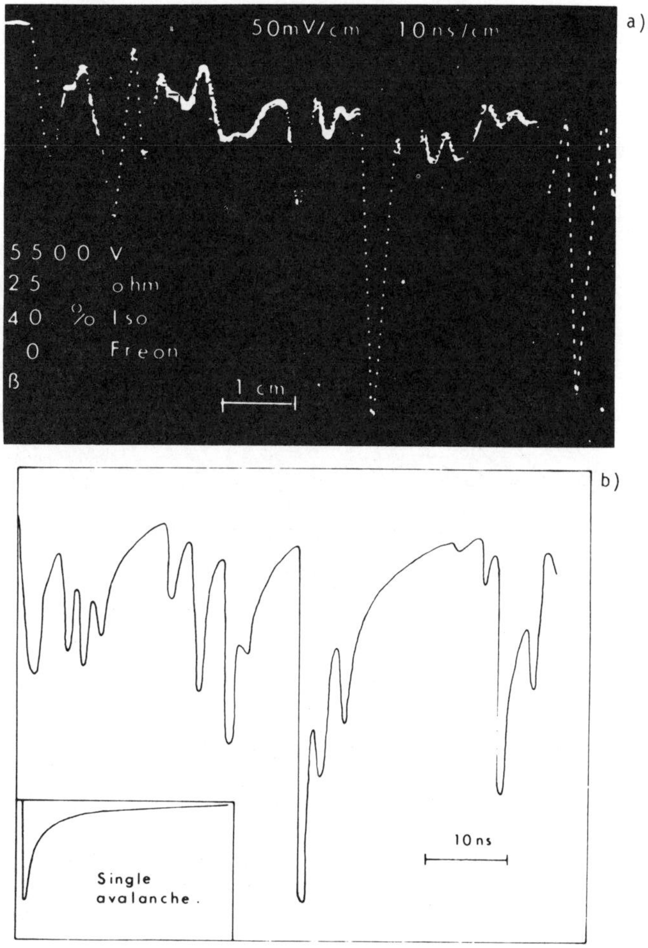

Fig. 5. (a) Oscilloscope display of proportional chamber pulse showing separate avalanches. (b) Pulse shape as simulated by computer [FI 75].

2.3. Planar drift chambers

A great reduction in cost is possible by using the experimental fact [CH 70a] that the time delay between the crossing of a charged particle through a proportional chamber and the creation of a pulse on the anode wire is related to the distance between particle trajectory and anode wire. This delay was found to be of the order of 20 nsec/mm, and if this time is measured for each anode wire with an accuracy of 4 nsec, a spatial resolution of 200 μm can be obtained.

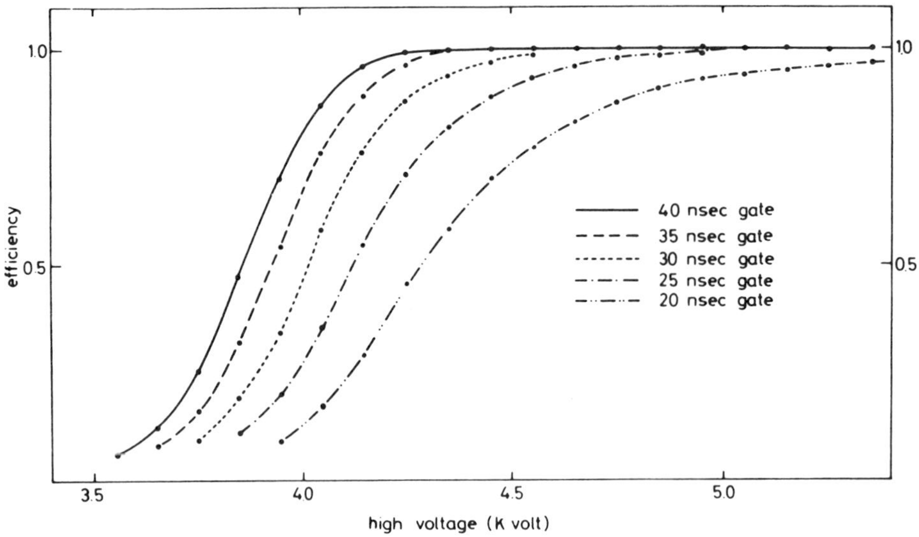

Fig. 6. Detection efficiency of a proportional chamber (2 mm wire distance, 2 × 6 mm gap) vs. high voltage for different gate opening times [SC 71].

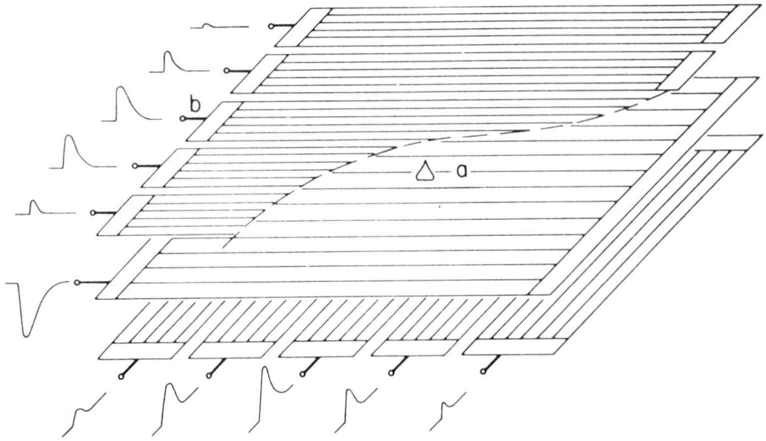

Fig. 7. Principle of cathode readout for proportional chambers. The center of gravity of the induced charges on cathode strips (b) determines the position of avalanche (a), [CH 78a].

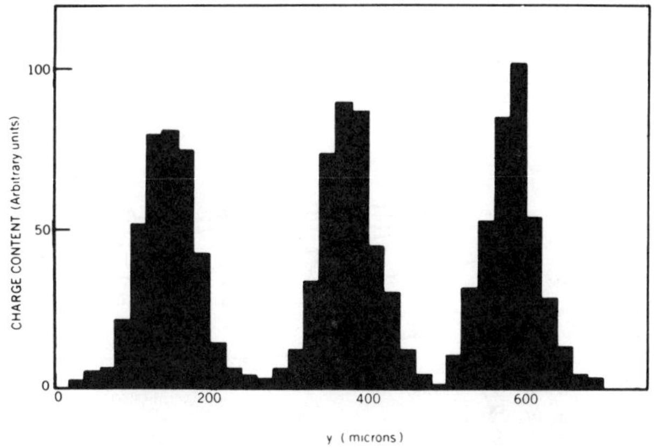

Fig. 8. Spatial resolution of a proportional chamber exposed to a 1.4 keV X-ray beam at 3 positions, 200 μ apart; plotted is total charge on cathode vs. center of gravity [CH 78a].

The invention of the drift chamber [CH 70a, WA 71] exploits this possibility. A scetch of the field configuration in one cell of a drift chamber is shown in fig. 9. The electrons from the primary ionization process drift in a low field (1000 V/cm) region into the high field amplification region around the anode wire, where avalanche formation occurs. Typical drift velocities for different gases are shown in fig. 10 for various argon–isobutane mixtures [BR 74]. For some of the mixtures, the drift velocity depends only mildly on the field strength, thus enabling a linear relation between distances and drift time even without constructing a perfectly constant field in the drift region. This is important because the requirement of a constant drift field necessitates the introduction of several field shaping wires per cell. An example for such a cell [MA 77] is shown in fig. 11, where the cell dimensions are 60×30 mm^2. The linearity between drift time and distance for this chamber is shown in fig. 12.

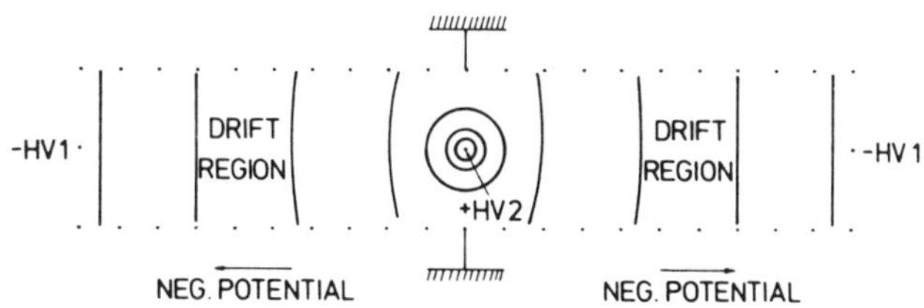

Fig. 9. Equipotential lines in a drift chamber cell.

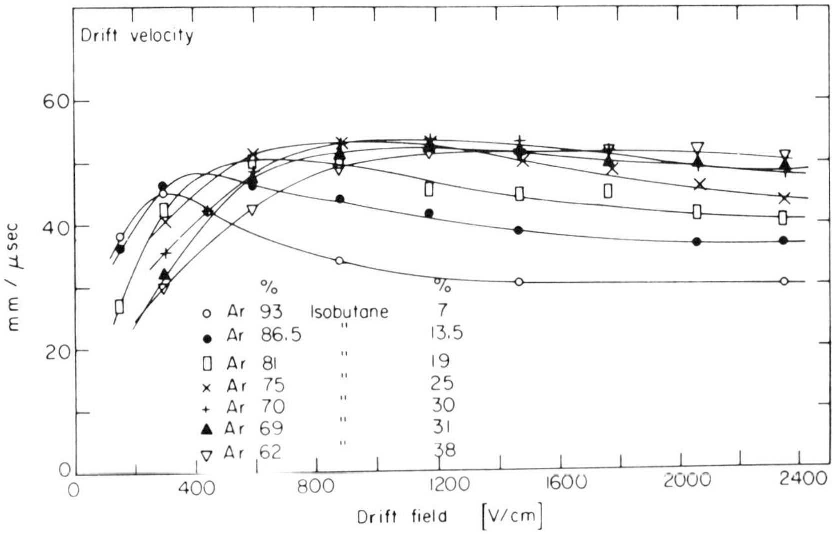

Fig. 10. Drift velocities in argon–isobutane mixtures [BR 74].

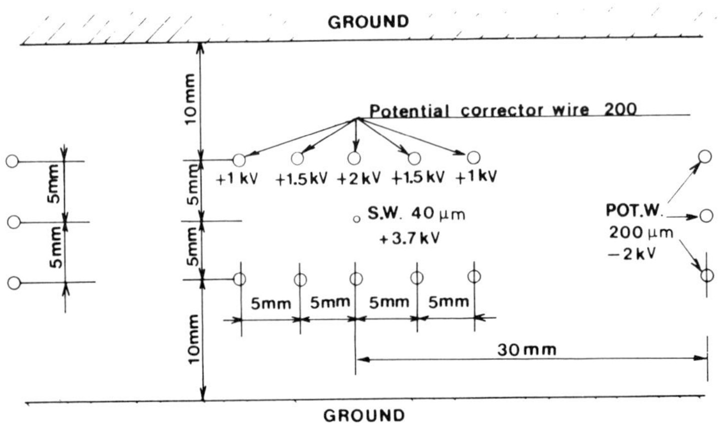

Fig. 11. Cell structure of large area drift chamber [MA 77].

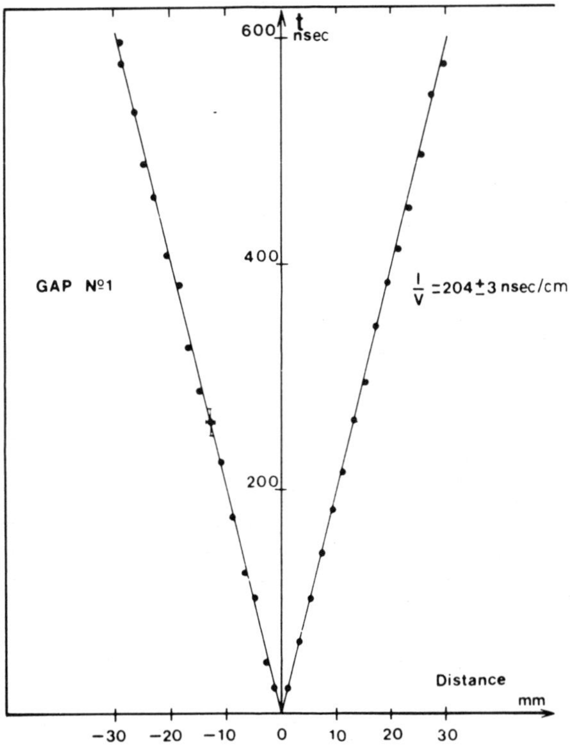

Fig. 12. Linear relation between drift time and position [MA 77].

2.4. Cylindrical wire chambers

For storage ring detectors, cylindrical geometries matched to solenoidal magnetic fields ($B_r = B_\varphi = 0, B_Z \neq 0$) have been widely used. These central detectors aim at the measurement of curvatures and initial directions of tracks emerging from the interaction point (fig. 13).

The first detectors of this kind used cylindrical layers of *proportional chambers* (see 2.2) or spark chambers (see 2.7) to determine the (r, φ) trajectory of the tracks (fig. 13). The wires are strung parallel to the B field, the E field is radial, but the displacement in φ of the avalanches due to the Lorentz force is small due to the small drift space to the anode wire.

For the next generation of central detectors, *cylindrical drift chambers* were used. Here the detector has up to 20 cylindrical layers of drift cells with electrical drift field in the $r\varphi$-plane (fig. 13). In order to save wires, the cells are open in the radial direction, but closed by at least 3 potential wires in the φ direction. Approximately half of the sense wires run exactly parallel to the B field, the others are inclined by a stereo angle (e.g. ±4°) relative to this axis in order to enable reconstruction of the z position of the tracks.

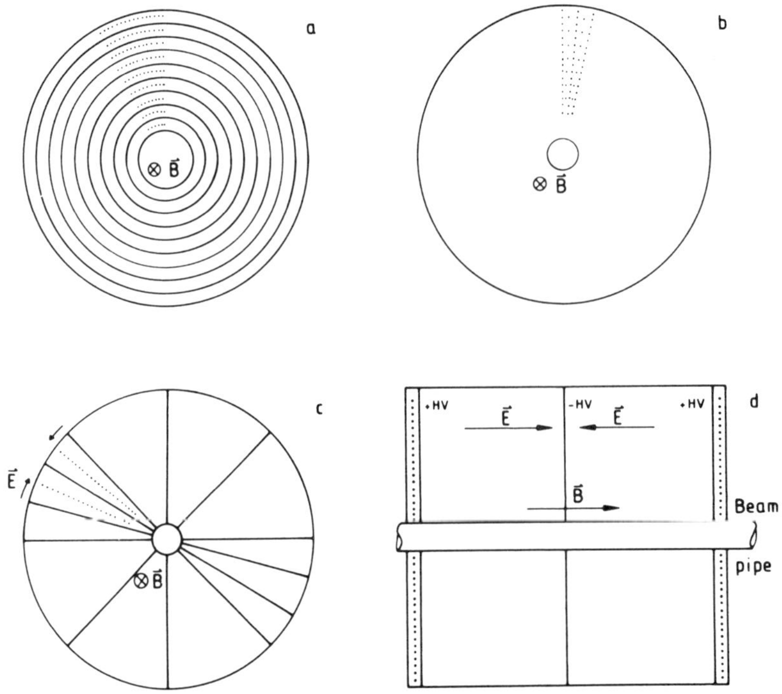

Fig. 13. Different types of cylindrical wire chambers; (a) proportional chambers, (b) cylindrical drift chambers, (c) pictorial drift chambers, (d) time projection chamber.

An example for such a central detector is the TASSO chamber [BO 80]. The drift cell (fig. 14) is radially open, the wires are 3.5 m long. The 15 chamber layers extend over 85 cm radial track length, 6 of these are equipped with $\pm 4°$ stereo wires. The resolution achieved was $\geq 200\ \mu m$ in the (r, φ) plane and 3–4 mm in the z direction. Fig. 15 shows the (r, φ) view of a hadronic event at 30 GeV c.m. energy recorded by the TASSO detector. Another example with a closed cell structure is the ARGUS drift chamber, shown in fig. 16 [WE 81]. Here the drift velocity is nearly radially symmetric around the sense wire, such that the geometrical position of all hits with a fixed drift time is a cylinder around the sense wire. This facilitates pattern recognition for tracks.

2.5. *Pictorial drift chambers*

This type of chamber records many ($\gtrsim 50$) three-dimensional points along the charged particle track. This measurement of true space points is very instrumental for the reconstruction of events with a high density of tracks. The first such pictorial chamber was built for the JADE detector at

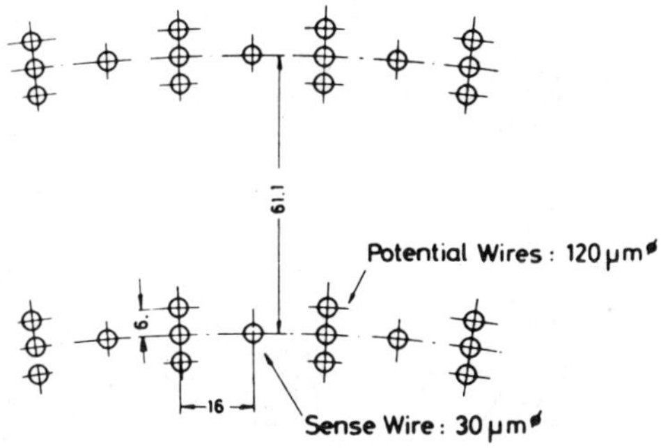

Fig. 14. Geometrical arrangement of wires in the TASSO wire detector [BO 80]. Dimensions are in mm.

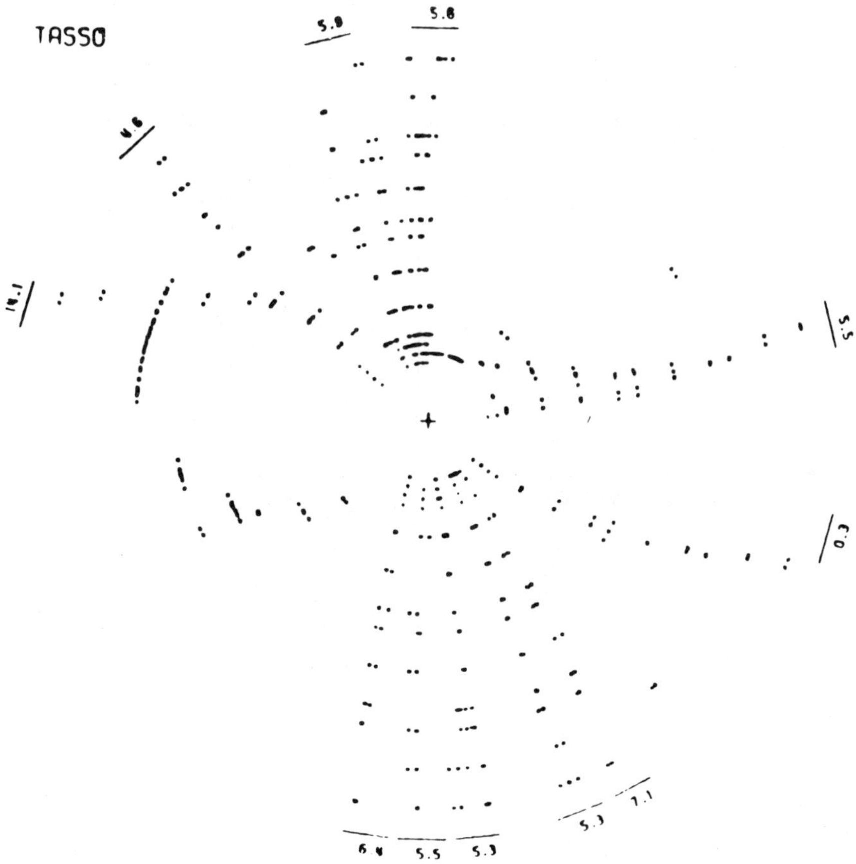

Fig. 15. Hadronic event produced by e^+e^- collision at 30 GeV c.m. energy in the TASSO wire detector, seen along the beam direction [BO 80].

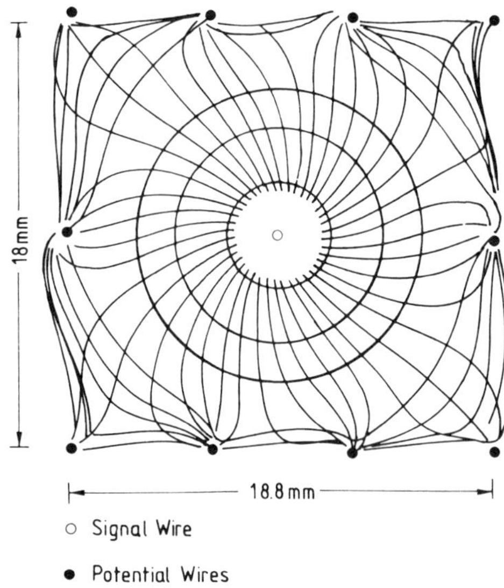

Fig. 16. Geometry of wires, lines of constant drift time and drift paths in one cell of the ARGUS detector for a 0.9 T magnetic field parallel to the wires [WE 81].

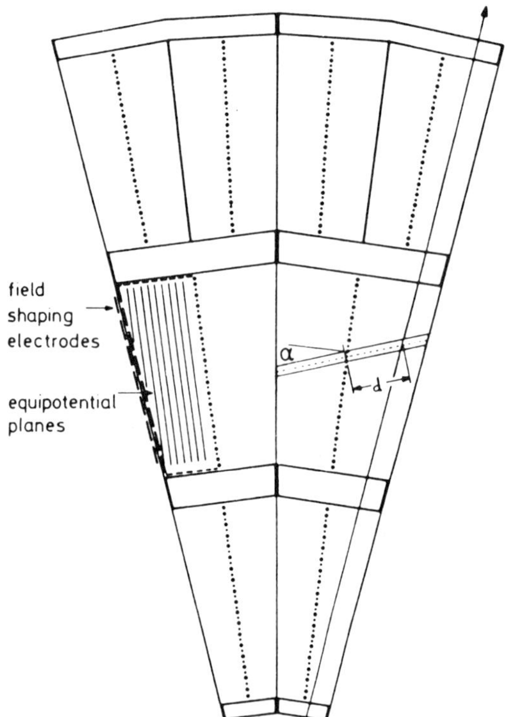

Fig. 17. Cross-section through two segments of the jet chamber of the JADE detector. Length of drift path d, Lorentz angle α [DR 80, WA 81b].

PETRA [BA 79, DR 80]. Here the cylindrical volume of the track detector is subdivided into 24 radial segments of 15° opening angle. In each segment (fig. 17), there are 64 sense wires parallel to the magnetic field, arranged in 4 cells of 16 wires, i.e. a total of 1536 sense wires of 234 cm length. The electric field is perpendicular to the sense wire plane, and therefore also to the magnetic field. Due to the longer drift length compared to normal cylindrical drift chambers, the effect of the Lorentz force $e\boldsymbol{v} \times \boldsymbol{B}$ becomes noticeable here.

It leads to a deviation of the direction of the drift velocity v_D from the direction of the electric field E by an angle α_L given approximately by the ratio of magnetic and electric forces, $\tan \alpha_L = k(E)\, v_D B/E$ [WA 81b], which in the JADE chamber at $B = 0.45$ T and 4 atm. pressure is 18.5°. The factor $k(E)$ depends on chamber gas and electric field [SC 80]. Due to the Lorentz force, the lines of equal drift time in the neighbourhood of the sense wire plane in the JADE chamber are rather complicated (fig. 18). Fig. 19 shows the (r, φ) projection of a two-jet event at 35 GeV c.m. energy. Up to 48 samplings per track are recorded and can be used for a measurement of the ionization energy loss. The z coordinates along the wires are obtained by charge division on the two ends of the wires with a precision of 1.6 cm.

The principle of the pictorial chamber was also used in the AFS [CO 81] and the UA1 [BA 80] central detectors (see 5.3).

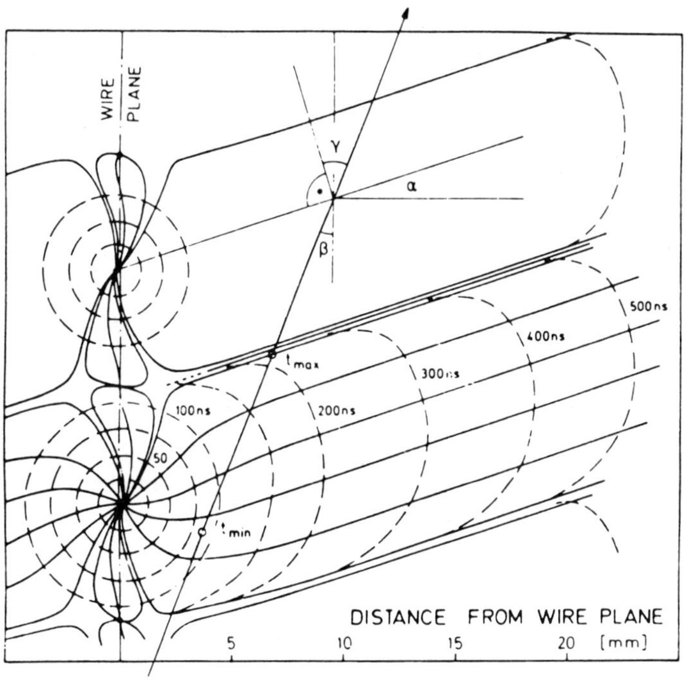

Fig. 18. Trajectories of drifting electrons (full lines) and lines of equal drift times (dashed) for the JADE field configuration [DR 80, WA 81b].

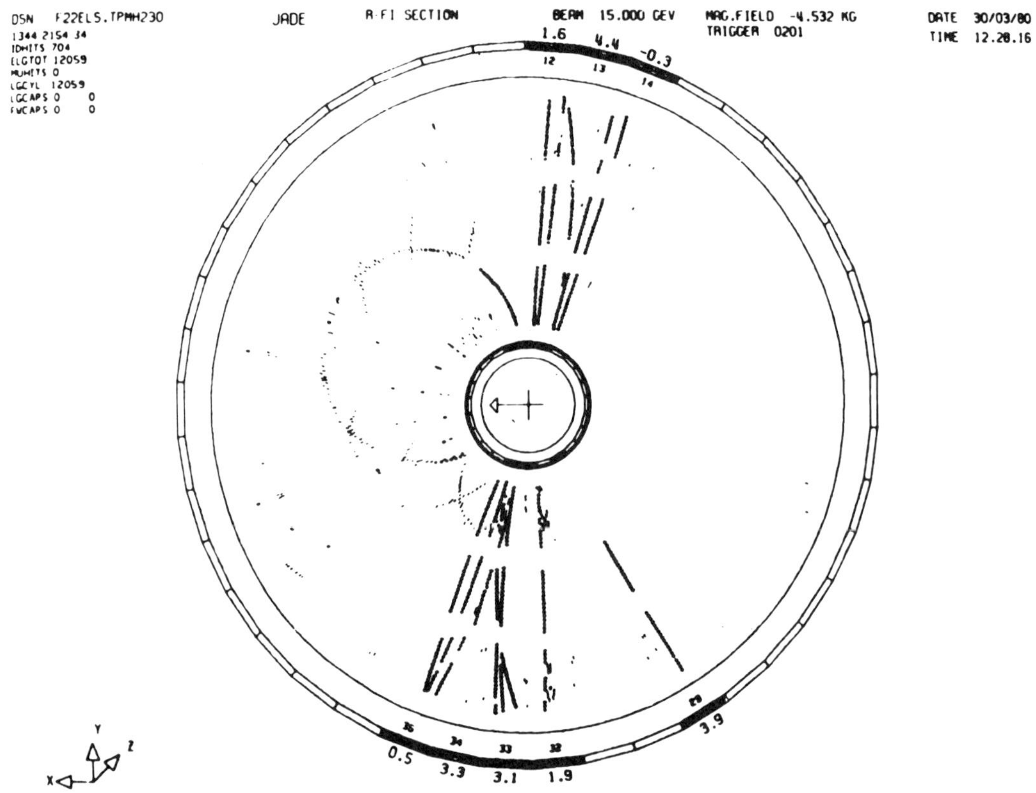

Fig. 19. Hadronic event produced by e^+e^- collision at 30 GeV c.m. energy in the jet chamber of the JADE detector. Hits assigned to a track are shown as crosses, remaining hits are dashes [DR 80].

2.6. *Time projection chamber*

An ingenious way of using the proportional and drift chamber principles for the central detector of a storage ring detector was proposed by Nygren [NY 74, NY 81]. Fig. 20 shows a large (1 m radius, 2 m length) cylindrical volume filled with argon–methane at 10 atmospheres. The two endcaps are equipped with one layer of multiwire proportional chambers subdivided into six sectors. Each of the sectors has 186 proportional signal wires for multiple ionization measurement and 15 wires with segmented cathode readout ("pads") for the spatial measurement of radius r and peripheral angle φ in the cylinder coordinates (fig. 21).

Most importantly here, the electric drift field (150 kV/m) is parallel to the magnetic field (1.5 T) of the solenoid used for magnetic analysis of tracks originating from the collisions in the center of the cylinder. $\boldsymbol{E} \times \boldsymbol{B}$ type forces vanish, and it is possible to have the ionization electrons drift over large distances to the end caps. Furthermore, the strong magnetic field reduces considerably

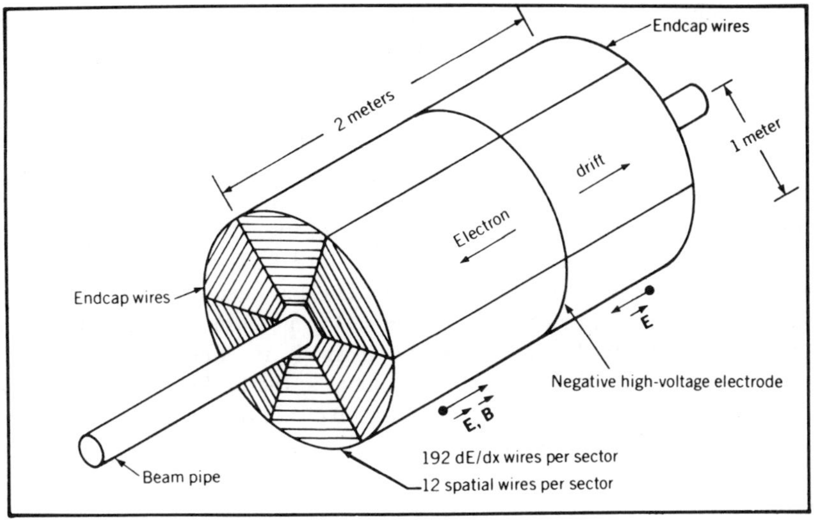

Fig. 20. Scetch of time projection chamber [MA 78].

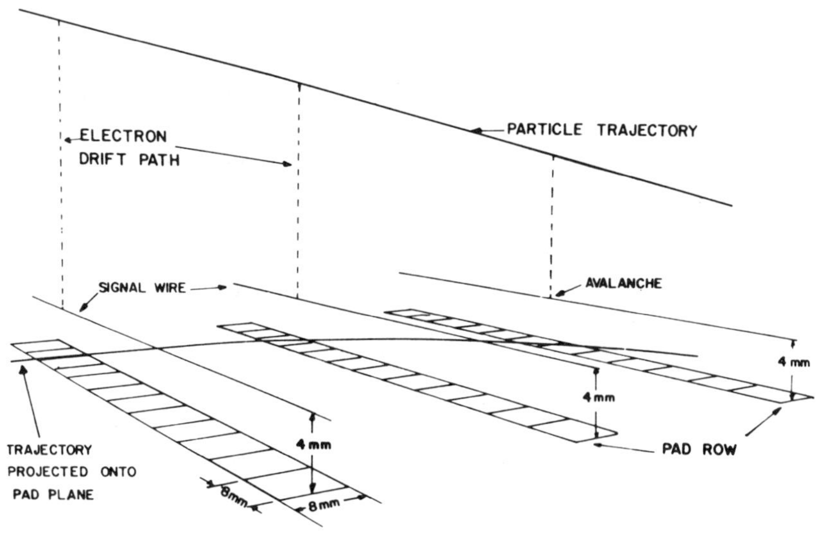

Fig. 21. Principle of cathode readout by pads used in TPC [FA 79].

(factor 10) the diffusion broadening of the track image on the end caps of the cylinder by causing helical movements of the electrons around the magnetic field lines.

The spatial reconstruction of the original tracks is obtained by measuring the two-dimensional images on the endcaps using the cathode plane readout and the center-of-gravity method and by measuring the drift time of each track segment parallel to the cylinder axis. In addition, the proportional chambers also measure the ionization energy loss of the track, thus enabling the separation of e, π, K and p by using dE/dx and momentum measurement.

Measurements on a small test chamber gave an (r, φ) resolution of less than 200 μm and a z resolution along the drift path of 0.2 mm [FA 79, NY 81]. Initial tests on the big chamber have not yet reached this precision. A similar and less ambitious TPC has been built and tested at TRIUMF [HA 81a]. Hexagonal endcaps with 12 sense wires per sector measure the space position of tracks. At atmospheric pressure for 80% Ar 20% CH_4 and with an electric field of 150 V/cm the ratio of electric field over pressure is 0.2 V/(cm Torr), the same as in the Berkeley TPC. The arrangement of the cathode pad readout shown in fig. 22 allows a precision of 120 μm along the anode wire. Preliminary initial tests gave a spatial resolution of 600 μm.

2.7. Bubble chambers

The bubble chamber [GL 52, GL 58, FR 55] consists of a pressure vessel filled with liquified gas around the boiling temperature. By an expansion mechanism, the pressure on the liquid is reduced for a short time (~ms). The liquid then becomes unstable against bubble formation, and if

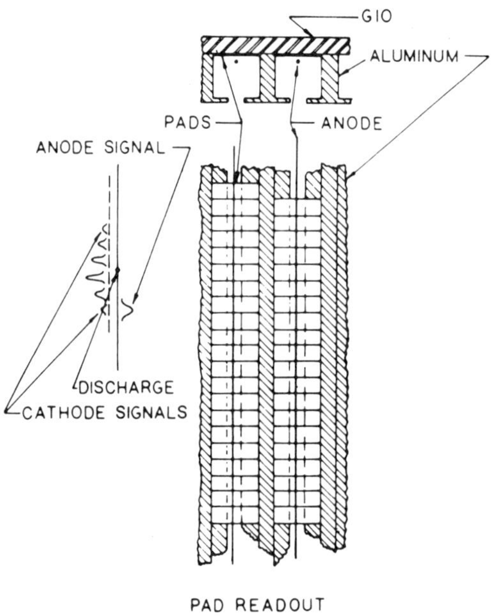

Fig. 22. Section of TRIUMF TPC showing pads and anode wires [HA 81a].

during the expansion time a charged particle forms a track of ionization in the liquid, bubble formation starts at this track. After a time left for bubble growth (μs–ms), the track is illuminated and photographed through windows in the vessel. With the chamber embedded in a magnetic field of up to 30 kG, the track curvature is used for momentum measurement, and the bubble density b can be used for a measurement of the velocity $v = \beta c$ and thus the mass of the particle [GL 58].

Chamber liquids range from hydrogen as a pure proton target, deuterium for study of interaction on neutrinos up to xenon for experiments which need a high conversion probability for γ rays. Table 1 gives some properties of chamber liquids [BE 77, HA 81b, WE 81b].

The bubble chamber is still unique in its capability of analyzing complicated events with many tracks and identifying those particles. A beautiful example for such a super-event is shown in fig. 23. However, the use of bubble chambers as isolated detectors has diminished because i) they cannot be employed at storage rings; ii) at high energy, showers are not contained in the chamber volume any more except if the chamber is a hybrid of calorimeter and bubble chamber techniques, which seems to be possible using liquid argon [HA 82]; iii) the lever arm for momentum measurement is not sufficient at high momenta. Future use will include small chambers with extremely high resolution, e.g. the 8 μm resolution obtained [DY 81, MO 80] by holographic readout in the BIBC chamber at CERN (fig. 24). Such chambers will be used in conjunction with large spectrometers for momentum measurement and identification of reaction products ("hybrid systems").

2.8. Streamer chambers

In a similar development streamer chambers are used as track sensitive targets. In this type of chamber electric fields above 50 kV/cm perpendicular to the track direction create an avalanche with gas amplification around 10^8 and light emission ("streamer"). The geometry of such a chamber is given in fig. 25. Very short high voltage pulses (few nsec) are required in order to keep the streamers short [SC 79, RI 74].

The excellent quality of streamer chambers presently in use can be seen in fig. 26, a picture from the NA5 experiment at CERN [NA 5]. The development of a very high resolution streamer chamber has been pioneered at Yale [DI 78] in order to measure the lifetimes of charmed par-

Table 1
Physical properties of bubble chamber liquid

Liquid	Temp. (K)	Pressure (Bar)	Density (g/cm^3)	Expansion ratio $\Delta V/V$ (%)	Rad. length X_0 (cm)	Absorption length λ (cm)
^{4}He	3.2	0.4	0.14	0.75	1027	
^{1}H	26	4.0	0.06	0.7	1000	887
^{2}D	30	4.5	0.14	0.6	900	403
^{20}Ne	36	7.7	1.02	0.5	27	89
^{131}Xe	252	26	2.3	2.5	3.9	
C_3H_8	333	21	0.43	3	110	176
CF_3Br	303	18	1.50	3	11	73
Ar	35	25	1.0	1.0	20	116

ticles, around 10^{-13} sec, corresponding to a flight path of 300 μm for a time dilatation factor $\gamma = 10$ which is typical for a particle of 30 GeV/c with a mass of 3 GeV. This chamber operates at 24 atmospheres, uses pulses of 0.5 nsec duration producing a field of 330 kV/cm, and a spatial resolution of 32 μm has been achieved [SA 80]. With this technique a good measurement of charmed particle lifetime seems feasible, competitive with similar experiments using nuclear emulsions.

2.9. Flash and spark chambers

Another gas discharge chamber is the flash chamber developed by Conversi et al. [CO 55, CO 78] and built in a similar way for the FMNN neutrino experiment at Fermilab [TA 78]. The chamber consists of an array of rectangular tubes made of polypropylene by extrusion. This array is placed between two metal electrodes and filled with a neon (90%)–helium (10%) mixture (fig. 27). A triggered HV pulse is applied on the electrodes generating a glow discharge in those cells where ionization has been induced by passing particles. This discharge can be recorded by photographing or by electronic readout. The flash chamber reaches an efficiency of 80%. Due to the extremely low cost, large volume calorimeters with fine grain sampling can be built.

The spark chamber has been used widely as a triggered track detector. A set of electrodes or massive plates is inserted in a noble gas volume (typically He/Ne) at atmospheric pressure. The plates are alternatingly connected to a pulsed HV supply or ground. After the passage of an ionizing particle, a HV pulse is triggered via a spark gap, causing a spark break-through at the place of the initial ionization but parallel to the electric field (fig. 28). For chamber gases at atmospheric pressure, the magnitude of the pulsed electric field has to be ~10–20 kV/cm in order to generate sparks.

The spark position is recorded optically or by magnetostriction. Here the electrodes of the chamber consist of wire planes, and a magnetostrictive wire (made of a Co–Ni–Fe-alloy) running across these wires picks up a signal induced by the spark current pulse on the HV or ground wires. The magnetostrictive wave travels along the Co–Ni–Fe-wire with a speed of 5000 m/s such that its arrival time at the end of the magnetostrictive wire measures the spark position. A precision of 200 μm is obtained. The large amount of charge in the spark plasma requires a long time (~2 ms) for clearing before the chamber can be triggered again [RI 74, AL 69].

2.10. Comparison of position detectors

The parameters to be compared are space and time resolution and rate of data acquisition. Table 2 gives typical values, where "dead time" for pulsed detectors means the time needed before a new trigger can be allowed to pulse the detector, and "sensitive time" is the time during which incoming particles are registered whether they are correlated or not with the event causing the trigger. Time overlay of different events can only be avoided if the mean time interval between events is large compared to this sensitive time.

Proportional and drift chambers are best suited for precise recording at high data rates, while the pulsed bubble and streamer chambers still have the potential for optimum space resolution

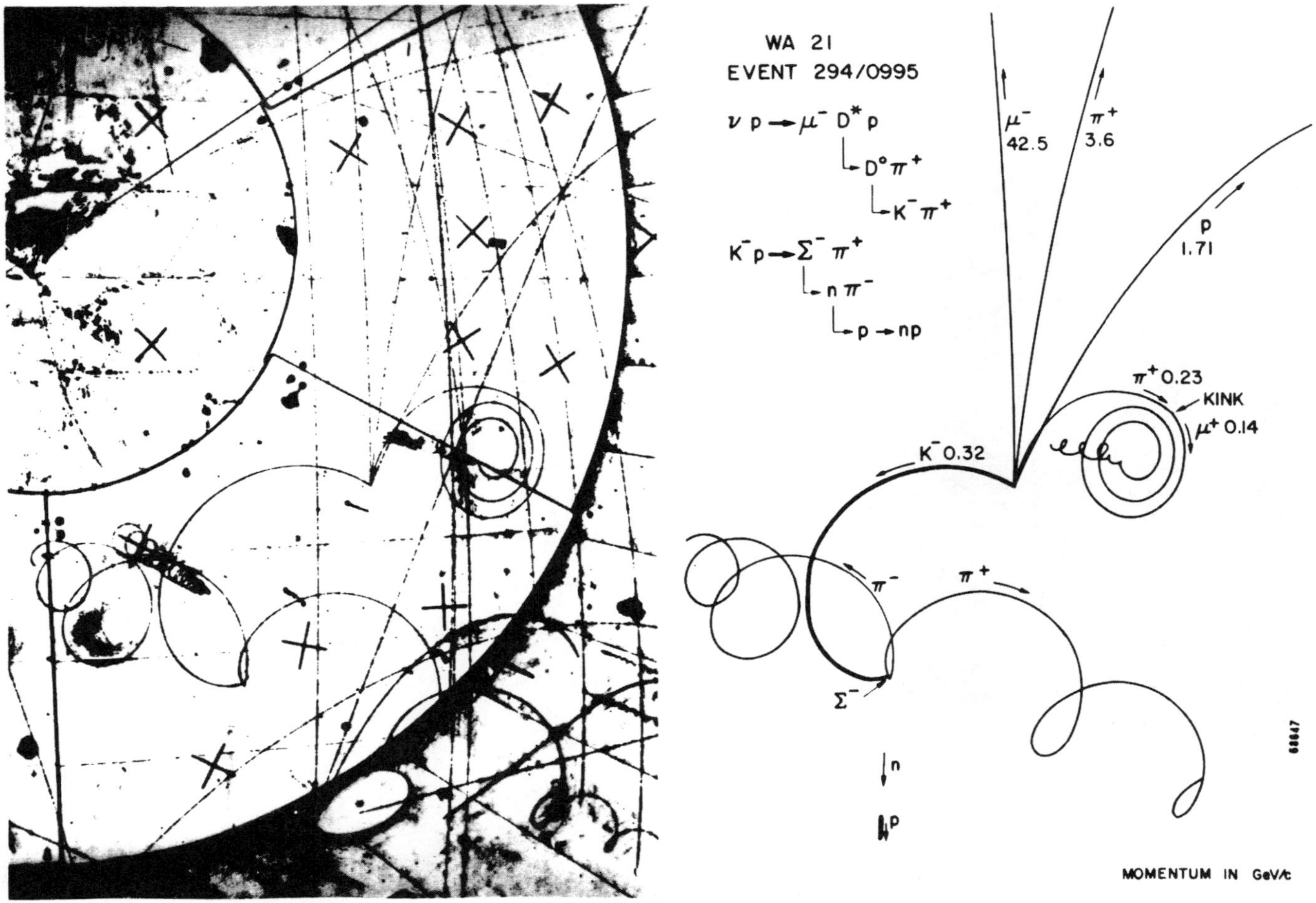

Fig. 23. Neutrino interaction in hydrogen bubble chamber BEBC [WA 21].

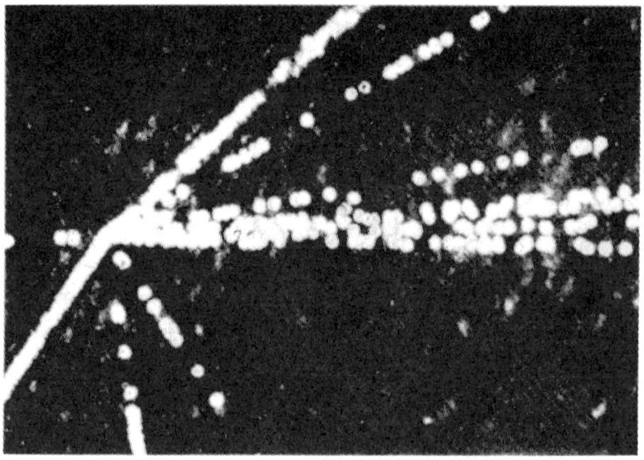

Fig. 24. Holographic photograph of a 15 GeV π^- interaction in a small freon bubble chamber BIBC. Bubble size is 8 μm [DY 81].

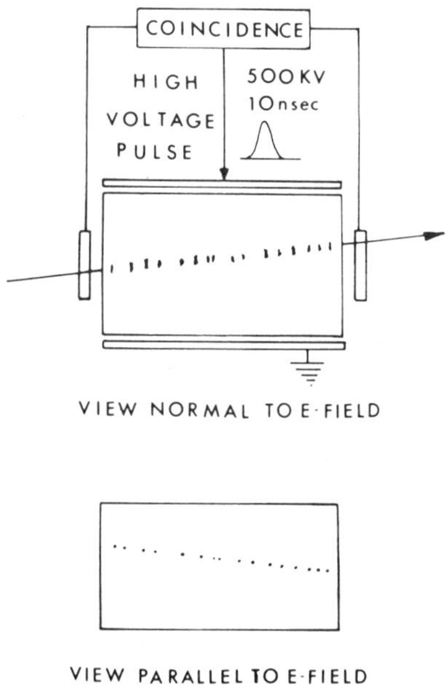

Fig. 25. Schematic view of streamer chamber [SC 79].

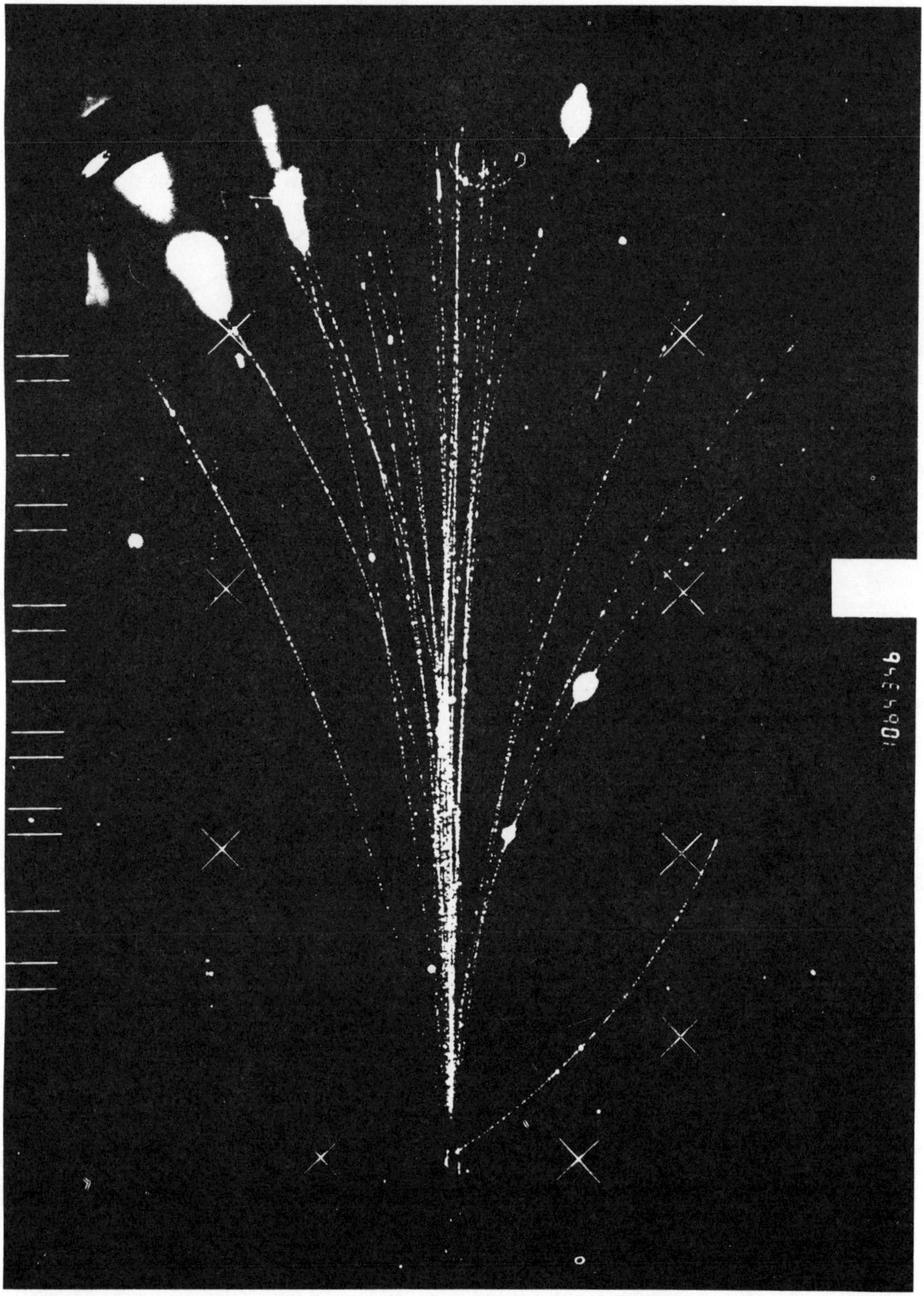

Fig. 26. Interaction of a 300 GeV π^- in a liquid hydrogen target as seen in a streamer chamber of dimensions $200 \times 120 \times 72\ cm^3$ [EC 80].

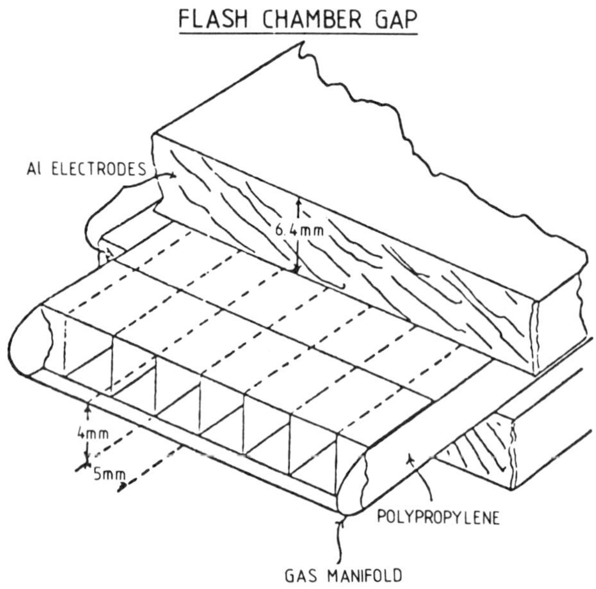

Fig. 27. Part of flash chamber made of extruded polypropylene [TA 78].

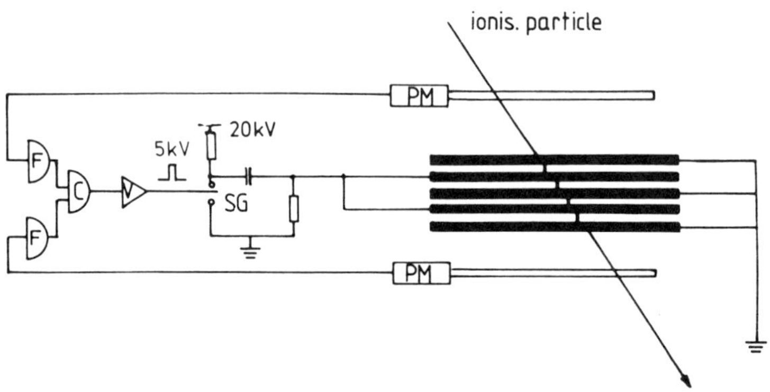

Fig. 28. Principle of spark chamber; PM photomultiplier, F pulse shaper, C coincidence unit, V amplifier, SG spark gap.

Table 2
Properties of position detectors

Type of chamber	Space resolutions (μm)		Deadtime (ns)	Sens. time (ns)	Readout time (ns)	Advantages
	normal	special				
Prop. chb.	700	100	–	50	10^3-10^4	time resolution
Drift chb.	200	50	–	500	10^3-10^4	space resolution
Bubble chb.	100	8	10^8	10^6	–	complex events
Streamer chb.	300	30	10^8	10^3	10^7	multiple tracks
Flash chb.	4000	2000	10^7	10^3	10^6	price
Spark chb.	200		10^7	10^3	10^6	simplicity

and multitrack analysis. The flash chamber, because of its low price and simple construction, may well find application in very large detectors using fine-grain calorimetry, e.g. for low-rate experiments like proton decay and neutrino experiments.

3. Time measurement

3.1. Photomultiplier

The main instrument for obtaining time information on a particle is the photomultiplier tube. Visible light from a scintillator liberates, by photoelectric effect, electrons from a photocathode made of alkali metals. For bialkali cathodes of Cs–K–Sb, the quantum efficiency reaches [VA 70] a maximum of 25% around 400 nm (fig. 29). In tubes of the linear-focussing type, the photoelectrons are then focussed on the first dynode consisting of materials like BeO or Mg–O–Cs. With secondary emission yields of 3–5 per incident electron, amplification of 10^8 for 14 dynode stages can be achieved. The risetime of the anode pulse is around 2 ns, the transit time is typically 40 ns.

The spread in the transit time ("time jitter") through the photomultiplier is mainly given by the variation of transit time of the photoelectron from the cathode to the first dynode. There are two effects, the broad velocity distribution of the photoelectrons and their different path lengths from different parts of the cathode to the first dynode. The photoelectron kinetic energy spectrum for bialkali cathodes illuminated by light of 400–430 nm wavelength extends [NA 70] from 0 to 1.8 eV peaking at 1.2 eV. For an electric field of $E = 150$ V/cm, the difference δ in transit time between a photoelectron initially at rest and another one of kinetic energy $T_k = 1.2$ eV is $\delta = (2mT_k)^{1/2}/(eE) \approx 0.2$ ns. The other effect contributing to the transit time jitter, the geometrical path length variation from cathode to first dynode, depends mainly on the cathode diameter. For a diameter of 44 mm, it is $\delta_2 = 0.25$ ns and 0.7 ns for the tubes XP2020 and XP2232 B, respectively [PH 78]. This seems to be the ultimate limitation in time resolution for conventional photomultipliers.

In microchannel plate multipliers, the paths of photoelectrons are straight channels (fig. 30). This reduces the transit time jitter by a factor of two compared to conventional multipliers; a jitter of 0.1 ns has been measured [LE 78b].

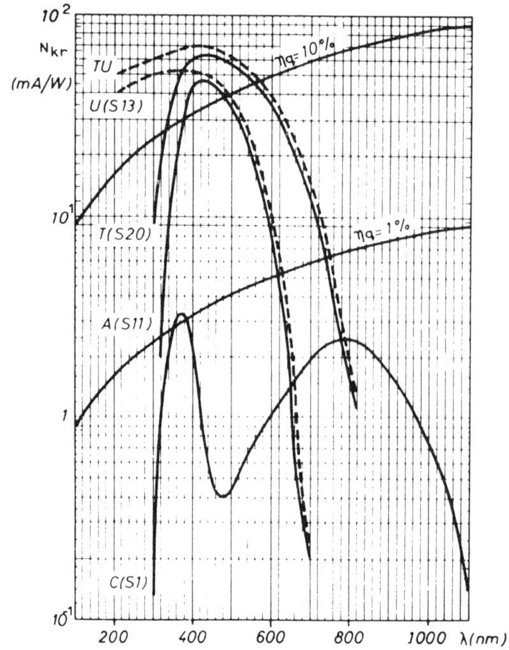

Fig. 29. Spectral sensitivity N_{Kr} (mA/W) and quantum efficiency η_q(%) for photocathodes; TU and U types have quartz window, others glass window [VA 70].

3.2. *Scintillators*

A scintillation counter has two functions: the conversion of the excitation caused by the ionizing particle into visible light and the transport of this light to the photocathode.

The mechanism of scintillation [BI 64] is completely different for anorganic crystal scintillators and for organic crystal, liquid or polymeric scintillators.

For *anorganic crystals* doped with activator centers, the energy level diagram looks qualitatively as shown in fig. 31. Ionizing particles produce free electrons, free holes and excitons. These move inside the crystal until they reach an activator center A, which they transform into an excited state A^* decaying to A with emission of light. The decay time of scintillation light is then given by the lifetime of the unstable state A^* and depends on temperature like $\exp(-E_1/kT)$, where E_1 is the excitation energy. Typical data for such crystals are given in table 3.

Organic scintillators, on the other hand, have very short decay times of order nanoseconds. The scintillation mechanism here is not a lattice effect, but proceeds through excitation of molecular levels which emit bands of UV light. The absorption length of this UV light in most transparent organic materials is short, of order mm, and the use of these scintillators is possible only through fluorescence excitation in a second molecule, called wavelength shifter. The emission of

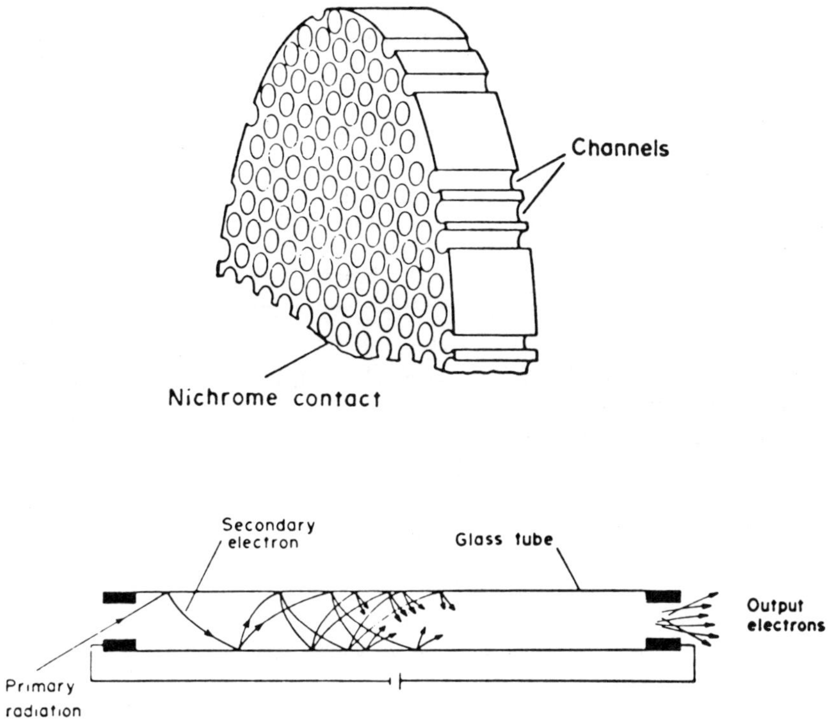

Fig. 30. Microchannel plate and principle of multiplication [DH 77].

this shifter material is usually chosen to be in the blue wavelength region detectable by photocathodes. These two active components in a scintillator can be solved in liquids or in a monomeric substance being polymerized subsequently. Two parameters determine the figure of merit for such a scintillator: the light yield and the absorption length in the scintillator.

Table 4 gives structure, wavelength of maximum emission and decay time for a few primary scintillators, as well as for two wavelength shifters [BE 71]. For polymerizing plastic scintillator, either aromatic compounds (styrol, vinyltoluene) or alifatic ones (acrylic glasses, "plexiglass") are

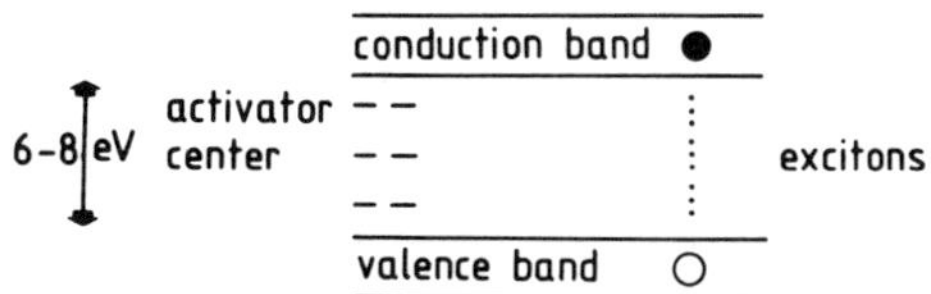

Fig. 31. Energy band structure in anorganic crystals.

Table 3
Properties of scintillating anorganic crystals

	NaJ (Tl)	LiJ (Eu)	CsJ (Tl)	$Bi_4Ge_3O_{12}$
Density (g/cm^3)	3.67	4.06	4.51	7.13
Melting point (°C)	650	450	620	
Decay time (μsec)	0.2	1.3	1	0.35
Pulse height for electrons	1.0	0.35	0.28	0.08
λ_{max} (emission) (nm)	410	470	550	480
Radiation length (cm)	2.5			1.12
Physical properties	hydroscopic	hygroscopic		non hygroscopic

used. The aromatic ones yield about twice as much light, but the alifatic ones are less expensive and much easier to handle mechanically.

In order to achieve a good energy resolution in large calorimeters, it is very important to obtain uniform response of a long but thin scintillator over its entire length even when the scintillator light is viewed by a photomultiplier from one end only. The observed attenuation of light from the far end is mainly due to the absorption of the short-wavelength part of the POPOP emission

Table 4
Organic scintillators and wavelength shifters

Primary scintillator	Structure	λ_{max} emis. (nm)	Decay time (ns)	Yield/ yield (NaJ)
Naphthalene	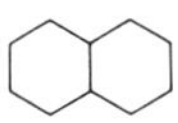	348	96	0.12
Anthracene	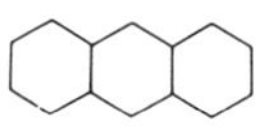	440	30	0.5
p-Terphenyl	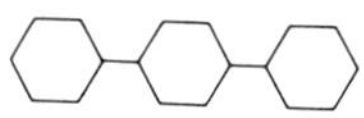	440	5	0.25
PBD	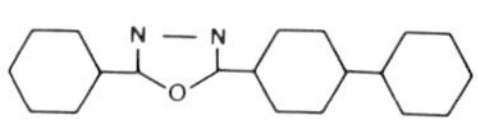	360	1.2	
Wavelength shifter				
POPOP	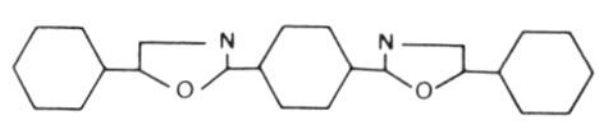	420	1.6	
bis-MSB	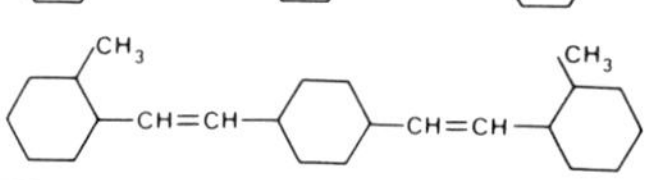	420	1.2	

spectrum, as shown [KL 81b] in fig. 32. In order to obtain a more uniform response it is therefore possible to filter out the short wavelength part. The effect of a filter at 430 nm can be seen in fig. 32: the light yield at the end of the scintillator near to the photomultiplier is diminished drastically, while the one from the far end is influenced much less.

Still, by using the filter, light is lost, and it is interesting therefore to search for an acrylic scintillator with long attenuation length and higher light yield than commercially available. One new mixture found [KL 81b] recently contains 3% naphthalene, 1% PBD and 0.01% bis-MSB. The attenuation curves for a scintillator of this material with size $1800 \times 150 \times 5$ mm^3 are shown in fig. 33. The attenuation length with black end and filter is $\lambda = 210$ cm, and the light yield at 160 cm from the photomultiplier side is 20% higher than the one for the commercial mixture plexiglas 1921 (1% naphthalene, 1% PBD, 0.01% POPOP). This new scintillator is used therefore for a new neutrino calorimeter of the CDHS collaboration (fig. 77).

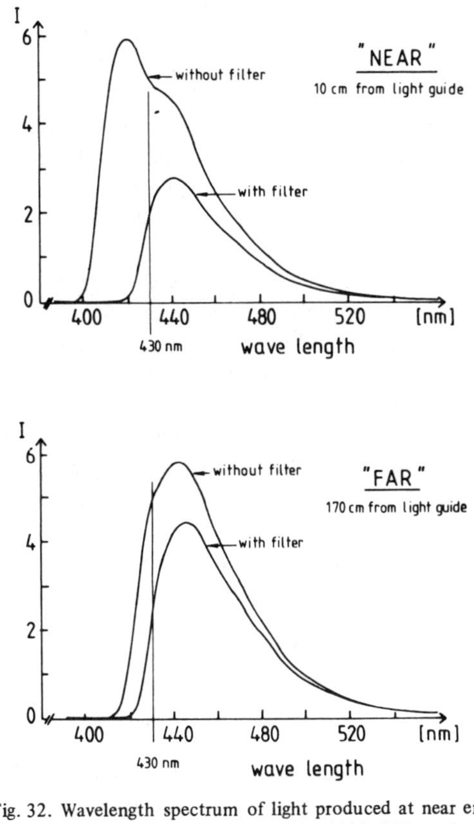

Fig. 32. Wavelength spectrum of light produced at near end or far end of scintillator (type plexiglas 1922, 180 cm long) with and without yellow filter at 430 nm [KL 81b].

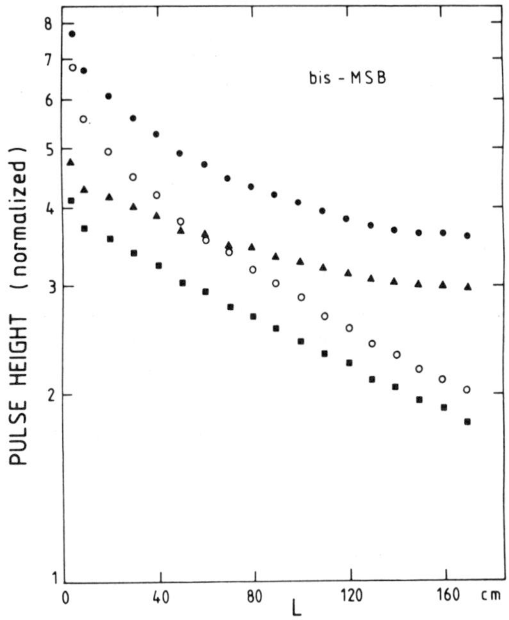

Fig. 33. Attenuation curves in a scintillator of dimensions $1800 \times 150 \times 5$ mm^3; without filter, reflecting end (full dots), black end (open circles); with yellow filter, reflecting end (triangles), black end (squares), [KL 81b].

Important developments of low cost scintillators have been done at Saclay [BO 81] for experiments at the proton–antiproton collider at CERN. Two new groups of scintillator material have been developed: i) the KSTI line based on polystyrene material which can be extruded between two polished rolls; it has 80–100% of the light output of the PVT-scintillator NE110, an attenuation length of 80 cm for sheets of 3 mm × 200 mm cross-section and a decay time of 3 ns, however careful handling is required as for other aromatic scintillators; ii) the Altustipe series based on polymethylmetacrylate (PMMA) with similar properties as Plexipop, i.e. 20–50% of the light output of NE110, and attenuation lengths of 1.0–1.5 m.

3.3. Light collection

The traditional way of collecting light from a scintillator is the adiabatic light guide. The (blue) scintillation light travels down the scintillator plate by multiple internal reflection. The usually rectangular radiating surface with cross-section F is imaged onto the photocathode surface f by means of bent transparent plastic rods or strips such that the radius of curvature of the rods is large compared to their thickness. In this way it can be avoided that light hits an internal surface under an angle larger than the one of total reflection. The amount of light reaching the photocathode is less than f/F due to Liouville's theorem.

The time resolution of direct-coupled scintillation counters comes from two sources: the transit time jitter of the photomultiplier and the time difference between different light paths in the scintillator and light guide. The latter contribution depends mainly on the scintillator dimensions and dominates for large ($\gtrsim$2m) counters, as shown by the data in fig. 34. For large counter systems, a resolution below σ_t = 200 ps has not been achieved.

An alternative method, due originally to Garwin [GA 60, SH 51], has been revived recently for applications in large-scale calorimetry [BA 78, SE 79]. The principle is shown in fig. 35: blue light from the wavelength shifter (e.g. POPOP) leaves the scintillator and enters, through an air gap, a second shifter bar. This rod is made of acrylic material doped with a molecule (BBQ, e.g.) absorbing the blue light and emitting isotropically green light (around 480 nm, see fig. 36). A part (10–15%) of the green light is catched by the shifter bar by internal reflection and reaches the photomultipliers looking at the end of the bar. The main problems in developing this technique were i) to find the appropriate shifting material matched to the POPOP emission spectrum and the photocathode spectral sensitivity; ii) to find a way of optimizing the self-absorption in the green bar.

These problems were solved [BA 78] by taking a 90 mg/ℓ concentration of BBQ in plexiglas 218. The product now is commercially available and has found wide application in large experiments.

The thickness of the green shifter bar needed for absorption of the POPOP light can be obtained from fig. 37, where the intensity of BBQ emission has been measured [KL 81a] as a function of thickness of the green bar. An absorption length of $\lambda = (5.2 \pm 0.2)$ mm is obtained for the BBQ concentration mentioned.

The shifter bar technique can be used to collect the light from very large scintillators with a few photomultipliers. One example is the CFR neutrino calorimeter [BA 78] with counters of 3 × 3 m viewed by 4 phototubes at the corners. These 4 pulseheights can be used to calculate the center of

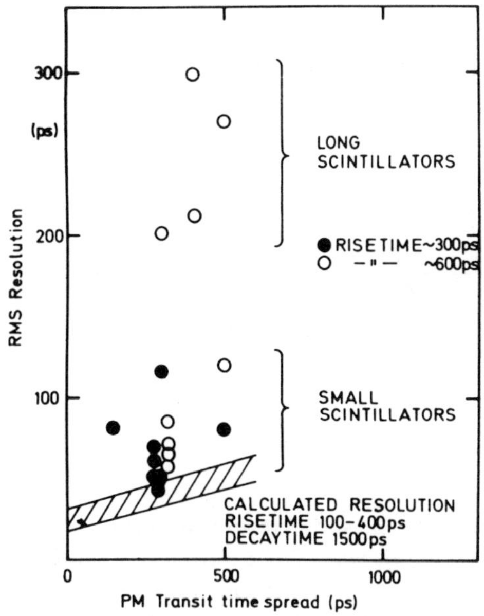

Fig. 34. Comparison of r.m.s. time resolutions of scintillation counters vs. the r.m.s. transit time spread in the photomultipliers used. Small scintillators with dimensions below 1 cm are compared to long scintillators (length ~2 m, thickness 2–5 cm, width 20–40 cm), [CA 81b].

Fig. 35. Principle of wavelength shifter bar technique.

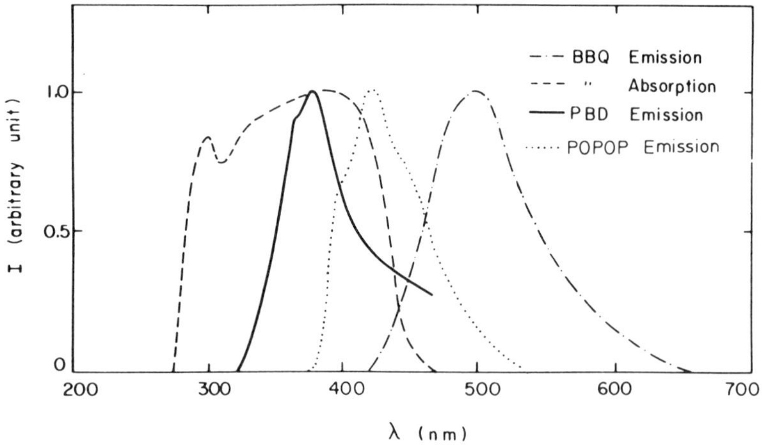

Fig. 36. Absorption and emission spectra of BBQ.

gravity of a shower of particles. Fig. 38 shows results of a measurement done with a $150 \times 300 \times 1.5$ cm^3 acrylic scintillator viewed in this way. The position of a shower with 100 equivalent particles can be reconstructed with an accuracy of $\sigma \sim 8$ cm [KL 81a]. This method, therefore, has the advantage of allowing to save photomultipliers, save mechanical work for light guides and permitting a measurement of the position of a shower of particles. It does not, however, permit to disentangle several showers.

One disadvantage of the wavelength shifter BBQ is its two-component decay time with lifetimes [KL 81a] of 18 ns and 620 ns, which causes timing difficulties when measuring pulseheights.

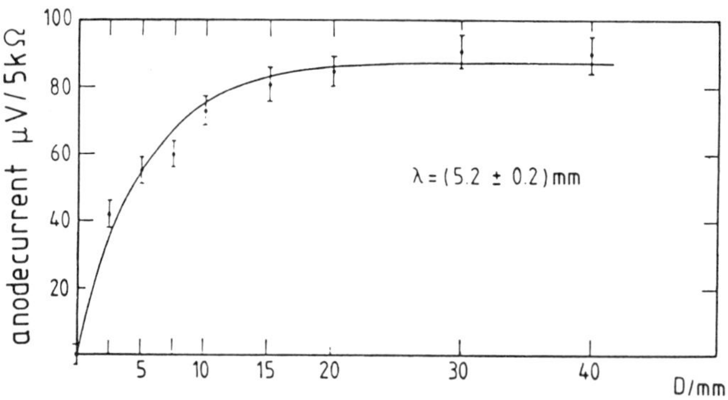

Fig. 37. Measurement of absorption length of POPOP light in BBQ [KL 81a].

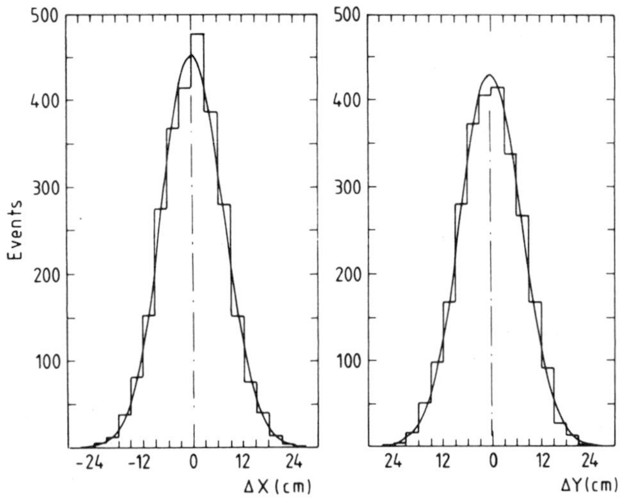

Fig. 38. Deviation of reconstructed position of a shower of 100 equiv. particles from real position; r.m.s. deviation $\sigma_x = (7.3 \pm 0.1)$ cm, $\sigma_y = (7.6 \pm 0.1)$ cm [KL 81a].

3.4. Planar spark counters

These counters consist of two planar electrodes generating an electric field above the static breakdown, i.e. at a ratio of field strength E to gas pressure p of $E/p \sim 30\text{–}60$ V/(cm Torr). The primary ionization of a passing charged particle develops into a spark, and the large current drawn by this spark can be recorded as a fast rising pulse. In counters with metallic electrodes [KE 48] the spark discharges the total capacity of the plates, leading to high temperatures and burning of the electrodes. The damaged surface gives spontaneous breakdowns at lower fields. A possibility of avoiding this deficiency consists in using material with high resistivity ($\sigma = 10^9 - 10^{10}$ Ωcm) for one of the electrodes [BA 56]. The spark then only discharges a small area of the condensor around the primary ionization, leading to a lower energy density in the spark. Fig. 39 shows a schematic of such a counter with copper strip readout on the semiconducting anode. The impedance of this strip line can be matched to the readout cable. Counters are operated with argon at 5–10 atmospheres, adding hydrocarbons (isobutane, ethane, 1.3-butadien) in order to absorb UV photons from the spark thus avoiding secondary sparking [BR 81].

The time jitter of the signal, δ, depends on the electric field E and the number of primary ions N, like $\delta \sim 1/(E\sqrt{N})$. Measured values [BR 81] are $\delta \sim 30\text{–}80$ ps for counters of 10×10 cm^2 area having detection efficiencies of $>95\%$. The distribution of the time difference between two parallel spark counters is not quite gaussian, showing broad tails (fig. 40).

Despite the excellent timing characteristics of these counters, wide application is not yet foreseeable because of the extreme difficulties in manufacturing and maintaining the high-quality surfaces.

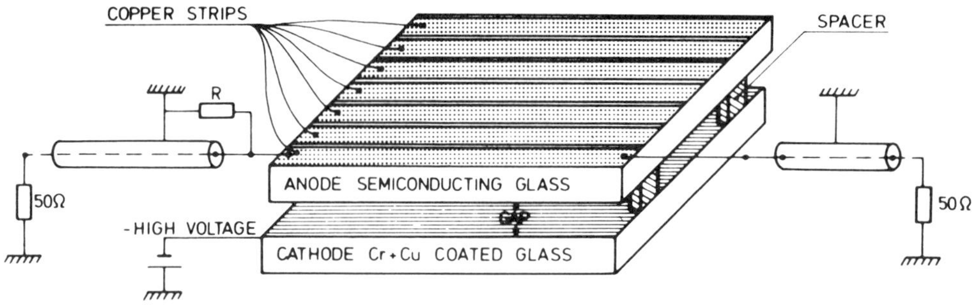

Fig. 39. Schematic view of a typical planar spark counter [BR 81].

4. Identification methods

4.1. Time-of-flight

The identification of charged particles through their flight time between two scintillation counters requires, for momenta above 1 GeV/c, very good time resolution and quite long flight path. The time difference between two particles with masses m_1 and m_2 is for a flight path L

$$\Delta t = L/(\beta_1 c) - L/(\beta_2 c) = (L/c)(\sqrt{1 + m_1^2 c^2/p^2} - \sqrt{1 + m_2^2 c^2/p^2}) \,,$$

which for $p^2 \gg m^2 c^2$ becomes $\Delta t \sim (m_1^2 - m_2^2)Lc/(2p^2)$. Fig. 41 shows flight time differences between pairs of charged particles. Using conventional scintillation counters (section 3.3) with a

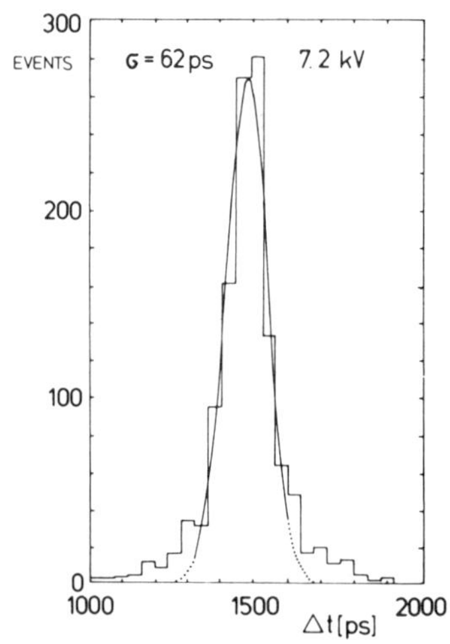

Fig. 40. Distribution of time difference of two planar spark counters. A gaussian is fitted to the histogram in the region where the curve is drawn as solid line [BR 81].

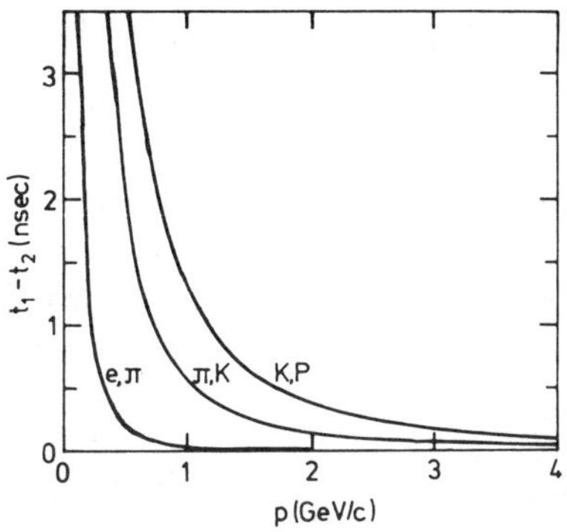

Fig. 41. Differences of time of flight $t_1 - t_2$ of particle pairs eπ, πK and Kp for a flight path of 1 m.

time resolution $\sigma_t = 300$ ps, π/K separation at the level of $4\sigma_t$ would require a flight path of 3m at 1 GeV/c and 12 m at 2 GeV/c. If parallel plate spark counters would come into operation, the required flight path would be reduced to 0.5 m or 2 m, respectively.

For this method of identification, therefore, at present a very long flight path is needed.

4.2. Cherenkov counters

Cherenkov radiation [CE 64] is electromagnetic radiation emitted by charged particles of velocity v traversing matter with refractive index n if $v > c/n$. The classical theory of the effect attributes this radiation to the asymmetric polarization of the medium in front of and behind the charged particle resulting in a net electric dipole moment varying with time. The radiation is emitted at an angle θ, where $\cos\theta = (ct/n)/(\beta ct) = 1/(\beta n)$. The threshold for Cherenkov-effect, $\beta > 1/n$, corresponds to a threshold in the γ factor of the particle,

$$\gamma > 1/\sqrt{1 - 1/n^2}\ .$$

Typical refractive indices and threshold values are given in table 5. Unfortunately there is a gap in

Table 5
Cherenkov radiators

Material	$n - 1$	γ (threshold)
Glass	0.46–0.75	1.22–1.37
Scintillator (toluene)	0.58	1.29
Plexiglass (acrylic)	0.48	1.36
Water	0.33	1.52
Aerogel	0.025–0.075	4.5–2.7
Pentane (STP)	1.7×10^{-3}	17.2
CO_2 (STP)	4.3×10^{-4}	34.1
He (STP)	3.3×10^{-5}	123

the refractive indices between the gases with highest index (pentane) and practical transparent liquids with lowest index. The development of silica–aerogel [CA 74] consisting of $n(SiO_2)$ + $2n(H_2O)$ has closed this gap and permits a velocity measurement in the range of $\gamma \sim 3$–5, where the specific ionization is nearly constant.

Large scale production of silica–aerogel with $n = 1.03$ and $n = 1.05$ in blocks of $18 \times 18 \times 3$ cm^3 is now possible [HE 81]. The Cherenkov light has been collected by cylindrical mirrors [CA 81a] or by a diffusor box [AR 81] behind the aerogel block (fig. 42) and recorded by photomultipliers. 6–12 photoelectrons have been obtained from a 15–18 cm long radiator of this material ($n = 1.03$).

From the relation $\cos\theta = 1/(\beta n)$ follows that the maximum Cherenkov angle becomes smaller if n approaches unity. The energy radiated per path length in the radiator is

$$\frac{dE}{ds} = \frac{2\pi\alpha h}{c^2} \int_{\beta n > 1} \left(1 - \frac{1}{\beta^2 n^2}\right) \nu \, d\nu ,$$

with α being the fine structure constant, $\alpha = 1/137$. This leads to the number of photons N emitted over a path length L in the wavelength interval λ_1 to λ_2

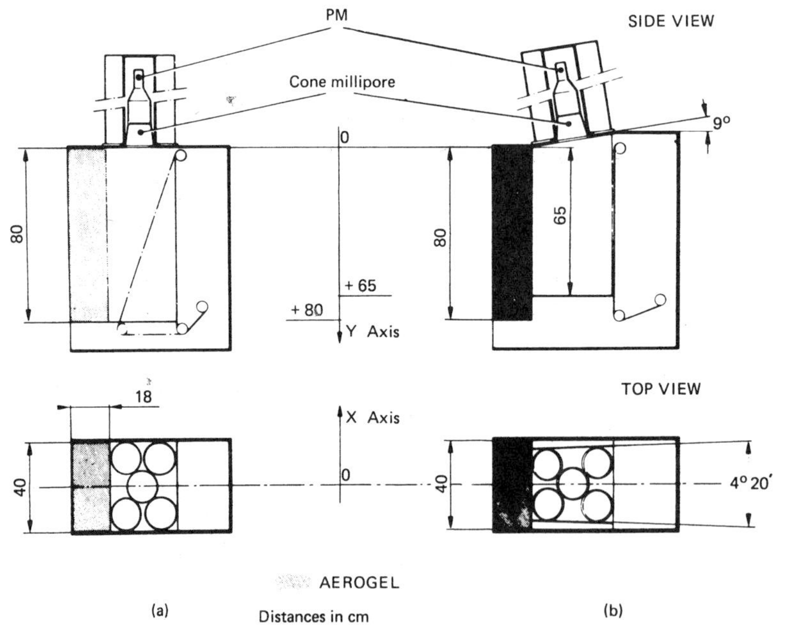

Fig. 42. Silica–aerogel counters used in EMC experiment. Particles incident from left on aerogel, diffusing box covered internally with millipore filter [AR 81].

$$N = 2\pi\alpha L \int_{\lambda_1}^{\lambda_2} d\lambda \, \sin^2\theta/\lambda^2 \ .$$

For a detector sensitive in the visible region $\lambda_1 = 400$ nm, $\lambda_2 = 700$ nm, this corresponds to $N/L = 490 \sin^2\theta$ photons/cm. Evidently, the detection of UV-light can increase this yield by a factor of 2–3.

The length of Cherenkov threshold detectors needed for separation of particles of momentum p increases as p^2: suppose two particles with masses m_1 and $m_2 > m_1$ have to be distinguished. Then the refractive index of the radiator can be chosen such that the heavier particle with mass m_2 does not yet radiate, or is just below threshold, $\beta_2^2 \simeq 1/n^2$, and $n^2 = \gamma_2^2/(\gamma_2^2 - 1)$. Then the amount of Cherenkov light from the particle with mass m_1 is proportional to

$$\sin^2\theta = 1 - 1/(\beta_1^2 n^2) \ ,$$

which for $\gamma \gg 1$ becomes

$$\sin^2\theta \simeq c^2(m_2^2 - m_1^2)/p^2 \ .$$

In a radiator of length L, detecting photons with a quantum efficiency of 20%, the number of photoelectrons is

$$P = 100\, Lc^2\, (m_2^2 - m_1^2)/(p^2 L_0) \ ,$$

where $L_0 = 1$ cm. In order to obtain $P = 10$ photoelectrons, a length

$$L/L_0 = p^2/((m_2^2 - m_1^2)\, c^2 \cdot 10)$$

is required in the optimistic case assuming that a radiator with exactly the refractive index required above can be found.

For practical purposes, one uses a combination of threshold counters with different refractive indices, as indicated in table 6. By using two or more of these counters, pions, kaons and protons can be identified in the momentum range given in fig. 43.

Apart from this utilization of the Cherenkov threshold, the angle of Cherenkov emission can also be measured in order to identify particles. The conical emission pattern around the radiating particle can be focussed into a ring-shaped image. An adjustable diaphragm at the focus transmits Cherenkov light emitted in a small angular range into a phototube. Changing the radius of the diaphragm allows a scan through regions of velocity. Differential gas Cherenkov counters [Li 73] correcting for chromatic dispersion in the radiator (DISC) have achieved velocity resolutions of $\Delta\beta/\beta \approx 10^{-7}$.

Since the length of these counters is limited to a few meters, there is a maximum momentum at which two kinds of particles can be separated (fig. 44). Separation of π and K mesons at several 100 GeV/c is possible with these devices. A velocity spectrum of charged hyperons in a short beam

Table 6
Possible choices of Cherenkov threshold counters [LE 81c]

Counter	Refractive index	Radiative medium	Radiator length	Counter length	Light yield
A	1.022	Aerogel	20 (cm)	50–100 (cm)	5–6 (e^-)
B	1.006	?(Aerogel)	?	50–100	?
C	1.00177	Neopentane	30	50	≃10
D	1.00049	(N_2O–CO_2) or Fr14	100	≃120	≃10
E	1.000135	(Ar–Ne) or H_2	185	≃200	≃ 5

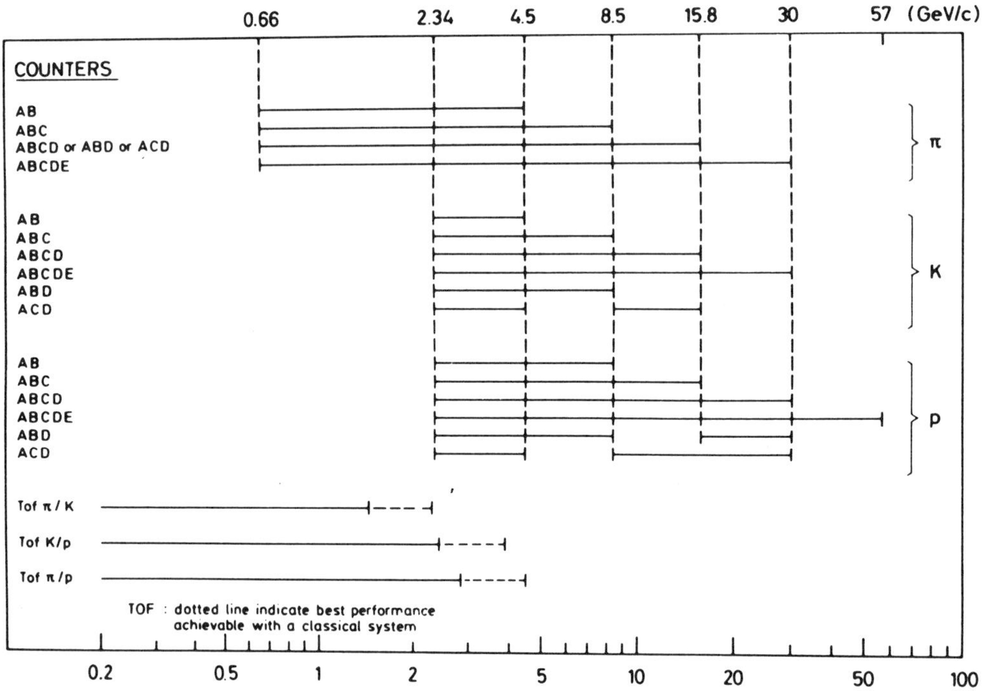

Fig. 43. Domain of particle identification for threshold Cherenkov counters as given in table 6 [LE 81c].

from an external proton target is shown in fig. 45, demonstrating separation of these hyperons at 15 GeV/c momentum.

An alternative to changing the radius of the diaphragm consists in changing the gas pressure and leaving the optical system in place.

While the DISC counters can only be used for particles parallel to the optical axis of the detector, a velocity measurement for diverging particles from an interaction region requires a different approach. Seguinot and Ypsilantis [SE 77] have proposed the idea of a Cherenkov ring imag-

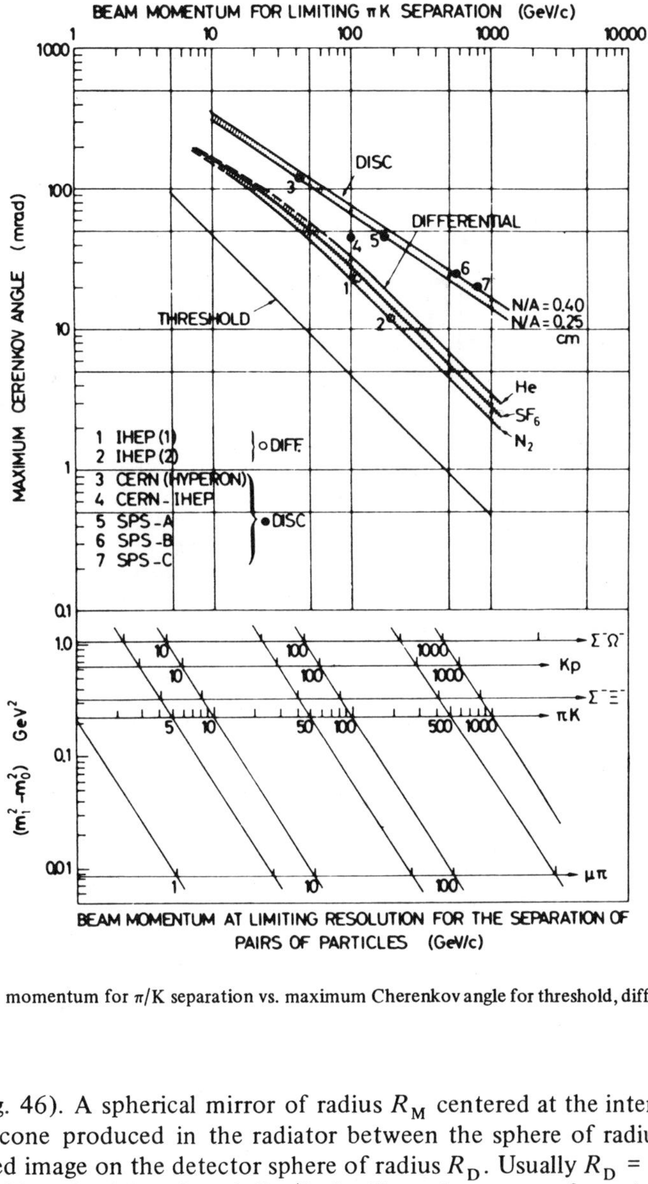

Fig. 44. Highest beam momentum for π/K separation vs. maximum Cherenkov angle for threshold, differential and DISC Cherenkov counters [LI 73].

ing detector (fig. 46). A spherical mirror of radius R_M centered at the interaction point focusses the Cherenkov cone produced in the radiator between the sphere of radius R_D and the mirror into a ring-shaped image on the detector sphere of radius R_D. Usually $R_D = R_M/2$.

Since the focal length of the mirror is $R_M/2$, the Cherenkov cones of opening angle $\theta_c = \arccos(1/(\beta n))$ emitted along the particle's path in the radiator are focussed onto a ring with radius r on the detector sphere. For $R_D = R_M/2$, the opening angle θ_D of this ring equals θ_c, in first approximation. The radius r of the ring image gives the Cherenkov angle via $\tan\theta_c = 2r/R$, and from this we

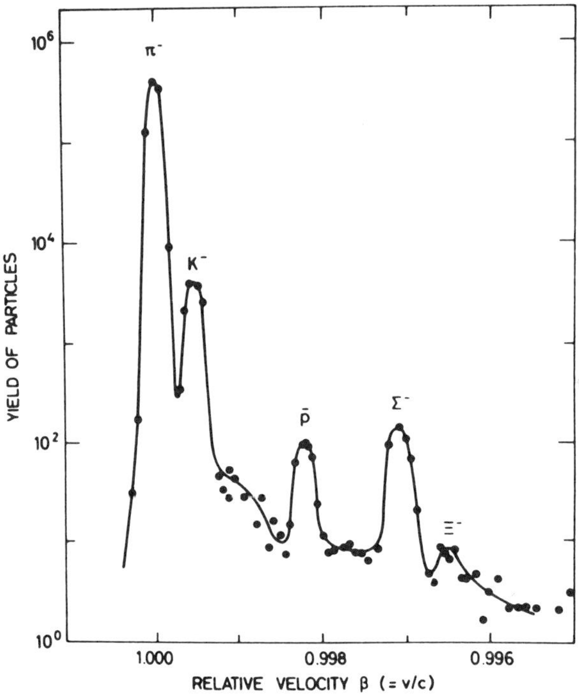

Fig. 45. Velocity distribution in a short hyperon beam selecting 15 GeV/c particles [LI 73].

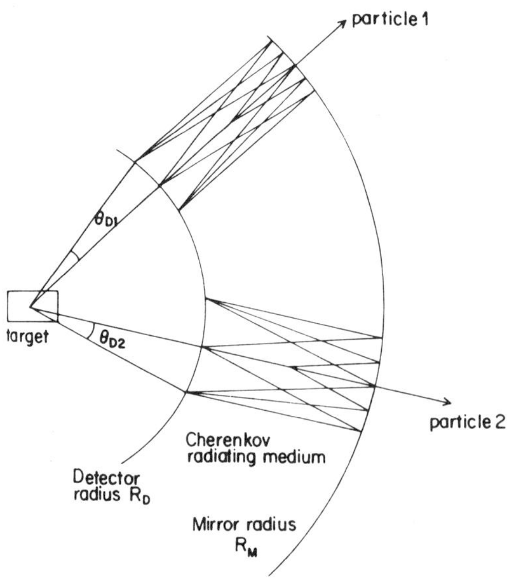

Fig. 46. Principle of ring imaging Cherenkov counter [SE 77].

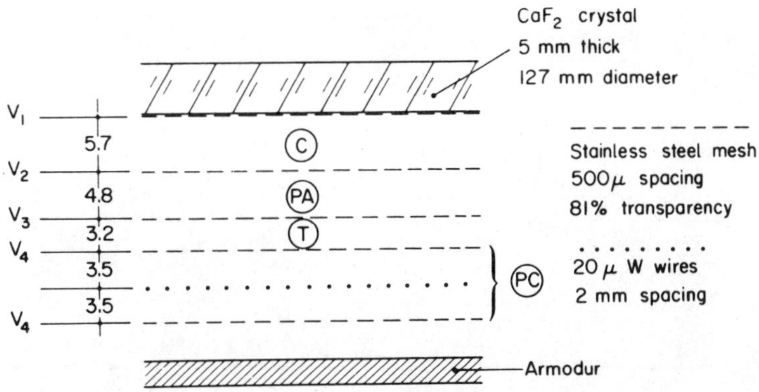

Fig. 47. Photon detector for ring imaging Cherenkov detector with CaF_2 window, three parallel-plane gaps: C for conversion, PA for amplification, T for transfer, and MWPC. Dimensions in mm [EK 81].

obtain the velocity $\beta = 1/(n \cos\theta_c)$. The relative error on β is $\Delta\beta/\beta = (\tan^2\theta_c\,(\Delta\theta_c)^2 + (\Delta n/n)^2)^{1/2}$. Neglecting the error from the uncertainty in n, one obtains $\Delta\gamma/\gamma = \gamma^2\beta^3 n \sin\theta_c\,\Delta\theta_c$, and the momentum of the particle $p = m\beta\gamma$ with error $\Delta p/p = \Delta\gamma/(\gamma\beta^2)$ [YP 81]. The critical point of the detector is the development of photoionization detectors. At present [EK 81], proportional chambers with an admixture of photosensitive triethylamine (TEA) are under study. Such a photon detector is shown in fig. 47. Behind the CaF_2 there are three gaps and a proportional chamber (PC). In gap C photons are converted by TEA to photoelectrons, gap PA serves for preamplification, gap T for transfer and PC for avalanche multiplication. Three photoelectrons per incident 10 GeV/c pion traversing 1 m of argon Cherenkov radiator at 1.2 atm. pressure have been observed. Development of such detectors is continuing. Amongst the possible improvements is the study of a new photosensitive gas [NA 72], Tetrakis-dimethylaminoethylene (TMAE), having a photo-ionization potential of 5.4 eV, lower than the one of TEA, 7.5 eV. Using this vapour, evidence for Cherenkov ring imaging has been obtained as shown in fig. 48, where ten events have been overlapped in the picture [SA 81].

4.3. Transition radiation detectors

If a charged particle traverses a medium with varying dielectric constant e.g. a periodic series of foils and air gaps, radiation is emitted from the interfaces between the two materials. This "transition radiation" (TR) was shown theoretically by Ginzburg and Frank [GI 46] to depend on the γ factor of the moving particle, thus permitting an identification of particles in the very high energy region ($\gamma > 1000$) where other methods fail.

The intensity of this radiation is expected theoretically to have a sharp forward peak at an angle $\theta \sim 1/\gamma$ and to be proportional to γ. If a periodic sandwich of many foils is used, interference effects [AR 75, FA 75] will produce a threshold effect in γ, such that the detector can be used for discriminating between particles of different mass.

Fig. 48. Evidence for Cherenkov ring image using multistep spark chamber with TMAE photosensitive gas. Ten events are overlapped in the picture; central spot is due to beam [SA 81].

Practical applications have followed the demonstration by Garibian [GA 73] that TR is emitted also in the X-ray region. Actual TR counters consist of a radiator followed by a proportional chamber for the detection of the X-rays emitted forward. Since the absorption of X-rays in the radiator material behaves as $Z^{3.5}$, the atomic number of the foils has to be as low as possible. In the pioneering work of Willis, Fabjan and co-workers [CO 77] the technology of thin lithium ($Z = 3$) foils has been mastered. As a counting gas for the X-ray detector, xenon ($Z = 54$) has been used.

The pulse height spectrum in a xenon chamber behind 1000 Li foils of 51 μ thickness is shown in fig. 49 together with a spectrum from a dummy radiator not producing TR. The pulseheight from TR can be clearly separated from the one from ionization loss only.

The increase of total radiated TR energy with γ is mainly due to an increase in the average X-ray energy, as shown by the measurements in different Li/Xe-detectors (fig. 50) using electrons with $\gamma \sim 2000$–6000. From these experiments we can conclude that i) TR detectors at the moment can be used for $\gamma > 1000$, i.e. for electrons above 0.5 GeV/c and pions above 140 GeV/c. ii) The extension of this method below $\gamma = 1000$ requires the detection of 1–5 keV X-rays.

Recently, Ludlam et al. [LU 81] have shown that an improvement in the separation of particles can be obtained by not only measuring the total energy deposited by TR quanta but counting ionization clusters along the track. The number of such clusters from an ionization particle track obeys a Poisson distribution, while the upper end of an energy loss curve has a very long tail. If therefore a charged particle below transition threshold has to be separated from a particle with TR, the region of overlap becomes much smaller for the cluster counting method. Fig. 51 shows the principle of method [LU 81], and fig. 52 the distributions in cluster number N and in deposited charge Q for 15 GeV pions and electrons [FA 81]. Also shown is the pion rejection vs. electron efficiency for a particular cluster threshold energy of 4 keV, obtained with a detector of 12 sets of 35 μ lithium foils, each one followed by a xenon proportional chamber. The detector has a total length of 66 cm and a thickness of 0.04 radiation lengths. For a 90% electron efficiency, a

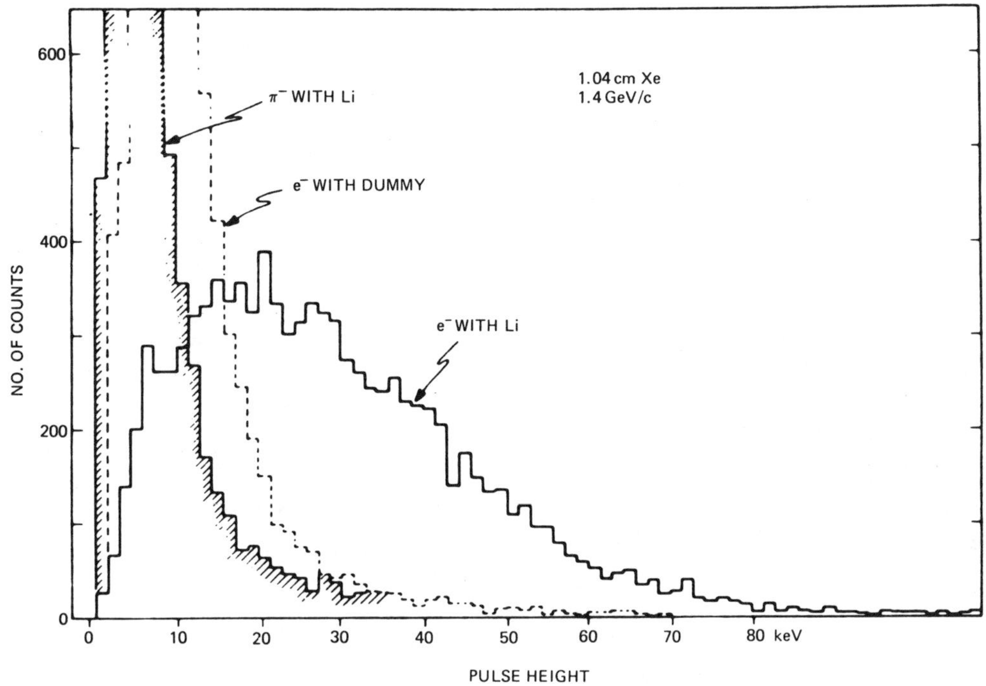

Fig. 49. Pulse height spectrum of transition radiation in Li foils detected by a xenon proportional chamber [FA 80].

pion rejection of 8×10^{-4} is obtained. The corresponding figure for a radiator made of pure carbon fibres of 7 μm thickness is 2×10^{-3}. With a similar detector of 132 cm length, a kaon rejection of 10^{-2} has been achieved for a 90% pion detection efficiency, as shown in the points labelled Exp.A in fig. 53. Also shown are measurements of Commichau et al. [CO 80] using only charge measurements (Exp.B). This detector nearly reaches the rejection obtained with the cluster method in Exp.A.

4.4. Multiple ionization measurement

Between the region ($\gamma > 1000$), where transition radiation can be utilized, and the medium and low energy domain, $\gamma < 100$, where Cherenkov counters and time-of-flight measurement are practical, there is a region of γ between 100 and 1000 where neither of these methods is applicable. Here a new kind of detector is provided by the exploitation of the relativistic rise of the ionization energy loss in this domain (see fig. 1). In gases, this energy loss rises by a factor of 1.5, and very precise measurement is required. Because of the Landau tail from knock-on electrons, the accuracy in the determination of the mean energy loss (or alternatively the most probable energy loss) does not improve considerably by increasing the thickness of the detector. However, the

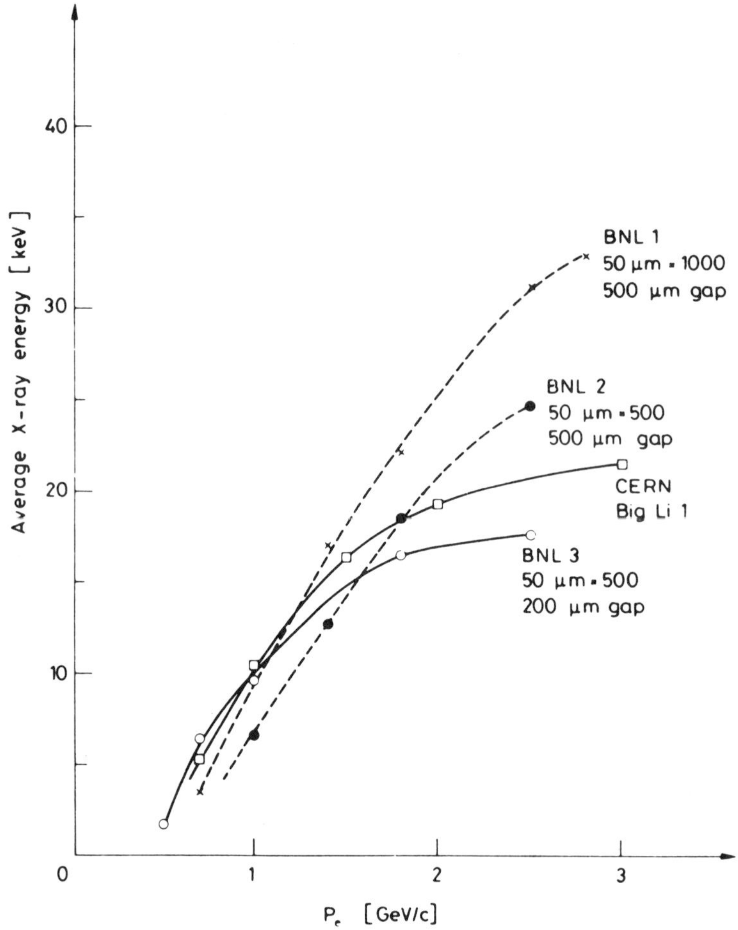

Fig. 50. Transition radiation measured in a xenon/CO_2-filled (80/20) proportional chamber (1.04 thick) behind radiators of different geometries traversed by electrons of momentum p_e [CO 77].

resolution increases if the energy loss is measured in many consecutive thin detectors and if the large pulse-heights from knock-on electrons occuring in some of the detectors are removed. This is done by taking the mean of the lowest 40–60% of ionization values. This sampling method with truncation reduces fluctuations in the mean and permits a measurement of energy loss precise enough in order to distinguish particles if their momentum is known. As can be seen from fig. 54, the ratio of most probable energy losses of pions and kaons at 100 GeV/c is 1.05, such that π/K separation at this energy requires a r.m.s. resolution of about 2%. Such a resolution can

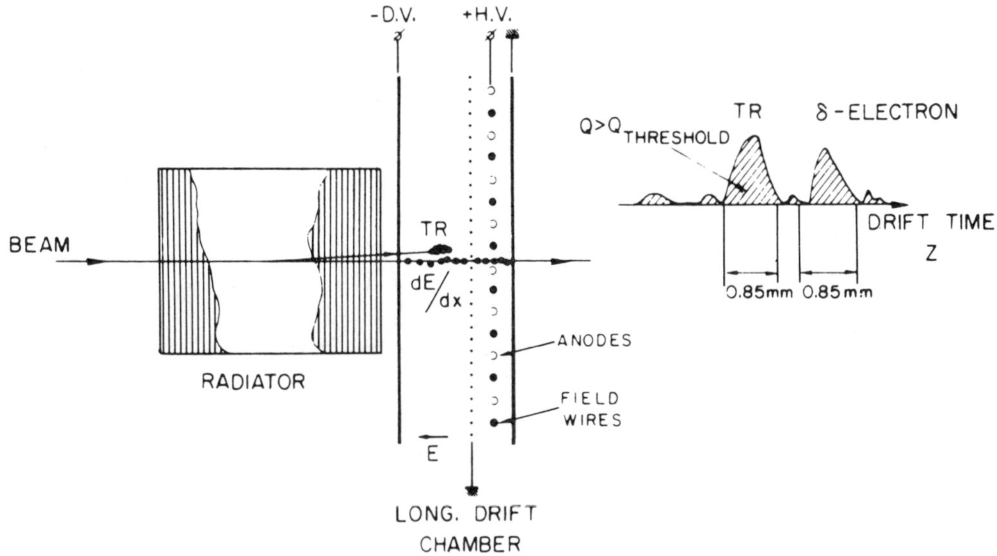

Fig. 51. Principle of detection of transition radiation by counting ionization clusters along the track. TR: transition radiation, D.V.: drift voltage [LU 81].

be achieved by using several hundred detectors with a total thickness of a few meters of gas. For 128 chambers, by measuring the average of the 40% smallest pulse-heights, a r.m.s. resolution of $\sigma = 2.5\%$ has been obtained for 50 GeV pions and protons [LE 78a].

The dependence of this resolution on pressure, detector length and number of samplings has been studied [AD 74]. The simplest statistical scaling law for the relative error on the energy loss measurement for 3 cm argon sampling is $\sigma_E \sim 5.6\,(L\,p)^{-1/2}\,\%$ with L being the detector length in meters and p the gas pressure in atmospheres. If one includes the sampling thickness, a graphical form of the relation is obtained (fig. 55). However, detailed measurements show that the increase in gas pressure does not improve the resolution as expected [LE 81a], (fig. 56). In addition, the density effect for the ionization leads to a considerable reduction of the relativistic rise (fig. 57). These two effects conspire to nearly delete the advantages of high pressure. In fig. 58 the ratio of distance D between the truncated mean energy losses of two particles and the resolution σ_E for π/p and e/π pairs is shown as a function of pressure. There is only a marginal gain in going from 1 to 2 atmospheres.

This result is at variance with measurements done in a JADE type test setup [WA 82], where 1 cm samples of 4 atm $Ar{-}CH_4$ give the resolution expected from fig. 55 for 4 cm samples at 1 atm.

However, on the basis of their results, Lehraus et al. [LE 81b] propose as a rule of thumb that an experimental detector will have a resolution equal to the one shown in fig. 55 but reading the graph for half of the actual sampling size.

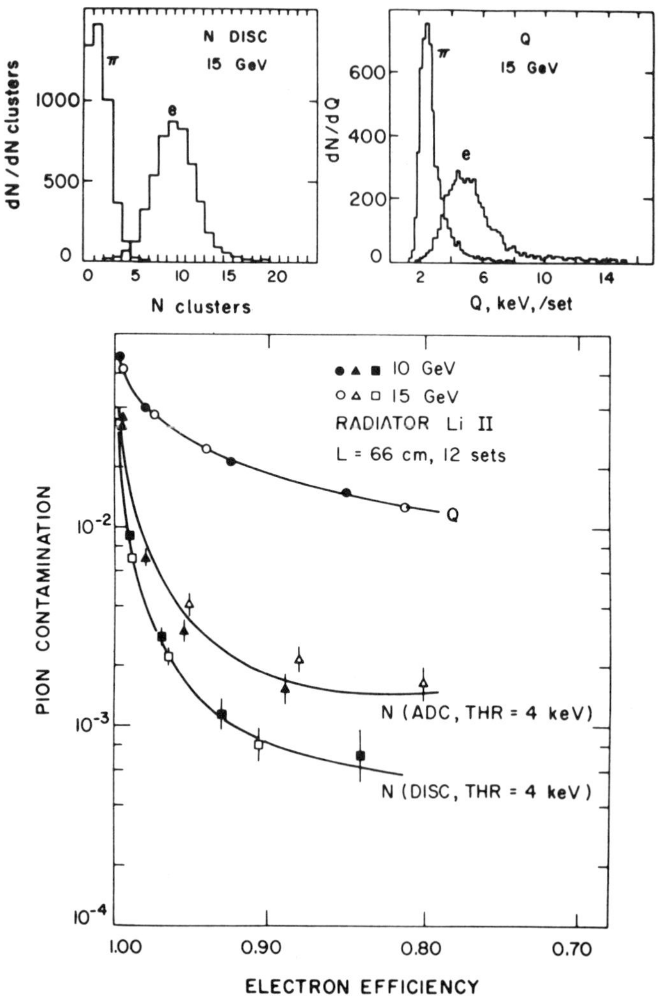

Fig. 52. Upper part: distribution in cluster number N and in deposited charge Q for pions and electrons. Lower part: π/e separation by threshold in Q or in N [FA 81].

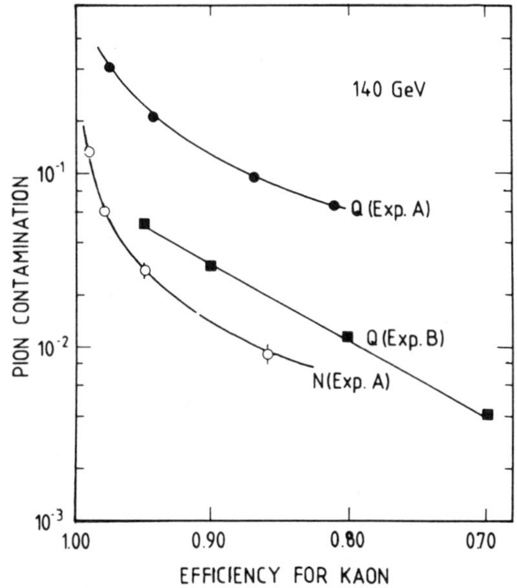

Fig. 53. π/K separation for a given efficiency of K meson detection at 140 GeV/c beam momentum. Exp.A [FA 81] has 24 radiators of carbon fibres followed by Xe chambers, with total length 132 cm, Exp.B [CO 80] has 20 sets of 5 μm mylar foil stacks and chambers, total length 147 cm. Q: charge discrimination, N: cluster counting.

4.5. Comparison of identification methods

The identification methods discussed above are usable in certain momentum domains: the time-of-flight measurement at low momenta, then threshold Cherenkov counters, DISC-Cherenkovs, multiple ionization measurement and, at ultrahigh momenta, transition radiation. The length required for π/K separation in these detectors is shown in fig. 59. Using a typical detector length of a fixed target experiment of 30 meters and a length of 3 meters for storage ring experiments, typical momentum ranges for π/K separation are calculated, as shown in table 7. It appears that the multiple ionization measurement is necessary for bridging the gap between threshold Cherenkov and transition radiation counters.

5. Energy measurement

5.1. Electron-photon shower counters

At energies well above 1 MeV, the ionization loss of fast ($\beta \simeq 1$) electrons is given by

$$-\left(\frac{\mathrm{d}E}{\mathrm{d}x}\right)_{\mathrm{ion}} = 4\pi N_0 \frac{Z}{A} r_e^2\, mc^2 \left[\ln(2mv^2\gamma^2/I) - 1\right] ,$$

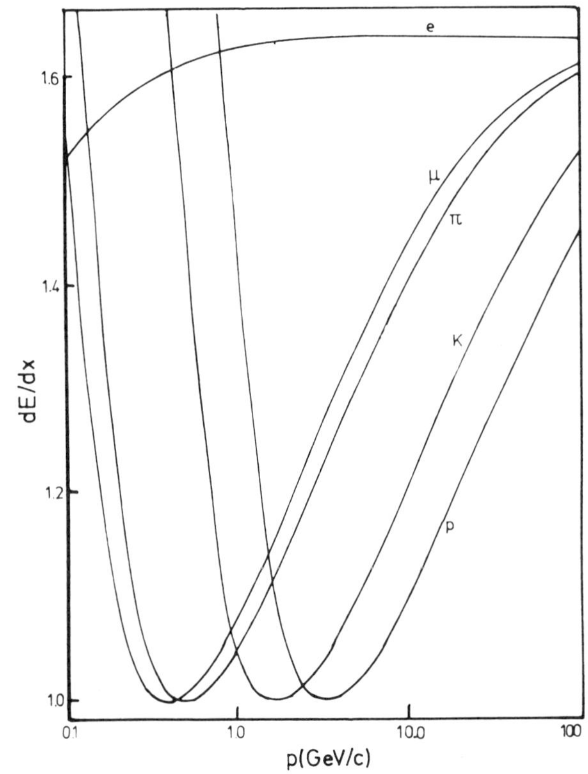

Fig. 54. Most probable energy loss in one cm of argon (80%)–methane (20%) mixture at STP, for electrons, muons, π and K mesons and protons [MA 78].

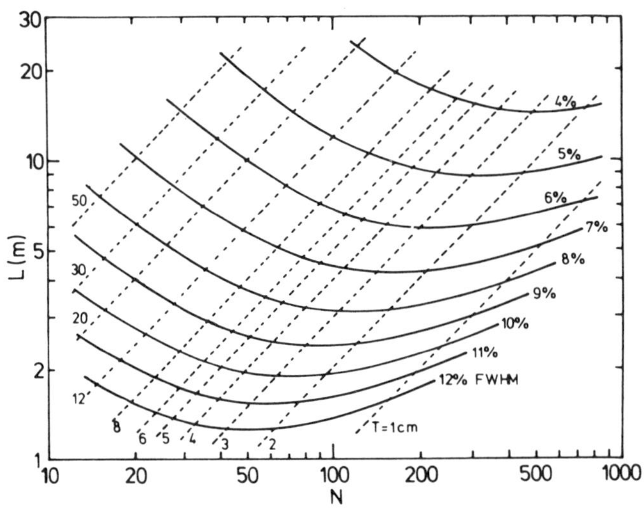

Fig. 55. Resolution expected in energy loss measurement as a function of number of samplings N and detector length L(m). $T = L/N$ is the thickness of one sampling detector [AD 74].

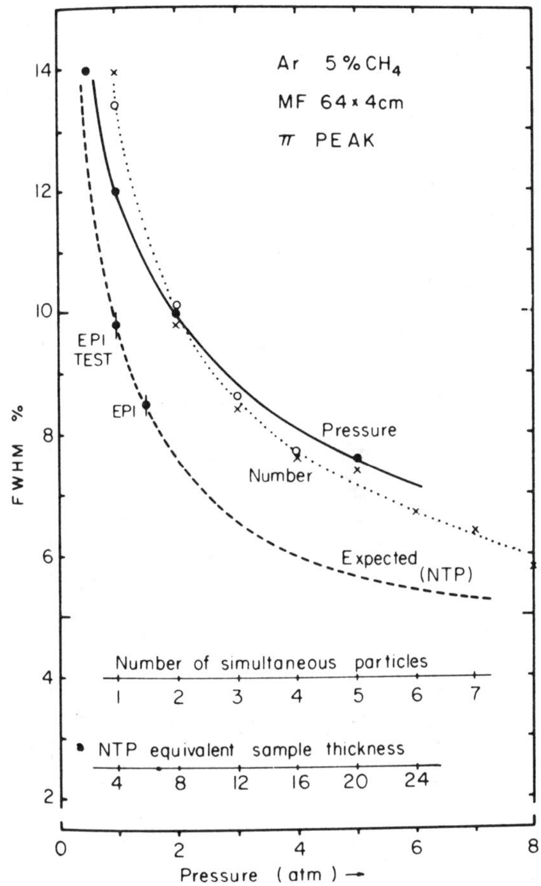

Fig. 56. d*E*/d*x* resolution obtained [LE 81a] by varying pressure ("Pressure"), by varying number of simultaneous particles ("Number") compared to expectation from EPI test [AD 74] at NTP without drift. Truncated mean of 64 samples 4 cm thick.

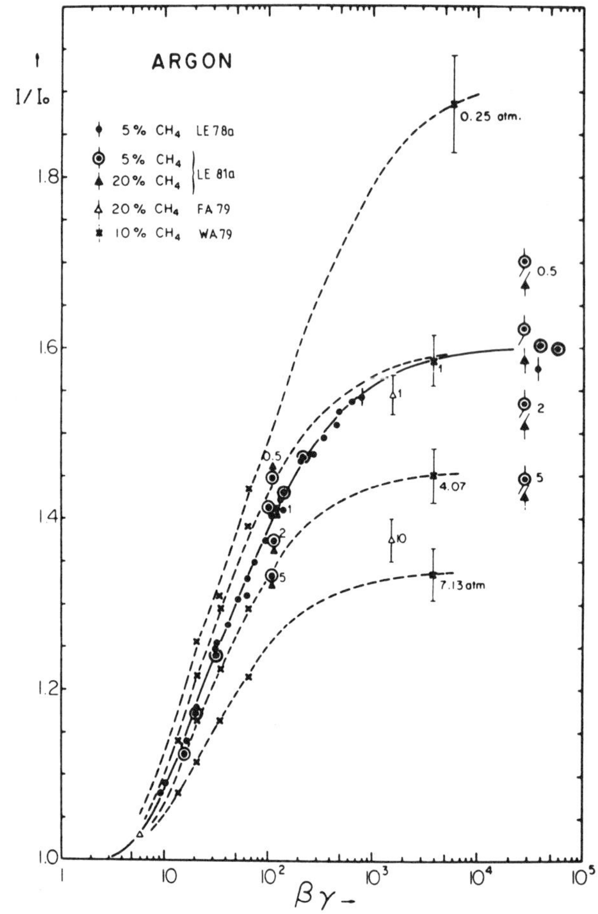

Fig. 57. Relativistic rise of ionization in argon–CH_4 mixtures. Numbers indicate pressure in atmospheres.

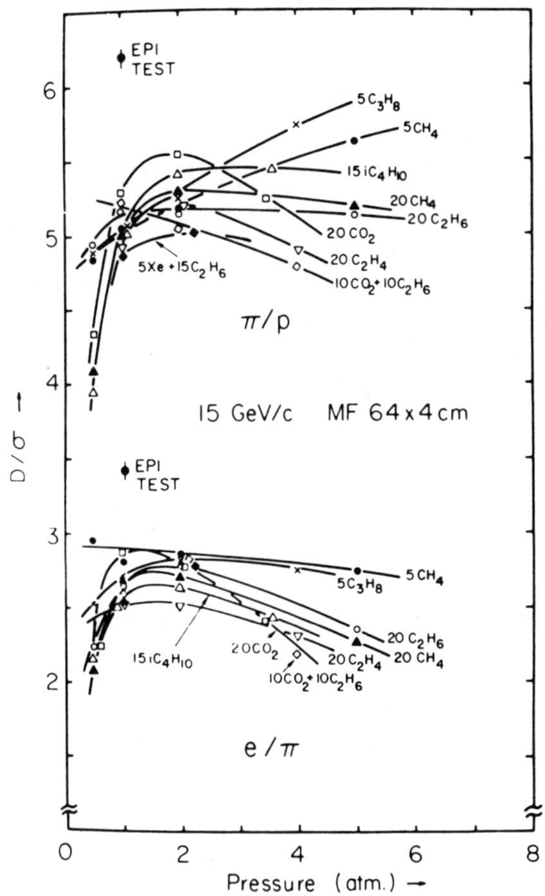

Fig. 58. Resolving power D/σ (see text) for separation of π/p or e/π vs. gas pressure for 15 GeV/c particles using dE/dx measurement in 64 × 4 cm samples [LE 81a].

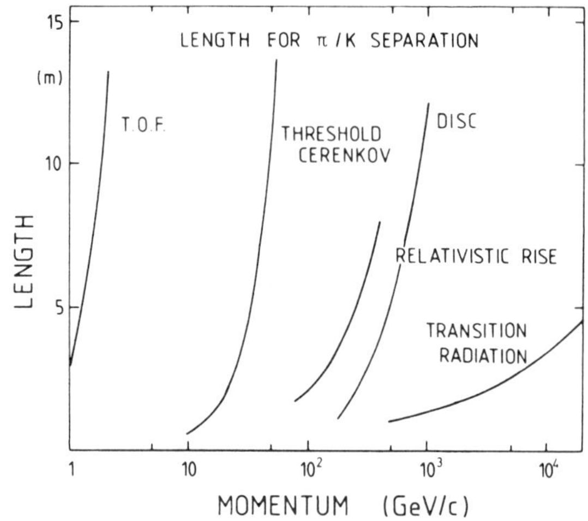

Fig. 59. Length of detectors needed for π/K separation by different identification methods vs. momentum.

with $r_e^2 = (e^2/mc^2)^2 \simeq (2.8\ \mathrm{fm})^2$, while the competing loss by bremsstrahlung takes the form

$$-\left(\frac{\mathrm{d}E}{\mathrm{d}x}\right)_{\mathrm{brem}} = 4\alpha\frac{N_0}{A}Z^2 r_e^2 E \ln\frac{183}{Z^{1/3}} =: \frac{E}{X_0},$$

where the "radiation length" X_0 is defined this way, and m is the electron mass.

While ionization dominates at low energies, bremsstrahlung takes over at high energies, and the ratio R of bremsstrahlung loss and ionization loss comes out to be $R \simeq ZE/550$, where E is measured in MeV. The energy at which this ratio becomes unity, the "critical energy" E_c therefore has the approximate value $E_c \sim 550/Z$ MeV which for lead is $E_c = 6.7$ MeV.

Table 7
Identification methods

Method	Domain for π/K separation		Requirements
	Fixed target geometry $L = 30$ m	Storage ring geometry $L = 3$ m	
Time-of-flight	$p < 4$ GeV/c	$p < 1$ GeV/c	$\sigma_t = 300$ ps
Threshold Cherenkov	$p < 80$ GeV/c	$p < 25$ GeV/c	10 photoelectrons
DISC-Cherenkov	$p < 2000$ GeV/c	–	achromatic gas counter
Ring imaging Cherenkov		$p < 65$ GeV/c	
Multiple ionization	$1.2 < p < 100$ GeV/c	$1.5 < p < 45$ GeV/c	$\sigma_E = 2.5\%$
Transition radiation	$\gamma > 1000$	$\gamma > 1000$	detection of >10 keV X-rays

The interaction of photons at high energy is also governed by the radiation length: the cross-section for pair creation

$$\sigma_{\text{pair}} = 4\alpha Z^2 r_e^2 [\tfrac{7}{9} \ln(183/Z^{1/3}) - \tfrac{1}{54}] ,$$

gives a probability P for pair creation in one radiation length

$$P = \sigma_{\text{pair}} \cdot \frac{N\rho}{A} \cdot \frac{X_0}{\rho} = \frac{7}{9} .$$

Table 8 gives radiation length and critical energy for some materials [PA 78].

The interaction of a high-energy photon or electron therefore leads to a cascade of electrons and photons; starting with a photon of energy E_0, after $1X_0$ we have 2 particles of average energy $E_0/2$, after nX_0 there are 2^n particles with mean energy $E_0/2^n$. The cascade stops approximately when the particles approach the critical energy, i.e. if $E_0/2^n = E_c$.

The number of generations up to the maximum therefore is $n = \ln(E_0/E_c)/\ln 2$, and the number of particles at the maximum $N_p = 2^n = E_0/E_c$. The total integral path length S of all electrons or positrons in the shower is approximately

$$S = \tfrac{2}{3} X_0 \sum_{\nu=1}^{n} 2^\nu + s_0 \tfrac{2}{3} N_p = (\tfrac{4}{3} X_0 + \tfrac{2}{3} s_0) E_0/E_c ,$$

where s_0 is the path length of electrons below the critical energy.

The path length S is proportional to the total energy E_0 if electrons and positrons can be detected until they come to rest. In practical detectors there is a minimum kinetic energy required for detection (cut-off energy E_k). This effect has the consequence [RO 52] that the visible path length becomes [AM 81]

$$S = F(z)\, X_0 E_0/E_c$$

with $F(z) \approx e^z (1 + z \ln(z/1.526))$ and $z = 4.58\, ZE_k/(AE_c)$.

Table 8
Radiation length and critical energy

Material	X_0 [g/cm²]	E_c [MeV]
H_2	63	340
Al	24	47
Ar	20	35
Fe	13.8	24
Pb	6.3	6.9
Leadglass SF 5	9.6	~11.8
Plexiglass	40.5	80
H_2O	36	93
NaJ (Tl)	9.5	12.5
$Bi_4Ge_3O_{12}$	8.0	~ 7

Including the effect of the cut-off energy into Monte Carlo [CR 62, NA 65, LO 75] calculations gives the following properties of electron–photon showers:

i) the number of particles at maximum N_p is proportional to the primary energy E_0,

ii) the total track length of electrons and positrons S is proportional to E_0,

iii) the depth at which the maximum occurs X_{max} increases logarithmically: $X_{max}/X_0 = \ln(E_0/E_c) - t$, where $t = 1.1$ for electrons and $t = 0.3$ for photons.

The longitudinal energy deposition in an electromagnetic shower can be seen in fig. 60 as measured [BA 70] for 6 GeV electrons. A useful parametrisation for this distribution is given by [LO 75]

$$(\mathrm{d}E/\mathrm{d}t) = E_0 A\, t^{\alpha}\, \mathrm{e}^{-\beta t} ,$$

where $t = X/X_0$ is the longitudinal depth X in units of X_0, and the parameters $\beta \approx 0.5$, $\alpha \approx \beta \cdot t_{max}$ and $A = \beta^{\alpha+1}/\Gamma(\alpha+1)$ vary logarithmically with energy. For proton energies around 1 GeV, the distribution can be approximated by $(\mathrm{d}E/\mathrm{d}t) = E_0\ 0.06\ t^2\ \mathrm{e}^{-t/2}$ for a lead converter.

The transverse dimension of a shower is determined by the multiple scattering of low energy electrons. It turns out that a useful unit for transverse shower distributions is the Molière unit $R_M = 21\ \mathrm{MeV} \cdot X_0/E_c$. As shown by the measurements [BA 70] in fig. 61, the distribution of shower energy in transverse (radial) bins scaled in R_M is independent of the material used, and 99% of the energy are inside the radius of 3 R_M.

The *energy resolution* of an idealized homogeneous detector of infinite dimensions is limited only by statistical fluctuations. For a cut-off energy of 0.5 MeV and a critical energy of 11.8 MeV a total track length of 176 cm/GeV and a resolution $\sigma(E)/E = 0.7\%/\sqrt{E(\mathrm{GeV})}$ have been computed [LO 75].

If the shower is not contained in the detector, the fluctuation of the energy leaking out gives a contribution to the resolution. As shown in [DI 80], longitudinal losses induce a larger degradation of the resolution than lateral ones. An estimate for this fluctuation due to longitudinal leakage is $\sigma(E) = (\mathrm{d}E/\mathrm{d}t)_{t_r} \cdot \sigma(t_{max})$, where t_r is the length of the detector and $\sigma(t_{max})$ the fluctuation of the position of the shower maximum. For photons of 1 GeV energy, $\sigma(t_{max}) \sim 1$ and $(\sigma(E)/E)_{leak} = 0.06\ t_r^2 \exp(-t_r/2)$. If the number of photoelectrons N_p detected per incident energy E_0 is limited, the fluctuation of this number gives an additional contribution to the resolution: $\sigma(E)/E = (N_p \cdot E_0)^{-0.5}$.

To these two sources of fluctuations, valid for homogeneous calorimeters, we have to add the sampling fluctuations if the shower calorimeter consists of a series of inactive absorber layers of thickness d interspersed with active detector layers ("sampling calorimeter"). If the detectors count only the number of particle traversals, N, the statistical fluctuation in N determines the contribution to the energy resolution. Since N depends on the total track length, $N = S/d = E_0 X_0 F(z)/(E_c d)$, we obtain [AM 81]

$$(\sigma(E)/E)_{sampl} = 1/\sqrt{N} = 3.2\% \sqrt{(550/ZF(z))}\ \sqrt{\frac{d/X_0}{E_0\ (\mathrm{GeV})}} .$$

In high Z materials, the lateral dimension of the showers is much larger than in those with low Z,

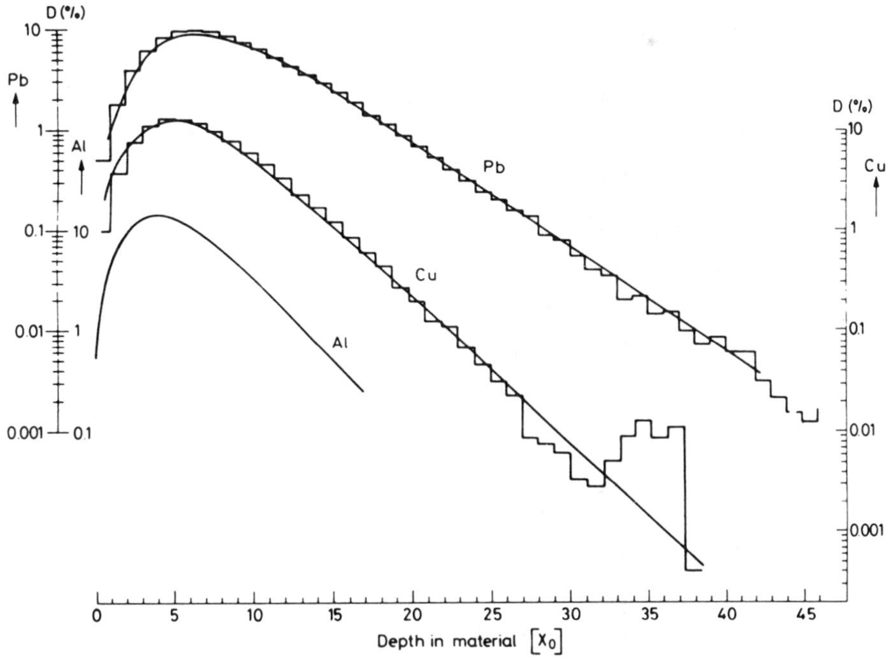

Fig. 60. Longitudinal distribution of energy deposition in a 6 GeV electron shower; measurements (line) and Monte Carlo calculation (histogram) [BA 70].

since the Molière unit in units of X_0, $R_M/X_0 = 21$ MeV$/E_c$, is larger for heavy materials. Consequently also the angle θ of electrons and positrons relative to the shower axis is larger [AM 81]. Those shower particles see a larger sampling thickness $d/\cos\theta$, and therefore a smaller number of traversals occurs, reducing further the energy resolution by a factor $\langle\cos\theta\rangle^{-1/2}$. A Monte Carlo calculation [FI 78] shows that the average $\langle\cos\theta\rangle = \cos(21\ \text{MeV}/(E_c \cdot \pi)) = 0.57$ for lead. From this calculation, the sampling fluctuation gives $\sigma(E)/E = 4.6\%/\sqrt{E(\text{GeV})}$ for 1 mm lead sampling thickness and $E_k = 0$.

Another large source of fluctuations enters if the sensitive layers of the calorimeter consist of a gas or a very thin layer of liquid argon ($\lesssim 2$ mm), used as proportional counters. Then low energy electrons moving at large angles relative to the shower axis induce large pulseheight fluctuations ("path length fluctuations"), and the Landau tail of the energy loss distribution also leads to a reduction of resolution. The computed effect of these two contributions on the energy resolution of a lead–argon calorimeter can be seen in fig. 62. The overall resolution is $18\%/\sqrt{E(\text{GeV})}$, more than twice the sampling fluctuation of $7\%/\sqrt{E(\text{GeV})}$.

Homogeneous shower counters

The best resolutions are obtained with anorganic scintillating crystals. NaJ(Tl) detectors with a diameter of $3R_M = 13$ cm and $15X_0 = 40$ cm length have yielded [PA 80] a resolution of

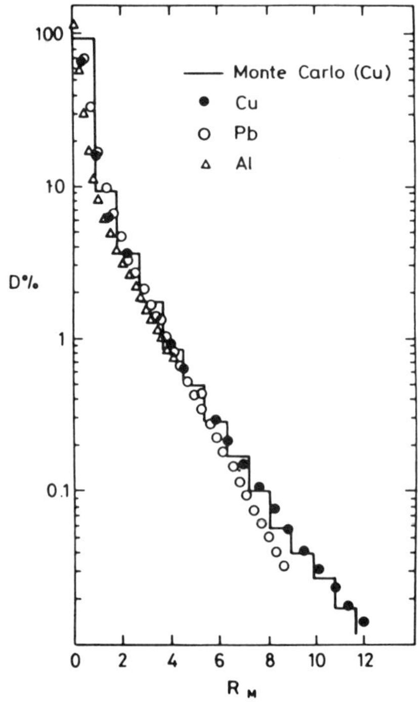

Fig. 61. Transverse distribution of energy deposition in a 6 GeV electron shower; data: points; Monte Carlo: histrogram; R_M = 21 MeV • X_0/E_c is the Molière unit [BA 70].

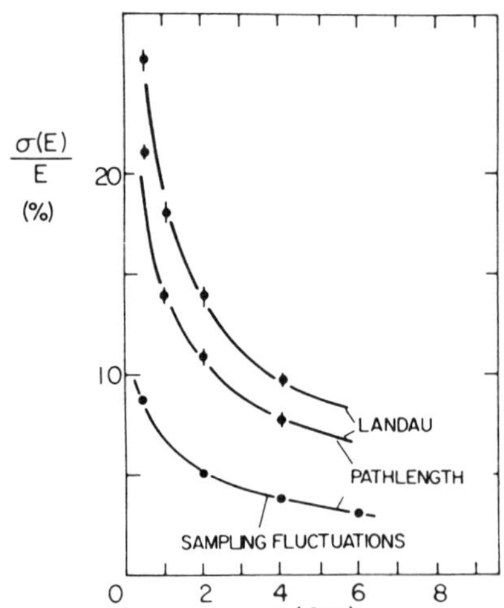

Fig. 62. Contributions of sampling, path length and Landau fluctuations to the energy resolution of a lead–gas quantameter [Fi 78].

$\sigma(E)/E = 2.8\%(E(\text{GeV}))^{-0.25}$ in a large scale application. For one $24X_0$ long counter $\sigma(E)/E = 0.9\%(E(\text{GeV}))^{-0.25}$ has been achieved [HU 72]. The new type of crystal, $BGO(B_4Ge_3O_{12})$ gives 8% of the light output of NaJ, and a resolution of $\sigma(E)/E = 2.5\%/\sqrt{E(\text{GeV})}$ [KO 81].

Lead glass counters detect the Cherenkov light of shower electrons, the resolution is limited by photoelectron statistics. A computation [PR 80], based on 1000 photoelectrons per GeV, gives $\sigma(E)/E = 0.006 + 0.03\ (\xi E)^{-0.5}$, ξ being the ratio of photocathode area and counter exit area. Actual measurements [BI 81] with 208 blocks of $36 \times 36 \times 420\ \text{mm}^3$ give a resolution of $\sigma(E)/E = 0.012 + 0.053/\sqrt{E(\text{GeV})}$ for $\xi = 0.35$, in agreement with the calculation.

Sampling shower detectors

The resolution of a lead-scintillator sandwich with 1 mm lead and 5 mm scintillator thickness for a total length of 12.5 radiation length is shown in fig. 63 versus incident energy [HO 79]. The values for $A = \sigma(E)/\sqrt{E}$ vary from 7% $\text{GeV}^{1/2}$ at 100 MeV to 9% $\text{GeV}^{1/2}$ at 5 GeV, in agreement with a calculated 5% $\text{GeV}^{1/2}$ from sampling fluctuations, 3–4% $\text{GeV}^{1/2}$ from photoelectron statistics, and 2–5% $\text{GeV}^{1/2}$ from leakage.

In lead–liquid argon calorimeters, the ionization is sampled in a proportional mode by the argon chambers defined by two lead plates as electrodes. Resolutions for 2 mm Pb plates and 3 mm liquid argon are $\sigma(E)/E = 12\%/\sqrt{E(\text{GeV})}$ [KA 81].

A summary on the energy resolution obtained with electron–photon shower counters is given in table 9.

Position resolution

The impact point of an electron or photon on an array of shower counters can be obtained by measuring the lateral distribution of energy in the shower. The precision of the position information increases with the number of cells hit by shower particles, and decreases with the cell size. In particular, the accuracy is best if the shower energy is shared equally between two adjacent cells. Binon et al. [BI 81] using cells of $36 \times 36 \times 420\ \text{mm}^3$, have obtained a position resolution

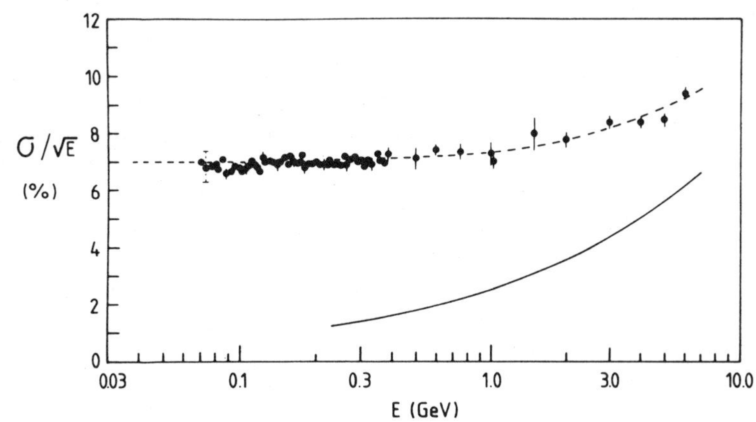

Fig. 63. The quantity $\sigma/\sqrt{E(\text{GeV})}$ vs. E for a lead scintillator sandwich with 1 mm lead plates. Full line is contribution of leakage [HO 79].

Table 9
Electromagnetic shower counters

Type	Sampling thickness (X_0)	Total thickness (X_0)	$\sigma(E)/\sqrt{E}$ %(GeV$^{1/2}$)	Spatial resol. Δx (mm)	Angul resol.	Lateral cell size (mm)	Group	Ref.
Na J	–	24	$0.9\,E^{1/4}$					[HU 72]
Na J	–	16	$2.8\,E^{1/4}$				Crystal ball	[PA 80, KI 79] [CH 78b]
Pb glass F8	–	17	$5.3 + 1.2\sqrt{E}$	1.3		36 × 36	IHEP	[BI 81]
Pb glass SF5	–	12.5	$\sqrt{6^2 + 2.5^2 E}$	6	10 mrad	80 × 104	JADE	[DR 80, BA 79]
Pb glass SF5	–	20	$\sqrt{6^2 + 0.5^2 E}$	2			NA 1	[NA 1]
Pb/scint	0.18	12.5	7–9	$11/\sqrt{E(\mathrm{GeV})}$		100 × 100	ARGUS	[HO 79]
Pb/scint.	0.21	13	9	$25/\sqrt{E(\mathrm{GeV})}$		200 × 250	LAPP–LAL	[SC 81]
Pb/LAR	0.36	13.5	10–12	5	5 mrad	{70 × 70 + strips 20 mm	TASSO	[KA 81]
Pb/LAR	0.26	21	10	4	4 mrad	23 × 23	CELLO	[BE 81]
Pb/LAR		14	11.5				Mark II	[DA 79]
Pb/PWC	0.5	12	16				Mark III	[HI 81]
Pb/prop. tube	1		28	13–18		pitch 7.7	NA 24	[BA 81]

of $\sigma_x = 1.3$ mm for 25 GeV electrons. For a lateral cell size $d > 30$ mm, an experimental increase of σ_x is calculated (fig. 64) by these authors. On the other hand, a variation of $\sigma_x \propto 1/\sqrt{E}$ has been found [AK 77], confirming the assumption that the spatial resolution depends mainly on the number of shower particles.

With lead-scintillator sandwiches of 10 × 10 cm^2 lateral dimensions [HO 79], the measured spatial resolution was $\sigma_x = 11\ \mathrm{mm}/\sqrt{E(\mathrm{GeV})}$.

5.2. Hadron calorimeters

The scale for the spatial development of a hadronic shower, the inelastic production of secondary hadrons, which again interact inelastically producing tertiary hadrons, and so on, is given by the nuclear absorption length λ. From the inelastic cross-section σ, $\lambda = A/(\sigma N_0 \rho)$ can be obtained. The experimental values of λ for materials usable for calorimetry are 77 g/cm^2(C), 135 g/cm^2(Fe), 210 g/cm^2(Pb) and 227 g/cm^2(U). Compared to the small values for the radiation length of high Z materials enabling the construction of correspondingly small shower counters, the size of hadronic showers is large; typical values for Fe calorimeters are 2 meters depth and 0.5 m transverse size. The need for such sizes is demonstrated by the measurements [HO 78b] on the longitudinal shower development shown in fig. 65, where the center of gravity, the length for 95% energy containment and the length, where the average particle number goes below one ("shower length") are displayed as a function of incident pion energy for a 5 cm Fe sampling calorimeter. A parametrization of L (95%) can be given: $L(95\%) = [9.4 \ln E(\mathrm{GeV}) + 39]$ cm Fe. In a similar way, fig. 66 gives lateral shower sizes for 95% energy containment.

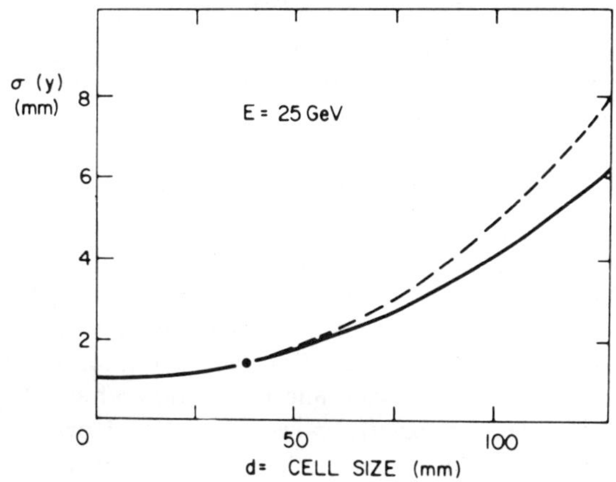

Fig. 64. Position r.m.s. resolution as a function of transverse block size d for an array of leadglass blocks. Full line: average over photon impact points across the block; dotted line: photon impinging on centre of block. Point: measured resolution for d = 36 mm [BI 81].

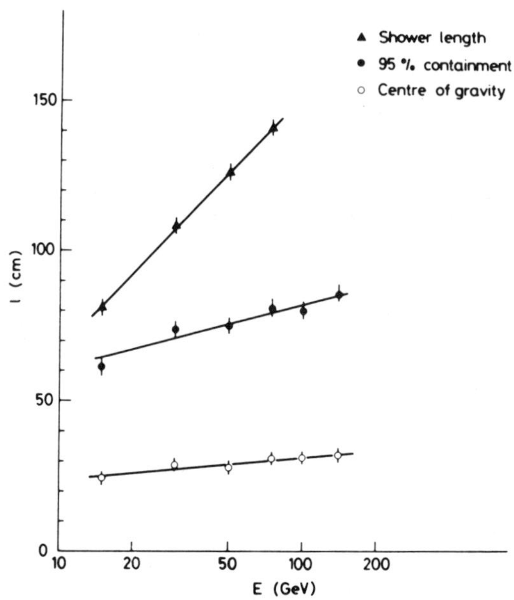

Fig. 65. Shower center of gravity in iron, length for 95% energy containment and length where average particle number goes below one as function of pion energy [HO 78b].

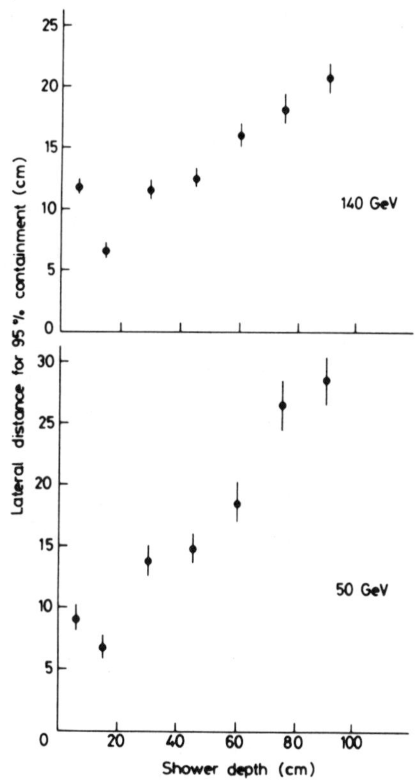

Fig. 66. Lateral dimension for 95% energy containment as a function of depth in iron [HO 78b].

Apart from the hadron shower particles leaking out longitudinally or laterally, the energy seen in a sampling calorimeter for hadrons is incomplete for several reasons:

i) there are particles escaping the calorimeter carrying away energy, like muons and neutrinos from pion decay (1% at 40 GeV),

ii) there is nuclear excitation and breakup resulting in low energy γ rays or heavy fragments, which do not reach the sensitive part of the sandwich (20–30% of total energy at 10 GeV).

This loss of visible energy, typically 30%, can be seen by comparing the light collected from electron- and hadron-induced showers in iron (fig. 67). Since in a hadronic shower the electromagnetic component can occasionally be dominant through energetic π^0 production, this loss induces a fluctuation in response which contributes significantly to the resolution.

On top of this fluctuation there is the sampling fluctuation which alone gives rise to a resolution about twice as large as in electromagnetic showers (see 4.1). However, the effects of the fluctuation in energy leakage and in the electromagnetic component of the hadronic shower are much larger here and lead to energy resolutions of about

$$\sigma(E)/E \sim (0.9 - 0.5)/\sqrt{E(\mathrm{GeV})}\ ,$$

if the thickness of material between the sampling devices ("sampling thickness") is below 5 cm of iron.

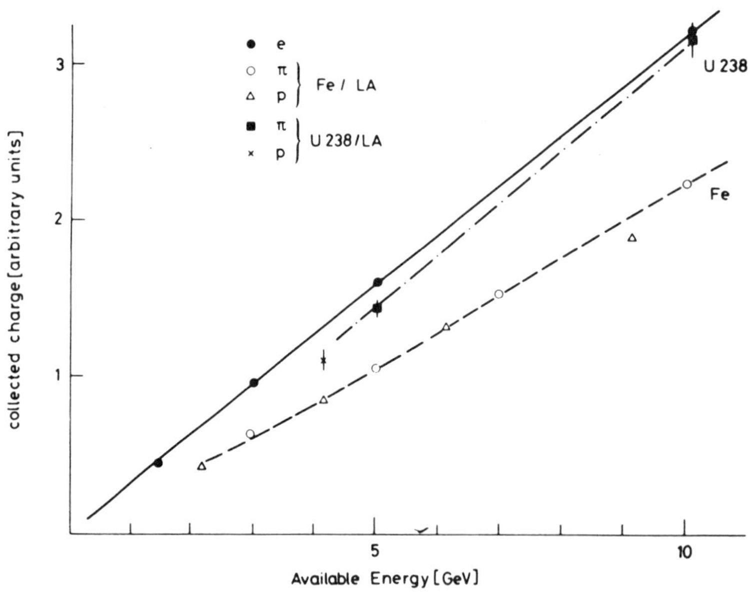

Fig. 67. Detected energy deposition in sampling calorimeters of Fe and U for electrons, pions and protons vs. particle energy [FA 77].

Two ways of improving this resolution have been invented and tried out successfully:

i) The loss of visible energy through the nuclear excitation and breakup mechanism can be nearly completely compensated by the energy release in nuclear fission of ^{238}U. Energetic photons from the fission contribute to the observed signal such that the pulseheight for hadron showers becomes nearly equal to the one for electromagnetic showers, as shown [FA 77] in fig. 67. The corresponding fluctuations disappear, and the energy resolution decreases by about a factor of two. Experimental results for U calorimeters are shown in fig. 68, they correspond to

$$\sigma(E)/E = 0.3/\sqrt{E(\mathrm{GeV})}\ ,$$

which is only 50% higher than the lower limit given by sampling fluctuations.

ii) Another method [DI 79, AB 81] reduces the fluctuation due to the electromagnetic component by weighting the response in individual counters. Electromagnetic parts of the shower are localized, therefore producing very large depositions in individual counters. If the measured response in one counter E_k is corrected downwards for large response, $E'_k = E_k(1 - CE_k)$, then the resulting resolution in the sum $\Sigma E'_k$ is markedly improved over the one in ΣE_k, as shown [AB 81] in fig. 69 for a 2.5 cm Fe sampling calorimeter exposed to 140 GeV/c pions. The resolution displayed in fig. 68 can be approximately described by

$$\sigma(E)/E = 0.58/\sqrt{E(\mathrm{GeV})}$$

between 10 and 140 GeV/c.

If the sampling thickness is larger, the sampling fluctuations increase and the resolution σ/E increases with d; fig. 70 gives some measurements.

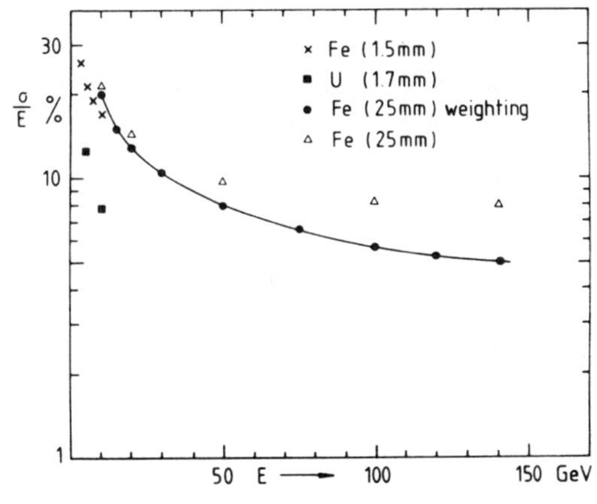

Fig. 68. r.m.s. energy resolutions obtained with hadronic sampling calorimeters; Fe (1.5 mm) and U (1.7 mm): [FA 77]; Fe (25 mm): [HO 78b] and [AB 81].

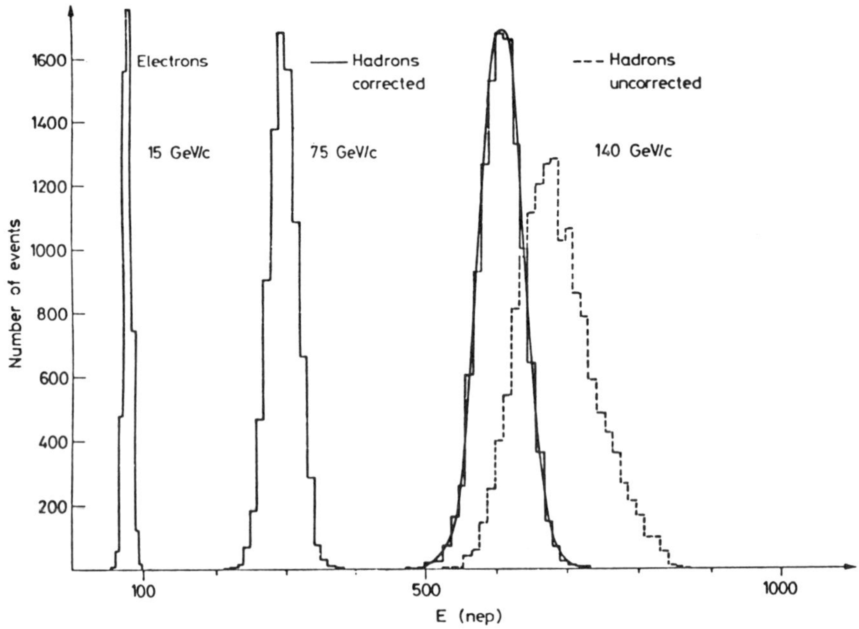

Fig. 69. Pulseheight spectra (in n.e.p.) for electrons and hadrons in a 2.5 cm sampling calorimeter [AB 81]

These data (obtained without the weighting procedure) can be parametrized using an empirical formula [AM 81], $\sigma(E)^2/E = 0.25 + (R')^2(4t/3)$, where t is the sampling thickness in units of X_0, $t = d/X_0$, and the parameter R' comes out to be 30–40%. It appears that a reduction of d below 2 cm of iron does not considerably improve the resolution any more, and that the limiting resolution for $d \to 0$ is around $0.5/\sqrt{E(\mathrm{GeV})}$.

The sampling of ionization in hadron calorimeters can be done by scintillators liquid argon ionization chambers, proportional chambers, or flash tubes. The choice between these detectors depends on the desired resolution, granularity and cost. For moderate-sized geometries, liquid argon and scintillators are used for best resolution. For very large fine grain calorimeters (νe scattering, proton decay), the proportional tubes or flash tubes give granularities down to 5 mm × 5 mm at a price which still allows the construction of multi-hundred ton calorimeters.

5.3. Monitoring of calorimeters

In a typical large-scale calorimeter there will be several thousand channels of analog pulseheight information which is converted to digits and registered. A severe problem with such a number of channels is their calibration and monitoring.

The calibration can be done by using suitable hadron beams and calibrating the response of the calorimeter, where for each sampling detector the pulseheight is measured in terms of minimum ionization deposited by high energy muons.

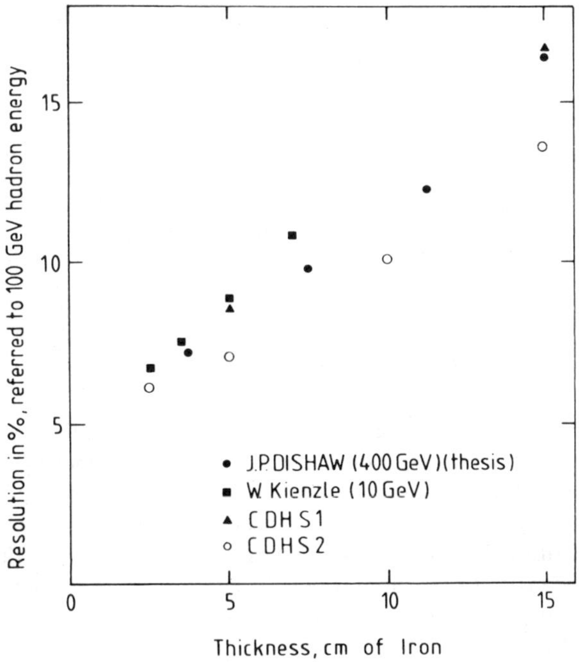

Fig. 70. r.m.s. hadronic energy resolution for iron calorimeters with different sampling thickness; J.P. Dishaw [DI 79]; CDHS 1 [HO 78b]; CDHS 2 [AB 81].

If there are not as many muons in each sampling detector as are needed for day-to-day monitoring, another source of calibrated pulseheights is needed. For liquid argon calorimeters, such a source is obtained by depositing a known amount of charge into the ion chamber. The same can be done for proportional chambers.

For scintillation counters, a novel kind of monitoring systems has been constructed recently [GR 80, EI 80]. The light source here is a pulsed nitrogen laser emitting at 337 nm. There are different schemes of distributing the light onto a few thousand counters.

In one of the systems [GR 80], built for the UA1 experiment, the laser light is injected into a rectangular box covered inside with a highly reflecting material (millipore). After many reflections inside the light is diffuse and leaves the box through quartz fibers of 200 μm diameter. This fiber has attenuations of $\sim$400 db/km in the UV. Each of the 8000 fibers is connected to a plexiglas prism glued to the center of the scintillator. The UV light pulse then produces scintillation light which reaches the photomultiplier and produces a digital pulseheight.

In another system [EI 80], designed for the improved CDHS detector, the laser beam passes through a filter, is widened up optically, and then illuminates a scintillator piece glued onto a plexiglas rod (fig. 71). The blue POPOP light emitted isotropically from the scintillator travels down the rod by internal reflection and is partially accepted by the 2304 fibers grouped into 144 bundles of 16, each bundle in one connector. The homogeneity of illumination of the fibers is

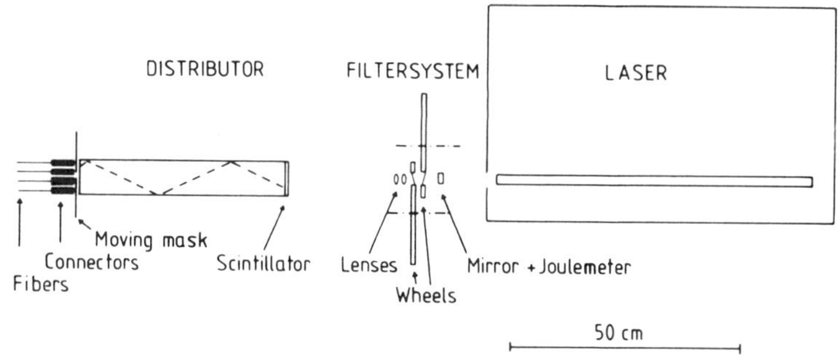

Fig. 71. Dortmund laser calibration system [EI 80].

within 1%. A mechanical mask moving across in front of the connectors permits one group of 192 fibers at a time to be illuminated. This is required by the number of ADC channels available. The transmission of the fibers of 200 μm diameter (QSF 200 A) is 180 db/km for the blue scintillator light, such that over a length of 25 m the attenuation is a factor of 2.8. Each fiber is connected to a light guide through a small (4 mm dia.) cylindrical rod. With this system, by exchanging filters of different density on a "filter wheel", the linearity of all tubes can be measured in a dynamic range from 1 to 2000 times minimum ionization. The absolute calibration is done by comparing one of the fiber outputs to the standard light from an α source embedded in a scintillator.

6. Momentum measurement

6.1. Magnet shapes for fixed target experiments

In a fixed target interaction, the reaction products are usually concentrated in a cone around the incident beam direction (z), because of their limited transverse momenta and the Lorentz boost for longitudinal momenta. If such a particle with momentum (P_x, P_y, P_z) traverses a homogeneous magnetic field $(0, B_y, 0)$, it receives a transverse momentum kick

$$\Delta P_x = -e \int B_y \, dz ,$$

which gives for a field integral of 10 kG·m a transverse momentum change of 0.3 GeV/c. The corresponding deflection of the particle is inversely proportional to its momentum, and a measurement of the projected angles in the (x,z) plane yields, in the simplest approximation, the momentum

$$P = e \int B_y \, dz/(\sin\theta_{in} - \sin\theta_{out}) .$$

If the magnetized volume is evacuated and the multiple scattering in the position detectors is neglected, the error in momentum, δP, comes from measurement error δx in the chambers alone

$$\delta P/P = 2(P/\Delta P_x)(\delta x/L)\,,$$

if the lever arm for the angle measurement before and after the magnet is L. For a field integral of 50 kG m, $\delta x = 0.3$ mm and $L = 3$ m this gives $\delta P/P \sim 1.3\%$ at 100 GeV/c.

These "air core" magnets come in different forms (fig. 72): H-magnets have symmetrical flux return yokes, C-magnets asymmetrical ones (and a less uniform field). The amount of iron in the flux return depends on the desired field strength in the air gap. For a cubic magnetized region, the volume of iron needed, V_{Fe}, relative to the magnetized air gap volume V_{Mag} for different field strength B in the gap is shown in fig. 73. If B has to reach the saturation field strength B_s, then $V_{\text{Fe}}/V_{\text{Mag}} \sim 3$, which is very uneconomical. More usual magnets have $B/B_s \sim \frac{1}{2}$ to $\frac{1}{3}$.

If the particles to be analyzed are high-energy muons, a more economical form of magnets are "iron core" magnets (ICM). Here the field lines stay completely within iron, either in the form of a toroid, where the field lines are circular around a central hole for the coils, or in a kind of H-magnet, where the central region is also filled with iron (fig. 72). The momentum resolution here is limited by multiple scattering of the muons in iron and the measurement error of the muon track. Multiple scattering results in a mean transverse momentum change of

$$\Delta p_{\text{T}}^{\text{MS}} = 21\,(\text{MeV}/c)\sqrt{L/X_0}\,,$$

where L is the length of iron traversed.

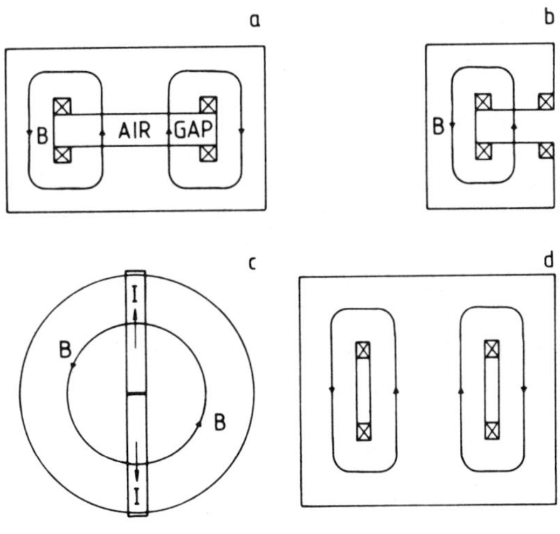

Fig. 72. Magnet shapes for fixed target experiments, (a) H-magnet, (b) C-magnet, (c) toroidal iron core magnet, (d) H-type iron core magnet.

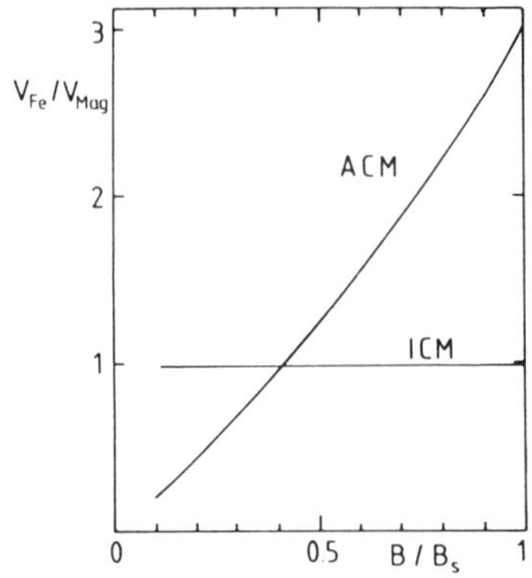

Fig. 73. Volume of iron V_{Fe} needed per magnetized volume used in experiment for iron core magnets (ICM) and air core magnets (ACM) vs. magnetic flux density B in units of saturation density B_s.

The momentum resolution is given by the ratio $\Delta p_T^{MS}/p_T$ and is therefore independent of the momentum. The resolution improves with the length of the ICM as $\sqrt{L}$, and is 12% at $L \sim 5$ m if the position measurement error is smaller than the error by multiple scattering. For high momenta ($P > 100$ GeV/c) this is not the case, since the momentum resolution from measurement error increases linearly with momentum. For a position error of 1 mm in drift chambers after every 75 cm of iron $(\Delta P/P)_{meas} \sim 12\%$ at 100 GeV/c, equal to the error by multiple scattering. The error $(\Delta P/P)_{meas}$ decreases with a 5/2 power of the magnet length L, because not only the bending power and the lever arm increase with L, but also the number of measurements along the track. Fig. 74 gives these two contributions to the momentum resolution for differenth length of iron core magnets.

6.2. Magnet shapes for storage ring experiments

Interaction rates in colliding beam experiments are notoriously low, and 4π solid angle coverage is very desirable. Various magnet geometries are imaginable (fig. 75).

i) Dipole magnet with two compensators to keep particles in orbit: uniform field, good analyzing power in the forward/backward direction, bad analyzing power for particles emitted parallel to the field lines; synchroton radiation of the beam is prohibitive for e^+e^- rings.

ii) Split-field dipole magnet: good resolution in the forward direction, very inhomogenous field at 90°, complicated track fitting procedure; synchroton radiation in e^+e^- rings; being used at the ISR.

iii) Toroid: the inner current sheet or copper bars have to be crossed by particles before momentum analysis, momentum resolution is affected for low energy particles, advantage: no field in beam region.

iv) Solenoid: no force on beam particles, good analyzing power for particles emitted at 90°; access to inner detector only through end caps.

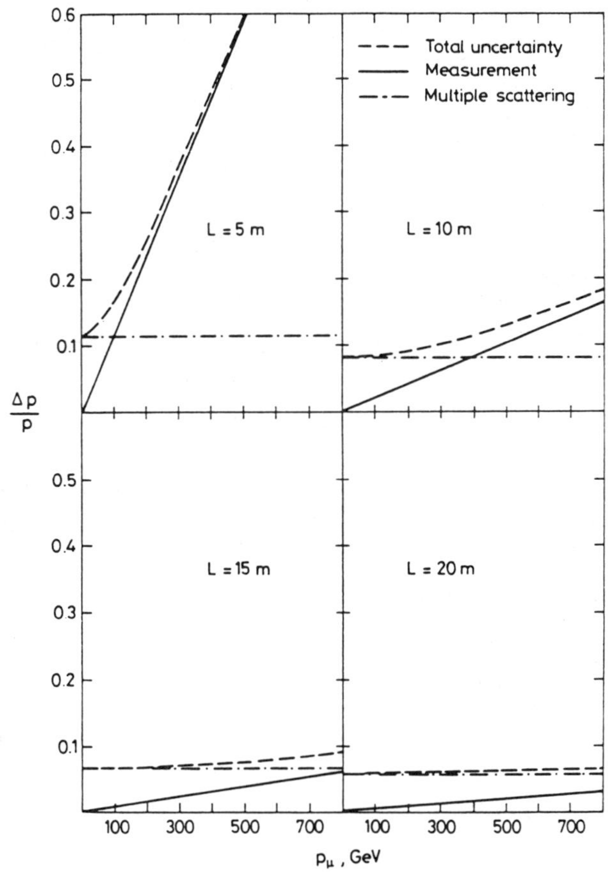

Fig. 74. Momentum resolution for iron core magnets for different iron length, L. Contributions from measurement and multiple scattering are given separately.

For proton (and antiproton) storage rings, split-field and dipole magnets are being used as well as solenoids, while for electron–positron rings the solenoid has been most widely chosen. In general, storage ring detectors resemble each other much more closely than the detectors at fixed target machines.

6.3. Central tracking detectors

For the solenoidal fields used at electron–positron storage rings, momentum measurement is usually done in a central tracking detector around the interaction point. This detector has cylindrical shape with cylinder coordinates; radius r, azimuthal angle φ, and z along the magnetic field

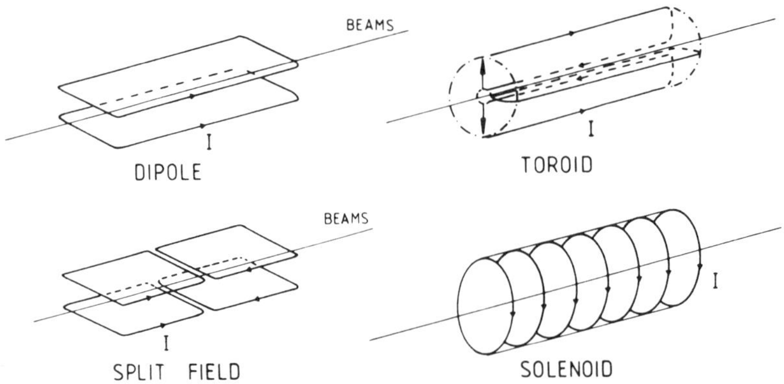

Fig. 75. Magnet shapes for storage ring experiments; lines are currents.

which is parallel to the cylinder axis. If the measurement error in the r, φ plane perpendicular to the field is $\sigma_{r\varphi}$, the momentum component in that plane p_T, is measured with an error [GL 63]

$$\left(\frac{\sigma_p}{p_T}\right)_m = (\sigma_{r\varphi} p_T/(0.3\, BL^2))\sqrt{720/(N+4)}\,,$$

where B is the flux density in Tesla, L the radial track length in meters and N the number of measured points along the track at uniform spacing.

In addition to this measurement error, there is the error due to multiple scattering

$$\left(\frac{\sigma_p}{p_T}\right)_{ms} = (0.05/BL)\sqrt{1.43\, L/X_0}\,.$$

For reconstructing the total momentum of the track, $p = p_T/\sin\theta$, also the polar angle in a plane containing the cylinder axis has to be measured.

The error coming from the measurement error in the z-coordinate, σ_z, is

$$(\sigma_\theta)_m = \frac{\sigma_z}{L}\sqrt{12(N-1)/(N(N+1))}$$

and the one from multiple scattering

$$(\sigma_\theta)_{ms} = \frac{0.015}{\sqrt{3}p}\sqrt{L/X_0}\,.$$

It appears from these relations that the momentum resolution improves with L^2 and B, but that an increase in the number N of measured points gives only an improvement with $\sqrt{N}$.

Table 10
Central tracking detector properties

Name	Ref.	Meas. track length radical L (cm)	Meas. track length axial z (cm)	Flux density (T)	No. sampl.	Gas press. (bar)	Sense wires	Spat. resol. $\sigma(r, \varphi)$ (μm)	Spat. resol. σ_z (mm)	Method of z meas.	Mom. resol. $\sigma_p/p^2\%$ calc.	Mom. resol. $\sigma_p/p^2\%$ meas.
TASSO	[BO 80]	85	330	0.5	15	1	2340	200	3–4	4° stereo		1.7
CELLO	[BE 81]	53	220	1.3	12	1	6432	170	0.44	cathodes		
CLEO	[ST 81]	75	190	0.5 (1.5)	17	1		250	5 (0.25)			5
MARK II	[DA 79]	104		0.4	16	1		200	4			1.9
JADE	[DR 80]	57	234	0.45	48	4	1536	180	16	charge div.		2.2
AFS	[CO 81]	60	128	0.5	42	1	3400	200	17	charge div.		
UA1	[BA 80]	112	250	0.7	~100	1	6100	drift: 250 μm	ch. div. 8–25 mm	charge div.		
TPC	[NY 81]	75	100	1.5	186	10	2232 +13824	$\leq$200	0.2	drift	1.0	
TRIUMF	[HA 81a]	54	69	0.9	12	1	144 +630	(600)	(0.6)	drift		

The three cylindrical drift chamber types described in sections 2.4, 2.5 and 2.6 have been used as central detectors. Some of their properties are listed in table 10.

In two of these detectors energy loss is measured in conjunction with the tracking information. The momentum resolution is in the range $\Delta p/p^2 \sim (1.5-5.0)\%$.

7. Realization of detector systems

Detectors of the kind described in the foregoing sections are assembled for specific experiments which nowadays reach big dimensions (~50 m length), large mass (up to 2000 tons) and huge complexity (up to 10^5 channels of analog and digital information). It is evident that there is a great variety and versatility of experiments at fixed target proton machines, because there are so many variations in experimental conditions possible: different incident particles, different targets, different energies, different interactions (weak, electromagnetic, strong). On the other hand, the higher c.m. energies at colliding beam machines and particularly the cleaner experimental conditions at electron–positron storage rings have enabled important discoveries with these machines (SPEAR and PEP at Stanford, DORIS and PETRA at Hamburg, CESR at Cornell). This evolution may continue with the proton–antiproton collider at CERN in 1981 and at FNAL in 1984, Isabelle at Brookhaven, the Large Electron Positron Machine (LEP) at CERN and the Electron Proton Projects at DESY (HERA) and at KEK (Tristan).

Out of the large number of detector systems in use or being built presently, I have chosen four examples.

7.1. A hadron beam detector

This experiment (NA 5 at CERN [NA 5]) uses a hybrid detector (fig. 76). A hadron beam from the SPS impinges on a hydrogen target, embedded in a large dipole magnet. Particles from the interaction vertex are bent by the field and analyzed in a streamer chamber. Large multiplicities of charged particles do not hinder the performance of the chamber (see fig. 26). The momentum measurement is further improved by the eight large (6 m wide) magnetostrictive spark chambers. A trigger on jets and their analysis is made possible by the hadron calorimeter [EC 77], whose angular range can be changed by moving it back and forth on rails. The experiment searches for jet structure in hadron–hadron reactions at high transverse momentum.

7.2. A neutrino detector

Since the neutrino nucleon total cross-section is only 10^{-36} cm^2 at 100 GeV neutrino energy, neutrino detectors have to be massive. This detector (fig. 77) of the CERN–Dortmund–Heidelberg–Saclay-Collaboration [HO 78a] uses the target weight of 1500 tons of iron, arranged in circular plates of 3.75 m diameter, for three other functions:

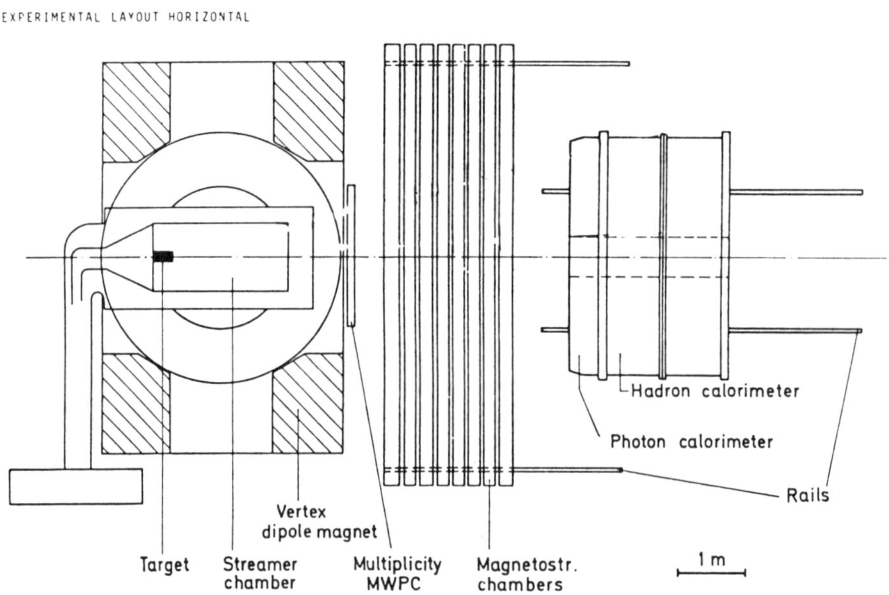

Fig. 76. Experimental layout of Na 5 hadron experiment [NA 5].

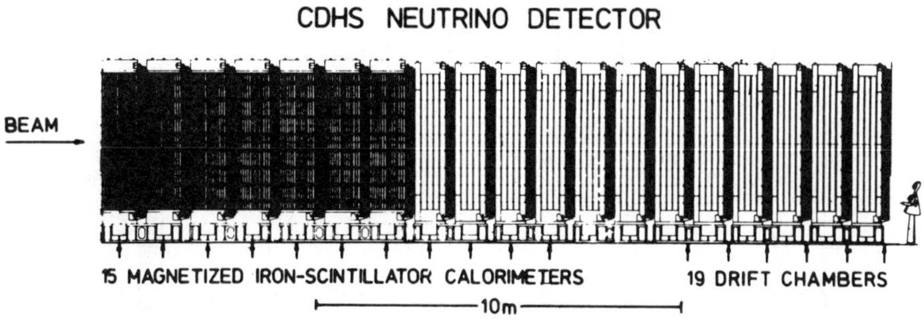

Fig. 77. Experimental layout of CDHS neutrino experiment [HO 78a].

i) 75 cm of iron thickness are combined to form a toroidal magnet,

ii) between the iron slabs of 5 cm (for 7 magnets) or 15 cm (for 8 magnets) 8 scintillators viewed by 16 photomultipliers sample the ionization energy deposited by hadronic showers,

iii) the range in iron enables identification of muons. Between two magnets, driftchambers [MA 77] with 3 planes of wires strung in 120° sequence measure the muon momentum to about 10%. The detector has recorded so far about 3×10^6 neutrino interactions, with zero, one, two, three and four muons.

For future experiments, a part of the detector is being rebuilt (fig. 78) with 2.5 cm iron plates and a finer scintillator mesh (15 cm wide strips in both directions).

7.3. A proton storage ring detector

For the proton–antiproton collider at CERN, one of the detectors is the UA 1 experiment [UA 1]. Fig. 79 shows a sketch of this enormous apparatus: a dipole field configuration with two compensator magnets was chosen. The central detector uses image readout similar to the TPC, but in a different geometry, because the magnetic field lines are perpendicular to the beam direction. The E and B fields are not parallel here, such that the improvement on diffusion broadening during drifting does not apply, and a drift space of 20 cm is used. From about 10000 wires pulse-heights and time information is read out to get a spatial rms resolution of 250 μm and a $\mathrm{d}E/\mathrm{d}x$ measurement. The central detector (radius 1.2 m, length 5.7 m) is surrounded by an electromagnetic calorimeter inside the magnetic field and by the hadronic calorimeter embedded in the pole pieces of the dipole magnet. The wavelength shifting technique is used for the light collection from the 8000 scintillators. The detector is covered on three sides by muon detectors (fig. 80).

According to the preliminary measurements [AL 82] the particle multiplicity in hadron–hadron collisions seems to rise logarithmically up to 540 GeV center-of-mass energy; the typical number of charged secondaries is around 25. Each of the events recorded by this detector will contain more than 10^5 bits of information. On-line data reduction will become very important for this enormous mass of information.

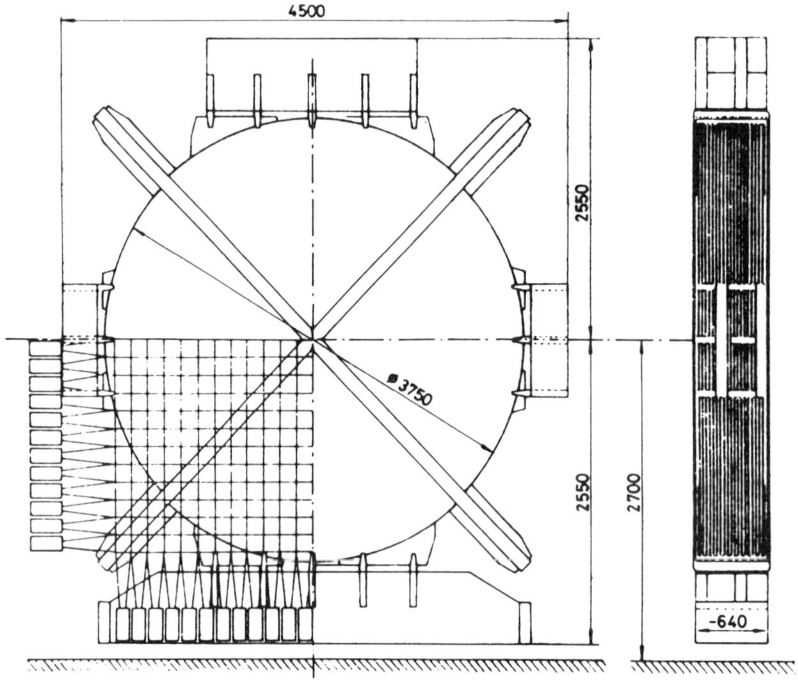

Fig. 78. New toroidal magnetic calorimeter of CDHSW collaboration.

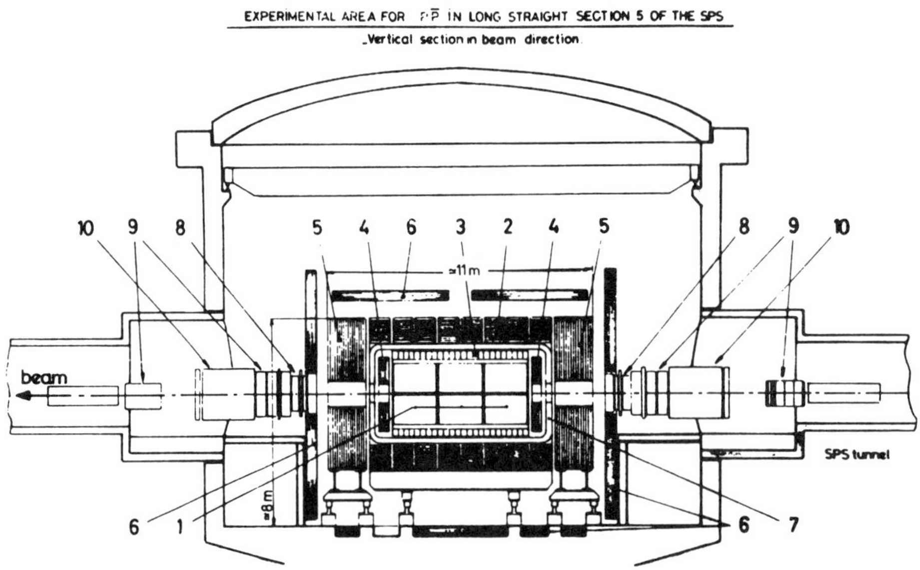

Fig. 79. Side view of UA 1 detector at antiproton-proton collider [UA 1]: 1. central detector with image readout, 2. large angle calorimeter and magnet yoke, 3. large angle shower counter, 4. end cap shower counter, 5. end cap calorimeter, 6. muon detector, 7. aluminium coil, 8. forward chambers, 9. forward shower counters and calorimeters, 10. compensator magnet.

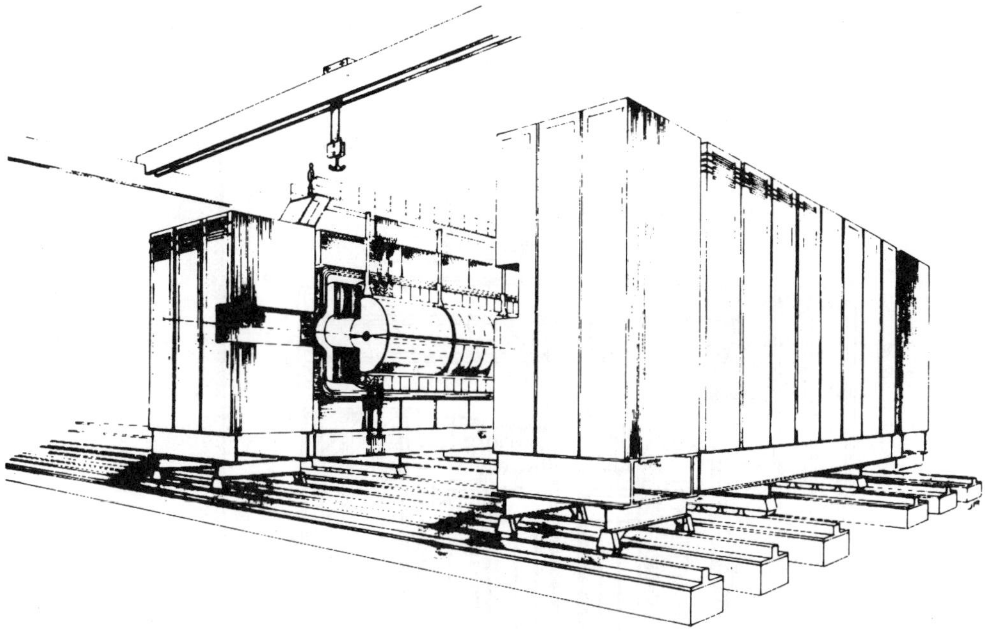

Fig. 80. Perspective view of UA 1 detector.

7.4. An electron–positron storage ring detector

One of the detectors at the PETRA storage ring is the JADE detector [BA 79, DR 80] (fig. 81). Around the interaction point, tracks are recorded in a central detector (section 6.3) of pictorial drift chambers (section 2.5) embedded in a 0.45 Tesla solenoidal field. Track coordinates in the (r, φ) plane are measured with an accuracy of 180 μm, and the momentum resolution is $\Delta p/p^2$ = 2.2% $(\mathrm{GeV}/c)^{-1}$. The central detector also measures $\mathrm{d}E/\mathrm{d}x$. The field is produced by a 7 cm thick aluminium coil, 3.5 m long, 2 m in diameter. Outside the coil, electron and photon energies are measured in an array of 2520 wedge-shaped lead glass counters, grouped in 30 rings of 84 elements. Together with the lead glass counters on the two endcaps, they cover 90% of the total solid angle. The magnetic flux of the solenoid returns through the iron covering the cylindrical detector from all sides as a rectangular box. The iron forms part of the muon absorber, of total thickness 785 g/cm^2 or six nuclear absorption lengths. Penetrating tracks are registered by 4 layers of planar drift chambers.

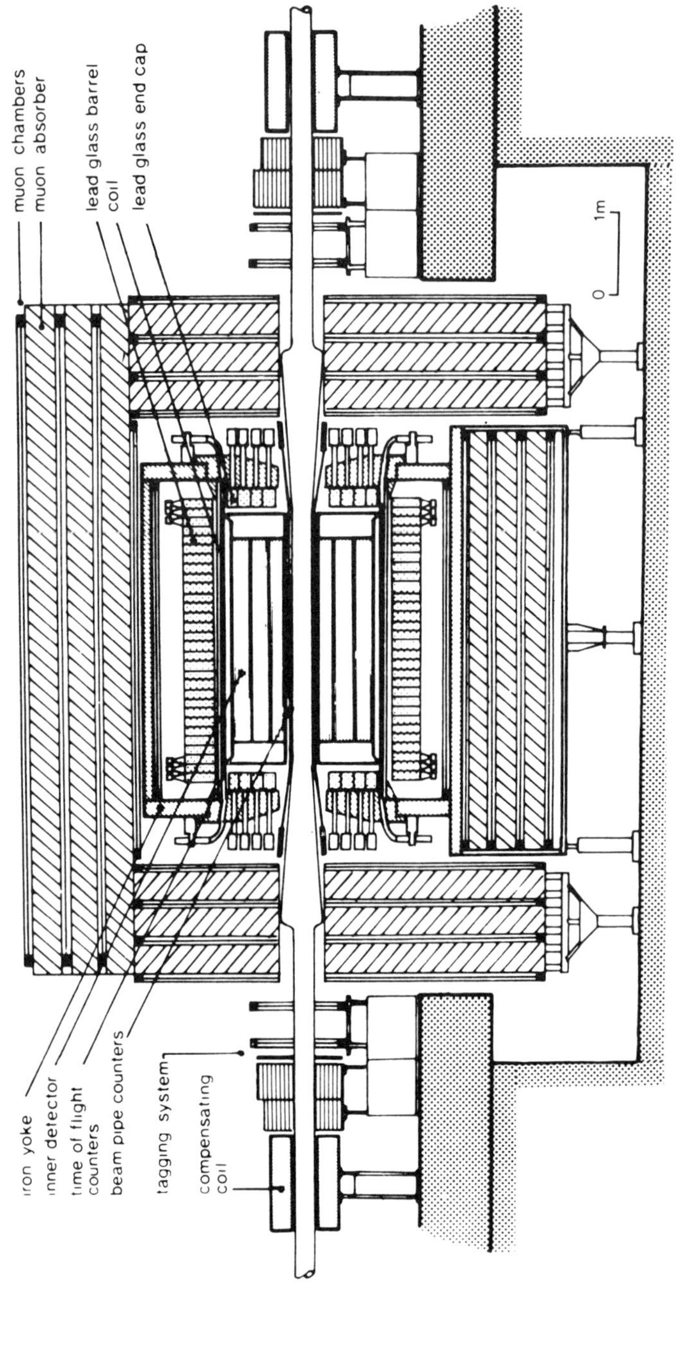

Fig. 81. Sectional view of the JADE detector [BA 79, DR 80].

8. Conclusion

During the last ten years, a lot of progress in detector technology has been achieved, and the discoveries during this exciting time of particle physics would not have been possible without it. Developments will go on: the precision of position measurement in large detectors may be improved by a calibration with nitrogen laser beams, with planar spark counters the precision of time-of-flight measurements could improve, the Cherenkov ring imaging technique may become usable, electromagnetic shower counters would shrink in size if BGO could be produced at reasonable cost, and data processing will become more efficient by using microprocessors and specialized 32-bit emulators.

It will be necessary and possible to construct for the new generation of accelerators and storage rings general purpose detector systems of the enormous size and complexity of, e.g., the UA 1 experiment. These experiments require large experimental teams for construction, maintenance, running and data analysis. It is an open question whether this is an unavoidable consequence of the physics questions at these accelerators and storage rings, or if there is still a reasonable chance for smaller specialized experiments to contribute significantly to the progress in elementary particle physics.

References

[AB 81] H. Abramowicz et al., Nucl. Instr. Meth. 180 (1981) 429.
[AD 74] M. Aderholz et al., Nucl. Instr. Meth. 118 (1974) 419.
[AK 77] G.A. Akapdjanov et al., Nucl. Instr. Meth. 140 (1977) 441.
[AL 69] O.C. Allkofer, Spark chambers (Thiemig Verlag, Munich, 1969).
[AL 80] W.W.M. Allison and J.H. Cobb, Ann. Rev. Nucl. Sci. 30 (1980) 253.
[AL 81] W.W.M. Allison, Phys. Scripta 23 (1981) 348.
[AL 82] K. Alpgard et al., (UA 5 Coll.), CERN-EP/81-153, subm. Phys. Lett. B.
[AM 81] U. Amaldi, Phys. Scripta 23 (1981) 409.
[AR 75] X. Artru et al., Phys. Rev. D12 (1975) 1289.
[AR 81] C. Arnault et al., Phys. Scripta 23 (1981) 710.
[BA 56] M.V. Babykin et al., Sov. Journ. of Atomic Energy IV (1956) 627.
[BA 70] G. Bathow et al., Nucl. Phys. B20 (1970) 592.
[BA 78] B. Barish et al., Very large area scintillation counters for hadron calorimetry, IEEE Trans. Nucl. Sci. NS 25 (1978) 532.
[BA 79] W. Bartel et al., Phys. Lett. 88B (1979) 171.
[BA 80] M. Barranco Luque et al., Nucl. Instr. Meth. 176 (1980) 175.
[BA 81] A. Bamberger, priv. comm. (1981).
[BE 30] H.A. Bethe, Annalen d. Physik 5 (1930) 325.
[BE 32] H.A. Bethe, Z. Physik 76 (1932) 293.
[BE 33] H.A. Bethe, Hdb. Physik 24 (1933) 518.
[BE 71] I.B. Berlman, Fluorescence Spectra of Aromatic Molecules (N.Y. and London, 1971).
[BE 77] BEBC Users Handbook, CERN (1977).
[BE 81] H.J. Behrend et al., Phys. Scripta 23 (1981) 610.
[BI 64] J.B. Birks, Theory and practice of scintillation counting (London, 1964).
[BI 81] F. Binon et al., Nucl. Instr. Meth. 188 (1981) 507.
[BL 50] O. Blunck and S. Leisegang, Z. Physik 128 (1950) 500.
[BL 81] W. Blum, priv. comm. (1981).
[BO 80] H. Borner et al., DESY 80/27 (1980).

[BO 81] H. Bourdinaud and J.C. Thevenin, Phys. Scripta 23 (1981) 534.
[BR 74] A. Breskin et al., Nucl. Instr. Meth. 119 (1974) 9.
[BR 81] W. Braunschweig, Phys. Scripta 23 (1981) 384.
[CA 74] M. Cantin et al., Nucl. Instr. Meth. 118 (1974) 177.
[CA 81a] P.J. Carlson et al., Phys. Scripta 23 (1981) 708.
[CA 81b] P.J. Carlson, Phys. Scripta 23 (1981) 393.
[CE 64] P.A. Cherenkov, I.M. Frank and I.E. Tamm, Nobel Lectures in Physics (New York, Elsevier, 1964).
[CH 68] G. Charpak et al., Nucl. Instr. Meth. 62 (1968) 262.
[CH 70a] G. Charpak et al., Nucl. Instr. Meth. 80 (1970) 13.
[CH 70b] G. Charpak, Ann. Rev. Nucl. Sci. 20 (1970) 195.
[CH 78a] G. Charpak et al., Nucl. Instr. Meth. 148 (1978) 471.
[CH 78b] Y. Chan et al., IEEE Trans. Nucl. Sci. NS 25 (1978) 333.
[CO 55] M. Conversi and A. Gozzini, Nuovo Cim. 2 (1955) 189.
[CO 75] J.H. Cobb, Ph.D. thesis, Univ. Oxford 1975.
[CO 76] J.H. Cobb et al., Nucl. Instr. Meth. 133 (1976) 315.
[CO 77] J.H. Cobb et al., Nucl. Instr. Meth. 140 (1977) 413.
[CO 78] M. Conversi and L. Federici, Nucl. Instr. Meth. 151 (1978) 93.
[CO 80] V. Commichau et al., Nucl. Instr. Meth. 176 (1980) 325.
[CO 81] D. Cockerill et al., Phys. Scripta 23 (1981) 649.
[CR 62] D.F. Crawford and H. Messel, Phys. Rev. 128 (1962) 352.
[CU 71] W. Cunitz et al., Nucl. Instr. Meth. 91 (1971) 211.
[DA 79] W. Davies-White et al., Nucl. Instr. Meth. 160 (1979) 227.
[DH 77] S. Dhawan and R. Majka, IEEE Trans. Nucl. Sci. NS 24 (1977) 270.
[DI 78] M. Dine et al., Fermilab proposal No. 490 (1978).
[DI 79] P. Dishaw, Limits on neutrino-like particles, thesis Stanford University 1979.
[DI 80] A.N. Diddens et al., Nucl. Instr. Meth. 178 (1980) 27.
[DR 80] H. Drumm et al., Nucl. Instr. Meth. 176 (1980) 333.
[DY 81] M. Dykes et al., Nucl. Instr. Meth. 179 (1981) 487.
[EC 77] V. Eckhardt et al., Nucl. Instr. Meth. 143 (1977) 235.
[EC 80] V. Eckhardt, MPI München, priv. comm. (1980).
[EI 80] F. Eisele, K. Kleinknecht, D. Pollmann and B. Renk, Dortmund University (1980).
[EK 81] T. Ekelöf et al., Phys. Scripta 23 (1981) 718.
[ER 77] V.C. Ermilova et al., Nucl. Instr. Meth. 145 (1977) 555.
[FA 75] C.W. Fabjan and W. Struczinski, Phys. Lett. 57B (1975) 484.
[FA 77] C.W. Fabjan et al., Nucl. Instr. Meth. 141 (1977) 61.
[FA 79] D. Fancher et al., Nucl. Instr. Meth. 161 (1979) 383.
[FA 80] C.W. Fabjan and H.G. Fischer, Rep. Progr. Phys. 43 (1980) 1003.
[FA 81] C.W. Fabjan et al., Nucl. Instr. Meth. 185 (1981) 119.
[FI 75] H.G. Fischer et al., Proc. Int. Meeting on Prop. and Drift Chambers, Dubna 1975 (JINR) report D 13-9164.
[FI 78] H.G. Fischer, Nucl. Instr. Meth. 156 (1978) 81.
[FL 81] G. Flügge et al., Phys. Scripta 23 (1981) 499.
[FR 55] W.B. Fretter, Ann. Rev. Nucl. Sci. 5 (1955) 145.
[GA 60] R.C. Garwin, Rev. Sci. Instr. 31 (1960) 1010.
[GA 73] G.M. Garibian, Proc. 5th Int. Conf. in Instrumentation for High Energy Physics, Frascati 1973, p. 329.
[GI 46] V.L. Ginzburg and I.M. Frank, IETP 16 (1946) 15.
[GL 52] D.A. Glaser, Phys. Rev. 87 (1952) 665; Phys. Rev. 91 (1953) 496.
[GL 58] D.A. Glaser, Hdb. Physik 45 (1958) 314.
[GL 63] R.L. Glückstern, Nucl. Instr. Meth. 24 (1963) 381.
[GR 80] G. Grayer and J. Homer (Rutherford Laboratory), priv. comm. (1980).
[HA 73] F. Harris et al., Nucl. Instr. Meth. 107 (1973) 413.
[HA 81a] C.K. Hargrove et al., Phys. Scripta 23 (1981) 668.
[HA 81b] G. Harigel et al., Nucl. Instr. Meth. 187 (1981) 363
[HA 82] G. Harigel, priv. comm. (1982).
[HE 81] S. Henning and L. Svensson, Phys. Scripta 23 (1981) 697.
[HI 81] D. Hitlin, Phys. Scripta 23 (1981) 634.

[HO 78a] M. Holder et al., Nucl. Instr. Meth. 148 (1978) 235.
[HO 78b] M. Holder et al., Nucl. Instr. Meth. 151 (1978) 69.
[HO 79] W. Hofmann et al., Nucl. Instr. Meth. 163 (1979) 77.
[HU 72] E.B. Hughes et al., Stanford Univ. Report No. 627 (1972).
[KA 81] V. Kadansky et al., Phys. Scripta 23 (1981) 680.
[KE 48] I.W. Keuffel, Phys. Rev. 73 (1948) 531; Rev. Sci. Instr. 20 (1949) 202.
[KE 79] G. Keil, Nucl. Instr. Meth. 83 (1970) 145; 87 (1970) 111.
[KI 79] I. Kirkbridge, IEEE Trans. Nucl. Sci. NS 26 (1979) 1535.
[KL 70] K. Kleinknecht et al., CERN NP Int. Report 70-18 (1970).
[KL 81a] P. Klasen et al., Nucl. Instr. Meth. 185 (1981) 67.
[KL 81b] F. Klawonn et al., A new type of acrylic scintillator, Nucl. Instr. Meth. (in press).
[KO 81] M. Kobayashi et al., Nucl. Instr. Meth. 189 (1981) 629.
[LA 44] L.D. Landau, J. Exp. Phys. (USSR) 8 (1944) 201.
[LE 78a] I. Lehraus et al., Nucl. Instr. Meth. 153 (1978) 347.
[LE 78b] B. Leskovar and C.C. Lo, IEEE Trans. Nucl. Sci. NS 25 (1978) 582.
[LE 81a] I. Lehraus et al., Phys. Scripta 23 (1981) 727.
[LE 81b] I. Lehraus et al., Preprint CERN/EF 81-14 (1981).
[LE 81c] P. Lecomte et al., Phys. Scripta 23 (1981) 377.
[LI 73] J. Litt and R. Meunier, Ann. Rev. Nucl. Sci. 23 (1973) 1.
[LO 61] L.B. Loeb, Basic Processes of Gaseous Electronics (U. of Calif. Press, Berkeley, 1961).
[LO 75] E. Longo and I. Sestili, Nucl. Instr. Meth. 128 (1975) 283.
[LO 81] E. Lorenz, MPI München, priv. comm. (1981).
[LU 81] T. Ludlam et al., Nucl. Instr. Meth. 180 (1981) 413.
[MA 69] H.D. Maccabee and D.G. Papworth, Phys. Lett. 30A (1969) 241.
[MA 77] G. Marel et al., Nucl. Instr. Meth. 141 (1977) 43.
[MA 78] J.N. Marx and D.R. Nygren, Physics today, Oct. 1978, p. 46.
[MO 80] L. Montanet, Proc. XXth Conf. on High En. Physics, Madison, Wisconsin, July 1980, p. 863.
[NA 1] Frascati–Milan–Pisa–Rome–Torino–Trieste Coll., S.R. Amendiola et al., CERN Proposal SPSC/74-15/P6.
[NA 5] Bari–Cracow–Liverpool–München(MPI)–Nijmegen-Coll., CERN Proposal SPSC/75-1/P 37.
[NA 65] H.H. Nagel, Z. Physik 186 (1965) 319.
[NA 70] R. Nathan and M. Mee, Phys. Sol. A2 (1970) 67.
[NA 72] Y. Nakato et al., Bull. Chem. Soc. Japan 45 (1972) 1299.
[NE 75] O.H. Nestor and C.N. Huang, IEEE Trans. NS 22 (1975) 68.
[NY 74] D.R. Nygren, LBL Int. Report, Feb. 1974.
[NY 81] D.R. Nygren, Phys. Scripta 23 (1981) 584.
[PA 78] Particle Data Group, Phys. Lett 75B (1981) 1.
[PA 80] R. Partridge et al., Phys. Rev. Lett. 44 (1980) 712.
[PH 78] Philips Data Handbook, Part 9 (1978).
[PR 80] Y.D. Prokoshkin, Proc. Second ICFA Workshop, Les Diablerets, Oct. 79, CERN Report (June 1980).
[RA 74] V. Radeka, IEEE Trans. Nucl. Sci. NS 21 (1974) 51.
[RI 74] P. Rice-Evans, Spark, Streamer, Proportional and Drift Chambers (London, 1974).
[RO 52] B. Rossi, High Energy Particles (Prentice Hall, N.Y., 1952).
[SA 80] J. Sandweiss, paper given at XXth Int. Conf. on High En. Physics, Madison, Wisconsin, July 1980.
[SA 81] F. Sauli, Phys. Scripta 23 (1981) 526.
[SC 71] P. Schilly et al., Nucl. Instr. Meth. 91 (1971) 221.
[SC 79] L.S. Schröder, Nucl. Instr. Meth. 162 (1979) 395.
[SC 80] B. Schmidt, Elektronendrift in Zählgasen unter dem Einfluss elektrischer und magnetischer Felder, Diplomarbeit Heidelberg 1980.
[SC 81] M.A. Schneegans et al., Preprint CERN-EP/81-37 (1981) subm. to Nucl. Instr. Meth.
[SE 77] J. Seguinot and T. Ypsilantis, Nucl. Instr. Meth. 142 (1977) 377.
[SE 79] W. Selove et al., Nucl. Instr. Meth. 161 (1979) 233.
[SH 51] W.A. Shurcliff, J. Opt. Soc. Am. 41 (1951) 209.
[ST 52] R.M. Sternheimer, Phys. Rev. 88 (1952) 851.
[ST 71] R.M. Sternheimer and R.F. Peierls, Phys. Rev. B3 (1971) 3681.
[ST 81] S. Stone, Phys. Scripta 23 (1981) 605.

[TA 78] F.E. Taylor et al., IEEE Trans. Nucl. Sci. NS 25 (1978) 312.
[TR 69] T. Trippe, CERN NP Int. Report 69-18 (1969).
[UA 1] Aachen–Annecy–Birmingham–CERN–London–Paris–Riverside–Rutherford–Saclay–Vienna-Coll. CERN Proposal SPSC/78-6, SPSC/P92 (1978).
[VA 70] Valvo Photomultiplier Book, Hamburg, April 1970.
[WA 21] Birmingham–Bonn–CERN–London–Munich–Oxford Coll. using BEBC at CERN. Exp. WA 21 (1979).
[WA 71] A.H. Walenta et al., Nucl. Instr. Meth. 92 (1971) 373.
[WA 79] A.H. Walenta et al., Nucl. Instr. Meth. 161 (1979) 45.
[WA 81a] A.H. Walenta, Phys. Scripta 23 (1981) 354.
[WA 81b] A. Wagner, Phys. Scripta 23 (1981) 446.
[WA 82] A. Wagner, priv. comm. (1982).
[WE 81a] D. Wegener, priv. comm. 1981; ARGUS.
[WE 81b] H. Wenninger, priv. comm. (1981).
[YP 81] T. Ypsilantis, Phys. Scripta 23 (1981) 371.

Reprinted, with permission of the author, from CERN Report 77-09 (1977).

PRINCIPLES OF OPERATION OF MULTIWIRE PROPORTIONAL AND DRIFT CHAMBERS

F. Sauli

1. INTRODUCTION

The first multiwire proportional chamber, in its modern conception, was constructed and operated by Charpak and his collaborators in the years 1967-68 [1]. It was soon recognized that the main properties of a multiwire proportional chamber, i.e. very good time resolution, good position accuracy and self-triggered operation, are very attractive for the use of the new device in high-energy physics experiments. Today, most fast detectors contain a large number of proportional chambers, and their use has spread to many different fields of applied research, such as X-ray and heavy ion astronomy, nuclear medicine, and protein crystallography[2]. In many respects, however, multiwire proportional chambers are still experimental devices, requiring continuous attention for good operation and sometimes reacting in unexpected ways to a change in the environmental conditions. Furthermore, in the fabrication and operation of a chamber people seem to use a mixture of competence, technical skill and magic rites, of the kind "I do not know why I'm doing this but somebody told me to do so".

In these notes I will try to illustrate the basic phenomena underlying the behaviour of a gas detector, with the hope that the reader will not only better understand the reasons for some irrational-seeming preferences (such as, for example, in the choice of the gas mixture), but will also be able better to design detectors for his specific experimental needs.

Most of the recent development on multiwire proportional chambers is due to the enthusiastic work of Georges Charpak who initiated me into this exciting field of applied research; these notes are dedicated to him.

A large number of the illustrations have been extracted from old textbooks and articles, and I would like to thank Claude Rigoni for her great skill and patience in making them suitable for reproduction.

2. DETECTION OF CHARGED PARTICLES

2.1 Generalities

A fast charged particle, traversing a gaseous or condensed medium, can interact with it in many ways. Of all possible interactions, however, only the electromagnetic one is generally used as a basis for detection, being many orders of magnitude more probable than strong or weak interactions and therefore leaving a "message" even in very thin samples of material. These notes are mainly concerned with the highly probable incoherent Coulomb interactions between the electromagnetic fields of the incoming charged particle and of the medium, resulting in both excitation and ionization of the atoms of the medium itself. The contribution of other electromagnetic processes (at least for particles heavier than electrons), such as bremsstrahlung, Čerenkov, and transition radiation, to the total energy loss is negligible in gas detectors and we will ignore them.

2.2 Energy loss due to electromagnetic interactions

An expression for the average differential energy loss (loss per unit length) due to Coulomb interactions has been obtained by Bethe and Bloch[3)] in the framework of relativistic quantum mechanics, and can be written as follows (in the electrostatic unit system):

$$\frac{dE}{dX} = -K\frac{Z}{A}\frac{\rho}{\beta^2}\left\{\ln\frac{2mc^2\beta^2E_M}{I^2(1-\beta^2)} - 2\beta^2\right\}, \qquad K = \frac{2\pi Nz^2e^4}{mc^2} \tag{1}$$

where N is the Avogadro number, m and e are the electron mass and charge, Z, A and ρ are the atomic number and mass, and the density of the medium, respectively, and I is its effective ionization potential; z is the charge and β the velocity (in units of the speed of light c) of the projectile. In the electrostatic unit system and expressing energies in MeV, $K = 0.154$ MeV g^{-1} cm^2 for unit charge projectiles. In the system used, the rest energy of the electron, mc^2, equals 0.511 MeV.

The quantity E_M represents the maximum energy transfer allowed in each interaction, and simple two-body relativistic kinematics gives

$$E_M = \frac{2mc^2\beta^2}{1-\beta^2} \quad . \tag{2}$$

For example, for 1 GeV/c protons ($\beta = 0.73$) $E_M = 1.2$ MeV. Not always, however, can this kinematical limit be used, particularly in the case of thin detectors where energy can escape in the form of δ-rays (see Section 2.4).

It is customary to substitute for the length X a reduced length x defined as Xρ and measured in g cm^{-2}. In this case, the reduced energy loss can be written as

$$\frac{dE}{dx} = \frac{1}{\rho}\frac{dE}{dX} \quad .$$

The value of the effective ionization potential I is in general the result of a measurement for each material; a rather good approximation is, however, $I = I_0 Z$ with $I_0 \simeq 12$ eV. Values of I_0 are given in Table 1 for several gases. For molecules and for gas mixtures, average values for Z, A, and I have to be taken.

Inspection of expression (1) shows that the differential energy loss depends only on the projectile velocity β, not on its mass. After a fast decrease dominated by the β^{-2} term, the energy loss reaches a constant value around $\beta \simeq 0.97$ and eventually slowly in-

Table 1

Properties of several gases used in proportional counters (from different sources, see the bibliography for this section). Energy loss and ion pairs per unit length are given at atmospheric pressure for minimum ionizing particles

Gas	Z	A	δ (g/cm^3)	E_{ex}	E_i (eV)	I_0	W_i	dE/dx (MeV/g cm^{-2})	dE/dx (keV/cm)	n_P (i.p./cm)[a]	n_T (i.p./cm)[a]
H_2	2	2	8.38×10^{-5}	10.8	15.9	15.4	37	4.03	0.34	5.2	9.2
He	2	4	1.66×10^{-4}	19.8	24.5	24.6	41	1.94	0.32	5.9	7.8
N_2	14	28	1.17×10^{-3}	8.1	16.7	15.5	35	1.68	1.96	(10)	56
O_2	16	32	1.33×10^{-3}	7.9	12.8	12.2	31	1.69	2.26	22	73
Ne	10	20.2	8.39×10^{-4}	16.6	21.5	21.6	36	1.68	1.41	12	39
Ar	18	39.9	1.66×10^{-3}	11.6	15.7	15.8	26	1.47	2.44	29.4	94
Kr	36	83.8	3.49×10^{-3}	10.0	13.9	14.0	24	1.32	4.60	(22)	192
Xe	54	131.3	5.49×10^{-3}	8.4	12.1	12.1	22	1.23	6.76	44	307
CO_2	22	44	1.86×10^{-3}	5.2	13.7	13.7	33	1.62	3.01	(34)	91
CH_4	10	16	6.70×10^{-4}		15.2	13.1	28	2.21	1.48	16	53
C_4H_{10}	34	58	2.42×10^{-3}		10.6	10.8	23	1.86	4.50	(46)	195

a) i.p. = ion pairs

creases for $\beta \to 1$ (relativistic rise). The region of constant loss is called the minimum ionizing region and corresponds to the more frequent case in high-energy physics.

When plotting dE/dX as a function of the projectile energy, see Fig. 1 (computed for air), one observes that at energies above a few hundred MeV all particles are at the minimum of ionization and therefore lose the same amount of energy per unit length; for most materials, dE/dx is equal to about 2 MeV g^{-1} cm^2 at the minimum. Computed values of the differential energy loss at the minimum are given in Table 1, together with other relevant parameters and for the gases commonly used in proportional counters.

It should be emphasized that, even for thin materials, the electromagnetic energy loss is the result of a small number of discrete interactions and, therefore, has the characteristics of a statistical average. The distribution, however, is not Gaussian for all cases where the energy loss ΔE is small compared to the total energy, as will be discussed below. Furthermore, a closer look at the interaction mechanism shows that individual events can be grouped in two classes: close collisions, with large energy transfers resulting in the liberation of electronic charges (ionizations), and distant collisions involving smaller energy transfers and resulting in both ionization and atomic excitation. Primary ionization and excitation share more or less equally the available energy loss, although secondary effects can increase the efficiency for one of the processes. In Fig. 2 [4)] the probability of the processes described is qualitatively shown as a function of the energy transfer, in a single event, for 100 keV incident electrons.

2.3 The relativistic rise of energy loss

At very high momenta, above 10 GeV/c or so, the logarithmic term in the Bethe-Block formula produces an increase of the energy loss; this may constitute a basis for particle identification at very high energies since, for a given momentum, the average energy loss will be slightly different for different masses. Polarization effects, not taken into account by expression (1), however, produce a saturation in the increase, that would otherwise indefinitely continue. In gases, saturation occurs around a few hundred GeV/c at a value which is about 50% above minimum ionizing; Fig. 3 shows a collection of experimental data for argon[5)]. For statistical reasons, discussed more fully later, an effective particle identification in the region of the relativistic rise requires a very large number of independent measurements of energy loss for each track, and stacks of multiwire proportional chambers have been succesfully used for this purpose[6)].

2.4 δ-ray production

In ionizing encounters, the ejected electron is liberated with an energy E that can assume any value, up to the maximum allowed E_M, as given by expression (2). An approximate

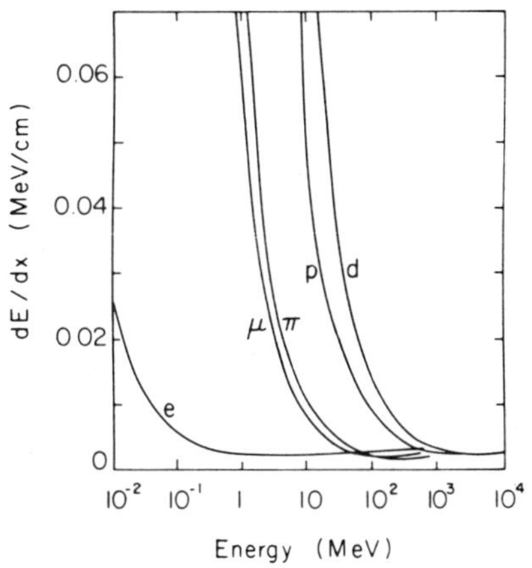

Fig. 1 Energy loss per unit length in air, as computed from expression (1), for different particles as a function of their energy. At energies above 1 GeV/c or so, all particles lose about the same amount of energy (minimum ionization plateau).

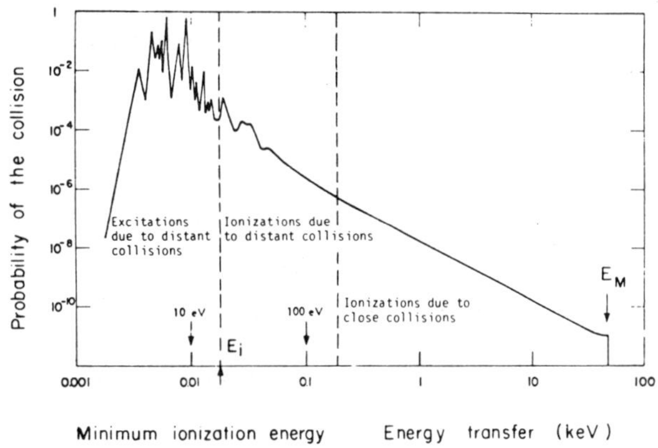

Fig. 2 Relative probability of different processes induced by fast (100 keV) electrons in water, as a function of the energy transfer in a collision[4]. The maximum kinematically allowed energy transfer, E_M = 50 keV in this case, is also shown.

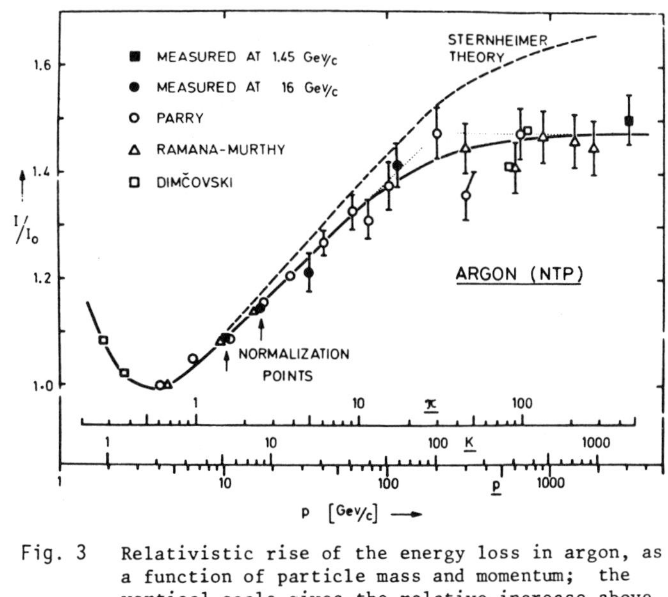

Fig. 3 Relativistic rise of the energy loss in argon, as a function of particle mass and momentum; the vertical scale gives the relative increase above the minimum of ionization[5]

expression for the probability of an electron receiving the energy E is given by[3)]

$$P(E) = K \frac{Z}{A} \frac{\rho}{\beta^2} \frac{X}{E^2} \tag{3}$$

that corresponds essentially to the first term in the Bethe-Bloch formula. More accurate expressions can be found elsewhere[7,8)], but for our needs the approximation (3) is sufficient. If the reduced thickness $x = X\delta$ is introduced, and given in g cm^{-2}, the expression can be conveniently rewritten as

$$P(E)dE = \frac{K}{\beta^2} \frac{Z}{A} \frac{x}{E^2} dE = W \frac{dE}{E^2} \quad . \tag{3'}$$

Electrons ejected with an energy above a few keV are normally called δ-rays, using an old emulsion terminology. Integration of expression (3′) allows one to obtain an expression for the number of δ electrons having an energy E_0 or larger:

$$N(E \geq E_0) = \int_{E_0}^{E_M} P(E)\ dE = W\left(\frac{1}{E_0} - \frac{1}{E_M}\right) \simeq \frac{W}{E_0} \quad ,$$

the last approximation being valid for $E_0 << E_M$.

As an example, Fig. 4 shows the number of electrons ejected with energy $E \gtrsim E_0$ by 1 GeV/c protons, as a function of E_0 in 1 cm of argon at normal conditions. There are, for example, about ten electrons emitted with energy above 15 eV, which is the ionization potential of argon (see Table 1); these considerations are very important for the understanding of secondary processes (see below). In Fig. 4, the maximum energy transfer for 1 GeV/c protons as projectiles is also shown, as given by expression (2).

The angle of emission of a δ electron of energy E is given in a free-electron approximation, by the expression[8)]

$$\cos^2 \theta = \frac{E}{E_M} \quad .$$

Therefore for minimum ionizing particles (E_M > 1 MeV), and up to energies of emission of several keV, δ rays are emitted perpendicularly to the incident track; however, multiple scattering in the medium quickly randomizes the direction of motion of δ electrons. In fact, typical electron molecule cross-sections, in the keV region, are around 10^{-16} cm^2 [3)], at atmospheric pressure, this corresponds to a mean free path between collisions of a few microns. For energies around a hundred electronvolts, the cross-section is increased by an order of magnitude or so owing to the high probability of inelastic collisions (see Section 4.7). Be-

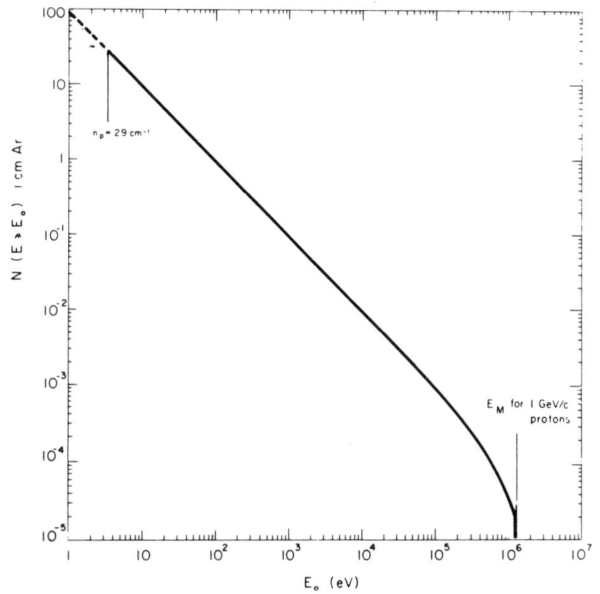

Fig. 4 Computed number of δ electrons ejected at an energy larger than or equal to E_0, as a function of E_0, in 1 cm of argon at normal conditions. The average number of primary ionizing collisions (29 per cm) and the maximum allowed energy transfer for 1 GeV/c protons are shown.

cause of the large mass difference between target and projectile, large momentum transfers (scattering angles) are highly probable, which means that in a few collisions any trace of the electron's original emission angle will be obliterated. A detailed discussion of the process can be found in Refs. 6 and 7.

2.5 Range of slow electrons

Depending on their energy, δ electrons will cover a certain distance in the gas, suffering elastic and inelastic scatters from the molecules. The total range R_T for an energy E, along the trajectory, can be calculated integrating the Bethe-Block formula over the length R_T and requiring the integral to equal the total available energy; however, it gives a bad representation of the distance effectively covered by an electron, because of the randomizing effect of the multiple collisions. It is customary to define a practical range R_p that appears to be two or three times smaller than the total range and in general is the result of an absorption measurement. For energies up to a few hundred keV, a rather good approximation for the practical range, in g cm^{-2}, is[8]

$$R_p = 0.71\ E^{1.72} \qquad (E \text{ in MeV}) \quad .$$

Figure 5 gives the range of electrons in argon, under normal conditions, as a function of energy. Combining the data of Figs. 4 and 5, one can deduce, for example, that in 1 cm of argon, one out of twenty minimum ionizing particles ejects a 3 keV electron having a practical range of 100 μm.

Since the average energy loss in argon is 2.5 keV/cm, these events will result in a much bigger pulse height, and also the centre of gravity of the detected charge, which is the best position information one can get, will be systematically displaced to one side of the original track. This obviously sets a limit to the best accuracy of position one can hope to obtain in a single gas counter operating at atmospheric pressure, somewhere between 20 and 30 μm. Actual measurements obtained so far with high accuracy drift chambers provide accuracies of localization of between 50 and 100 μm [9,10]. Notice that an increase in

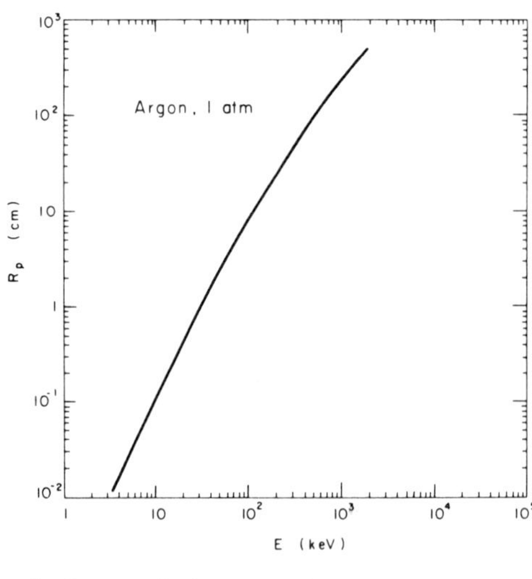

Fig. 5 Range of electrons in argon, at normal conditions as a function of energy, deduced from measurement in light materials[8]

the gas density or pressure does not obviously improve the accuracy since, although the range of electrons will decrease, the number produced at any given energy will increase.

Also, from Fig. 5 one can deduce that electrons produced with energies above 30 keV have a range larger than 1 cm of argon, and will escape detection from the 1 cm thick layer. In this case use of the Bethe-Block formula is not completely justified, since the maximum energy transfer E_M cannot in fact be dissipated in the detector.

2.6 Energy loss distribution

The fact that in thin materials the total energy loss is given by a small amount of interactions, each one with a very wide range of possible energy transfers, determines a characteristic shape for the energy loss distribution. In a classical formulation due to Landau, the energy-loss distribution in thin media is written as[11)]

$$f(\lambda) = \frac{1}{\sqrt{2\pi}} e^{-\frac{1}{2}(\lambda + e^{-\lambda})} \quad ,$$

where the reduced energy variable λ represents the normalized deviation from the most probable energy loss $(\Delta E)_{mp}$:

$$\lambda = \frac{\Delta E - (\Delta E)_{mp}}{\xi} \quad , \qquad \text{where } \xi = K \frac{Z}{A} \frac{\rho}{\beta^2} X \quad ,$$

ΔE being the actual loss and ξ the average energy loss given by the first term in the Bethe-Block formula. Figure 6 shows the characteristic shape of the Landau distribution, and indicates the meaning of the average and of the most probable energy losses. Notice the long tail at very large energy losses, corresponding to events where one or more energetic δ electrons have been produced. The energy resolution of thin counters for fast particles is therefore very poor; increasing the thickness of the detector is of no help, since the number of energetic electrons will increase (unless, of course, one gets close to total absorption where Gaussian statistics again dominate).

The large fluctuation in energy loss for individual events has several important practical consequences. First, when designing the amplification electronics for a gaseous detector one has to take into account the large dynamic range of the signals. Secondly, a single measurement of a track contains very little information about the average energy loss; when trying to identify particles in the relativistic rise region from the small increase of dE/dx, one is obliged to sample each track as much as several hundred times. This requires the development of relatively cheap, large surface multiwire proportional chambers.

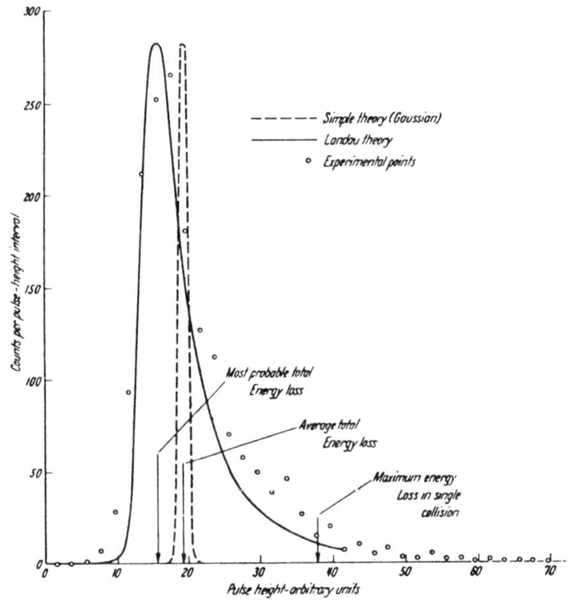

Fig. 6 (From Franzen and Cochran, see bibliography for Section 5.) Comparison of the experimental energy loss distribution, in an argon-carbon dioxide counter, with the distribution computed using a simple Gaussian theory and the Landau expression.

2.7 Primary and total ionization

We are now in a position to understand in detail the process of energy loss by ionization of a charged particle. On the passage of the particle, a discrete number of primary ionizing collisions takes place which liberate electron-ion pairs in the medium. The electron ejected can have enough energy (larger than the ionization potential of the medium) to further ionize, producing secondary ion pairs; the sum of the two contributions is called total ionization. Both the primary and the total ionization have been measured for most gases, although not always in the minimum ionizing region. The total number of ion pairs can be conveniently expressed by

$$n_T = \frac{\Delta E}{W_i} ,$$

where ΔE is the total energy loss in the gas volume considered, and W_i is the effective average energy to produce one pair. Values of W_i for different gases are given in Table 1. No simple expression exists for the number of primary ion pairs; in the table, experimental

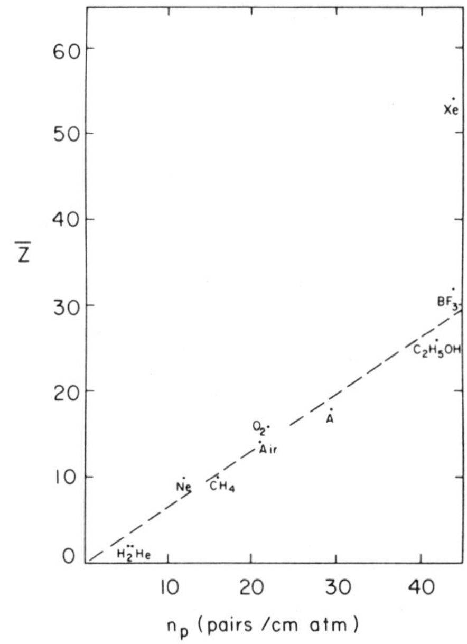

Fig. 7

Primary ionizing events produced by fast particles per unit length at normal conditions for several gases, as a function of their average atomic number (from Table 1). With the exception of xenon, all experimental points lie around a straight line. The plot may be used to estimate the number of primary pairs in other gases.

values of n_p for minimum ionizing particles are given, per unit length and at normal conditions. As Fig. 7 shows, n_p is roughly linearly dependent on the average atomic number of the gas, and the figure can be used to estimate the number of primary pairs for other molecules (an exception is xenon). This has been done, for example, to obtain W_i for carbon dioxide (CO_2) and isobutane (C_4H_{10}) (values in brackets in Table 1).

The reader should be warned that the values of W_i and of n_p listed in the table correspond just to a reasonable average over results of different experimenters[12]; values different from the averages by as much as 20-30% can be found.

A simple composition law can be used for gas mixtures; as an example, let us compute the number of primary and total ion pairs produced in a 1 cm thick 70-30 mixture of argon-isobutane, at normal conditions:

$$n_T = \frac{2440}{26}\,0.7 + \frac{4500}{23}\,0.3 \quad = 124 \text{ pairs/cm}$$

$$n_p = 29.4 \times 0.7 + 46 \times 0.3 = 34 \text{ pairs/cm} \quad .$$

In other words, we see that the average distance between primary interactions is around

300 μm at normal conditions, and that each primary produces about 2.5 secondaries on the average.

2.8 Statistics of ion-pair production

The primary ionization encounters, being a small number of independent events, follow Poisson-like statistics; if n is the average number of primary interactions ($n = n_p$), the actual number k in one event will have a probability

$$P_k^n = \frac{n^k}{k!} e^{-n} \quad .$$

The inefficiency of a perfect detector is therefore given by

$$1 - \varepsilon = P_0^n = e^{-n} \quad . \tag{4}$$

In the gas mixture previously considered, the inefficiency will be around 10^{-15} for a 1 cm thick detector ($n = 34$), and 3.3% for a 1 mm thick detector ($n = 3.4$). As a matter of fact, it is from inefficiency measurements in low-pressure proportional counters that the values of n_p have been deduced. When k ion pairs are produced in a given event, simple probabilistic considerations provide the space distribution of each pair j ($1 \le j \le k$) along a normalized coordinate x ($0 \le x \le 1$):

$$D_j^k(x) = \frac{k!}{(k-j)!(j-1)!} (1-x)^{k-j} x^{j-1}$$

and the general expression for the space distribution of the pair j , when n is the average number produced, is obtained as follows:

$$A_j^n(x) = \sum_{k=j}^{\infty} P_k^n D_j^k(x) = \frac{x^{j-1}}{(j-1)!} n^j e^{-nx} \quad .$$

Consider, in particular, the distribution of the pair closer to one end of the detection volume,

$$A_1^n(x) = n e^{-nx} \quad , \tag{5}$$

which is represented in Fig. 8, for $n = 34$, as a function of the coordinate across a 10 mm thick detector. If the time of detection is the time of arrival of the closest electron at one end of the gap, as is often the case, the statistics of ion-pair production set an obvious limit to the time resolution of the detector. A scale of time is also given in the figure, for a collection velocity of 5 cm/μsec typical of many gases; the FWHM of the distri-

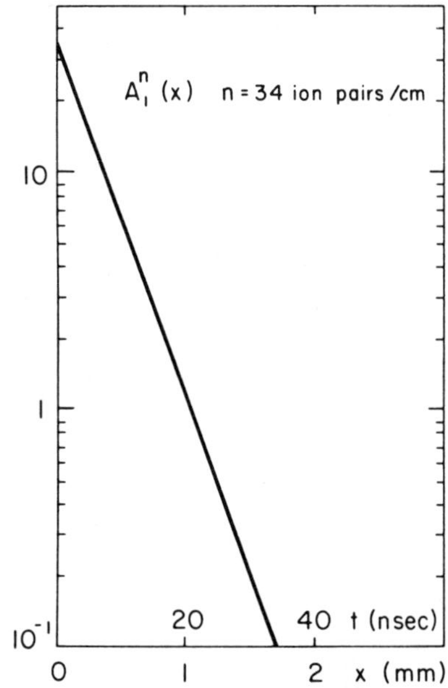

Fig. 8

Statistics of primary ion pair production: probability of finding the closest pair at a distance x from one electrode in a counter, in argon-isobutane 70-30. The corresponding electron minimum collection time is shown, for a typical drift velocity of electrons of 5 cm/μsec.

bution is about 5 nsec. There is no hope of improving this time resolution in a gas counter, unless some averaging over the time of arrival of all electrons is realized.

3. DETECTION OF PHOTONS

3.1 Different processes of absorption

As for charged particles, an electromagnetic interaction allows the detection of photons; in this case, however, the interaction is a single localized event. The probability of absorption can be written in terms of the cross-section σ, and the attenuation of a beam of photons traversing a thickness X of a medium having N molecules per unit volume is given by

$$I = I_0 e^{-\sigma NX} = I_0 e^{-\mu x} \quad , \tag{6}$$

where μ is the mass attenuation coefficient (normally expressed in $cm^2\ g^{-1}$) and $x = \rho X$ is the reduced thickness of the medium, see Section 2.2. One can also define an absorption mean free path $\lambda = (\mu\rho)^{-1}$ and in this case Eq. (6) is rewritten as

$$I = I_0 e^{-X/\lambda} \quad . \tag{7}$$

Depending on photon energy, the interaction can follow different mechanisms. At low energies, up to several keV, the dominant process is photoelectric conversion; then Compton scattering takes over, up to energies of a few hundred keV, and at even higher energies electron-positron pair production is the most probable process (Fig. 9).

3.2 Photoelectric absorption

We will consider this process in some detail for two reasons. Firstly, it is common practice for laboratory testing of proportional chambers to use X-ray emitting isotopes producing energy losses in the few keV region (as can be seen from Table 1, this is equal to the typical energy loss of a minimum ionizing particle in 1 cm of gas). Secondly, multi-

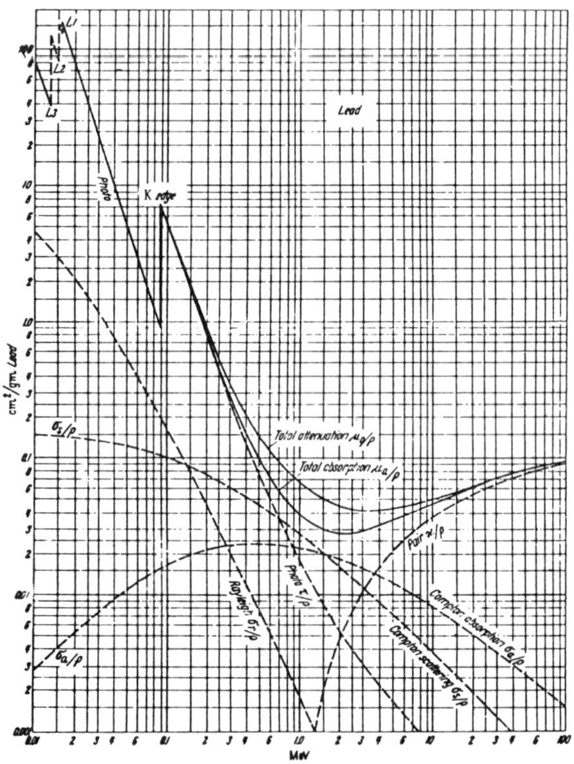

Fig. 9 Cross-section for photon absorption in lead, as a function of energy, showing the relative contribution of different processes[16])

wire proportional chambers are used as soft X-ray detectors in many applications, such as, for example, to study transition radiation, in crystal diffraction experiments, and in X-ray astronomy. A clear understanding of the detailed process of energy loss by the photoelectric effect is necessary in all these cases.

Photoelectric absorption is a quantum process involving one or more transitions in the electron shells of a molecule. Denoting by E_j the energy of a shell j, photoelectric absorption in the shell can take place only for photon energies $E_\gamma \geq E_j$ and, at a given energy, the contributions of all levels having $E_j < E_\gamma$ add up. The absorption is a maximum at the edge, and then very rapidly decreases with energy.

The binding energy of a given shell increases rapidly with atomic number, as shown in Fig. 10; precise values can be found in many textbooks (see the bibliography for this section). Figure 11, drawn using the tables of Grodstein[13], gives the photoelectric absorption coefficient as defined in the previous section, for several elements. Approximate values of the absorption edges are also shown. At a given energy, light elements have the smallest coefficient and, except for the heavier noble gases, at energies above a few keV K-edge ab-

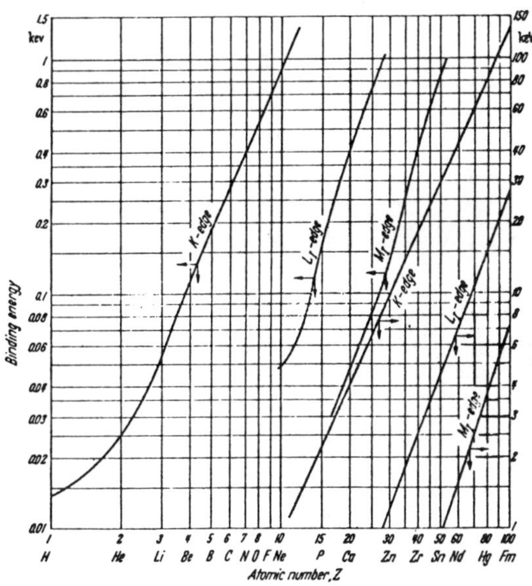

Fig. 10 Binding energy of K, L and M electron shells, as a function of atomic number[16]

sorption dominates. For a gas mixture or for complex molecules one can assume a simple composition rule to be valid; if p_1, p_2, ..., p_n are the percentages, in weight, of atoms 1, 2, ..., n in the mixture, the absorption coefficient of the compounds can be written as

$$\mu_{1,2\ldots n} = p_1\mu_1 + p_2\mu_2 + \ldots + p_n\mu_n \quad .$$

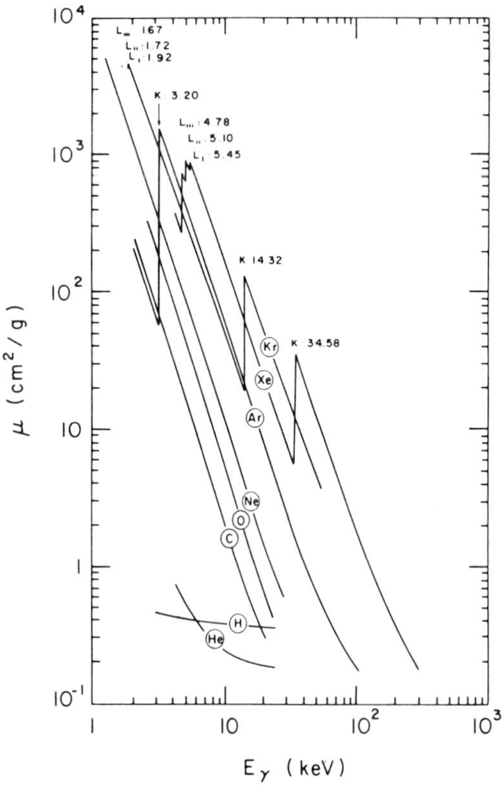

Fig. 11 Absorption coefficient versus energy of photons, in several gases used in proportional counters. Carbon is included, to allow the calculation of absorption in hydrocarbons. The figure has been drawn using the tabulations of Ref. 13.

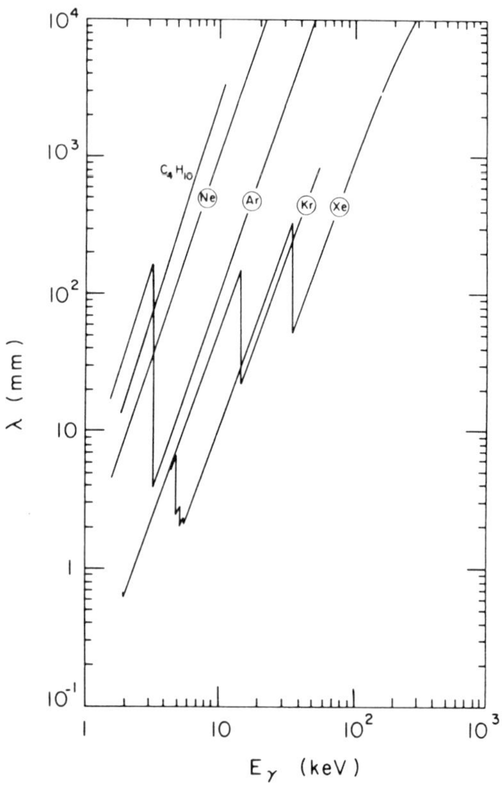

Fig. 12 Mean free path for absorption in several gases at normal conditions, for photons of energy E_γ (from the tabulation in Ref. 13)

In particular, the organic vapours used as quenchers in proportional counters have a very small absorption coefficient, since they mainly contain light elements.

Using the definition (7) one can compute the mean free path for absorption in gases under normal conditions, as shown in Fig. 12. Notice, for example, that the absorption in argon increases by almost two orders of magnitude when the photon energy increases over the K-shell edge of 3.20 keV.

Absorption of a photon of energy E_γ in a shell of energy E_j results in the emission of a photoelectron of energy $E_e = E_\gamma - E_j$; the excited molecule can then return to its ground state mainly through two competing mechanisms:

- fluorescence, i.e. the transition of an electron from an energy shell $E_i < E_j$ into the j-shell, with the emission of a photon of energy $E_j - E_i$;
- radiationless transition, or Auger effect, which is an internal rearrangement involving several electrons from the lower energy shells, with the emission of an electron of energy very close to E_j.

The fraction of de-excitations producing the emission of a photon is called fluorescence yield. For the K-shell, the fluorescence yield increases with the atomic number as shown in Fig. 13 [14]. In argon, for example, about 15% of the photoelectric absorptions are followed by the emission of a photon, while in 85% of the events two electrons, of energy $E_\gamma - E_K$ and slightly smaller than E_K, respectively, are produced. The secondary photon, emitted at an energy just below the K-edge, has of course a very long mean free path for absorption and can therefore escape from the volume of detection. This produces the characteristic escape peak of argon, at an energy $E_\gamma - E_K$. In detectors where localization of the conversion point is required, the emission of a long-range photon can introduce a large error if the position is estimated with a centre-of-gravity method as is often the case. A quantitative discussion on this point can be found in Bateman et al.[15] for xenon-filled counters.

The primary photoelectron is emitted in a preferential direction, depending on its energy, as shown in Fig. 14; up to a few tens of keV, the direction of emission is roughly perpendicular to the incoming photon direction. However, as already discussed in Section 2.4, multiple scattering quickly randomizes the motion of the heavily ionizing photoelectron. The range of electrons in gases was also discussed in Section 2.5.

3.3 Compton scattering

When the photon energy rises well above the highest atomic energy level, Compton scattering begins to be the dominant process. The incident photon, with energy $h\nu_0$, is scattered by a quasi-free electron by an angle θ, and takes a new energy $h\nu'$ such that

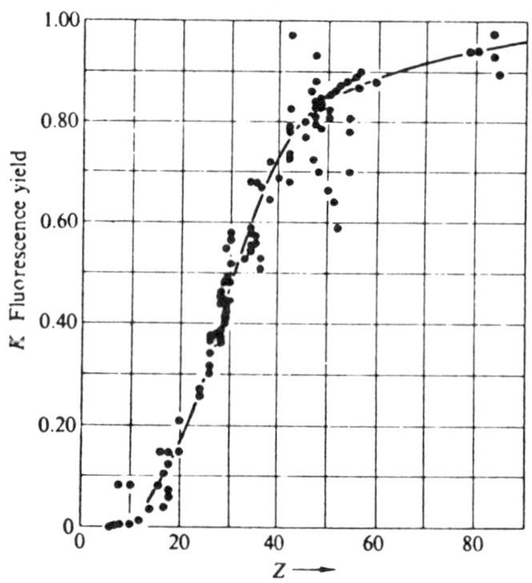

Fig. 13 Fluorescence yield of the K shell i.e. the fraction of de-excitations producing the emission of a photon, versus atomic number[14])

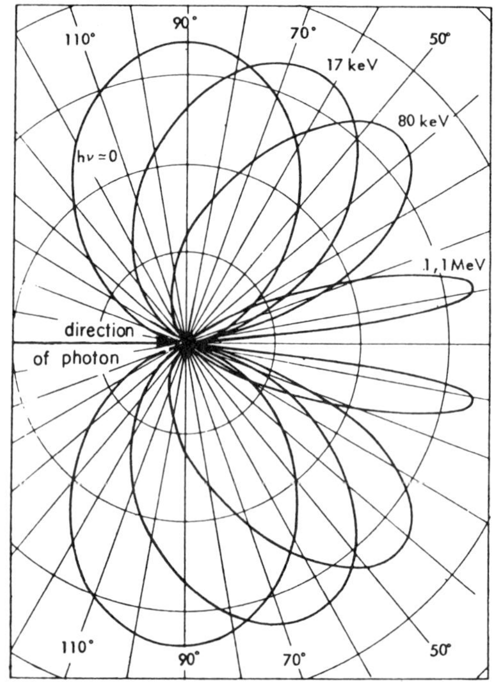

Fig. 14 Probability of emission of a photoelectron at a given angle in respect to the incoming photon direction, as a function of the quantum energy[4])

$$\frac{1}{h\nu'} - \frac{1}{h\nu_0} = \frac{1}{mc^2}(1 - \cos\theta) \quad .$$

The energy of the scattered electron is $h\nu_0 - h\nu'$. Quantum-mechanical theory provides the differential cross-section and the angular distributions for this effect (see the bibliography for this section). Figure 15 shows the computed absorption coefficient for Compton scattering (coherent and incoherent), as from Ref. 13; Figs. 16a and 16b [16]) show correlations between the energies and angles of diffusion, while Fig. 17 is a polar plot of the cross-section for different photon energies[16]).

In a thin gas counter, it is unlikely for the scattered photon to be absorbed again; therefore, the energy deposit depends on the (unknown) angle of scattering. No energy resolution is therefore obtained in the Compton region.

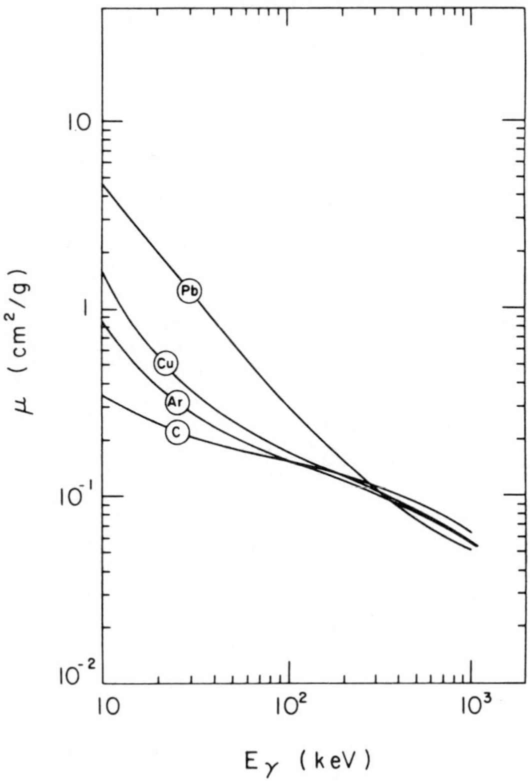

Fig. 15 Absorption coefficient for Compton scattering (coherent and incoherent) as a function of photon energy, for several elements[16)]

3.4 Pair production

Electron-positron pair production can take place at photon energies above the threshold of 1.02 MeV (corresponding to two electron masses); Figure 18 shows the absorption coefficient due to pair production for several elements[13)], while Fig. 19 gives the relative energy-sharing between e^+ and e^-. In the figure, the ordinate is a quantity proportional to the differential cross-section for the process, increasing with photon energy (given in terms of the electron rest mass). Because of the very small value of the absorption coefficient, proportional chambers are used in this energy region only to detect the pairs pro-

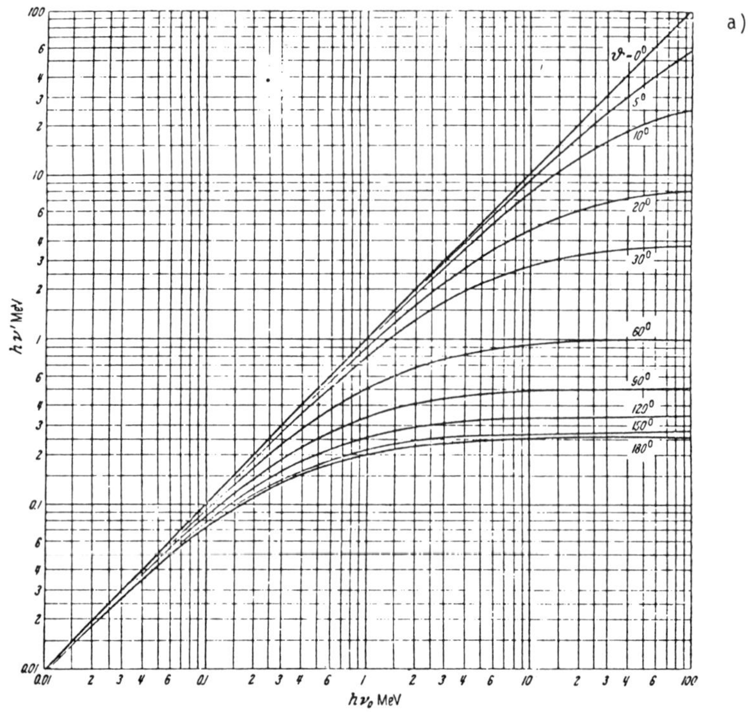

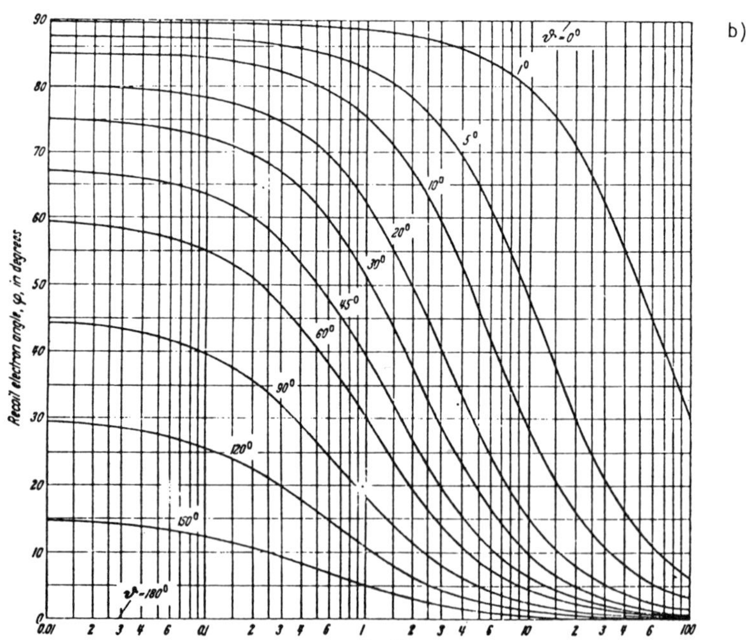

Fig. 16 Computed correlation (a) between energy $h\nu'$ and angle θ of the Compton scattered photon, and (b) between the recoil electron angle ϕ and θ, both as a function of the incident photon energy $h\nu_0$ [16])

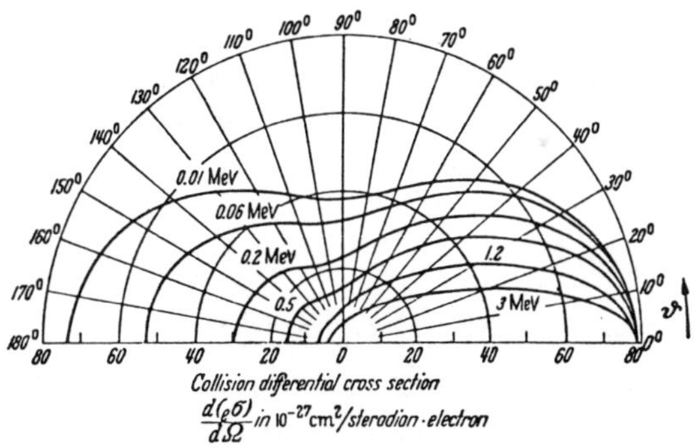

Fig. 17 Polar plot of the Compton cross-section as a function of the photon scattering angle θ, for several incident photon energies[16])

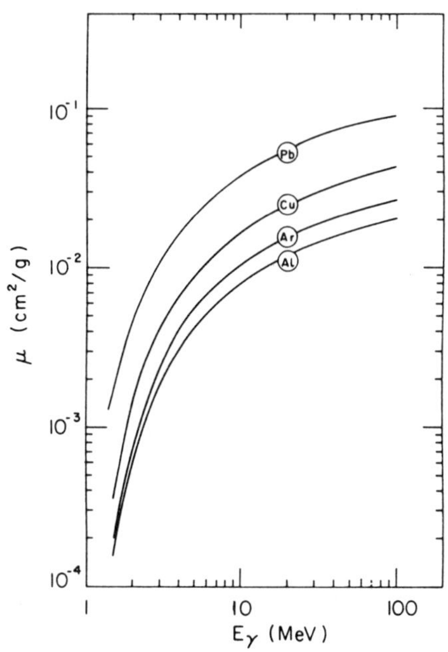

Fig. 18 Absorption coefficient for e^+e^- pair production in several materials, as a function of the incoming photon energy (drawn from the Tables of Ref. 13)

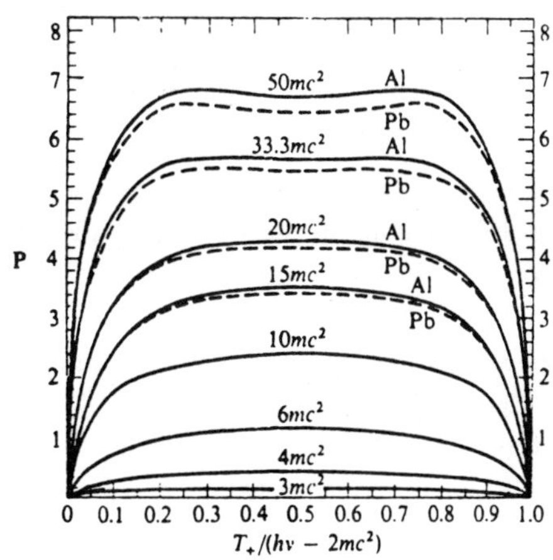

Fig. 19 Relative energy sharing between e^+ and e^- in pair production, for several incoming energies (given in terms of the electron rest energy mc^2 = 0.511 MeV). The ordinate is proportional to the cross-section for the process, as given in the previous curve. (From N.A. Dyson, see bibliography for this section.)

duced in a layer of heavy material placed in front of the chamber. Shower counters are either constructed in this way sandwiching chambers with conversion plates, or by using high density drift chambers with conversion cells of heavy material in the gas volume[17].

4. DRIFT AND DIFFUSION OF CHARGES IN GASES

4.1 Ion and electron diffusion without electric fields

Charges produced by an ionizing event quickly lose their energy in multiple collisions with the gas molecules and assume the average thermal energy distribution of the gas. Simple kinetic theory of gases provides the average value of the thermal energy, $\varepsilon_T = (3/2)\ kT \simeq 0.04$ eV at normal conditions, and the Maxwellian probability distribution of the energies

$$F(\varepsilon) = C \sqrt{\varepsilon}\, e^{-(\varepsilon/kT)} \quad . \tag{8}$$

In the absence of other effects, a localized distribution of charges diffuses by multiple collisions following a Gaussian law:

$$\frac{dN}{N} = \frac{1}{\sqrt{4\pi Dt}}\, e^{-(x^2/4Dt)}\, dx \quad , \tag{9}$$

where dN/N is the fraction of charges found in the element dx at a distance x from the origin, and after a time t; D denotes the diffusion coefficient. The root mean square (r.m.s.) of the distribution, or standard deviation, is given by

$$\sigma_x = \sqrt{2Dt} \quad \text{or} \quad \sigma_V = \sqrt{6Dt} \quad , \tag{10}$$

respectively, for a linear and a volume diffusion. As an example, Fig. 20 [18] shows the space distribution of ions produced in air, at normal conditions, after different time intervals; values of the diffusion coefficient of several ions in their own gas are given in Table 2. During diffusion, ions collide with the gas molecules; the mean free path, average velocity, and time between collisions are also given in the table. Electrons move much faster than ions (because of their small mass), their average thermal velocity being about 10^7 cm/sec; also, because of their negligible size, their mean free path is classically four times longer than that of ions in a like gas.

A positive ion can be neutralized recombining in the gas volume with a negative charge carrier, either an electron or a negative ion, or extracting an electron at the walls. A process of charge transfer is also possible with a molecule of its own gas or with molecules of another kind having lower ionization potentials. We will reconsider this very important effect in the next section.

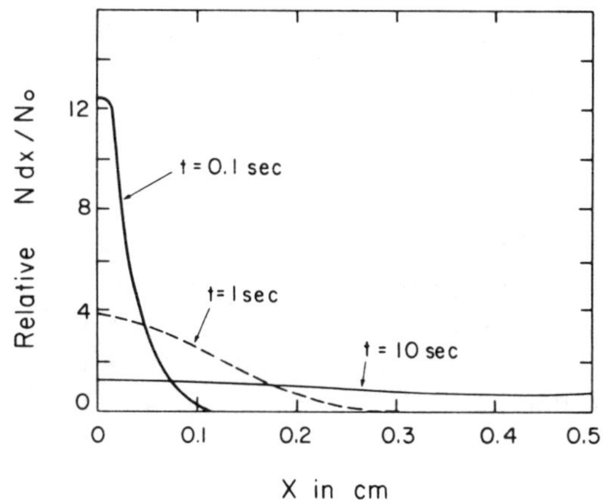

Fig. 20 Space distribution of ions produced in air, at normal conditions, after different time intervals[18]

Electrons, instead, can be neutralized by an ion, can be attached to a molecule having electron affinity (electro-negative), or can be absorbed in the walls. The probability of attachment h is essentially zero for all noble gases and hydrogen, while it assumes finite values for other gases (see Table 3). In the table we have also shown, taking into account the previous data, the average attachment time $t = (hN)^{-1}$, if N is the number of collisions per unit time. One can see, for example, that in oxygen the average time it takes for a thermal electron to be attached is 140 nsec. The attachment coefficient is a strong function of the electric field, as will be shown later.

Table 2

Classical mean free path, velocity, diffusion coefficients, and mobility for molecules, under normal conditions[18-21]

Gas	λ (cm)	u (cm/sec)	D^+ (cm^2/sec)	μ^+ ($cm^2\ sec^{-1}\ V^{-1}$)
H_2	1.8×10^{-5}	2×10^5	0.34	13.0
He	2.8×10^{-5}	1.4×10^5	0.26	10.2
Ar	1.0×10^{-5}	4.4×10^4	0.04	1.7
O_2	1.0×10^{-5}	5.0×10^4	0.06	2.2
H_2O	1.0×10^{-5}	7.1×10^4	0.02	0.7

Table 3

Coefficient, number of collisions,
and average time for electron attachment
in several gases under normal conditions[12,18,21]

Gas	h	N (sec^{-1})	t (sec)
CO_2	6.2×10^{-9}	2.2×10^{11}	0.71×10^{-3}
O_2	2.5×10^{-5}	2.1×10^{11}	1.9×10^{-7}
H_2O	2.5×10^{-5}	2.8×10^{11}	1.4×10^{-7}
$C\ell$	4.8×10^{-4}	4.5×10^{11}	4.7×10^{-9}

4.2 Mobility of ions

When an electric field is applied across the gas volume, a net movement of the ions along the field direction is observed. The average velocity of this slow motion (not to be confused with the instant ion velocity) is called drift velocity w^+, and it is found to be linearly proportional to the reduced field E/P up to very high fields, P being the gas pressure. It is therefore convenient to define a quantity μ^+, called mobility, as

$$\mu^+ = \frac{w^+}{E} . \tag{11}$$

The value of the mobility is specific to each ion moving in a given gas. A constant mobility is the direct consequence of the fact that, up to very high fields, the average energy of ions is almost unmodified; we will see that this is not the case for the electrons.

A classical argument allows one to obtain the following relationship between mobility and diffusion coefficient:

$$\frac{D^+}{\mu^+} = \frac{kT}{e} . \tag{12}$$

Values of the mobility and diffusion coefficient for ions moving in a like gas were given in Table 2, while Table 4 gives the mobility of several ions drifting in gases commonly used in proportional and drift chambers[20].

In a mixture of gases G_1, G_2, ..., G_n, the mobility μ_i^+ of the ion G_i^+ is given by the relationship (Blanc's law):

Table 4

Experimental mobilities of several ions in different gases, at normal conditions[20]

Gas	Ions	Mobility ($cm^2 V^{-1} sec^{-1}$)
Ar	$(OCH_3)_2CH_2^+$	1.51
$IsoC_4H_{10}$	$(OCH_3)_2CH_2^+$	0.55
$(OCH_3)_2CH_2$	$(OCH_3)_2CH_2^+$	0.26
Ar	$IsoC_4H_{10}^+$	1.56
$IsoC_4H_{10}$	$IsoC_4H_{10}^+$	0.61
Ar	CH_4^+	1.87
CH_4	CH_4^+	2.26
Ar	CO_2^+	1.72
CO_2	CO_2^+	1.09

$$\frac{1}{\mu_i^+} = \sum_{j=1}^{n} \frac{p_j}{\mu_{ij}^+} , \tag{13}$$

where p_j is the volume concentration of gas j in the mixture, and μ_{ij}^+ the mobility of ion G_i^+ in gas G_j. In gas mixtures, however, a very effective process of charge transfer takes place, and very quickly removes all ions except the ones with the lower ionization potential. Depending on the nature of the ions, and on the difference in the ionization potential (small differences increase the charge-transfer probability), it takes between 100 and 1000 collisions for an ion to transfer its charge to a molecule having a lower ionization potential. Since mean free paths for collision are of the order of 10^{-5} cm under normal conditions, see Table 2, one can assume that after a drift length between $10^{-3}\ p^{-1}$ and $10^{-2}\ p^{-1}$ cm, where p is the percentage of the lowest ionization potential molecules, the charge-exchange mechanism will have left migrating only one kind of ion. Figure 21 [20] shows the measurement of ion mobility in the mixtures argon-isobutane and argon-isobutane-methylal. As shown by the relationship (13), for a given kind of ion drifting in a gas mixture, the inverse of the mobility depends linearly on the mixture's specific weight; lines of equal slope, therefore, represent migration of the same kind of ions. In the figure, curve F represents the mobility

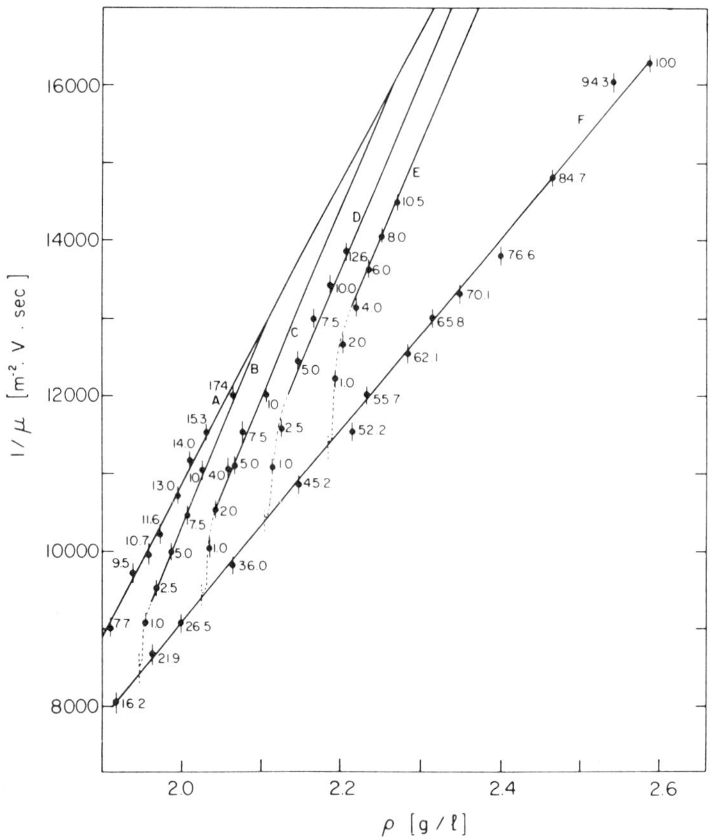

Fig. 21 Inverse mobility of ions migrating in argon-isobutane mixtures (curve F), in argon-methylal (curve A) and argon-isobutane-methylal (B = 80% argon, C = 70% argon, D = 60% argon, E = 50% argon). The numbers close to the experimental points represent the methylal concentration (curves A to E) or the isobutane concentration (curve F)[20].

of isobutane ions in variable argon-isobutane mixtures, while curves B, C, D, E represent the measured mobility of methylal ions in several argon-isobutane-methylal mixtures; the fraction of methylal in each measurement is written close to the experimental points. Clearly in the range of electric fields considered (a few hundred to a thousand volts/cm) and for 1 cm of drift, if methylal is present in the mixture by more than 3-4% the exchange mechanism is fully efficient and only methylal ions are found to migrate. The relevance of this mechanism in the operation of proportional counters will be discussed in Section 5.3.

Ions migrating in a time t over a length x diffuse with a probability distribution expressed by Eq. (9), and the standard deviation is given by [introducing expressions (12) and (11) in (10)]:

$$\sigma_x = \sqrt{2Dt} = \sqrt{\frac{2kTw^+t}{eE}} = \sqrt{\frac{2kTx}{eE}} .$$

Therefore the r.m.s. linear diffusion is independent of the nature of the ions and the gas; the variation of σ_x with the electric field, at 1 atm, is shown in Fig. 22, as well as the equivalent time dispersions σ_t for several gas mixtures[20], and 1 cm drift.

4.3 Drift of electrons

A simple theory of mobility can be constructed following the same lines as for positive ions; it was found very early on, however, that except for very low fields the mobility of electrons is not constant. In fact, due to their small mass, electrons can substantially in-

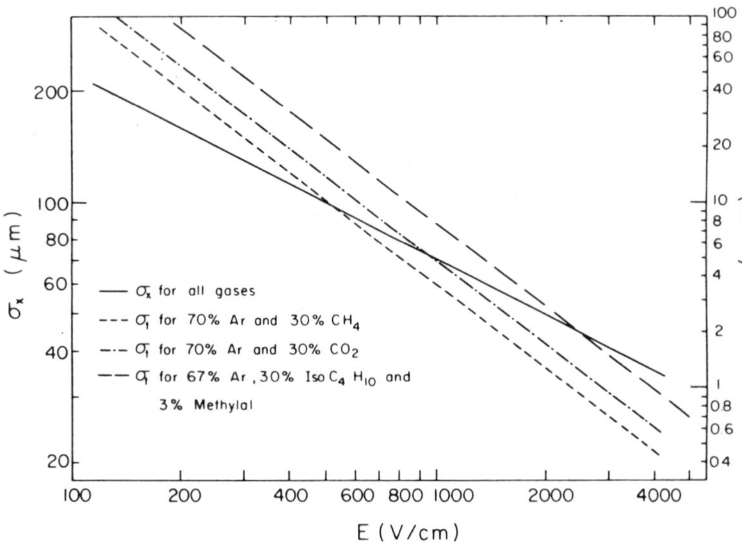

Fig. 22 Positive ion diffusion in space (σ_x) and in time (σ_t) for a drift length of 1 cm, at normal conditions as a function of electric field[20]. Notice that σ_x is the same for all gases.

crease their energy between collisions with the gas molecules under the influence of an electric field. In a simple formulation, due to Townsend[21] one can write the drift velocity as

$$w = \frac{e}{2m} E\tau , \tag{14}$$

where τ is the mean time between collisions, in general a function of the electric field E. It has been found that the collision cross-section, and therefore τ, varies for some gases very strongly with E, going through maxima and minima (Ramsauer effect). This is a consequence of the fact that the electron wavelength approaches those of the electron shells of the molecule, and complex quantum-mechanical processes take place. The energy distribution will therefore change from the original Maxwellian shape [as given by Eq. (8)] and the average energy can exceed the thermal value by several orders of magnitude, at high fields. As an example, Fig. 23 shows the energy distribution of electrons in helium at several values of the electric field[18]. Figure 24 gives instead a typical Ramsauer cross-section measurement for argon[22]; it appears that the addition of even very small fractions of another gas to pure argon can, by slightly modifying the average energy, dramatically change the drift properties, as illustrated in Fig. 25 [22]. We have collected from different sources[9,12,22-24], in Figs. 26 to 30 the measured drift velocities for several pure gases

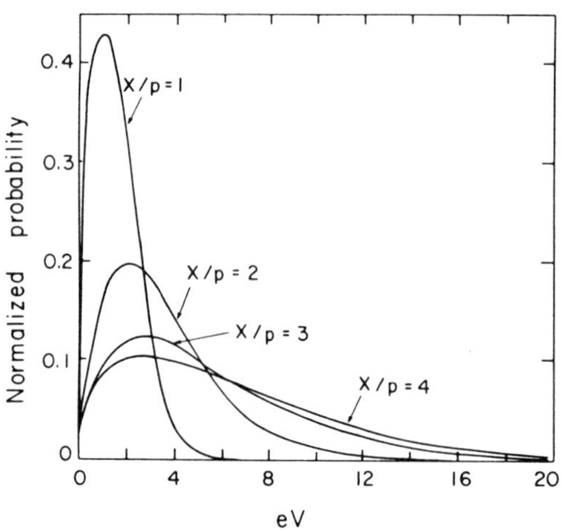

Fig. 23

Computed energy distribution of electrons in helium at different field values (X/p = 1 means 760 V/cm at atmospheric pressure)[18]

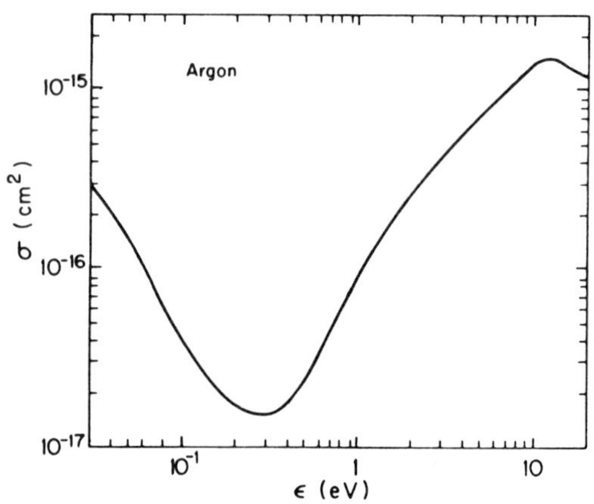

Fig. 24

Ramsauer cross-section for electrons in argon as a function of their energy[22]

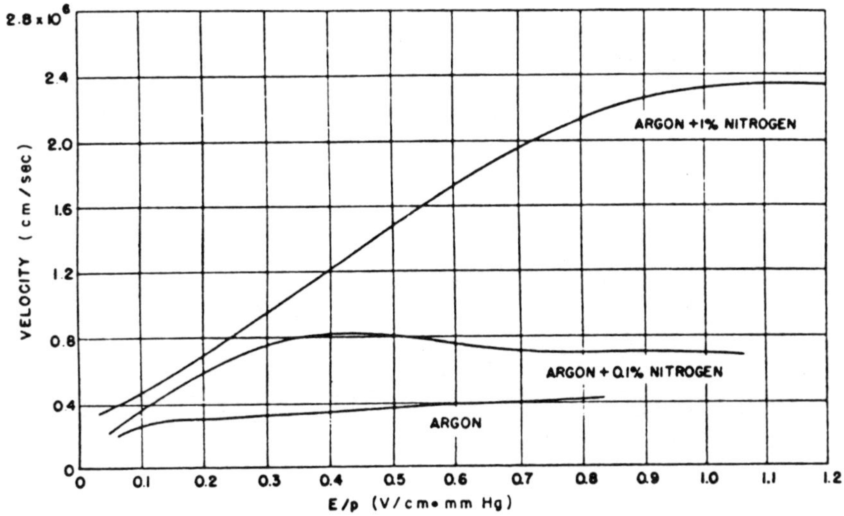

Fig. 25 Drift velocity of electrons in pure argon, and in argon with small added quantities of nitrogen. The very large effect on the velocity for small additions is apparent[22]).

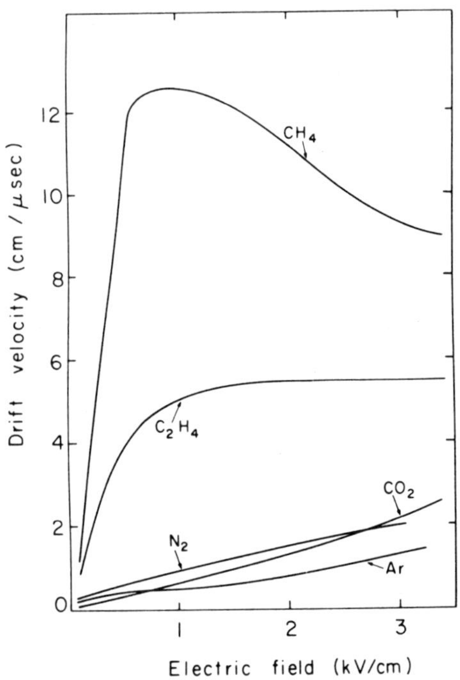

Fig. 26 Drift velocity of electrons in several gases at normal conditions[12,22,23])

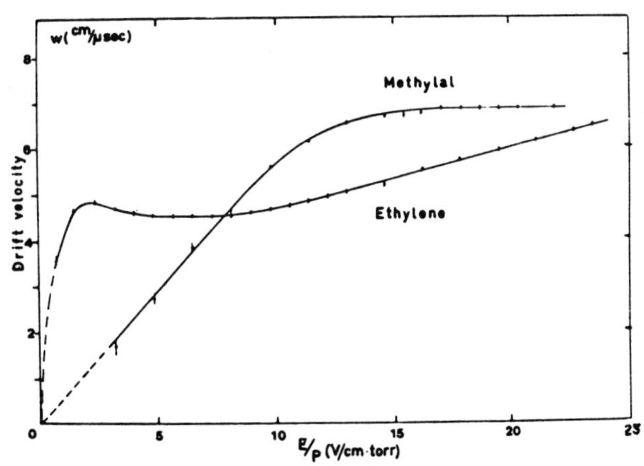

Fig. 27 Drift velocity of electrons in methylal [$(OCH_3)_2CH_2$] and in ethylene (C_2H_4) [24])

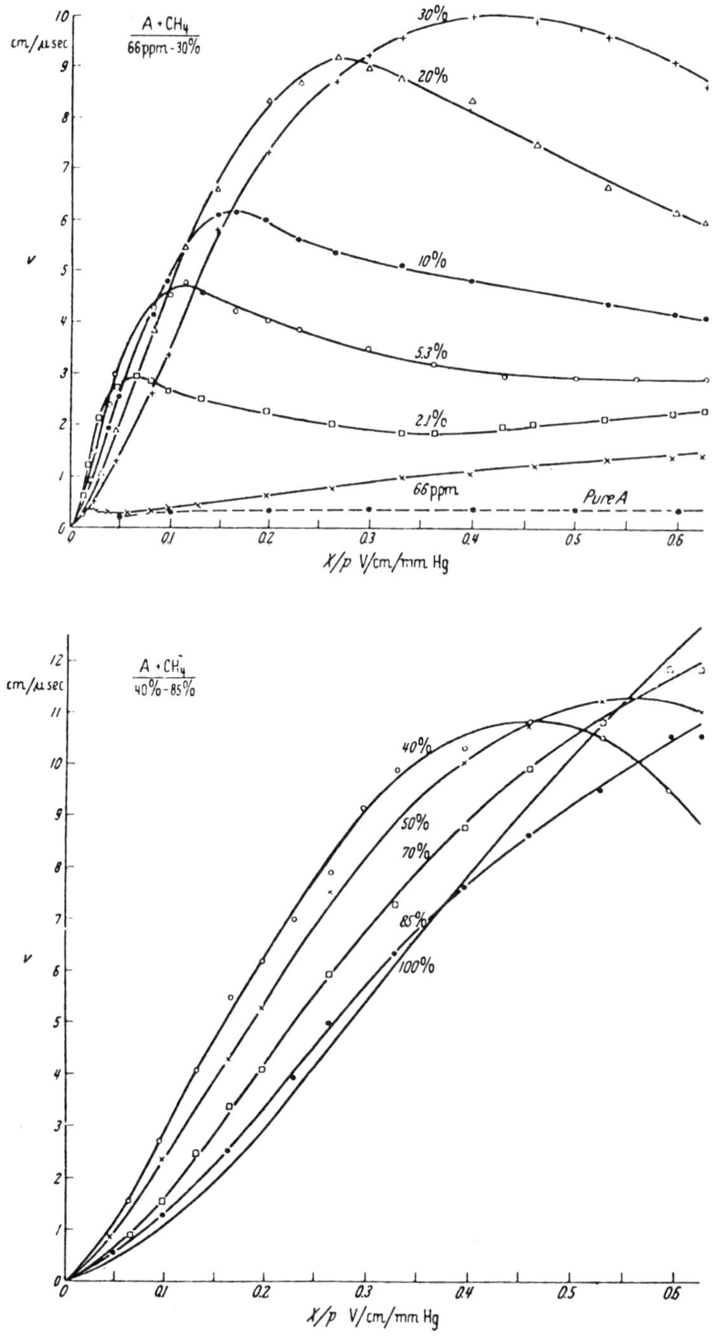

Fig. 28 Drift velocity of electrons in several argon-methane mixtures[12)]

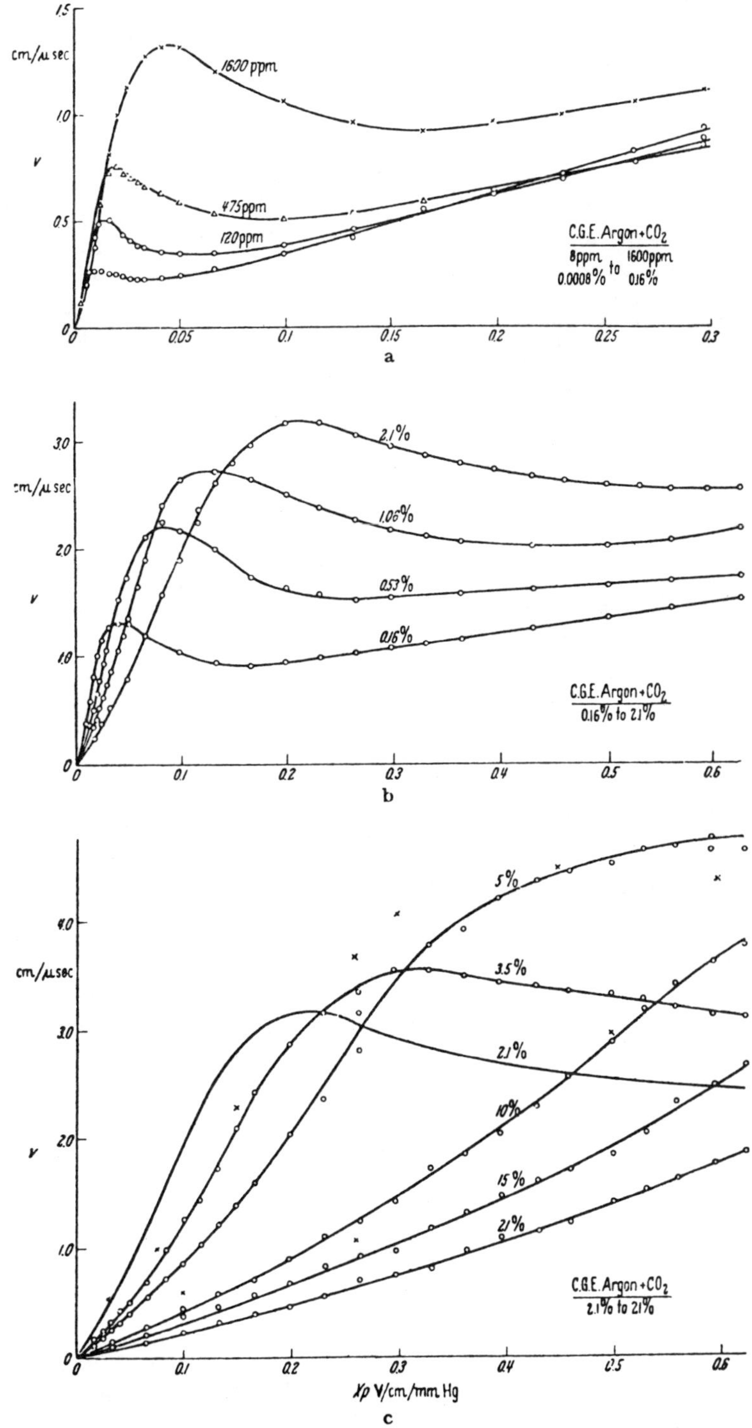

Fig. 29 Drift velocity of electrons in several argon-carbon dioxide mixtures[12)]

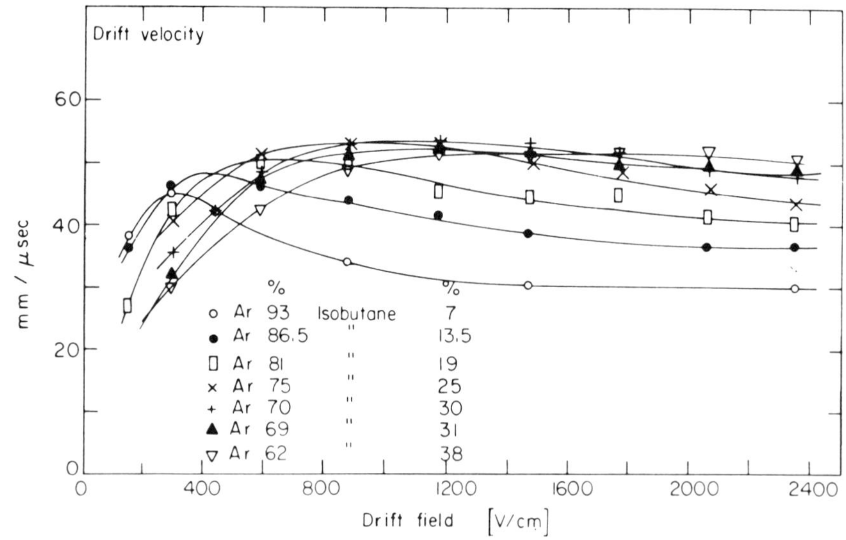

Fig. 30 Drift velocity of electrons in argon-isobutane mixtures, at normal conditions[9])

and gas mixtures of interest in proportional chambers. Depending on the source, the abscissa is given in terms of the electric field at normal conditions, or of the reduced field E/P; to obtain in the last case the equivalent field at 1 atm one has to multiply the scale by 760. At high fields, typical values of w around 5 cm/μsec are obtained; use of Table 2 shows that, under similar conditions, ions are roughly a thousand times slower.

A rigorous theory of electron drift in gases exists; it has recently been reviewed for the study of drift chamber performances[25,26]). Here, we will only summarize the main results. Under rather broad assumptions, and for fields such that only a negligible fraction of the electrons get enough energy to experience ionizing collisions, one can deduce the following expression for the energy distribution:

$$F(\varepsilon) = C\sqrt{\varepsilon}\,\exp\left(-\int\frac{3\Lambda(\varepsilon)\varepsilon d\varepsilon}{[eE\lambda(\varepsilon)]^2 + 3\varepsilon kT\Lambda(\varepsilon)}\right), \tag{15}$$

where the mean free path between collisions, $\lambda(\varepsilon)$, is given by

$$\lambda(\varepsilon) = \frac{1}{N\sigma(\varepsilon)} \quad , \tag{16}$$

N being the number of molecules per unit volume. At the temperature T and pressure P, N is given by

$$N = 2.69 \times 10^{19} \frac{P}{760} \frac{273}{T} \text{ molecules/cm}^3$$

and the cross-section $\sigma(\varepsilon)$ is deduced from the Ramsauer curve of the gas considered. In Eq. (15), $\Lambda(\varepsilon)$ is the fraction of energy lost on each impact, or inelasticity; in other words the amount of energy spent in processes like rotational and vibrational excitations. If the elastic and inelastic cross-sections are known, $F(\varepsilon)$ can be computed and the drift velocity and diffusion coefficient are given by

$$w(E) = -\frac{2}{3}\frac{eE}{m} \int \varepsilon\lambda(\varepsilon) \frac{\partial[F(\varepsilon)\, u^{-1}]}{\partial\varepsilon} d\varepsilon \tag{17}$$

$$D(E) = \int \frac{1}{3} u\lambda(\varepsilon)F(\varepsilon)d\varepsilon \quad , \tag{18}$$

where $u = \sqrt{2\varepsilon/m}$ is the instant velocity of electrons of energy ε.

Simple rules hold for gas mixtures

$$\sigma(\varepsilon) = \sum p_i\sigma_i(\varepsilon) \quad \text{and} \quad \sigma(\varepsilon)\Lambda(\varepsilon) = \sum p_i\sigma_i(\varepsilon)\Lambda_i(\varepsilon)$$

with obvious meaning. It is customary to define a characteristic energy ε_k as follows:

$$\varepsilon_k = \frac{eE\,D(E)}{w(E)} \quad . \tag{19}$$

Figures 31 and 32 show the computed and measured drift velocity and the characteristic energy for argon on carbon dioxide[26], while Figs. 33 and 34 give the dependence of ε_k on the electric field, at normal conditions, for several pure gases and gas mixtures[25].

4.4 Diffusion of electrons

During the drift in electric fields, electrons diffuse following a Gaussian distribution like formula (9); the change in the energy distribution due to the electric field does, of

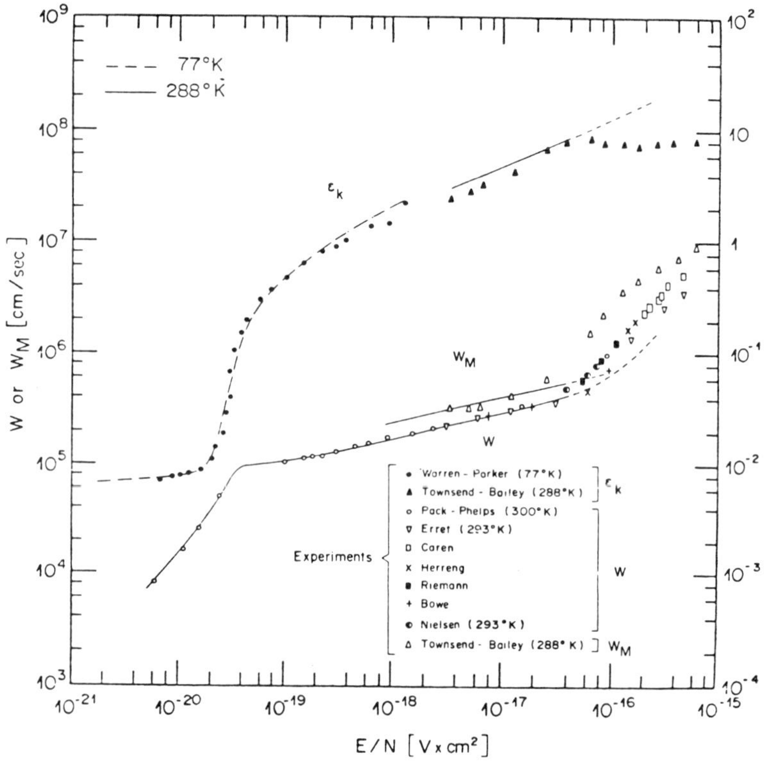

Fig. 31 Comparison of measured and computed drift velocities and characteristic energy for argon[26])

course, result in a coefficient of diffusion dependent on E, as shown by Eq. (18). Figure 35 gives, as a function of the electric field, the computed value of the standard deviation of space diffusion σ_x, as defined by formulae (10) and for 1 cm of drift[25]). The thermal limit is also shown, corresponding to a fictitious gas where the energy of electrons is not increased by the presence of the field; carbon dioxide, because of its very low characteristic energy, is very close to the thermal limit. For a 75-25 mixture of argon-isobutane, very close to the one often used in proportional and drift chambers, $\sigma_x \simeq 200\ \mu m$ independently of E. In drift chambers, one obtains the space coordinates of ionizing tracks from the measurement of the drift time, in a more or less uniform field, of the electron swarm. A small diffusion coefficient leads of course to a better accuracy;

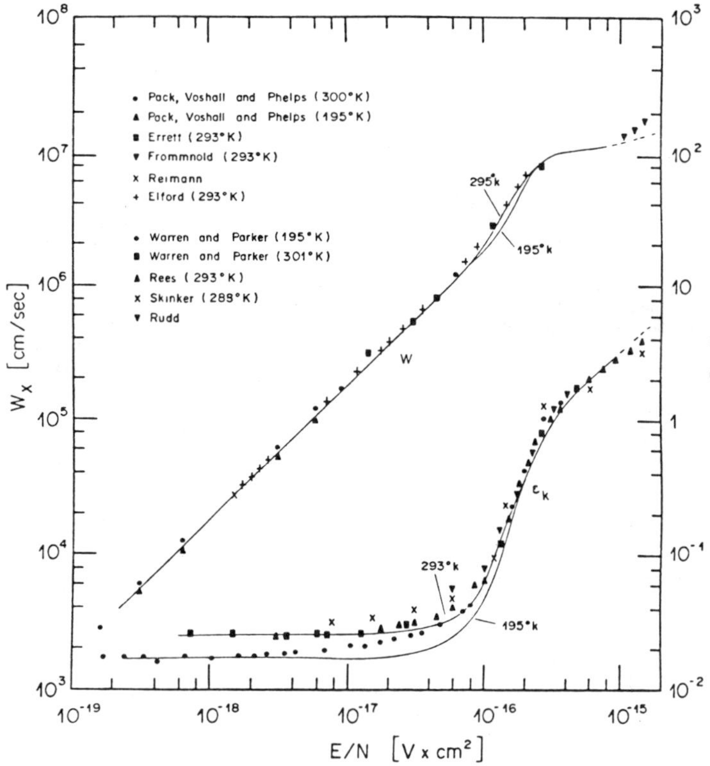

Fig. 32 Comparison of measured and computed drift velocities and characteristic energy for carbon dioxide[26)]

the choice of carbon dioxide is, however, often forbidden by the poor quenching properties of this gas in proportional counters (see Section 5.3) and the quoted argon-isobutane mixture is often preferred.

Notice that the limiting accuracy with which one can localize the drifting swarm is not directly given by σ_x, but by its variance, depending on the number of electrons necessary to trigger the time-measuring device. For example, if the average time of the n drifting electrons is measured, the limiting accuracy will be $\sigma_x/\sqrt{n}$. A general expression, which can be written for the case where k electrons out of n are necessary to trigger the detecting electronics[26)], is

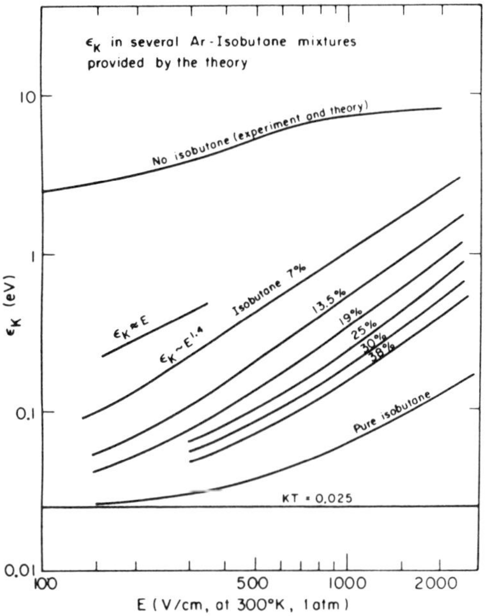

Fig. 33 Computed dependence of the characteristic energy on electric field for several argon-isobutane mixtures[25)]

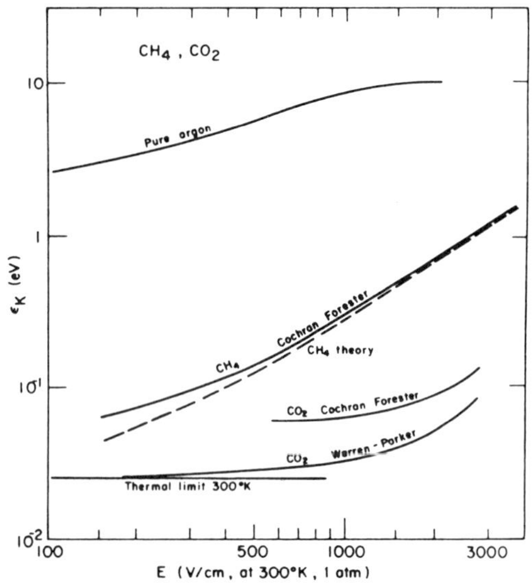

Fig. 34 Computed dependence of the characteristic energy on electric field for pure argon, carbon dioxide and methane, from different sources[25)]

$$\sigma_k^2 = \frac{\sigma_x^2}{2 \ln n} \sum_{i=k}^{n} \frac{1}{i^2} \; .$$

For k = 1, the expression can be seen to reduce to

$$\sigma_1 = \frac{\pi}{2\sqrt{3}\, \ln n}\, \sigma_x \; , \qquad (20)$$

for example, for n = 100 (typical value for 1 cm of gas), $\sigma_1 = 0.4\ \sigma_x$. We will discuss this point further in the section devoted to high-accuracy drift chambers.

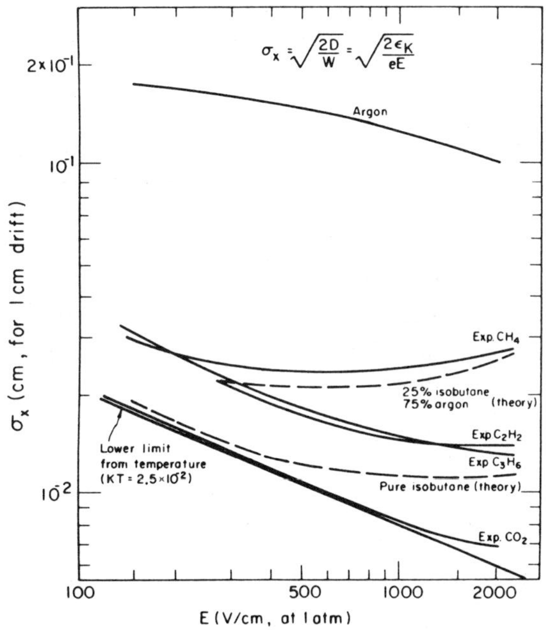

Fig. 35 Computed and experimental dependence of the standard deviation of electron diffusion from the electric field for 1 cm drift, in several gases at normal conditions[25]

4.5 Drift of electrons in magnetic fields

The presence of a magnetic field modifies the drift properties of a swarm of electrons. The Lorentz force applied to each moving charge transforms the small segment of motion between two collisions into circular trajectories, and also modifies the energy distribution. The net effect is a reduction of the drift velocity, at least at low electric fields, and a movement of the swarm along a line different from a field line. In the case of a movement in a constant electric and magnetic field, the swarm will drift along a straight line at an angle α_H with the field lines, and with a velocity $w_H \neq w$. The same simple theory that gives the expression (14) allows one to write the effect of a magnetic field H applied in a direction perpendicular to the electric field[21] as follows:

$$w_H = \frac{w}{\sqrt{1 + \omega^2\tau^2}} , \qquad \omega = \frac{eH}{m}$$

$$\tan \alpha_H = \omega\tau \quad .$$

Substituting the value of τ obtained from Eq. (14), one gets an approximation which is rather good for low electric fields, as shown in Fig. 36, where experimental points are compared with calculation, for the standard gas mixture used in high-accuracy drift chambers[9], i.e. 67.2% argon, 30.3% isobutane and 2.5% methylal. At higher fields, the presence of the magnetic field modifies the energy distribution of electrons and a more rigorous analysis is necessary, similar to the one sketched in the previous section[25,26]. We will present here some experimental measurements of w_H and α_H, in the quoted gas mixture,

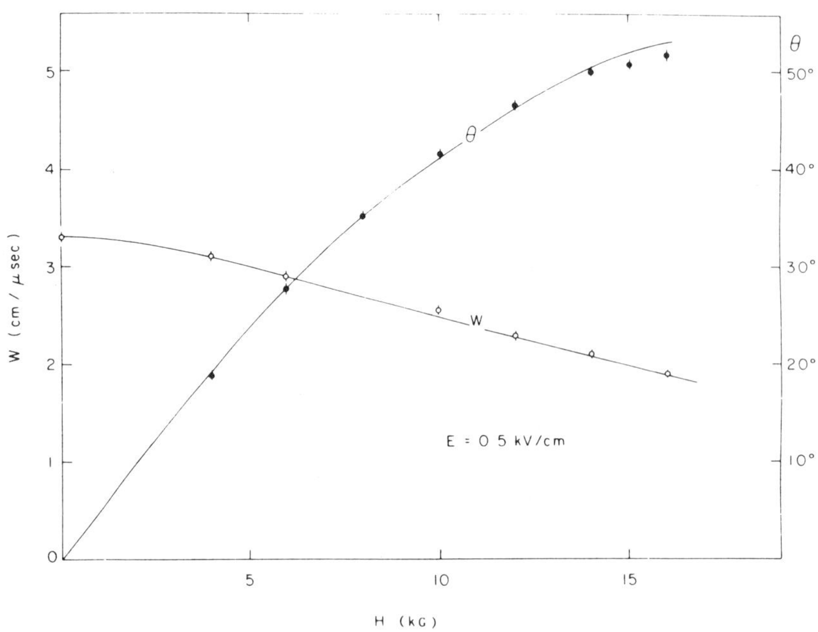

Fig. 36 Measured dependence of electron drift velocity and drift angle from the magnetic field for a low value of electric field (500 V/cm), in argon-isobutane-methylal (67.2%, 30.3%, 2.5%, respectively). The curves represent the predictions of an approximated model[9].

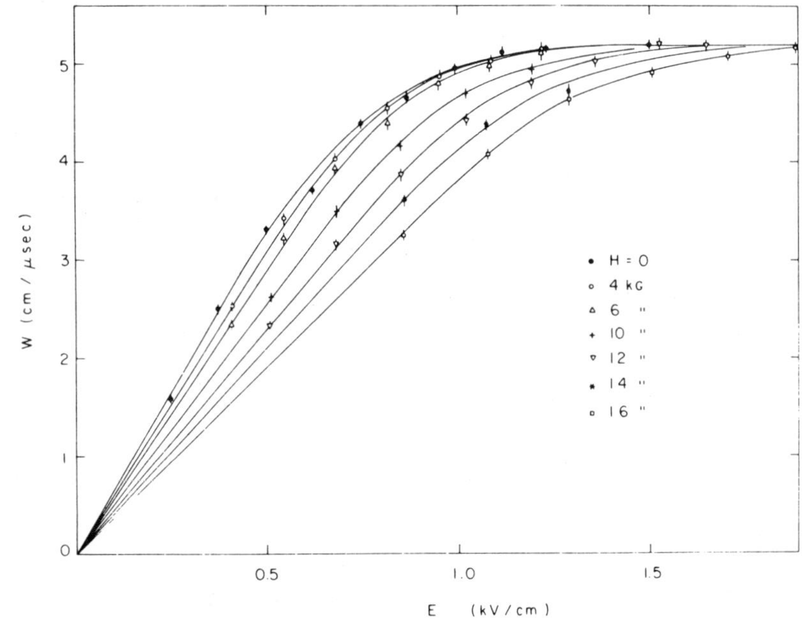

Fig. 37 Measured drift velocity of electrons, in the direction of the motion, as a function of electric field for several values of the magnetic field (perpendicular to the plane of drift)[9].

from Ref. 9. Figure 37 shows that the drift velocity, although reduced at low electric fields, tends to reach the same saturation value for all values of magnetic field; Figure 38 gives instead the angle of drift which appears to follow almost a linear dependence on H for large electric fields. The deflection of drifting electrons has to be taken into account, especially when operating drift chambers close to or inside strong magnetic fields. Notice also, from Fig. 38, that the use of a heavier gas (xenon replacing argon) reduces the deflection of the drifting swarm.

4.6 Effect of electronegative gases

Addition to an inert gas of even small quantities of electronegative products sensibly modifies the drift properties due to electron capture. The attachment coefficients h for several gases, in the absence of electric field, were given in Table 3; the cross-section for electron capture varies, however, with the electron energy, as shown in Fig. 39 for oxygen. Figures 40 and 41 show instead the attachment coefficient as a function of the reduced field E/p for air and chlorine in argon; similar curves for many other gases can be found in Refs. 18 and 22.

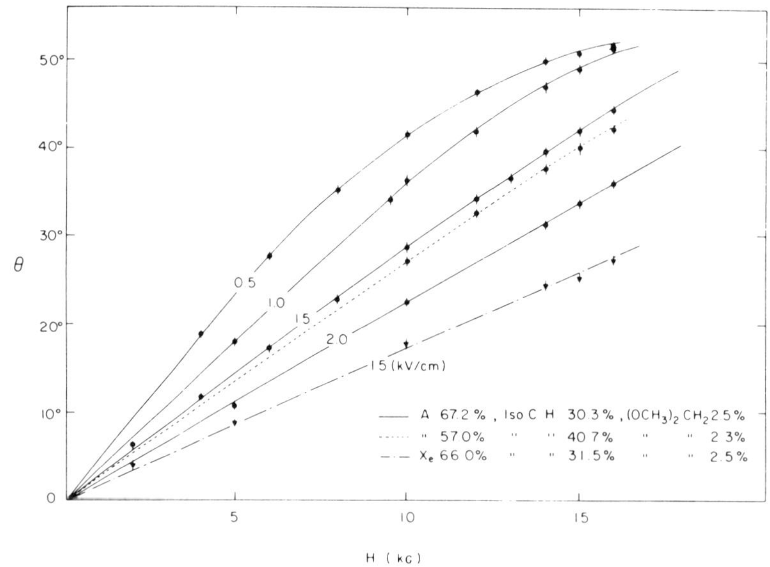

Fig. 38 Measured drift angle (angle between the electric field and the drift directions) as a function of electric and magnetic field strength[9].

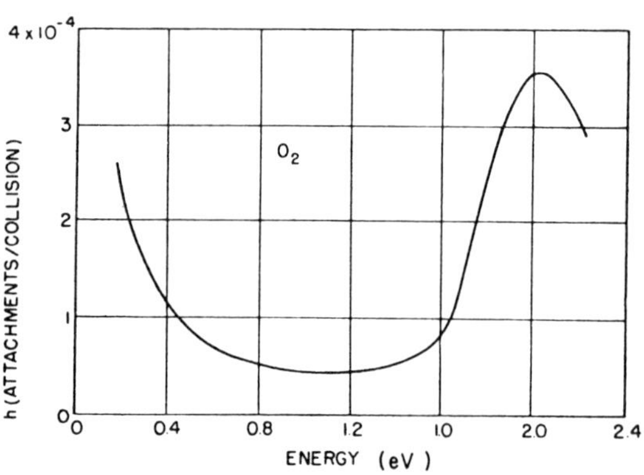

Fig. 39 Attachment coefficient for electrons in oxygen, as a function of electron energy[22].

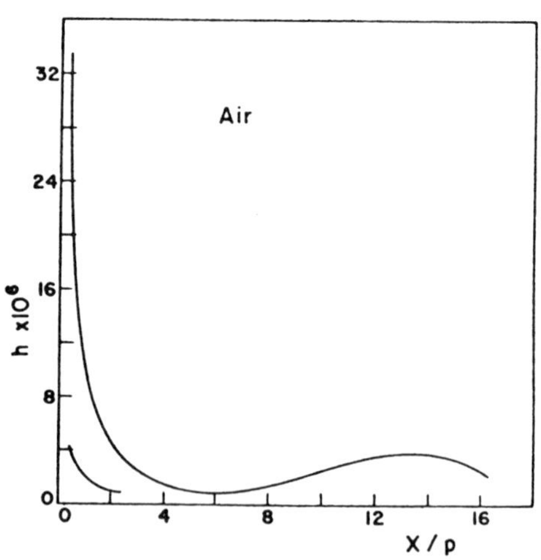

Fig. 40 Attachment coefficient for electrons in air as a function of the reduced electric field[22].

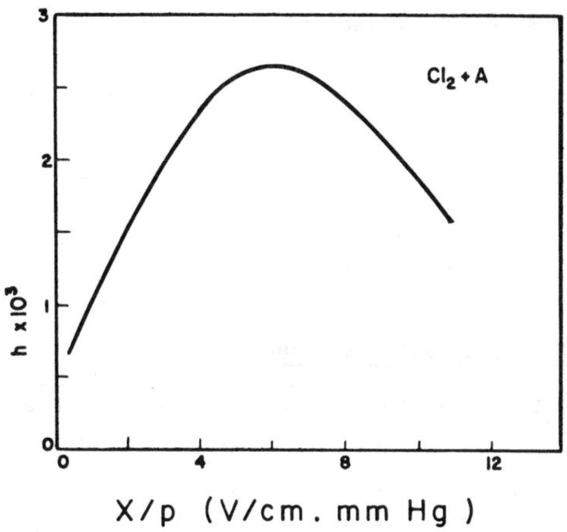

Fig. 41 Attachment coefficient for electrons in an argon-chlorine mixture[22].

The presence of electronegative polluants, mostly oxygen and water, in proportional counters reduces the detected pulse height because of electron capture. Let p be the fraction of electronegative polluant in the gas, λ and u the electron mean free path and instant velocity, respectively, and w the drift velocity. The number of collisions per unit time of an electron with the electronegative molecules is therefore up/λ and the probability of attachment

$$\frac{hup}{\lambda w} = \frac{1}{\lambda_c} \quad , \tag{21}$$

where λ_c is the mean free path for capture. Substituting in Eq. (21) the expression (16) and the definition of u , one gets

$$\lambda_c = \sqrt{\frac{m}{2\varepsilon}} \frac{w}{Nhp\sigma(\varepsilon)} \quad .$$

The loss of electrons in a swarm drifting in constant fields across a distance x is given by

$$\frac{n}{n_0} = e^{-x/\lambda_c} \quad . \tag{22}$$

As an example, let us consider the effect of air pollution in pure argon, at normal conditions. We shall assume that the presence of the pollution does not modify the energy distribution of electrons in argon. (This is, as we have seen, a rather naîve assumption!) From previous figures one gets, at E = 500 V/cm: $\varepsilon_k = 6$ eV $= 1.9 \times 10^{-12}$ erg, $w = 4 \times 10^5$ cm/sec, $\sigma(\varepsilon_k) = 5 \times 10^{-16}$ cm^2, and for air $h = 2 \times 10^{-5}$. From Eq. (21), therefore, $p\lambda_c \simeq 2.5 \times 10^{-2}$ cm. A 1% pollution of air in argon, therefore, will remove about 33% of the migrating electrons, per cm of drift, due to electron capture.

4.7 High electric fields: excitation and ionization

Increasing the electric field above a few kV per cm, more and more electrons can receive enough energy between two collisions to produce inelastic phenomena, excitation of various kinds, and ionization. Even a simplified description of the energy dependence of inelastic processes would exceed the purpose of these notes; here, we will only summarize some phenomenological aspects which are relevant for understanding the operation of proportional counters. A molecule can have many characteristic modes of excitation, increasing in number and complexity for polyatomic molecules. In particular, noble gases can only be excited through photon absorption or emission, while weakly-bound polyatomic molecules, for example the hydrocarbons used in proportional counters as a quencher, have radiationless transitions of a rotational and vibrational nature. Addition of an organic vapour to noble gases will therefore allow the dissipation of a good fraction of energy in radiationless transition, and, as will be discussed later, this is essential for high gain and stable operation of proportional counters.

When the energy of an electron increases over the first ionization potential of the gas E_i (see Table 1), the result of the impact can be an ion pair, while the primary electron continues its trip. The probability of ionization is rapidly increasing above threshold and has a maximum, for most gases, around 100 eV (see Fig. 42). Approximate curves showing the fraction of energy going into different processes, as a function of the reduced field E/P, are given in Fig. 43 for argon, nitrogen and hydrogen[18)]. In the figure curves labelled El represent the elastic impacts, EV the vibrational excitations, EE the excitation leading to photon emission, and I the ionizations.

Consider now a single electron drifting in a strong electric field; at a given time, it will have an energy ε with a probability given by the appropriate energy distribution function $F(\varepsilon)$. When, following the statistical fluctuations in the energy increase between collisions, the electron gains an energy in excess of the ionization potential, an ionization encounter may occur. The mean free path for ionization is defined as the average distance an electron has to travel before getting a chance to become involved in an ionizing collision. The inverse of the mean free path for ionization, α, is called the first Townsend

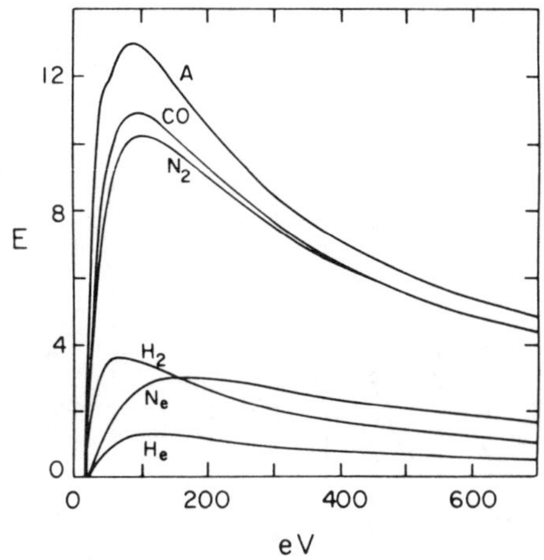

Fig. 42 Probability of ionization by the impact of electrons, as a function of their energy, in several gases[22])

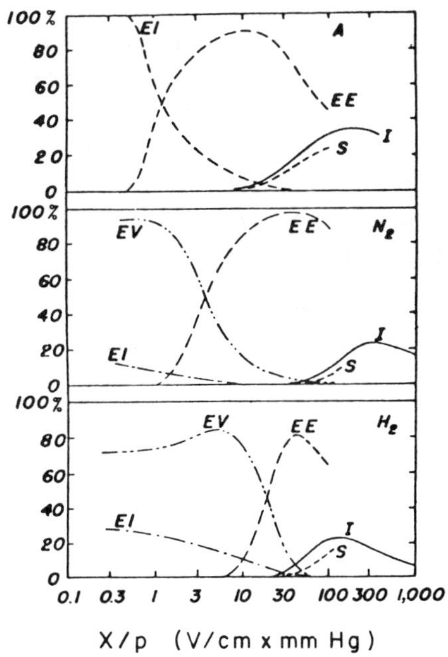

Fig. 43

Approximate curves showing the fraction of energy going into different processes in argon, nitrogen and hydrogen as a function of the reduced electric field[18]). In the figure, El represents the elastic impacts, EV the vibrational excitations, EE the excitations leading to photon emission and I the ionizations.

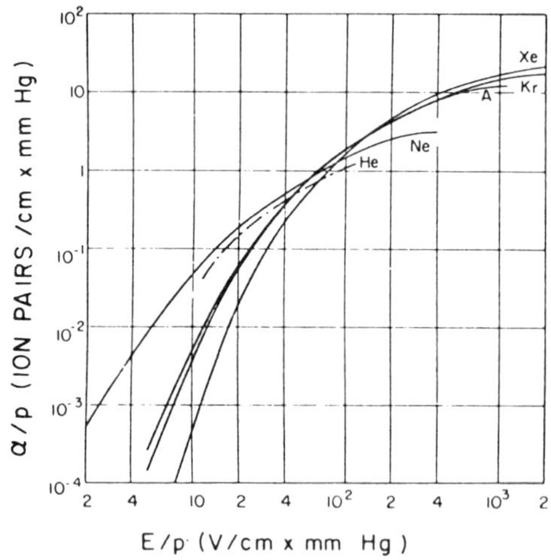

Fig. 44 First Townsend coefficient as a function of the reduced electric field, for noble gases[22])

coefficient and represents the number of ion pairs produced per unit length of drift. Values of α/P as a function of the reduced electric field E/P are given in Fig. 44, while Fig. 45 shows the dependence of α/P on the energy of the electrons ε.

The process of ionization by collision is the basis of the avalanche multiplication in proportional counters. Consider an electron liberated in a region of uniform electric field. After a mean free path α^{-1}, one electron-ion pair will be produced, and two electrons will continue the drift to generate, again after one mean free path, two other ion pairs and so on. If n is the number of electrons at a given position, after a path dx, the increase in the number will be

$$dn = n\alpha \, dx$$

and, by integration

$$n = n_0 \, e^{\alpha x} \quad \text{or} \quad M = \frac{n}{n_0} = e^{\alpha x} \quad . \tag{23}$$

M represents the multiplication factor. In the general case of a non-uniform electric field, $\alpha = \alpha(x)$, Eq. (23) has to be modified in the following way:

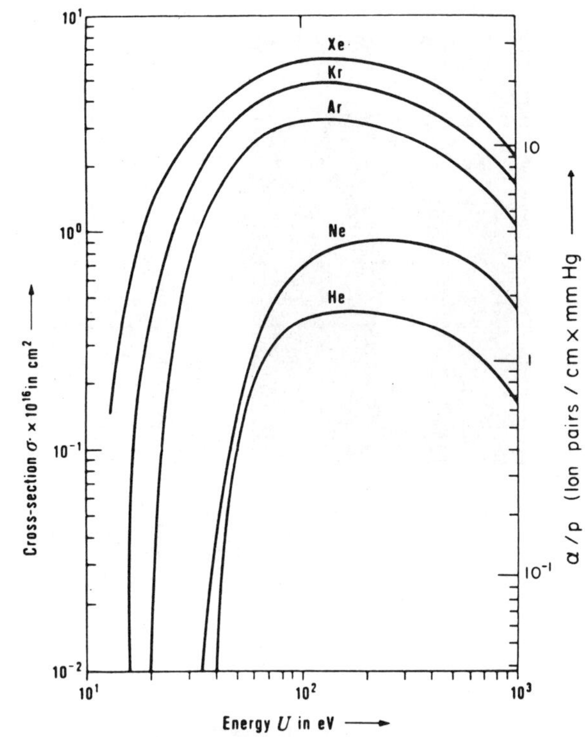

Fig. 45 Cross-section and first Townsend coefficient as a function of electron energy, for noble gases[28])

$$M = \exp\left[\int_{x_1}^{x_2} \alpha(x)\, dx\right] . \tag{24}$$

If one remembers now the big difference in the drift velocity of ions and electrons -- about a factor of thousand -- and the diffusion of migrating charges in the gas, the following picture of an avalanche multiplication appears (see Fig. 46): at a given instant, all electrons are situated in the front of a drop-like distribution of charges, with a tail of positive ions behind, decreasing in number and lateral extension; half of the total ions are contained in the front part, since they have just been produced in the last mean free

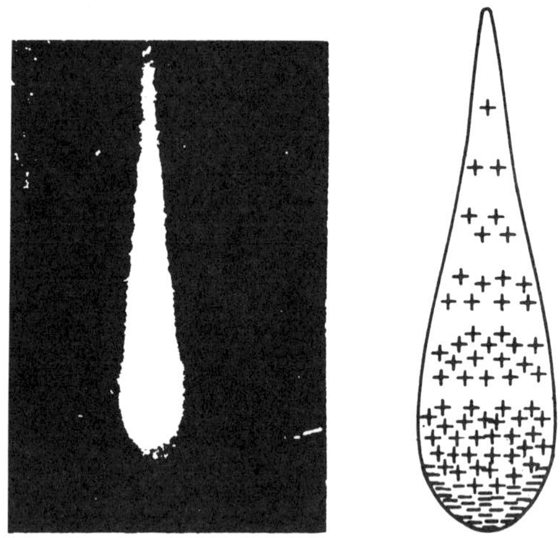

Fig. 46 Drop-like shape of an avalanche, showing the positive ions left behind the fast electron front. The photograph shows the actual avalanche shape, as made visible in a cloud chamber by droplets condensing around ions[18].

path. Knowing the dependence of the Townsend coefficient on the electric field, one can compute the multiplication factor for any field geometry. Many approximated analytic expressions exist for α, valid in different regions of E; for a summary see Ref. 25. We will mention here a simple approximation, due to Korff[27] and valid for low values of α

$$\frac{\alpha}{P} = A\,e^{-BP/E} ,$$

where A and B are constants as given in Table 5. In the same region one can assume the coefficient to be linearly dependent on the energy of the electrons

$$\alpha = kN\varepsilon , \tag{25}$$

N being the number of molecules per unit volume; the values of k are given in Table 5. The limits of the approximation can readily be estimated by inspection of Figs. 44 and 45. Knowing the dependence of α on the electric field, the multiplication factor can be computed for any field configuration; an example will be given in Section 5.1. No simple rules can be given for the behaviour of the first Townsend coefficient in gas mixtures. As a general trend, addition to a noble gas of a polyatomic gas or vapour increases the value

Table 5

Parameters appearing in Korff's approximated expression for the first Townsend coefficient α [27]

Gas	A (cm^{-1} Torr)	B (V cm^{-1} Torr)	k (cm^2 V^{-1})
He	3	34	0.11×10^{-17}
Ne	4	100	0.14×10^{-17}
Ar	14	180	1.81×10^{-17}
Xe	26	350	
CO_2	20	466	

of the field necessary to obtain a given value of α; Figure 47 shows the effect of adding several vapours to argon[28]. An exponential approximation is still possible, at least for small values of α.

The multiplication factor cannot be increased at will. Secondary processes, like photon emission inducing the generation of avalanches spread over the gas volume, and space-

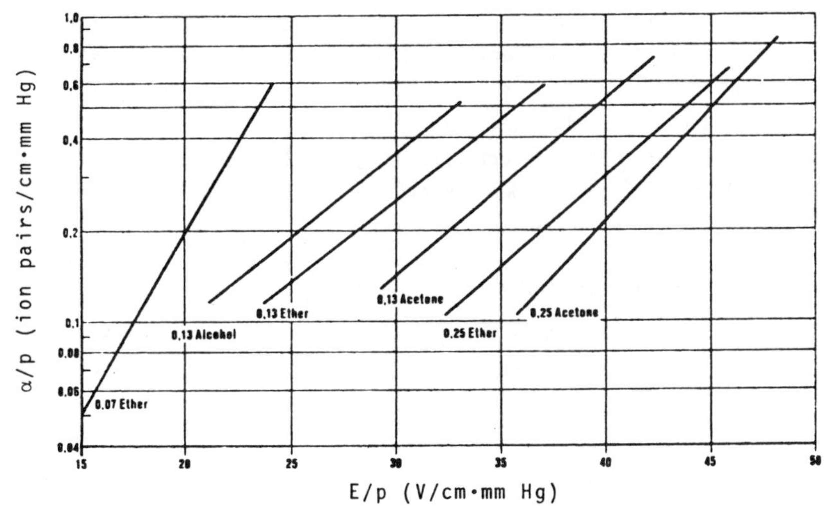

Fig. 47 First Townsend coefficient as a function of the reduced electric field, for several vapours added to argon[28].

charge deformation of the electric field (which is strongly increased near the front of the avalanche), eventually result in a spark breakdown. A phenomenological limit for multiplication before breakdown is given by the Raether condition

$$\alpha x \sim 20\,, \tag{26}$$

or $M \sim 10^8$; the statistical distribution of the energy of electrons, and therefore of M, in general does not allow one to operate at average gains above $\sim 10^6$ if one wants to avoid breakdowns. Notice also that by increasing the gap thickness, the Raether condition will be met at decreasing values of α; in other words, for a given field strength, the breakdown probability increases with the gap thickness.

5. PROPORTIONAL COUNTERS

5.1 Basic operation

Consider a thin layer of gas, for example 1 cm of argon at normal conditions, between two flat electrodes. A minimum ionizing particle will release (see Section 2.7) about 120 ion pairs; if this charge is collected at one electrode, the detected signal will be

$$V = \frac{ne}{C}\,,$$

which, for n = 120 and a typical system capacitance C = 10 pF, gives $V \sim 2\ \mu V$; this is so far below any possibility of detection. If, however, a strong electric field is applied between the electrodes, avalanche multiplication can occur, boosting the signal amplitude by several orders of magnitude. This kind of parallel plane detector suffers, however, from severe limitations. The detected signal, to start with, depends on the avalanche length, i.e. on the point where the original charge has been produced: no proportionality can be obtained between the energy deposit and the detected signal. Also, because of the uniform value of α over the thickness, the Raether condition (26) is readily met for some electrons in the energy distribution: only moderate gains can be obtained before breakdown, except under very special conditions[29]. A cylindrical coaxial geometry allows one to overcome the quoted limitations (Fig. 48). A thin metal wire is stretched on the axis of a conducting cylinder and insulated from it so that a difference of potential can be applied between the electrodes. The polarity is chosen so that the central wire is positive in respect to the outer cylinder. The electric field in the system is a maximum at the surface of the anode wire and rapidly decreases, as r^{-1}, towards the cathode; using thin wires, very high values of the field can be obtained close to the anode. The operation of the counter will then be as follows. In most of the region where the charges are produced by the primary interaction processes, the electric field only makes electrons drift towards

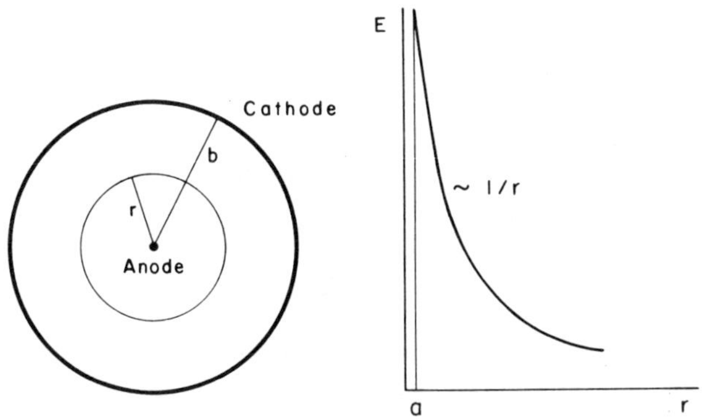

Fig. 48 The coaxial cylindrical proportional counter, and the shape of the electric field around the thin anode. Only very close to the anode the field grows high enough to allow avalanche multiplication.

the anode and, of course, positive ions towards the cathode. But very close to the anode, normally at a few wire radii, the field gets strong enough so that multiplication starts; a typical drop-like avalanche develops with all electrons in the front and ions behind. Because of lateral diffusion and the small radius of the anode, the avalanche surrounds the wire as shown in Fig. 49 [30]; electrons are collected and positive ions (half of them produced in the last mean free path) begin to drift towards the cathode.

Figure 50 shows how the detected charge depends on the potential difference V_0 between anode and cathode. At very low voltages, charges begin to be collected but recombination is still the dominant process; then full collection begins and the counter is said to operate in the ionization chamber mode. At a certain voltage, called threshold voltage V_T, the electric field close to the surface of the anode is large enough to begin the process of multiplication. Increasing V_0 above V_T, gains in excess of 10^4 can be obtained, still having the detected charge proportional, through a multiplication factor M, to the original deposited charge. At even higher voltages, however, this proportionality is gradually lost, as a consequence of the electric field distortions due to the large space charge built around the anode. This region of limited proportionality eventually ends in a region of saturated gain, where the same signal is detected independently of the original ionizing event. Proceeding even further, the photon emission process outlined in Section 4.7 begins to propagate avalanches in the counter, and the full length of the anode wire is surrounded by a sheath of electrons and ions: this is the typical Geiger-Müller operation.

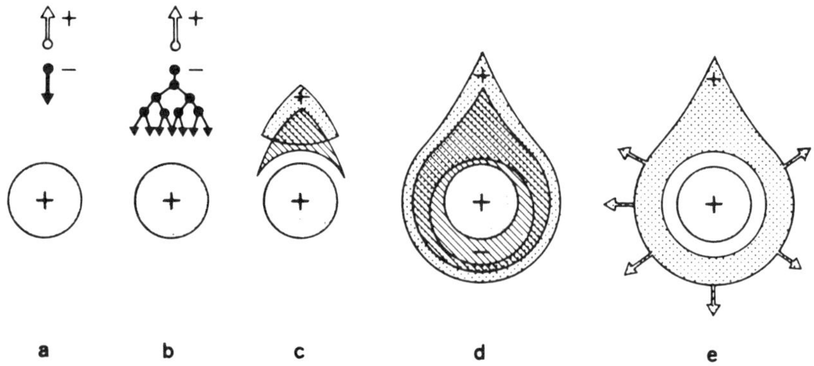

Fig. 49 Time development of an avalanche in a proportional counter[30]. A single primary electron proceeds towards the anode, in regions of increasingly high fields, experiencing ionizing collisions; due to the lateral diffusion, a drop-like avalanche, surrounding the wire, develops. Electrons are collected in a very short time (1 nsec or so) and a cloud of positive ions is left, slowly migrating towards the cathode.

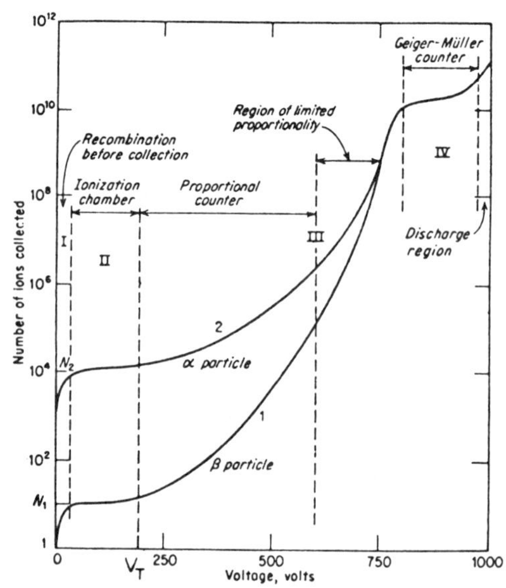

Fig. 50 Gain-voltage characteristics for a proportional counter, showing the different regions of operation (from W. Price, see bibliography for Sections 2 and 3).

In the development of multiwire proportional chambers, in order to reduce the cost and complexity of the electronics, very often one tends to work at the highest possible gains, without entering into the Geiger-Müller region which would introduce prohibitive dead-times for the counter (see later), i.e. either in the semiproportional or in the saturated mode. The maximum amplification is limited, of course, by discharge or spark breakdown; the maximum gain that can be obtained before breakdown depends on the gas used (see later).

Let a and b then be the radii of anode and cathode; at a distance r from the centre, the electric field and potential can be written as

$$E(r) = \frac{CV_0}{2\pi\varepsilon_0}\frac{1}{r}$$

$$V(r) = \frac{CV_0}{2\pi\varepsilon_0}\ln\frac{r}{a}, \qquad \text{where } C = \frac{2\pi\varepsilon_0}{\ln(b/a)}, \tag{27}$$

$V_0 = V(b)$ is the over-all potential difference and $V(a) = 0$; C is the capacitance per unit length of the system and ε_0 is the dielectric constant (for gases $\varepsilon_0 \simeq 8.85$ pF/m). Following Rose and Korff[27,31] we will compute the multiplication factor of a proportional counter, within the limits of the approximation (25). If $1/\alpha$ is the mean free path for ionization, the average energy ε obtained by an electron from the electric field between collisions is E/α; using the explicit value of E and expression (25) one gets

$$\varepsilon = \sqrt{\frac{CV_0}{2\pi\varepsilon_0 kN}\frac{1}{r}}$$

and therefore

$$\alpha(r) = \sqrt{\frac{kNCV_0}{2\pi\varepsilon_0}\frac{1}{r}}\ . \tag{28}$$

The multiplication factor can then be obtained from the definition (24). To fix the limits of integration, we will assume that avalanche multiplication begins at a distance r_c from the centre, where the electric field exceeds a critical value E_c:

$$M = \exp\left[\int_a^{r_c} \alpha(r)\, dr\right]\ . \tag{29}$$

Recalling the definition of threshold voltage V_T, one can write

$$E_c = \frac{CV_T}{2\pi\varepsilon_0}\frac{1}{a} \quad \text{and} \quad \frac{r_c}{a} = \frac{V_0}{V_T} \quad . \tag{30}$$

Substituting Eq. (28) in Eq. (29), integrating and using Eq. (30) one gets to the two alternative expressions:

$$M = \exp\left[2\sqrt{\frac{kNCV_0 a}{2\pi\varepsilon_0}}\left(\sqrt{\frac{V_0}{V_T}} - 1\right)\right] \tag{31}$$

or

$$M = \exp\left[\sqrt{2kNE_c}\; a\sqrt{\frac{V_0}{V_T}}\left(\sqrt{\frac{V_0}{V_T}} - 1\right)\right] \quad . \tag{31'}$$

For $V_0 >> V_T$, the gain is seen to depend exponentially on the charge per unit length $Q = CV_0$;

$$M = K e^{CV_0} , \tag{32}$$

having introduced expression (30) in (31) and approximating the term in parenthesis to $(V_0/V_T)^{1/2}$. Once the threshold voltage has been determined, using the value of k given in Table 5, the multiplication factor can be computed and compared with the measurements; Figure 51 (from Ref. 31) shows that the agreement is excellent at least for moderate gains. For higher values above 10^4 , the approximation used for α is not justified; in Ref. 25 a summary of calculations by different authors is given. For a qualitative understanding of the proportional counters' operation, however, either expressions (31), (31') or (32) are quite sufficient.

5.2 Time development of the signal

We can now analyse in detail the time development of the avalanche and of the detected signal. As shown in Fig. 49, the whole process begins at a few wire radii, i.e. typically at less than 50 μm from the anode surface. Taking a typical value of 5 cm/μsec for the drift velocity of electrons in this region, it appears that the whole process of multiplication will take place in less than 1 nsec: at that instant, electrons have been collected on the anode and the positive ion sheath will drift towards the cathode at decreasing velocity. The detected signal, negative on the anode and positive on the cathode, is the consequence of the change in energy of the system due to the movement of charges. Simple electrostatic considerations show that if a charge Q is moved by dr, in a system of total capacitance ℓC (ℓ is the length of the counter), the induced signal is

$$dv = \frac{Q}{\ell CV_0}\frac{dV}{dr}\,dr \quad . \tag{33}$$

Electrons in the avalanche are produced very close to the anode (half of them in the last mean free path); therefore their contribution to the total signal will be very small:

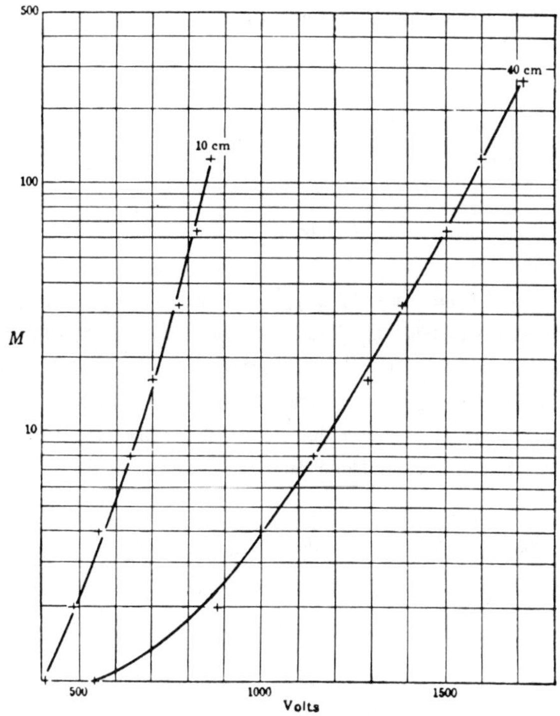

Fig. 51 Comparison between the computed [through expression (31)] and the measured multiplication factor in argon-filled proportional counters[31]. The two curves refer to the indicated gas pressures.

positive ions, instead, drift across the counter and generate most of the signal. Assuming that all charges are produced at a distance λ from the wire, the electron and ion contributions to the signal on the anode will be, respectively,

$$v^- = -\frac{Q}{\ell C V_0}\int_a^{a+\lambda}\frac{dV}{dr}\,dr = -\frac{Q}{2\pi\varepsilon_0\ell}\ln\frac{a+\lambda}{a}$$

and

$$v^+ = \frac{Q}{\ell C V_0}\int_{a+\lambda}^{b}\frac{dV}{dr}\,dr = -\frac{Q}{2\pi\varepsilon_0\ell}\ln\frac{b}{a+\lambda} \quad .$$

The total maximum signal induced on the anode is seen to be

$$v = v^+ + v^- = -\frac{Q}{2\pi\varepsilon_0 \ell} \ln \frac{b}{a} = -\frac{Q}{\ell C}$$

and the ratio of the two contributions is

$$\frac{v^-}{v^+} = \frac{\ln (a + \lambda) - \ln a}{\ln b - \ln (a + \lambda)} .$$

Typical values for a counter are a = 10 μm, λ = 1 μm, and b = 10 mm; substituting in the previous expression one finds that the electron contribution to the signal is about 1% of the total. It is therefore, in general, neglected for all practical purposes. The time development of the signal can easily be computed assuming that ions leaving the surface of the wire with constant mobility are the only contribution. In this case, integration of formula (33) gives for the signal induced on the anode

$$v(t) = -\int_0^t dv = -\frac{Q}{2\pi\varepsilon_0 \ell} \ln \frac{r(t)}{a} . \tag{34}$$

From the definition of mobility, Eq. (11), it follows that

$$\frac{dr}{dt} = \mu^+ \frac{E}{P} = \frac{\mu^+ C V_0}{2\pi\varepsilon_0 P} \frac{1}{r}$$

and therefore

$$\int_a^r r\, dr = \frac{\mu^+ C V_0}{2\pi\varepsilon_0 P} \int_0^t dt \quad \text{or} \quad r(t) = \left(a^2 + \frac{\mu^+ C V_0}{\pi\varepsilon_0 P} t\right)^{\frac{1}{2}} .$$

Substituting in Eq. (34) one gets

$$v(t) = -\frac{Q}{4\pi\varepsilon_0 \ell} \ln \left(1 + \frac{\mu^+ C V_0}{\pi\varepsilon_0 P a^2} t\right) = -\frac{Q}{4\pi\varepsilon_0 \ell} \ln \left(1 + \frac{t}{t_0}\right) . \tag{35}$$

The total drift time of the ions, T, is obtained from the condition r(T) = b, and is

$$T = \frac{\pi\varepsilon_0 P(b^2 - a^2)}{\mu^+ C V_0} \tag{36}$$

and it is easily seen that $v(T) = -Q/\ell C$ as it should be. As an example, let us consider an argon-filled counter under normal conditions, and with a = 10 μm, b = 8 mm; it follows from Eqs. (27) that C = 8 pF/m. From Table 2, $\mu^+ = 1.7\ \text{cm}^2\ \text{sec}^{-1}\ \text{V}^{-1}\ \text{atm}^{-1}$; for a typical operational voltage V_0 = 3 kV, one gets T = 550 μsec. The time growth of the signal is very fast at the beginning, as shown in Fig. 52; from Eq. (35) one can see that

$$v\left(\frac{a}{b} T\right) \simeq \frac{Q}{2\ell C} \quad ,$$

hence half of the signal is developed after one thousandth of the total time, about 700 nsec in the example. It is therefore normal practice to terminate the counter with a resistor R, such that the signal is differentiated with a time constant τ = RC; very short pulses can be obtained using low impedance terminations, with the aim of increasing the rate capability of the counter (see below). The figure shows some examples of pulse shape obtained by differentiation. At the limit for R → 0, one speaks rather of a current signal than a voltage; this is given by

$$i(t) = \ell C \frac{dv(t)}{dt} = - \frac{QC}{4\pi\varepsilon_0} \frac{1}{t_0 + t} \quad .$$

The current is maximum for t = 0:

$$i_{max} = i(0) = - \frac{\mu^+ QC^2 V_0}{4\pi^2 \varepsilon_0^2 a^2 P} \quad .$$

Substituting the numerical values of the previous example, and assuming $Q = 10^6$e, (where e

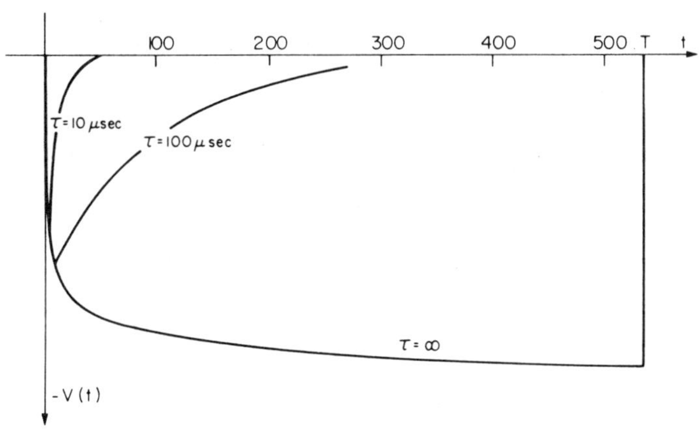

Fig. 52 Time development of the pulse in a proportional counter; T is the total drift time of positive ions from anode to cathode. The pulse shape obtained with several differentiation time constants is also shown.

is the electron charge) one gets $i_{max} \simeq 13\ \mu A$, which is a typical value for the operation of multiwire proportional chambers.

5.3 Choice of the gas filling

Since avalanche multiplication occurs in all gases, virtually any gas or gas mixture can be used in a proportional counter. In most cases, however, the specific experimental requirements restrict the choice to several families of compounds; low working voltage, high gain operation, good proportionality, high rate capabilities, long lifetime, fast recovery, etc., are examples of sometimes conflicting requirements. In what follows, we will briefly outline the main properties of different gases in the behaviour of the proportional counter; a more detailed discussion can be found in the bibliography for this section (in particular in the book by Korff).

Comparison of Figs. 44 and 47 shows that avalanche multiplication occurs in noble gases at much lower fields than in complex molecules: this is a consequence of the many non-ionizing energy dissipation modes available in polyatomic molecules. Therefore, convenience of operation suggests the use of a noble gas as the main component; addition of other components, for the reasons to be discussed below, will of course slightly increase the threshold voltage. The choice within the family of noble gases is then dictated, at least for the detection of minimum ionizing particles, by a high specific ionization; with reference to Table 1, and disregarding for economic reasons the expensive xenon or krypton, the choice falls naturally on argon. An argon-operated counter, however, does not allow gains in excess of 10^3-10^4 without entering into a permanent discharge operation; this is for the following reasons. During the avalanche process, excited and ionized atoms are formed. The excited noble gases can return to the ground state only through a radiative process, and the minimum energy of the emitted photon (11.6 eV for argon) is well above the ionization potential of any metal constituting the cathode (7.7 eV for copper). Photoelectrons can therefore be extracted from the cathode, and initiate a new avalanche very soon after the primary.

Argon ions, on the other hand, migrate to the cathode and are there neutralized extracting an electron; the balance of energy is either radiated as a photon, or by secondary emission, i.e. extraction of another electron from the metal surface. Both processes result in a delayed spurious avalanche: even for moderate gains, the probability of the processes discussed is high enough to induce a permanent régime of discharge.

Polyatomic molecules have a very different behaviour, especially when they contain more than four atoms. The large amount of non-radiative excited states (rotational and vibrational) allows the absorption of photons in a wide energy range: for methane, for example, absorption is very efficient in the range 7.9 to 14.5 eV, which covers the range of energy of photons emitted by argon. This is a common property of most organic compounds in the hydrocarbon and alcohol families, and of several inorganic compounds like freons, CO_2, BF_3 and others. The molecules dissipate the excess energy either by elastic collisions, or by dissociation into simpler radicals. The same behaviour is observed when a polyatomic ionized molecule neutralizes at the cathode: secondary emission is very unlikely. In the neutralization, radicals recombine either into simpler molecules (dissociation) or forming larger complexes (polymerization). Even small amounts of a polyatomic quencher added to a noble gas changes completely the operation of a counter, because of the lower ionization potential that results in a very efficient ion exchange (see Section 4.2). Good photon absorption and suppression of the secondary emission allows gains in excess of 10^6 to be obtained before discharge.

The quenching efficiency of a polyatomic gas increases with the number of atoms in the molecule; isobutane (C_4H_{10}) is very often used for high-gain stable operation. Secondary emission has been observed, although with low probability, for simpler molecules like carbon dioxide, which may therefore occasionally produce discharge.

Addition of small quantities of electronegative gases, like freons (CF_3Br in particular) or ethyl bromide (C_2H_5Br) allows one to obtain the highest possible gains before Geiger-Müller discharge, i.e. saturated operation (see Section 5.1). Apart from their additional photon-quenching capability, the electronegative gases capture free electrons forming negative ions that cannot induce avalanches (at least in the field values normally met in a proportional counter). If the mean free path for electron capture is shorter than the distance from anode to cathode, electrons liberated at the cathode by any of the described processes will have very little probability of reaching the anode, and gains around 10^7 can be safely obtained before discharge or breakdown. To preserve detection efficiency, however, only limited amounts of electronegative gases can be used. Unfortunately, the use of polyatomic organic gases can have a dramatic consequence on the lifetime of counters, when high fluxes of radiation are detected. To start with, the dissociation process, which is at the basis of the quenching action, quickly consumes the available molecules in a sealed counter. For a gain of 10^6 and assuming there are 100 ion pairs detected in each event, about 10^8 molecules are dissociated in each event. In a typical 10 cm^3 counter operated at atmospheric pressure in a 90-10 mixture of noble gas and quencher, there are about 10^{19} polyatomic molecules available and therefore a sealed counter is expected to change its operational characteristics substantially after about 10^{10} counts. As we will see later, multiwire proportional chambers are normally operated in an open flow of gas, essentially because the necessity for large surfaces of detection with thin gas windows does not allow efficient sealing, and therefore gas consumption is not a problem. However, we have seen that some products of molecular recombination are liquid or solid polymers. These products will deposit on cathodes and anodes, depending on their affinity, and substantially modify the operation of the counter after integral fluxes of radiation around 10^7-10^8 counts per cm^2. The following process takes place (Malter effect)[32]. When a thin layer of insulator develops on the cathode, as a result of the deposit of polymers, positive ions created in further avalanches deposit on the outer side of the layer and only slowly leak through the insulator to be neutralized on the cathode. When the detected radiation flux grows above a threshold value (10^2 or 10^3 counts per second per cm^2) the production rate of the ions exceeds the leakage rate and very quickly a high density of charge develops across the thin layer. The dipole electric field can be so high as to extract electrons from the cathode and through the insulator: a régime of permanent discharge is therefore induced, even if the original source of radiation is removed. Temporary suppression of the counter voltage stops the discharge; the counter, however, remains damaged and an exposure to lower and lower radiation fluxes will start the process again. Only complete cleaning can regenerate a damaged counter.

Flows of 10^8 counts per cm^2 are very quickly met in high-energy beams having typical intensities of 10^6/sec cm^2. Use of non-polymerizing quenchers, like alcohols, aldehydes, and acetates, was soon recognized to strongly suppress the ageing effect; however, their low vapour pressure as compared with hydrocarbons does not allow one to obtain an efficient quenching against photoionization and secondary emission (for example, isopropylic alcohol has a vapour pressure, at 20°, of 30 Torr). A solution to the dilemma has been found by taking advantage of the ion exchange mechanism already mentioned several times. If a non-polymerizing agent is chosen having its ionization potential lower than those of the other constituents of the gas mixture, addition of even a small quantity of the new quencher will modify the nature of the ions neutralized at the cathode into a non-polymerizing species. Propylic alcohol $[C_3H_7OH]$ and methylal $[(OCH_3)_2CH_2]$ are often used, having ionization potentials of 10.1 and 10.0 eV, respectively. Integrated rates in excess of 10^{10} counts per cm^2 have been measured, without alteration of the counter properties[33].

Use of a single inorganic quenching gas, like carbon dioxide, would of course avoid the ageing effect; the quoted instability of operation at high gains of argon-carbon dioxide counters, however, limits their use. Incidentally, radiation damage of carbon dioxide counters has been reported, probably as a result of a small amount of polluting agents in the gas[34,35]. Use of gas mixtures of three or four components may appear to be a nuisance, but in practice the advantages of having a high gain, stable operation in a large multiwire proportional chamber complex widely counteract the mixing problems.

5.4 Space-charge gain limitation

The growth in the avalanche process of a positive ion sheath around the wire has as a consequence the local reduction of the electric field; the normal field is completely restored only when all ions have been neutralized at the cathode (i.e. after several hundred μsec). When the counter is operated in the proportional or semiproportional mode, the extension of the avalanche along the wire is very small, between 0.1 and 0.5 mm, and therefore the field modification is confined to a small region of the counter. In Geiger-Müller operation, on the other hand, the avalanches spread all along the wire and the field in the whole counter is distorted: for several hundred μsec no further detection is possible.

When a counter is operated in a proportional mode, the effect of a uniformly distributed flow of radiation is to reduce the average gain. Figures 53 and 54 show the pulse-height reduction due to space charge, as measured in a drift chamber for ^{55}Fe 5.9 keV X-rays and for minimum ionizing particles[9]. The incident fluxes are given in counts per second per millimetre of anode wire; the corresponding surface rate can be obtained taking into account the counter geometry. Following the formulation of Hendricks[36], one can compute the approximate change in the anodic potential V_0 due to a flux R of ionizing events, each producing nMe charges in the avalanche,

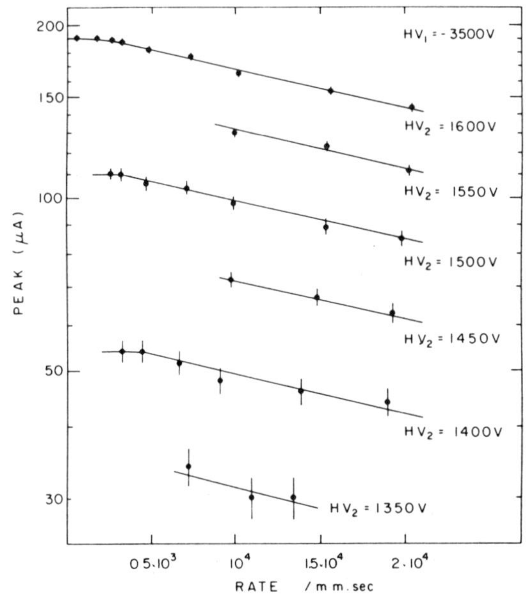

Fig. 53 Rate dependence of the peak current on 5.9 keV X-rays in a drift proportional chamber, for several anodic potentials[9]

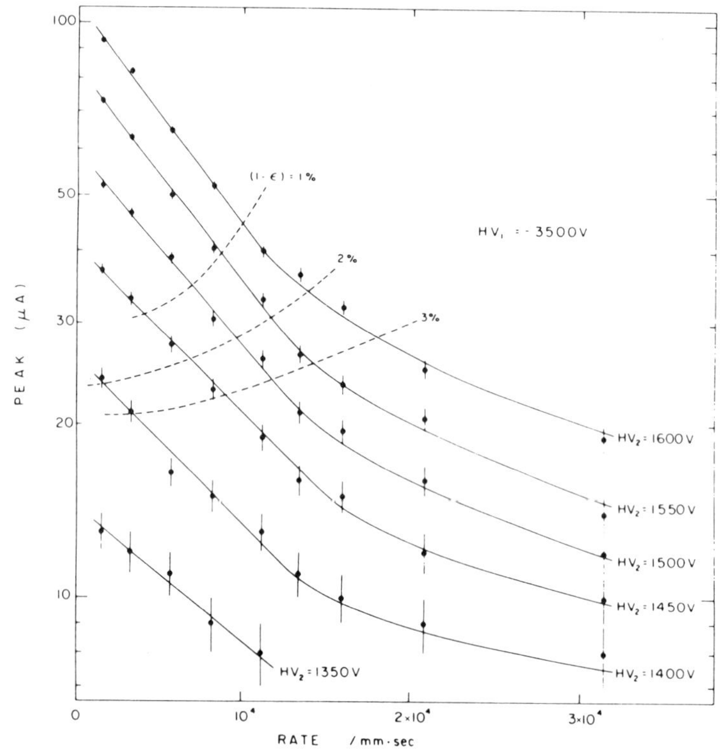

Fig. 54 Rate dependence of the peak current for minimum ionizing particles, in a drift chamber. The dashed curves show the equal inefficiency intercept, for a fixed detection threshold (5 μA) [9].

$$\Delta V = \frac{nMeRT}{4\pi^2\varepsilon_0} = KMR \quad , \qquad (37)$$

where T is the total drift time of positive ions, as given by Eq. (36). Introducing Eq. (37) into the dependence of the multiplication factor on the voltage, one can obtain the corresponding change in the gain. This could be done using expression (31); however, it appears that in the region of gain values we are normally concerned with (10^6 or more), the simpler expression for M given by formula (32) can be used; in this case $M_0 = K e^{CV_0}$ and $M = M_0 e^{-\Delta V}$. Substituting this in Eq. (37) one gets

$$\Delta V\, e^{\Delta V} = KM_0R \quad ,$$

which, for small variations of ΔV ($e^{\Delta V} \simeq 1$) gives $\Delta V = KM_0R$ and

$$M = M_0\, e^{-KM_0R} \quad .$$

The exponential decrease in the multiplication factor with the rate is precisely what has been observed (see Figs. 53 and 54), at least for moderate rates. At higher rates, $e^{\Delta V} > 1$ and therefore $\Delta V < KM_0R$; then the decrease of the gain with rate is less than exponential as shown by the data. A practical consequence of a gain decreasing with the rate is, of course, loss of efficiency when operating at fixed detection thresholds, as is the case in multiwire proportional chambers. We will come back to this point in Section 6.

6. MULTIWIRE PROPORTIONAL CHAMBERS

6.1 Principles of operation

Proportional counters have been and are widely used whenever measurement of energy loss of radiation is required. The space localization capability of a counter is, however, limited to the determination that a particle has or has not traversed the counter's volume. Stacking of many independent counters is possible, but is not very attractive mechanically. There was a vague belief that multiwire structures in the same gas volume would not properly work, because the large capacitance existing between parallel non-screened wires would cause the signal to spread, by capacitive coupling, in all wires, therefore frustrating any localization attempt in the structure. It was the merit of Charpak and collaborators to recognize that the positive induced signals in all electrodes surrounding the anode interested by an avalanche largely compensate the negative signals produced by capacitive coupling; these authors operated in 1967-68 the first effective multiwire proportional chambers[1)], which comprised a set of anode wires closely spaced, all at the same potential, each wire acting as an independent counter.

A multiwire proportional chamber consists essentially of a set of thin, parallel and equally spaced anode wires, symmetrically sandwiched between two cathode planes; Fig. 55 gives a schematic cross-section of the structure. For proper operation, the gap ℓ is normally three or four times larger than the wire spacing s. When a negative potential is applied to the cathodes, the anodes being grounded, an electric field develops as indicated by the equipotentials and field lines in Fig. 56 and in a magnified view around the anodes, in Fig. 57 [37)]. Suppose now that charges are liberated in the gas volume by an ionizing event; as in a proportional counter, conditions are set such that electrons will drift along field lines until they approach the high field region, very close to the anode wires, where

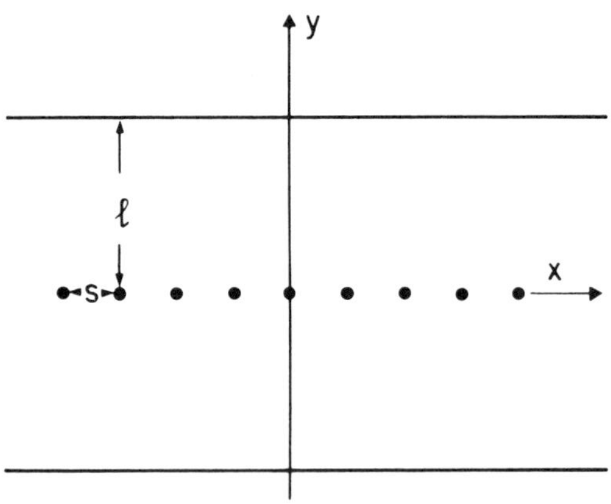

Fig. 55 Principle of construction and definition of parameters in a multiwire proportional chamber. A set of parallel anode wires is mounted symmetrically between two cathode planes (wires or foils).

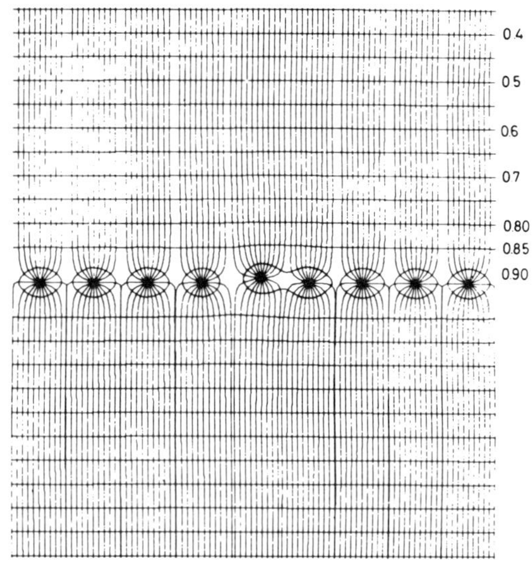

Fig. 56 Electric field equipotentials and field lines in a multiwire proportional chamber. The effect on the field of a small displacement of one wire is also shown[37].

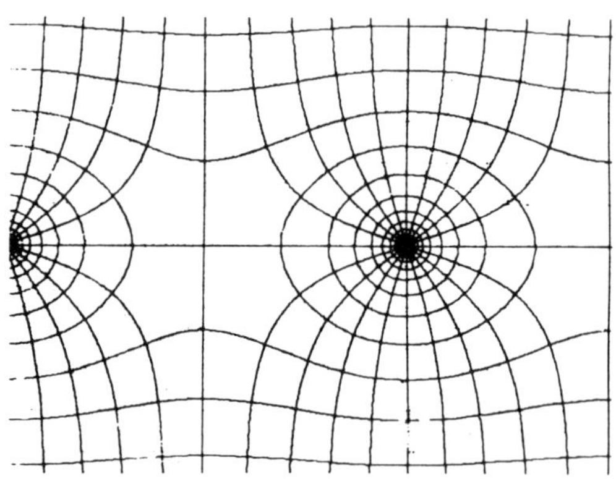

Fig. 57 Enlarged view of the field around the anode wires (wire spacing 2 mm, wire diameter 20 μm) [37]

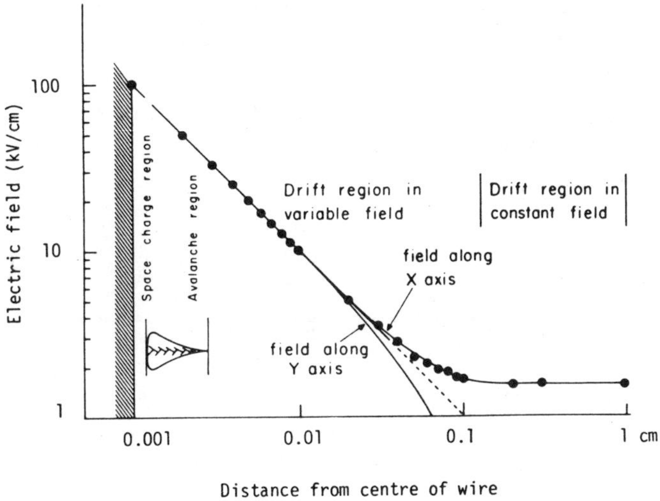

Fig. 58 Variation of the electric field along the axis perpendicular to the wire plane and centred on one wire in a multiwire proportional chamber (x), and along the direction parallel to the wire plane (y) [38]

avalanche multiplication occurs; Fig. 58 [38] shows the variation of the electric field along a direction perpendicular to the wire plane, for a typical multiwire proportional chamber.

The analytic expression for the electric field can be obtained by standard electrostatic algorithms, and is given in many textbooks[39,40]. An approximated expression has been given by Erskine[37], who also computed the field deformations due to wire displacements.

With the definitions of Fig. 55 and requiring $V(a) = V_0$, $V(\ell) = 0$, one gets

$$V(x,y) = \frac{CV_0}{4\pi\varepsilon_0}\left\{\frac{2\pi\ell}{s} - \ln\left[4\left(\sin^2\frac{\pi x}{s} + \sinh^2\frac{\pi y}{s}\right)\right]\right\}$$

$$E(x,y) = \frac{CV_0}{2\varepsilon_0 s}\left(1 + \mathrm{tg}^2\frac{\pi x}{s}\,\mathrm{tgh}^2\frac{\pi y}{s}\right)^{1/2}\left(\mathrm{tg}^2\frac{\pi x}{s} + \mathrm{tgh}^2\frac{\pi y}{s}\right)^{-1/2} \tag{38}$$

and the capacitance per unit length is given by

$$C = \frac{2\pi\varepsilon_0}{(\pi\ell/s) - \ln(2\pi a/s)}\,, \tag{39}$$

where a is the anode wire radius. Notice that, since $a \ll s$, the value given by Eq. (39) is always smaller than the capacity of the plane condenser with the same surface $(2\varepsilon_0 s/\ell)$. Computed values of C are given in Table 6 for several typical geometries; in general, one can see that the capacitance is quickly decreasing with the wire spacing, while it does not depend very much on the wire diameter. Along the symmetry lines x = 0 and y = 0, the electric field can be written as

$$E_y = E(0,y) = \frac{CV_0}{2\varepsilon_0 s} \coth \frac{\pi y}{s}$$

$$E_x = E(x,0) = \frac{CV_0}{2\varepsilon_0 s} \operatorname{cotg} \frac{\pi x}{s} .$$

It is also instructive to consider the following approximations:

$$\text{for } y \ll s: \qquad E(x,y) \simeq \frac{CV_0}{2\pi\varepsilon_0} \frac{1}{r} , \qquad r \simeq (x^2 + y^2)^{1/2} ; \tag{40}$$

$$\text{for } y \geq s: \qquad \coth \frac{\pi y}{s} \simeq 1 , \qquad E_y = \frac{CV_0}{2\varepsilon_0 s} . \tag{41}$$

Equation (40) shows that the field is radial around the anode, with an expression identical to that of a cylindrical proportional counter, Eq. (27). One can therefore use the main results obtained in Sections 5.1 and 5.2 to discuss the operational characteristics of a multiwire proportional chamber, provided that the correct value for the capacity per unit length is used.

6.2 Choice of geometrical parameters

The accuracy of localization in a multiwire proportional chamber is obviously determined by the anode wire spacing; spacings of less than 2 mm are, however, increasingly difficult to operate. One can understand the reasons by inspection of the expressions (40) and (41) and of the approximate multiplication factor, expression (32). For a fixed wire diameter, to obtain a given gain one has to keep the charge per unit length CV_0 constant, i.e. increase V_0 when s (and therefore C) is decreased. For example, going from 2 to 1 mm spacing V_0 has to be almost doubled (see Table 6). At the same time, however, the electric field in the drift region is also doubled; the chance that some drifting electrons meet the Raether condition (26) is strongly increased. Practical experience has shown that, if 2 mm wire spacings are possible, 1 mm wire spacings are rather hard to operate for surfaces larger than 100 cm^2 or so. Decreasing the wire diameters helps, but there are obvious mechanical and electrostatic limitations (see Section 6.4). Notice that scaling down all geometrical parameters (i.e. the distance and diameter of the wires, and the gap) is not sufficient to preserve good operation: in fact, the mean free path for ionization remains invariant, unless the gas pres-

Table 6

Capacitance per unit length, in pF/m, for several proportional chambers' geometries

ℓ (mm)	2a (μm)	s (mm) 1	2	3	5
8	10	1.94	3.33	4.30	5.51
	20	2.00	3.47	4.55	5.92
	30	2.02	3.56	4.70	6.19
4	10	3.47	5.33	6.36	7.34
	20	3.63	5.71	6.91	8.10
	30	3.73	5.96	7.28	8.58

sure is correspondingly increased. Work in this direction has been carried out to obtain good accuracies over small surfaces[41].

Let us now investigate the influence of the wire diameter, on a given geometry. Figure 59 shows how the gain varies, as a function of the ratio V_0/V_T, for an s = 2 mm, ℓ = 8 mm chamber with several anode radii; the approximate expression (31) has been used, with the correct value for C as given by formula (39) for each geometry and for argon ($k = 1.8 \times 10^{-21}$ m^2 V^{-1}). Clearly, although in principle any wire diameter allows one to obtain any gain, the steeper slope obtained for large diameters means a much more critical operation. We can represent the effect of all mechanical and electrical tolerances (see the next section) as a widening of the lines drawn in the figure into narrow bands; for a given V_0/V_T, the multiplication factor will vary, across the chamber, between two extremes. The steeper the slope of a given band, the more likely it is that one section of a chamber will begin to discharge when another section is not yet properly amplifying. Obviously, however, thick anode wires are easier to handle than thin ones and a compromise has to be found; diameters around 10 μm are a practical limit, while that of 20 μm is more frequently used.

6.3 Dependence of the gain on mechanical tolerances

The gain of a chamber at a given operational voltage depends on the detailed shape and value of the electric field in the multiplication region and can therefore change along a wire or from wire to wire as a consequence of mechanical variations. The maximum tolerable

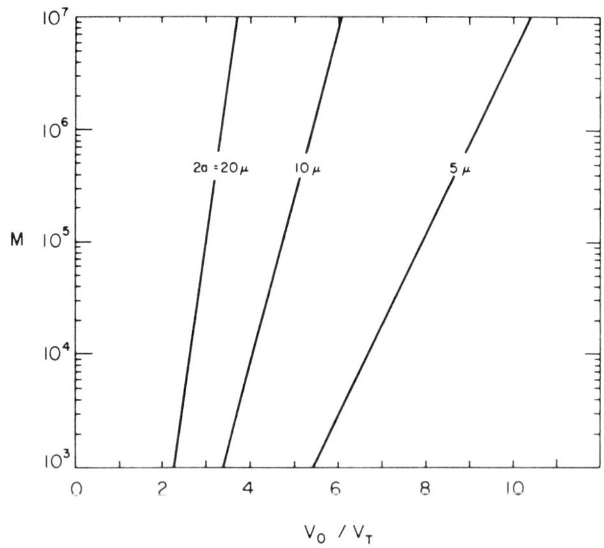

Fig. 59 Dependence of the multiplication factor on the operational voltage, relative to the threshold voltage, in a 2 mm spacing multiwire proportional chamber with several wire diameters [computed from expression (31)].

differences in gain depend, of course, on the specific application: in pulse-height measuring devices, requirements are much more severe than in threshold-operated chambers.

A detailed analysis of the gain variations due to several kinds of mechanical tolerances can be found in Erskine[37] and Dimčowski[42]. Here we would like to give only a qualitative formulation, based on the approximated expression (32) for the multiplication factor. Let $Q = CV_0$ be the charge per unit length of the wires; differentiation of Eq. (32) gives

$$\frac{\Delta M}{M} = \ln M \frac{\Delta Q}{Q} . \tag{42}$$

The problem of gain variation is therefore reduced to the calculation of the change in the wires' charge.

Recalling the expression for the capacitance per unit length C in a proportional chamber, Eq. (39), we can compute the effect of a uniform change in the wire radius Δa and in the gap $\Delta \ell$

$$\frac{\Delta Q}{Q} = \frac{C}{2\pi\varepsilon_0} \frac{\Delta a}{a}$$

$$\frac{\Delta Q}{Q} = \frac{C\ell}{2\pi\varepsilon_0 s} \frac{\Delta \ell}{\ell} \; . \qquad (43)$$

Consider, for example, a typical $\ell = 8$ mm, $s = 2$ mm chamber with $2a = 20$ μm operating at a gain around 10^6. The gain variation will be

$$\frac{\Delta M}{M} \simeq 3 \frac{\Delta a}{a} \quad \text{and} \quad \frac{\Delta M}{M} \simeq 12 \frac{\Delta \ell}{\ell} \; .$$

Typical diameter variations around 1% have been measured on standard 20 μm wires, which result in a 3% change in gain, while a 0.1 mm difference in the gap length results in about 15% gain change.

The effect of a displacement of one wire in the wire plane is also a change in the charge, in the displaced wire as well as in its neighbours. Erskine[37] has computed the relative charge modification for a displacement of a wire (wire 0) both in the x and in the y directions (see Fig. 55 for the definition of the reference system). The results for a typical chamber geometry are shown in Figs. 60 a and b. It appears that a 0.1 mm displacement

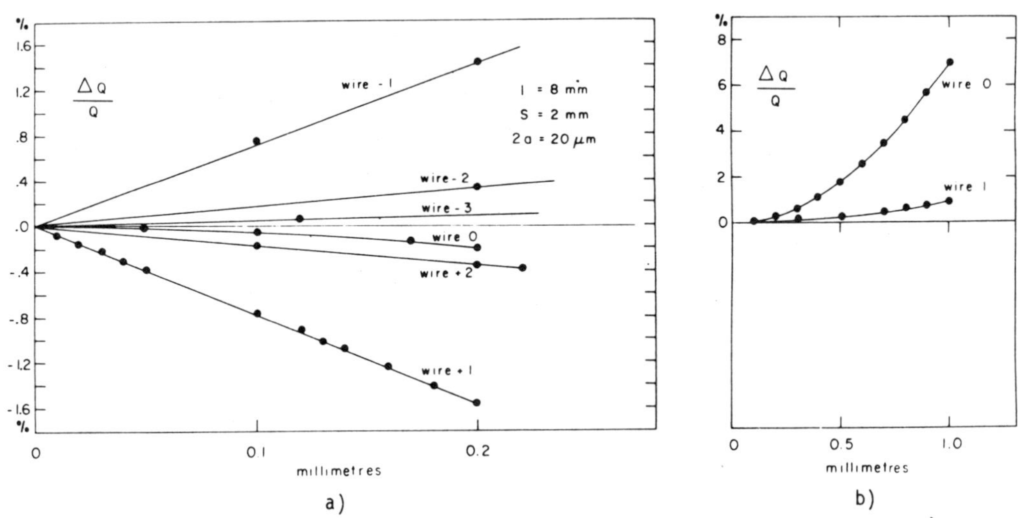

Fig. 60 Relative change in the charge per unit length due to wire displacements[37]. In (a), wire 0 is displaced in the x direction (see Fig. 55), in (b) in the y direction. The change in the charge of the neighbouring wires is also shown.

of a wire in the wire plane results in a 1% change in the charge of the two adjacent wires, and from Eq. (42) this means, at gains around 10^6, more than a 10% change! This is obviously the most critical parameter in chambers' construction, if a good energy resolution is required.

Expression (42) can also be used to estimate the variation of the multiplication factor with the operational voltage V_0; a 1% change in V_0 results in an increase of about 15% in M at gains around 10^6. Practical experience shows that, with normal mechanical tolerances and for medium-size chambers, an over-all gain variation around 30-40% can be expected, and this has to be taken into account when designing the detection electronics.

6.4 Electrostatic forces

In a structure like that of a multiwire proportional chamber, the anode wires are not in a stable equilibrium condition when a difference of potential is applied. In fact, assuming that one wire is slightly displaced from the middle plane, it will be attracted more to the side of the displacement and less to the other, and the movement would continue indefinitely if there was no restoring force (the mechanical tension on the wires). It has been observed in large size chambers (1 m or more) that, above a certain value of the operational voltage, the wires are unstable and the new equilibrium has all wires alternatively displaced up and down, as shown in Fig. 61. Following Trippe[43] we can compute the critical length of a chamber for wire stability. Assuming a radial field as given by Eq. (40), not modified by the small displacement δ, the force between two equal linear charges CV_0 at a distance r, per unit length, is

$$F(r) = \frac{(CV_0)^2}{2\pi\varepsilon_0}\frac{1}{r} ,$$

and, from Fig. 61 and approximating the tangents to the arcs, the total force on a given wire, per unit length, in the direction normal to the wire plane, is

$$\sum F_\perp \simeq 2\,\frac{(CV_0)^2}{2\pi\varepsilon_0}\left\{\frac{1}{s}\,\frac{2\delta}{s} + \frac{1}{3s}\,\frac{2\delta}{3s} + \ldots\right\} = \frac{(CV_0)^2\pi}{4\varepsilon_0}\,\frac{\delta}{s^2} .$$

If T is now the mechanical tension of the wire, the restoring force, in the direction perpendicular to the wire plane and per unit length, is

$$R = T\,\frac{d^2\delta}{dx^2} ,$$

where $\delta = \delta(x)$ is the displacement of the wire along its length, with the conditions $\delta(0) = \delta(L) = 0$ if L is the total wire length. For equilibrium, therefore, one must have $R = -\sum F_\perp$, or

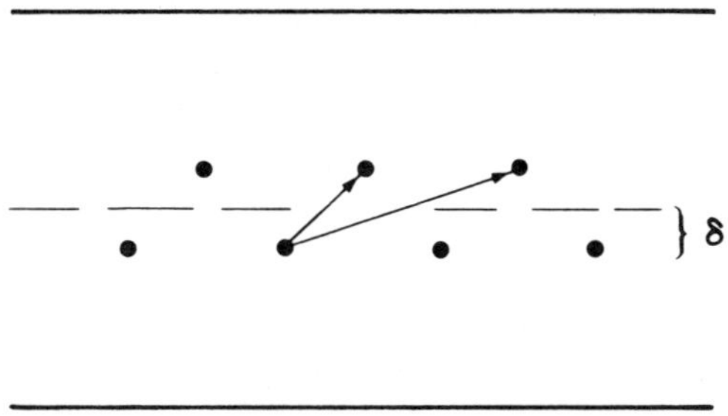

Fig. 61 Electrostatic instability in multiwire proportional chambers. The wires are shown alternatively displaced by a quantity δ from the central plane.

$$T \frac{d^2\delta}{dx^2} = - \frac{(CV_0)^2 \pi}{4\varepsilon_0} \frac{\delta}{s^2} .$$

The equation has the solution

$$\delta(x) = \delta_0 \sin\left(\frac{CV_0}{2s} \sqrt{\frac{\pi}{\varepsilon_0 T}}\, x \right)$$

and the boundary condition $\delta(L) = 0$ means

$$\frac{CV_0}{2s} \sqrt{\frac{\pi}{\varepsilon_0 T}}\, L = \pi \quad \text{or} \quad T = \frac{1}{4\pi\varepsilon_0} \left(\frac{CV_0 L}{s} \right)^2 .$$

For tensions larger than this, no solution is possible other than $\delta(x) \equiv 0$, which means that the wires remain stable. The required stability condition is, therefore,

$$T \geq T_C = \frac{1}{4\pi\varepsilon_0} \left(\frac{CV_0 L}{s} \right)^2$$

or, for a given maximum tension T_M allowed by the elasticity module of a given wire, the critical stability length is

$$L_C = \frac{s}{CV_0} \sqrt{4\pi\varepsilon_0 T_M} . \qquad (44)$$

Table 7 gives for tungsten wires of several diameters the maximum mechanical tension that can be applied before inelastic deformation. As an example, in an $s = 2$ mm, $\ell = 8$ mm multiwire proportional chamber with 20 μm wires, the critical length is about 85 cm for an operational voltage $V_0 = 5$ kV (normal for this geometry, see Section 6.5). When larger sizes are necessary, some kind of mechanical support has to be foreseen for the wires, at intervals shorter than L_C [44,45]. Using thinner wires, if on the one hand one gains in performance as shown above, on the other hand one needs much more closely spaced supports and this is a severe mechanical complication and a source of localized inefficiencies.

Table 7

Maximum mechanical tension (in newtons) for tungsten wires, as a function of their diameter

2a (μm)	T_M (N)
5	0.04
10	0.16
20	0.65
30	1.45

Another consequence of electrostatic forces in a multiwire chamber is the over-all attraction of the outer electrodes towards the anode plane, and therefore an inflection of the cathode planes with a reduction of the gap width in the centre of a large chamber. As we have seen in the previous section, the multiplication factor is very sensitive to the gap width and this can be a problem in large size chambers. Taking into account all charge distributions in the multiwire chamber structure, one can compute the over-all electrostatic force per unit surface, or pressure, on each cathode [46]; the calculation is, however, rather tedious. We can reach essentially the same result using the approximated expression (41) for the field in the drift volume. The field at the surface of the cathode conductors is then

$$E_S = \frac{E_y}{2} = \frac{CV_0}{4\varepsilon_0 s} . \tag{45}$$

Simple charge balance shows that the average charge per unit surface on each cathode is $CV_0/2s$ [this can also be deduced from expression (41) if one remembers the basic electrostatic relationship $E = \sigma/\varepsilon_0$]. The electrostatic pressure on each cathode is, therefore,

$$p = \frac{C^2 V_0^2}{8\varepsilon_0 s^2} . \tag{46}$$

The limits of validity of this expression can be found in the requirement that the field (and therefore the charge distribution) on the surface of the cathodes is constant; from Fig. 56, one can see that this assumption is essentially true for chambers having $\ell \gg s$.

The maximum inwards deflection of a square foil of surface H^2, stretched with a linear tension T, and subject to a pressure p, is given by

$$\Delta y = \frac{p}{T}\frac{H^2}{8} . \tag{47}$$

Combining Eqs. (46) and (47), one cane deduce the maximum deflection for a given chamber geometry, or the maximum size for a chamber if a limit is set for the deflection. In case of large surfaces, a mechanical gap-restoring spacer may be necessary.

As an example, let us compute the maximum gap reduction in an s = 2 mm, 8 mm gap chamber with H = 3 m operating at 4.5 kV; from Table 6 one obtains

$$Q = CV_0 = 1.5 \times 10^{-8}\ C/m$$

and, from Eq. (46), $p = 0.8\ N/m^2$. The yield strength of aluminium alloys is around $2 \times 10^8\ N/m^2$, and for a 20 μm thick foil this gives a maximum stretching tension T of $4 \times 10^3\ N/m$. Introducing this value in Eq. (47), one can see that the maximum deflection is about 220 μm, which implies a gain increase in the centre of the chamber of about 35% at gains around 10^6 (see Section 6.3), in general not tolerable for correct operation. A mechanical gap-restoring device or spacer is necessary; several kind of spacers have been developed[47,48] but they are, of course, a source of inefficiency.

6.5 General operational characteristics: proportional and semi-proportional

A large variety of gases and gas mixtures is currently used in multiwire proportional chambers. Except for special applications (for example, high densities or very good proportionality) all mixtures are more or less equivalent, in terms of performances, at least for moderate gains. Exceptionally large gains are, however, possible only in several specific mixtures containing an electronegative controlled impurity.

A general discussion on gases for proportional counters was given in Section 5.3; here we will present only a collection of experimental observations. A stable proportional or semi-proportional operation can be achieved in mixtures of argon or xenon with carbon dioxide, methane, isobutane, ethylene, ethane, etc; gains above 10^5 can be obtained before breakdown. For a typical energy loss of 6 keV this implies (see Table 1) a charge signal around 3 pC, or 300 mV on a 10 pF load. A collection of measured average pulse heights on ^{55}Fe X-rays on a very large impedance (100 kΩ) is shown in Figs. 62 and 63 [49], for mixtures of argon-carbon dioxide and argon-isobutane, respectively. In Fig. 64 the actual pulse-height distributions are shown as measured under identical conditions for 5.9 keV X-rays (Fig. 64a) and for minimum ionizing particles (Fig. 64b) in a 2 × 8 mm gap chamber, operating with a 60-40 argon-isobutane mixture. As from the considerations of Section 2 and from Table 1, in the quoted geometry the average energy loss of minimum ionizing particles is also around 6 keV, the characteristic Landau distribution of pulse height is, however, observed. In both cases, the horizontal scale is about 1 μA/div., or 1 mV/div. on a 1 kΩ load. Full efficiency of detection for minimum ionizing particles can be obtained with an electronic threshold around one tenth of the peak amplitude, i.e. about 0.5 mV on 1 kΩ; this is a typically adopted value

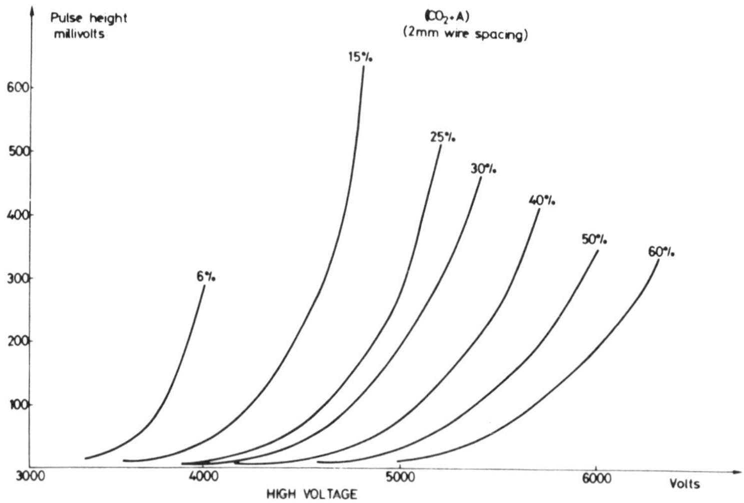

Fig. 62 Peak pulse height on ^{55}Fe 5.9 keV X-rays, in a 2 mm wire spacing chamber for several argon-carbon dioxide mixtures (on 100 kΩ)[49].

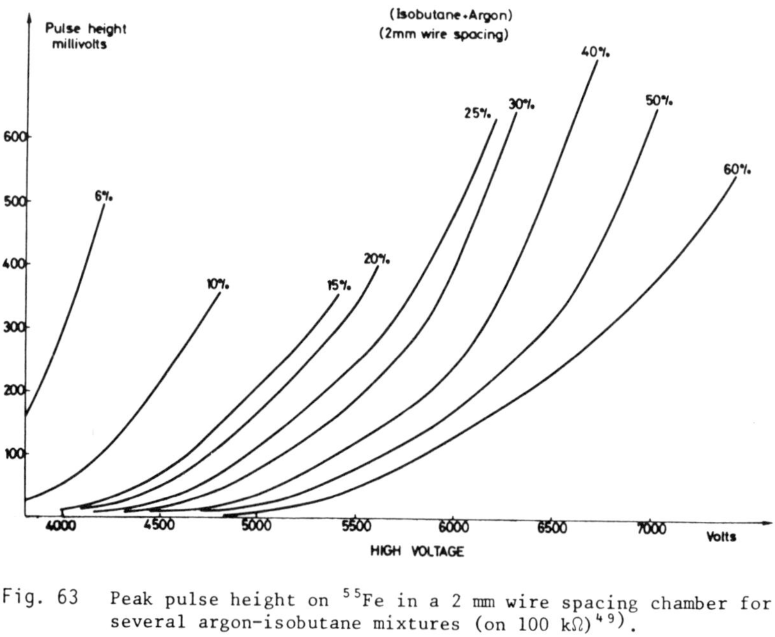

Fig. 63 Peak pulse height on ^{55}Fe in a 2 mm wire spacing chamber for several argon-isobutane mixtures (on 100 kΩ)[49].

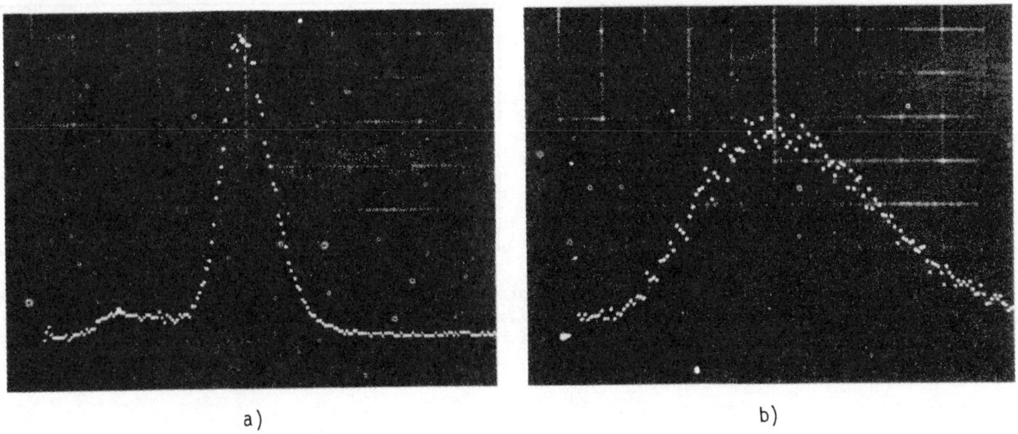

Fig. 64 Pulse-height distribution in the proportional region for ^{55}Fe 5.9 keV X-rays (a) and for minimum ionizing particles (b) in a standard ℓ = 8 mm, s = 2 mm wire chamber. The horizontal scale corresponds to about 1 mV/div. (on 1 kΩ).

for operation in the porportional region. Notice also in Fig. 64a the characteristic 3.2 keV escape peak of argon.

Figure 65 shows the approximate limits of operation of a standard 8 mm gap, 2 mm spacing chamber for argon-isobutane and argon-carbon dioxide mixtures; the beginning of the plateau is, of course, determined by the detection electronics' threshold, 0.5 mV in this case.

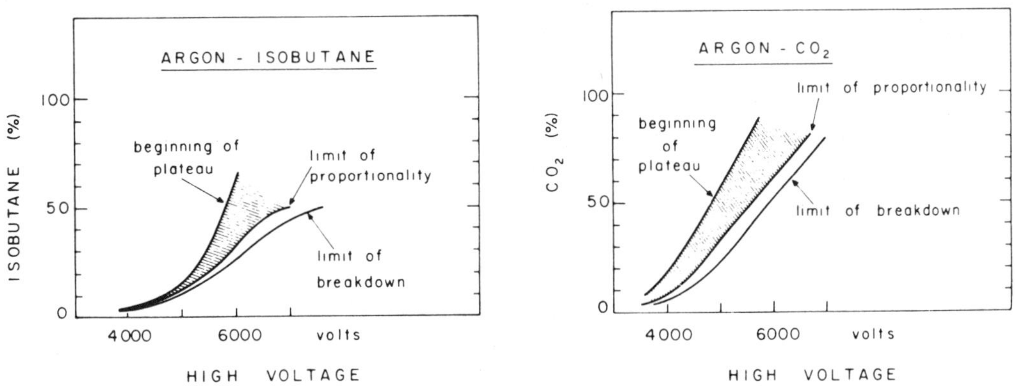

Fig. 65 Approximate limits of operation of an 8 mm gap, 2 mm wire spacing chamber, as a function of isobutane and carbon dioxide content in argon[49].

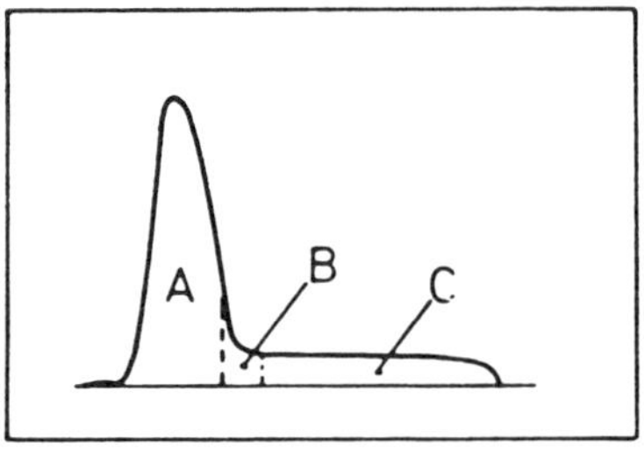

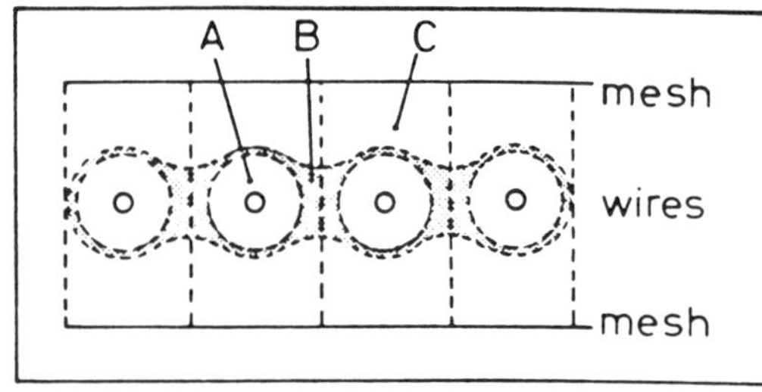

Fig. 66 Timing properties of a multiwire proportional chamber[38]. Depending on where the original charge has been deposited, one can distinguish a "fast" region A, a "drift" region C, and an intermediate region B contributing to the tail in the fast region time distribution.

The timing properties of a proportional chamber are determined by the collection time of the electrons produced by the ionizing track; the peculiar structure of the electric field around the wires allows the separation of three regions, indicated as A, B, and C in Fig. 66. Electrons produced in region A are quickly collected (typical drift velocities in this region of high fields are above 5 cm/μsec); tracks crossing the low field region B, however, will produce a characteristic tail in the time distribution. Electrons produced in region C, on the other hand, smoothly drift to the anode where they are amplified and collected. The time resolution of a chamber is defined as the minimum gate width necessary on the detection electronics for full efficiency; it is of around 30 nsec for a 2 mm spacing chamber. Figure 67 shows the typical time distribution observed in such a chamber, when all wires are connected together (meaning that each track crosses region A or B of at least one wire), while Fig. 68 shows the same spectrum for a single wire and an inclined beam: the long uniform tail in this case corresponds to tracks crossing region C of the considered wire. When detecting tracks not perpendicular to the chamber, the number of wires hit on each track (or cluster size) will obviously depend on the time gate on the associated electronics. If the gate length is the minimum allowed by the requirement of full efficiency (around 30 nsec), the cluster size will be of one or two wires in a ratio depending on the angle; if, one the other hand, the gate length corresponds to the maximum drift time from region C (about 200 nsec for an 8 mm gap), the cluster size will correspond to the maximum allowed by geometry. Figure 69 shows the measured average cluster size as a function of the angle of incidence of the tracks ($\alpha = 0°$ means tracks perpendicular to the wires plane) for a large gate width. A detailed discussion on cluster size and efficiency can be found in Fischer et al.[50].

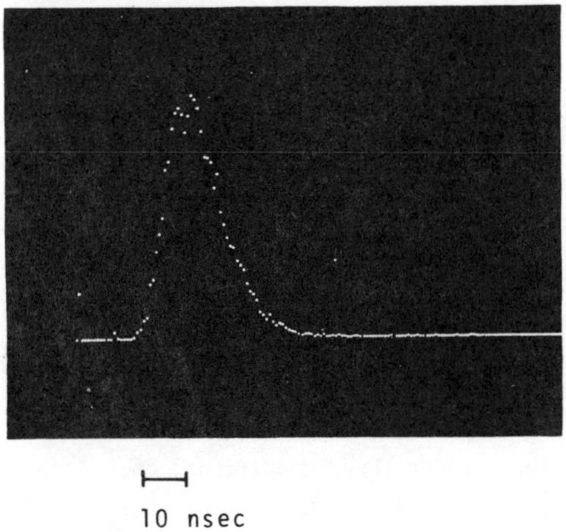

Fig. 67 Typical time distribution measurement in a chamber when all wires are connected together (total OR)

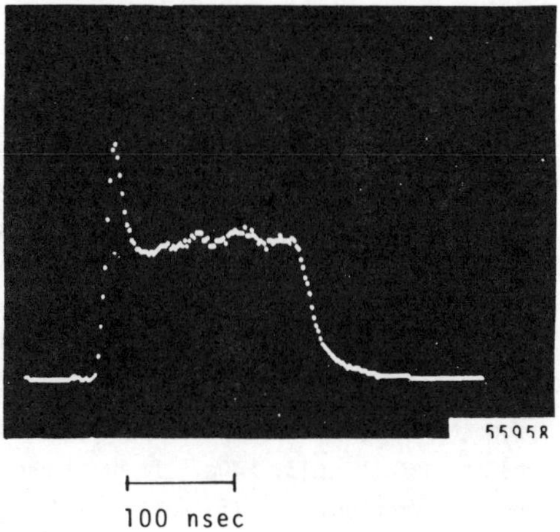

Fig. 68 Typical time distribution measured on a single wire for an inclined beam of minimum ionizing electrons.

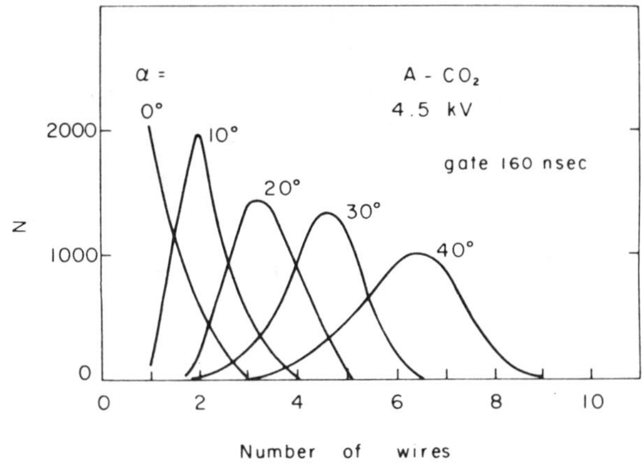

Fig. 69 Average cluster size as a function of the incidence angle (0° means a track perpendicular to the wire planes) for a large time gate width[49])

6.6 Saturated amplification region

Addition to a proportional gas mixture of small quantities of electronegative vapours, like freon (CF_3Br) or ethylbromide (C_2H_5Br) allows the multiplication factor to be pushed to values as high as 10^7 before breakdown, at the same time obtaining a saturated gain condition, i.e. a pulse-height distribution which is entirely independent of the amount of charge lost in the ionizing event. This particular behaviour was first noticed by Charpak and collaborators[49)] in the so-called "magic gas", argon-isobutane-freon in the volume proportions 70-29.6-0.4. The appearance of saturation in these conditions is illustrated by Fig. 70, where the pulse-height spectra for minimum ionizing electrons and 5.9 keV photoelectrons are compared at increasingly high operational voltages. In Fig. 70a, the amplification is still proportional (the lower peak corresponds to fast electrons), in Fig. 70b saturation appears and it is full in Fig. 70c. It has been proved that, under these conditions, one single photoelectron provides the full pulse height[51)]. Notice that the horizontal scales in the three pictures are not the smae, owing to the large increase in the multiplication factor. In Fig. 70c, the peak pulse height corresponds to about 50 μA, or 50 mV on 1 kΩ; thresholds of detection around 5-10 μA are sufficient for full efficiency. Notice also, comparing Figs. 70c and 64b, the reduced dynamic range of saturated pulses: this is greatly reducing the overload recovery time problems of the electronics.

Figure 71 shows a typical efficiency plateau for minimum ionizing particles, of a multiwire proportional chamber operated with magic gas; the average wire noise rate, i.e. the counting rate in the absence of radiation, is also indicated.

The amount of electronegative gas that can be used in a chamber is obviously limited by the requirement of full efficiency; roughly speaking, the mean free path for electron capture λ_c should not be smaller than half the wire spacing. Figure 72 shows the measured efficiency for a 2 mm wire spacing chamber, at increasingly high concentrations of freon; under reasonable assumptions on the detection and capture mechanisms, the experimental points are well approximated by a calculation that assumes $\lambda_c^{-1} = 1.5\ p\ mm^{-1}$, where p is the percentage of freon.

An important consequence of the presence of an electronegative gas in the mixture is a different behaviour in the cluster size versus gate length; electrons produced in the drift region (region C of Fig. 66) have a very small probability of reaching the anodes, and the cluster size is limited even for long gates, as illustrated in Fig. 73.

6.7 Limited Geiger and full Geiger operation

A peculiar mode of operation has been observed in proportional chambers having thick wires widely spaced[52)]. At sufficiently high voltages, and using a reduced concentration of organic quenchers (for example a 90-10 argon-isobutane mixture), a transition is observed

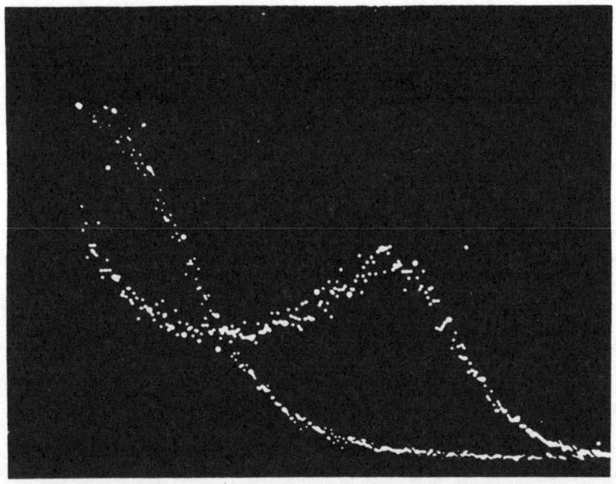

a) HV = 4100 V

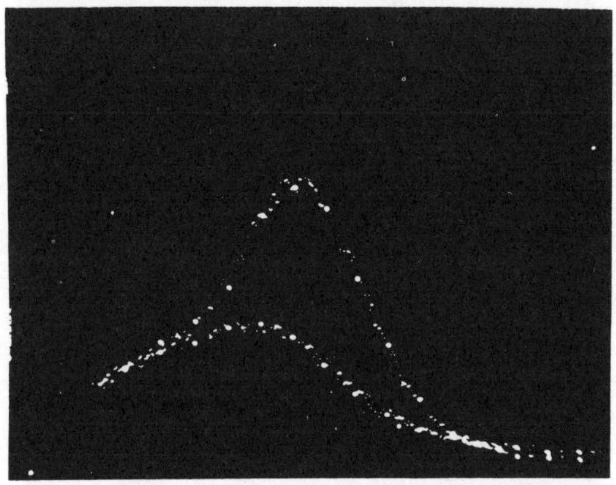

b) HV = 4300 V

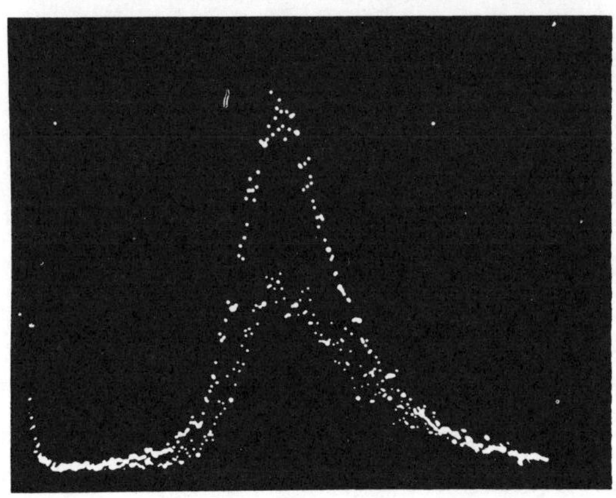

c) HV = 4500 V

Fig. 70 Pulse-height spectra measured in a 5 mm gap, 2 mm wire spacing multiwire chamber for ^{55}Fe X-rays and minimum ionizing electrons at increasing anodic voltages, in "magic" gas, showing the pulse-height saturation effect[38]. Notice that the horizontal scales in the three pictures do not coincide (the average pulse height is increasing by more than one order of magnitude from (a) to (c).

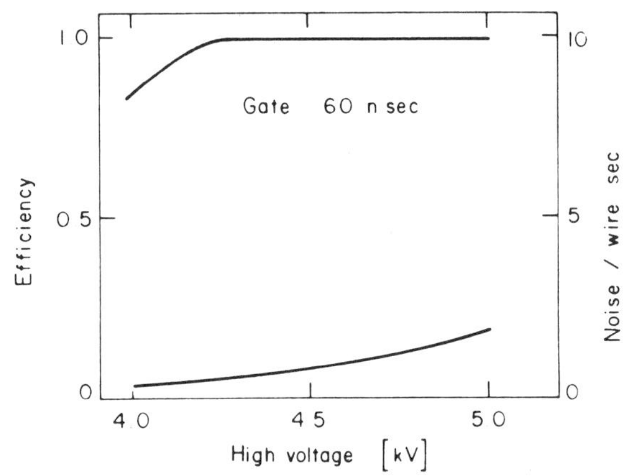

Fig. 71 Typical efficiency plateau and noise rate versus high voltage for an 8 mm gap, 2 mm wire spacing chamber, operating with magic gas and 5 mV (on 1 kΩ) detection thresholds[45])

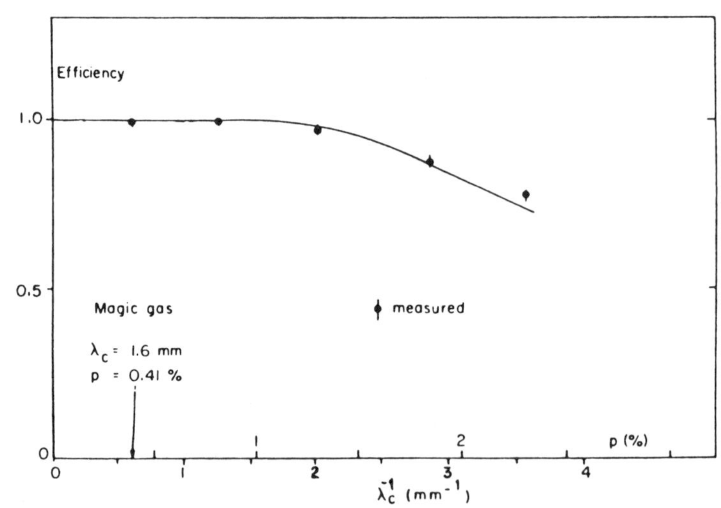

Fig. 72 Average efficiency for minimum ionizing particles, as a function of freon content in a 2 mm wire spacing chamber[51])

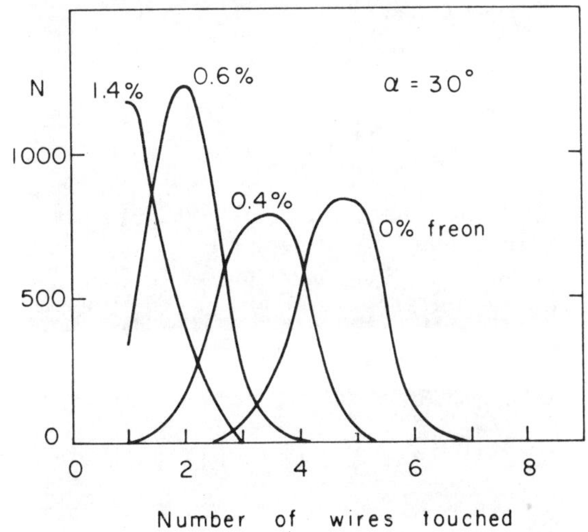

Fig. 73 Cluster size for 30° tracks, for a large time gate, as a function of freon content[49]. Reduction of the sensitive zone around the wires results in a reduced cluster size.

from the normal proportional régime to a damped Geiger propagation, limited in spatial extension to 10 mm or so along the wire. Although very attractive because of the remarkable signal pulse height that can be obtained (30-40 mV on 50 Ω, see Fig. 74), this mode of operation has a severe rate limitation in the long time it takes the positive ion sheath to clear the activated anode wire section. For an 8 mm gap chamber, the assumption that after each count a 10 mm long section of the wire is dead for about 300 μsec provides a good agreement with the experimentally measured efficiency.

Full Geiger propagation has been observed in multiwire proportional chambers when only a small percentage of quencher, like methylal or ethyl bromide, is added to pure argon: again, very long dead-times are obtained. On the other hand, a measurement of the propagation time of the Geiger streamer along the wires can be used to provide two-dimensional images of the conversion points[53].

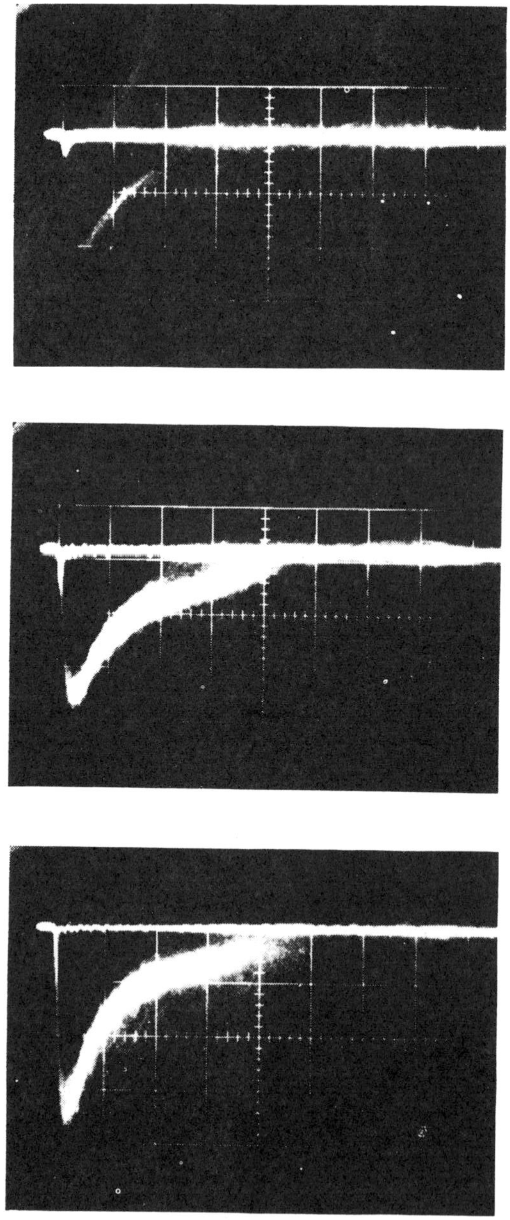

Fig. 74 Transition between the proportional and the limited Geiger operation in a multiwire chamber as seen directly on a 50 Ω termination[52]. Horizontal and vertical scales are, respectively, 200 nsec/div. and 10 mV/div. In (a) the small peak at the left shows the detection of 5.9 keV X-rays still in the proportional (or semiproportional) region, while limited Geiger pulses begin to appear; Figs. 74b and c, obtained at increasingly high voltages, show that all detected pulses enter the limited Geiger mode. The time extension of the signal (less than a μsec) proves the limited extension of the Geiger streamer (around 1 cm).

6.8 Rate effects and ageing

We have already discussed in Section 5.4 the effect on the gain of a positive space charge built up at high rates. Since the gain reduction is a localized effect, extending perhaps one or two gap lengths around the hot spot in a chamber, substantial modification in the distributions of measured particles can be produced if the rate limit is locally exceeded. Measurements of the average pulse-height reduction as a function of rate were given in Figs. 53 and 54 for a specific geometry and gas mixture; of course, the effect depends on the gap and ion mobility, which however cannot vary over a very wide range. Inspection of the figures shows clearly that, at fixed threshold of detection, one will start losing efficiency when the lower part of the pulse-height spectrum distribution decreases below threshold. High chamber gains and low thresholds are therefore recommended for higher rate full efficiency operation. It appears in practice, however, that at any given chamber gain the threshold value cannot safely be reduced below one tenth or one twentieth of the average pulse height (essentially because of noise due to micro-discharges), and therefore almost identical efficiencies of detection versus rate have been measured in a large variety of gases and conditions. We will quote here a measurement realized in a multiwire proportional chamber having 1 mm wire spacing, and operating in magic gas [Fig. 75 [54)]], and a similar measurement obtained in a drift chamber having 50 mm wire spacing and operating in argon-isobutane [Fig. 76 [9)]]. The efficiency loss per unit length of the anode wire at high rates is about the same; obviously, the different surface acceptance in the two cases (50 to 1) reduces correspondingly the tolerable surface rate. Small wire spacings are therefore necessary for operation at high rates, although the spatial extension of the positive charges may create some degree of interdependence between wires.

Another effect related to high counting rates is the polymerization of some quenchers or impurities with the appearance of secondary discharge (see also Section 5.3). A multiwire chamber -- in which, either owing to improper cleaning or to the deposit of polymers due to long exposures to radiation, the secondary discharge mechanism is active -- has a very characteristic behaviour. Normally operating at low counting rates, it manifests an increasing and sustained background rate (or dark current) when exposed to higher radiation fluxes even for a short time, in the damaged regions. Rapid switch off and on of the operational voltage, however, restores the original low-noise condition by capacitive removal of the positive ion sheath on the insulating layers, which is responsible for the secondary sustained discharge. The appearance of radiation damage can be quantitatively measured monitoring the singles counting rates on a variable intensity radioactive source, as a function of the voltage. Figure 77 shows such a measurement using a ^{55}Fe X-ray source before and after a long irradiation (around 10^7 counts/cm^2) [33)]. The background rate is clearly increased, and appears at lower and lower voltages for increasing rates of the source. A shift to higher voltages

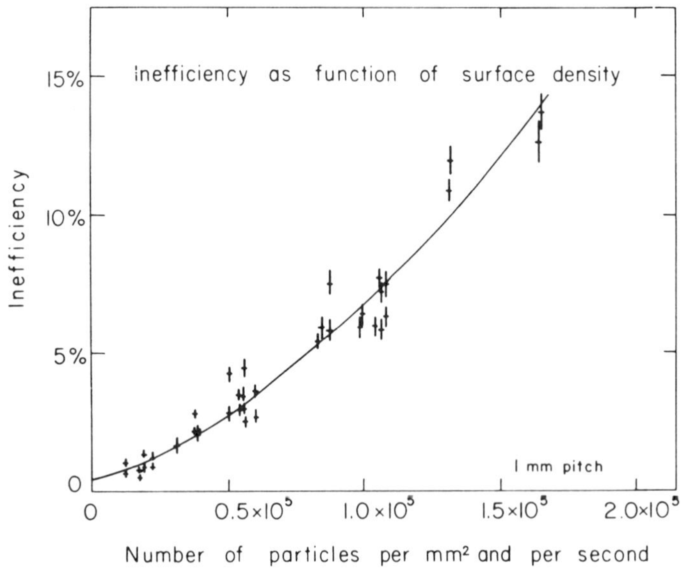

Fig. 75 Space-charge effect on chamber efficiency. Measured inefficiency on a s = 1 mm chamber, operating in magic gas, as a function of beam intensity[54].

of the lower edge of the plateau also appears, either due to the field reduction induced by the charge dipole created on the cathodes over the thin insulator film, or to an increase on the anode wire diameter due to conductive deposits (a thin carbonaceous deposit is normally found covering the anodes of a damaged chamber).

Use of non-polymerizing additives having an ionization potential lower than any other constituent in the gas mixture eliminates, or at least displaces by several orders of magnitude, the integral flux capability of a proportional chamber. In Fig. 78 [33] the efficiency plateau on minimum ionizing particles, the background rate, and the singles rate on ^{55}Fe are measured in a chamber operating with magic gas and methylal (72% argon, 23.5% isobutane, 0.5% freon and 4% methylal) before and after a total irradiation of 3.3×10^{10} counts/cm^2. No change is observed in the main operational parameters. Total exposures exceeding 10^{12}/cm^2 have been reported, without detectable ageing effects.

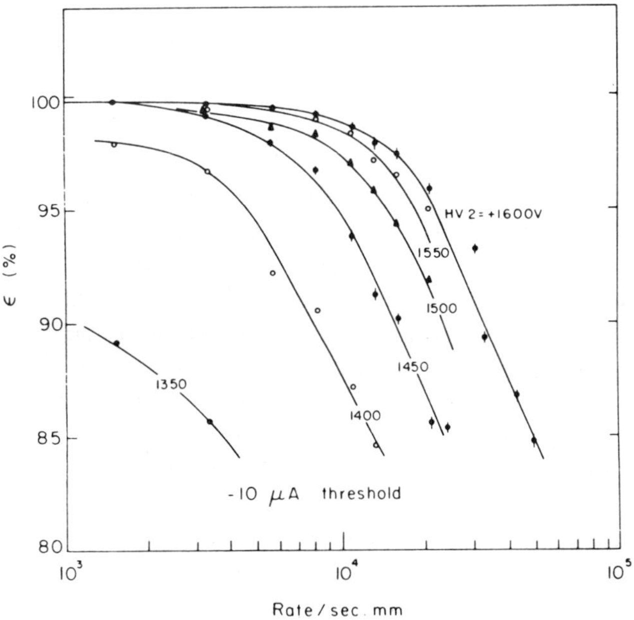

Fig. 76 Space-charge effect measured in a drift chamber[9]. To present the results in a plot independent from the wire spacing, the horizontal scale has been given as rate per unit length of the anodic wire.

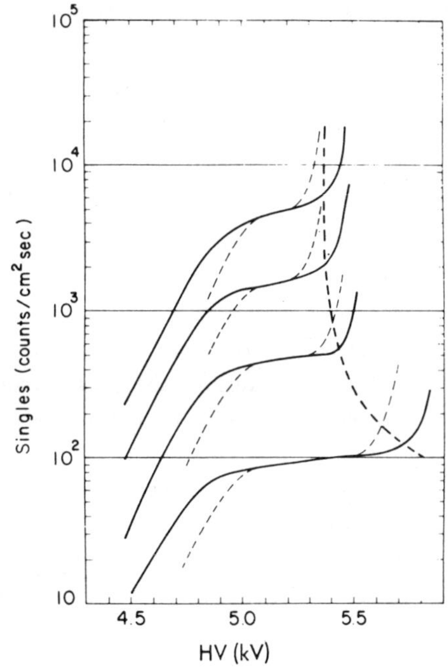

Fig. 77 Ageing effect in a chamber operating with organic gas mixtures (argon-isobutane 70-30)[33]. The singles counting rate has been measured in a chamber for increasing intensities of ^{55}Fe X-rays, as a function of high voltage, before and after an irradiation of about 10^7 counts/cm^2. After the irradiation, the average pulse height is reduced, for a given voltage, and the discharge knee appears at lower voltages.

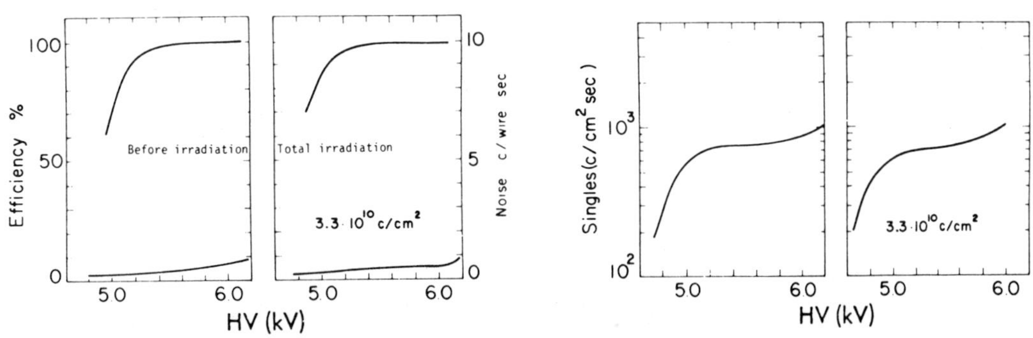

Fig. 78 Effect of a long irradiation in a chamber operating with 4% methylal added to magic gas[33]. At left, the effect on efficiency for minimum ionizing particles, [illegible] the noise, and at right the effect on ^{55}Fe singles rate.

6.9 Mechanics and electronics

Just for completeness, I shall give here some hints on the mechanical construction of a multiwire chamber, as well as a basic description of the attached electronics; the reader should, however, refer to the quoted literature to obtain more details.

The basic problem of chamber construction is to support, on suitable frames, a succession of foils or wire planes constituting the electrodes, within a given set of mechanical and electrical tolerances, and to make the whole structure gas tight. Several methods of construction have been developed; we shall mention here two representative examples. The first, and more frequently used, consists in fabricating a set of self-supporting insulating frames, normally of extruded or machined fibre-glass, one per electrode. The wire electrodes are in general soldered on printed circuits, which are an integral part of the frames; the chamber is then mounted either gluing together the required number of frames (in case of small chambers) or assembled with pins and screws traversing the frames, the gas tightness being guaranteed by suitable rubber O-rings embedded in the frames, and by two thin gas windows (mylar or equivalent) on the two outer surfaces. Figure 79 shows an assembled three-gap

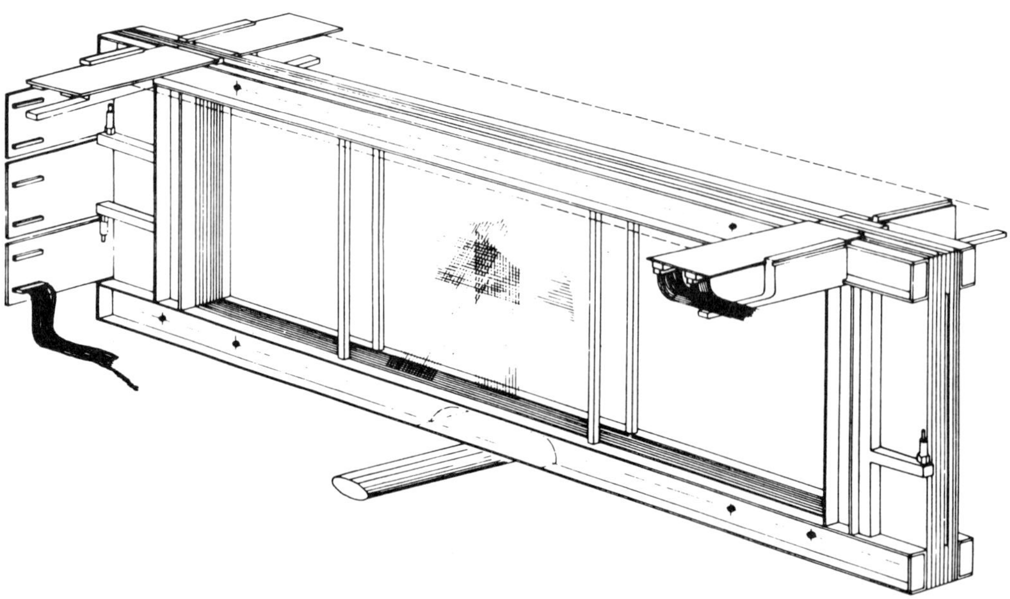

Fig. 79 Schematic view of an assembled three-coordinate medium-size multiwire proportional chamber[45].

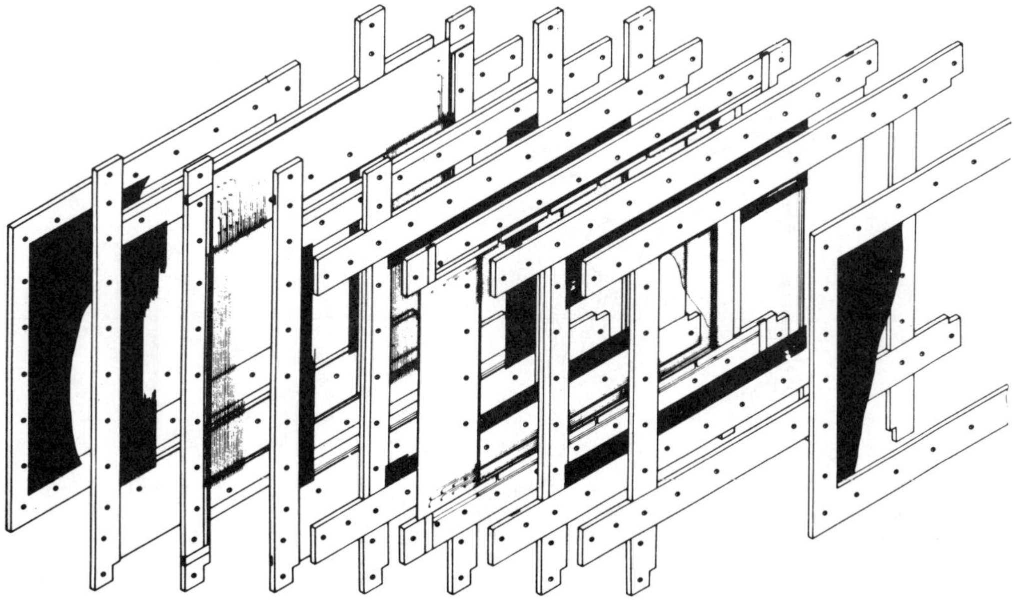

Fig. 80 Exploded view of a 2-coordinates multiwire chamber, showing the number and the complexity of the required parts[55].

chamber of this kind[45] and Fig. 80 [55] an exploded view of a similar chamber; the complexity and the large number of parts required is apparent. The reader will find elsewhere[44,45,55-61], a selection of papers devoted to a detailed description of this method of construction.

A drawback of the described technique lies in the rather unfavourable ratio between the active and the total surface (for a 1 m^2 chamber it is typically around 70%). In case the total available detection area is limited, as in a spectrometer magnet, the loss is considerable. For this reason, a different construction principle has been developed, based on the use of metallized self-supporting honeycomb or expanded polyurethane planes that constitute both the cathode plane and the chamber support[47]. A sketch of such self-supporting chambers is shown in Fig. 81 for a two-gap element; notice the very thin (1 cm or less) passive frame on the edges, on which the anode wires are stretched and soldered and the particular shape of one side, intended to contain the vacuum tubes at the CERN Intersecting Storage Rings. The perspective cut shows also the fishbone structure of the cathode plane strips, used to obtain an ambiguity-resolving third coordinate. The obvious advantages of large active area and

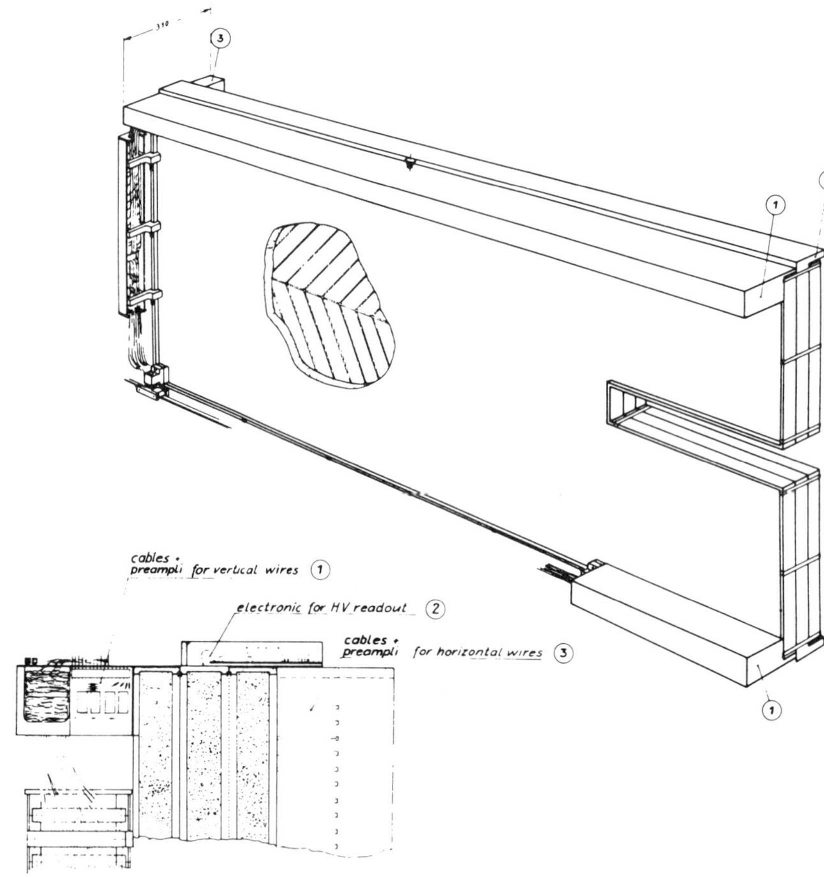

Fig. 81 Principle of construction of a self-supporting multiwire proportional chamber[47]. The basic building elements are foam or honeycomb plates, coated with insulating or conducting thin sheets as shown in the cross-section.

ease of construction are, on the other hand, counterbalanced by a rather substantial increase in the chamber thickness in the active area (0.6 g/cm^2 against a few mg/cm^2 for a conventional construction).

As already mentioned in Section 6.4, when the size of a chamber exceeds a square metre or so, electrostatic wire instabilities and over-all electrode deflections appear, which spoil the chamber's operation unless suitable mechanical support lines are used to balance the electrostatic forces. Supports must obviously be insulators if they are in contact with the anodes and, because they have a dielectric rigidity very different from that of a gas, a substantial field modification is produced that spoils the efficiency over a large area (a centimetre or so around the support). Several solutions have been developed in which an insulated wire or a conductive strip, close to the anode wires but not in contact with them, is raised to a potential high enough to, at least partially, restore the field and therefore the efficiency of detection. In Fig. 82 an example is given of a vinyl-insulated support line and of the corresponding local efficiency measurement[45]; the photograph in Fig. 83 [48] shows a corrugated thin kapton strip, with a printed conductor on one side, which corrects both the anode wire instability and the gap squeezing, with essentially the same efficiency reduction as the previous support line.

As far as the electronics is concerned, I will again refer the reader to the abundant literature existing on the subject; almost every group using proportional chambers has developed its own electronic circuit [see the previously quoted references for the chamber's construction, and also Lindsay et al.[62]]. Completely assembled systems are commercially available[63,64]. The basic principle of a single-wire electronic channel is shown in Fig. 84. The signal from one anode wire is amplified, discriminated, and shaped to a logical level; typical discrimination levels are between 0.5 mV and 5 mV on 1 kΩ (0.5 to 5 μA), depending on chamber gain and performances (see Section 6.5). The pulse is then delayed by either a passive or an active delay element (cable, delay line, or one-shot monostable), so that the experimenter can select, using a logic gate after the delaying element, only events considered good by fast trigger electronics. In the trigger may participate external devices (such as scintillation counters, particle identifiers, etc.) as well as the signals from the wires or group of wires of the chambers, available through the fast OR output. For accepted events, the wire pattern stored in the memory elements (one per wire) is then read out in a sequential way into a computer.

We shall only mention here that several alternatives to the one channel-per-wire electronics have been developed, based either on delay line read-outs[65-67] or on other analogic methods[68,69].

7. DRIFT CHAMBERS

7.1 Principles of operation

The possibility of measuring the electrons' drift time to get information about the spatial coordinates of an ionizing event was recognized in the very early works on multiwire proportional chambers[1]. In its basic form, a single-cell drift chamber consists of a region of moderate electric field, followed by a proportional counter (Fig. 85). Suitable field shaping electrodes, wires or strips, as shown in the figure, allow one to obtain the desired

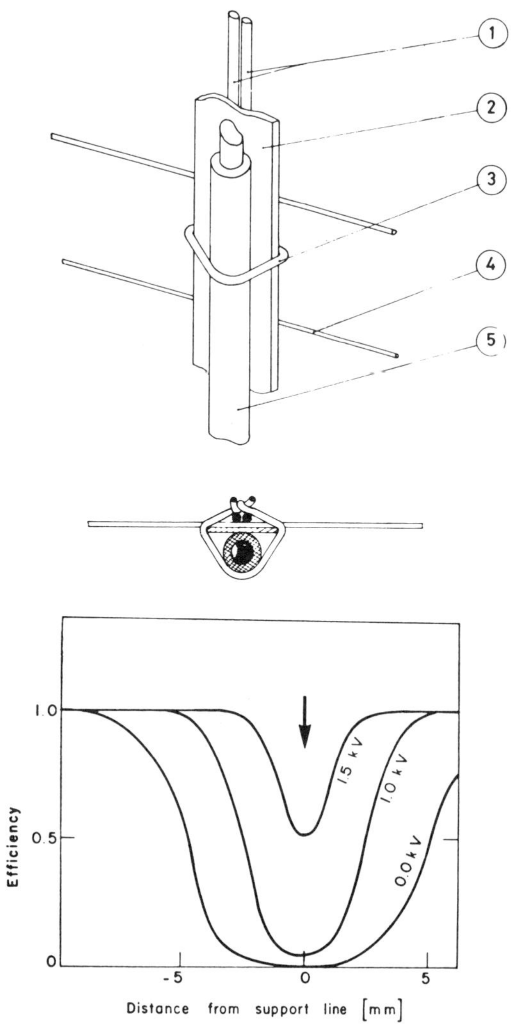

Fig. 82

One of the several support lines developed to avoid electrostatic instability in proportional chambers, together with a measurement of efficiency across the region perturbed by the line[45]. A correction potential given to the central conductor greatly reduces the extension of the inefficient region, as shown by the family of curves.

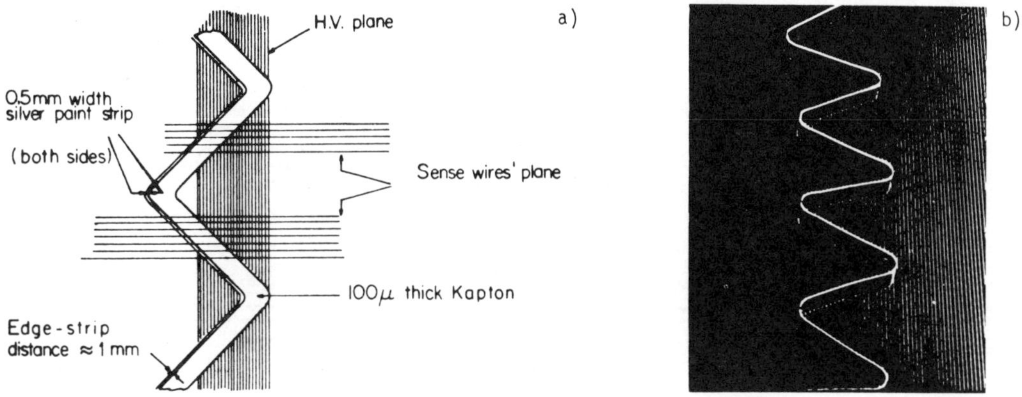

Fig. 83 Example of a mechanical device used both to avoid gap compression (due to the anode-cathode planes attraction) and anode wire instabilities, with a potential-correcting conductor on the anodic side[48]

electrical configuration. Electrons produced at time t_0 by the incoming charged particle migrate against the electric field with velocity w, and reach the anode wire where avalanche multiplication occurs at a time t_1. The coordinate of the track, in respect to the anode wire, is therefore given by

$$x = \int_{t_0}^{t_1} w \, dt , \tag{48}$$

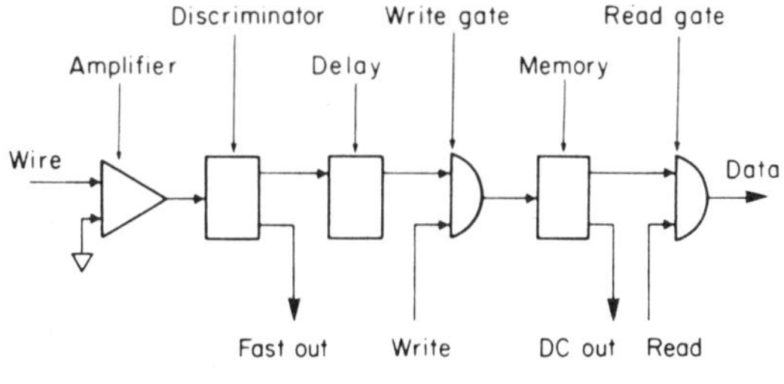

Fig. 84 Basic scheme of the electronics required on each wire in a proportional chamber

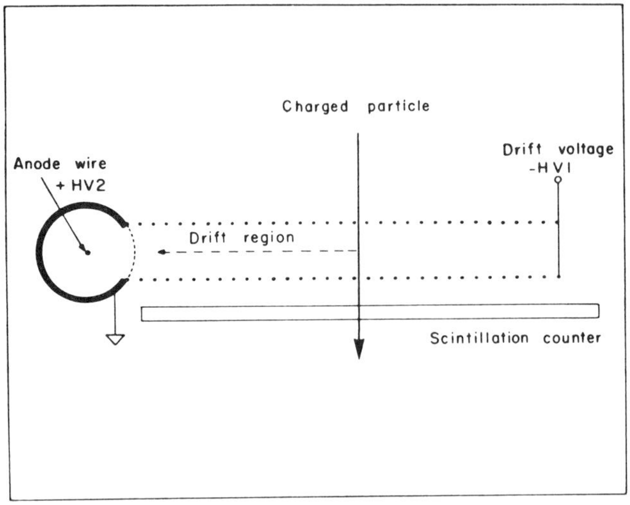

Fig. 85 Principle of operation of a single-cell drift chamber. A set of cathode wires, at suitable potentials, generate in the drift space a region of uniform field. The electrons produced by an ionizing event migrate to one end of the cell, where avalanche multiplication occurs in a single wire proportional counter. The coordinate is then proportional to the time of drift (the time reference being given by an external scintillation counter).

which reduces, for a constant drift velocity, to $x = (t_1 - t_0)w$. It is obviously very convenient to have a linear space-time relationship, and this can be obtained in structures with uniform electric field. If a large surface of detection is required, however, a simple structure like the one of Fig. 85 leads to uncomfortably large working voltages and very long drift times; nevertheless, chambers of this kind having as much as 50 cm drift lengths have been operated, with an over-all drift voltage around 50 kV and maximum drift time (or memory) of 7 μsec [70]. For even larger surfaces, or in cases where shorter memory times are necessary because of the expected particle rates, a multicell structure can be used; in this case, since the region of the anode wire becomes necessarily part of the active volume, it is not possible to obtain a constant drift field all across the cell.

In principle, a structure identical to the one of a multiwire proportional chamber can be used to realize a multiwire drift chamber; however, the low field region between the anode wires would result in a strong non-linearity of the space-time relationship, especially for large wire spacings. A modification of the original proportional chamber structure allows

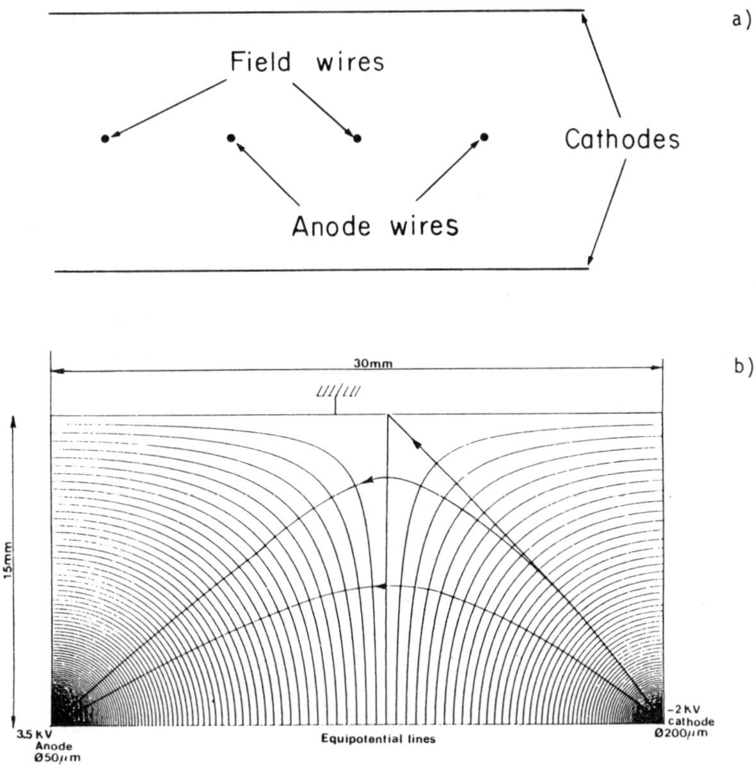

Fig. 86 Principle of the multiwire drift chambers with uniform cathode planes: (a) the basic geometry and (b) the electric field equipotentials in a chamber having 2 × 15 mm gap and 60 mm between anode wires[75].

the elimination of low field regions in the central plane, as shown in Figs. 86 a and b. The anode wires are alternated with thick field-shaping cathode wires that reinforce the electric field in the critical region. Chambers of this design with anode wire spacing of 1 cm were the first operational drift chambers[71], and were built, with wider wire spacings, up to sizes of about 4 × 4 m^2 [72]. Other designs, similar in principle to the one described, have been developed, which allow a simpler construction of large surface, mechanically very stiff, drift chambers; Fig. 87 shows one example[73]. Thin aluminium profiles (I-beams), insulated from the cathode planes and kept at a negative potential, serve both the purpose of mechanical spacers and field-reinforcing electrodes. The cathodes are grounded, while the anode wires are maintained at a positive potential to collect and amplify the electrons.

The major limitation of the structures represented in Figs. 86 and 87 lies in the fact that, in order to obtain a relatively uniform drift field, the ratio of the gap length to the wire spacing has to be maintained close to unity. For typical convenient wire spacings (5 to 10 cm) this implies rather thick chambers, and therefore a reduced packaging density. Moreover, it takes a long time to collect at the anode all the electrons produced by a track, and therefore multitrack capability per wire is excluded. These considerations have led to the development of the structure shown in Fig. 88 [9,74]. Two sets of parallel cathode wires are connected to increasingly high negative potentials on both sides starting from the centre of a basic cell; the anode wire is maintained at a positive potential, and two field wires, at the potential of the adjacent cathode wires, sharpen the transition from one cell to the next. The equipotential lines are shown on the figure, for a typical choice of operational voltages; a uniform field drift region is produced in most of the cell. Small gap-to-anode wire spacing ratios can therefore be implemented; typical values of 6 mm and 50 mm have been used for the gap and the anode spacing, respectively. Notice that, since the cathode planes

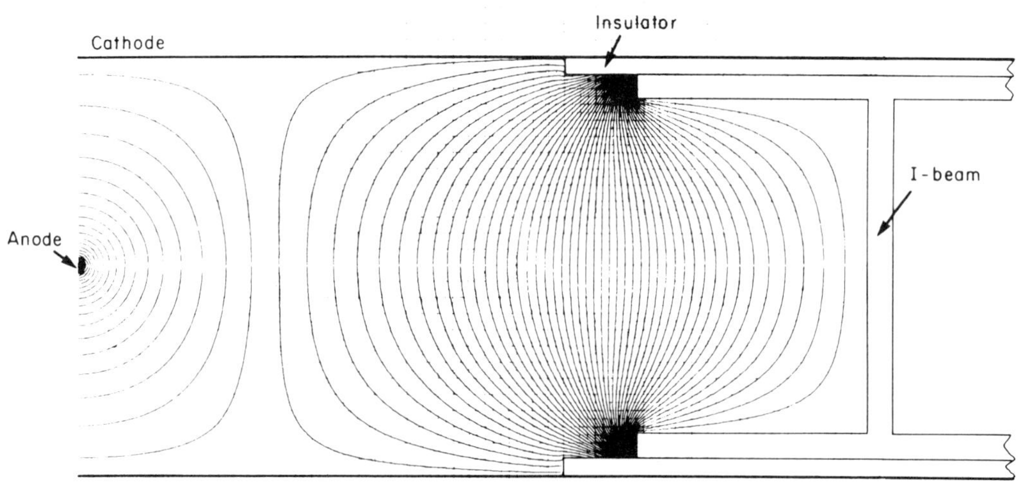

Fig. 87 Equipotentials computed in the Harvard-MIT drift chambers[73]. The field wire has been replaced by an I-shaped profile, insulated from the cathode planes, and serving both the purpose of field reinforcing element and of mechanical stiffener. The gap is 26 mm, the anode wire spacing about 10 cm.

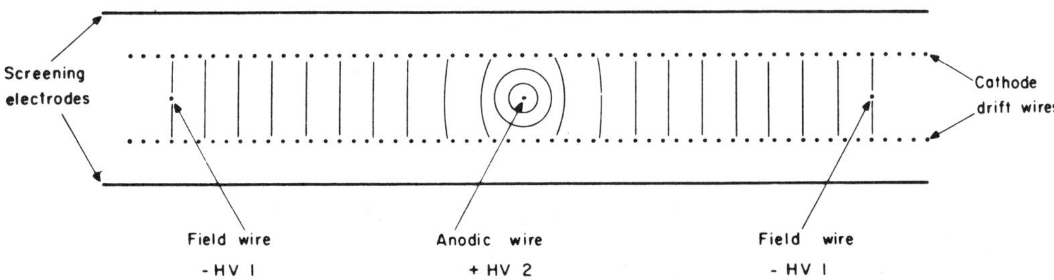

Fig. 88 Principle of construction of the adjustable field multiwire drift chambers[9]. Cathode wires are connected to uniformly decreasing potentials, starting from ground in front of the anode. Field wires reinforce the field in the transition region to the next cell.

are not equipotentials, some field lines escape from the structure; subsidiary grounded screening electrodes, as shown in the figure, guarantee the immunity of the drift field from external perturbations.

Other structures, intermediate between the ones described, have been developed; Fig. 89 shows, for example, a modification of the geometry of Fig. 86, with the introduction of several additional field-shaping wires that allow reduction of the long collection time on a track by limiting the effective volume of detection[75]. Figure 90 shows instead a scheme where flat electrodes on both sides of the anodes replace the field-shaping wires[76].

7.2 Space-time correlation and intrinsic accuracy

The ultimate accuracy that can be obtained in a drift chamber depends both on the good knowledge of the space-time relationship and on the diffusion properties of electrons in gases. From the considerations of Section 4.4 it appears that, for most gases commonly used in proportional counters, intrinsic accuracies (due to diffusion) below 100 μm are possible for minimum ionizing particles. The space-time correlation, however, may not be known at this level of accuracy, especially for large chambers where various mechanical tolerances can locally modify the electric field structure. Very good results in terms of stability of operation and reproducibility have been obtained combining the good electric field characteristics of the structure shown in Fig. 88 and the saturated drift velocity peculiarity obtained in selected gas mixtures. Inspection of Fig. 30 shows, for example, that in a 70-30 mixture of argon-isobutane the electrons' drift velocity is roughly constant at fields exceeding 1 kV/cm. If a chamber is designed in such a way as to avoid regions of field lower

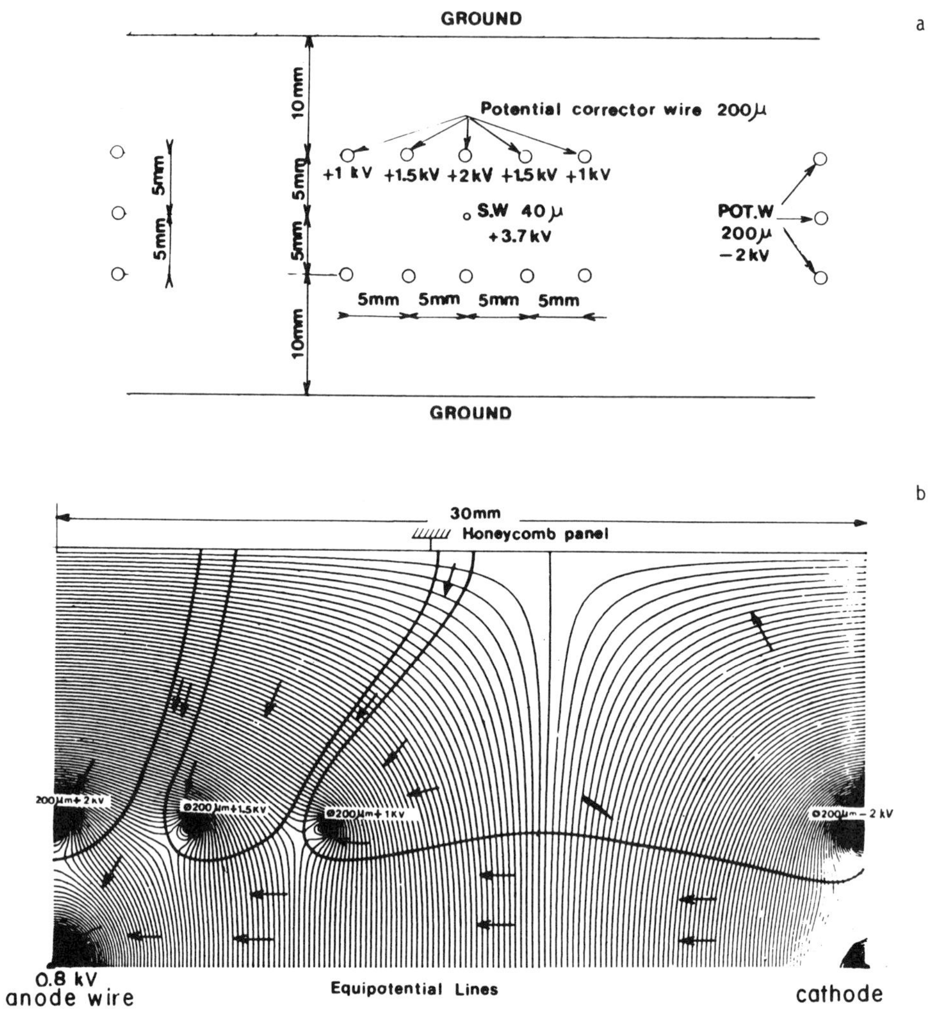

Fig. 89 (a) Geometry of the Saclay chambers, and (b) electric field equipotential in one quarter of the structure[75]. The role of the intermediate potential wires is to limit the accepted fraction of ionizing track so as to have a more or less uniform response along the drift cell.

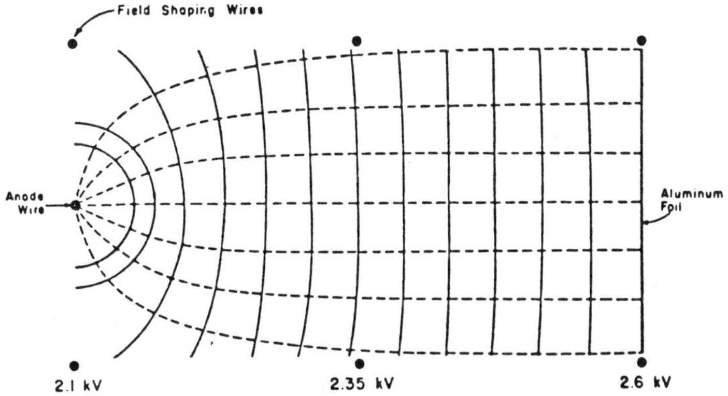

Fig. 90 The drift chambers developed at Fermilab with a thin aluminium foil separating adjacent cells[76].

than the quoted value, the sensitivity of the response to local field variation is strongly reduced.

We shall describe in what follows the main results obtained with a drift chamber of the kind depicted in Fig. 88, with 6 mm cathode plane separation and 50 mm cell size, operating in a gas mixture of argon, isobutane and methylal in the proportions, respectively, 67.2%, 30.3% and 2.5% [9]. The dependence of the drift velocity on electric field for this mixture was given in Fig. 37 *). Similar sets of measurements exist for all other drift chamber structures and can be found in the corresponding quoted references.

A simple way of measuring the space-time relationship in a drift chamber is to record its time spectrum on a uniformly distributed beam. In fact,

$$\frac{dN}{dt} = \frac{dN}{ds}\frac{ds}{dt} = k\ w(t)\ . \tag{52}$$

Therefore the time spectrum represents the drift velocity as a function of time of drift, and its integral the space-time relationship. An example of this kind of measurement is

*) Notice, comparing Figs. 30 and 37, that the addition of methylal to an argon-isobutane mixture (necessary for the reasons illustrated in Sections 5.3 and 6.8) increases the saturation voltage to about 1.2 kV/cm.

given in Fig. 91. Obviously, the limitation of the method lies in the accuracy with which a uniform beam can be produced over a large surface; for more accurate measurements, several methods of mechanical or electronic scanning have been used. Figure 92 shows the result obtained in a chamber of the kind illustrated in Fig. 88 [9]; the space-time correlation has been measured for several angles of incidence of the minimum ionizing beam, and it is strictly linear, within the measurement errors (±50 μm) over all the cell for tracks perpendicular to the chamber plane ($\theta_V = 0°$) with a slope of 5.20 ± 0.02 cm/μsec. For a beam inclined in the plane perpendicular to the wires, the correlation is modified by the fact that the electrons having the shortest time of drift are not those produced in the middle plane of the drift cell. However, a simple two-straight-lines approximation is possible to the measurement, as shown in the figure, under the assumption that the shortest distance to an inclined track is radial around the anode wire and follows the cathode planes thereafter, i.e. as expressed by

$$\begin{cases} x = \dfrac{w}{\cos\theta_V}\, t & \text{for } 0 \le t \le \dfrac{g}{w\sin\theta_V} \\[2ex] x = wt + \dfrac{g}{\sin\theta_V}\left(\dfrac{1}{\cos\theta_V} - 1\right) & \text{for } t \ge \dfrac{g}{w\sin\theta_V} \end{cases} \tag{53}$$

where g is the total gap width. No deviation for linearity is observed for a beam inclined

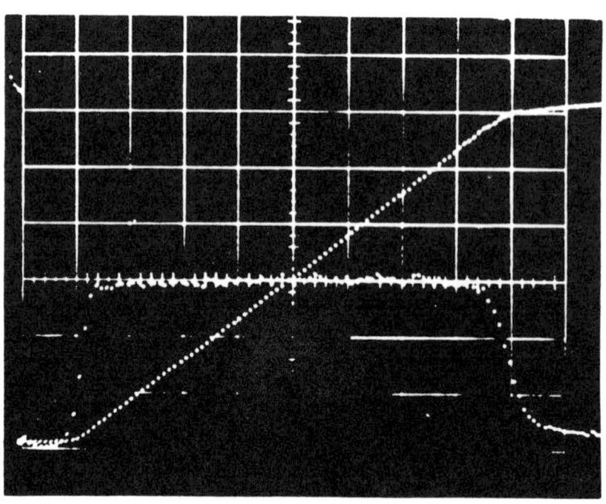

Fig. 91 Example of time spectrum and its integral in a uniform beam, representing respectively the drift velocity and the space-time relationship. The measurement has been obtained by the author with a chamber of the kind described in Fig. 88.

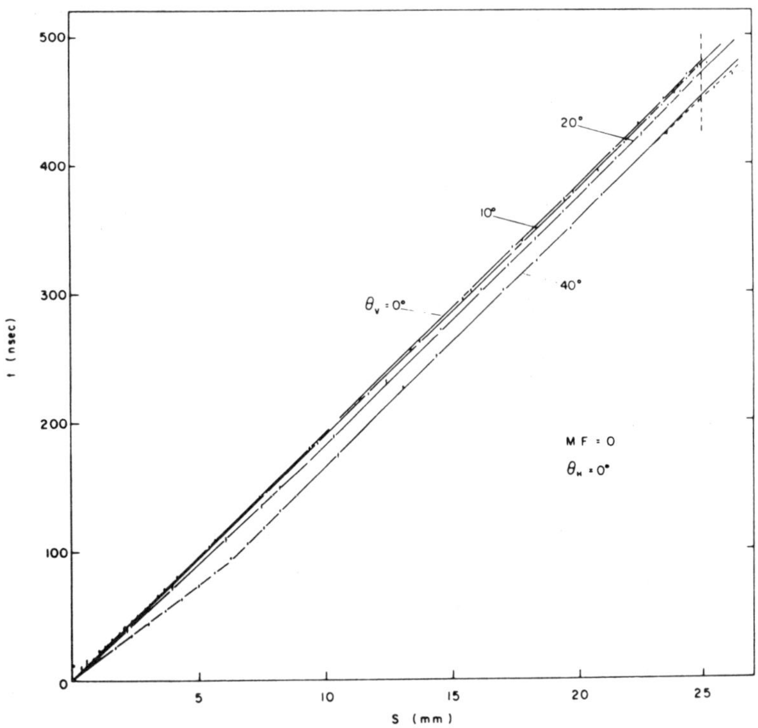

Fig. 92 Measured and computed space-time relationship for the chamber in Fig. 88, as a function of the minimum ionizing beam angle of incidence[9]

in a plane parallel to the wires. Normally, of course, the angle of incidence of the tracks is not known *a priori*; an iterative procedure using the data from a set of chambers is then followed , computing an approximated incidence angle under the assumption of a linear correlation. The procedure is, in general, very quickly converging. In structures where the electric field is less uniform, the space-time relationship is obviously more or less deviating from linearity especially if regions of field are met low enough not to allow drift velocity saturation. As an example, Fig. 93 shows the measured correlation in a very large (3.6 × × 3.6 m^2) chamber with the structure shown in Fig. 86 and 10 cm between anode wires[72].

The intrinsic accuracy of a chamber can be estimated by the usual method of measuring the same track in a set of equal chambers, and computing the standard deviation of the difference, in a given chamber, between the measured and fitted coordinate. One of the best

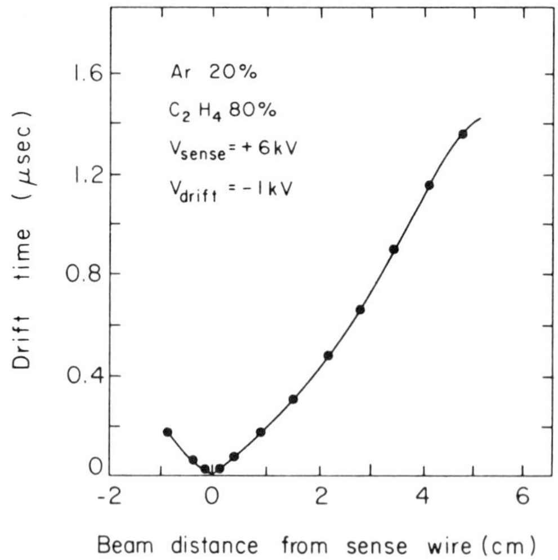

Fig. 93 Space-time relationship for the chamber illustrated in Fig. 86, with 10 cm wire spacing[72]. Due to the large electric field variability, the correlation is not linear.

results obtained so far is presented in Fig. 94 [10], which gives the accuracy as a function of the drift distance in a chamber like the one in Fig. 88, but with 42 mm anode wire distance. The result can be decomposed into three main contributions: a square root dependence on the distance of drift, due to electron diffusion, a constant electronics spread estimated to correspond to about 40 μm as from the figure, and a contribution of the primary electrons' production statistics, particularly important close to the anode wire (see the discussion in Section 2.8).

As far as the proportional gain is concerned, drift chambers are, in general, easier to operate than multiwire proportional chambers, being essentially isolated proportional counters; all considerations on multiplication factors and gas choice, developed in Sections 5 and 6, apply as well with the necessary modifications due to the geometry. Exceedingly long efficiency plateaux (about 30% of the working voltage) have been obtained with rather high discrimination thresholds, 5 to 15 μA [77]. The gas purity is, of course, of primary importance in a drift chamber, especially if long drift spaces are used; the effect of electronegative gas pollution has been discussed in Section 4.6. Common practice has shown that commercial grade purities are sufficiently good for moderate drift lengths (a few cm), but that the gas

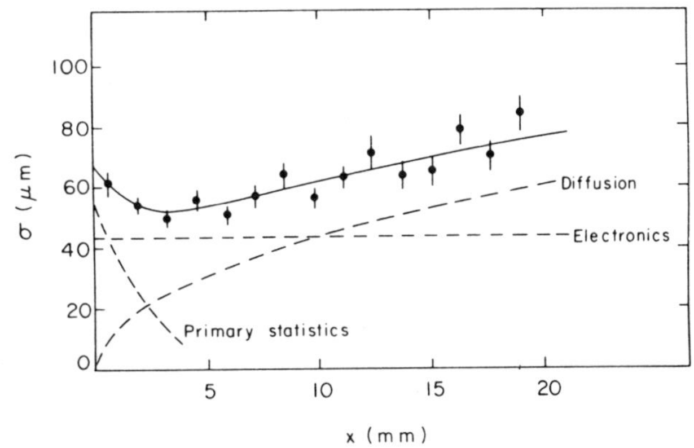

Fig. 94 Measured intrinsic accuracy in the drift chamber of Fig. 88, as a function of drift space[10]. The experimental results have been decomposed into three contributions: a constant electronics dispersion, a physical diffusion term function of the square root of the drift space, and a contribution of the primary ion pair statistics.

tightness of the chamber and of the tubing has to be carefully checked. In some cases, a gas monitoring device at the output of a chamber or a chamber system is advisable.

7.3 Stability of operation

It should be emphasized that the intrinsic accuracy given in Fig. 94 is the result of a local measurement, normally realized in a short run and essentially independent of the detailed space-time correlation. To make use of the quoted accuracies in an actual coordinate measurement, one has, of course, to know precisely the space-time correlation or the drift velocity $w = w(t)$. For a given chamber geometry, the major factors that can influence the drift velocity are: the electric field strength and direction, the atmospheric pressure, the gas composition and temperature, the presence of external factors modifying the drift properties (electric or magnetic stray fields), and the mechanical imperfections. Alghough it is, in principle, possible to take all these factors into account by proper calibration or monitoring, for a realistic system it is more reasonable to set definite limits to the tolerable variations, as a function of the desired final accuracy. The choice of a drift-velocity saturating gas obviously decreases or eliminates the dependence on the reduced electric field E/P. It appears also [25,26] that for several gases and gas mixtures the temperature dependence of w is reduced at high fields (see Fig. 95a); for the particular mixture used by the

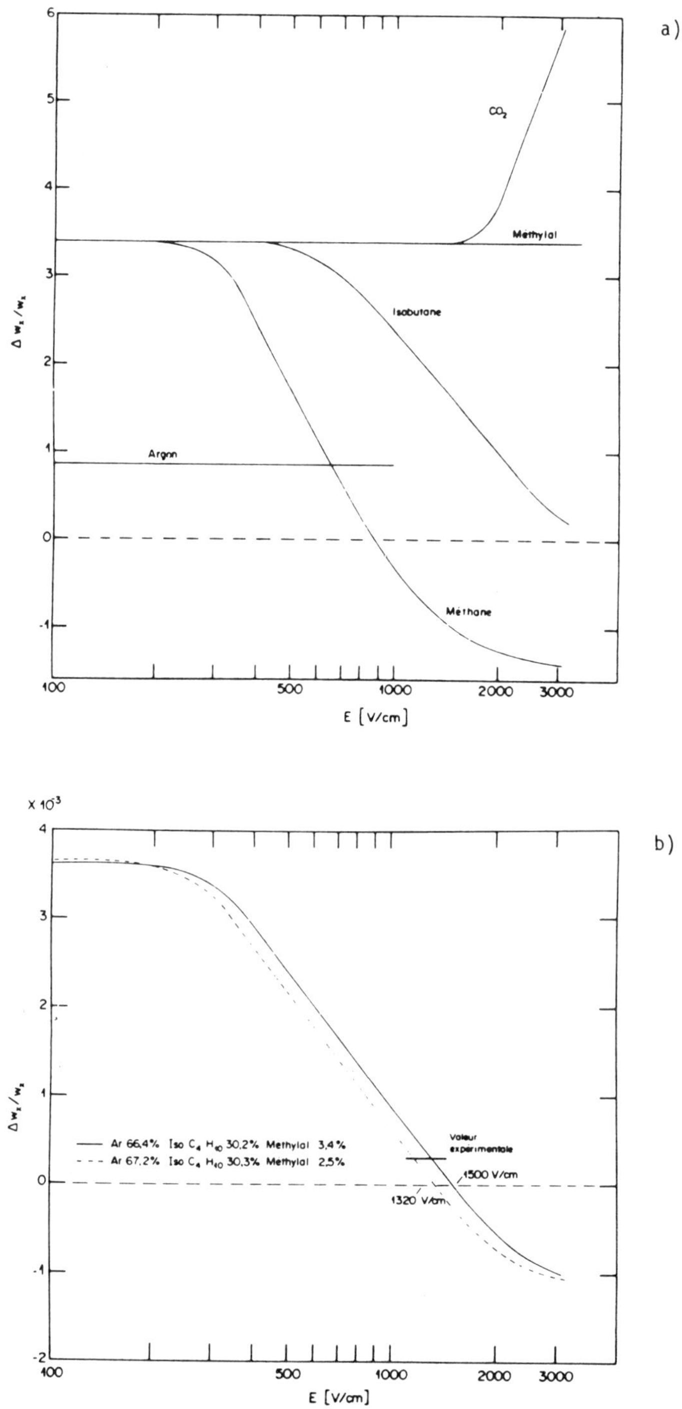

Fig. 95 Computed relative dependence of the drift velocity on the temperature, at 20°C and 1 atm (a) for several gases and (b) for the particular mixture used in high-accuracy drift chambers[26])

authors of Ref. 9, the calculation gives a value very close to the experimentally measured one, i.e. $\Delta w/w = 3 \times 10^{-4}$ per °C at a field around 1.4 kV/cm (Fig. 95b). At low fields, and for the same mixture, the relative variation is about one order of magnitude larger. The dependence on gas composition also reduces at saturation, and has been measured in the quoted gas mixture to be $\Delta w/w = 1.2 \times 10^{-3}$ for 1% change in the gas[77]. Using the measured relative variations, one can see that to maintain a ±50 μm stability over 25 mm of drift ($\Delta w/w = \pm 2$‰) a maximum temperature variation of ±7°C, and a maximum gas composition change of ±1.6% can be tolerated.

Mechanical tolerances and electrostatic deformations contribute, of course, directly to the limiting accuracy, and may moreover produce electric field distortions. For large size chambers one should also take into account the thermal expansion of the materials.

7.4 Behaviour of drift chambers in magnetic field

We have seen in Section 4.5 that the presence of a magnetic field other than parallel to the drift direction modifies both the drift velocity and the angle of the electron swarm. In some cases this can be a perturbing factor, if a chamber is situated too close to a stray field, in others instead installation and proper operation of a chamber inside a magnet is desired. Several structures have been studied that permit the use of a drift chamber in strong magnetic fields. Figure 96, for example, shows the computed electron trajectories in

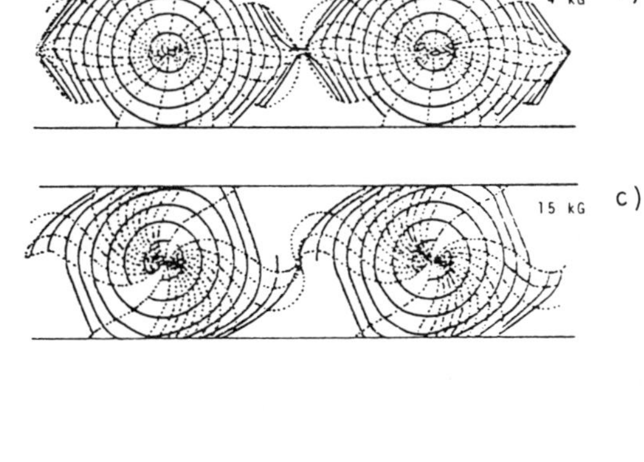

Fig. 96

Computed electron trajectories in a chamber like the one of Fig. 86, for several values of the magnetic field, parallel to the wires[78]

a structure similar to the one of Fig. 86, at increasingly high values of the magnetic field parallel to the sense wires, and Fig. 97 the corresponding time-space relationships[78]. Obviously, precise knowledge of the magnetic field strength and direction is necessary for each position in the chamber and a rather complex parametrization or tabulation for the correlation must be used. In the case of a uniform magnetic field, the electric field in a structure like the one of Fig. 88 can be modified to compensate for the angle of drift. By a suitable choice of the drift voltage connections, the equipotential surfaces in the chamber can be tilted by an angle corresponding to the expected angle of drift (see Fig. 98), the fine adjustment being then possible by a modification of the field strength (see Fig. 38) still maintaining the saturation properties of the velocity. The result of such a procedure can be seen in Fig. 99, where the space-time relationship has been measured for several angles of incidence and in a magnetic field, parallel to the wires, of 10 kG [9]. The equipotentials' tilt angle was, in this case, about 29°; the similarity of behaviour to the case of no magnetic field, Fig. 92, is apparent. In more complex geometries of magnetic field, or for non-uniform fields, a simple formulation of the space-time relationship is, of course, not possible; corrections are then done by using a set of measured 'rift velocities and angles, like the ones shown in Figs. 37 and 38. Denoting by $w_{\parallel}$ the projection of the drift velocity

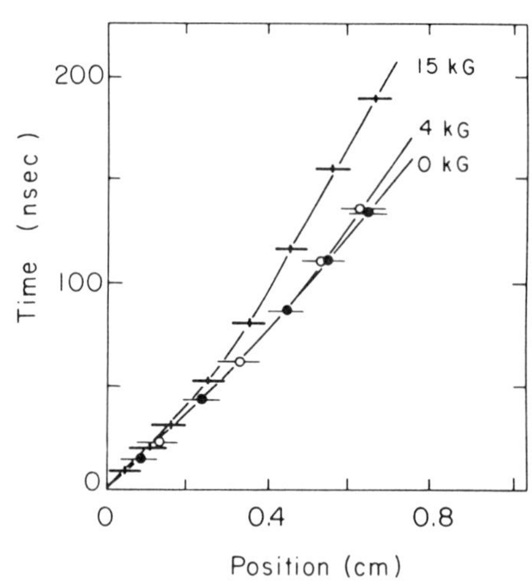

Fig. 97 Computed space-time relationship for normal tracks in the geometry of Fig. 96, at increasing values of the magnetic field[78]

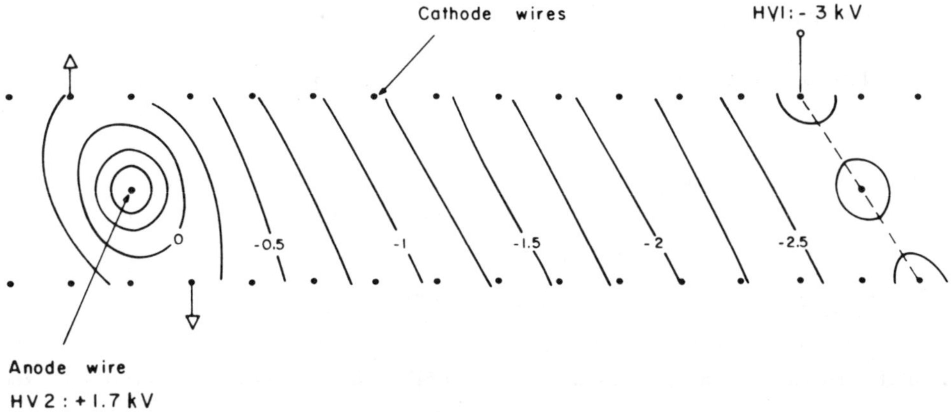

Fig. 98 Modification of the electric field equipotentials in the structure of Fig. 88, to allow operation in strong magnetic fields (parallel to the wires)[77]

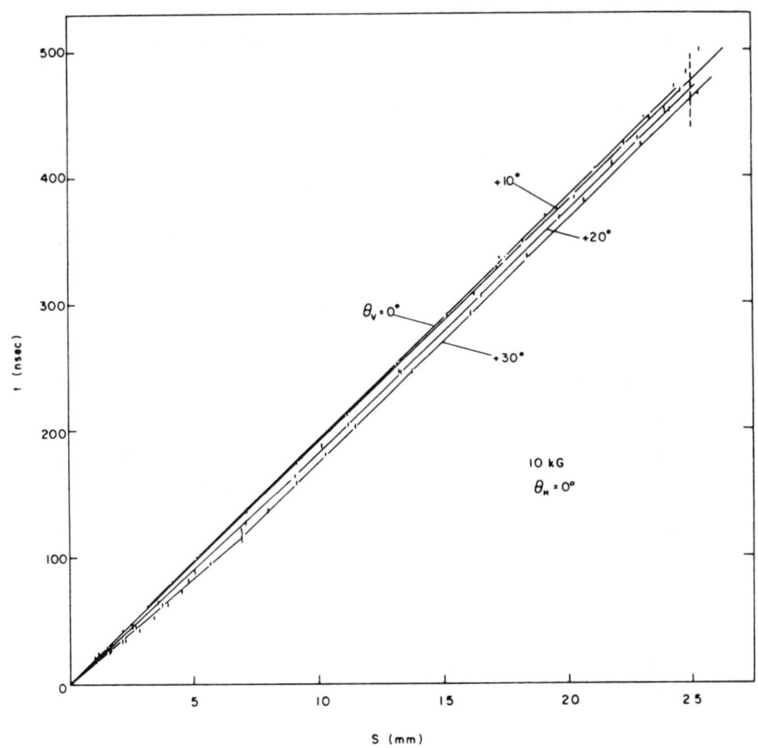

Fig. 99 Space-time relationship measured, for several angles of incidence, in the chamber of Fig. 98 at 10 kG [9]. The similarity with the results of Fig. 92 is apparent.

in the plane of the chamber, or in other words the slope of the space-time correlation, the following variance has been measured[9] (around 10 kG): $\Delta w_{\parallel}/w_{\parallel} = 3 \times 10^{-3}$ per kG. To maintain a ±50 μm stability, one can therefore tolerate a variation of the axial magnetic field of ±660 G; larger variations imply the use of a modified slope. About the same tolerance has been found for weak stray fields in a chamber operating with normal geometry.

7.5 Mechanical construction and associated electronics

As for the multiwire proportional chambers, only a brief mention will be given here about the mechanical construction of drift chambers, and the reader is referred to the quoted literature for more information. Basically, the same techniques as those developed for the construction of proportional chambers have been used for drift chambers, except for high accuracy chambers where the general mechanical tolerances and the anode wires' positioning, in particular, have to satisfy more stringent criteria.

Calculation of the electrostatic forces in a chamber with a structure like the one in Fig. 86 is relatively straightforward and shows that anodic instabilities do not appear for any reasonable length of the wire[72]; on the other hand, the two cathode planes are attracted inwards and the over-all movement can be estimated following the methods outlined in Section 6.4. The same is not true for the configuration depicted in Fig. 88, since the cathode wires are not equipotentials and one cannot assume a uniform charge distribution. Programs have been written to compute the charge on each wire and the relative forces[79,80]. It appears that the more critical point is in the region of the field wire; since the adjacent cathode wires have a charge equal in sign and rather large, they receive a strong outward force. A gap-restoring strip, like the one depicted in Fig. 83, but inserted between the cathodes and the screening electrodes every 50 cm along the wires have proved to be sufficient to compensate the electrostatic forces. Being outside the active volume of the chamber, the strip has no influence on the behaviour of the drift chamber and does not need a field-restoring conductor.

The problem of amplifying and shaping the signals in a drift chamber is also similar to the one encountered in multiwire proportional chambers, except for the fact that one wants, in general, to obtain a better time resolution (implying wider bandwidth and smaller slewing for the amplifier). To measure the time of drift, then, it is possible to use standard time-to-digital converters working in the range of 500 or 1000 nsec, with a channel resolution around 1 nsec (for the highest accuracy) or worse. However, several dedicated electronic circuits have been developed for drift chambers either cheaper than the standard time-to-digital converters, or allowing multihit-per-wire capability and faster conversion and read-out times[81-87]. Some circuits are already commercially available. A very convenient mode of operation has been incorporated in almost all dedicated drift time digitizers, which allows

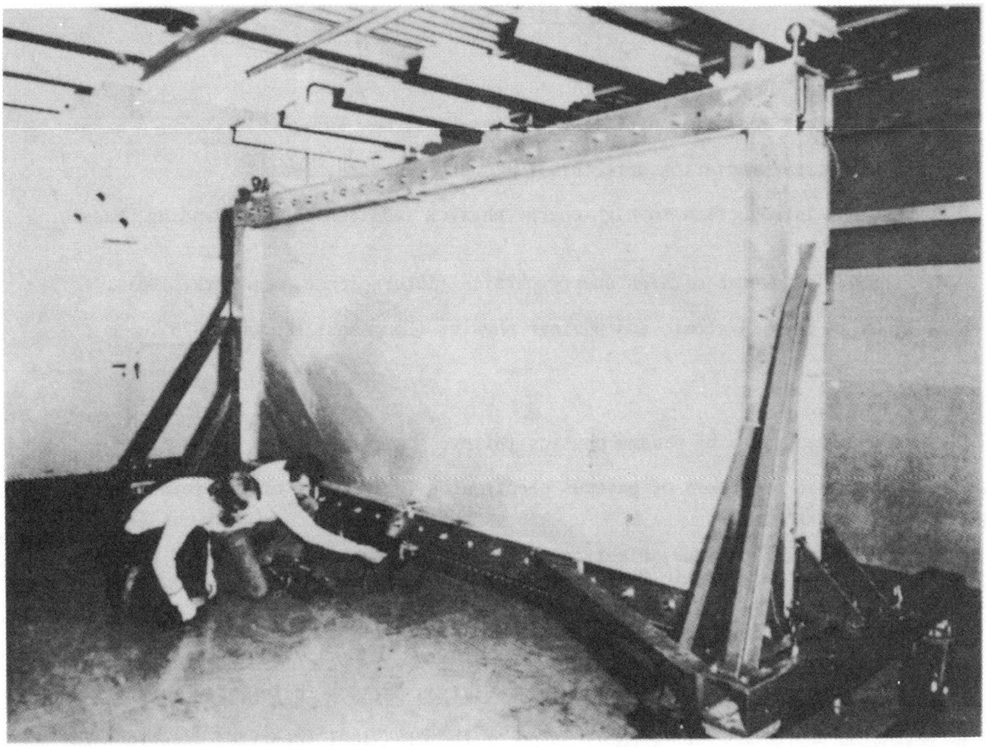

Fig. 100 Large drift chamber (about $4 \times 2.5\ m^2$) constructed at CERN for the Ω spectrometer lever arm[88]. The module has four independent coordinates, similar in construction to the one depicted in Fig. 88.

the provision of the time reference, or time zero, as a late stop instead of as an early start (as required in conventional units). Several hundred nsec are therefore available to the experimenter for making up a trigger selection, using external devices like scintillation and Čerenkov counters, before permanently storing the drift chamber data.

Several large drift chamber systems are already or are close to being operational, including many thousands of drift wires, at CERN and elsewhere. Figure 100 shows a high accuracy drift chamber module developed at CERN[88]; three such chambers approximately $4 \times 2.5\ m^2$, have been constructed for the Omega magnetic spectrometer.

BASIC BIBLIOGRAPHY

Sections 2 and 3

W. PRICE, Nuclear radiation detection (McGraw-Hill, New York, 1958).

J. SHARPE, Nuclear radiation detectors (Methuen, New York, 1964).

D.H. PERKINS, Introduction to high energy physics (Addison-Wesley, Reading, Mass., 1972).

D.M. RITSON, Techniques of high energy physics (Interscience, New York, 1961).

N.A. DYSON, X-rays in atomic and nuclear physics (Longmans, London, 1973).

Section 4

S.C. BROWN, Basic data of plasma physics (Wiley, New York, 1959).

L.B. LOEB, Basic processes of gaseous electronics (University of California Press, Berkeley, 1961).

R.N. VARNEY and L.H. FISHER, Electron swarms, *in* Methods of experimental physics, (Ed. L. Marton) (Academic Press, New York, 1968), Vol. 7B p. 29.

R.H. HEALEY and J.W. REED, The behaviour of slow electrons in gases (Amalgamated Wireless, Sidney, 1941).

L.G. CHRISTOPHOROU, Atomic and molecular radiation physics (Wiley, New York, 1971).

V. PALLADINO and B. SADOULET, Application of the classical theory of electrons in gases to multiwire proportional chambers and drift chambers, Report LBL-3013 (1974).

Section 5

S.A. KORFF, Electron and nuclear counters (Van Nostrand, New York, 1955).

W. FRANZEN and L.W. COCHRAN, Pulse ionization chambers and proportional counters, *in* Nuclear instruments and their uses (Ed. A.H. Snell) (Wiley, New York, 1956).

S.C. CURRAN and J.D. CRAGGS, Counting tubes, theory and applications (Butterworths, London, 1949).

Sections 6 and 7

G. CHARPAK, Evolution of the automatic spark chambers, Annu. Rev. Nuclear Sci. 20, 195 (1970).

F. SAULI, Principes de fonctionnement et d'utilisation des chambres proportionnelles multifils *dans* Méthodes expérimentales en physique nucléaire, Ecole d'Eté de Physique théorique, Les Houches, semptembre 1974 (SETAR, Ivry).

P. RICE-EVANS, Spark, streamer, proportional and drift chambers (Richelieu, London, 1974).

REFERENCES

1) G. Charpak, R. Bouclier, T. Bressani, J. Favier and Č. Zupančič, Nuclear Instrum. Methods 62, 235 (1968).

2) See, for example, G. Charpak and F. Sauli, An interesting fall-out of high-energy physics techniques; the imaging of X-rays at various energies for biomedical applications, Proc. Conf. on Computer-Assisted Scanning, Padova, 21-24 April 1976 (Istituto di Fisica, Univ. di Padova, 1976) p. 592.

3) See, for example, D. Ritson, Techniques of high-energy physics (Interscience, New York, 1961).

4) U. Amaldi, Fisica delle radiazioni (Boringhieri, Torino, 1971).

5) D. Jeanne, P. Lazeyras, I. Lehraus, R. Matthewson, W. Tejessy and M. Aderholz, Nuclear Instrum. Methods 111, 287 (1973).

6) M. Aderholz, P. Lazeyras, I. Lehraus, R. Matthewson and W. Tejessy, Nuclear Instrum. Methods 118, 419 (1974).

7) H.L. Brandt and B. Peters, Phys. Rev. 74, 1828 (1948).

8) E.J. Kobetich and R. Katz, Phys. Rev. 170, 391 (1968).

9) A. Breskin, G. Charpak, F. Sauli, M. Atkinson and G. Schultz, Nuclear Instrum. Methods 124, 189 (1975).

10) N.A. Filatova, T.S. Nigmanov, V.P. Pugachevich, V.D. Riabtsov, M.D. Shafranov, E.N. Tsyganov, D.V. Uralsky, A.S. Vodopianov, F. Sauli and M. Atac, Study of drift chambers system for K^-e scattering experiment at Fermi National Accelerator Laboratory, Nuclear Instrum. Methods 143, 17 (1977).

11) See, for example, J.E. Moyal, Phil. Mag. 46, 263 (1955).

12) H.W. Fulbright, Ionization chambers in nuclear physics, *in* Encyclopedia of Physics (Ed. S. Flügge) (Springer-Verlag, Berlin, 1958), p. 1.

13) G. White Grodstein, National Bureau of Standard Circ. 583 (NBS, Washington, 1957). R.T. McGinnies, National Bureau of Standards, Supplement to Circ. 583 (NBS, Washington, 1959).

14) C.D. Brogles, D.A. Thomas and S.K. Haynes, Phys. Rev. 89, 715 (1953).

15) J.E. Bateman, M.W. Waters and R.E. Jones, Nuclear Instrum. Methods 135, 235 (1976).

16) R.D. Evans, Compton effect, *in* Handbuch der Physik (Ed. J. Flügge) (Springer-Verlag, Berlin, 1958), Vol. 34, p. 218.

17) A.P. Jeavons, G. Charpak and R.J. Stubbs, Nuclear Instrum. Methods 124, 491 (1975).

18) L.B. Loeb, Basic processes of gaseous electronics (University of California Press, Berkeley, 1961).

19) S.C. Curran and J.D. Craggs, Counting tubes, theory and applications (Butterworths, London, 1949).

20) G. Schultz, G. Charpak and F. Sauli, Mobilities of positive ions in some gas mixtures used in proportional and drift chambers, Rev. Physique Appliquée 12, 67 (1977).

21) J. Townsend, Electrons in gases (Hutchinson, London, 1947).

22) S.C. Brown, Basic data of plasma physics (MIT Press, Cambridge, Mass., 1959).

23) W.N. English and G.C. Hanna, Canad. J. Phys. 31, 768 (1953).

24) A. Breskin, Nuclear Instrum. Methods 141, 505 (1977).

25) V. Palladino and B. Sadoulet, Nuclear Instrum. Methods 128, 323 (1975).

26) G. Schultz, Etude d'un détecteur de particules à très haute précision spatiale (chambre à drift) Thèse, Université L. Pasteur de Strasbourg (1976); also CERN EP Internal Report 76-19 (1976).

27) S.A. Korff, Electrons and nuclear counters (Van Nostrand, New York, 1946).

28) O.K. Allkofer, Spark chambers (Theimig, München, 1969).

29) R. Bouclier, G. Charpak and F. Sauli, Parallel-plate proportional counters for relativistic particles, CERN NP Internal Report 71-72 (1971).

30) G. Charpak, Filet à particules, Découverte (février 1972), p. 9.

31) H. Staub, Detection methods, *in* Experimental Nuclear Physics (Ed. E. Segré) (Wiley, New Nork, 1953), Vol. 1, p. 1.

32) L. Malter, Phys. Rev. 50, 48 (1936).

33) G. Charpak, H.G. Fisher, C.R. Grühn, A. Minten, F. Sauli, G. Plch and G. Flügge, Nuclear Instrum. Methods 99, 279 (1972).

34) A.J.F. Den Boggende, A.C. Brinkman and W. de Graaff, J. Sci. Instrum. (J. Phys. E) 2, 701 (1962).

35) W.S. Bawdeker, IEEE Trans. Nuclear Sci. NS-22, 282 (1975).

36) R.W. Hendricks, Rev. Sci. Instrum. 40, 1216 (1969).

37) G.A. Erskine, Nuclear Instrum. Methods 105, 565 (1972).

38) G. Charpak, Annu. Rev. Nuclear Sci. 20, 195 (1970).

39) P. Morse and H. Feshbach, Methods of theoretical physics (McGraw Hill, New York, 1953).

40) E. Durand, Electrostatique (Masson, Paris, 1964) Vol. 1.

41) P.A. Souder, J. Sandweiss and D.A. Disco, Nuclear Instrum. Methods 109, 237 (1973).

42) Z. Dimčowski, Calculation of some factors governing multiwire proportional chambers, CERN NP Internal report 70-16 (1970).

43) T. Trippe, Minimum tension requirements for Charpak chambers' wires, CERN NP Internal Report 69-18 (1969).

44) P. Schilly, P. Steffen, J. Steinberger, T. Trippe, F. Vannucci, H. Wahl, K. Kleinknecht and V. Lüth, Nuclear Instrum. Methods 91, 221 (1971).

45) G. Charpak, G. Fischer, A. Minten, L. Naumann, F. Sauli, G. Flügge, Ch. Gottfried and P. Tirler, Nuclear Instrum. Methods 97, 377 (1971).

46) A. Michelini and K. Zankel, A first design study of a large area proportional counter, CERN Internal Note OM/SPS/75/5 (1975).

47) R. Bouclier, G. Charpak, E. Chesi, L. Dumps, H.G. Fischer, H.J. Hilke, P.G. Innocenti, G. Maurin, A. Minten, L. Naumann, F. Piuz, J.C. Santiard and O. Ullaland, Nuclear Instrum. Methods 115, 235 (1974).

48) S. Majewski and F. Sauli, Support lines and beam killers for large size multiwire proportional chambers, CERN NP Internal Report 75-14 (1975).

49) R. Bouclier, G. Charpak, Z. Dimčowski, G. Fischer, F. Sauli, G. Coignet and G. Flügge, Nuclear Instrum. Methods 88, 149 (1970).

50) H. Fischer, F. Piuz, W.G. Schwille, G. Sinapsius and O. Ullaland, Proc. Internat. Meeting on Proportional and Drift Chambers, Angle measurement and space resolution in proportional chambers, Dubna, 1975.

51) M. Breidenbach, F. Sauli and R. Tirler, Nuclear Instrum. Methods 108, 23 (1973).

52) S. Brehin, A. Diamant Berger, G. Marel, G. Tarte, R. Turlay, G. Charpak and F. Sauli, Nuclear Instrum. Methods 123, 225 (1975).

53) G. Charpak and F. Sauli, Nuclear Instrum. Methods 96, 363 (1971).

54) B. Sadoulet and B. Makowski, Space charge effects in multiwire proportional counters, CERN D PH II/PHYS 73-3 (1973).

55) S. Declay, J. Duchon, H. Louvel, J.P. Datry, J. Seguinot, P. Baillon, C. Bricman, M. Ferro-Luzzi, J.M. Perreau and T. Ypsilantis, K^-p and K^-n final states in $\bar{K}d$ collisions between 1.2 and 2.2 GeV/c, submitted to Nuclear Phys. B, May 1976.

56) S. Parker, R. Jones, J. Kadyk, M.L. Stevenson, T. Katsure, V.Z. Peterson and D. Yount, Nuclear Instrum. Methods 97, 181 (1971).

57) E. Bloom, R.L.A. Cottrell, G. Johnson, C. Prescott, R. Siemann and S. Stein, Nuclear Instrum. Methods 39, 259 (1972).

58) R. Lanza and N. Hopkins, Nuclear Instrum. Methods 102, 131 (1972).

59) J. Buchanan, L. Gulson, N. Galitzsch, E.V. Hungerford, G.S. Mutchler, R. Persson, M.L. Scott, J. Windish and G.C. Phillips, Nuclear Instrum. Methods 99, 159 (1972).

60) K.B. Burns, B.R. Grummon, T.A. Nunamaker, L.W. Who and S.C. Wright, Nuclear Instrum. Methods 106, 171 (1973).

61) L. Baksay, A. Böhm, H. Foeth, A. Staude, W. Lockman, T. Meyer, J. Rander, P. Schlein, R. Webb, M. Bozzo, R. Ellis, B. Naroska, C. Rubbia and P. Strolin, Nuclear Instrum. Methods 133, 219 (1976).

62) J. Lindsay, Ch. Millerin, J.C. Tarlé, H. Verweij and H. Wendler, A general-purpose amplifier and read-out system for multiwire proportional chambers, CERN 74-12 (1974).

63) LeCroy Model 2720 and LD604 (LeCroy Research Systems Co., New York).

64) FILAS, 8-channel MOS circuit for proportional chambers (Société pour l'Etude et la Fabrication de Circuits intégrés spéciaux, Grenoble, France).

65) R. Grove, K. Lee, V. Perez-Mendez and J. Sperinde, Nuclear Instrum. Methods 89, 257 (1970).

66) D.M. Lee, J.E. Sobottka and H.A. Thiessen, Los Alamos Report LA-4968-MS (1972).

67) F. Bradamante, S. Connetti, C. Daum, G. Fidecaro, M. Fidecaro, M. Giorgi, A. Penzo, L. Piemontese, A. Prokofiev, M. Renevey, P. Schiavon and A. Vascotto, Electric delay-line read-out for multiwire proportional chambers *in* Proc. Internat. Conf. on Instrumentation for High-Energy Physics, Frascati, 1973 (Lab. Naz. CNEN, Frascati, 1973).

68) J. Hough and R.W.P. Drever, Nuclear Instrum. Methods 103, 365 (1972).

69) H. Foeth, R. Hammarström and C. Rubbia, Nuclear Instrum. Methods 109, 521 (1973).

70) J. Saudinos, J.C. Duchazeaubeneix, C. Laspalles and R. Chaminade, Nuclear Instrum. Methods 11, 77 (1973).

71) A.H. Walenta, J. Heintze and B. Schürlein, Nuclear Instrum. Methods 92, 373 (1971).

72) D.C. Cheng, W.A. Kozanecki, R.L. Piccioni, C. Rubbia, R.L. Sulak, H.J. Weedon and J. Wittaker, Nuclear Instrum. Methods 117, 157 (1974).

73) Experiment R209 at CERN (H. Newman, private communication).

74) G. Charpak, F. Sauli and W. Duinker, Nuclear Instrum. Methods 108, 413 (1973).

75) G. Marel, P. Bloch, S. Brehin, B. Devaux, A.M. Diamant-Berger, C. Leschevin, J. Maillard, Y. Malbequi, H. Martin, A. Patoux, J. Pelle, J. Plancoulaine, G. Tarte and R. Turlay, Large planar drift chambers, Nuclear Instrum. Methods 141, 43 (1977).

76) M. Atac and W.E. Taylor, Nuclear Instrum. Methods 120, 147 (1974).

77) A. Breskin, G. Charpak, B. Gabioud, F. Sauli, N. Trautner, W. Duinker and G. Schultz, Nuclear Instrum. Methods 119, 9 (1974).

78) B. Sadoulet and A. Litke, Nuclear Instrum. Methods 129, 349 (1975).

79) F. Bourgeois and J.P. Dufey, Programmes de simulation des chambres à drift en champs magnétique, CERN NP Internal Report 73-11 (1973).

80) A. Wylie, Some calculations of drift chamber electric fields, CERN NP Internal Report 74-7, September 1974.

81) B. Schurlein, W. Farr, H.W. Siebert and A.H. Walenta, Nuclear Instrum. Methods 114, 587 (1974).

82) E. Schuller, A two-clock sensing system, a new time-to-digital converter, CERN NP Internal Report 73-15 (1973).

83) H. Verweij, Electronics for Drift Chambers, IEEE Trans. Nuclear Sci. NS-22 437 (1975). See also: DTR, Drift Time Recorder, CERN Type EP247.

84) W. Sippach, Drift chambers electronics, NEVIS Internal Report (1975).

85) C. Rubbia, New electronics for drift chambers, CERN NP Internal Report 75-10 (1975).

86) M. Atac and T. Droege, Notes on drift chambers time digitizer meeting, Fermilab Int. Note TM-553 (April 1975).

87) LeCroy Model 2770A, Drift chambers digitizer.

88) A. Dalluge and M. Jeanrenaud, private communication at CERN.

HIGH-RESOLUTION ELECTRONIC PARTICLE DETECTORS

by G. Charpak and F. Sauli

1. INTRODUCTION

High-energy physics currently presents a heavy demand on high-accuracy particle detectors, in the μm or tens of μm range. As is almost always the case in particle-detector development, this demand from physics stimulates the conception of new ideas or, more often, the perfection of well-known techniques whose potential properties for the highest possible accuracy were not fully exploited. Very illustrative is the spectacular rejuvenation of the bubble chamber, which has stopped developing in the direction of mammoth dimensions to become a live target with an accuracy of 5 μm, thanks to the use of holographic methods for the retrieval of the bubble position. A similar development is observed for streamer chambers, where the combination of high pressure and refined read-out methods has been used to improve greatly the localization accuracies of the device. In the two cases the interest in these developments has been fostered by the need to obtain clear information on the interaction vertices where short-lived particles are produced and decay over path lengths of μm or tens of μm. While these detectors—which provide an almost complete picture of the most complex vertices—have gained a new popularity with the research on charmed particles, a considerable effort has been undertaken in the development of electronic detectors capable of withstanding high fluxes to bring them to the same level of accuracy. The study of short-lived particles is not the only goal in these searches. The improvement of accuracy in the measurement of momenta of charged particles in detectors being used around high-energy colliders would result in a serious reduction of size in some expensive systems. The most important developments—those permitting a realistic projection in the near future for the design of experiments—are based on solid-state or gaseous electronics and are the subject of this review.

2. THE PHYSICAL MESSAGE

Charged particles, traversing a gaseous or condensed medium, release energy in various ways. Of all interactions, however, only the electromagnetic one provides a physical message that can be exploited for detection in thin layers of matter, since it is several orders of magnitude more probable than nuclear or weak collisions. The interaction between the electromagnetic fields of the particle and of the medium can result, for energy transfers exceeding several eV, in the excitation or ionization of the molecules; the ionization message is the one exploited in most position-sensitive detectors.

The probability of a given collision is a fast decreasing function of the energy transfer involved. As an example, Figure 1 gives the computed number of ionizing collisions in argon for fast singly charged particles, as a function of the energy of the ejected electron (1). Typically, 80% of all ionization encounters lead to the production of an electron with energy below the average that is needed for further ionizations (around 25 eV in argon). Despite their low probability, however, large-energy transfer encounters whose outcome is an energetic electron capable of ionizing the medium (a delta electron) dominate the statistics of the energy-loss processes. Indeed in 1 cm of argon at normal temperature and pressure (NTP) there is a 5% probability of emission of a 2-keV electron, which doubles the energy loss (2 keV/cm on an average) and, with its 150 μm of range, offsets the ionization trail.

The total number of ion pairs released in the medium can be evaluated by dividing the fractional average energy loss by the average energy per ion pair, w. This phenomenological quantity depends on the nature and energy of the ionizing radiation, and equals roughly twice the lowest ionization potential of the medium. Values of w measured under different conditions and in various materials are given, for example, by Franzen & Cochran (2) and Christophorou (3).

The localization accuracy in detectors is often dominated by the physical extension of the ionization trail in the medium due to the nonresolved, heavily ionizing, slow electrons ejected in the primary encounter at energies in the keV region. Precise range-energy curves can be computed; however, because of the very large scattering angles suffered by low-energy electrons in their collisions with molecules, the integrated range does not represent well the physical extension of the released ionization. It is customary to define a practical range as the extrapolation to zero intensity in an absorption curve; the ratio between integral path length and practical range is energy-dependent, and is about a factor of 2 around 10 keV. Figure

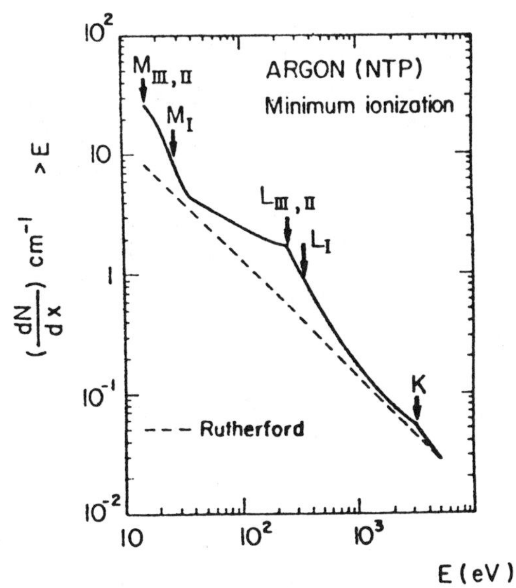

Figure 1 Collisions per unit length, in argon at NTP, with an energy transfer $\geq E$ as a function of the energy (1).

complications arise in the calculation of energy loss for thin layers of condensed materials because of the delta electrons escaping from the layer. Still an excellent agreement can be obtained with experimental measurements, as shown in Figure 4 (7).

2 (4) shows values of practical range measured in solids between 0.1 and 10 keV and given in g/cm^2; the range in cm can be computed by dividing these values by the corresponding density. Although data are scarce for gases in this energy domain, it is assumed here that the average values from Figure 2 can be used.

Analytical and statistical methods have recently been developed to allow a detailed description of the energy-loss process of fast charged particles (1, 5, 6). Although accurately computing the number and distribution of the ionization clusters, the quoted works unfortunately do not include an estimate of the physical track width. Figure 3 (6) shows the remarkable agreement between computed and experimental values of differential energy loss for fast particles in 1 cm of argon at NTP. Additional

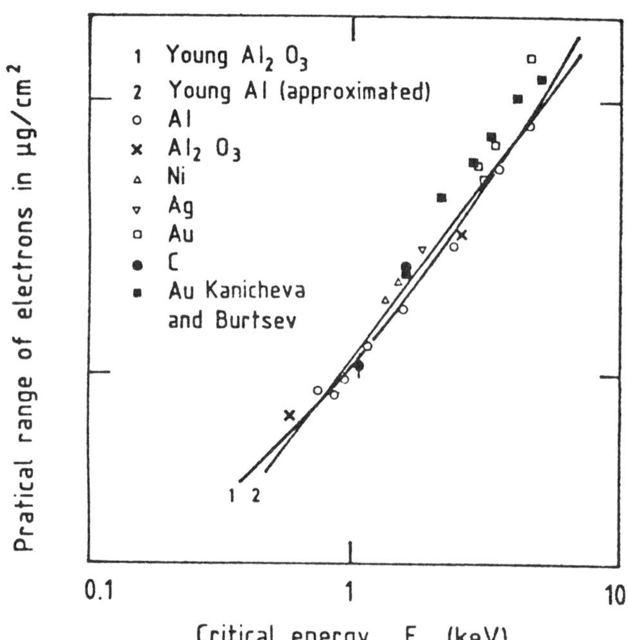

Figure 2 Practical range-energy relation for slow electrons (4).

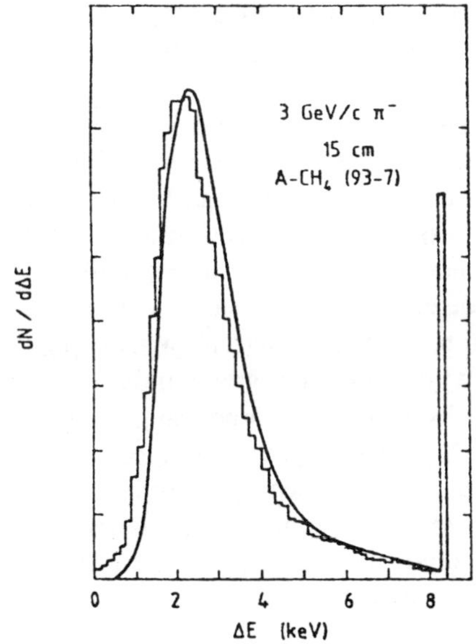

Figure 3 Differential energy loss of minimum ionizing particles in argon at NTP compared with a model calculation (6).

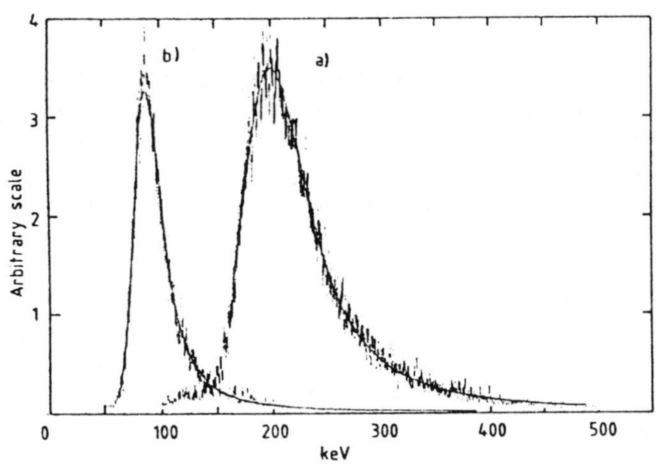

Figure 4 Experimental and computed differential energy loss in silicon (7) for 0.736 (*a*) and 115 GeV/*c* (*b*) protons.

Soft x-ray generators or sources are often used to check efficiency and localization accuracies of detectors, since they are easily collimated. The interaction of protons with matter consists in a single localized encounter generally involving large transfers. In the energy domain we are concerned with, a few keV to a few tens of keV, the main absorption process is due to photo-ionization with the emission of an electron, accompanied either by a lower-energy photon or by an Auger electron. The range in the medium of the emitted electron or electrons is one of the factors determining the achievable localization accuracy of the detector. Reconversion—within the active volume—of a fluorescence photon, if not resolved, may also spoil localization. For example, a 5.9-keV x-ray converts in argon mainly on the K shell (3.2 keV); the 2.7-keV photoelectron has a practical range of 30 $\mu g/cm^2$ or 200 μm at NTP. With 88% probability, another electron of energy around 3 keV (250 μm range) is ejected; in the remaining cases, a 3.2-keV fluorescence photon is produced with a mean absorption length of 40 mm. This illustrates the physical limits in the attainable localization accuracy for soft x-rays using gaseous detectors.

3. GASEOUS DETECTORS

3.1 *Introduction*

Position-sensitive gaseous detectors based on the multiwire proportional chamber (MWPC) are widely used in particle physics. Submillimeter localization accuracies in the detection of ionizing radiation are routinely achieved and there have been many attempts to further improve the accuracy by using suitable geometries and operational conditions. Among the resolution-limiting factors, the major ones are the width of the physical message due to emission of long-range delta electrons in the gas and to the diffusion of charges before collection. In drift chambers localization accuracies around 20 μm rms have been achieved, about the same as intrinsically allowed by the center-of-gravity avalanche localization method.

For energetic charged particles, not suffering too much from Coulomb scattering, multiple sampling can be realized to improve localization accuracy, using large stacks of detectors or thick gas regions with many independent collecting electrodes (as in the time projection chamber). In the absence of correlations, one expects the accuracy to improve as the square root of the number of samples. At the limit, each individual electron or cluster in the ionization trail could be located, as attempted in the time expansion chamber; replacement of the multiwire detection element with a parallel-plate chamber, single or multistep, may further improve the issue by removing the dispersive effects of discrete anode wires.

Both the physical ionization width and the diffusion can be reduced by increasing the gas pressure. However, while operation of gas proportional chambers at pressures above several hundred atmospheres has been demonstrated, implementation of large, multielectrode detectors at high pressures is difficult and limited in application by the use of thick containment vessels. This has encouraged experimenters to turn their attention to liquid or solid ionization devices; however, since electron transport is observed over distances of interest only in ultrapure condensed noble gases, the necessity of using cryogenic containment and purification, for the time being, seriously limits the usefulness of such an approach. Recently in the field of calorimetry research, several room-temperature liquids capable of reasonable electron transport were found and may change the issue in the near future.

3.2 *Drift and Diffusion of Charges in Gases*

Charges released in a gas by the ionizing encounters diffuse uniformly in the absence of external fields, having multiple elastic collisions with the molecules until they eventually get neutralized by mutual collision, capture, or collision at the walls. However, in the presence of an applied electric field, a global motion in the direction of the field superimposes on the random thermal motion; the average displacement of the swarm per unit time is called drift velocity.

For positive ions, and up to high values of field (several kV/cm · atm) the drift velocity increases linearly with the field; the ratio $\mu = w/E$ is called mobility. The average energy of ions remains in this region equal to its thermal value kT. Detailed theories of ion motion and experimental values of mobility can be found, for example, in McDaniel & Mason (8).

The much lighter electrons, on the other hand, can easily be accelerated between collisions with the molecules by the external field, reaching average energies far exceeding the thermal one even at moderate fields. The behavior of drift velocity and diffusion depends very strongly on the gas nature, through the detailed structure of the elastic and inelastic electron cross section of molecules. Because of the very different values and energy dependence of the cross section in different gases, addition of even very small quantities of one species to another may substantially modify the drift properties; often, in mixtures of noble gases with organic quenchers (which increase the maximum gain attainable by proportional amplification) a maximum or saturated value of drift velocity is reached at fields around 1 kV/cm · atm, typically of several cm/μs.

The general requirements for a gas mixture optimal for obtaining high resolutions depend on the detector design. Very high drift velocity is desired whenever collection time plays a role, while a slow gas is preferred if a measurement of drift time is performed to provide space coordinates, in order to ease the electronics requirements. In this case, it is also desirable to use a gas with velocity saturation at high fields, as this generally reduces the dependence of drift velocity on variations of field, temperature, pressure, and gas composition. For example, Figure 5 shows the field dependence of drift velocity and of its variance with temperature, for an argon-methane 90-10 mixture at NTP (9). Electron diffusion, setting an ultimate limit to the obtainable accuracy, has to be kept small, thus favoring the choice of "cool" gas mixtures (where the energy of electrons remains close to thermal at high fields); since diffusion is inversely proportional or the square root of the gas density, increasing the gas pressure improves accuracy. Needless to say, one has to preserve the other requirements of the detectors (for example, proportional or semiproportional avalanche multiplication, small electron

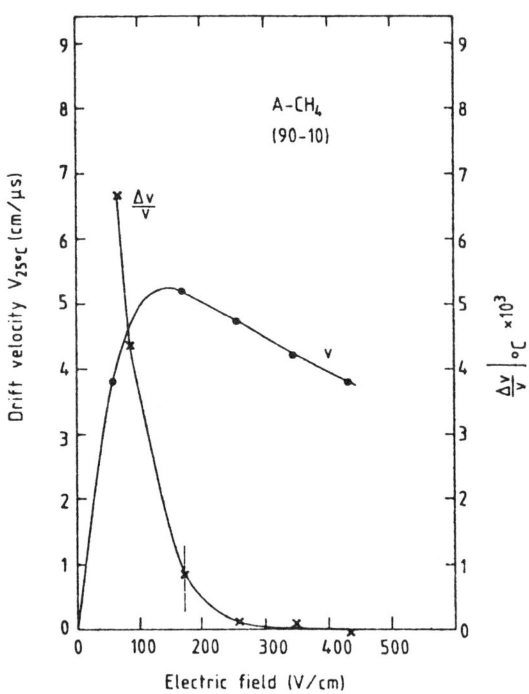

Figure 5 Electron drift velocity and its temperature dependence, as a function of electric field, in argon–methane (90-10) at NTP (9).

attachment, and so on). Extensive compilations of cross sections and drift velocities in pure gases and in some mixtures can be found in the literature (3, 10–14). Theories and experimental methods are reviewed in the quoted references and elsewhere (15–17).

Recently, and specifically in the development of gas detectors, electron transport theories have been reviewed by several authors and applied to the gas mixtures commonly used in MWPCs and drift chambers (18–21). The agreement between theory and experiment is remarkable, and the quoted calculation methods can be used to tailor a gas mixture to specific needs or to compute its drift properties, e.g. in the presence of magnetic fields.

In the case of a spherically symmetric, Gaussian diffusion (this assumption is not verified for most gases at large fields, see below), the dispersion with time of an originally point-like charge distribution is described by a field-dependent diffusion coefficient D such that

$$\sigma_x = \sqrt{2Dt} \qquad 1.$$

represents the rms of the swarm along any direction x after a time t. It is customary to define a quantity named characteristic energy ε_K as the ratio between diffusion coefficient and mobility; one can write then

$$\sigma_x = (2\varepsilon_K x/eE)^{1/2} = (2\varepsilon_K x/EP^{-1})^{1/2}P^{-1/2}, \qquad 2.$$

where the last expression shows explicitly the pressure dependence of space diffusion at given values of the reduced field EP^{-1}.

For an ideal gas, where the energy distribution of electrons remains thermal at any value of the electric field, the characteristic energy equals its classic value kT and the space diffusion is given by

$$\sigma_x = (2kTx/eE)^{1/2}, \qquad 3.$$

a quantity often called the thermal limit.

In "hot" gases, where their average energy is greatly increased at high fields, electrons can have diffusions that are orders of magnitude larger than the thermal limit [see Figure 6, computed for 1 cm of drift in gases at NTP (18)]. In carbon dioxide, on the other hand, electrons remain thermal up to rather high fields; this is partly true also for mixtures containing a large fraction of CO_2. It has been found recently that dimethylether $(CH3)_2O$ presents thermal diffusion properties at even higher fields, see Figures 7*a* and 7*b* (22); at the same time, this gas has a low drift velocity, which makes it interesting in some applications.

As mentioned, diffusion is not always symmetric, especially at high fields, but tends to be smaller in the direction of drift. One has therefore to consider a transverse diffusion coefficient D_T (as given by the previous

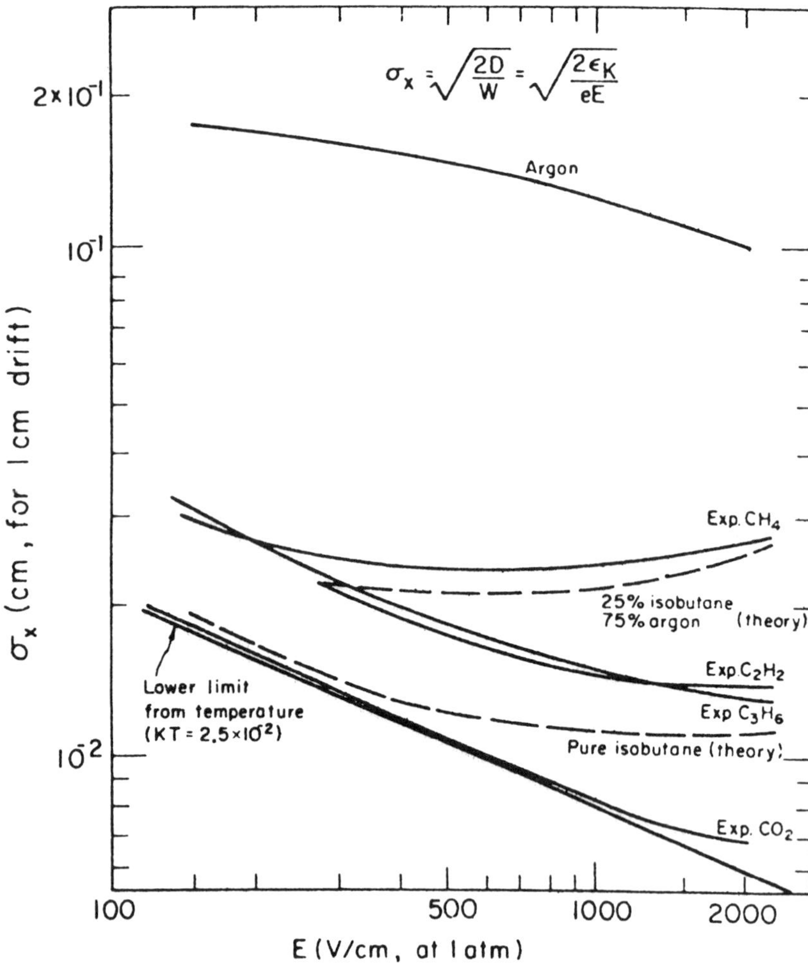

Figure 6 Single-electron diffusion rms for 1 cm drift, as a function of field for various gases at NTP (18).

considerations), and a longitudinal one D_L. Although a detailed theory has been developed on the subject (23) the above-mentioned calculation methods have so far not been extended to include the effect, and provide only the transverse values. Longitudinal diffusions, on the other hand, have been measured in several gas mixtures commonly used in proportional chambers (24–26); Figure 8 is a compilation from the quoted sources, while Figure 9 shows the comparison of computed transverse and measured longitudinal diffusions for a particular gas mixture (24).

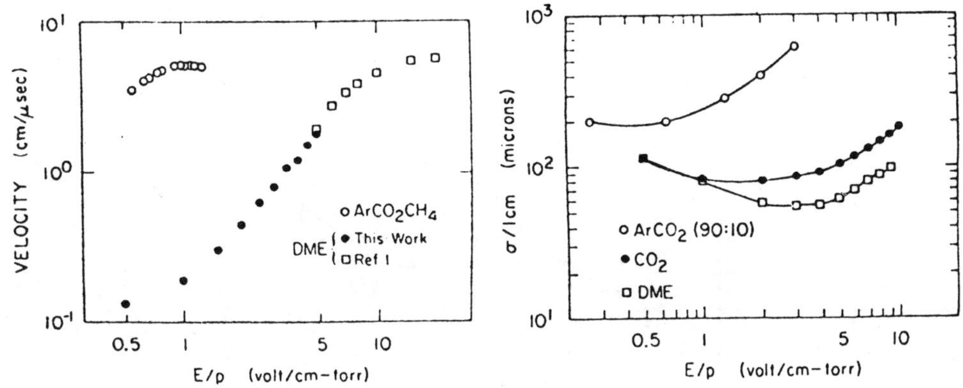

Figure 7 Drift velocity and diffusion, for 1 cm drift, in dimethylether (DME) (22).

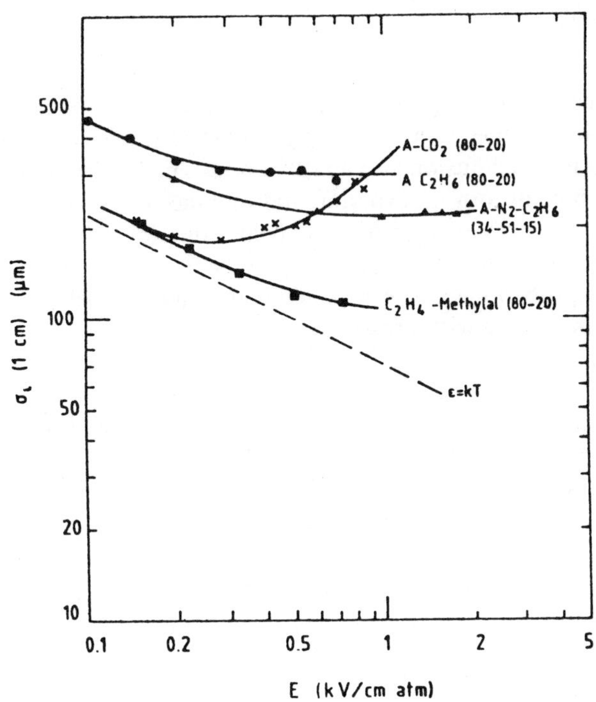

Figure 8 Experimental measurements of single-electron longitudinal diffusion for 1 cm drift, in various gases at NTP (24–26).

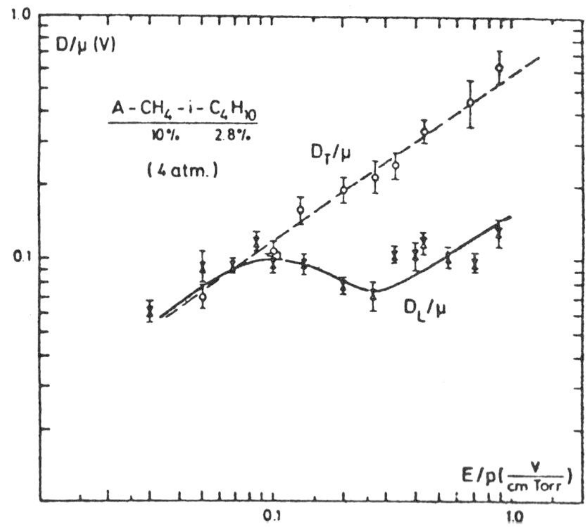

Figure 9 Comparison between transverse and longitudinal diffusion coefficients in a mixture of argon, methane, and isobutane (24).

Other effects may appear to partly invalidate the previous considerations. As an example, Figure 10 (27) gives the dependence on electric field of the longitudinal diffusion measured for a 16-mm drift length in pure CO_2 at various pressures in an experimental high-accuracy drift chamber. At each pressure, there is an optimum field value for minimum diffusion. Notice, however, that the results shown in the figure violate the usual E/P scaling law, probably because of residual attachment of electrons to CO_2 molecules, increasing with pressure.

Any search for high-resolution detectors has to take into account the dispersive role of diffusion; the quoted values, however, provide the expected localization accuracy for single electrons. Very often the physical message consists instead of many electrons, localized or scattered along an ionization trail. Calculation of the resulting resolution depends then on the way the detector handles the message; for the ideal case of a localized cluster consisting of N electrons (a soft x-ray conversion in high-pressure gases), and if one can somehow determine the center of gravity of the detected swarm, the expected accuracy will be $\sigma_x(N)^{-1/2}$.

For more complex topologies of energy loss and electric fields, as is the case for drift chambers in the detection of charged tracks, the limiting resolution can either be computed using Monte Carlo simulations, or be directly measured (see the next sections).

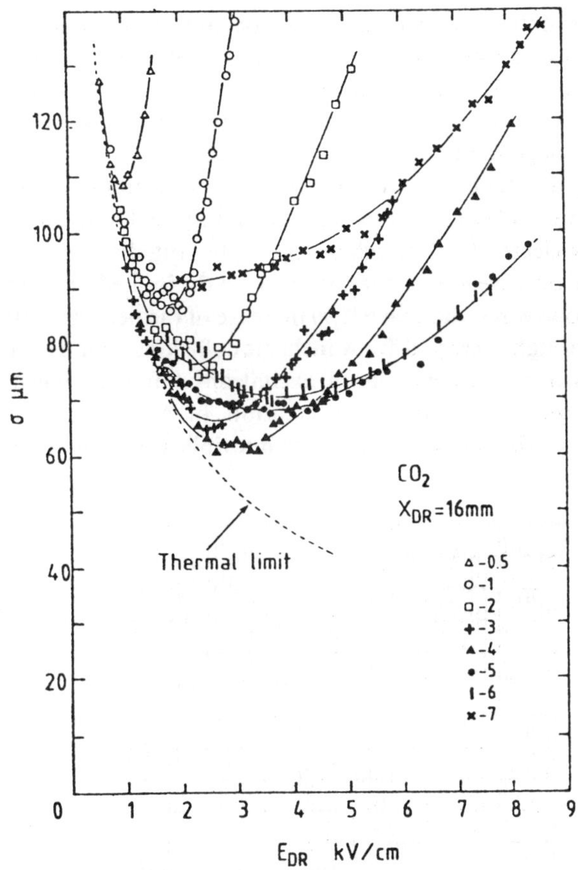

Figure 10 Experimental single-electron diffusion width (rms for 1 cm drift), in carbon dioxide at various pressures in atmospheres (27).

3.3 *The Multiwire Proportional Chamber*

The MWPC was conceived fifteen years ago as a fast, high-resolution detector of charged particles and soft x-rays [Charpak et al (28)]. Since then, an impressive amount of work has been done to improve the general operational characteristics of the device, as documented in many review articles (9, 29–34).

In its basic design, the MWPC consists of a set of parallel, evenly spaced, thin anode wires between two equally spaced cathode planes. Typical values for the anode wire spacing range between 1 and 5 mm, the anode to cathode distance (the gap) being 5 to 10 mm. The operation gets increasingly difficult at smaller wire spacing, a fact that has generally prevented taking this obvious direction for obtaining higher resolutions. The chamber is filled with a gas mixture suitable for charge collection and proportional multiplication, in a range of pressures between a few Torr and tens of atm, depending on experimental needs.

Applying a potential difference between anode and cathodes, field lines and equipotentials develop as shown in Figure 11. In most of the gas volume, the electric field is such as to simply drift charges: electrons to the anodes, and positive ions to the cathodes. In the immediate surroundings of the thin anode wires, the sharp $(1/r)$ increase of the field strength imparts to electrons enough energy to allow inelastic collisions with the gas molecules, causing, as a consequence, both excitations (often followed by photon emission) and ionizations. The creation of electron-ion pairs in the collisions is at the origin of avalanche multiplication, and is exploited as a

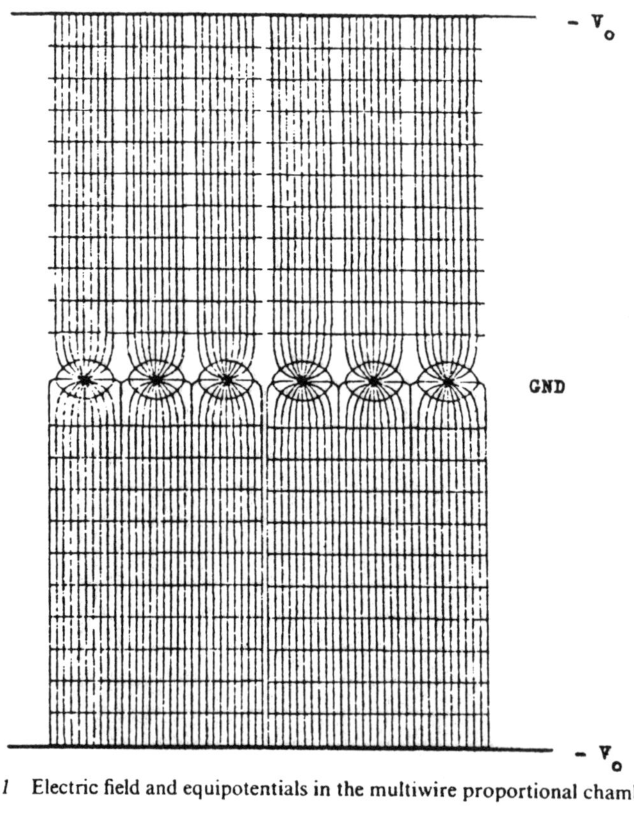

Figure 11 Electric field and equipotentials in the multiwire proportional chamber (29).

signal amplification mechanism in gas counters; excitations and subsequent photon emissions participate in the avalanche spread processes, and can be detected directly by optical means.

The typical avalanche growth around a thin wire is depicted in Figure 12. Approaching the anode wire, electrons multiply in cascade, leaving behind an exponentially increasing number of positive ions. Both mechanical and photon-mediated diffusion contribute to the spread of the growing avalanche, which may completely surround the anode; the whole process, which begins a few wire radii from the anode, is over after a fraction of a nanosecond, leaving the cloud of positive ions receding from the anode at decreasing speed as a consequence of the decrease in field strength. This motion is responsible for the largest fraction of charge signals detected on the anode induced by the avalanche and on all surrounding electrodes; a measurement of the charge profile induced on the cathodes, suitably segmented, allows bi-dimensional localization of the ionizing event (35) to be achieved. This is the center-of-gravity method, which allows attainment of the highest localization accuracies in MWPCs (see Figure 13); if Y_i is the measured charge on the strip of central coordinate y_i, the average position of the avalanche along the direction y is estimated as

$$\bar{y} = \sum Y_i y_i / \sum Y_i. \qquad 4.$$

An example of induced charge profile, recorded in a MWPC with a 5-mm gap and 5-mm wide cathode strips, is shown in Figure 14. It has a FWHM of about 10 mm, twice the anode to cathode distance.

The computed center of gravity of induced pulses has a continuous

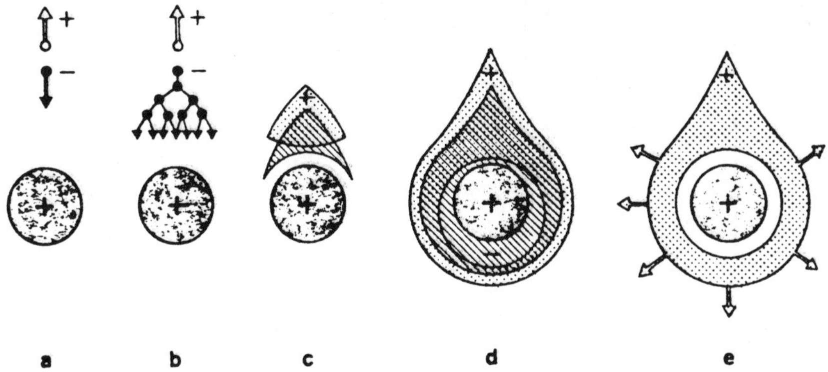

Figure 12 Schematic picture of the time development of the avalanche around the anode wire of a proportional counter (29).

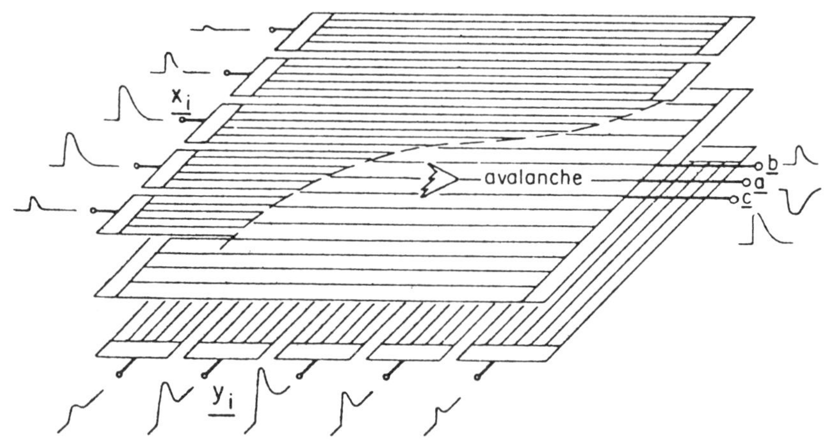

Figure 13 Principle of the center-of-gravity localization method by read-out of the induced charges on cathode planes (43).

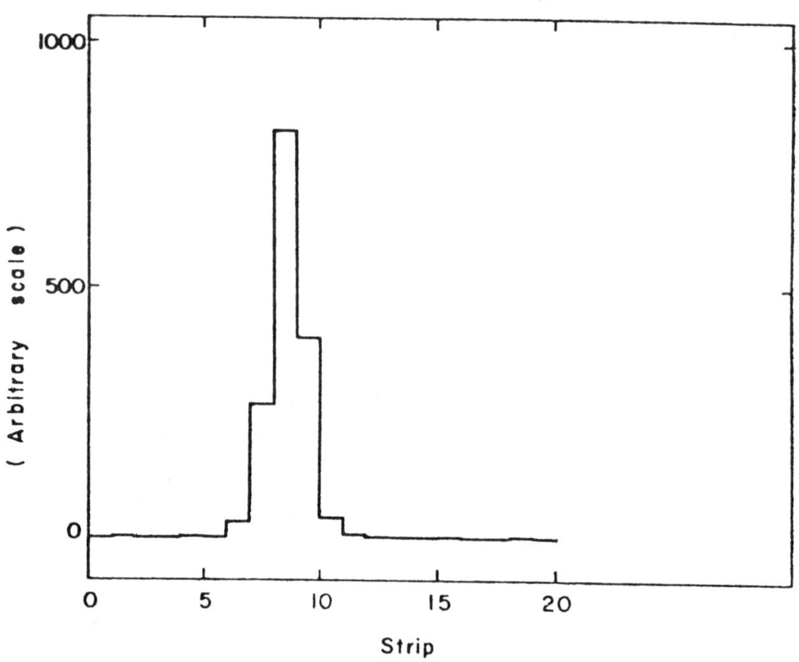

Figure 14 Typical charge profile induced on cathode strips by an avalanche; strip width and MWPC gap are 5 mm (43).

dependence on the position of the original ionization only in one direction, that of the anode wires. For the perpendicular coordinate the dependence reflects the quantizing effect of the anode wires; should the avalanche symmetrically surround the anodes, the localization accuracy in this direction would be limited to the wire spacing. As we see below, this is not necessarily the case and indeed, with a proper choice of operating conditions and reconstruction algorithms, almost equal localization accuracies can be obtained in both directions, at least in soft x-ray detection.

A precise knowledge of the induced charge distribution is necessary to obtain the highest resolutions. Indeed, while the distribution for a localized avalanche is intrinsically symmetric (and therefore its weighted average is the best estimate of the real position), for practical reasons the charge is integrated over cathode strips or pads with finite width, thus introducing a nonlinear correlation between the real position and the computed center of gravity. The distribution of induced charge on the cathode planes has been computed by many authors with various assumptions; the results of some model calculations are compared with experimental points in Figure 15 (36). Model D (37), overestimating the width, consists in a simple solid-angle dependence of charges induced by a point charge at an anode position, neglecting all other electrodes. Curve C (38) assumes a continuous anode plane at a very small distance from the avalanche; curve B (39) considers the presence of both cathode planes but neglects the adjacent anodes. Curve A is the result of a very detailed and rigorous electrostatic calculation (40). While this obviously provides the best match with the experiment, the simplified expressions of the other models are often used for practical reasons. The effect of finite sampling width has been analyzed by many authors (36–41); in particular, Gatti et al (40) estimated theoretically both the integral and differential dispersions due to finite sampling size, to mutual capacitance between strips and from strips to ground, and to amplifier noise. Figure 16 shows one of the results, the figure of merit of avalanche localization, as a function of the ratio a/D ($2a$ being the cathode strip width and D the anode to cathode distance or gap). The result is given for three positions of the avalanche, in front of the center of a strip ($\lambda = 0$), between two strips ($\lambda = a/D$), or in an intermediate position. The result depends on the anode wire spacings ($b = s/D$ in the figure), but obviously there is an optimum in the region around $a/D = 0.5$, which is for the cathode strip width equal to the MWPC gap, as is already well known experimentally (42). This is a natural choice for a high-accuracy detector; a smaller strip width increases the number of samples, but this advantage is spoiled by the decrease of detected charge on each strip, with the resulting increase of both cross-talk and noise effects.

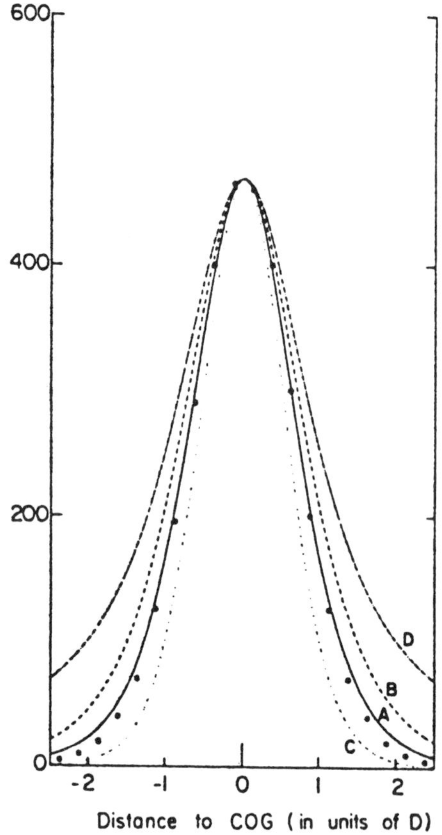

Figure 15 Experimental (*full points*) and computed cathode charge profile, according to various models (36).

The simplest way to reduce the dispersive effects of noise and finite sampling size is the subtraction of a bias level to individual strip amplitudes before averaging (43):

$$\bar{y} = \sum(Y_i - B)y_i / \sum(Y_i - B). \qquad 5.$$

A good choice of the constant B appears to be a fraction of the total amplitude for each event:

$$B = b\sum Y_i. \qquad 6.$$

Figure 17 (43) shows the dependence on b of the localization accuracy in the direction of the anode wires, measured for a minimum ionizing particle

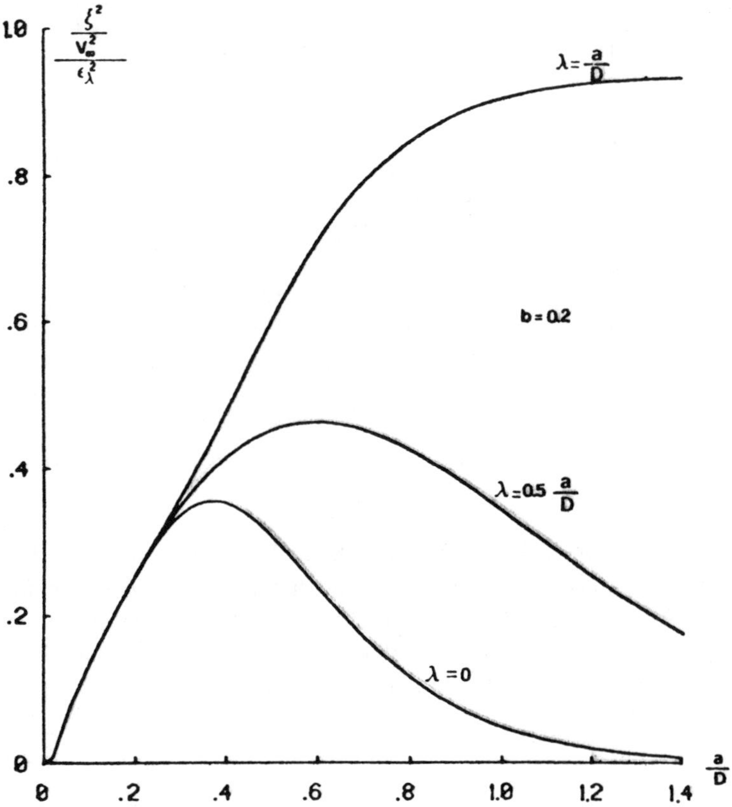

Figure 16 Figure of merit for avalanche localization, as a function of the strip width/gap ratio, for different avalanche relative positions (40).

beam perpendicular to the MWPC plane. The choice of the value of b is not very critical, and produces consistently around 40 μm rms for the localization accuracy. Figure 18 (36) shows the systematic nonlinearity error in units of the half gap D, as a function of avalanche position, for several choices of the bias constant b [computed with the model of (40)]. The error appears as a periodic modulation, with a wavelength corresponding to the cathode-strip width. For the particular case of large cathode strips or pads, such as used in the time projection chambers, simple algorithms have been developed to restore the linearity of the response in reconstruction using a functional weighted average of the detected charges.

Mutual capacitance between strips and electronics cross-talk, producing

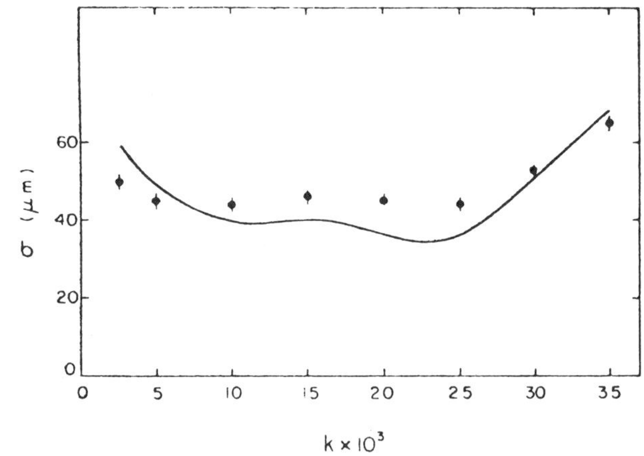

Figure 17 Influence of the choice of the bias constant (Equation 6) on localization accuracy (43).

a reinjection of charge in adjacent channels, may cause considerable distortions if not taken into account, as shown for example by Piuz et al (36). As noted by these authors, systematic effects of this kind may be masked by the experimental procedure that often involves the use of three identical, aligned chambers in a perpendicular beam.

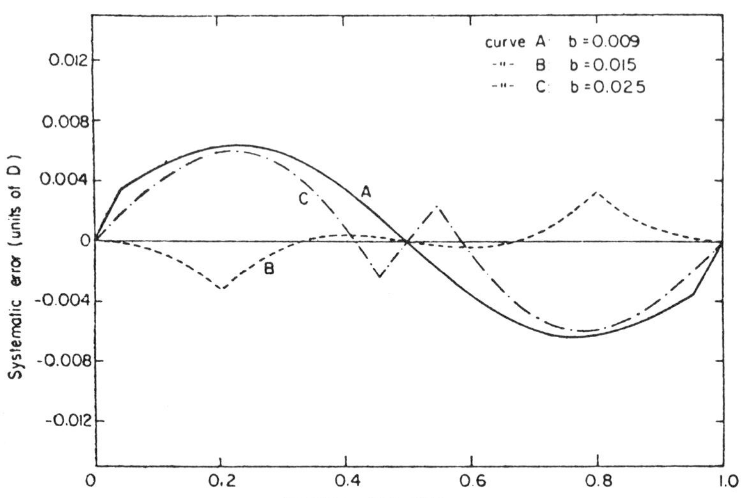

Figure 18 Systematic nonlinearity error as a function of avalanche position, for several choices of the bias constant (36).

Gain differences between channels, mostly due to the electronic amplifiers, also cause local nonlinearities. From Equation 4, and for a typical induced charge distribution such as that shown in Figure 14, one can estimate that a 10% gain difference in one strip results in a position error corresponding to 2% of the strip width (40 μm in the example). Rather straightforward calibration procedures can be used to determine the relative gains of all channels, applying then the corresponding scale correction at the analysis level. A convenient method of calibration consists in uniformly irradiating the detector with a monochromatic soft x-ray source (such as ^{55}Fe producing a 5.9-keV line), and recording the complete charge profile in each event. For each channel, a plot is then constructed containing the recorded charge only for the events in which the computed center of gravity lies within the size of the corresponding strip: this guarantees a good enough energy resolution to allow the determination of the average relative gain with an accuracy of a few percent.

Localization accuracies of 35 μm rms have been reported in the detection of soft x-rays (44, 45). Figure 19 (45) shows the reconstructed center-of-gravity distributions measured for three positions of a collimated 1.5-keV x-ray beam, 200 μm apart; the localization accuracy (in the direction of the anode wires) is 35 μm rms. The MWPC used for the measurement had an 8-mm gap, 3-mm cathode strips, and was operated in a xenon-isobutane mixture in order to reduce the photoelectron range in the gas (see Section 2.1). Figure 20, from the same reference, shows the linearity of the correlation between real and computed positions of the source.

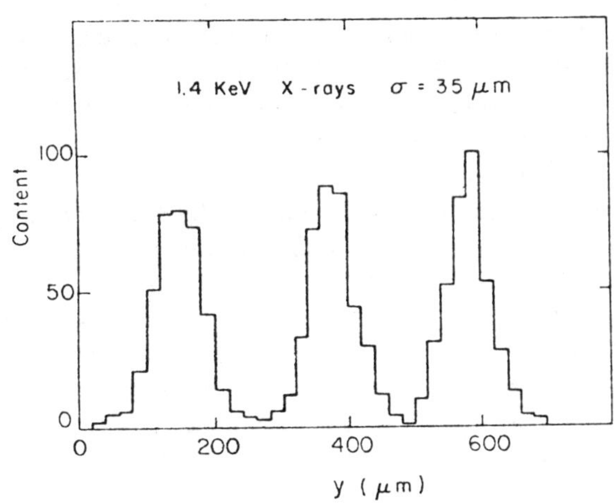

Figure 19 Position resolution, measured with the center-of-gravity method, in the direction parallel to the anode wires for three positions, 200 μm apart, of a collimated source (45).

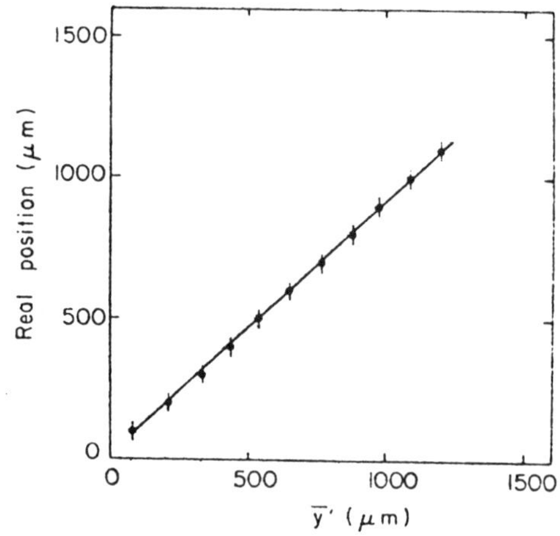

Figure 20 Linearity of the correlation between the computed center of gravity and the real position, in the direction of the anode wires (45).

Positioning accuracy is limited in the detection of charged particles by the physical extension and by the non-uniform density of the ionized trail, both due to emission of delta electrons. The second effect is largely dominant for tracks nonperpendicular to the MWPC plane. This is illustrated in Figure 21 (43). The localization accuracy for fast, charged particles is given for four incidence angles ($\theta = 0°$ meaning a beam perpendicular to the MWPC), and is shown in each plot for several intervals of actual energy loss. Resolution tends to deteriorate for large energy losses at each angle, and moreover the average accuracy goes from 50 μm rms at $\theta = 0°$ to 300 μm at 30°. For large losses, the resolution distribution has considerable non-Gaussian tails; this is very apparent when the experimental conditions allow for very high intrinsic resolutions, as shown for example in Figure 22 (46).

Should the avalanche process symmetrically surround the anode wire, the localization accuracy of a MWPC would be necessarily limited in one direction by the wire spacing. It was realized, however, rather early that such is not always the case and that the induced charge distribution reflects the original position of the (localized) ionization between wires (47). Detailed measurements have shown indeed that at moderate proportional gains and in well-quenched gases the avalanche grows along the field lines of approach of multiplying electrons with a rather narrow angular spread (44, 45, 48–52). This can be exploited for obtaining localization accuracies far better than the wire spacing.

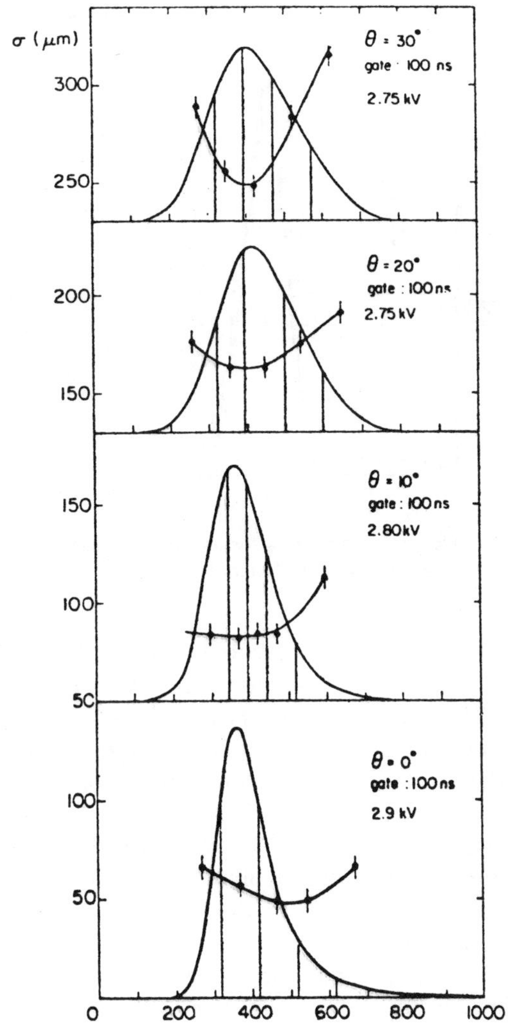

Figure 21 Experimental accuracy in the detection of minimum ionizing particles at different angles of incidence. The accuracy is computed for different slices in the energy loss as shown (43).

The angular spread of avalanches around the anode has been measured both directly in special counters and indirectly by observation of the induced charge ratios on cathodes. The spread depends strongly on the counter gain and, to a smaller extent, on the gas mixture and anode

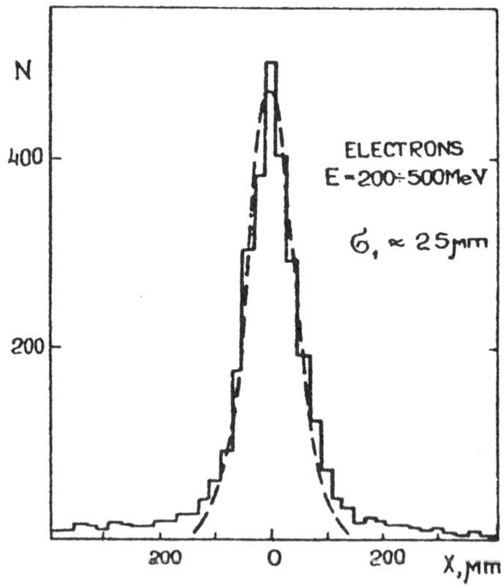

Figure 22 Localization accuracy distribution for an electron beam perpendicular to the chamber, showing the extended tails at large deviations (46).

diameter. Figure 23 (49) shows the measured spread, FWHM, as a function of avalanche size in argon-methane mixtures and for several anode wire diameters. Figure 24 instead gives the measured correlation between real position and center of gravity, computed from charges induced on cathode strips parallel to the anode wire, for a collimated 5.9-keV x-ray source (45). A 2-mm displacement of the source is reflected by a change of about 150 μm in the center of gravity. Despite this scale compression, using a plot such as the one in the figure one can deconvolute the measurements to achieve along the direction perpendicular to the anode wires almost the same accuracy as for the crossed coordinate. This is apparent in Figure 25, a bi-dimensional plot obtained by irradiating a MWPC with soft x-rays through a mask with letters cut out (45); the real size of the image is 4 $\times$ 2 mm. A single anode wire was hit, located horizontally in the center of the figure, and the appropriate transformation was applied to restore the vertical scale.

The sensitivity of the induced charge distributions to the position and angular spread of the avalanche has been computed in detail (53); Figure 26 is an example of time dependence of the charge induced on two adjacent cathode strips, one of which is facing the anode, for several average angular locations of the avalanche (90° meaning an average avalanche growth perpendicular to the anode plane). The apparent displacement of the center

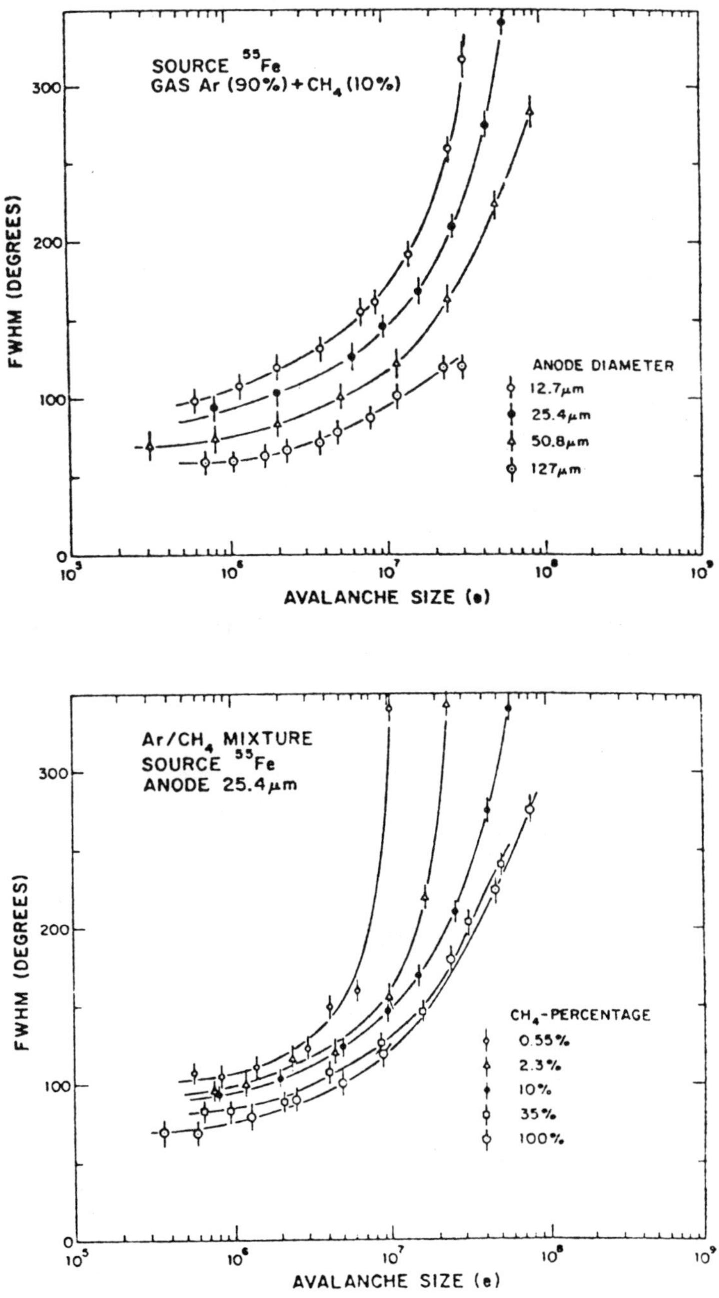

Figure 23 Avalanche width measured in argon–methane mixtures as a function of gain and for various wire diameters (48).

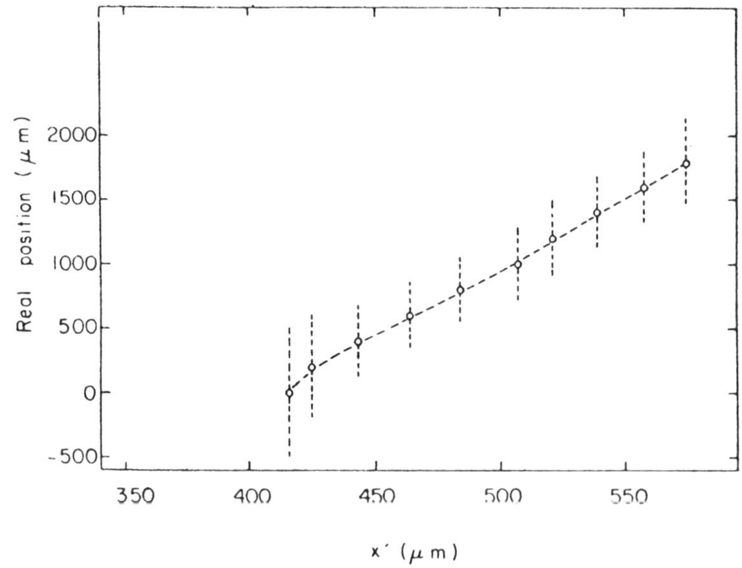

Figure 24 Real and measured coordinates in the direction perpendicular to the anode wires (47).

of gravity is clearly time dependent, as experimental results obtained with different shaping constants in the amplifiers have shown. In particular, the asymmetry tends to vanish for very short constants, a reflection of the screening effect of the anode wire.

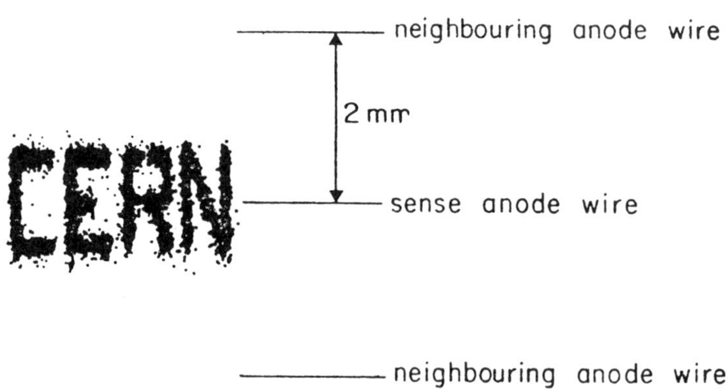

Figure 25 Bi-dimensional image obtained with a high-accuracy MWPC with cathode read-out using a collimator with letters cut out; the size of the mask is 4 × 2 mm^2 (45).

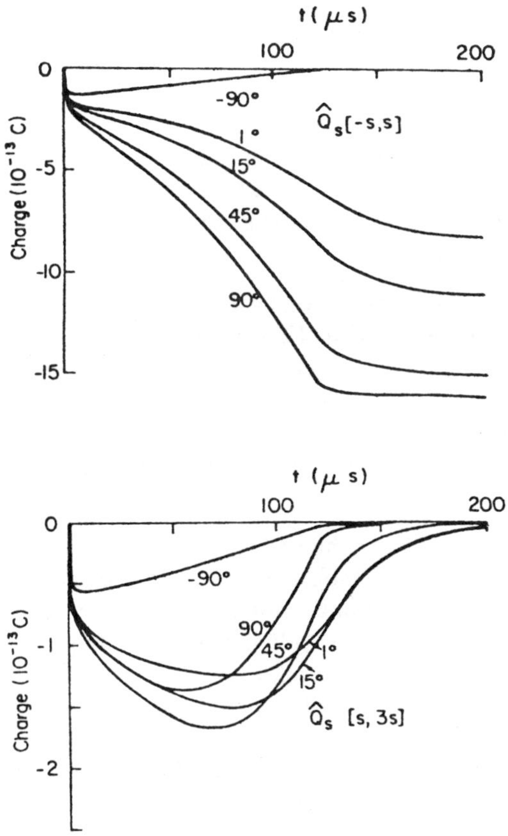

Figure 26 Computed time dependence of the charge induced on adjacent cathode strips for several average avalanche positions (53).

3.4 *Drift Chambers*

As indicated in Section 2.1, electrons released in a gas drift and diffuse under the action of an electric field. A measurement of the time lag between the ionizing encounter and detection of a charge signal at the anodes in a suitable structure provides the average distance, along the field lines, of the ionization, as pointed out already in the early works on the MWPC (35). Of course, this can be done only if the interaction time is known. The first detectors specifically designed to exploit the drift process for localization were operated by Saudinos et al (54, 55) and Walenta et al (56); with various improvements, the drift chamber is still a very popular detector in high-energy physics because of the rather good position accuracy that can be obtained over large surfaces at moderate cost. Large-volume imaging detectors, such as the time projection chamber (57–59), have been developed by fully exploiting the drift properties in gases. For a review of existing drift chamber systems see the references quoted in the previous section; we are concerned here only with the high-resolution detectors.

The localization accuracy that can be achieved in a drift is limited mainly by the following effects:

1. the physical extension and statistical spread of the original ionization mostly due to energetic delta electron emission;
2. the value and stability of the electron drift velocity as well as the detailed knowledge of the drifting field, both in direction and strength; and
3. the dispersion due to diffusion, which depends on the gas nature and pressure and on the field strength, and increases with the drifting distance.

A structure optimized to achieve high resolutions in detecting charged tracks is shown in Figure 27 (60). The closely spaced cathode wires receive from an external network a uniformly decreasing potential, from ground (for the wire facing the anode) to a high negative value in front of the field wires; the electric field is then uniform across most of the cell, with the exception of the region around the anode. This design, together with the choice of a gas mixture exhibiting velocity saturation, allows one to obtain a very stable operation (61, 62). Figure 28 (63) shows an example of positioning accuracy obtained with a set of small drift chambers detecting a beam of minimum ionizing particles perpendicular to the wire planes, as a function of drift distance. Various contributions to the dispersion are also indicated, and the decrease in accuracy, due to diffusion, is apparent; in the argon-isobutane mixture used for this measurement, the single electron diffusion is about 200 μm for 1 cm of drift (see Figures 6 and 8).

One way to reduce the dispersion caused by diffusion and electronics sampling errors is to use a gas in which electrons remain thermal and with small drift velocity up to very high electric fields. Villa (22), operating with dimethylether at atmospheric pressure, measured a localization accuracy of 16 μm rms for minimum ionizing tracks drifting over about 1 cm (see also Figure 7).

The most obvious way to improve localization accuracy is to increase the gas pressure (see Equation 2). This both reduces diffusion and increases the primary ionization yield. Figure 29 shows measurements in a drift chamber at increasing pressures of a propane-ethylene gas mixture (64); Figure 30 is

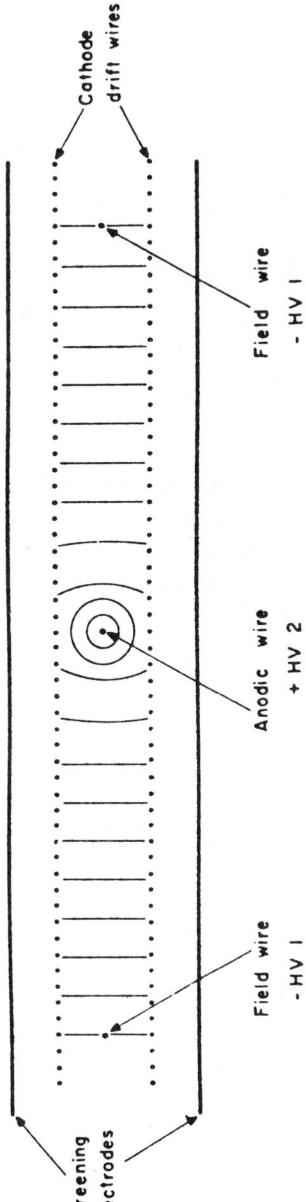

Figure 27 Principle of construction of the high-precision drift chamber (60).

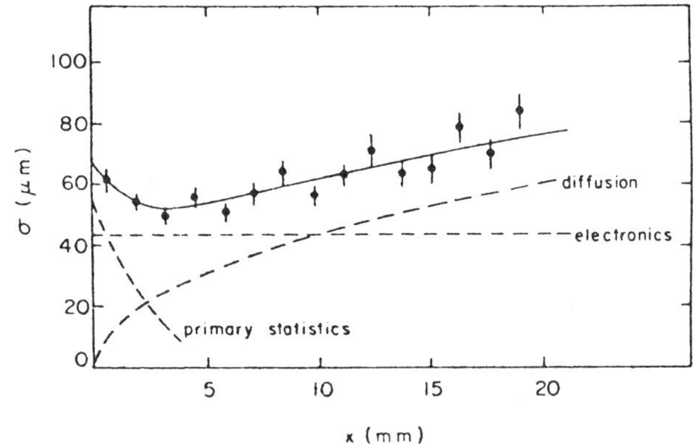

Figure 28 Experimental localization accuracy measured with a drift chamber as a function of the distance of tracks to the anode wire (63).

the result obtained in xenon at 20 atm (65). Figure 31, from the same work, shows the actual accuracy distribution plot for a 20-mm drift distance; the standard deviation of the distribution is 16 μm, but one should notice the non-Gaussian tails due to long-range delta electrons. They contain a non-negligible fraction of the events—about 6%. This information does not appear in plots such as the ones in Figures 28 and 29, and should be taken into proper account when designing high-resolution detectors; redundancy

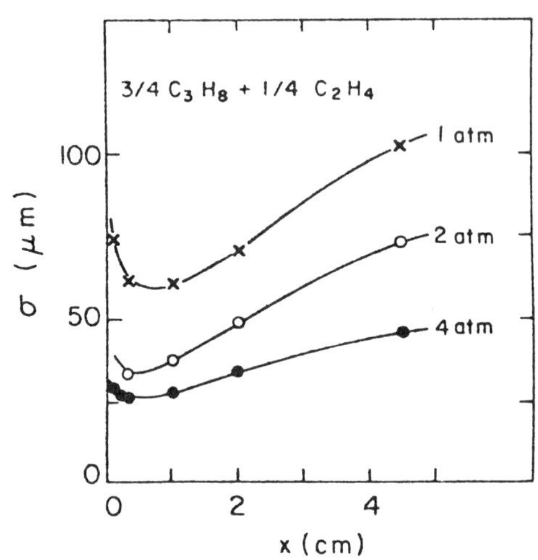

Figure 29 Influence of gas pressure on localization accuracy of a drift chamber (64).

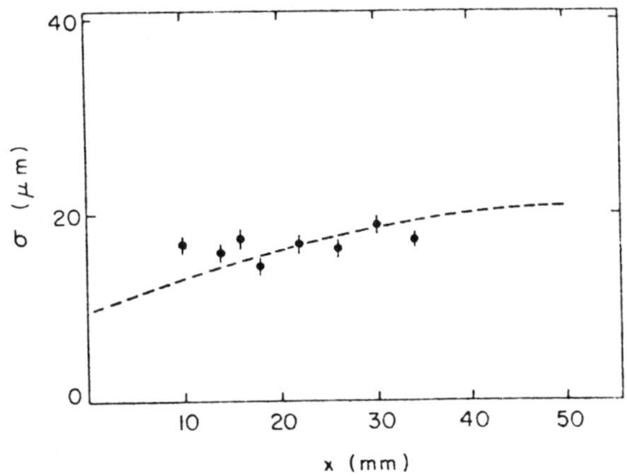

Figure 30 Localization accuracy measured with a scintillating drift chamber operated in xenon at 20 atm (65).

in the number of recorded points for each track allows one to disregard the measurements exceeding a given confidence level in the fit. For this measurement, the authors used the so-called scintillating drift chamber, a device in which charges are detected by a photomultiplier through the light

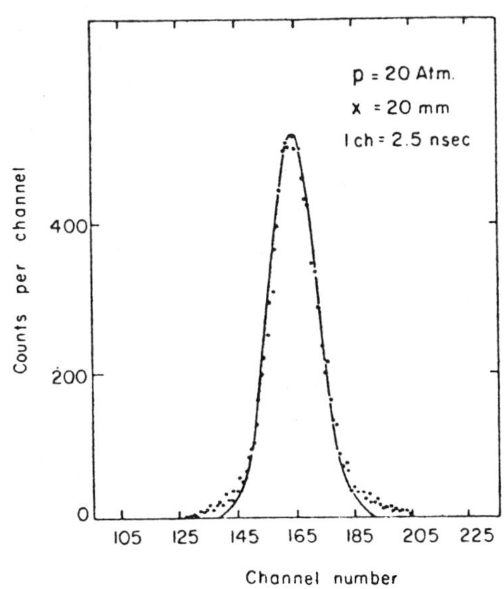

Figure 31 Accuracy distribution in a high-pressure drift chamber, for 20 mm of drift; the tails due to delta electrons are apparent (65).

emitted by secondary gas scintillation in the high field around the anodes (66, 67). The intrinsically faster response of a photomultiplier, as compared to electronic amplifiers, and the absence of space-charge distortion (light emission can be obtained without large charge multiplication) encourage this approach for high-resolution, high-rate detectors.

Various ways have been proposed for coupling scintillating gas detectors to photomultipliers using scintillating rods, wavelength shifters, or optical filters (68), but no operational system based on this approach has emerged so far. Operation of drift chambers at ultra-high pressures (up to 200 atm) has been investigated (69); recent results (27) suggest, however, that the $1/P$ decrease in diffusion may not continue in some gases at high pressures, probably because of enhanced electron attachment and detachment (see also Figure 10).

In the devices described so far, tracks were mostly perpendicular to the drift direction, a particularly favorable case. This is not so in the large-volume drift detectors such as the time projection chamber (57–59), the jet chamber (24), and the imaging chamber (70). In these devices ionized trails form within the gas volume, as a result of external or internal interactions, and a suitable electric field drifts the tracks to a single plane of detection, generally a modified MWPC having field wires interleaved with anode wires. Track segments are continuously sampled and recorded electronically, using systems of analog memories [charge-coupled devices (CCDs)] or fast flash encoders. The time information is then used to compute the coordinate in the drift direction. In the time projection chamber the coordinates in the direction of the anode wires are determined by recording the induced charges on rows of cathode pads and computing the center of gravity by one of the algorithms discussed in the previous section. In other imaging devices this coordinate is obtained either by current division on the anodes or by assembling different modules at a small stereo angle; although less accurate than the center of gravity, these methods allow a more flexible construction geometry. Most of the detectors described are mounted inside a large magnet to allow momentum analysis.

Va'vra (71, 72) recently analyzed both experimentally and theoretically, in great detail, the physical and geometrical factors limiting the attainable accuracy in a drift chamber, in view of building a high-resolution vertex detector. His microjet chamber prototype design is shown in Figure 32; it is essentially a miniature drift chamber with thin anode wires (7.8 μm) operated at high pressures, with tracks developing parallel to the wire plane. Figure 33, generated by a Monte Carlo simulation program, illustrates the different sources of dispersion at detection. Because of the electric field structure, segments of track at equal distance from the anodes

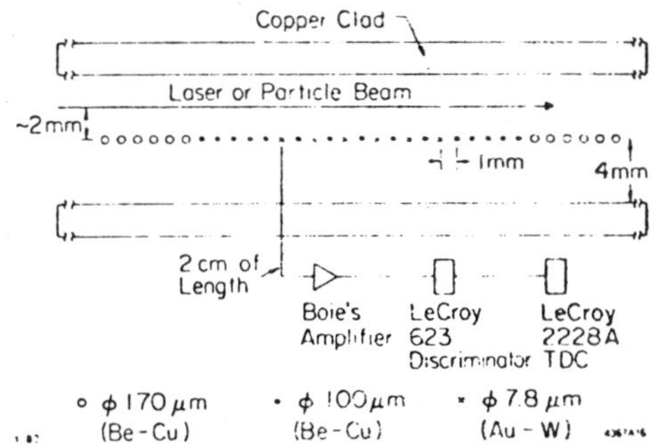

Figure 32 Principle of construction of the microjet chamber (71).

reach the wires with a characteristic (and unavoidable) U-shaped distribution. In the absence of clustering, i.e. local increases of ionization due to delta electrons, the distortion does not necessarily introduce a localization error, since the average time of recorded charge corresponds to the average track distance. A large cluster, on the other hand, may largely offset the time measurement because of the local increase in detected signal. Diffusion helps to reduce the cluster effects by smearing the charge, but of course represents another source of dispersion. The author extensively describes the best strategy for determining the timing of track segments, reaching the conclusion that first electron timing (obtained using large chamber gains and low discrimination thresholds) is the simplest and most accurate method. Figure 34 shows a comparison between computed and measured accuracies, as a function of distance of drift; the result was obtained by operating the microjet prototype in argon-methane 90–10 at 6.1 atm.

To overcome the intrinsic limitation in resolution due to the described clustering effect, various chamber geometries have been proposed, either focusing or otherwise limiting in extension the accepted segments of ionized tracks (27, 73); Figure 35 shows the design of the precision drift imager (27) operated at high pressures and in which only an 0.8-mm long segment of track contributes to the detected signal to each wire. Although of course the same performance could be obtained with a conventional drift chamber having small wire spacing, operation of such a detector at high pressures would require very thin wire diameters with the associated safety and reliability problems. Figure 36 shows the accuracy measured with the detector in a high-energy hadron beam, using as gas filling a carbon dioxide-isobutane mixture at 4 atm; Figure 37 shows instead the measured accuracy for a fixed drift distance, as a function of pressure, in pure CO_2. As already mentioned, the resolution does not improve indefinitely with

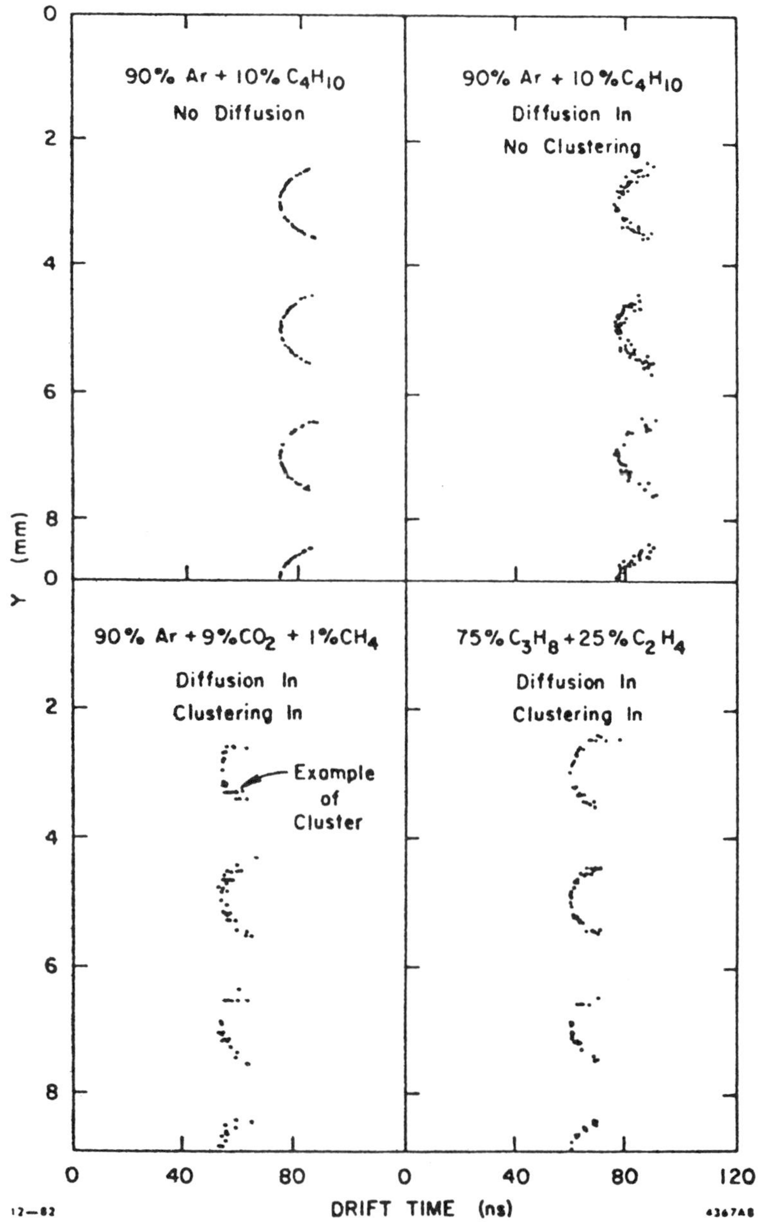

Figure 33 Computed time-shape of detected electrons in the microjet chamber, under various assumptions (72).

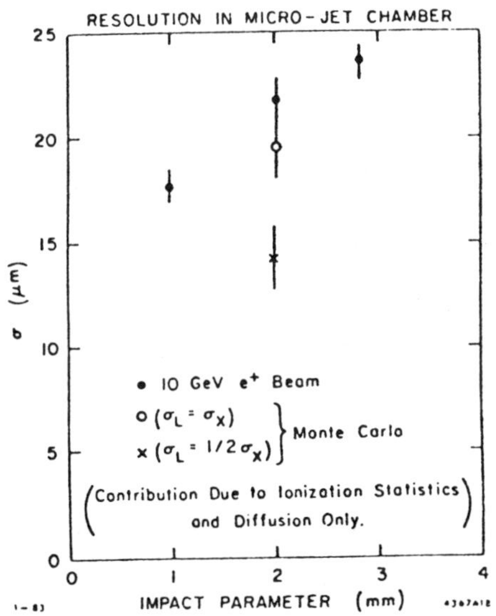

Figure 34 Computed and measured resolution in the microjet chamber (72).

increasing pressures, as would be expected from Equation 2, but an optimum is reached around 4–5 atm. The decrease in accuracy above this value could either be due to charge asymmetries produced by clustering (the probability of obtaining energetic electrons in a given length increases with gas density) or, as suggested by the authors, to residual electronegative pollutants in the gas, which attach and lose electrons, thus producing delayed afterpulses at the wires. The low diffusion and narrow sampling in the described detectors results also in a very good two-track resolution;

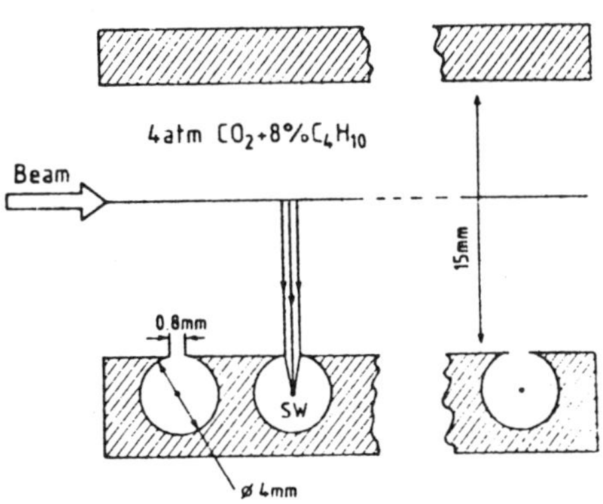

Figure 35 Principle of construction of the high-accuracy drift imager (27).

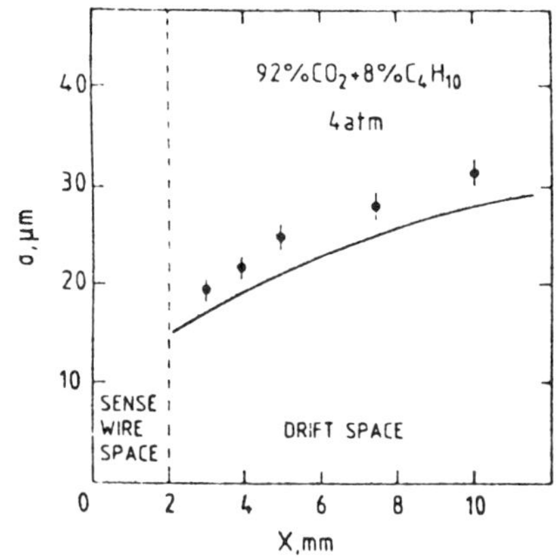

Figure 36 Localization accuracy measured with the drift imager operated at high pressure (27).

100% two-track separation has been obtained above 700 μm of distance, a rather remarkable result (73).

A different approach for obtaining high resolutions has been proposed by Walenta with the time expansion chamber, see Figure 38 (74). As in a standard time projection chamber, the drift region is separated from the amplifying section by a grid. In this case, however, the gas mixture and the field strength in the drift region are chosen such as to obtain a very low drift velocity (typically 0.5 cm/μs), while in the amplification region a normal fast collection is restored. A relatively coarse time measurement is then sufficient to obtain good spatial resolutions. Using a recording electronics system of fast flash ADCs, with 10-ns binning (corresponding in the drift region to a sampling length of 50 μm) and computing the center of gravity of

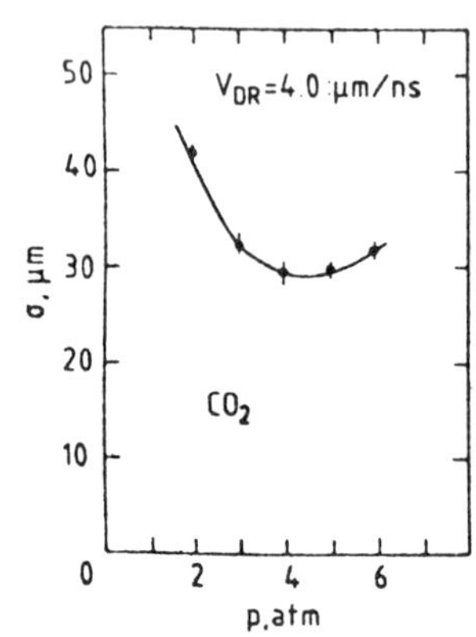

Figure 37 Best localization accuracy measured in the drift imager chamber as a function of gas pressure (27).

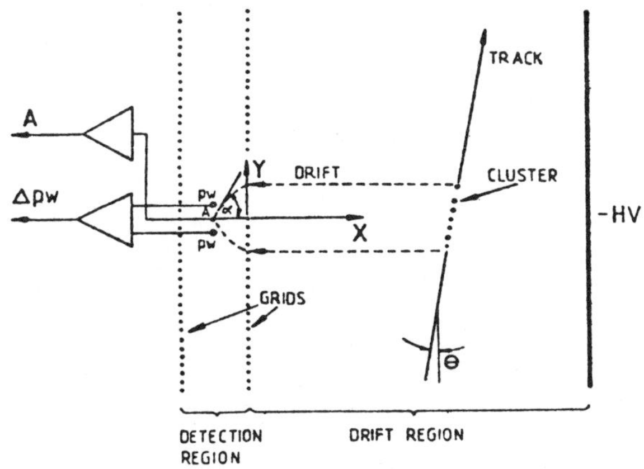

Figure 38 Principle of construction of the time expansion chamber (74).

the recorded time information one can reach resolutions that are mostly diffusion-limited. Figure 39 shows the accuracy measured with a time expansion chamber, operated in a mixture of CO_2 and isobutane in the ratio 80–20 and detecting tracks parallel to the wire plane (75).

The chamber geometry and operating conditions can be optimized in order to attempt a measurement of drift-time for each individual electron or localized cluster in the track segments; such a piece of information contains the full history of the ionizing event and, in principle, allows one to extract the best estimate of its coordinates. With a typical 20–30-ns duration for each detected pulse, obtainable using fast, low-impedance amplifiers, clusters with an average distance of 200–300 μm can be resolved (34). Obviously, in order for the clusters to maintain individuality the electron diffusion has to be kept to a value well below the average cluster separation; whence the need for using a gas mixture having low diffusion and low drift velocity.

For inclined tracks a good localization along the drift direction is obviously not sufficient to locate the cluster, because of the difference in the drift lengths depending on the transverse position. To obviate this problem, two field wires on each side of the anode are connected to a difference amplifier, providing additional information on the direction of approach of each cluster (see the discussion in the previous section on the avalanche localization).

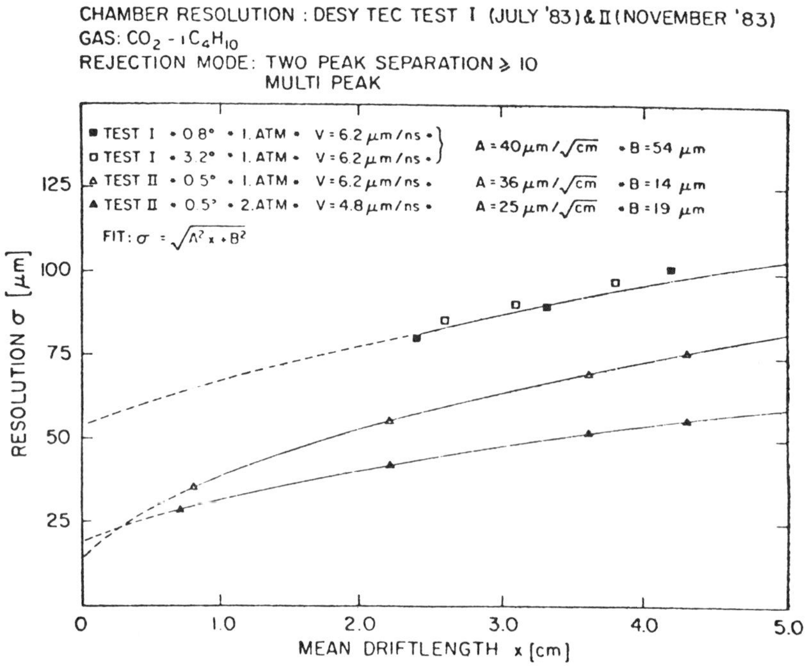

Figure 39 Localization accuracy measured with the time expansion chamber, for tracks parallel to the wire plane (75).

3.5 *Parallel-Plate and Multistep Chambers*

The finite distance between anode wires has a distorting effect on localization along the coordinate perpendicular to the wire, because of the different length of field lines. Although the center of gravity of the induced charge reflects, to some extent, the original position of ionization, this is only true at moderate proportional gains (see Section 3.2), which seems a major limitation to the possibility of locating individual clusters of one or a few electrons, as attempted in the time expansion chamber. Except for tracks parallel to the wire planes, the different trajectory and therefore drift-time for clusters at equal distance from a wire, if not corrected for, is obviously a source of large errors. In the presence of external strong magnetic fields, the deflection of the electron swarm introduces also large dispersions.

A device that, in principle, allows these problems to be avoided is the parallel-plate avalanche chamber, as it has a uniform electric field in the multiplication region. Generally realized with thin meshes or metal foils, parallel-plate chambers can be built using printed-circuit boards, or otherwise patterned electrodes, for localization. Proportionality and imaging capability can be obtained by separating with a mesh the drift from the multiplication regions, much in the same way as in normal time projection chambers.

In the absence of other effects (space charge, gap distortions, etc.) the overall charge gain in a gap of thickness d is given by [see, for example, Raether (76)]

$$Q = Q_0 \exp(\alpha d), \qquad 7.$$

where α is the (constant) first Townsend coefficient pertinent to the gas and field conditions. Quite differently from the MWPC case, the fast electron signal represents a considerable fraction of the total induced charge, and is given by

$$Q_e = Q_0 \exp(\alpha d)/\alpha d \qquad 8.$$

or about 10% of the total signal at gains around 10^5 (as compared with 1% in MWPCs). Moreover, since the ions move in the uniform field with constant velocity, a simple differentiation at the amplifiers' input removes their contribution leaving a very fast detected signal. This makes the parallel-plate chamber a rather promising device to obtain good time and multitrack resolutions in time projection chambers (77–79).

Although widely used in nuclear physics as a detector of heavily ionizing projectiles (80–85), the parallel-plate chamber has not been much considered so far in high-energy physics because of the difficulty of obtaining large and uniform gains over extended surfaces. The discovery, however, that two or more parallel-plate amplifying elements can be combined in the same structure can modify the picture [Charpak & Sauli (86)].

Figure 40 shows the simplest two-step chamber design. Charges produced by ionizing encounters in the upper region drift into a region of very high field, the preamplification region, separated from the previous one by a thin mesh. Here, avalanche multiplication occurs; a fraction of the electrons generated in the avalanche escape into the following, lower field transfer region and continue toward the second element of amplification, another parallel-plate chamber in the drawing. With typical amplification factors in both multiplying gaps of 2×10^3, and 20% transfer efficiency, overall stable gains around 10^6 are attained, quite sufficient for single-electron detection. Using a gated two-step device (87), the authors detected minimum ionizing tracks perpendicular to the electrodes with a localization accuracy of about 150 μm and a two-track separation of about 1.5 mm (Figure 41). The single-electron imaging capability of multistep devices has also been demonstrated (88–90) in Cherenkov ring imaging, a

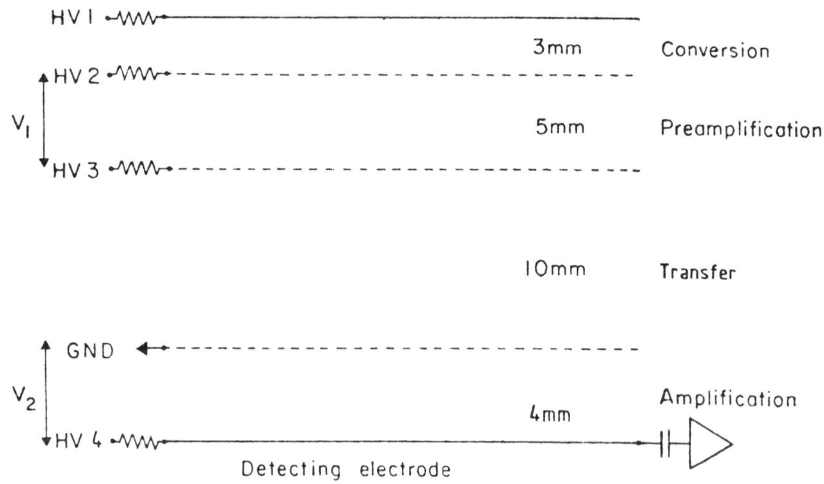

Figure 40 Principle of construction of the multistep avalanche chamber (86).

technique devised for particle identification through the detection and localization of photons radiated by fast charged particles in suitable media.

It appears that the presence of a transfer low-field region, originally thought to help in suppressing photon and ion feedback between the two amplifying elements, may not be necessary to obtain large gains, at least in

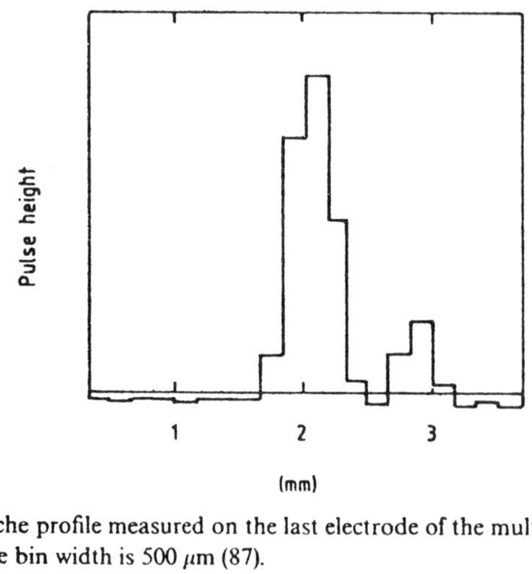

Figure 41 Avalanche profile measured on the last electrode of the multistep chamber, for a two-track event; the bin width is 500 μm (87).

well-quenched gases. Peisert (77) demonstrated that a double-step structure, where the multiplying gaps follow each other, has a stable gain at least 20 times that of each gap individually, for reasons that are not very clear. The detector structure is shown in Figure 42; a region of drift is followed by two multiplying gaps, 4 and 1 mm thick, realized using crossed wire meshes or thick parallel wires at a small pitch; setting the operational conditions such as to have a multiplication factor of at least 20 or 30 in the last gap, a very fast electron signal can be detected. The last anode, realized with a printed-circuit board, contains the read-out structure, which is made of an alternance of strips, 500 μm apart, and rows of 1 mm^2 pads in a geometry rather similar to that of the time projection chamber except for the size of the elements. Figure 43, from the same reference, shows that the electron avalanche measured on the strips of the last electrode has a FWHM of 1 mm, just the right value to obtain a good localization by center-of-gravity measurement on the strips sharing the charge; using a low-impedance amplifier, with fast time constant, only the electron signal with 8-ns full width is detected (see Figure 44), with the consequent very good intrinsic two-track resolution in the drift direction (typical drift times are around 20 ns/mm).

For the reasons indicated, we expect the parallel-plate structure, simple or multistep, to take an important part in the future detector development because of its intrinsic superior characteristics and simplicity of construction.

3.6 *Condensed Noble-Gas Ionization Detectors*

Both diffusion and physical track width are reduced by increasing the density of the medium, a fact naturally leading to the search for suitable liquids or solids as detectors. Most liquids have, however, very large attachment coefficients, leading to a quick loss of drifting electrons, the exception being liquified noble gases with exceedingly low impurity levels.

More than a decade ago Derenzo et al (91) demonstrated electron drift and detection in a multistrip ionization chamber filled with liquid argon.

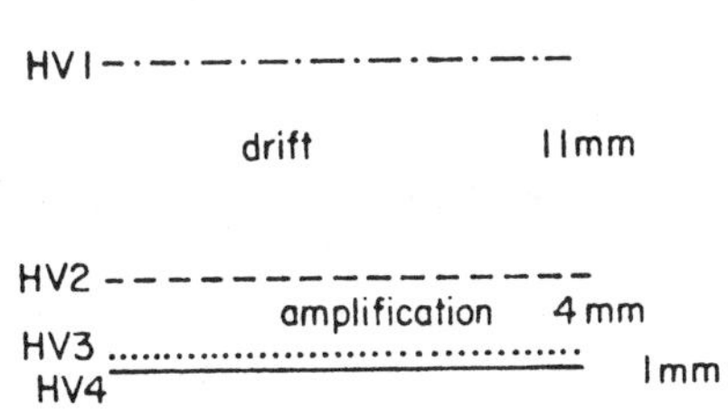

Figure 42 A two-step avalanche chamber as end-cap detector of a time projection chamber (77).

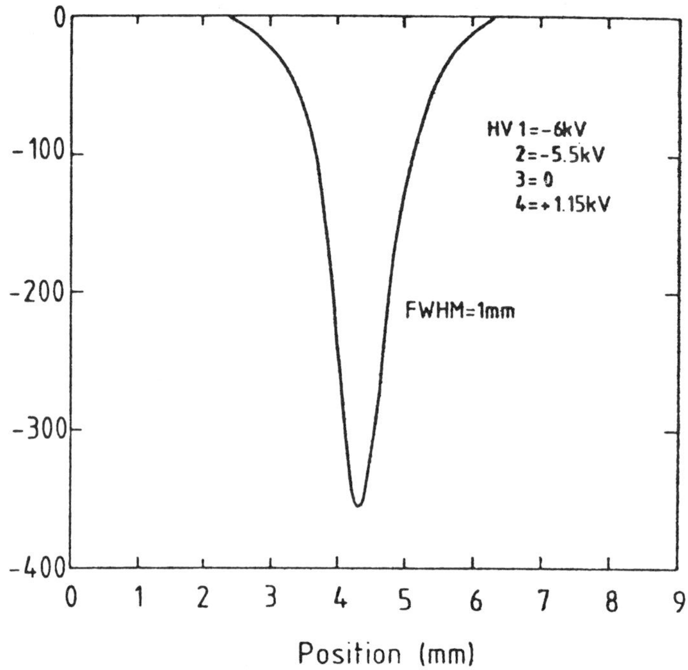

Figure 43 Avalanche profile measured in the two-step chamber shown in Figure 42 (77).

Attempts to obtain proportional amplification around thin wires in the same conditions were, however, only partly successful, preventing at the time the use of the device as a detector of minimum ionizing particles. With the present availability of inexpensive charge-sensitive amplifiers with a typical noise of 0.2 fC rms (92–94) there is no real need for amplification within the detector, as minimum ionizing particles lose about 1.5 fC/mm.

A liquid-argon microstrip ionization chamber with resolutions better than 10 μm has been operated by Deithers et al (95). It consists of a 2-mm wide gap with a flat cathode and an anode made with vacuum-deposited metal strips, 10-μm wide at 20-μm intervals. The active area of the detector is about 3×0.4 mm^2, each individual strip being read out with a low-noise charge amplifier connected to analog-to-digital converters.

Because of diffusion and of the large gap, several strips (8 to 10) collect charge in each event, and the tracks' coordinates is computed by the usual center of gravity method. A stack of several aligned identical detectors allow the measurement of the localization accuracy by a straight-line fitting

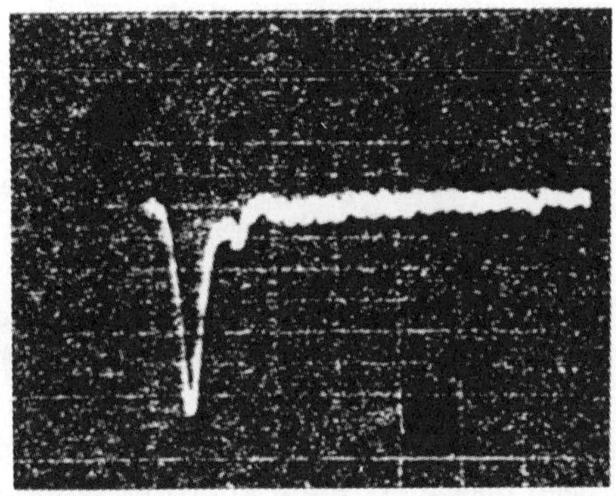

Figure 44 Typical signal detected with the two-step chamber using a fast current amplifier: only the fast electron component is seen (77).

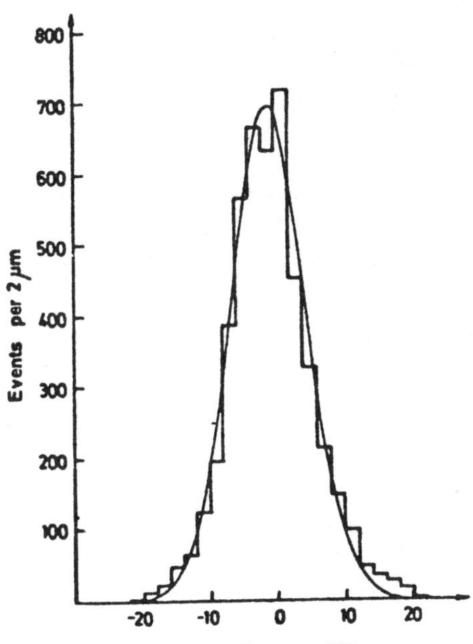

Figure 45 Localization accuracy measured with a stack of microstrip ionization chambers operated in liquid argon (95).

algorithm; Figure 45 shows the residuals' distribution, having a width of 8.5 μm rms.

In the described detector electrons are collected over a few millimeters and the requirements on gas purity are not very stringent (0.3 ppm residual oxygen is quoted by the authors). It was proposed some time ago (96) to use liquid argon in a detector with the time projection chamber geometry, having several tens of centimeters drift length; in this case the required impurities should be reduced to the ppb level.

A 50-liter liquid-argon time projection chamber has been operated successfully (97–99); see Figure 46. A large cryogenic vessel contains the actual detector, consisting of a drift volume of variable thickness; a grid separates the volume from the segmented anode electrode. After extensive purification, an attentuation length for electrons of about 30 cm at a field of 1 kV/cm has been measured, quite appropriate for track detection. The pulse profile within one unit of drift time and for a number of adjacent strips is shown in Figure 47, for two cosmic-ray events. Because of the modest

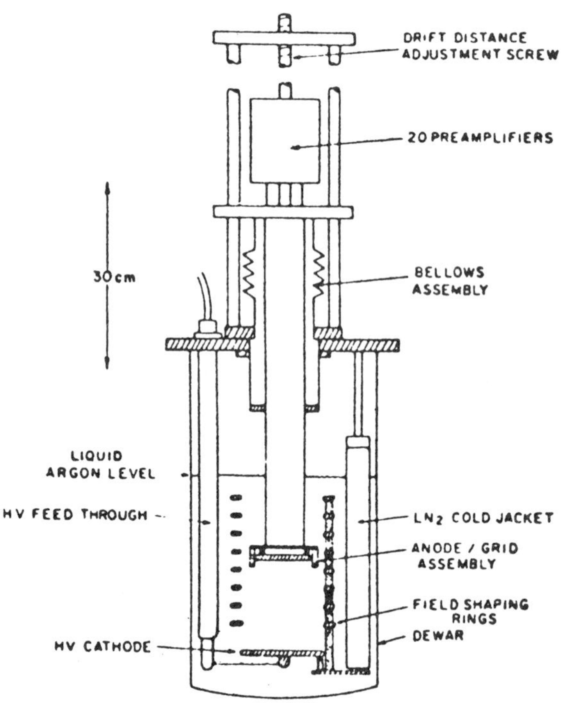

Figure 46 Schematics of the liquid-argon time projection chamber assembly (97).

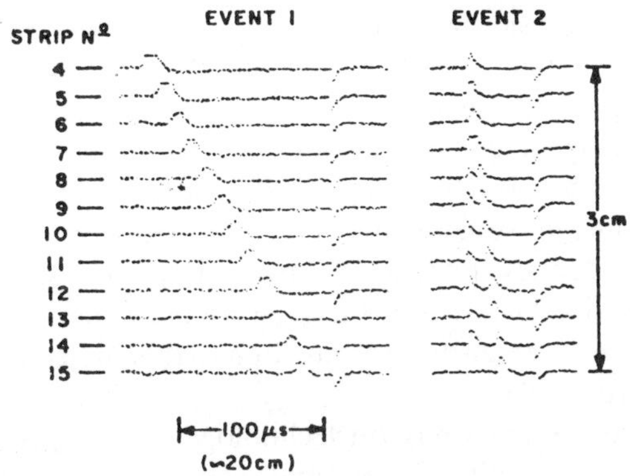

Figure 47 Pulse-height profile as a function of drift time, recorded in the liquid-argon time projection chamber, for a single- and double-track event (99).

sampling frequency of the waveform digitizers employed (1 MHz, corresponding to 2-mm samplings), the data are far from reaching the ultimate localization accuracy that should be permitted by the reduced diffusion in the condensed medium.

Various methods of obtaining bi-dimensional localization on the detector plane have been investigated by the authors, as mosaics of pads and interleaved strips. Because of the small detected charge, however, some of these methods may turn out to be impractical. Further progress is also certainly required in the amplifiers' design to increase the signal-to-noise ratio and the bandwidth.

Even a small amplification factor would be rewarding in this kind of game. Because of the difficulty of obtaining gains in the liquid phase, Dolgoshein et al (100) successfully attempted to drift tracks, produced within the condensed medium, into an upper layer of gas where a conventional MWPC would work. This very challenging approach does not seem to have been pursued further.

4. SOLID-STATE DETECTORS

4.1 *Introduction*

While most of the basic phenomena underlying the gaseous detectors are often well understood by the users, this is not the case with solid-state detectors. The reason is not connected with the complexity of the phenomena in semiconductors but with the fact that most experimental groups in particle physics have had to build their own gaseous detectors and thus became familiar with the underlying physics. Solid-state detectors require high-technology devices built by specialists and appear as black boxes with unchangeable characteristics. The situation is changing since solid-state devices so clearly present perfect features for the solution of pressing problems that some teams of high-energy physicists have started to play with these devices and there is a strong pressure in some laboratories to acquire expensive facilities of the type used by the manufacturers of integrated circuits. An extensive introduction to this subject can be found in recent reviews (101–104).

The solid-state detectors rely on technologies developed for the semiconductor industry. While silicon and germanium present favorable features for particle detection, only silicon allows operation at normal temperature. This is important for devices aiming at the localization of charged particles where the material of the cryogenic tanks would be unacceptable in many cases.

Silicon has four valence electrons. In the crystalline structure each atom is equidistant from the other atoms. Each valence electron is coupled to an electron of a neighboring atom in a covalent bond. In a pure crystal at absolute zero all the electrons are bound and cannot conduct electricity. The valence band is filled. It is separated from the conduction band by an energy gap of 1.1 eV for silicon and 0.7 eV for germanium.

At higher temperatures the thermal energy can break valence bonds, thus allowing electrons to reach the conduction band and become current carriers (thermal generation). The vacancy left behind behaves like a current carrier of opposite sign. An electron from a neighboring atom can fill the hole by breaking its own valence bond and jump over to the first atom. It appears as if the hole has moved. At room temperature the thermal energy is sufficient for a large number of free electrons, of the order of 10^{11} cm^{-3}, to exist in a perfectly pure crystal ("intrinsic semiconductor").

In most applications a controlled amount of impurities is added to the pure crystals ("extrinsic semiconductors"). The impurities are of two types: an element that has three valence electrons (p-type) and one that has five (n-type). The impurity atoms replace semiconductor atoms in the lattice. With a p-type impurity one of the four covalent bonds will not be formed. This makes it easy for an electron in a bond in the neighboring semiconductor atoms to fill this vacant bond, leaving a hole behind it. Common p-type impurities are boron, gallium, and indium. With an n-type impurity the fifth electron is in excess and can easily be removed, becoming free. Typical impurities are antimony, phosphorus, and arsenic.

In intrinsic semiconductors the number of holes equals the number of free electrons (with no applied voltage). In extrinsic semiconductors, at room temperature, the number of current carriers is usually much greater than the number that would be present in the undoped material, but the product of the concentration of electrons n and holes p is a constant $n \cdot p = N_c N_v \exp(-E_g/kT)$, where E_g is the band gap, k is Boltzmann's constant, N_c and N_v are the numbers of allowed energy levels in the conduction band and the valence band, respectively; typically, at 300 K, $N_c = 2.8 \times 10^{19}$ and $N_v = 1.04 \times 10^{19}$, for silicon.

The most important material for charged-particle detection is silicon doped with an n-type impurity. Most of the current is carried by free electrons in the conduction band. They are called the majority carriers. With p-type impurities the holes are majority carriers. Table 1 shows some intrinsic properties of silicon and germanium, at room temperature. It illustrates the advantages of silicon for operation at room temperature.

4.2 *The Depletion Region in Rectifying Structures*

Particle detection exploits the properties of rectifying structures made of the junction of materials of different types, for instance an n-type and a p-type, which we take as examples.

The charges in an unbiased p–n junction are shown schematically in Figure 48. The silicon atoms are not shown. The holes and electrons are free to move. There will be a current generated by the diffusion of holes from the p region to the n region and vice-versa. The electrons and holes recombine. The current stops because each of the regions was initially neutral and the departure of the majority carrier leaves a charge, produced by the fixed ions in the lattice. The p region becomes negative and the n region positive. This build-up of space charge prevents further diffusion. It is only near the junction that the current carriers have disappeared. This region is not electrically neutral; it is called the "space-charge region" or the "depletion region."

Table 1 Some intrinsic properties of silicon and germanium at room temperature

Property	Ge	Si	Unit
Electron mobility	4×10^3	1.4×10^3	$cm^2/V \cdot s$
Hole mobility	1.9×10^3	500	$cm^2/V \cdot s$
Intrinsic resistivity	65	2×10^5	$\Omega \cdot cm$
Energy gap	0.7	1.1	eV

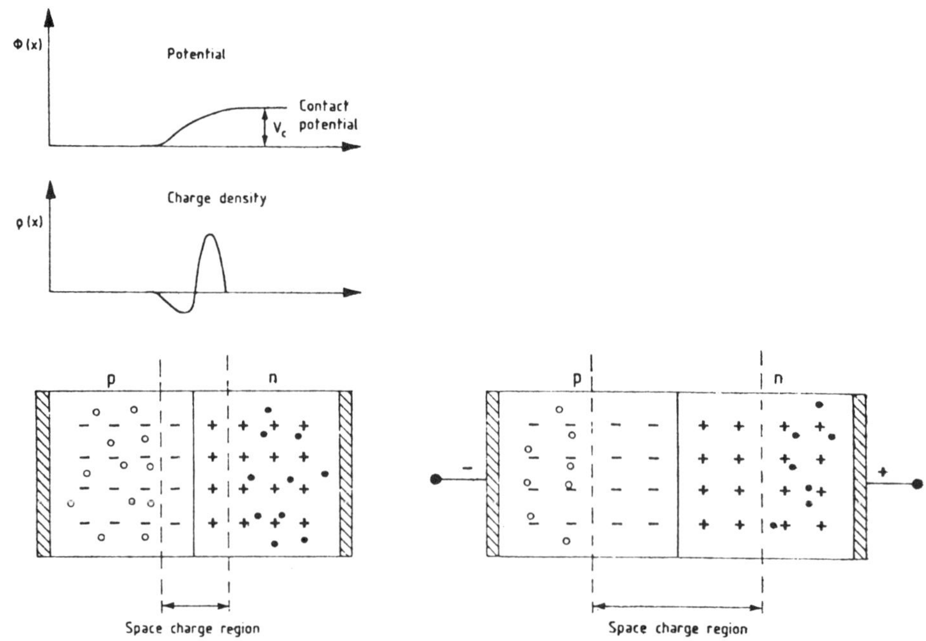

Figure 48 Schematic view of a p-n junction; − and + represent the fixed ionized acceptor and donor atoms. ○ and ● represent the mobile charge carrying holes and electrons. The substrate Si or Ge atoms are not represented. (*a*) No applied external potential. The holes and electrons recombine in the space-charge or depletion region; the extension of this region is prevented by the potential produced by the fixed ions. (*b*) External potential in the reverse direction. More atoms are "uncovered" by the field. The depletion region is extended to a thickness depending on the external field strength.

If a voltage source is applied across the junction in such a way as to pull away the majority carriers from the junction, more fixed donor and acceptor ions are "uncovered," more space charge is built up, and the potential barrier between the two regions is increased; the only current flowing in such a structure is due to minority carriers and is called the saturation current of the rectifying structure.

When ionizing radiations liberate charges in this depletion region the strong internal electric field drifts them away and produces a detectable current. This current has to be larger than the noise that is due to the fluctuations of the saturation current. This is where silicon presents its main advantage over germanium, at room temperature, because of its high resistivity and its larger energy gap.

The art of construction of the detecting silicon diodes for high-accuracy detectors in high-energy physics consists in having a depletion layer of sufficient thickness for an efficient detection of minimum ionizing particles and a minimum thickness of the dead region necessary to produce the rectifying structure and carry the currents. The materials of the two types can have very different impurity concentrations leading to different resistivities ρ_p and ρ_n; if $\rho_p \ll \rho_n$ the active region extends primarily into the n-type side of the device and is referred to as a p^+n structure. In a p^+n device the charge mobility is essentially the mobility of the electrons, which can be an advantageous feature for the speed of the device. A depleted region can also be obtained by the junction of an n-type crystal with a metal, such as a thin layer of gold (Schottky diode). Silicon has the advantage of growing a natural oxide that serves as an insulator, as a passivation layer, and also as a diffusion mask. The various types of detecting diodes are described in (101, 105). Figure 49*a* shows a practical structure with the various manufacturing steps only to give an idea of the complexity of the construction.

The width X_D of the depletion layer depends on the doping density and the applied bias voltage V_B:

$$X_D = [2\varepsilon\mu_e\rho(V_0 + V_B)]^{1/2}. \qquad 9.$$

V_0 is the voltage created by the space charge in the junction, ε is the permittivity ($\varepsilon = 1.054 \times 10^{-12}$ F/cm), μ_e is the electron mobility.

The resistivity of n-type silicon can be 20 kΩ·cm and a depletion layer of 1 mm thickness can be obtained with $V_B = 300$ V. A depletion layer X_D has a capacitance, per unit area, of $C_D = \varepsilon/X_D$. Above a certain voltage the capacitance becomes constant and the complete thickness of silicon is depleted, the electrical field extending from the rectifying contact to the ohmic rear contact. This is the situation reached in most detectors, which then behave like a solid ionization chamber.

The typical detecting thicknesses vary between 20 μm in charge-coupled devices (CCDs) and a few hundred μm in silicon detectors. With liberated charges of 80 electron–hole pairs per μm and, as we discuss below, with structures allowing for an energy sharing between several electrodes, the signals to detect can vary between a few hundred electron charges to a thousand times more. The various sources of noise, intrinsic to the solid-state detector, vary from a few tens of electron charges to a few thousands in silicon detectors and can relatively easily be made negligible compared to the signals. We refer the reader to the articles discussing the various sources of noise (101–105) and the matching of the external read-out electronics. The main source of noise is due to the statistical fluctuations in the number

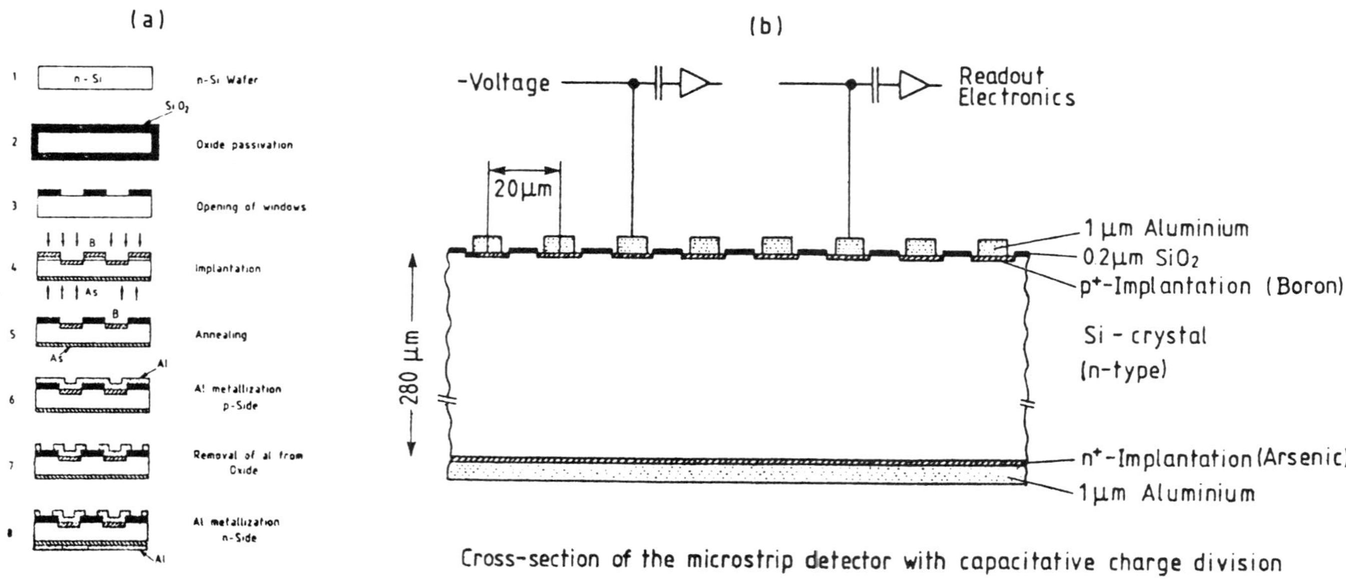

Figure 49 Successive steps of the manufacturing process of passivated ion-implanted silicon detectors. (*a*) The manufacturing steps. (*b*) Details of the final product (103).

of carriers, leading to changes in conductivity, and it is proportional to the current flow:

$$\langle i^2 \rangle = 2qI\Delta f, \quad 10.$$

where I is the dark current, q is the charge of the carrier, and Δf is the bandwidth. Optimizing the electronics as discussed by Radeka (in 101), and making the capacitance C_d of the detector equal to the input capacitance C_a of the amplifier, leads to an equivalent noise charge ENC,

$$\mathrm{ENC} \simeq 8kTC_d/f_T\lambda, \quad 11.$$

where f_T is the frequency for unity gain of the amplifier and λ is the time constant of the filter. For typical values of $f_T = 1$ GHz and $\lambda = 100$ ns, ENC $\simeq 3 \times 10^2$ [$C_d^{1/2}$, C_d being in pF] rms electron charges. For $C_d = 10^{-2}$ pF, which is more than the capacitance of the detecting elements of a CCD, ENC is only $30e$; for $C_d = 1$ pF, ENC is equal to $300e$; and even for parasitic capacitances of 100 pF, ENC $\simeq 1000e$. With present amplifiers, where the input noise is of the order of $1000e$, the contribution of the solid-state detector is not the dominant factor. And it is easy, with the various structures, to obtain charge signals from minimum ionizing particles larger than the noise. In many applications, however, where high accuracy is desirable, the thickness is also limited by the errors introduced by multiple scattering. This contribution is very dependent upon the geometry of the experiment. With a radiation length of 9 cm, the most promising silicon detectors, with a thickness of about 300 μm, do not contribute much more than gaseous detectors if they are operated in vacuum.

The finite thickness leads also to another source of error. While the initial column of charges produced by the ionizing particle is estimated to have an extension of less than 1 μm, it broadens by diffusion when the charges are drifting to the electrodes. For silicon and a drift length of 1 mm, the diffusion leads to a spread of about 30 μm (FWHM).

4.3 *Position-Sensitive Detectors*

There are many concepts for position sensing with silicon detectors, of which we mention five:

1. Charge transfer from the whole detector with current division to localize the initial charge deposit. The energy loss of minimum ionizing particles in thin layers of silicon, say 300 μm, liberates about 25,000 charges. For a surface of a few cm^2, with a capacitance around 100 pF and a high-resistivity electrode, the total noise is equivalent to 50 keV. This does not allow accuracies in the micrometer range to be reached over a length of

the order of cm, and the method is used only for particles of a few MeV stopping in the detector where 1% accuracy is possible.
2. Arrays of narrow strips, read out individually or using interpolation methods to reduce the number of strips.
3. Charge-coupled devices where a mosaic of small read-out elements is used to localize the ionizing event.
4. Charge storage in P–I–N diodes.
5. Silicon drift chambers.

We describe these different approaches, which are, at present, at quite different levels of development.

4.4 *Microstrip Detectors*

The breakthrough for the high accuracy was the demonstration in 1983 by Hyams et al (105) that resolutions of 5 μm rms can be achieved with strips 20-μm wide, read out every 60 μm, while a resolution of 8 μm was achieved with a read-out of 120 μm, exploiting capacitive charge division between adjacent strips. A large-scale experiment has confirmed the operational character of this device (105–108). Figure 49*b* shows the detailed structure of the detector. The sensitive area of the counter is a rectangle of 24 mm × 36 mm with 1200 strips of 12 μm × 36 mm and 20-μm pitch.

A relativistic particle of charge one traversing the detector produces ~25,000 electron-hole pairs, which are collected within <10 ns at the electrodes. The intermediate strips are kept at the potential of the read-out strip. The impedance between read-out strips is much greater than the input impedance of the electronics in order to avoid cross-talk. The interstrip capacitance is greater than the strip to ground capacitance. The charge collected at intermediate strips can be divided among the neighboring read-out strips. The position of the impinging particles is found by computing the center of gravity of the collected charges.

Figure 50 shows the strip layout of the detector. The results obtained with such a detector are illustrated by Figure 51, which shows that a resolution of 4.5 μm can be achieved. A two-particle separation of ~ 120 μm has been obtained. Detectors of this type have now been employed for several years in an experiment and constitute one of the best candidates for detectors in this range of accuracy. It should be mentioned that the width of the charge distribution arriving at the strip is approximately 6 μm and that one would expect the charges to be collected by no more than 2 strips. However 1/3 of all clusters originating from a single particle involve at least 3 strips. This effect, spoiling the particle separation, is so far unexplained. The counting rate achieved was 10^6 per second per strip, limited only by the electronics.

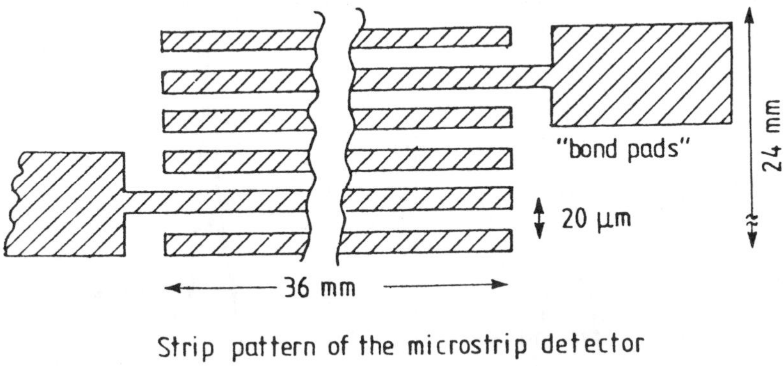

Figure 50 Strip pattern of a microstrip detector (105).

Belau et al (106) have studied the behavior of such a detector as a function of applied voltage and magnetic field. Their results are summarized in Figure 51. A simple model permits the prediction of the observed spatial resolution. A magnetic field perpendicular to the direction of the drifting charges and parallel to the read-out strip increases the average number of the collecting strips, as expected from the effect of the Lorentz force. A field of 1.68 T shifts the measured coordinate by an amount of the order of 10 μm and increases the width of the collected charge from 5 to about 12 μm. The accuracy of 5 μm is not considered the lowest limit by Belau et al. Fluctuations in the center of gravity of the deposited charge, of the order of only 1 μm, are expected from delta electrons. These authors calculate that with a read-out pitch of 20 μm an accuracy of 2.8 μm could be reached, with the diode arrangement on the strip side. This has to be compared with the observed resolution of 4.5 μm, with the read-out of 60 μm, while their model gave 3.6 μm. With a read-out pitch of 20 μm and a diode arrangement opposite to the strips, their model gives an accuracy of 0.8 μm, i.e. the limit of accuracy where the delta electrons enter into the game.

The understanding of the microstrip silicon detectors and the experience acquired in their operation give support to the estimate that they are at present good candidates for accurate measurements of relativistic particles in the range of 5 μm and probably down to 2 μm.

It should be stressed that the possibility of individual read-out of strips spaced by 20 μm, as demonstrated by Belau et al (106), which leads to the same accuracy as that reached so far by capacitive read-out, is an attractive solution for the future. The shaped pulse width can be of only 50 ns at the basis, against 800 ns with current division, thus permitting higher rates. This fast response is of value for applications where the microstrip chambers are used for the triggering on preselected multiplicities. This

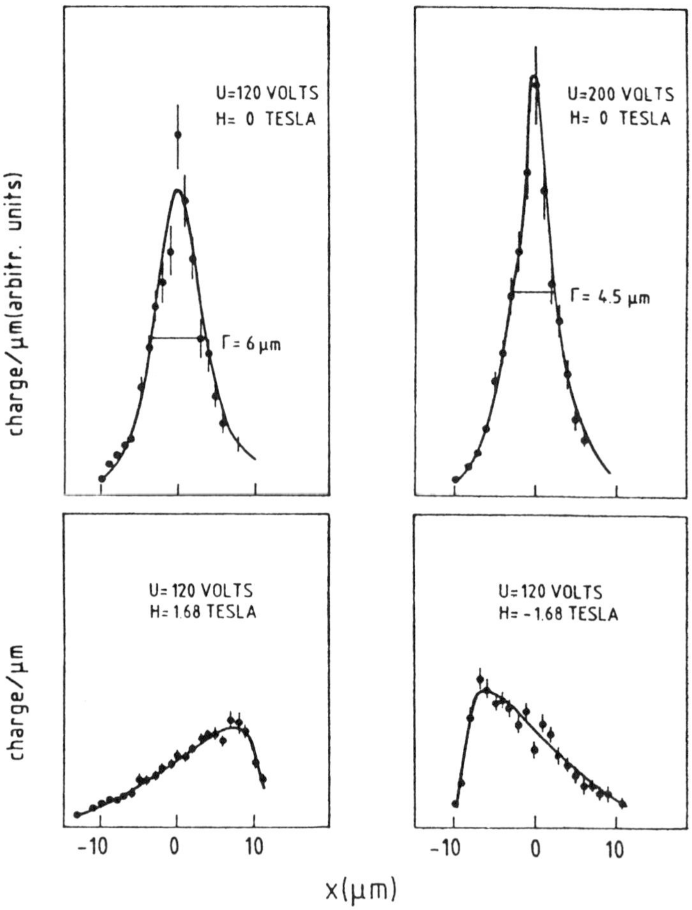

Figure 51 Position resolution of a microstrip detector, as a function of voltage and magnetic field: U = applied voltage, H = magnetic field. The magnetic field is parallel to the strip (106).

is permitted by the resolution, close to 20 keV (FWHM), for the energy loss of minimum ionizing particles (109). The two particles' resolution can be 40 μm, which is also a very attractive feature for high-multiplicity reactions.

Large-scale integration of the associated electronics on the same chips as the microstrips would make a considerable difference in the prospects for a generalization in the use of this technique. The present ratio of the area of the electrons to the area of the detector is 300, while with a suitable integration it could possibly reach one. Such a development could permit the construction of the moderate surfaces needed for the vertex detectors close to the interaction regions at storage rings. Figure 52 shows a detector planned for LEP, with only a partial integration of the electronics (108).

The commercial cost of a microstrip detector is only $1000 for an area of about 5 cm^2. The main cost is in the associated electronics and large-scale integration of the whole device could make it very attractive. Present technology could permit the construction of chips of 7×7 cm^2.

The radiation damage could be a worry for some applications. It is very dependent upon the nature of the radiation involved. Sizeable changes take place at silicon surface barriers irradiated on the front surface by fast electrons at above 10^{14} cm^{-2}, 10^{13} cm^{-2} for protons, 10^{11} cm^{-2} for α particles, about 3×10^8 cm^{-2} for fission fragments and x-ray doses of about 10^4 Gy (103).

Passivated, ion-implanted, silicon junction detectors were exposed to doses of high-energy protons at CERN up to fluences of 8.3×10^{13} particles cm^{-2}, and a considerable increase in leakage current was

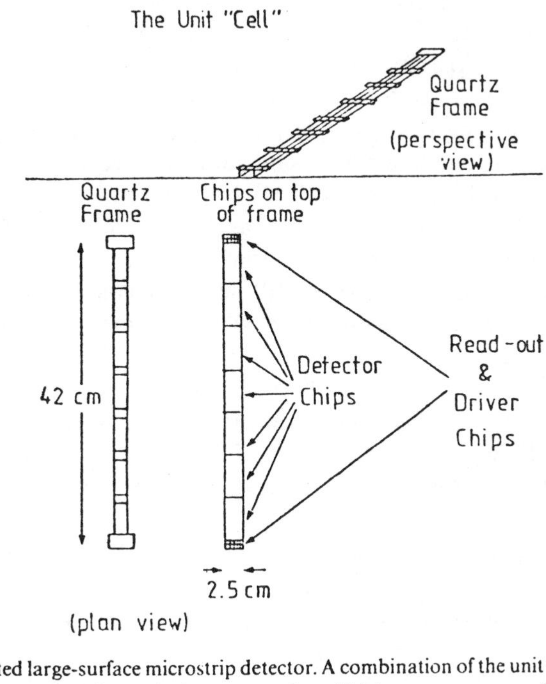

Figure 52 Projected large-surface microstrip detector. A combination of the unit "cell" made of microstrip detectors should permit the construction of high-accuracy vertex detectors for colliding beams. Part of the electronics is integrated on the detecting chips (108).

observed. Nevertheless full charge collection can be maintained by linear increase of the electric field with the fluence (110). Also, for extreme cases, different methods of manufacture or choice of the silicon resistivity offer possibilities to decrease the sensitivity to radiation damage.

4.5 *Charge-Coupled Devices*

The CCD is, like the silicon diode, a concept of semiconductor electronics. However, it does not rely mainly on the modulation of electrical currents but on the manipulation of information. This information is in the form of electrons stored in potential wells, located in very small volumes of a few μm^2 of apparent surface, constituting a matrix of tens of thousands of elements per silicon chip.

Charge-coupling is the collective transfer of all the charge stored in a potential well to the adjacent similar element, by external manipulation of voltages. The quantity of the stored charge is conserved in this transfer and can represent information.

One of the main applications of this device is the imaging of light in the visible or near infrared region. The light liberates electrons in the silicon elements and these electrons are trapped in the wells, the charge being proportional to the amount of light. The transfer of the charges to external read-out elements is done very simply by manipulating collectively the potentials of a few electrodes, which are part of each storage element. The interest of this device for television cameras, and its extension to particle detection, can be understood from some typical characteristics: (*a*) The quantum efficiency can be as high as 80%. (*b*) The low capacity for each storage element permits a very low intrinsic noise of the transferable charges. Overall rms noise of 10 electrons can be achieved. Excellent pictures are obtained with CCDs where charges as small as about 100 electrons each are transferred through a centimeter of silicon.

This is not the only application of CCDs. They can be used as analogue memory elements. The charges are then introduced through diodes and gates and they then operate as discrete delay lines, with the timing properties controlled by the clocking potential applied to the transferring electrodes.

Before discussing the use of CCDs for particle detection (111, 112) a summary description of their physical structure may be appropriate. This device illustrates beautifully the huge potential for high-accuracy particle detection that exists in solid-state electronics.

Figure 53 shows the structure of one type of CCD used by Damerell et al.[1] The CCD is fabricated on an epitaxial p-type layer (dopant concen-

[1] Charge Coupled Device P 8600, English Electric Valve Co. Chelmsford, England.

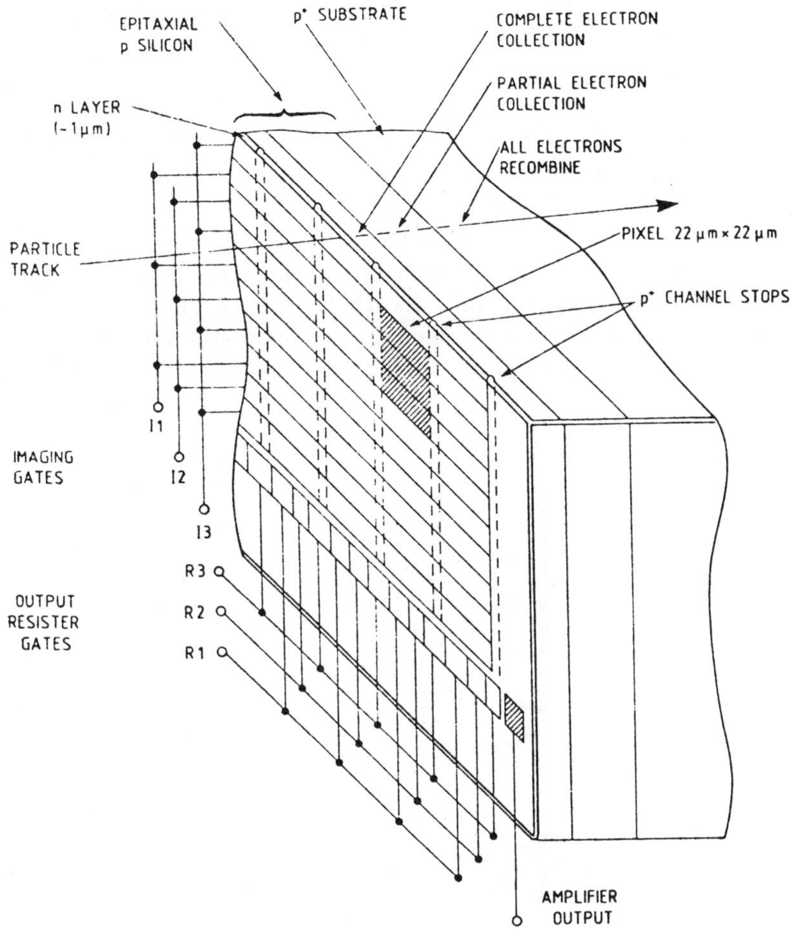

Figure 53 A CCD structure. A corner of the CCD enlarged to show details of the pixel (storage element) structure (111).

tration 5×10^{14} cm^{-3}) of thickness 25 μm. This layer is grown on a p^+ substrate (dopant 5×10^{18} cm^{-3}), which is inactive from the point of view of particle detection. The surface of the epitaxial silicon is converted, by ion implantation, to an n-type, with dopant concentration of 10^{16} cm^{-3} to a depth of approximately 1 μm. Above a thin oxide layer are deposited transparent electrodes, called the imaging gates, which are insulated from one another and form the substrate. Holding the I_2 gates high (~ 10 V) and the I_1 and I_3 low (0 V) creates a matrix of potential wells (pixels) with minima near the upper interface (buried channel), defined in the x direction by the I_2 gates and in the y direction by narrow p^+ implants. The CCD has 576 × 385 such pixels in a sensitive area of 12.7 mm × 8.5 mm. With an interval between pixel centers of 22 μm in the x and y directions, the CCD is essentially a very precise rectangular matrix of potential wells that act to trap electrons. Figure 54 shows a typical distribution of potentials in a CCD slightly different from that represented in Figure 53 and the first investigated by Damerell et al (111, 112).

At room temperature the potential wells rapidly fill with electrons, owing to thermal generation. This is reduced to a negligible level at moderately low temperatures, which require the CCD to be kept in a cryostat in thermal contact with a bath of liquid nitrogen. In the applications to optical imaging the read-out of the stored image is obtained by shifting the stored

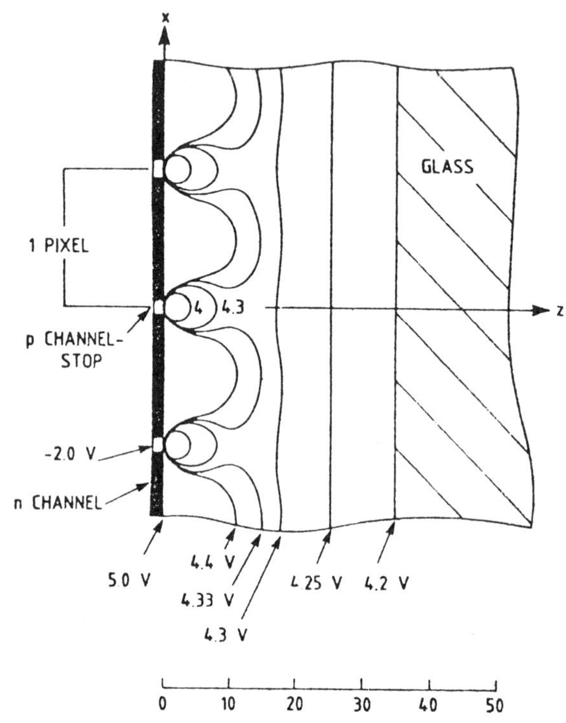

Figure 54 Map of the electrostatic potentials in the CCD substrate when all gate voltages are held to zero (109). The electrons generated in the depletion region diffuse to the n channel. Their drift is limited in the x direction by the p^+ stops. In the y direction the spread among several channels improves the accuracy of the centroid-finding methods. The application of a pulse of a few volts to the gates I_2 freezes the charges inside the pixels.

electrons in the y direction, from I_2 to I_3, I_3 to I_1, I_1 to I_2 (for a three-phase CCD) [Figure 55, from (113)]. At this step the charges have moved by one pixel in the x direction. Those charges that were in the bottom row are shifted to the R register, which provides 385 R triplets to read out the first I triplet, by moving the charges, on the same principle, in the orthogonal direction, sideways. The charge in each pixel is sensed at the output by an analog circuit. Once the first row has been read out, there follows another I triplet and 385 R triplets as the second row is read. This system has been invented to read out an optical image where most of the pixels have a stored bucket and for speeds of about 25 images per second.

When the particle ionizes the depletion region, the initial column of liberated electrons, of about 1 μm thickness, starts diffusing toward the region of minimum potential, which is part of the depletion region closest to the I gates. It is collected in a time of the order of 30 ns, during which it diffuses transversely by about 10 μm. The diffusion is stopped in the x direction by the p^+ stops.

In the y direction the electrons would continue to diffuse laterally if all the I gates were at the same potential. The initial idea of the first investigators of

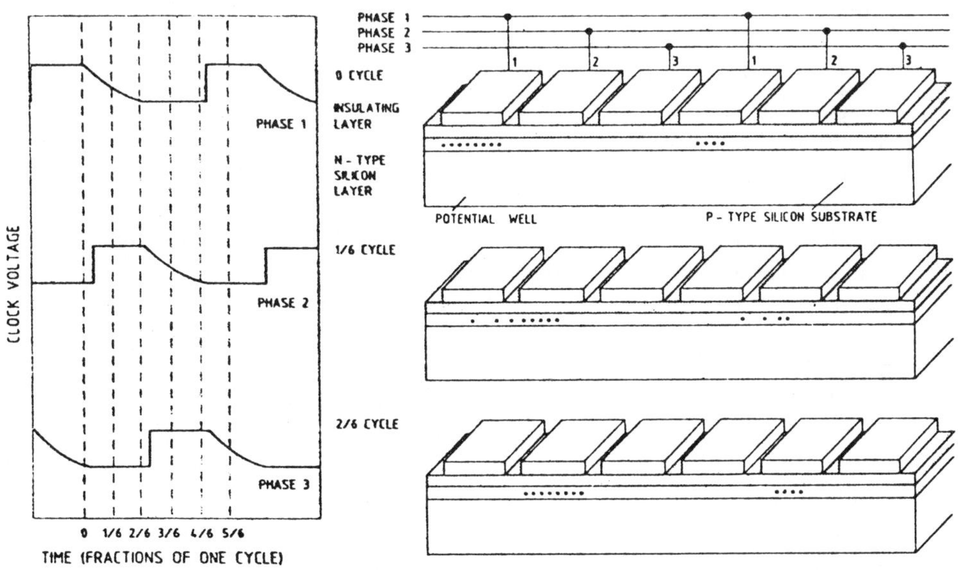

Figure 55 The read-out of a three-phase CCD. A full cycle of the clock voltages represented at the left shifts the electrons by one pixel (3 strips). The electrons in the end row are moved to the R registers (Figure 53), where the electron buckets are moved sideways, on the same principle, and their contents are measured, one after the other (113).

the CCD was to apply a pulsed voltage on every third I gate, say I_2, 100 ns after having obtained a signal from auxiliary counters detecting an interesting event. This would have frozen the electrons in a matrix of potential wells which could then only be read out by the conventional potential manipulation of the I gates. Because of the diffusion the electrons would be distributed among several elements of this matrix in the y direction, thus permitting an accurate measurement of the charge centroid. In order to avoid beam tracks being registered during the read-out procedure, a fast kicker magnet switching off the beam was foreseen.

The experience gained during the study of this method has led to another approach. The I_2 gates are polarized as soon as the accelerator is active and the potential matrix is then permanent. The events are clocked out continuously, at a rate of 3 MHz, to the R registers. For the most remote events this leads to a clearing time of a few hundred μs during which accidental beam tracks traverse the detectors. However, by having a second device back to back with the first one, a precise geometrical relation is imposed on every track. The great number of pixels, close to 0.25×10^6, compensates for the lack of time resolution, and at a beam rate of 10^5/s there is no difficulty in tagging the beam tracks superimposed on a good event.

The sensitivity of the system is very high. A minimum ionizing particle generates 1300 electrons while the noise is only a few tens of electrons. The pulse-height distribution from x-rays of 5.9 keV shows a 12% (FWHM) resolution (Figure 56*a*). Figure 56*b* shows the pulse-height distribution from minimum ionizing particles for which an efficiency of $98 \pm 2\%$ is reached with a spatial accuracy of 4.3 μm and 6.1 μm in two orthogonal directions and a two-track resolution of 40 μm in space (with only 2% overlap). The authors claim that a spatial precision better than 2 μm can be achieved by fabricating the CCDs on high-resistivity silicon. It should be mentioned that in astronomy, CCDs have permitted the tracking of star images with an accuracy of 0.2 μm, thus illustrating the considerable usefulness of this device. The main advantage of CCDs over silicon microstrip detectors is their better two-track resolution and the two-dimensional read-out from every plane. They are much more limited in counting rate.

This work illustrates beautifully the fact that semiconductor technology has the potential properties for high-accuracy, high-resolution detection of charged particles. However, because of the considerable investment in money required for the development of this technology, physicists have so far been forced to use commercially available chips, even if they do not

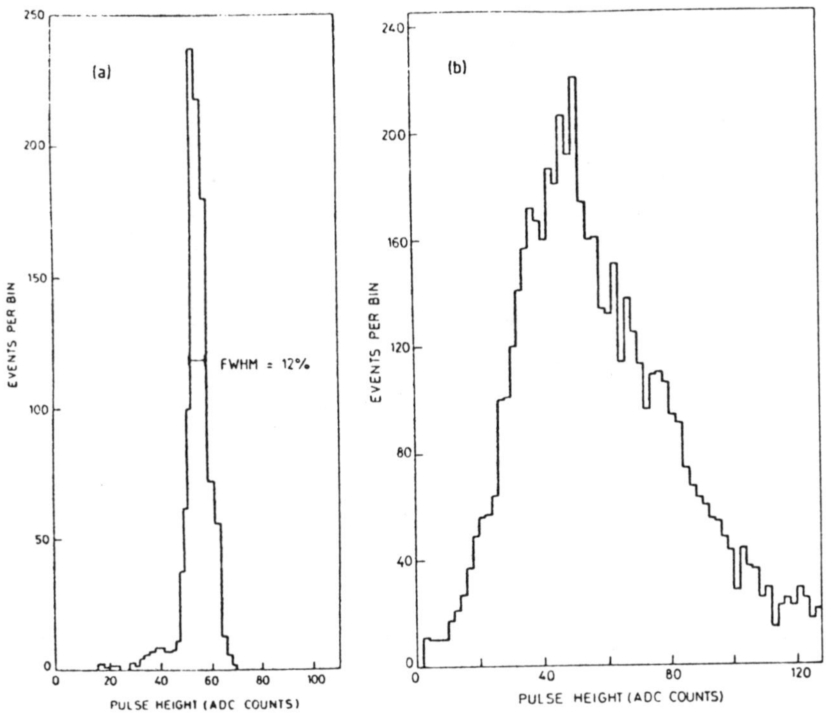

Figure 56 Energy resolution of a CCD. (*a*) Pulses from 5.9-keV x rays. (*b*) Energy loss of a minimum ionizing particle (111).

perfectly fit their needs. The time is ripe for the elaboration of specific systems.

4.6 *Detection by Charge Storage in Shallow Levels in Semiconductors*

In 1978 it was observed in so-called P–I–N diodes, which in our notation is a pnn^+ structure, that charges liberated by ionizing particles can be stored and subsequently released by the application of an electric field (Figure 57) [Shepard, in (101)]. This was interpreted as being due to the capture of the carriers by the ionized impurities. The storage charge could be released by applying an electric field of a few V/μm. Figure 58 illustrates the process. The *I* region is silicon-doped with an n-type impurity. The trapped charges are tunnelled out from their bound levels by the applied electrical field (Figure 58*b*). It was established that at 4.2 K the trapping lifetime is $\gg 10^5$ s.

Under the influence of an electric field at 2 V/μm the charge is extracted in

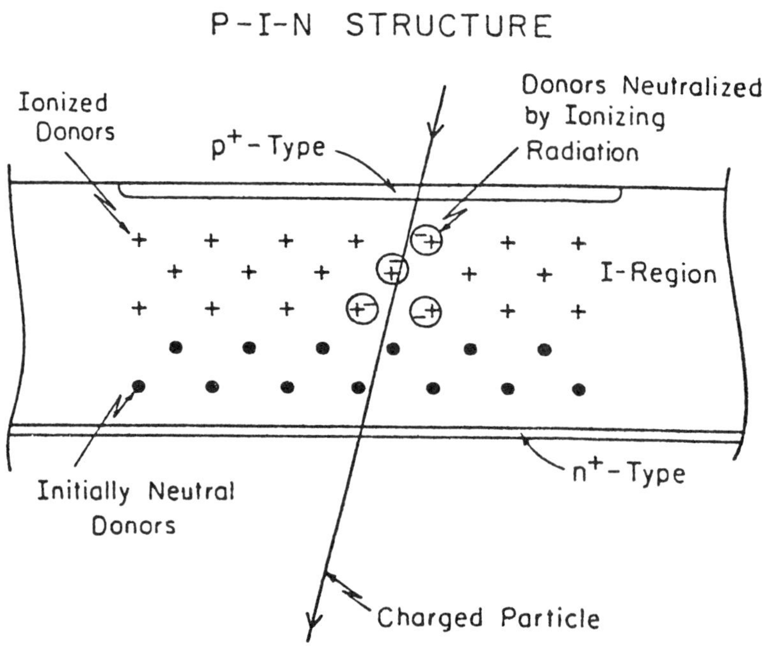

Figure 57 Schematic view of P–I–N structure suitable for cryogenic charge storage produced by ionizing radiation. [From Shepard, in (101) page 73.]

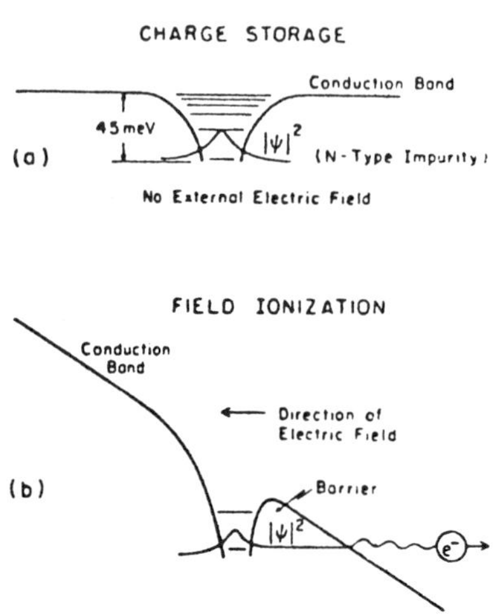

Figure 58 Schematic view of the potential energy levels and a ground-state wave function near a donor impurity site in (*a*) the absence of an external electric field and (*b*) the presence of an electric field of sufficient strength to produce field ionization by quantum mechanical tunnelling. [From Shepard, in (101) page 73.]

times of about 10 ps from phosphorus impurities in silicon. The tests were performed in commercial diodes with a thickness of 50 μm for the *I* region doped with phosphorus at a concentration of $\sim 10^{13}/\text{cm}^3$. The authors estimate that the capture occurs within a characteristic distance of the primary ionizing particle of only 1 μm, which sets the ultimate accuracy limit.

The preliminary test permitted successfully trapping and releasing of ionization electrons, the best results being obtained with a diode of 5×10^{-4} cm^{-2}. The authors extrapolate these observations into the conception of structures capable of exploiting the intrinsic properties into the few μm range of accuracies. The read-out relies on a strip structure, of 10 μm width on the n^+ side, in a diode, of a total width 0.5 mm. Each strip surface is only 10 times smaller than the tested commercial diode, and the calculations show that the accuracy should be about 3 μm in one dimension, with a perfect separation between signals from minimum ionizing particles and from the noise.

4.7 *The Silicon Drift Chamber*

The silicon drift chamber, introduced by Gatti & Rehak (114), relies on the idea that a flat pnp structure can be depleted from the side over a long distance (Figure 59*a*).

Until complete depletion is reached, the undepleted region around the median plane is a conductor at the potential of the side electrode. Thus the whole volume of the wafer is depleted at the voltage that would be required to deplete a wafer of half this thickness constructed as a conventional planar diode.

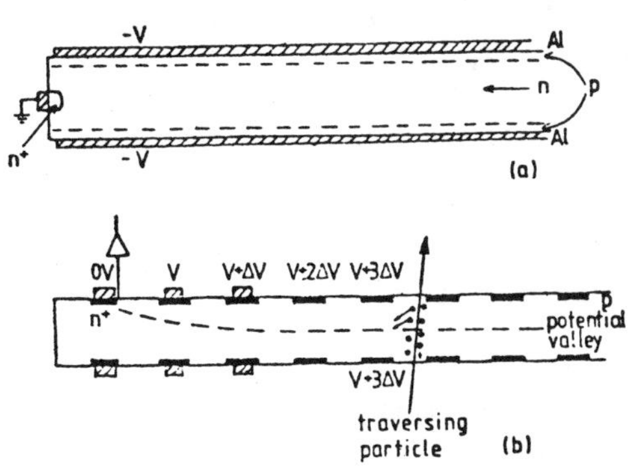

Figure 59 The silicon drift chamber (114). (*a*) Principle. The whole volume is depleted except the median plane, which acts as a conductor at the potential of the side electrode. (*b*) Practical scheme with a sidewise superimposed drift field.

The positive charges left after free electrons are removed by depletion give rise to a potential minimum in the median plane. Figure 59*b* shows how one may construct a practical drift chamber by introducing strip electrodes to give a transverse field. Electrons liberated by a charged track will drop to the potential minimum and then drift to the positive electrode; collection times in the range 100 ns to 10 μs can be envisaged. Preliminary tests of this device have been reported; it can aim at the same accuracy as the microstrip detectors, with much simpler electronics and mechanics.

5. CONCLUSION

We have reviewed the efforts being undertaken to improve the accuracy recently achieved in tracking high-energy particles. The great flexibility and relative ease of construction of gaseous detectors has encouraged a sizeable effort in this field. It seems, however, that in practical cases it will be hard to achieve accuracies better than 30–40 μm. This is compensated partly by the fact that multisampling along a track is easy.

Solid-state detectors offer considerable potentialities and two satisfactory structures emerge: (*a*) the multistrip detectors, which have demonstrated a one-dimensional accuracy of 5 μm (rms) with 120 μm multitrack resolution; and (*b*) the CCDs, which offer an accuracy of 5 μm, with a multiparticle resolution of 40 μm. The first device permits high counting rates and has been tested up to 10^6 counts/s on a single strip. The second has considerable limitations connected with its read-out system but has been used successfully up to about 10^5 counts/s $\cdot$ cm^2. These two devices can still be improved. The other approaches that we mentioned illustrate the considerable potentialities of solid-state electronics and may well produce new practical devices.

ACKNOWLEDGMENT

The authors are grateful to Drs. C. Damerell, B. D. Hyams, and P. G. Rancoita, for illuminating comments on the last section of this article.

Literature Cited

1. Lapique, F., Piuz, F. 1980. *Nucl. Instrum. Methods* 175:247
2. Franzen, W., Cochran, L. W. 1971. In *Nuclear Instruments and Their Use*, ed. A. H. Snell, p. 1. New York: Wiley
3. Christophorou, L. G. 1971. *Atomic and Molecular Radiation Physics*. New York: Wiley
4. Kanter, H. 1961. *Phys. Rev.* 121:461
5. Talman, R. 1979. *Nucl. Instrum. Methods* 159:198
6. Allison, W. W. M., Cobb, J. H. 1980. *Ann. Rev. Nucl. Part. Sci.* 30:253
7. Hancock, S., James, F., Movchet, J., Rancoita, P. G., Van Rossum, L. 1983. *Phys. Rev. A* 28:615

8. McDaniel, E. W., Mason, E. A. 1973. *The Mobility and Diffusion of Ions in Gases*. New York: Wiley
9. Sauli, F. 1983. In *Techniques and Concepts of High Energy Physics*, ed. T. Ferbel, Vol. 11, p. 301. New York: Plenum
10. Brown, S. C. 1959. *Basic Data of Plasma Physics*. New York: Wiley
11. Brown, S. C. 1966. *Basic Data of Plasma Physics*. Cambridge, Mass: MIT Press
12. Varney, R. N., Fisher, L. H. 1968. In *Methods of Experimental Physics*, ed. L. Marton. New York: Academic
13. Huxley, L. G., Crompton, R. W. 1974. *The Diffusion and Drift of Electronics in Gases*. New York: Wiley
14. Fehlmann, J., Viertel, G. 1983. *Compilation of Data for Drift Chamber Operation*. Zurich: ETH
15. Loeb, L. B. 1961. *Basic Processes of Gaseous Electronics*. Berkeley: Univ. Calif. Press
16. McDaniel, E. W. 1964. *Collision Phenomena in Ionized Gases*. New York: Wiley
17. Massey, H. S. W., Burhop, E. H. S. 1969. *Electronic and Ionic Impact Phenomena*. Oxford: The Univ. Press
18. Palladino, V., Sadoulet, B. 1975. *Nucl. Instrum. Methods* 128:323
19. Schultz, G., Gresser, J. 1978. *Nucl. Instrum. Methods* 151:413
20. Mathieson, E., el Hakeem, N. 1979. *Nucl. Instrum. Methods* 158:489
21. Ramanantsizehena, P. R. 1979. Thesis and Rep. *No. CRN-HE 79-13*. Strasbourg: Univ. Louis-Pasteur
22. Villa, F. 1983. *Nucl. Instrum. Methods* 217:273
23. Parker, J. H., Lowke, J. J. 1969. *Phys. Rev.* 181:302
24. Drumm, H., Ganz, B., Heintze, J., Heinzelmann, G., Heuer, R. D., et al. 1980. *Nucl. Instrum. Methods* 176:333
25. Piuz, F. 1982. *Nucl. Instrum. Methods* 205:425
26. Farilla, A., Sauli, F., Ropelewski, L. 1983. Search for a low drift velocity, low diffusion gas. Unpublished
27. Bobkov, S., Cherniatin, V., Dolgoshein, B., Evgrafov, G., Kalinovsky, A., et al. 1983. Preprint CERN-EP/83-81. Submitted to *Nucl. Instrum. Methods*
28. Charpak, G., Bouclier, R., Bressani, T., Favier, J., Zupančič, Č. 1968. *Nucl. Instrum. Methods* 62:235
29. Charpak, G. 1970. *Ann. Rev. Nucl. Sci.* 20:195
30. Rice-Evans, P. 1974. *Spark, Streamer, Proportional, and Drift Chambers*. London: Richelieu
31. Sauli, F. 1977. *Rep. CERN 77-09*. Geneva: CERN
32. Charpak, G., Sauli, F. 1979. *Nucl. Instrum. Methods* 162:405
33. Fabjan, C. W., Fischer, H. G. 1980. *Rep. Prog. Phys.* 43:1003
34. Walenta, A. H. 1983. *Nucl. Instrum. Methods* 217:65
35. Charpak, G., Rahm, D., Steiner, H. 1970. *Nucl. Instrum. Methods* 80:13
36. Piuz, F., Roosen, R., Timmermans, J. 1982. *Nucl. Instrum. Methods* 196:451
37. Fischer, H. G., Pech, J. 1972. *Nucl. Instrum. Methods* 100:515
38. Lee, D. E., Sobottka, S. E., Thiessen, H. A. 1972. *Nucl. Instrum. Methods* 104:179
39. Endo, I., Kawamoto, Y., Mizuno, Y., Ohsugi, T., Taniguchi, T., Takeshita, T. 1981. *Nucl. Instrum. Methods* 188:51
40. Gatti, E., Longoni, A., Okuno, H., Semenza, P. 1979. *Nucl. Instrum. Methods* 163:83
41. Fancher, D. L., Schaffer, A. C. 1979. *IEEE Trans. Nucl. Sci. NS*-26:150
42. Breskin, A., Charpak, G., Demierre, C., Majewski, S., Policarpo, A., et al. 1977. *Nucl. Instrum. Methods* 143:29
43. Charpak, G., Melchart, G., Petersen, G., Sauli, F. 1979. *Nucl. Instrum. Methods* 167:455
44. Charpak, G., Petersen, G., Policarpo, A., Sauli, F. 1978. *IEEE Trans. Nucl. Sci. NS*-25:122
45. Charpak, G., Petersen, G., Policarpo, A., Sauli, F. 1978. *Nucl. Instrum. Methods* 148:471
46. Bondar, A. E., Onuchin, A. P., Penin, V. S., Telnov, V. I. 1983. *Nucl. Instrum. Methods* 207:379
47. Charpak, G., Sauli, F. 1973. *Nucl. Instrum. Methods* 113:381
48. Fisher, J., Okuno, H., Walenta, A. H. 1978. *Nucl. Instrum. Methods* 151:451
49. Okuno, K., Fisher, J., Radeka, V., Walenta, A. H. 1979. *IEEE Trans. Nucl. Sci. NS*-26:160
50. Harris, T. J., Mathieson, E. 1978. *Nucl. Instrum. Methods* 154:183
51. Mathieson, E., Harris, T. J. 1979. *Nucl. Instrum. Methods* 159:483
52. Kochelev, N. I., Telnov, V. I. 1978. *Nucl. Instrum. Methods* 154:407
53. Erskine, G. A. 1982. *Nucl. Instrum. Methods* 198:325
54. Saudinos, J. 1970. *Proc. Topical Seminar on Interactions of Elementary Particles on Nuclei*, p. 313. Trieste: INFN
55. Saudinos, J., Duchazeaubeneix, J. C., Laspalles, C., Chaminade, R. 1973. *Nucl. Instrum. Methods* 111:77
56. Walenta, A. H., Heintze, J., Schürlein, B. 1971. *Nucl. Instrum. Methods* 92:373

57. Allison, W. W. M., Brooks, C. B., Bunch, J. N., Cobb, J. H., Lloyd, J. L., Pleming, R. W. 1974. *Nucl. Instrum. Methods* 119:499
58. Clark, A. R., et al. 1976. Proposal for a PEP facility based on the Time Projection Chamber, PEP-4. Stanford, Calif: SLAC
59. Nygren, D., Marx, J. 1978. *Phys. Today* 31, No. 10
60. Charpak, G., Sauli, F., Duinker, W. 1973. *Nucl. Instrum. Methods* 108:413
61. Breskin, A., Charpak, G., Gabioud, B., Sauli, F., Trautner, N., et al. 1974. *Nucl. Instrum. Methods* 119:9
62. Breskin, A., Charpak, G., Sauli, F., Atkinson, M., Schultz, G. 1975. *Nucl. Instrum. Methods* 124:189
63. Filatova, N., Nigmanov, T., Pugachevich, V., Riabtsov, V., Shafranov, M., et al. 1977. *Nucl. Instrum. Methods* 143:17
64. Farr, W., Heintze, J., Hellenbrand, K. H., Walenta, A. H. 1978. *Nucl. Instrum. Methods* 156:283
65. Baskakov, V. Y., Chernjatin, V. K., Dolgoshein, B. A., Lebedenko, V. N., Romanjuk, A. S., et al. 1979. *Nucl. Instrum. Methods* 158:129
66. Charpak, G., Majewski, S., Sauli, F. 1975. *Nucl. Instrum. Methods* 126:381
67. Mathis, K. D., Simon, M., Henkel, M. 1983. *Siegen Univ. Preprint SI-83-19*
68. Anderson, D., Charpak, G. 1982. *Nucl. Instrum. Methods* 201:527
69. Chernyatin, V. K., Dolgoshein, B. A., Golutvin, I. A., Kaftanov, V. S., Kalinovskii, A. N., et al. 1979. *Proc. Second ICFA Workshop, Les Diablerets, 1979*, p. 320. Geneva: CERN
70. Barranco Luque, M., Calvetti, M., Dumps, L., Girard, C., Hoffmann, H., et al. 1980. *Nucl. Instrum. Methods* 176:175
71. Va'vra, J. 1983. *Nucl. Instrum. Methods* 217:322
72. Va'vra, J. 1983. *Stanford preprint SLAC-PUB-3131* (Presented at the 2nd Pisa Meeting on Advanced Detectors, Castiglione, 1983). Stanford, Calif: SLAC
73. Bettoni, D., Dolgoshein, B., Evans, M., Fabjan, C. W., Hoffman, H., et al. 1983. In *Proc. Int. Conf. on High Energy Physics, Brighton, 1983*, p. 424
74. Walenta, A. H. 1979. *IEEE Trans. Nucl. Sci. NS*-26:73
75. Vertex chamber group (Aachen-Siegen-Zurich), LEP experiment L3, 1983 (private communication from G. Viertel)
76. Raether, H. 1964. *Electron Avalanches and Breakdown in Gases*. London: Butterworth
77. Peisert, A. 1983. *Nucl. Instrum. Methods* 217:229
78. Hilke, H. 1983. *Nucl. Instrum. Methods* 217:189
79. Peisert, A., Charpak, G., Sauli, F., Viezzoli, G. 1984. *IEEE Trans. Nucl. Sci. NS*-31:125
80. Hempel, G., Hopkins, F., Schatz, G. 1975. *Nucl. Instrum. Methods* 131:445
81. Jared, R. C., Glaessel, P., Hunter, J. B., Moretto, L. G. 1978. *Nucl. Instrum. Methods* 150:597
82. Harrach, D. V., Specht, H. J. 1979. *Nucl. Instrum. Methods* 164:477
83. Breskin, A., Zwang, N. 1977. *Nucl. Instrum. Methods* 146:461
84. Van der Plicht, J. 1980. *Nucl. Instrum. Methods* 171:43
85. Fabris, D., Fortuna, G., Gramegna, F., Prete, G., Viesti, G. 1983. *Nucl. Instrum. Methods* 216:167
86. Charpak, G., Sauli, F. 1978. *Phys. Lett.* 78B:523
87. Breskin, A., Charpak, G., Majewski, S., Melchart, G., Peisert, A., et al. 1979. *Nucl. Instrum. Methods* 161:79
88. Bouclier, R., Charpak, G., Cattai, A., Million, G., Peisert, A., et al. 1983. *Nucl. Instrum. Methods* 205:403
89. Adams, M., et al. 1983. *Nucl. Instrum. Methods* 217:237
90. Charpak, G., Sauli, F. 1983. *Preprint CERN-EP/83-128* (Presented at 2nd Pisa Meet. on Adv. Detectors, Castiglione, 1983). Geneva: CERN
91. Derenzo, S. E., Kirschbaum, A. R., Eberhard, P. H., Ross, R. R., Solmitz, F. T. 1974. *Nucl. Instrum. Methods* 122:319
92. Willis, W. J., Radeka, V. 1974. *Nucl. Instrum. Methods* 120:221
93. Karlovac, N., Mayhugh, T. L. 1977. *IEEE Trans. Nucl. Sci. NS*-24:327
94. Gatti, E., Hrisoho, A., Manfredi, P. F. 1983. *IEEE Trans. Nucl. Sci. NS*-30:319
95. Deithers, K., Donat, A., Lanius, K., Leiste, R., Roeser, U., et al. 1981. *Nucl. Instrum. Methods* 180:145
96. Rubbia, C. 1977. *Preprint CERN-EP/77-8*. Geneva: CERN
97. Chen, H. H., Doe, P. J. 1981. *IEEE Trans. Nucl. Sci. NS*-28:454
98. Doe, P. J., Mahler, H. J., Chen, H. H. 1982. *Nucl. Instrum. Methods* 199:639
99. Mahler, H. J., Doe, P. J., Chen, H. H. 1983. *IEEE Trans. Nucl. Sci. NS*-30:86
100. Dolgoshein, B. A., Kruglov, A. A.,

Lebedenko, V. N., Miroshnichenko, V. P., Rodionov, B. U. 1973. *Sov. J. Part. Nucl.* 4:70

101. Ferbel, T., ed. 1982. *Proc. Fermilab Workshop on Silicon Detectors for High-Energy Physics, Batavia, 1981*. Batavia: FNAL
102. Stefanini, A. 1983. *Miniaturization of High-Energy Physics Detectors*, Ettore Majorana International Science Series. New York: Plenum
103. Rancoita, P. G., Seidman, A. 1982. *Riv. Nuovo Cimento* 5: No. 7
104. Heijne, H. M. 1983. *Rep. CERN 83-06*. Geneva: CERN
105. Hyams, B., Kötz, U., Belau, E., Klanner, R., Lutz, G., et al. 1983. *Nucl. Instrum. Methods* 205:99
106. Belau, E., Klanner, R., Lutz, G., Neugebauer, E., Seebrunner, H. J., et al. 1983. *Nucl. Instrum. Methods* 214:253
107. Hyams, B., Kötz, U. 1983. *Nucl. Instrum. Methods* 205:9
108. Hyams, B., private communication
109. Rancoita, P. G. 1984. *J. Phys. G* 10: No. 3, 299
110. Borgeaud, P., McEwen, J. G., Rancoita, P. G., Seidman, A. 1983. *Nucl. Instrum. Methods* 211:363
111. Damerell, C. J. S., Farley, F. J. M., Gillman, A. R., Wickens, F. J. 1981. *Nucl. Instrum. Methods* 185:33
112. Bailey, R., Damerell, C. J. S., English, R. L., Gillman, A. R., Lintern, A. L., et al. 1982. *Nucl. Instrum. Methods* 213:201
113. Amelio, G. F. 1974. *Sci. Am.* 230:22
114. Gatti, E., Rehak, P. 1983. *Brookhaven preprint BNL 33523* (Presented at 2nd Pisa Meeting on Advanced Detectors, Castiglione, 1983). Brookhaven, NY: BNL

Reprinted, with permission, from *Techniques and Concepts of High Energy Physics-III*, T. Ferbel, ed. (Plenum Pub. Corp., 1985).

CALORIMETRY IN HIGH-ENERGY PHYSICS

by C. Fabjan

1. INTRODUCTION

Much of our present knowledge about elementary particles has been established through a continuing refinement of techniques for measuring the trajectories of individual charged particles. Only in recent years has a different class of detectors—calorimeters—been widely employed, but these have already greatly influenced the scope of experiments.

Conceptually, a calorimeter is a block of matter which intercepts the primary particle, and is of sufficient thickness to cause it to interact and deposit all its energy inside the detector volume in the subsequent cascade or 'shower' of increasingly lower-energy particles. Eventually, most of the incident energy is dissipated and appears in the form of heat. Some (usually a very small) fraction of the deposited energy is detectable in the form of a more practical signal (e.g. scintillation light, Cherenkov light, or ionization charge), which is proportional to the initial energy.

The first large-scale detectors of this type were used in cosmic-ray studies [1]. Interest in calorimeters grew during the late 1960's and early 1970's in view of the new accelerators at CERN [the Intersecting Storage Rings (ISR) and the Super Proton Synchrotron (SPS)] and at the Fermi National Accelerator Laboratory (FNAL), with their greatly changed experimental directions and requirements [2]. One consequence of the new fixed-target accelerators was the advent of intense, high-energy neutrino beams with the need for very massive detectors to study their interactions. This detector development was paralleled by the rapid growth of analog signal-processing techniques: during the last decade the typical number of analog signal-channels of nuclear spectroscopy quality has increased from about 10 to 10^4 in high-energy physics experiments.

Calorimeters offer many attractive capabilities, supplementing or replacing information obtained with magnetic spectrometers:

1) They are sensitive to charged and neutral particles.
2) The 'energy degradation' through the development of the particle cascade is a statistical process, and the average number $\langle N \rangle$ of secondary particles is proportional to the energy of the incident particle. In principle, the uncertainty in the energy measurement is governed by statistical fluctuations of N, and hence the relative energy resolution σ/E improves as $1/\sqrt{\langle N \rangle} \sim E^{1/2}$.
3) The length of the detector scales logarithmically with particle energy E, whereas for magnetic spectrometers the size scales with momentum p as $p^{1/2}$, for a given relative momentum resolution $\Delta p/p$.
4) With segmented detectors, information on the shower development allows precise measurements of the position and angle of the incident particle.
5) Their different response to electrons, muons, and hadrons can be exploited for particle identification.
6) Their fast time response allows operation at high particle rates, and the pattern of energy deposition can be used for rapid on-line event selection.

In these notes we comment first on the principal features of detectors designed to measure the energy of photons and electrons, the 'electromagnetic shower detectors' (ESD). The underlying physics has been understood for many years, and such detectors were the main components in many experiments—some of which were credited with important discoveries. Recent developments have been emphasizing precision measurements of energy and position in large arrays.

In the subsequent section the physics of 'hadronic calorimeters' is reviewed. Progress during the last decade contributed to an understanding of the physics of this technique and to a steadily growing range of applications.

The final section concentrates on the technical issues of information processing from calorimeters. We can only select representative examples from the numerous and ingenious methods devised to extract the energy information. We end with a discussion on the state-of-the-art Monte Carlo simulation of electromagnetic and hadronic showers.

These notes follow an earlier review [3], emphasize recent developments, update the bibliography, but do not supersede other excellent introductions to this field [4, 5].

2. ELECTROMAGNETIC SHOWER DETECTORS

2.1 Energy Loss Mechanism

The contributions of the various energy loss mechanisms as a function of particle energy are given in Fig. 1 for electrons and positrons and in Fig. 2 for photons [6]. Above approximately 1 GeV, the principal processes—bremsstrahlung for electrons and positrons, pair production for photons—become energy independent. It is through a succession of these energy loss mechanisms that the electromagnetic cascade (EMC) is propagated, until the energy of the charged secondaries has been degraded to the regime dominated by ionization loss. Within this description, the combined energy loss of the cascade particles in the detector equals the energy of the incident electron or photon. The measurable signal—excitation or ionization of the medium—can be considered as the

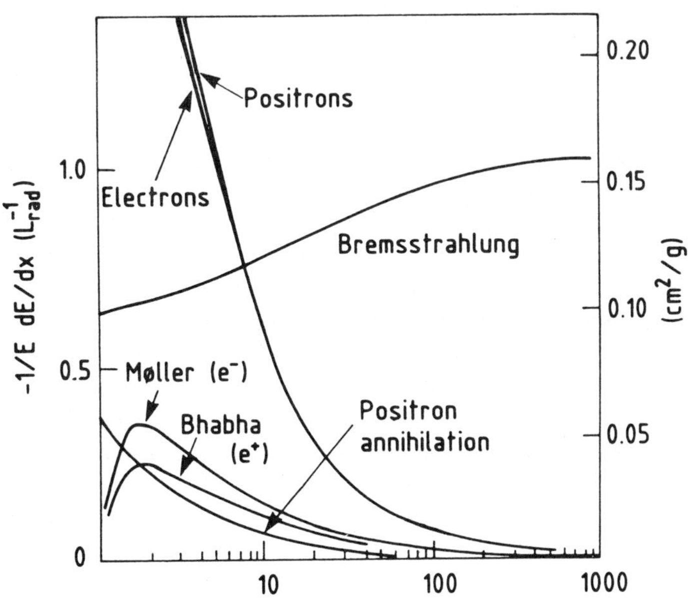

Fig. 1: Fractional energy loss per radiation length (left ordinate) and per g/cm^2 (right ordinate) in lead as a function of electron or positron energy. (Review of Particle Properties, April 1982 edition).

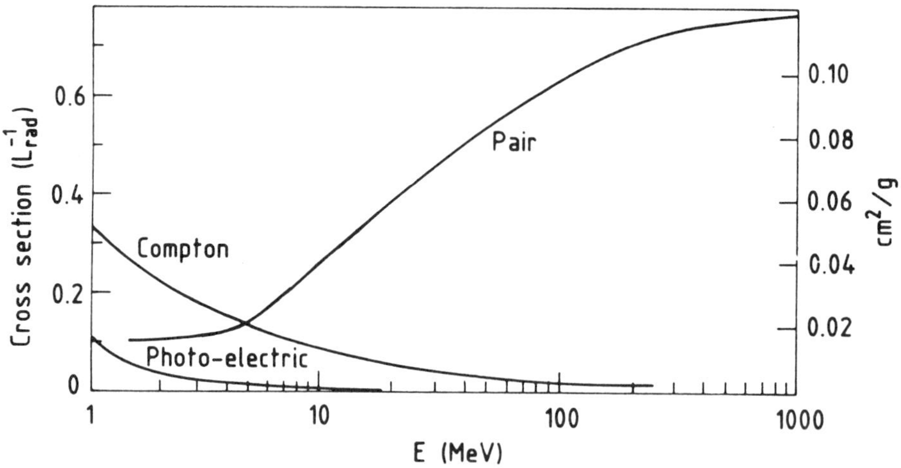

Fig. 2: Photon cross-section σ in lead as a function of photon energy. The intensity of photons can be expressed as $I = I_0 \exp(-\sigma x)$, where x is the path length in radiation lengths. (Review of Particle Properties, April 1980 edition).

sum of the signals from the track segments of the positrons and electrons. Naively one might therefore expect that this signal should be equivalent to that produced by muons traversing the detector and whose combined track length equals that of the track segments of the EMC. This picture ignores finer points concerning low-probability photo-nuclear interactions, non-linear response of the ionizing medium as a function of ionization density or the detailed response to the very low energy e^+'s or e^-'s of the last generation of the shower. It emphasizes, however, the concept of 'track length T': a calorimeter is a useful device, because in the process of cascade formation the total track length T required for absorption is broken up into a 'tree' of many individual segments.

The electromagnetic cascading is fully described by quantum electrodynamics (QED) [7], and depends essentially on the density of electrons in the absorber medium. For this reason it is possible to describe the characteristic longitudinal dimensions of the high-energy EMC (E > 1 GeV) in a material-independent way, using the 'radiation length X_0'. The energy loss ΔE *by radiation* in length Δx can then be written

$$(\Delta E)_{\text{radiation}} = -E\,(\Delta x / X_0)$$

and the numerical value is well approximated by the following expression:

$$X_0\,[\text{g/cm}^2] \simeq 180\,A/Z^2 \text{ (to better than 20\% for } \simeq Z > 13).$$

Whilst the high-energy part of the EMC is governed by the value of X_0, the low-energy tail of the shower is characterized by the 'critical energy ϵ' of the medium. It is defined as the energy loss *by collisions* of electrons or positrons of energy ϵ in the medium in one radiation length, i.e.

$$(dE)_{collision} = -\epsilon(dx/X_0), \qquad \text{where } \epsilon\ (\text{MeV}) \simeq 550 \times Z^{-1}$$

(accurate to better than 10% for $Z > 13$). This value of ϵ coincides approximately with that value of the electron energy below which the ionization energy loss starts to dominate the energy loss by bremsstrahlung. The critical energy ϵ is seen to define the dividing line between shower multiplication and the subsequent dissipation of the shower energy through excitation and ionization.

A rigorous, analytical description of the longitudinal shower profile has been given by Rossi [8] based on the following assumptions ('Rossi's approximation B'), and the most useful results are given in Table 1:

i) the cross-section for ionization is energy independent, i.e. $dE/dx = -\epsilon/X_0$;
ii) multiple scattering is neglected and the EMC is treated one-dimensionally;
iii) Compton scattering is neglected.

The characteristic longitudinal EMC profile is shown in Fig. 3 for four very different materials and demonstrates the 'longitudinal scaling in radiation length'. A convenient analytical description of the profile has been given in the form [9]

$$dE/dt = E_0\, b^{\alpha+1}/\Gamma(\alpha+1) t^{\alpha} e^{-bt}\ ;\ t = x/X_0\ ,\ \alpha = bt_{max}, \text{ and } b \simeq 0.5\ .$$

The transverse shower properties, which are not described within the framework of Rossi's 'approximation B' can also be easily understood qualitatively. In the early, most

Table 1: EMC Quantities Evaluated with Rossi's Approximation B
($y = E/\epsilon$; T measured in units of X_0)

	Incident electron	Incident photons
Peak of shower, t_{max}	$1.0 \times (\ln y - 1)$	$1.0 \times (\ln y - 0.5)$
Centre of gravity, t_{med}	$t_{max} + 1.4$	$t_{max} + 1.7$
Number e^+ and e^- at peak	$0.3\, y \times (\ln y - 0.37)^{-1/2}$	$0.3\, y \times (\ln y - 0.31)^{-1/2}$
Total track length T	y	y

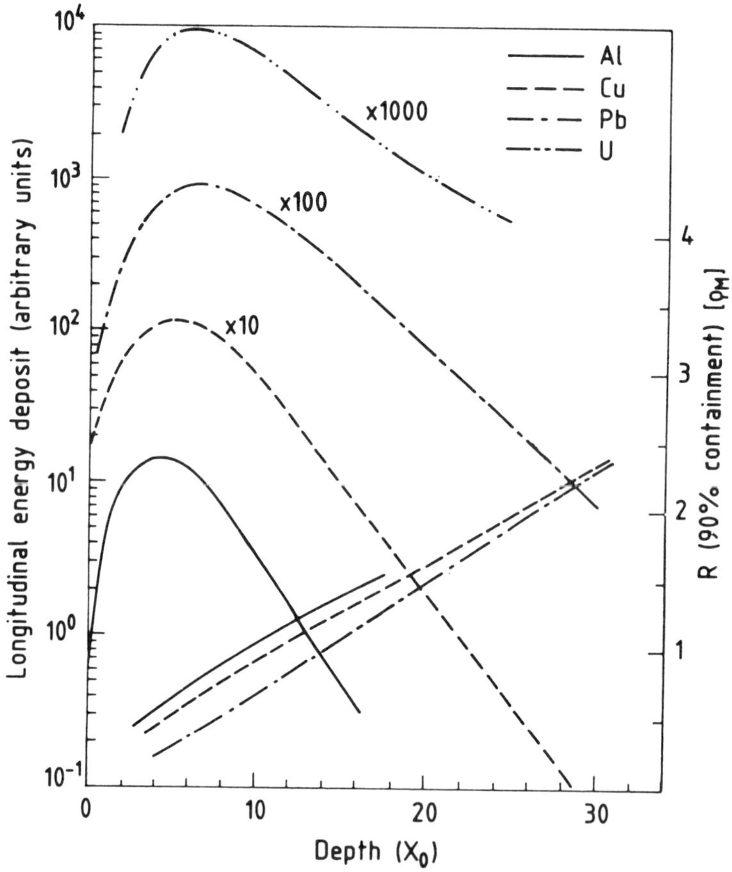

Fig. 3: Longitudinal shower development (left ordinate) of 6 GeV/c electrons in four very different materials, showing the scaling in units of radiation lengths X_0. On the right ordinate the shower radius for 90% containment of the shower is given as a function of the shower depth. In the later development of the cascade, the radial shower dimensions scale with the Molière radius $\varrho_M \sim 7A/Z$. [Al, Cu, and Pb, adapted from G. Bathow et al., Nucl. Phys. B20:592 (1970). Uranium data from G. Barbiellini et al., Ref. [127].

energetic part of the cascade the lateral spread is characterized by both the typical angle for bremsstrahlung emission, $\theta_{\text{brems}} \sim p_e/m_e$, and multiple scattering in the absorber. This latter process increasingly influences the lateral spread with decreasing energy of the shower particles and causes a gradual widening of the shower. For the purpose of total energy measurement, the EMC occupies a cylinder of radius R

$$R \approx 2\varrho_M ; \qquad \varrho_M = 21X_0/\epsilon \simeq 7A/Z \ [\text{g cm}^{-2}],$$

ϱ_M being the 'Molière Radius', which describes the average lateral deflection of electrons of energy ϵ after traversing one radiation length. In Fig. 4, the transverse shower profile as a function of depth clearly exhibits the rather pronounced central and energetic core surrounded by a low-energy 'halo'.

2.2 Limits on Energy Resolution of EMCs

In the discussion in the previous section we have represented the shower by a total track length T, which could be expressed as $T(X_0) = E_{particle}(\text{MeV})/\epsilon(\text{MeV})$. The 'detectable' track length T_d, i.e. the equivalent track length which corresponds to the

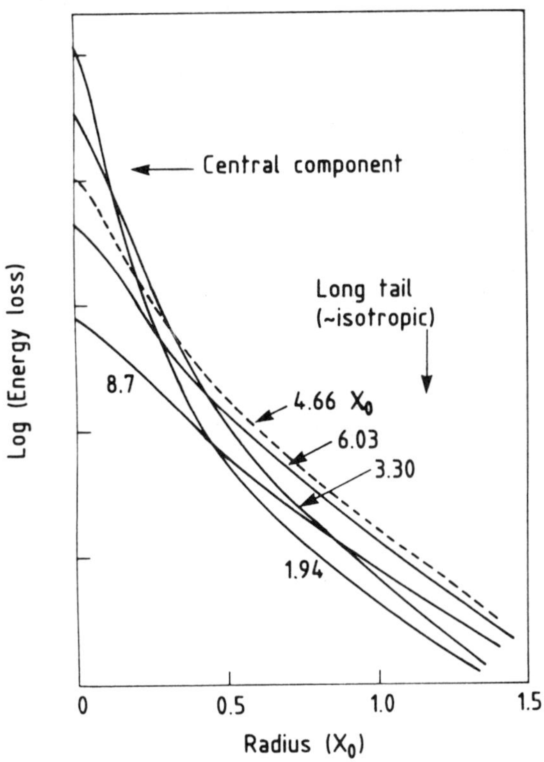

Fig. 4: Radial shower profile of 1 GeV electrons in aluminium; a pronounced central core, surrounded by 'halo', gradually widens with increasing depths of the shower [17].

measured signal in a particular detector, will in general be shorter, $T_d \leq T$, as practical devices are only sensitive to the cascade particles above a certain threshold energy η. The fractional reduction in visible track length as a function of η/ϵ [4] is indicated in Fig. 5. The dotted lines are the result of an analytic calculation [8] for $E \gg \eta$, and $F(\xi)$ is given by $F(\xi) = [1+\xi \ln(\xi/1.53)] \exp \xi$ ($\xi = 2.29\ \eta/\epsilon$). The points are obtained by Monte Carlo calculations [9–11].

The average detectable track length $\langle T_d \rangle$ is given by $\langle T_d \rangle (X_0) = F(\xi) \cdot E/\epsilon$ and calorimetric energy measurements are possible because $\langle T_d \rangle \propto E$ for any reasonable value of ϵ. The *resolution* of the energy measurement is determined by the *fluctuations* in the shower propagation. The intrinsic component of the resolution is caused by the fluctuations in T_d. This represents the lower bound on the energy resolution and may be qualitatively estimated in the following way: the maximum number of track segments $N_{max} = E/\eta$ hence $\sigma(E)/E \geq \sigma(N_{max})/N_{max}$. In a lead-glass shower counter for which $\eta \sim 0.7$ MeV, one estimates for a 1 GeV shower, $N_{max} = 1000/0.7 = 1.5 \times 10^3$, implying an energy resolution at the level of one to two percent, somewhat higher than the level computed by detailed Monte Carlo calculations [9].

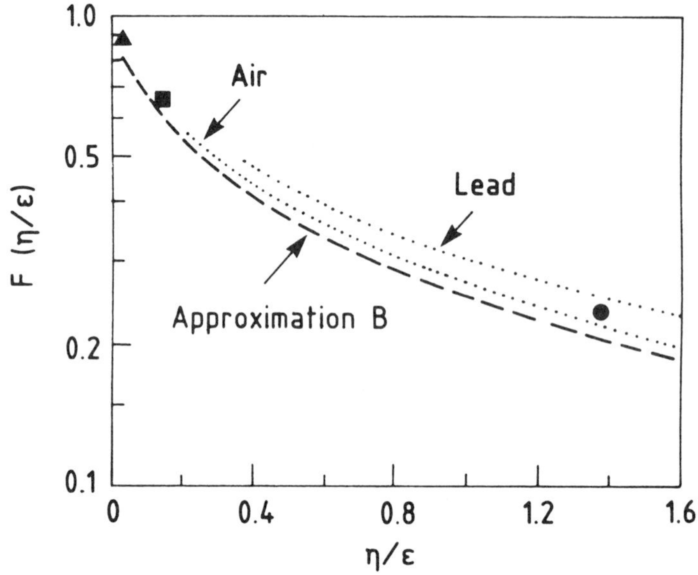

Fig. 5: Fraction F of the total track length which is seen on the average in a fully contained electromagnetic shower. The dotted lines represent analytical calculations and the points represent Monte Carlo results. ▲: lead glass; ■ and ●: lead sampling devices (see Ref. [4]).

In practical detectors, however, usually a number of additional components must be considered, which may conspire to affect the resolution. One important instrumental contribution to the energy resolution comes from incomplete containment of the showers ('energy leakage'), as can be seen from Fig. 6. Available information on average longitudinal containment (both experimental and calculational) may be parametrized as

$$L(98\%)_{av} \simeq t_{max} + 4\,\lambda_{att},$$

where L(98%) gives the length for 98% longitudinal containment. The quantity λ_{att} characterizes the slow, exponential decay of the shower after the shower maximum (see Fig. 3) expressed as $\exp(-t/\lambda_{att})$. The values of λ_{att} are found to be rather energy independent, but material dependent and close in value to the mean free path of photons that have minimum attenuation in a given material. Experimental values cluster around $\lambda_{att}\,[X_0] \simeq 3.4 \pm 0.5X_0$. This estimate is in reasonable agreement with other parametrizations [12], e.g. $L(98\%) \simeq 2.5\,t_{max}$ for E in the 10 to 1000 GeV range. The effect of longitudinal leakage on the energy resolution is consistent with the parametrization

$$\sigma(E)/E \simeq [\sigma(E)/E]_{f=0} \times [1 + 2\sqrt{E(\mathrm{GeV})} \times f]$$

for values f of the fractional energy loss through leakage, $f < 0.2$ and $E < 100$ GeV. One notes that longitudinal leakage is more critical than transverse leakage due to the fact that fluctuations about the average longitudinal loss are much larger than for transverse leakage.

Homogeneous detectors have always played a very important role as e.m. calorimeters. Historically, it was NaI that was used as one of the first calorimetric detectors, still unsurpassed in energy resolution. For energies $E \simeq 1$ GeV, one obtains a value close to the intrinsic limit; at higher energies, the full statistical gain $\sim 1/\sqrt{E}$ is not obtained, even for very carefully tuned instruments [13] for which the resolution is quoted as $\sigma(E)/E \simeq 0.009 \times E^{-1/4}$ (GeV). Such a behaviour is characteristic of contributions other than those governed by statistical processes, such as non-uniformity in the signal collection, energy leakage, etc. A second, very frequently used homogeneous detector is lead glass, i.e. glass loaded with 50–60% PbO. The EMC is sampled by the production of Cherenkov light emitted from the relativistic electron-positron pairs. These detectors are therefore characterized by a relatively low light yield—typically 1000 photoelectrons per GeV are measured—and a relatively large cut-off energy η. These two effects combined give a resolution of

$$\sigma_{tot} = (\sigma_\eta^2 + \sigma_{ph}^2)^{1/2} \simeq (0.020^2 + 0.032^2)^{1/2} = 3.8\% \text{ at } 1\text{ GeV},$$

in agreement with the best values reported, of $\gtrsim 4\%$. Recent developments of new

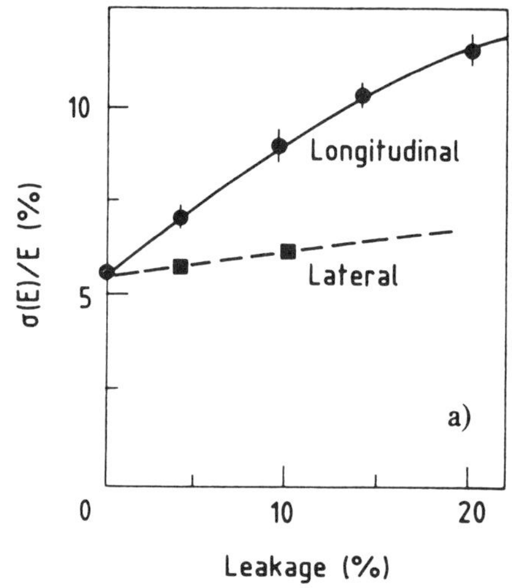

Fig. 6a: Effects of longitudinal and lateral losses on the energy resolution as measured for electrons in the CHARM neutrino calorimeter [23].

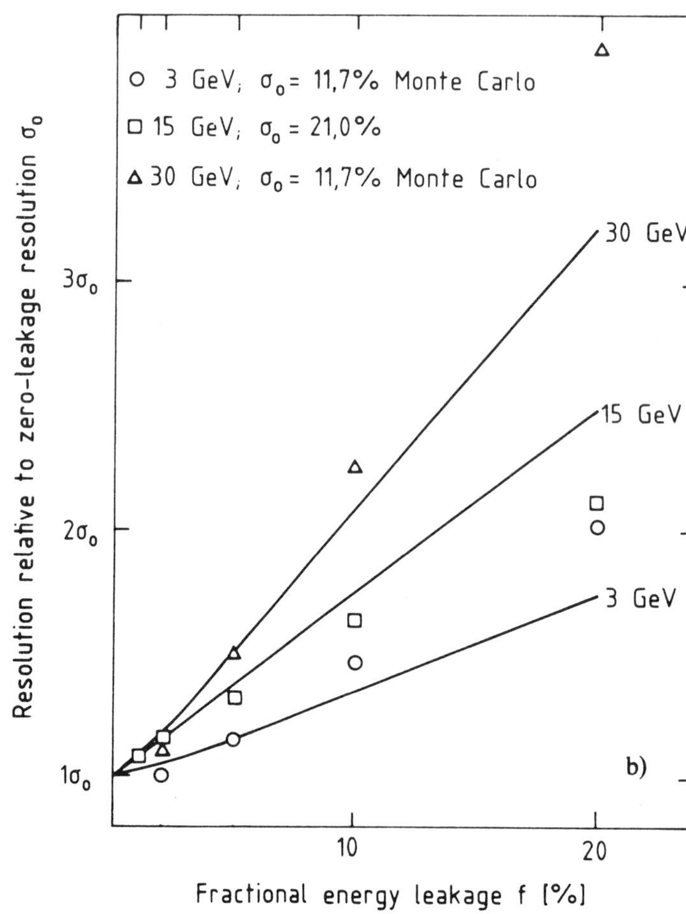

Fig. 6b: Deterioration of the zero-leakage resolution σ_0 as a function of fractional energy leakage f for three different electron energies [21, 23].

scintillating crystals have stimulated interest in such homogeneous detectors, which promise exceedingly good performance in the 1 to 100 GeV regime (see the discussion in Section 5).

2.3 Energy Resolution in 'Sampling' Calorimeters

'Sampling' calorimeters are devices in which the functions of energy degradation and energy measurement are separated in alternating layers of different substances. This allows a considerably greater freedom in the optimization of detectors for certain specific applications. The choice of a 'passive' absorber—typically plates made of Fe, Cu, or Pb, each ranging in thickness from a fraction of X_0 to a few X_0—makes it possible to build rather compact devices, and it permits optimization for a specific experimental requirement such as electron/pion discrimination or position measurement. Independently of the choice of the absorber, the readout method may be selected for best uniformity of signal collection, high spatial subdivision, rate capability or other criteria. The disadvantage is that only a fraction of the total energy of the EMC is 'sampled' in the active planes, resulting in additional *'sampling' fluctuations* of the energy determination.

These general comments apply to both electromagnetic and hadron calorimeters. The following discussion of sampling fluctuations is specifically valid for the measurement with e.m. calorimeters, for which sampling fluctuations have been rather carefully studied. Today we know that they depend on the characteristics of both the passive and the active medium (in particular, thickness and density) and that several effects contribute to the 'total' sampling fluctuation.

The *'intrinsic sampling' fluctuations* express the statistical fluctuations in the number of e^+e^- pairs traversing the active signal planes and can be estimated in the spirit of approximation B. The number N_x of crossings is ($\eta = 0$)

$$N_x = T \text{ (total track length)}/d \text{ (distance between active plates)} ,$$

where $T = E/\epsilon$ and hence $N_x = E/\epsilon d = E/\Delta E$, ΔE being the energy loss per unit cell. The contribution to the energy resolution is

$$\sigma(E)/E_{\text{sampling}} = \sigma(N_x)/N_x = 1/\sqrt{N_x} = 3.2\% \, [\Delta E \text{ (MeV)}/E \text{ (GeV)}]^{1/2} .$$

This expression has to be regarded as a lower bound on the sampling fluctuations for the following reasons:

- tracks originate from pair-produced particles and therefore the number of independent gap crossings would be only $N_x/2$ for totally correlated production;

- approximation B ignores multiple scattering, which increases the effective distance d to $d = d/\langle\cos\theta\rangle$, where the characteristic multiple scattering angle θ is given by $\langle\cos\theta\rangle \sim \cos(21\ \text{MeV}/\epsilon\pi)$ [4];

- for $\eta \neq 0$, $T_d = F(\xi)T$.

Considering these effects, the contribution of sampling fluctuations to the energy resolution is evaluated as

$$[\sigma(E)/E]_{\text{sampling}} \gtrsim 3.2\% \,\{\Delta E\,(\text{MeV})/[F(\xi) \times \cos(21/\epsilon\pi)\,E\,(\text{GeV})]\}^{1/2}.$$

This expression does not include possible additional effects due to *'Landau' fluctuations* of the energy deposit in the active signal planes, which can be estimated to contribute

$$[\sigma(E)/E]_{\text{Landau}} \simeq 3/[\sqrt{N_x} \times \ln(1.3 \times 10^4\,\delta)],$$

where δ (MeV) gives the energy loss per active detector plane. Such additional fluctuations are small for energy losses δ of a few MeV (e.g. a few millimetres of scintillator), but may become comparable to the 'intrinsic' sampling fluctuations for very thin detectors, e.g. gaseous detectors with δ in the keV range. In addition to these 'Landau' fluctuations there is a further source of errors which also depends on the density of the active medium, *'path-length' fluctuation:* low-energy electrons may be multiply scattered into the plane of the active detector and then travel distances much larger than, for example, the gap thickness in gaseous detector planes, depositing considerably more energy compared to that deposited under perpendicular traversal. This effect is quantitatively less significant in dense active layers, because the range of the low energy electrons is comparable to the thickness of these layers. Moreover, increased multiple scattering in dense detector planes will also tend to reduce this effect relative to light absorbers. From Fig. 7 it can be seen that path-length fluctuations may contribute as much as Landau fluctuations to the resolution [14] in detectors with gaseous readouts.

Concluding this section, we compare in Table 2 the measured performance of some characteristic sampling devices, and compare it with the estimated contributions using the formulae given here. The energy resolution is seen to be rather well described by these estimates, provided that instrumental effects (such as calibration errors, photon statistics, leakage, etc.) do not dominate.

It is interesting to note that the path length and the Landau fluctuations are not negligible even in the case of dense active readout gaps, if these are very thin (e.g. measurements with the W/Si calorimeter).

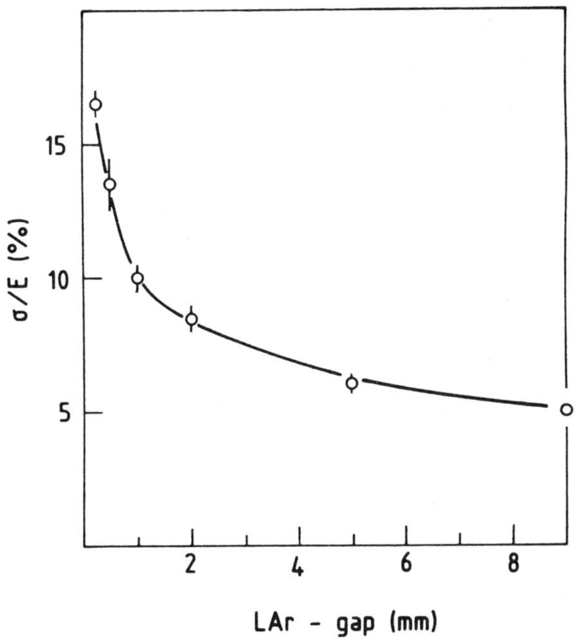

Fig.7a: Energy resolution versus thickness of the active liquid-argon layer for 1 GeV electrons in an iron/argon sampling calorimeter.

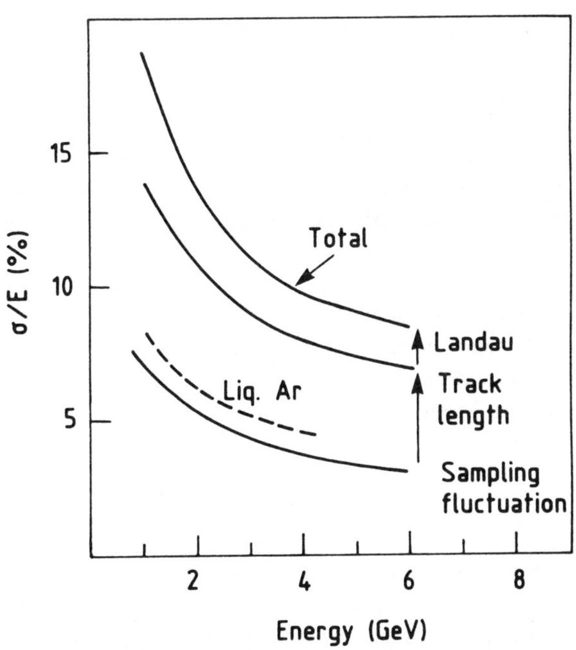

Fig. 7b: Contributions of sampling, path length, and Landau fluctuations to the energy resolution of a lead/MWPC sampling calorimeter. The latter contribute approximately equally (~ 12% at E = 1 GeV), and combined quadratically with the sampling fluctuations (~ 7%) they account for the overall resolution of ~ $18\%/\sqrt{E}$ [14].

Table 2: Measured and Estimated Performance of Electromagnetic Sampling Calorimeters

Device passive/active (mm)	Al/scint. 89/30	Fe/LAr 1.5/2.0	Cu/scint. 5/2.5	W/Si detector 7.0/0.2	$Pb/Ar/CO_2$ at NTP 2.0/10.0	U/scint. 1.6/2.5
Energy resolution measured at 1 GeV(%)	20	7.5	13.0	25.0	$\leqslant$ 20.0	11.0
η (MeV)	3.0	0.7 (?)	0.7 (?)	0.7 (?)	$\leqslant$ 0.6 (?)	0.7 (?)
$F(\xi)^{-1/2}$	1.16	1.10	1.10	1.18	1.18	1.20
$\langle\cos\theta\rangle^{-1/2}$	1.00	1.03	1.03	1.27	1.36	1.51
σ^{sample}	23	4.8	9.2	19.1	8.2	10.6
σ_{Landau}	3.8	1.0	1.0	4.5	8.70	1
$\sigma_{\text{path length}}$		5.7	6	17.5	13.0	6 (?)
$\sigma_{\text{estimated}}$	23	7.5	10.0	25.9	17.7	12.2
Note	a	b, c	c, d, e	c, f	c, g	c, e, h

a) A.N. Diddens et al., Nucl. Instrum. Methods 178:27 (1980).
b) C.W. Fabjan et al., Nucl. Instrum. Methods 141: 61 (1977).
c) Path-length fluctuations estimated from H.G. Fischer, Nucl. Instrum. Methods 156:81 (1978).
d) O. Botner, Phys. Scripta 23:555 (1981).
e) Difference consistent with photon statistics.
f) G. Barbiellini et al., Nucl. Instrum. Methods 235:55 (1983).
g) J.A. Appel, Fermilab FN-380 (1982).
h) R. Carosi et al., Nucl. Instrum. Methods 219:311 (1984).

2.4 'Transition' Effects

The concept of total track length T has been repeatedly used to estimate properties of the EMC. In particular, it suggests equating the measurable signal in a specific calorimeter with the energy deposit of penetrating particles such as muons of equivalent path length, or in terms of the number of equivalent particles n_{ep}. One equivalent particle (1 ep) is defined as the most probable detected energy, $(dE/dx)^{mp}$ of a penetrating muon

$$1\ ep = (dE/dx)^{mp}_{visible,\ \mu}.$$

The expected number n_{ep} for electrons with kinetic energy E^{e}_{kin} would be

$$n^{el}_{ep}(\text{expected}) = E^{el}_{kin}/1\ ep.$$

Experimentally, yet, one always observes:

$$n^{el}_{ep}(\text{visible}) < n^{el}_{ep}(\text{expected}) \quad \text{or} \quad \text{'}e/\mu < 1\text{'}.$$

A summary of some representative measurements is given in Table 3.

The calibration of an EMC with muons is one way of establishing an absolute energy scale. If carried out in a reproducible and consistent way, it would allow us to compare, on an absolute energy scale, the electron response of different calorimeters—a crucial ingredient also in the understanding of hadronic calorimeters (see Section 3). As an example, the energy scale for muons quoted in Table 3 is based on the most probable energy loss evaluated for the total thickness of the device, applying the energy loss formula [15] for the appropriate muon momentum, including the non-negligible relativistic rise. The energy loss in the active medium was estimated to follow the ratio of the respective mass of the passive and active materials. Table 3 shows that the discrepancies from the 'naïve' expectations are substantial, with some indication that the response depends on the sampling thickness and the atomic number of the active and passive materials.

These discrepancies have been repeatedly attributed to 'transition effects' at the boundary between the different layers [16–18], often characterized by very different critical energies and hence different collision losses per radiation length. One expects that at a boundary from high Z to low Z (e.g. Fe to scintillator), the increased collision losses in the low-Z substance will reduce the electron flux, in agreement with measurements [17, 18] and some recent Monte Carlo calculations [19]. Apart from local disturbances of the EMC, multiple scattering tends to increase the effective path length in the high-Z absorber relative to the low-Z readout and this mechanism may also suppress the electron response relative to muons, for increasing Z. Furthermore, a considerable fraction of the energy is deposited by the last generation of the cascade, consisting of low-energy particles, and saturation in the response of readout substances (which occur in scintillator or liquid argon) will further suppress the measured response relative to muons.

Table 3: Average Calorimeter Response for Pions and Muons[*)]
Relative to Electrons [38]

Type of particle (energy)	Sampling calorimeter structure $\frac{\text{scintillator}}{\text{liquid argon}}$		
	with Fe (Cu)	with Pb	with ^{238}U
Electrons (10 GeV/c)	1	1	1
	1	1	1
Pions (10 GeV/c)	0.63 ± 0.03 [a,b]	0.68 ± 0.04 [b]	0.93 ± 0.03 [b,d]
	0.7 [c]	not yet measured	1.0 ± 0.05 [c]
Muons (~ 10 GeV/c)	1.15 [b]	1.26 [b]	1.29 [b,d]
	1.1	1.4 [e,f]	1.65 [g]

*) See text for definition of muon response.

NB: Errors of typically 10% have to be assumed for figures quoted without error.

a) A. Beer et al., Nucl. Instrum. Methods 224:360 (1984).

b) O. Botner, Phys. Scripta 23:555 (1981).

c) C.W. Fabjan et al., Nucl. Instrum. Methods 141:61 (1977).

d) T. Akesson et al., Properties of a fine-sampling uranium-copper scintillator hadron calorimeter, submitted to Nucl. Instrum. Methods (1985).

e) J. Cobb et al., Nucl. Instrum. Methods 158:93 (1979).
A. Lankford, CERN-EP Internal Report 78-3 (1978).
C. Kourkoumelis, CERN Report 77-06 (1977).

f) P. Steffen (NA31 Collaboration, CERN), private communication.

g) C.W. Fabjan and W. Willis, unpublished note on measurements reported in c).

It may be concluded that for a more refined understanding of sampling detectors it will be important to calibrate carefully the electron response on an absolute energy scale with a reproducible standard, as provided e.g. by muons.

2.5 Spatial Resolution

In subsection 2.1 we described in general the physical processes contributing to the shower propagation and its characteristic dimension. Typical angles for bremsstrahlung emission and multiple scattering depend on the energy of the shower particles and hence alter the transverse shower profile as a function of longitudinal depth inside the shower. Before the shower maximum, typically more than 90% of the energy is contained in a cylinder of radius $r \simeq 0.5X_0$, whereas the radius for 90% containment of the total energy is $r \simeq 2\varrho_M$. For the localization of the impact point of a photon it is therefore advantageous to probe the shower in the early part before the shower maximum. In principle, given sufficiently fine-grained instrumental resolution, the localization σ_x of the centre of gravity of the transverse distribution is determined by signal/noise considerations and, therefore, should improve with increasing particle energy E as $\sigma_x \sim E^{-1/2}$, which is confirmed experimentally [20], reaching sub-millimetre accuracy for 100 GeV showers [21]. If position resolution is the principal criterion, one may achieve very high spatial subdivision using multiwire proportional chamber (MWPC) techniques [22] allowing localization at the millimetre level.

Somewhat different criteria apply if both good position and energy resolution are required, e.g. for the determination of the invariant mass of particles such as π^0's, $J/\psi \to e^+e^-$, etc. In this case the centres of gravity of the complete showers have to be determined—frequently with the constraint of minimizing the sharing of energy between the neighbouring showers—; even then, excellent spatial resolutions of the order of 1 mm have been reported [20], e.g. in an array of lead-glass blocks of 35 × 35 mm cross-section.

Given simultaneous information on the transverse and longitudinal shower development, the *direction* of a shower and hence the angle of incidence of the particle may be reconstructed. As an example, for the CHARM neutrino calorimeter [23] an angular resolution of $\sigma(\theta_e)$ (mrad) $= 20/E^{1/2} + 560/E$ (E in GeV) was measured; for a FNAL neutrino calorimeter [24] the following result is quoted:

$$\sigma(\theta_e)\ (\text{mrad}) = 3.5 + 53/E\ (\text{GeV}).$$

3. HADRONIC SHOWER DETECTORS

3.1 General properties

Conceptually, the energy measurement of hadronic showers is analogous to that of EMCs, but the much greater variety and complexity of the hadronic processes propagating the hadronic cascade (HC) complicate the detailed understanding. No simple analytical description of hadronic showers exists, but the elementary processes are well studied.

Typical of hadronic interactions is multiple particle production with transverse momentum $\langle p_T \rangle \simeq 0.35$ GeV/c, for which about half of the available energy is consumed (the inelasticity $K \simeq 0.5$). The remainder of the energy is carried by fast, forward-going (leading) particles. The secondaries are mostly pions and nucleons, and their multiplicity is only weakly energy-dependent. The characteristic stages in the HC development are summarized in Table 4. Two specific features have been identified as the principal physics limitations to the energy resolution of hadronic calorimeters:

i) A considerable part of the secondaries are π^0's, which will propagate electromagnetically without any further nuclear interactions; the average fraction converted into $f_{\pi^0} \simeq 0.1 \ln E$ (GeV), for energies E in the range of a few to several hundred GeV. The size of the π^0 component is largely determined by the production in the first interaction, and event-by-event fluctuations about the average value are, therefore, important.
ii) A sizeable amount of the available energy is converted into excitation or break-up of the nuclei, of which only a fraction will result in detectable ('visible') energy.

The two processes, intimately correlated, may lead, for a given entering hadron, to a very different shower composition, which has a very different detectable response. Together they impose the intrinsic limitation on the performance of hadronic calorimeters.

Table 4 gives some indications of the relative importance of these competing processes. Considerable insight has been gained from very detailed Monte Carlo calculations, which in their most ambitious form aim to simulate the full nuclear and particle physics aspects of the hadronic cascade based on the measured cross-sections of the elementary processes (see also Section 6) [25]. Examples showing the energy dependence of the principal effects are given in Fig. 8. It should be noted that these various processes contribute in varying degrees to the visible energy of the HC, and that a considerable fraction—such as nuclear binding energy, muons, and neutrinos—will be lost in the form of 'invisible' or undetectable energy.

Table 4: Characteristic Properties of the Hadronic Cascade

Reaction	Properties	Influence on energy resolution	Characteristic time (s)	Characteristic length (g/cm^2)
Hadron production	Multiplicity $\simeq A^{0.1} \ln s$ Inelasticity $\simeq 1/2$	π^0/π^+ ratio Binding energy loss.	10^{-22}	Abs. length $\lambda \simeq 35A^{1/3}$
Nuclear de-excitation	Evaporation energy $\simeq$ 10% Binding energy $\simeq$ 10% Fast neutrons $\simeq$ 40% Fast protons $\simeq$ 40%	Binding energy loss. Poor or different response to n, charged particles, and γ's.	10^{-18}–10^{-13}	Fast neutrons $\lambda_n \simeq 100$ Fast protons $\lambda_p \simeq 20$
Pion and muon decays	Fractional energy of μ's and ν's $\simeq$ 5%	Loss of ν's	10^{-8}–10^{-6}	$\gg \lambda$
Decay of c, b particles produced in multi-TeV cascades	Fractional energy of μ's and ν's at percent level	Loss of ν's. Tails in resolution function.	10^{-12}–10^{-10}	$\ll \lambda$

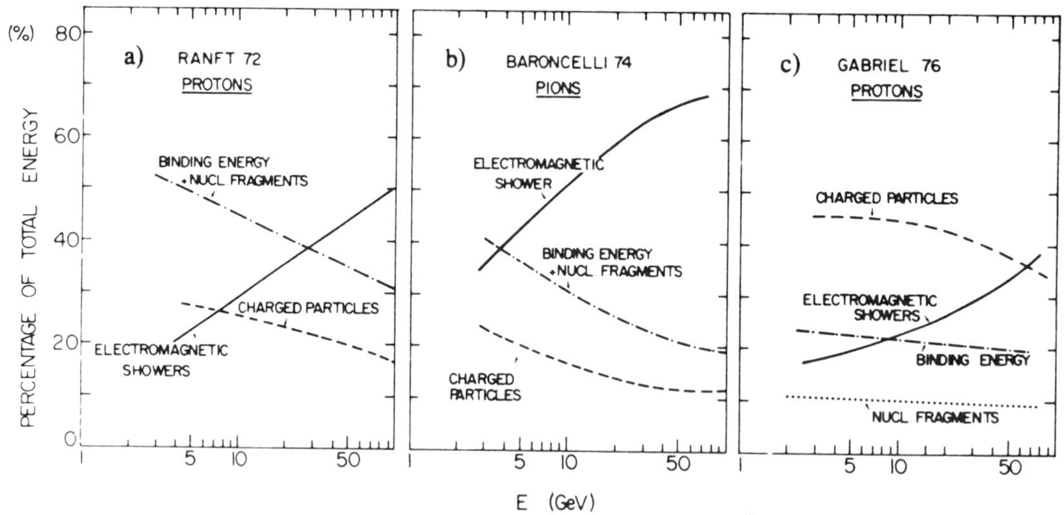

Fig. 8: Relative contributions of the most important processes to the energy dissipated by hadronic showers, as evaluated from three representative Monte Carlo calculations [25].

Average longitudinal and transverse distributions (Fig. 9) are useful estimates of the characteristic dimensions for near-complete shower containment. The average longitudinal distribution exhibits 'scaling in units of absorption length λ'. The transverse distributions depend—as in the case of EMCs—on the longitudinal depth: the core of the shower is rather narrow (FWHM from 0.1 to 0.5λ), increasing with shower depth. The highly energetic, very collimated core is surrounded by lower-energy particles, which extend a considerable distance away from the shower axis, such that for 95% containment a cylinder of radius $R \sim 1\lambda$ is required.

Experimental data are consistent with the following parametrization:

a) the shower maximum, measured from the face of the calorimeter, is given by

$$t_{max}(\lambda) \sim 0.2 \ln E\,(\mathrm{GeV}) + 0.7;$$

it occurs at a smaller depth in high-Z materials due to the smaller ratio of X_0/λ.

b) The longitudinal dimension required for almost full containment is approximated by

$$L_{0.95}(\lambda) \simeq t_{max} + 2.5\,\lambda_{att},$$

again measured from the face of the calorimeter. The quantity λ_{att} describes the exponential decay of the shower beyond t_{max} and increases with energy approximately

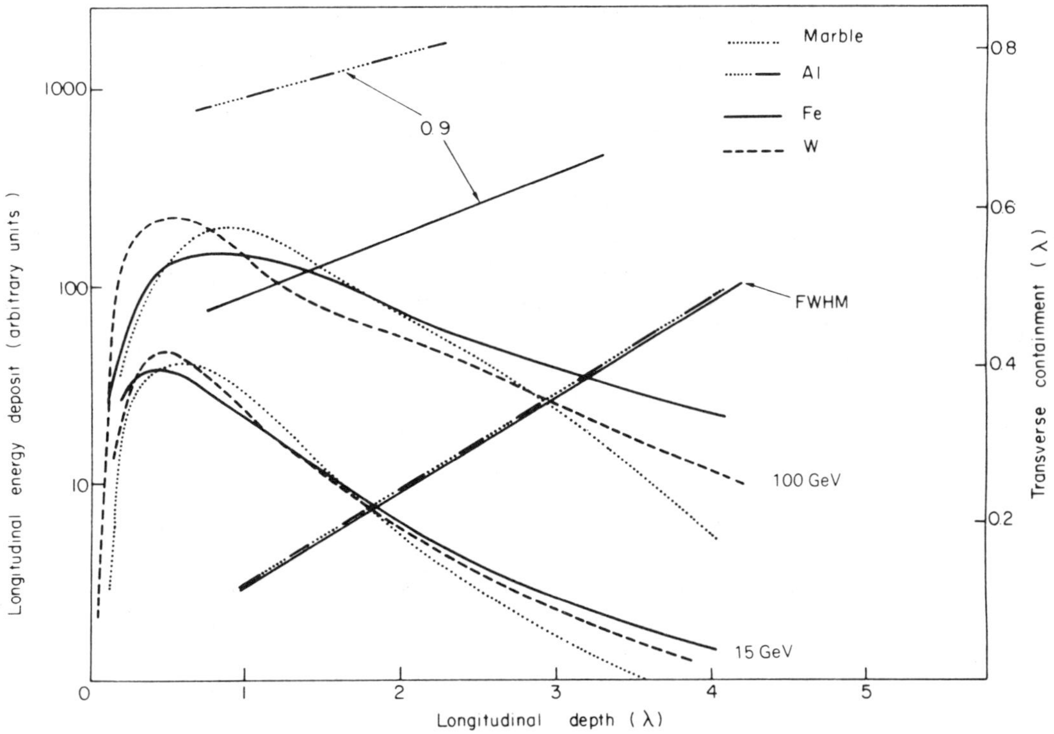

Fig. 9: Longitudinal shower development (left ordinate) induced by hadrons in different materials, showing approximate scaling in absorption length λ. The shower distributions are measured from the vertex of the shower and are therefore more peaked than those measured with respect to the face of the calorimeter. For the transverse distributions as a function of shower depth, scaling in λ is found for the narrow core (FWHM) of the showers. The radius of the cylinder for 90% lateral containment is much larger and does not scale in λ. [10 GeV/c π's: B. Friend et al., Nucl. Instrum. Methods 136:505 (1976)]. Note that marble and aluminium have almost identical absorption and radiation lengths [Marble: M. Jonker et al., Nucl. Instrum. Methods 200:183 (1982); Fe: M. Holder et al., Nucl. Instrum. Methods 151:69 (1978); W: D.L. Cheshire et al., Nucl. Instrum. Methods 141:219 (1977)].

as $\lambda_{att} \simeq \lambda[E\ (\mathrm{GeV})]^{0.13}$, with an indication of a weaker energy dependence for high-Z absorbers. The expression for $L_{0.95}$ describes available data in the energy range of a few GeV to a few hundred GeV to within 10%.

c) The transverse radius R of the 95%-containment cylinder is very approximately $R_{0.95} \lesssim 1\lambda$; it does not scale with λ and is smaller in high-Z substances.

d) A useful parametrization of the longitudinal shower development is

$$dE/ds = K[wt^a e^{-bt} + (1-w)\ell^c e^{-d\ell}],$$

where t is the depth, starting from the shower origin, in radiation lengths, and ℓ is the same depth in units of absorption lengths. The parameters a,b,c,d are fits to the data and are given a logarithmic energy dependence. Crude shower fluctuations may be simulated by i) randomly varying the depth of the shower origin; ii) smearing the incident particle energy to simulate the calorimeter energy resolution; iii) randomly varying the length of the shower by scaling the values of t and l [26].

Although the total depth needed for near-complete absorption increases only logarithmically with energy, it does require, for example, about 8λ to contain, on the average, more than 95% of a 350 GeV pion.

3.2 Intrinsic Energy Resolution

In the previous subsection we indicated that the fluctuations in the HC development, producing a range of different particles—from π^0's to slow neutrons, muons, and neutrinos—with vastly different detection characteristics, are the principal limitations to the energy resolution. These fluctuations have been found to be large—of the order of 50% at 1 GeV—in strong contrast with the measurement of e.m. calorimeters, where the *intrinsic* fluctuations of the visible track length are less than 1% at 1 GeV. This understanding of hadronic cascades emerged from studies in which the various possible contributions could be individually identified and measured [27]. The dominant influence of the nuclear processes manifests itself also in the shape of the response function (Fig. 10) and is corroborated by detailed Monte Carlo estimates. Available experimental evidence indicates that the intrinsic hadronic energy resolution is

$$\sigma(E)/E|_{\text{intrinsic}} \simeq 0.45/\sqrt{E}\ (\text{GeV}) .$$

This relation describes devices made from materials covering almost the complete periodic table from aluminium [23] to lead [28]. Only in hydrogen are these nuclear effects absent, but they are already sizeable in hydrogen-rich absorbers (e.g. scintillators): one measurement in an homogeneous liquid-scintillator calorimeter is reported [29] for which the quoted energy resolution is also consistent with the above-quoted value. The sole known exception from this rule is given by uranium-238, for reasons that are explained in the next subsection.

The level of these nuclear effects and, more generally, the level of 'invisible' energy is sensitively measured by comparing the response of a calorimeter for electrons and hadrons at the same 'available' energy, which is the kinetic energy of electrons and

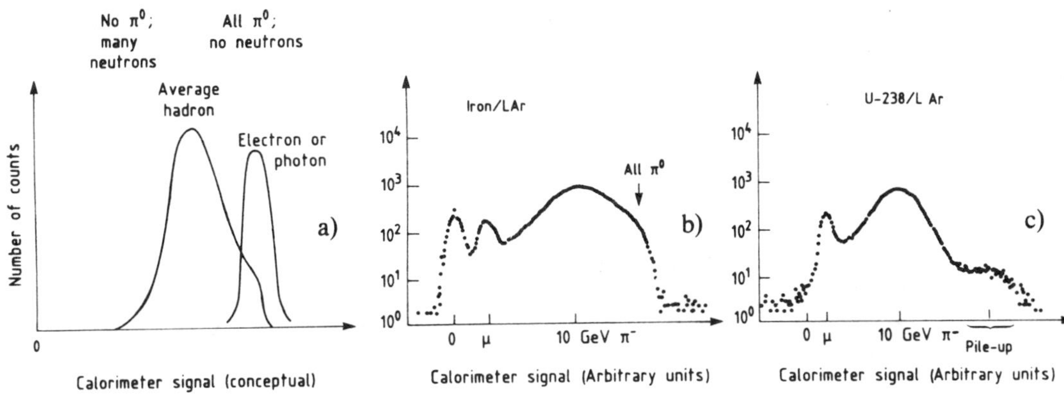

Fig. 10: Calorimeter response for 10 GeV/c pions.
a) Conceptually, the fluctuations are dominated by the nature of the first inelastic interaction. On the average, a certain number of π^0's, charged pions, nuclear fragments, and slow protons and neutrons will be produced. In one extreme case, no π^0's will be produced—only charged pions, neutrons, etc. (low-energy side of response function). At the other extreme the reaction products are mostly π^0's, and the energy deposit will be very similar to that of electrons or photons of equivalent energy.
b) The measured response function is shown for a calorimeter using iron as absorber. The logarithmic ordinate exposes a break in the resolution function corresponding to the case of mostly electromagnetic propagation.
c) The response for a U-238 sampling calorimeter is shown. The nuclear losses are effectively compensated, leading to a response that is nearly equal for charged and neutral pions. The essentially Gaussian response function is also an indication of this uniform response. (The muon peak sets the energy scale corresponding to 'one equivalent particle').

nucleons, the total energy for mesons, and the total energy plus the rest mass for antinucleons. A summary of some representative data is shown in Fig. 11. Two features deserve a comment: all calorimeters except those made from uranium show a visible energy of approximately 70% relative to electrons, which slowly increases owing to the rise in the electromagnetic component at higher energies. On the other hand, with energies decreasing below E ~ 1.5 GeV, the nature of the hadronic cascade changes: to a larger measure, the energy is degraded by ionization alone, with the hadron response approaching that of muons and being above that of electrons (see subsection 2.4). In this low-energy limit all calorimeters, including those using uranium as a degrader, are expected to give similar responses. This interpretation is confirmed by the fact that in this

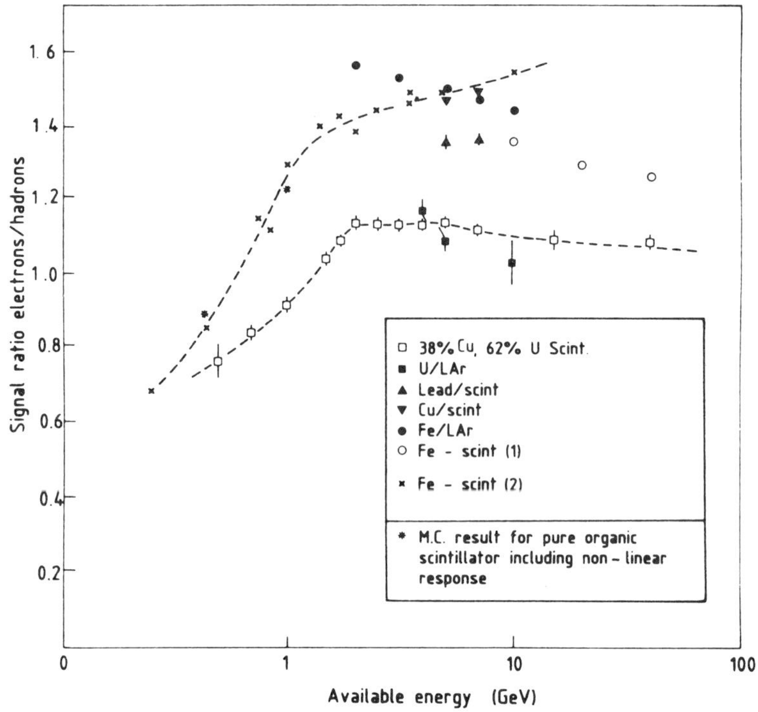

Fig. 11: The ratio of electromagnetic to hadronic energy response as a function of energy for different calorimeter systems: 38% Cu, 62% U/scint. [32]; U/LAr [37]; Cu/scint. [88]; Fe/LAr [37]; Fe/scint. (1) [35]; Fe/scint. (2) [31].

low-energy regime the relative resolution improves, $\sigma/E < 0.45/\sqrt{E}$ [30–32], as well as by quantitative Monte Carlo estimates [32].

In summary, the response of hadrons relative to electrons is a sensitive probe of the level of nuclear effects. Typical values are $e/h \simeq 1.4$ for most materials. Strongly correlated with this average suppression are fluctuations in the response; these are due to large fluctuations of the electromagnetic component. The intrinsic resolution of hadron calorimeters is, for these reasons, limited to $\sigma(E)/E_{intrinsic} \simeq 0.45/\sqrt{E}$ (GeV), unless event-to-event fluctuations in the electromagnetic component of hadron cascades are somehow corrected for. This applies likewise to *homogeneous* and *sampling* devices.

3.3 Compensating Fluctuations

We have emphasized that the relative response between electrons and pions is a sensitive measure of the level of nuclear interactions. An improvement in the energy resolution would be expected if the response of the electromagnetic cascade were identical compared with the purely hadronic one, i.e. if devices with an e/h ratio equal to one were available. Alternatively, given sufficiently detailed information on the individual hadronically induced shower, one would be able to assess the relative components and apply suitable corrections to improve the energy resolution. Both approaches have been explored and are described here.

Several suggestions have been made for monitoring the level of the electromagnetic component event-by-event. One suggestion was to use, as an indicator, the Cherenkov light from relativistic particles dominantly produced by e^+e^- pairs [33], but Monte Carlo estimates [33] for practical devices suggest that it is difficult to obtain a very useful correlation and to improve the resolution significantly. Another suggestion was to monitor the level of the nuclear component by associating heavily ionizing particles with the 'late' component of the hadronic shower [34]. If a calorimeter is instrumented so as to provide detailed longitudinal information, then some useful compensation on a shower-by-shower basis is possible for the electromagnetic/hadronic fluctuations. The most successful attempt was made by the CERN–Dortmund–Heidelberg–Saclay (CDHS) collaboration using the longitudinal information from their relatively fine-grained neutrino calorimeter [35, 36]. This was done by a weighting algorithm applied to the individual longitudinal measurements relative to the total energy measured. In Fig. 12 the unweighted and weighted results for the energy resolution are presented. Firstly, it can be seen that the raw results show a marked deviation from the expected $E^{-1/2}$ dependence. This cannot be ascribed to instrumental effects given the reported result for the electron resolution, which is well described by an $E^{-1/2}$ law. Secondly, the weighting algorithm improves the resolution, particularly at the highest energies, to a level that would be expected from extrapolating the low-energy resolution according to an $E^{-1/2}$ law. In subsection 3.5 we comment further about this deviation from the naïve $1/\sqrt{E}$ behaviour of hadronic calorimeters at high energies.

The more direct cure for these fluctuations would be to equalize the response for electrons and hadrons. In principle, equalizations of these differences, which are at the 30% to 40% level, may be accomplished in two ways: either by decreasing the electron response—typically 20% to 40% lower relative to a minimum-ionizing particle calibration (subsection 2.2); or by boosting the hadronic signal. This latter aspect is being exploited by using uranium-238 as the energy degrader [37]. In that material (and probably also to a lesser extent in thorium) some of the normally invisible energy

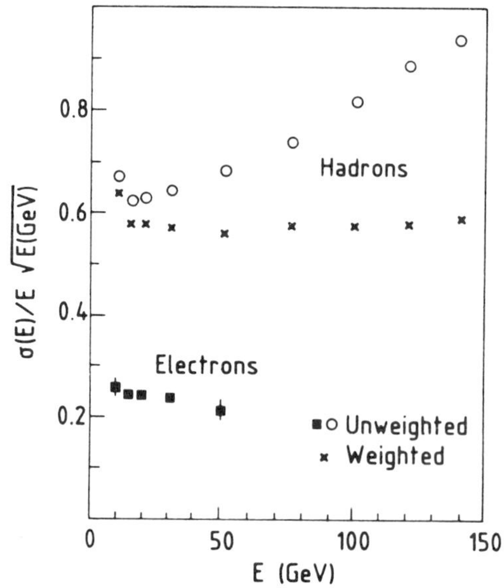

Fig. 12: Energy resolution measured in the CDHS neutrino calorimeter (2.5 cm Fe/scint.) versus the energy of the incident particle. Note that the uncorrected energy resolution for hadrons does not improve as $1/\sqrt{E}$. With a weighting procedure to reduce the large fluctuation due to the e.m. component the resolution is improved, and is consistent with the $1/\sqrt{E}$ scaling up to the highest energies measured [35].

expended in the nuclear break-up leads to neutron-induced fission, which in turn produces detectable energy in the calorimeter. It can be estimated that on the average 40 fissions are induced per GeV of energy deposited, which altogether liberate about 10 GeV of fission energy. Only a very small fraction (300 to 400 MeV) needs to be detected in order to compensate the nuclear deficit; this could be done either by the few-MeV γ-component or through the fission neutrons liberated in the fission process. Which component and what fraction of the fission contributions are measured depends on the nature of the active sampler. One may achieve essentially complete compensation not only on average but also event-by-event, because the intrinsic resolution is measured to be [32, 37]

$$\sigma(E)^{\text{uranium}}_{\text{intrinsic}} \simeq 0.22/\sqrt{E}\ (\text{GeV}) .$$

The fundamental importance of equalizing the hadronic and electromagnetic response should again be emphasized. The latter sensitively depends on the details of the low-energy part of the EMC and hence critically on the material and the sampling frequency. It would appear that this is one further contribution to the tuning of the e/h ratio. Hence for hadron showers, the level of visible compensation is expected to be affected not only by the choice of the passive absorber but also by the response of the active readout to densely ionizing particles (from the HC) and to the electromagnetic component. We do not yet have a complete set of measurements, but Table 3 attempts to organize the available information [38].

3.4 Instrumental Effects to the Energy Resolution

Most hadronic calorimeters are 'sampling' detectors, using preferentially rather dense passive absorbers to reduce the linear dimensions of the instrument. As a consequence, sampling fluctuations of statistical origin analogous to the case of electromagnetic sampling fluctuations (section 2.3) may contribute to the energy resolution, although, for the sampling of the HC we do not have a similar detailed description. Available measurements (see Fig. 13 with the quoted references) are consistent with a parametrization of the form

$$\sigma(E)/E/\text{hadron-sampling} \simeq 0.09\,[\Delta E\,(\text{MeV})/E\,(\text{GeV})]^{1/2}\,.$$

The quantity ΔE expresses the energy loss per unit sampling cell for minimum ionizing particles. Hadronic sampling fluctuations are approximately twice as large as the electromagnetic sampling fluctuations for the same detector; unlike the e.m. case however, where sampling is the predominant contribution to the resolution, sampling in hadronic detectors can be made small relative to the large intrinsic component, and energy resolution need not be sacrificed in hadronic sampling calorimeters.

Energy leakage due to partial shower containment will not only degrade the energy resolution, but will also give rise to very asymmetric resolution functions with low-energy tails. Calorimetric experiments, which emphasize measurements such as neutrino detection based on missing energy or hadronic high-p_T jet production have therefore particularly stringent requirements to achieve very close to 100% containment. Again, as already noted for the measurement of e.m. calorimeters, longitudinal fluctuations are larger than transverse fluctuations and hence longitudinal leakage is more critical to the performance. For values of fractional leakage $f \leq 0.3$ the degradation of the energy resolution follows approximately the expression

$$\sigma(E)/E \simeq [\sigma(E)/E]_{f=0} \times (1+4f)\,,$$

with the effect being somewhat more pronounced at higher energies for a given fractional energy leakage.

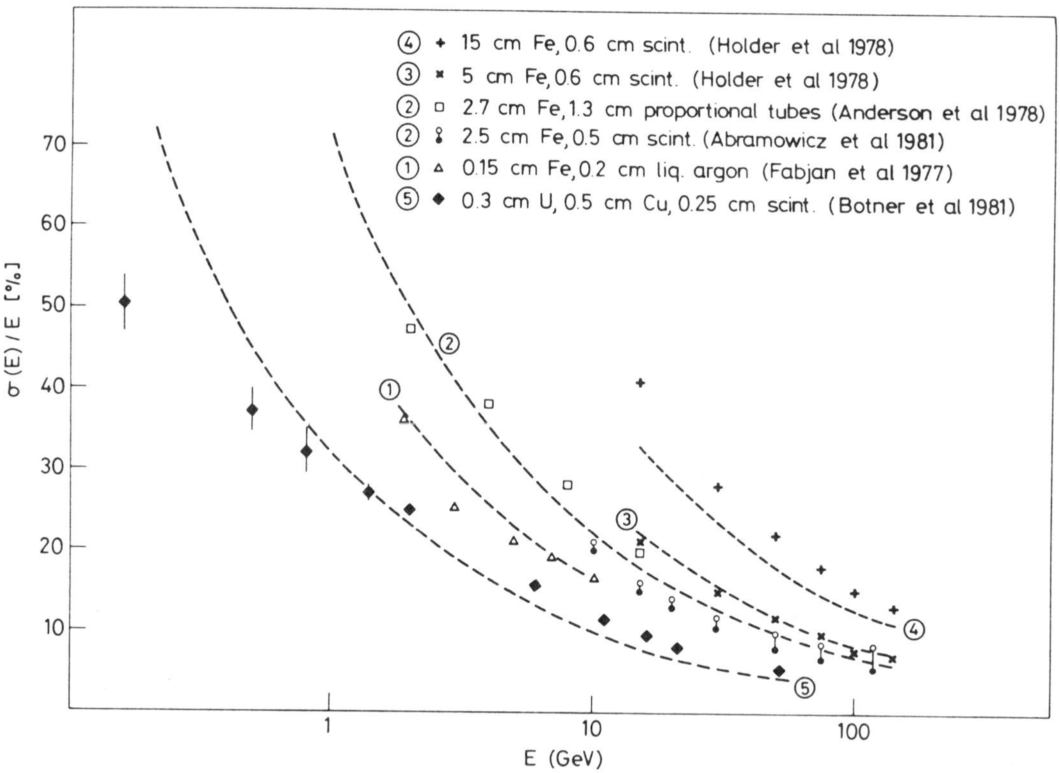

Fig. 13: Energy resolution for hadrons measured with iron and uranium sampling calorimeters. Curves 1–4 are calculated with the values for intrinsic and sampling fluctuations as given in Table 5. For the data on curve 2 [35], the open circles are the raw data; the solid circles are the results of the off-line analysis, using the longitudinal shower information to correct for fluctuations in the electromagnetic/hadronic energy ratio. For curve 5 the intrinsic fluctuation is assumed to be $0.2/\sqrt{E}$, and does not take account of the 35% (in units of λ) admixture of Cu. Below 1 GeV the resolution improves over the expected value and indicates the influence of mechanisms such as ranging and reduced nuclear effects [32, 88]. The data labelled 3 and 4 are by M. Holder et al., Nucl. Instrum. Methods 151:69 (1978), the open squares refer to R.L. Anderson et al., IEEE Trans. Nucl. Sci. NS-25:1-340 (1978).

3.5 Calorimetric Energy Resolution of Jets

Increasingly, the physics emphasis is shifting from the measurement of single particles to the analysis of jets of hadrons considered as the principal manifestation of quarks and gluons. This trend is expected to be pursued at the future multi-TeV hadron colliders, where the spectroscopy of particles in the 100 to 1000 GeV range will largely be done through the invariant mass determination of multijet systems [39]. There are two distinct contributions to the resolution of this invariant mass determination. The first effect is associated with the physics of jet production. Jets, unlike single particles, are not unambiguously defined objects, but have to be defined operationally by a 'jet algorithm' (Fig. 14). For example, hadron-initiated jets are produced together with particles originating from peripheral interactions; multijets may partially coalesce. The second contribution to the mass resolution depends on the calorimeter performance itself, and in

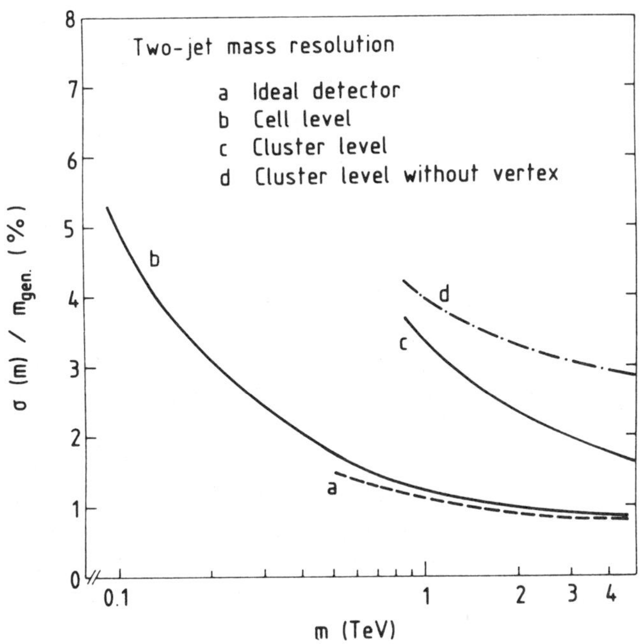

Fig. 14: Two-jet mass resolution as a function of the two-jet mass and for various assumptions on the detector performance. The ideal detector measures the jet mass at the individual particle level; at the cell level the energy information from each cell ($\Delta\phi \times \Delta\eta = 5° \times 0.05$) is considered. At the 'cluster level' certain pattern recognition criteria are introduced. Precise knowledge of the event vertex is important [39].

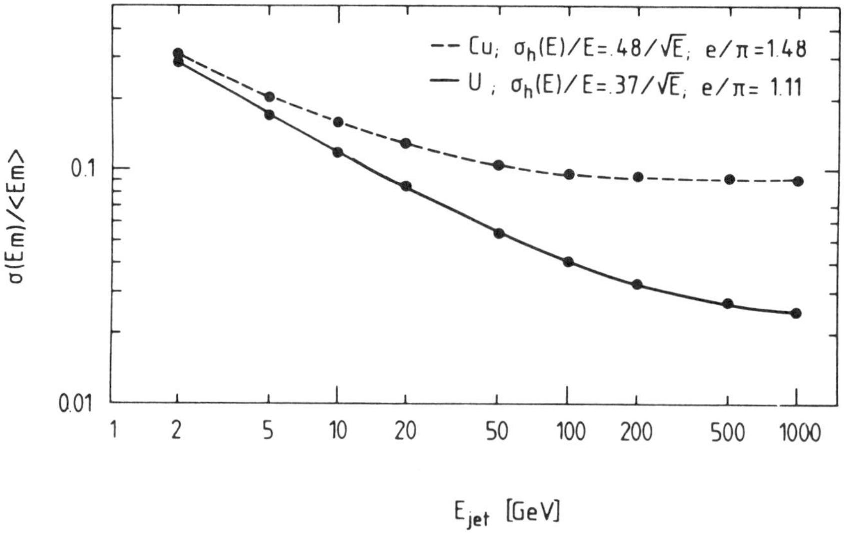

Fig. 15: Jet resolution for an 'infinitely' thick, 4π calorimeter, assuming a Feyman–Field–like fragmentation function. The advantage of a (nearly) compensated calorimeter is particularly evident at very large jet energies. For calorimeters with e/h very different from one, the resolution ceases to improve as $E^{-1/2}$ in the high-energy limit [39].

particular on the momentum response to different particles (Fig. 11). This is seen conceptually in Fig. 15 for two different calorimeters, which have rather comparable nominal resolutions but a very different response to electrons and pions. For very low energies of the jets ($E_{jet} \leq 5$ GeV), the jet resolution is dominated by the very non-linear response to low-momentum particles, and is similar for both calorimeters. At very high energies, the performance is dominated by the relative electron/hadron response. In particular, for the e/h = 1.48 calorimeter it is estimated that the energy resolution levels at approximately 10%. Qualitatively, such a strong influence is expected, because a large fraction of the jet energy is carried by only a few high-momentum particles; fluctuations in the charged hadron/π^0 ratio of these leading particles at sufficiently high energies will dominate over the simple statistical $E^{-1/2}$ improvement.

A similar argument should also be valid for single, very energetic hadrons, which after the first inelastic interaction in a calorimeter will be similar to a jet of particles of comparable energy. Therefore, hadron calorimeters with e/h $\neq$ 1 are expected to show an intrinsic energy resolution $\sigma(E)/E \sim c \times E^{-\alpha}$, with $\alpha < 1/2$ for large energies (E > 50 GeV). This behaviour is consistent with the careful analysis (Refs. [35, 36], and

Table 5: Principal Contributions to Energy Resolution in Electromagnetic and Hadronic Calorimeters

Mechanisms (add in quadrature)	Electromagnetic showers	Hadronic showers
Intrinsic shower fluctuations	Track-length fluctuations: $\sigma/E \gtrsim 0.005/\sqrt{E}$ (GeV).	Fluctuations in the energy loss: $\sigma/E \simeq 0.45/\sqrt{E}$ (GeV). Scaling weaker than $1/\sqrt{E}$ for high energies. With compensation for nuclear effects: $\sigma/E \simeq 0.22/\sqrt{E}$ (GeV).
Sampling fluctuations	$\sigma/E \simeq 0.04\sqrt{\Delta E/E}$. Nature of readout may augment sampling fluctuations.	$\sigma/E \simeq 0.09\sqrt{\Delta E/E}$
Instrumental effects	Noise and pedestal width: $\sigma/E \sim 1/E$ - determine minimum detectable signal; - limit low-energy performance. Calibration errors and non-uniformities: $\sigma/E \sim$ constant and therefore limits high-energy performance.	
Incomplete containment of shower	$\sigma/E \sim E^{-\alpha}, \alpha < 1/2$ (see subsec. 2.2, resp. 3.4). For leakage fraction $\geq$ few %: non-linear response and non-Gaussian 'tail'.	

a) ΔE = energy loss of a minimum ionizing particle in one sampling layer, measured in MeV; E = total energy, measured in GeV.

Fig. 12) discussed in the previous subsection. Table 5 summarizes the contributions to the energy resolution for both electromagnetic and hadronic calorimeters.

3.6 Spatial Resolution for Hadronic Showers

This discussion follows closely the related comments on e.m. calorimeters (subsection 2.5). Hadron showers are found to consist of a narrow core surrounded by a 'halo' of particles extending to several times the dimensions of the core. Consequently, somewhat different criteria apply to the measurement of the position and to considerations of shower separation. Measurements on the spatial resolution of the impact point [23, 32] may be parametrized approximately in the form

$$\sigma(\text{vertex})\,(\text{cm}) \simeq \langle\lambda\rangle/[4\sqrt{E}\,(\text{GeV})]\,.$$

In compact calorimeters, where the average interaction length may be as low as $\langle\lambda\rangle \leq 20$ cm, spatial resolutions in the range of a few centimetres at 1 GeV are achievable. The influence of the transverse segmentation has also been studied [40] and the following dependence can be derived:

$$\sigma(\text{vertex}) \simeq \sigma_0\,(\text{vertex})\,\exp\,(2d)\,,$$

where the segmentation d is expressed in units of absorption length and σ_0 refers to the intrinsic vertex resolution in the absence of instrumental effects due to finite segmentation. This expression suggests that the improvement becomes rather modest if the lateral segmentation is increased beyond $d(\lambda) \leq 0.1$, even disregarding other aspects such as photon statistics or noise.

Finally, the angular resolution of hadron showers has been carefully studied for several calorimeters used to investigate neutrino scattering. The limitations again stem from fluctuations in the $\pi^{\pm}/\pi^{0}$ composition of the HC, because of their (usually) very different spatial shower developments. These effects were purposely minimized in the case of the CERN-Hamburg-Amsterdam-Rome-Moscow (CHARM) neutrino calorimeter with the choice of marble as the passive absorber material in which EMCs and HCs have approximately the same dimensions [$3X_0$ (cm) $\sim \lambda$ (cm)]. An angular resolution of

$$\sigma(\theta)_{\text{hadron}}\,(\text{mrad}) \simeq 160/\sqrt{E}\,(\text{GeV}) + 560/E\,(\text{GeV})$$

is reported [23], and a similar result was obtained with a detector constructed for the same purpose at FNAL [24].

4. PARTICLE IDENTIFICATION

With hadronic calorimeters it is possible to identify a class of particles which are not always easily identified by other methods, and which may be particularly interesting for very topical physics studies, as summarized in Table 6.

In the following, we discuss in some detail the identification of electrons, muons, and neutrinos.

Table 6: Particle Identification with Calorimeters

Particle produced	Calorimeter technique	Comment
Electron, e	Charged particle initiating the electromagnetic shower	Background from charge exchange $\pi^{\pm} N \to \pi^0 + X$ in calorimeter; π discrimination of ~ 10–1000 possible
Photon, γ	Neutral particle initiating the electromagnetic shower	Background from photons from meson decays
$\pi^0, \eta, \ldots \to \gamma\gamma$ $\varrho, \phi, J/\psi, \Upsilon, \ldots$ $\to e^+e^-$	Invariant mass obtained from measurement of energy and angle	Classical application for electromagnetic calorimeters;
Protons, deuterons, tritons, ... and their antiparticles	Comparison of visible energy E_{vis} in calorimeter with momentum of particle	$E_{vis}^{b(\bar{b})} = (\vec{p}_b^{\,2} + m_b^2)^{1/2} - (+)\, m_b$ Protons (antiprotons) identified up to 4 (5) GeV/c; deuterons (antideuterons) correspondingly higher
(Anti)neutrino	Visible energy E_{vis} in calorimeter compared with missing momentum	Important tool for $e^+e^- \to \nu(\bar{\nu}) + X$ and at CERN Collider (FNAL $p\bar{p}$ collider, $pp(p\bar{p}) \to \nu(\bar{\nu}) + X$
Muon	Particle interacting only electromagnetically (range). E_{vis} compared to $\vec{p}$.	Background from non-interacting pions
Neutron or $K_L^0(\bar{n}, \bar{K}_L^0)$	Neutral particle initiating hadronic shower	Some discrimination perhaps possible based on detailed (longitudinal) shower information

4.1 Discrimination between Electrons (Photons) and Hadrons

The discrimination is based on the difference in the shower profiles, accentuated in materials with very different radiation and absorption lengths. One finds approximately

$$\lambda\,(g/cm^2)/X_0\,(g/cm^2) \sim 35A^{1/3}\, Z^2/180A \sim 0.12Z^{4/3},$$

which explains that heavy materials (lead, tungsten, or uranium) are best suited for electron–hadron discrimination.

The principal physics limitation is imposed by the charge exchange reaction $\pi^- p \to \pi^0 n$ (or $\pi^+ n \to \pi^0 p$), which may, under unfavourable circumstances, simulate an electromagnetic shower, closely matching the energy of the incident pion. For pion energies in the few-GeV range, the cross-section for this process is at the one percent level of the total inelastic cross-section, and decreases logarithmically with energy [41]. Typical values for pion discrimination are of the order of 1 in 10^2 in the 1 to 10 GeV region and of 1 in 10^3 or more for particles energies beyond 100 GeV [42]. Considerably better performance (close to 10^3 pion rejection for few-GeV particles) is reported for instruments with very fine longitudinal subdivision [43], which helps to recognize the hadronic origin of a charge-exchange-dominated cascade. Only relatively small further improvements (a factor of 3 to 5) are obtained if transverse shower profile information is available; this is because of the very high degree of correlation between transverse and longitudinal profile [44, 45]. The quoted values apply to electron–hadron discrimination based on shower shape analysis only. If, in addition, energy information can be used, for example knowledge of the electron and pion momenta from magnetic spectroscopy, a further improvement in the rejection of typically one order of magnitude is obtained.

4.2 Muon Identification

Several calorimetric methods exist for discriminating between muons and hadrons or electrons, all based on the very large differences of energy deposit.

i) Calorimeters with fine longitudinal subdivision: such calorimeters, typically many tens of absorption lengths long, have been used predominantly in experiments on incident neutrinos. Energetic muons are very clearly recognized as isolated, minimum-ionizing tracks, frequently ranging far beyond the tracks from hadronic showers.
ii) Muon penetration through active or passive absorbers: the absorbers or calorimeters are deep enough to contain the hadrons adequately and to reduce the 'punch through' probability P of pions ($P \sim e^{-d/\langle\lambda\rangle}$). The observed path length d is measured in units of 'detectable' absorption lengths $\langle\lambda\rangle$, which is found to agree closely with tabulated values [46]. The detailed rejection power against hadrons depends critically on the experimental precautions taken and may be improved by
a) reducing the background from pion and K decay before the calorimeter; the active 'beam dump' experiments have refined this method [47];
b) measuring the muon momentum after the calorimeter in, for example, magnetized iron [48] or a precision magnetic spectrometer [49]; momentum matching of muon candidates before and after the absorber may further improve the rejection [50];
c) correlating the direction of the particle before and behind the absorber. The applicability of the method is limited by multiple scattering of the muons in the absorber (Fig. 16), and accidental overlap with nearby tracks before the absorber.

Very good muon identification will be of increasing importance for experimentation at the storage rings under construction [the FNAL 2 TeV $p\bar{p}$ collider, the CERN Large Electron–Positron Storage Ring (LEP)], or *a fortiori* at those being discussed [the

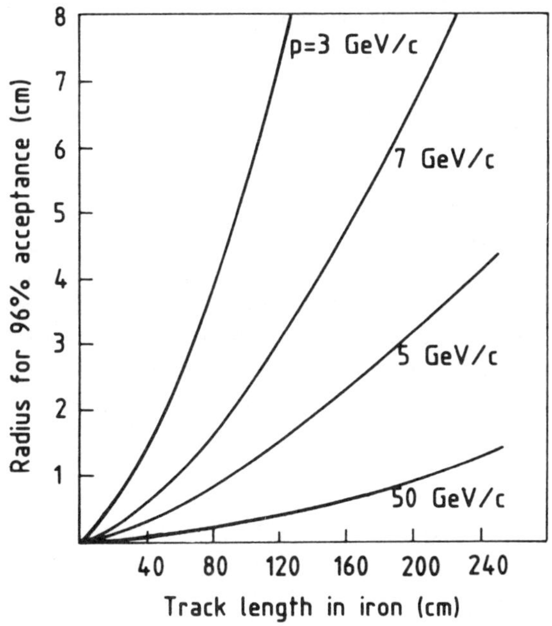

Fig. 16: Radius of 96% acceptance circle for multiply scattered muons as a function of muon track length in iron and of muon momentum. [H. Burmeister et al., CERN/TCL/Int. 74-7 (1974)].

Superconducting Super Collider (SSC) in the USA, the Large Hadron Collider (LHC) in the LEP tunnel]. The very high particle density will make the identification of electrons inside jets extremely difficult, leaving possibly only the muon as a charged lepton signature. In addition, accurate momentum measurement of the muon will be inevitable for those experiments which will increasingly exploit very good total energy measurement, for which, of course, the muon momentum has to be included [50, 51].

4.3 Neutrino Identification

The recent discovery of W production based on 'missing momentum' analysis [52] has reminded us of the power of such information. Two related methods can be used:

i) total energy measurement can be accomplished provided 4π calorimetric coverage in the c.m. system is available for all particles (charged, neutrals, muons). This can be practically achieved at e^+e^- storage rings (although 4π hadron calorimetry is not the forte of the forthcoming LEP experiments) or in a fixed-target environment [50]. Neutrino production is implied whenever the measured energy is lower than the total available energy and incompatible with the resolution function of the detector. Total energy measurement does not work well at a hadron collider, such as the CERN $p\bar{p}$ Collider, because a significant fraction of the total energy is always produced at very small angles relative to the incident beams, making a calorimetric measurement impractical. Fortunately, help is provided by

ii) a missing transverse momentum measurement. In this method, clearly related to method (i), the production of a neutrino is signalled by $\Sigma p_{T,i} \neq 0$ to a degree which is incompatible with the detector resolution. Very good missing-momentum resolution at the level of $\sigma(p_{T,miss})/p_{total} \gtrsim 0.3/\sqrt{E}$ has been estimated [53]. This is to be contrasted with the actual performance of a much cruder device for which a $\sigma(p_{T,miss}) \sim 0.7\sqrt{p_T}$ (GeV) is quoted, which is still adequate for a range of striking experimental results [52].

5. SIGNAL READOUT TECHNIQUES FOR CALORIMETERS

During the last 10 years, considerable effort has been devoted to calorimeter instrumentation with a view to developing readout techniques which optimally match an experimental application. The principal goal is to develop methods which will minimize the instrumental effects, relative to the intrinsic performance, caused by the physics of the detectors. Modern beam facilities with their steadily increasing particle energies impose ever more taxing criteria:

- the response must be linear as a function of the particle energy, frequently over a very large dynamic range; only for the exceptional case of energy measurement on isolated particles is a non-linear response acceptable, because it could be remedied (in principle) by calibration;
- the 'noise' or the non-uniformity of the readout system (photoelectron statistics, equivalent noise charge of preamplifiers) must not dominate the energy resolution;
- the readout system must have a rate capability adapted to the observed interaction rate;
- provision must be made for adequate longitudinal and transverse segmentation;
- the absolute and relative energy response must be monitored and maintained with sufficient accuracy;
- other operational characteristics, such as sensitivity to magnetic fields, radiation and temperature, have to be considered.

We have already emphasized the fundamental distinction between homogeneously active or sampling devices. The former have been used in many practical applications for the measurement of electromagnetic showers, whilst the latter represent the only really practical form of hadronic calorimeters. The instrumentation used can be categorized as 'light-collecting' devices measuring scintillator or Cherenkov light or as 'charge-collecting' methods, operating in the ion chamber, proportional, streamer, or saturated Geiger modes. In the following discussion, only recent developments or novel applications are emphasized, as witnessed by the choice of a very restricted number of references from the vast amount of literature.

5.1 Homogeneous Calorimeters

Some of the active readout materials have a density that is high enough for them to be used as homogeneously sensitive calorimeters. The properties of most frequently used materials are summarized in Table 7 [54–79].

Among recent noteworthy developments we find:
- a programme to use BGO crystals (amongst the presently known, optically transparent materials, the one with the shortest radiation length) for large 4π photon calorimeters with excellent space and energy resolution [49, 61];
- materials with considerably improved radiation resistance [64, 65, 74, 75];
- the use of BaF_2 crystals, coupled to very fast UV-sensitive light detectors for very high rate applications [73, 74].

5.2 Readout systems for sampling calorimeters

A great and very diversified number of readout systems have been developed, reflecting the desire to tailor the systems performance to a physics application.

5.2.1 Light-collecting sampling calorimeters. The renaissance of such calorimeters started with the introduction of cheap 'plastic scintillators' and elegant light-readout techniques using 'wavelength shifters' (WLSs) to replace the technique of scintillator plates individually coupled to a lightguide [80] (Fig. 17a). The principle is indicated in Fig.17b [81–89]. Scintillation light crosses an air gap and enters the WLS, where it is absorbed and subsequently re-emitted at longer wavelengths; a fraction of this 'wavelength shifted' light is then internally reflected to the light detector. This scheme avoids complicated and costly optical contacts between the scintillators and the light collectors, and minimizes dead spaces. A variety of scintillators have been developed for the large calorimeter facilities. They are based on polymethyl methacrylate (PMMA) [90] or polystyrene [30] as the matrix for the primary scintillating agent. The light yield is close to that of more conventional organic scintillators (usually based on a polyvinyl toluene solvent) if certain aromatic compounds, e.g. up to about 20% naphthalene, are added. These new scintillators are more easily mass-produced, hence cheaper, and have

Table 7: Properties and Performances of Homogeneous e.m. Shower Detectors

Detector type	NaI(Tl)	CsI(Tl)	BaF_2	$Bi_4Ge_3O_{12}$	Scintillating glass	Lead glass 55% PbO + 45% SiO_2	$Tl(HCO_2)$-liquid 'Helicon'	Liquid argon
Radiation length (cm)	2.59	1.86	2.1	1.12	~ 4	2.36	~ 1.9	14
Density (g/cm^3)	3.7	4.51	4.9	7.13	~ 3.5	4.08	~ 4.3	1.4
Detection mechanism	Scintillation	Scintillation	Scintillation (20% around 210 nm, 80% around 310 nm)	Scintillation	Scintillation	Cherenkov light	Cherenkov light	Ionization charge
Energy resolution (E in GeV)	$\sim 0.015\ E^{-1/2} < 1$ $< 0.015\ E^{-1/4} > 1$	Comparable to NaI(Tl)	Comparable to NaI(Tl)	Comparable to NaI(Tl)	$\sim 0.002\ E^{-1/2}$	$\sim 0.04\ E^{-1/2}$	Comparable to lead glass	$\gtrsim 0.02\ E^{-1/2}$
Principal limitation to $\sigma(E)$	Shower fluctuations optically non-uniform	Similar to NaI(Tl)	Light collection non-uniformities	Similar to NaI(Tl)	Photon statistics	Photon statistics	Photon statistics	Effect of shower fluctuation on electron collection
Signal[a)] (photo-el/GeV)	$\sim 10^7$	$\sim 5 \times 10^6$	$\sim 10^6$	$\sim 10^6$	Few $\times 10^3$	10^3	$\leq 10^3$ (?)	$\leq 2 \times 10^6$
Characteristic time (ns)	250	900	0.6 ; 300	350	~ 70	~ 20	~ 20	≥ 100
Rad. damage at appr. dose[b)] (Gy)	$\leqslant 10$	$\leqslant 10$	$\sim 10^5$	~ 10	$\sim 10^4$	$\sim 10^2$	$\geq 10^4$	Not measured; expected to be very large
Mechanical stability	Hygroscopic, fragile	Very good	Good	Good	Very good	Very good	Toxic liquid	Cryogenic liquid
References	[54, 55, 57]	[72]	[73, 74]	[61–63]	[64, 65]	[66–69]	[75]	[76–79]

a) Values are approximate, and depend on spectral matching between light source and photon detector.
b) Values are guidelines only and very substantially depending on experiment and measuring conditions.

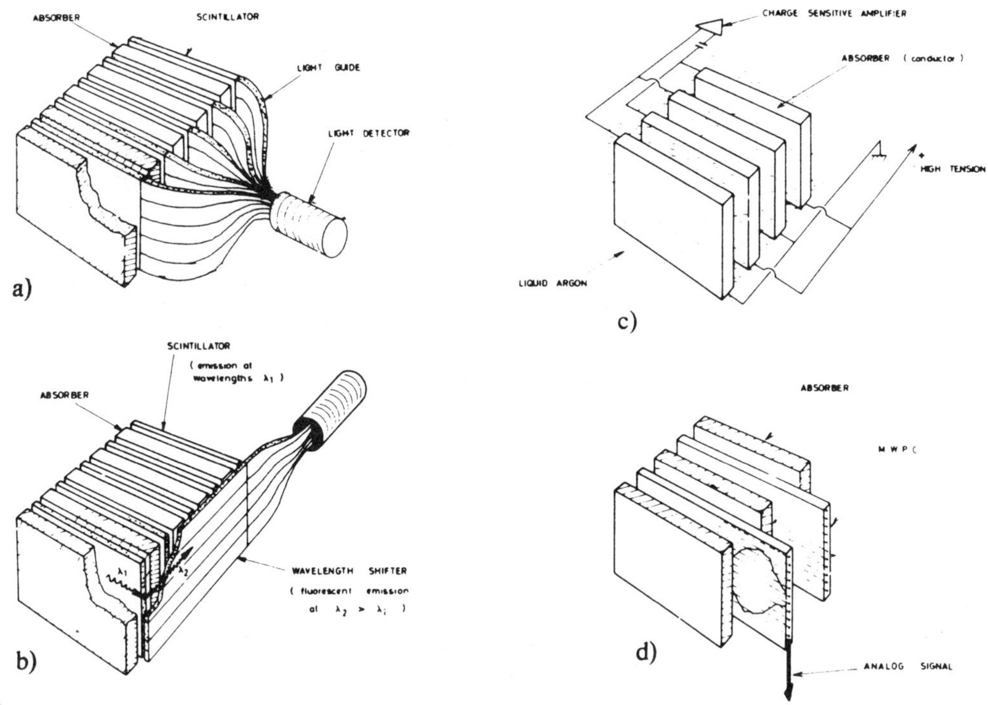

Fig. 17: Schematic representation for frequently used calorimeter readout techniques: a) Plates of scintillator optically coupled individually to a photomultiplier. b) Plates of scintillator read out by photon absorption and conversion in a wavelength shifter plate. c) Charge produced in an electron-transporting medium (e.g. liquefied or high-pressure argon) collected at electrodes, which may also function as the passive absorber plates. d) Charge produced in a proportional gas and amplified internally on suitable readout wires (proportional or saturated gas amplification).

superior mechanical properties. Some of the limitations of the WLS method may be removed after further development: better spectral matching between the scintillator emission and the WLS absorption, and also between the WLS emission and the photocathode sensitivity, will increase the number of detected photons, which is marginal in present systems. Related developments might result in the use of thinner yet more uniform WLSs; increased granularity might be achieved with WLSs having spatially different spectral sensitivities [87], or with very thin foils of WLS [91]. Potentially the most promising developments concern scintillators. They are still rather inefficient (only a few percent of the energy loss is converted into visible photons), and reduced saturation of the response to densely ionizing nuclear fragments should improve the energy resolution of hadron calorimeters (see subsection 3.3). The scintillator properties are important for the energy resolution of calorimeters, and need to be carefully investigated and specified when comparing various seemingly equivalent calorimeters.

Interest in high granularity and very compact readouts has recently led to 'double wavelength-shifter' applications [92–95] (Fig. 18). Although the second shifting reduces the number of photons by a further factor of ~ 5, the compression of the light into a very small cross-section light-pipe is attractive for several reasons, such as small insensitive zones and the possibility to use light detectors with small active areas. Such schemes favour the light registration with vacuum photodiodes [56, 70, 96] or silicon

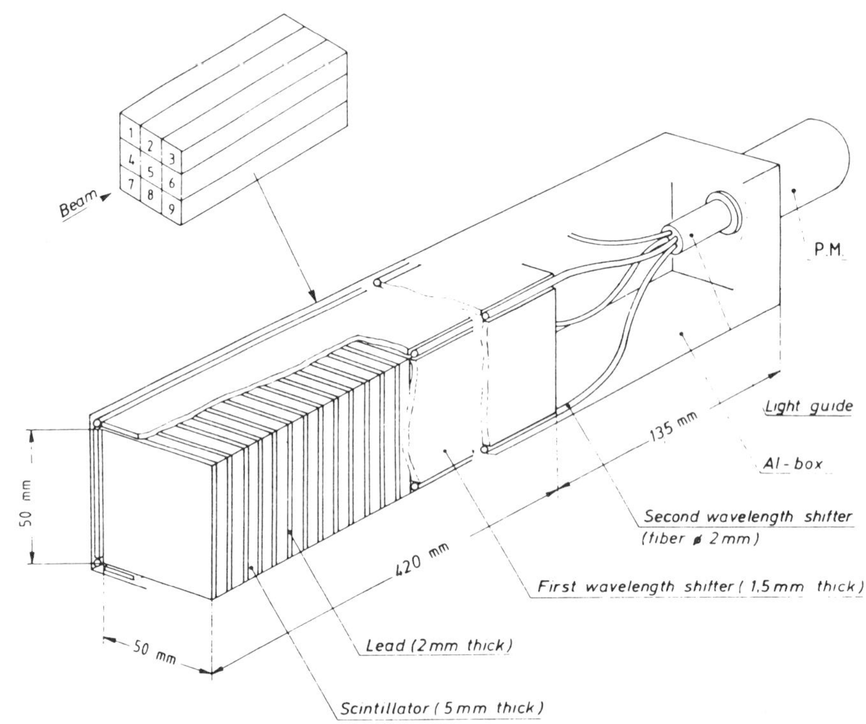

Fig. 18: Schematic view of a 'tower' of a sampling calorimeter array, using double-wavelength shifting techniques for the light measurement. The thin plates of the first WLS cover all four sides of the tower, whereas the second WLS, in the form of a fibre, registers the light emerging from the first shifter and guides it to the photomultiplier [92].

photodiodes [59, 93]. These devices—lacking an internal charge amplification mechanism—are operated with low-noise charge-sensitive preamplifiers, and are therefore more stable compared with photomultipliers; furthermore, they are insensitive to the commonly used levels of magnetic fields (vacuum diodes with some restrictions). These light detectors are therefore particularly attractive for the photon calorimeters of storage ring detectors, which most frequently are operated inside magnetic spectrometer fields [49, 56, 59, 60, 70].

Another line of study concerns the innovative use of scintillators. Very long, narrow, Teflon tubes filled with liquid scintillator have been used for a large photon calorimeter [97]. The Teflon tubes define the spatial granularity of the active element and guide the light through total internal reflection to photomultipliers. A logical refinement consists in using small scintillating fibres [98] embedded in a metal matrix. With this technique a photon detector was constructed with the very short average radiation length of $X_0 = 14.5$ mm and an energy resolution of $\sigma_\gamma/E = 0.11/\sqrt{E}$—the modest man's BGO [99].

In recent years considerable effort was devoted to minimizing two disadvantages of the scintillator readouts; namely, the inherent non-uniformity in the light collection, and the difficulty of energy calibration.

The principal source of non-uniform light collection is not primarily the attenuation of light propagating in thin scintillator sheets, but is usually due to the light collection geometry. The non-uniform response, measured for example by scanning the active surface with a monochromatic electron beam, is at a level of $\pm 5\%$ in finely tuned instruments [32], and one representative example is shown in Fig. 19. Such a level of non-uniform response will of course influence the energy resolution for electrons with $p \geq 10$ GeV/c, if no correction for the impact point is applied. Hadron showers are much less affected, if the geometric extension of the non-uniformities is comparable to or smaller than the shower size. Such problems may be considerably aggravated if the usually sufficiently high transmission of the scintillator is affected, e.g. by radiation damage [32], surface cracking, or for other reasons. Plastic scintillators will show radiation damage after exposure to less than 100 Gy, if in contact with air [100], whilst toluene-based scintillator may sustain approximately 10 times more radiation [101, 102] before its usefulness is severely limited. Closely connected with these problems of light collection is the strategy of relative and absolute energy calibration. For precision applications, it is necessary to expose each calorimeter cell at least once to some kind of particle beam in order to establish an absolute calibration, which subsequently has to be transferred and maintained with some kind of absolutely stable light source. This light source is usually an external reference lamp, whose output is distributed to the calorimeter cells [103]. In some cases, the light produced by internal radioactive sources [32] has served this purpose.

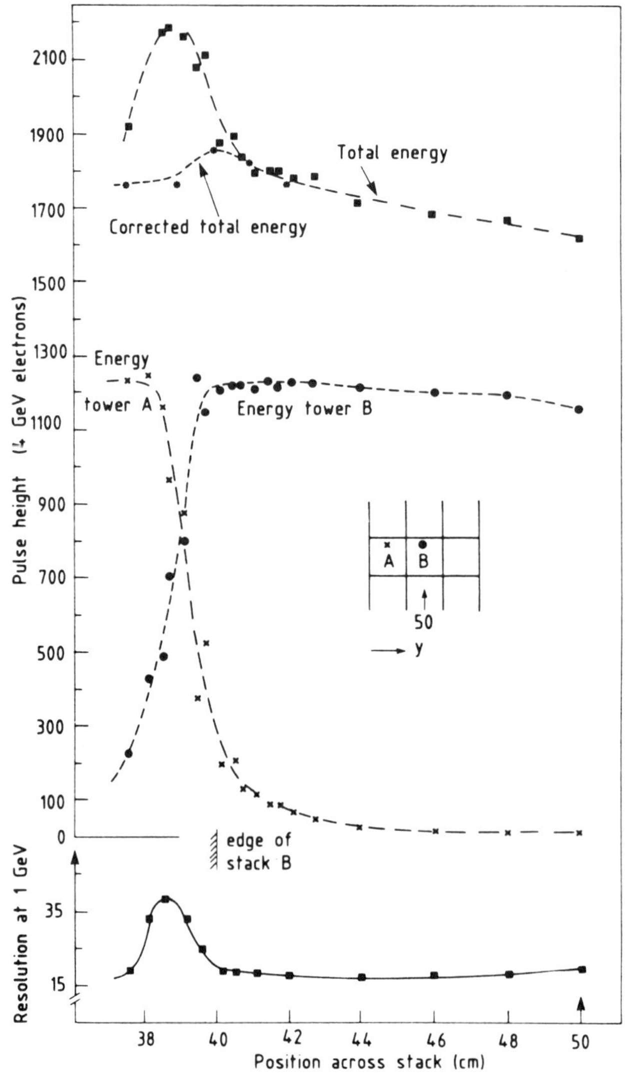

Fig. 19. The worst-case optical non-uniformity obtained by scanning with a 4 GeV/c electron beam under normal incidence across the gap between neighbouring stacks. The approximately 15% non-uniformity can be further reduced with a correction algorithm using the two signals in each tower, A and B (short-dashed curve) [32].

The limitations outlined here become a major concern for applications where very high energy deposits could, in principle, be measured at the one percent accuracy level [50, 95, 104] and correspondingly benefit the quality of the physics data.

Increasingly, therefore, the experimental teams are evaluating alternative solutions, as discussed in the next subsection.

5.2.2. Charge collection readout. The ionization charge produced by the passage of the charged particles of the shower may be collected from solids, liquids, or gases. Solids [105] and liquids can only be used in an ionization chamber mode with no internal amplification. The best known and, to date, the only practical example is based on the use of liquid argon [106]. In specific cases, liquid xenon may be used [107–110]. The use of room-temperature liquids has also been repeatedly advocated [95, 111], but with increasing operating temperature the tolerable level of impurities decreases strongly. If gas is used as the active sampling medium, internal amplification to various degrees is usually exploited: proportional chambers or tubes provide a signal proportional to the energy loss. At higher gas gain, with devices operating in a controlled streamer or Geiger mode, the measured signal is related to the number of shower particles which traverse the active medium ('digital readout') [112]. The conceptual arrangements are shown in Fig. 20.

The principal advantages common to all these charge collection methods are seen in the ease of segmentation of the readout and the capability to operate in magnetic fields. Some features specific to the various types are:

a) operation in the ionization mode, i.e. liquid-argon calorimeters, provides the best control of systematic effects [44, 113–116];
b) gas proportional devices offer a wide variety of relatively inexpensive construction methods [117];
c) digital operation, in the Geiger or streamer mode, allows for very simple and cheap signal-processing electronics [112, 118, 119].

The ion chamber technique is the preferred solution whenever optimum performance is at a premium because it excels in the following points, all of which have been realized in practice in devices using liquid argon as the active medium:

- uniformity of response at the fractional percent level;
- ease in fine-grained segmentation with a minimum of insensitive area;
- excellent long-term operating stability (radiation damage absent in liquid argon; response of active medium controllable).

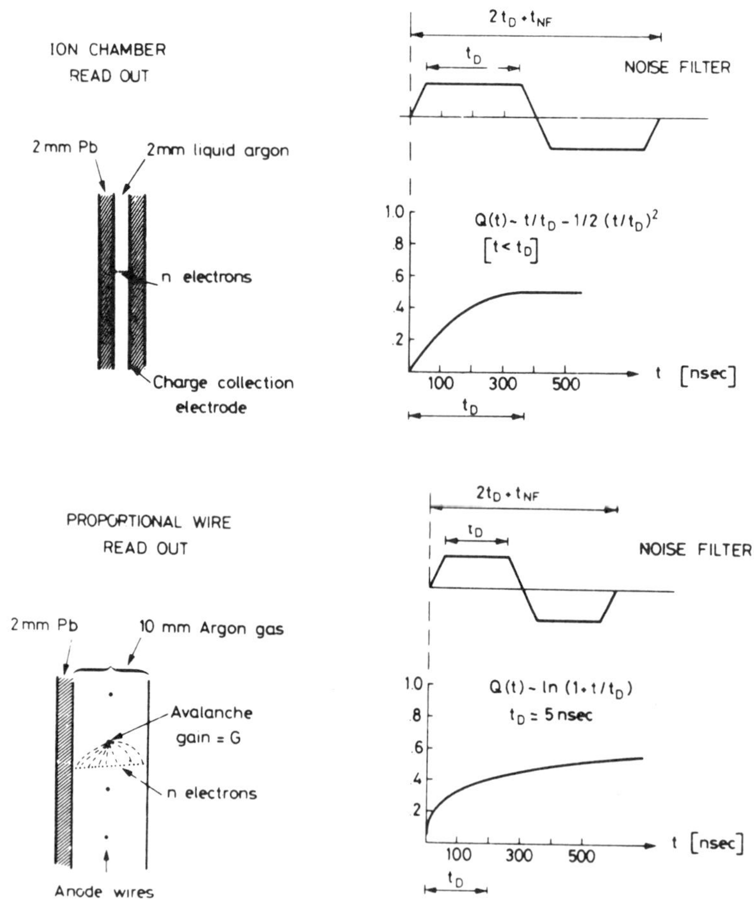

Fig. 20: Charge collection in a single sampling layer for a) a liquid-argon calorimeter with ion chamber readout; b) gas proportional wire readout. The signal charge Q(t) is shown as a function of time, and t_D is the time required for all ionization electrons to be collected. For each case a bipolar noise-filter weighting function is indicated (see text).

Experimentation at today's fixed-target or storage ring facilities often justifies the use of such high-performance calorimeters, and as a consequence the liquid-argon ion chamber technique has been adopted for several large facilities, both in existence [44, 113–116] or planned [51, 120, 121]. One noteworthy exception is made by the teams preparing the four LEP detectors—they have opted for scintillator or proportional chamber readout solutions.

For many years, several liquid substances have been known, in which electrons may be drifted over large (several centimetres) distances, given adequate electric fields and levels of purity [122]. Generally, it may be said that operation at cryogenic temperatures, $T < 100$ K, eases the purification problem. Liquid-argon calorimeters, for example, may be operated with levels of $\sim$ 1 ppm O_2, but the requirements are already considerably more severe for liquid xenon ($\sim$ 10 ppb O_2 tolerable) and develop into a major engineering difficulty for room-temperature liquids [95, 123]. In practical applications one has to weigh the complexity of a cryogenic detector (difficult access to components inside, extra space for the cryostat, increased mechanical engineering problems) against the difficulties arising from impurity control (large multistage purification plant, use of ultra-high vacuum techniques and components in the construction) [123]. It was thought that some of the lighter room-temperature liquids [such as tetramethylsilane (TMS)] might be intrinsically more advantageous for hadron calorimeter readouts, because one expected relatively small saturation for densely ionizing particles (e.g. recoil protons) [124]. Recent measurements [95, 125] however, have indicated that in practical electric drift fields such an advantage may not exist.

The very low cost of sufficiently pure liquid argon and the relative ease of maintaining it in operating conditions suggest its use in large-mass detectors for neutrino-scattering or proton-decay experiments [77–79, 126]. As with all rare-event detectors, the possibility of very large drifts of the ionization [time projection chamber (TPC)] have been studied with the aim of reducing the number of electronic channels to an 'acceptable' level [76, 77].

A recent extension of the ion chamber techniques to the use of solid dielectrics has met with considerable success [127], the signal being measured with Si surface barrier detectors in sampling calorimeter detectors. This scheme combines the advantages of a room-temperature ion chamber readout, with the attractive feature that these Si detectors are usually extremely thin, less than 500 μm, permitting the realization of detectors with $X_0 \leq 4$ mm, and hence offering the ultimate localization of showers.

For ion chambers and proportional wire readout, the measured signal typically amounts to a few picocoulombs of charge per GeV of shower energy. Since sampling calorimeters are inherently devices having a large capacitance, the optimum charge measurement requires careful consideration of the relationship between signal, noise, resolving time, and detector size. A detailed noise analysis gives [128]

$$ENC_{opt} = k \times 10^6 \, (C_D/t_{NF})^{1/2},$$

the 'equivalent noise charge' ENC being the input signal level which gives the same output as the electronic noise. The parameter k is proportional to the r.m.s. thermal noise of the input field-effect transistor; for practical detector arrangements a value of k $\simeq$ 5 is realized. [C_D is the detector capacitance (in μF), and t_{NF} is the noise filter-time (in ns)]. This relation determines the fundamental lower limit to the noise, which is achieved with optimal capacitance matching between the detector and amplifier. The noise figure grows with increasing detector capacitance, but can be reduced at the expense of augmenting the resolving time.

Charge collection in gases, usually followed by some degree of internal amplification, forms the basis of another important category of calorimeter readouts [117]. The method lends itself naturally to highly segmented construction, of particular value for the topological analysis of the energy deposit (e.g. γ/hadron discrimination, muon identification). The technique has profited from the diversified developments in gaseous position detectors [129] over the last fifteen years: the versatility of arranging readout anode wires combined with the ease of gain control have produced a great variety of different solutions tailored to the specific requirements of an experiment.

With these types of detectors, spatial segmentation and localization can be easily implemented. This may be achieved in a projective geometry using strips, or in a 'tower' arrangement, e.g. by measuring the signal charge induced on a pattern of cathode pads [130, 131]. The tower arrangement is mandatory for reducing ambiguities and confusion in multiparticle events.

Another technique, currently being pursued, aims at achieving a very high degree of spatial segmentation, and uses a TPC method, the so-called 'drift-collection' calorimeter [132]. The ionization produced by the charged particles of a cascade is drifted over very long distances and collected on a relatively small number of proportional wire planes, equipped with a two-dimensional readout, while the third shower coordinate is determined by a drift-time measurement. The conceptual configuration is shown in Fig. 21. An example of the pictorial quality of shower reconstruction expected with this technique is given in Fig. 22.

Gas sampling calorimetry in a digital mode (using Geiger, streamer, or flash-tube techniques) is a means of simplifying the signal processing circuitry, and offers an expedient method for achieving a high degree of segmentation. Flash chambers consist of an array of tubular cells filled with a mixture of about 96% neon + 4% helium; a pulsed high-voltage is applied across each cell after an external event trigger. In the presence of ionization charge, a signal-producing plasma discharge propagates over the full length of the cell. In one such array for a large neutrino experiment at Fermilab, 608 flash-chamber planes with a total of some 400,000 cells are sandwiched between absorber layers of sand and steel shot [24]. The pattern of struck cells in each plane is read out by sensing induced signals on a pair of magnetostrictive delay lines. Wire readout planes may also be operated in the 'streamer mode', where the charge gain is controlled to cover

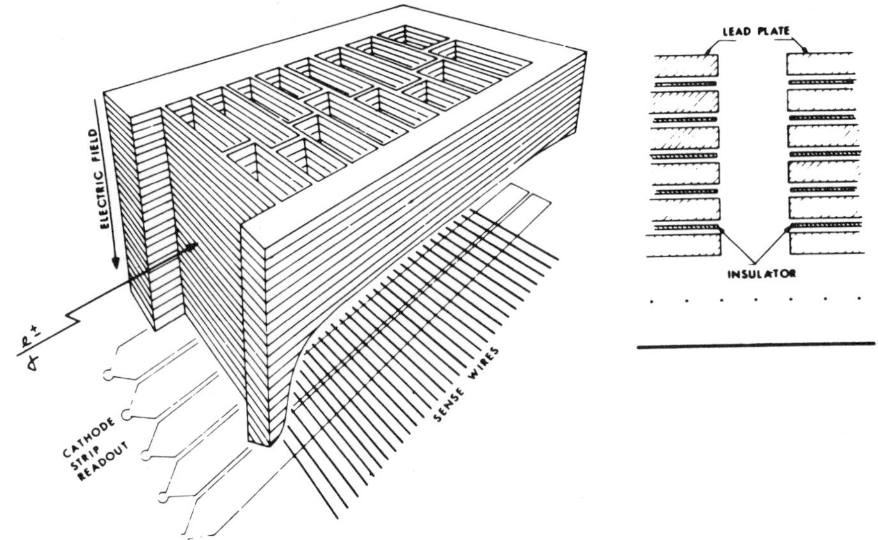

Fig. 21: Geometrical arrangement for the high-density drift calorimeter. Cavities between absorber plates allow the drifting of ionization electrons over long distances on to MWPC-type detectors. Very high spatial granularity can be achieved at the cost of mechanical complexity and rate capability. A very uniform magnetic guidance field parallel to the drift direction is usually required [132].

only a segment around the particle impact point [133, 134]. Common to these saturated gas-gain readout schemes are the following properties:

- the energy resolution is in principle better than in a proportional gain system, because Landau fluctuations are reduced or suppressed;
- the mechanical tolerances of the readout system are less stringent than for proportional systems;
- these readouts are inherently non-linear. One charged particle causes an insensitive region along the struck wire, which prevents other nearby tracks from being registered. Typically, non-linearities become measurable above ~ 10 GeV.

Usually, care is taken to limit the geometrical extension of the discharge region, e.g. with various mechanical discontinuities (beads, nylon wires, etc.). A calorimeter operated in this mode gave an energy resolution of $\sigma \simeq 14\%/\sqrt{E}$ for electron energies up to 5 GeV [135]. This is better than is normally achieved for gas sampling calorimeters, reflecting the absence of Landau and path-length fluctuations. At higher energies the

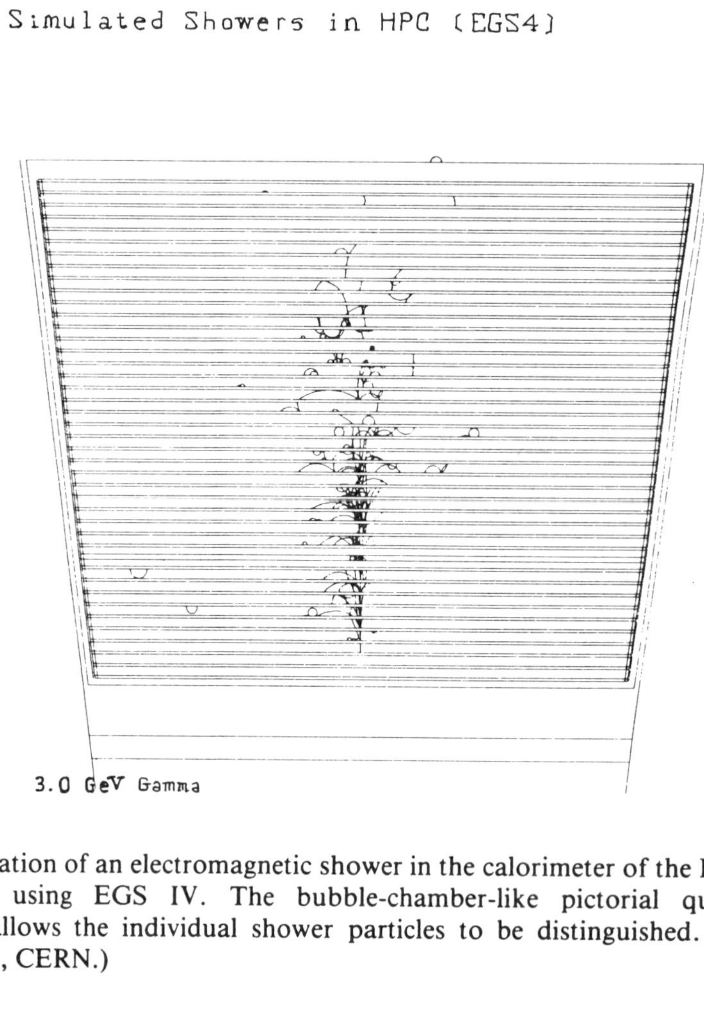

Fig. 22: Simulation of an electromagnetic shower in the calorimeter of the LEP DELPHI collaboration using EGS IV. The bubble-chamber-like pictorial quality of the information allows the individual shower particles to be distinguished. (Courtesy of H. Burmeister, CERN.)

calorimeter showed saturation effects due to the increasing probability of multiple hits over the geometrical extension of the discharge region.

The *rate capability* of calorimeters is an important parameter for high-rate fixed-target experiments or the planned hadron colliders (Table 8) [136]. Several different time constants characterize the readout:

1) 'occupation' time specifies the length of time during which the physical signal produced by the particle is present in the detector (pulse length);

Table 8: Temporal Response of Readout Systems

Calorimeter system	Occupation time (ns)	Pulse width (integration) (ns)	Timing resolution σ (ns)	Radiation resistance
Metal/scintillator with WLS readout	50	50	2.5 for few-GeV deposit; better at higher energy, if not limited by PM	Depends on scintillator, dose rate, environment; $> 10^3$ Gy appear achievable [102].
^{238}U/scintillator with fast WLS readout	~ 100	100	2.5	As above
Metal/fast scintillator with fast WLS readout	$\leqslant 20$	$\leqslant 20$	< 2 (?)	
Metal/proportional or saturated gas-gain readout	50–100	100–200 bipolar shaping	$\leqslant 10$	Adequate for chamber; lifetime of on-detector electronics may be a limitation; readout elements need to be shielded from U radioactivity.
^{238}U/proportional or saturated gas-gain readout	$\geqslant 100$	$\geqslant 200$ bipolar shaping	$\leqslant 10$	
Metal/LAr ion chamber	~ 200 per 1 mm gap	$2\lambda = 400$ bipolar shaping	~ 2 for few-GeV deposit	Lifetime of on-detector electronics may be a limitation
Metal/LAr–CH_4 ion chamber	~ 100 per 1mm gap	$2\lambda \leq 200$	$\lesssim 2$ for few-GeV deposit	NB: shorter pulse width (2λ) possible the expense of signal/noise $\simeq \lambda^{\alpha}$, at $\alpha \sim 1$.

2) 'integration' time corresponds to the externally (e.g. electronically) chosen time defining the bandwidth of the signal processing system;

3) 'time resolution' specifies the precision with which the impact time of a particle may be determined.

For scintillator-based methods the signal duration is typically about 20–50 ns but, with special care, signals of about 10 ns length have been achieved [94].

For liquid-argon devices, the charge-collection time is ~ 200 ns/mm gap; it may be reduced by a factor of 2 by adding ~ 1% of methane. Sometimes it is acceptable not to integrate over the full signal length, entailing a reduction in the signal-to-noise ratio. For liquid-argon devices, the noise is proportional to $(\tau_{\text{integ}})^{-1/2}$ and the signal is almost proportional to $(\tau_{\text{integ}})^{1/2}$, if $\tau_{\text{integ}} < \tau_{\text{tot. coll.}}$. A very short integration time is sometimes chosen ('clipping' of a signal) to enable very fast trigger decisions; only if the event information is of interest is the signal processed with a longer integration time, in order to obtain an optimum signal-to-noise ratio.

The time resolution achievable with calorimeters may help to associate the calorimeter information with different events, separated by a time interval much shorter than the integration time. For scintillator and liquid-argon calorimeters, time resolutions of 2–3 ns have been measured for few-GeV energy deposits [44, 137].

The ultimate rate limitation is, however, determined by the physics to be studied with a calorimeter. In collisions involving hadrons in the initial state, reactions which occur with very different cross-sections may be characterized by very different event topologies. A typical example is the production of several high-p_T jets in a pp collision, the cross-section of which is very small in comparison with the total inelastic cross-section. In such a case, several events may be recorded within the occupation time of the detector without serious effects (Fig. 23). This figure suggests that calorimeters may still be very useful even if, during the occupation time of an interesting event, several other events produce energy deposits and are recorded in the detector [32, 39, 138].

5.2.3. Bolometric Readout. Repeatedly, particles have been detected through the temperature rise in a calorimeter, caused by the absorption of the particles. Such experiments contributed decisively to our understanding of radioactivity and to the concept of the neutrino [139]. Already 50 years ago it was recognized that owing to the large reduction in heat capacity of many materials at cryogenic temperatures, very sensitive instruments were permitted [140]. More recently, interest has refocused on cryogenic calorimetry, operated in the temperature range of 1 mK to ~ 1 K. Such detectors [141–143] are considered as possible particle detectors offering an energy

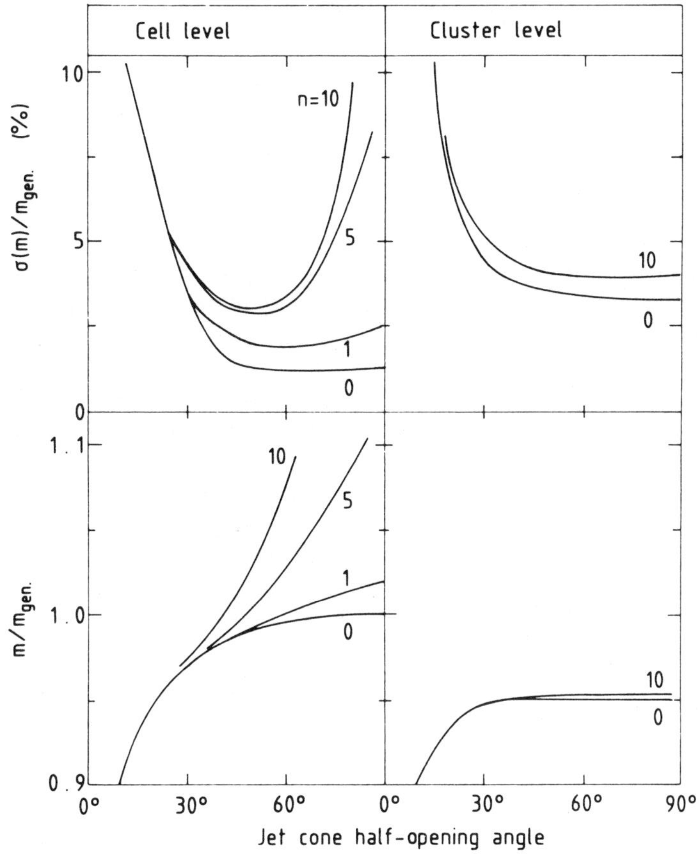

Fig. 23: Two-jet mass resolution (upper part) and reconstructed mass value (lower part) for a 1 TeV system at the cell and at the cluster level as a function of the opening angle between the jets, and the number n of additional accidental minimum-bias events [39] (see Fig. 14 for definition of cell and cluster level).

resolution of ≤ 10 eV and having a time response in the 100 μsec to 1 msec range. Possible experiments include the search for double-beta decay or the measurement of the end point of tritium beta decay. More ambitiously, multiton silicon detectors operated at mK temperatures are considered as a solar neutrino observatory to measure the spectrum of solar neutrinos and to probe for neutrino mass differences at the 10^{-6} eV level. Cryogenic calorimetry promises far more precise energy measurements than any of the other 'standard' techniques and allows us to contemplate some of the most challenging and fundamental experiments in particle physics.

6. SYSTEM ASPECTS OF CALORIMETERS

6.1 General Scaling Laws of the Calorimeter Dimensions

A number of parameters which may be external to the calorimeter design, e.g. the dimensions of a charged-particle spectrometer, may ultimately determine the size of a calorimeter.

If, however, global optimization of an experiment is attempted, one should aim for the most compact calorimeter layout that may lead to the achievement of the physics goals of the detector.

The distance D of a calorimeter from the interaction vertex, and hence its necessary size, is determined by the achievable useful segmentation and the characteristic angular dimension θ required to be resolved in the measurement. The useful segmentation d is determined by the shower dimensions, approximately $d \sim 2\varrho_M$ for electromagnetic detectors and $d \sim \lambda$ for hadronic ones. The characteristic angular dimension θ may be the minimum angular separation of the photons from π^0 decay, or the typical angle between the energetic particles in a hadron jet.

For correctly designed detector systems, the calorimeter dimensions are determined by the angular topology and size of the showers to be measured. The minimum detector distance is then $D \geq d/\mathrm{tg}\,\theta$, and the required calorimeter volume V is found to be

$$V \propto D^2 \times L \text{ (depth of calorimeter required for total absorption)}$$

whence

$$V \propto \varrho_M^2 \times X_0 \text{ for e.m. detectors,}$$

and

$$V \propto \lambda^2 \times \lambda \text{ for hadronic detectors.}$$

The third-power dependence of the calorimeter volume on shower dimensions implies that it may be economically advantageous to select very compact calorimeter designs, even if the price per unit volume is very high. Table 9 explains why sometimes only the most expensive materials (per cm^3) can be afforded.

6.2. Calorimeter Systems for Physics Applications

6.2.1 Neutrino physics and nucleon stability. Detectors for these studies share some common features: the event rate is low and is proportional to the total instrumented mass; the physics requires a fine-grained readout system, which permits detailed

Table 9: Characteristic Dimensions and Price Comparisons for e.m. Calorimeters

Material / Quantity	NaI	BGO	U/Si sampling calorimeter
X_0 (mm)	26	11	4
ϱ_M (mm)	44	23	11
Reference volume (cm^3) for 95% containment of ~ 5 GeV electrons	1600	180	15
Approx. price/Ref. vol. (arbitrary units)	1	1	0.1

three-dimensional pattern reconstruction. Bubble chambers have therefore been used extensively but are limited by the long analysis time per event. The present generation of neutrino detectors makes extensive use of wire-chamber techniques to approach the intrinsic spatial and angular resolution of calorimeters. New 'visual' electronic techniques are being developed for proton decay experiments, for which very massive detectors with a high density of signal channels are required.

A neutrino detector, even when exposed to present-day intense neutrino beams, must have a very large mass (typically hundreds of tons), and its volume must be uniformly sensitive to the signature that an interaction has occurred and to the characteristics of the reaction products. These requirements explain the modular construction typical of modern electronic neutrino detectors. The general form of the interaction is $\nu_{(\mu,e)}$ + nucleon = $\ell_{(\mu,e)}$ + X, where ℓ is a charged lepton (charged-current interaction) or a neutrino (neutral-current interaction), and X represents the hadronic system. The signature for a neutrino interaction is the sudden appearance, in the detector, of a large amount of energy in a few absorber layers. If the scattered lepton is a muon, it leaves in each layer the characteristic signal of a single minimum-ionizing particle. In some detectors the absorber layers are magnetized iron, which makes possible a determination of the muon momentum from the curvature of its trajectory. For a detailed study of neutral-current interactions, a very fine grained subdivision of the calorimeter system is required for measuring the energy and direction of the hadronic system X and for reconstructing the 'missing' transverse momentum of the final-state neutrino.

Clearly the scope and sensitivity of neutrino experiments would be much improved if even the most massive detectors could resolve final state particles with the reliability and precision typical of a bubble chamber. With this goal in mind, some schemes are currently being investigated that use drift chamber methods with large volumes of liquid argon [79], which provides a visual quality characteristic of homogeneously sensitive detectors. In another case [144] a cylindrical detector, 3.5 m in diameter and 35 m long, is foreseen, containing about 100 tons of argon gas at 150 atm pressure, with ionization electrons collected on planes of anode wires. The idea has also been advanced [145] of using compressed mixtures of more common gases (air or freon), large liquid-argon TPCs [78], or room-temperature liquid hydrocarbons [146] to detect the ions that migrate away from charged-particle tracks (positive and/or negative ions produced by electron attachment).

Detectors to search for the decay of nucleons are also characterized by a very large instrumented mass, varying from a hundred to several thousand tons. Current theoretical estimates place the proton lifetime at around $\tau_p \simeq 3 \times 10^{29 \pm 1.5}$ y compared with present experimental limits [147] of $\tau_p \geq 10^{32}$, reached by detectors of at least 1000 tons mass, or approximately $> 10^{33}$ nucleons. These detectors have to be instrumented to search sensitively for some of the expected decay modes such as $p \to e^+ + h^0$, where h^0 is a neutral meson (π^0, η, ϱ^0, ω^0). The signature of such a decay is clear, provided a detector is sufficiently subdivided to recognize the back-to-back decay into a lepton and a hadron with the relatively low energy deposit of about 1 GeV. The sensitivity is limited by the flux of muons and neutrinos originating from atmospheric showers. Only muons can be shielded by placing the experiments deep underground in mine shafts or in road tunnels beneath high mountains. The ν_μ-induced rate simulating nucleon decay is estimated at $\sim 10^{-2}$ events per ton per year if energy deposition alone is measured. If complete event reconstruction is possible, an experimental limit of $\tau \geq 10^{33}$ y may be reached [147].

The most massive calorimetric detectors to date have been conceived to explore the very high energy cosmic-ray spectrum. The detector volumes needed are so large that only the sea water [148] and air [149] are available in sufficiently large quantities. The interaction provoked by cosmic-ray particles is detected through the Cherenkov radiation emitted in the ensuing particle cascade in the case of the deep underwater array. In the case of the atmospheric detector it is the light from the excited N_2 molecules which is detected and measured with great ingenuity. The 'air calorimeter' represents the largest ($V \simeq 10^3$ km^3) and most massive ($W \sim 10^9$ tons) calorimeter conceived to date; already it has produced evidence about the cosmic-ray flux at $E > 10^{20}$ eV.

6.2.2. Calorimeter facilities for storage rings. At hadron machines the studies focus on reactions that are characterized by a large transverse energy (E_T) flow, as a signal for an inelastic interaction between the nucleon constituents. The signature appears in many different characteristic event structures and may therefore be efficiently selected with hadron calorimeters : examples are single high-p_T particles, 'jets' of particles, or events exhibiting large E_T, irrespective of their detailed structure. Topical applications include invariant mass studies of multijet events, often in conjunction with electrons, muons, or neutrinos (missing E_T). The power of this approach has been demonstrated by the results obtained in recent years at the CERN $p\bar{p}$ Collider [52], which in turn have led the UA1 and UA2 collaborations to proceed with major upgradings or replacements of their calorimeter detectors [95, 150]. This central role of calorimetry in exploratory hadronic physics programmes [39, 151, 152] is also recognized in the planning for the second detector facility for the FNAL Tevatron. The group is planning a 400 ton uranium/liquid-argon hadron calorimeter, which is expected to be the most advanced calorimeter facility in use during the coming years [51].

At electron–positron colliders, electromagnetic detectors are frequently used to measure the dominant fraction of neutral particles, the π^0's. They are also the ideal tool for detecting electrons, which may signal decays of particles with one or more heavy (c, b, ...) quarks. Unique investigations of cc and bb quark spectroscopy were accomplished with high-resolution NaI shower detectors [153].

For the physics programmes at LEP [154–156] and at the SLAC Linear Collider [120, 157] extensive use of hadron calorimetry will be made. At e^+e^- machines the event topology—production of particles at relatively large angles, with a total energy equal to the centre-of-mass (c.m.) collision energy, favours the experimental technique of total energy measurement. For future e^+e^- physics this method will be important because

- the fraction of neutron and K_L^0 production, measurable only with hadronic calorimeters, increases with energy [158];
- a large and most interesting fraction of events will contain neutrinos in the final state; missing energy and momentum analysis provides the sole handle for such reactions;
- a considerable fraction of events will show good momentum balance but large missing energy—these may be two-photon events or, above the Z^0 pole, events on the radiative tail; total energy will provide the cleanest signature;
- hadronic calorimeters will be the most powerful tool for measuring the reaction $e^+e^- \rightarrow W^+W^-$, either in channels where each W decays hadronically (a total of four jets) or through leptonic decay channels;

- most importantly, a measurement of energy topology is a powerful way of unravelling very rare and unexpected physics phenomena [159].

The technically most difficult requirements for the calorimetry will be imposed by the physics programme at HERA (DESY, Hamburg) [160]. Owing to the very asymmetric energy of the beams (800 GeV protons on 30 GeV electrons), the jets of particles fragmenting from the scattered quarks will have to be measured with the greatest possible energy and *angular* precision in a geometry similar to that of a fixed-target experiment. This implies that the detectors will have a very asymmetrical arrangement, a very large dynamic range, and very high granularity. Innovative developments [93, 127] signal the HERA groups' anticipation of this challenge.

At e^+e^- machines the performance with respect to the energy and space resolution of a well-designed calorimeter is matched to the physics programme foreseen at LEP and at the SLC. Consider, as an example, a 100 GeV multijet event of which the total energy can be measured with an accuracy of $\sigma \simeq 3$ 5 GeV and the total momentum balance checked at a level of $\sigma = 3$ GeV/c. These are intrinsic performance figures, disregarding possible instrumental effects (see Section 3). At hadron colliders, however, a further serious difficulty arises from the convolution of the energy response function with the steeply falling p_T distribution of hadronically produced secondaries [161–164]. As a consequence, the measured energy deposit E' in the detector will originate predominantly from incident particles with energy $E < E'$; count rates and trigger rates are higher than the true physics rates. The result can be devastating for detectors with poor energy resolution or a non-Gaussian response function ('tails' in energy resolution), introducing large errors in the deconvolution. The problem is compounded if the calorimeter response is different for charged and neutral pions (see Section 3): without adequate precautions, these detectors would preferentially select π^0's, making the use of calorimeters marginal for general trigger applications.

The trigger capability is a unique and perhaps the most important requirement of hadron calorimeters employed at hadron machines. For satisfactory operation, one needs uniform response irrespective of event topology and particle composition; good energy resolution at the trigger level to minimize effects of the response function; and adequate granularity for the selection of specific event topologies. For high selectivity, rather complex analogue computations are required, as may be seen from the examples in Table 10 [165]. A tabular summary on calorimeter facilities may be found in Ref. [3].

6.3 Monte Carlo Simulations

The development of calorimeters from crudely instrumented hadron absorbers to finely tuned precision instruments owes much to the development of a number of simulation codes. The relatively simple physics governing the electromagnetic showers

Table 10: Triggering with Hadron Calorimeters

Experiment	Trigger
Single-particle inclusive distribution; correlations	Localized energy deposit in spatial coincidence with matching track; several thresholds used concurrently.
Jet studies	Extended (~ 1 sr) energy deposit; several thresholds and multiplicities.
Inclusive leptons, multileptons	Electromagnetic deposit in spatial coincidence with matching track; several thresholds and multiplicities.
Heavy flavour jets; correlations	Various combinations of above triggers.

has facilitated their Monte Carlo simulation. Today, one program has emerged as the world-wide standard for simulating e.m. calorimeters [166]. It has progressed through several improvement stages up to the currently used version, EGS IV, which allows one to follow the shower history, tracking electron pairs down to zero kinetic energy and photons down to ~ 100 keV. It has successfully passed many very detailed tests, including the perhaps ultimate one, that of simulating absolutely the response of electrons relative to muons [167].

The physics and consequently its simulation are considerably more complex for hadronic showers. Over the last decade several programs have been developed, the aim of which is to simulate fairly accurately the detailed particle production of a hadronic cascade [25, 168, 169]. As an example, the flow chart of one of the most detailed simulations is shown in Fig. 24. The attentive reader will no doubt realize that even the most faithful physics simulation of the hadronic process will not guarantee unconditional success. Already the uncertainties associated with the sampling medium (relative response to minimum and heavily ionizing particles) and the complexities of the nuclear interactions are too large to make an *ab initio* calculation possible at present. These programs therefore do require careful tuning against many different measurements, before they become a reliable guide for designing new facilities. The results of a sample calculation are shown in Fig. 25 and give an impression of the power of this code. The status of these shower calculations has recently been extensively discussed [170].

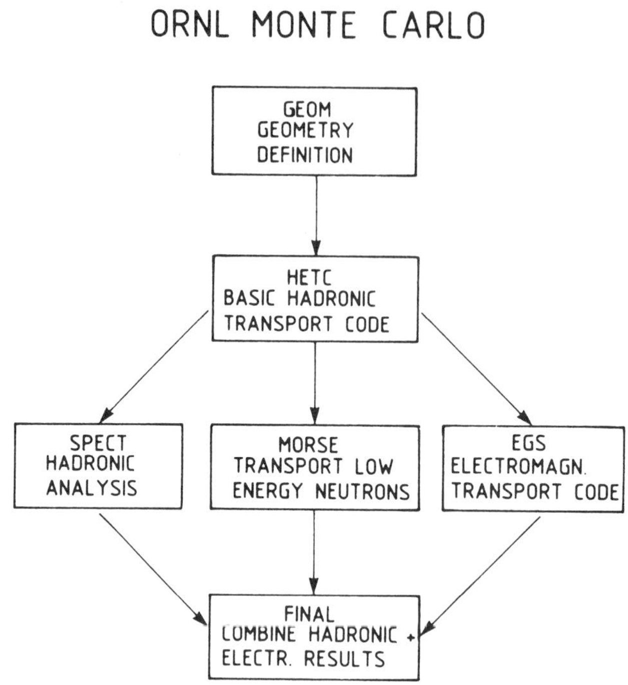

Fig. 24: Flow chart of the hadronic cascade Monte Carlo [T.A. Gabriel et al., Nucl. Instrum. Methods 195:461 (1982)].

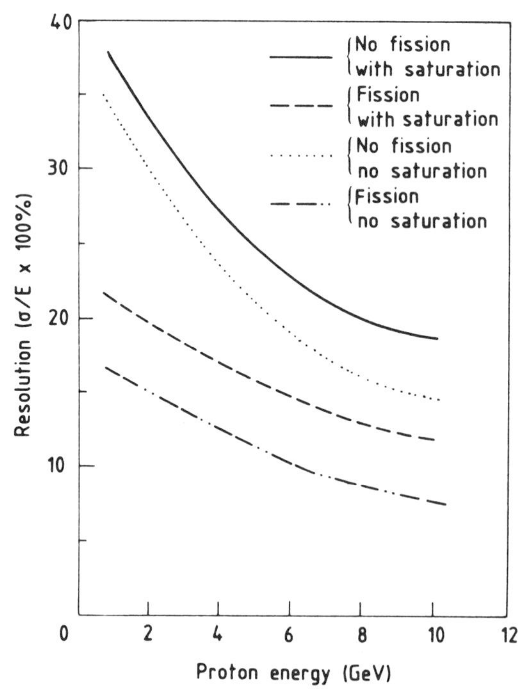

Fig. 25: Results of a 'Gedanken-experiment' using HETC to study the importance of the fission contribution and its manifestation is different readouts, with and without saturation [T. Gabriel et al., Nucl. Instrum. Methods. 164:609 (1979)].

7. OUTLOOK

The use of calorimetric methods in high-energy physics began with rather specialized applications, which capitalized on some unique features not attainable with other techniques: electromagnetic shower detectors for electron and photon measurements; neutrino detectors; and muon identifiers.

The evolution towards a more general detection technique—similar in scope to magnetic momentum analysis—had to wait for extensive instrumental developments driven by strong physics motivations. The application of this technique to the hadron storage rings (which were inaugurated in the early 1970's) was therefore delayed, but it has shaped the detectors for the CERN $p\bar{p}$ Collider and has proved to be of major importance for the detector facilities currently in preparation. At the same time, physics studies have evolved along lines where measurements based on 'classic' magnetic momentum analysis are supplemented or replaced by analyses which are based on precise global measurements of event structure, frequently requiring extraordinary trigger selectivity, and which are much more suitable for calorimetric detection. During the next decade, experiments will rely increasingly on these more global studies, with properties averaged over groups of particles and with the distinction between individual particles blurred, unless they carry some very specific information. This role of calorimetry in both present and planned physics programmes is summarized in Table 11; the impact of instrumental advances and technology is outlined in Table 12.

Despite recent conceptual and technical progress, a number of questions deserve further attention:

1) What is the precise energy dependence of hadronic energy resolution?
2) What improvement in energy, position, and angular resolution could be obtained with complete information on the individual shower distributions? With such information, could we minimize the effect of increased longitudinal leakage on the energy resolution?
3) What contributes to the measured energy resolution $\sigma(E) \simeq 0.2 \times E^{-1/2}$ of a fission-compensated hadron calorimeter?
4) Is there a cure for the low-energy non-linearities?
5) Can we understand hadronic sampling fluctuations to the same degree as we understand electromagnetic ones?
6) Can particle identification and separation be improved if more detailed shower information is available?
7) Are there any advantages in mixing different absorber materials, or in changing the sampling step inside a calorimeter?
8) Can we tailor the signal response of the readout to improve the calorimeter response?

Table 11: Future Role of Absorptive Spectroscopy

Source of particles	Physics emphasis	Calorimeter properties	Technical implications
pp ($p\bar{p}$) collider	Rare processes: high-p_T lepton, photon production; manifestations of heavy quarks $W^{\pm}$, Z^0, ...	~ 4π coverage with e.m. and hadronic detection; high trigger selectivity	Approach intrinsic resolution in multicell device; control of inhomogeneities, stability
e^+e^- collider	Complex high-multiplicity final states (multi-jets, electrons in jet, neutrinos)	Precision measurements of total visible energy and momentum	Very high granularity; particle identification
ep colliders (secondary beams, $p \geq 1$ TeV/c)	High-multiplicity final states; strong emphasis on global features	Calorimeter becomes primary spectrometer element	High granularity, high rate operation
Penetrating cosmic radiation; proton decay	Detailed final-state analysis of events with extremely low rate	Potentially largest detector systems (≥ 10,000 tons) with fine-grain readout	Ultra-low-cost instrumentation

Table 12: Interdependence of Detector Physics and Technologies for Calorimeters

Principles	Electronics	Mode of operation
Improved understanding of limitation to energy resolution in hadron calorimeters →	Gain stability at ~ 1% gain monitoring at 0.1% of ~ 10^5 analogue channels	
↓	↑	
Calorimeters replace magnetic spectrometers at high energies →	Operational systems of ~ 10^5 light- or charge-measuring channels →	Very high spatial resolution for electromagnetic and hadronic showers
	↓	↓
Particle identification through energy deposit pattern (μ, e, γ, π^0, n, $\bar{p}$, K^0_L, ν)	Cheap, fast ADC for high-level fast trigger decisions →	Helps or allows pattern recognition of very complex events (jets, ...)

The very diverse applications of calorimetric techniques will ensure continued study of these and the many technical questions connected with the signal processing of calorimeter information. There can never be a unique solution, but there should always be a search for the most suitable method. We hope that the information provided in this review will be useful for attaining this goal.

ACKNOWLEDGEMENTS

I wish to express my appreciation to Professor T. Ferbel who, as director of the 'Advanced Institute', has developed a School, which by its size, its students, its style, and its 'ambiance' appears optimally conducive to the pursuance of a variety of very topical themes.

R. Wigmans helped me considerably with a critical reading of these notes. C. Comby, L. Karen-Alun, M. Mazerand, M.-S. Vascotto, and K. Wakley efficiently and cheerfully converted my manuscript into printable form.

REFERENCES

[1] V.S. Murzin, Progress in elementary particle and cosmic-ray physics, J.G. Wilson and I.A. Wouthuysen, eds., North Holland Publ. Co., Amsterdam (1967), Vol. 9, p. 247.
[2] M. Atac, ed., Proc. Calorimeter Workshop, Batavia, 1975, FNAL, Batavia, Ill. (1975).
[3] C.W. Fabjan and T. Ludlam, Ann. Rev. Nucl. Part. Sci. 32:335 (1982).
[4] U. Amaldi, Phys. Scripta 23:409 (1981).
[5] S. Iwata, Nagoya University report DPNU–3–79 (1979).
[6] H. Messel and D.F. Crawford, Electron-photon shower distribution: Function tables for lead, copper and air absorbers, Pergamon Press, London (1970).
[7] Y.S. Tsai, Rev. Mod. Phys. 46:815 (1974).
[8] B. Rossi, High-energy particles, Prentice Hall, New York (1964).
[9] E. Longo and I. Sestili, Nucl. Instrum. Methods 128:283 (1975).
[10] H.H. Nagel, Z. Phys. 186:319 (1965).
[11] D. F. Crawford and H. Messel, Phys. Rev. 128:2352 (1962).
[12] Yu.D. Prokoshkin, Proc. Second ICFA Workshop on Possibilities and Limitations of Accelerators and Detectors, Les Diablerets, 1979, U. Amaldi, ed., CERN, Geneva (1980), p. 405.
[13] E.B. Hughes et al., IEEE Trans. Nucl. Sci. NS–19: 126 (1972).

[14] H.G. Fischer, Nucl. Instrum. Methods 156:81 (1978).
[15] R.W. Sternheimer et al., Phys. Rev. B3:3681 (1971).
[16] K. Pinkau, Phys. Rev. B139:1548 (1965).
[17] T. Yuda, Nucl. Instrum. Methods 73:301 (1969).
[18] C.J. Crannel, Phys. Rev. 182:1435 (1969).
[19] T. Kondo et al., A simulation of electromagnetic showers in iron–lead and uranium–liquid argon calorimeters using the EGS, and its implication for e/h ratios in hadron calorimetry, contributed paper to the Summer Study on the Design and Utilization of the Superconducting Super Collider, Snowmass, Colo. (1984).
[20] G.A. Akopdjanov et al., Nucl. Instrum. Methods 146:441 (1977).
S.R. Amendolia et al., Pisa 80–4.
R. Rameika et al., Measurement of electromagnetic shower position and size with a saturated avalanche tube hodoscope and a fine grained scintillator hodoscope, to be published in Nucl. Instrum. Methods.
[21] T. Kondo and K. Niwa, Electromagnetic shower size and containment at high energies, contributed paper to the Summer Study on the Design and Utilization of the Superconducting Super Collider, Snowmass, Colo. (1984).
[22] E. Gabathuler et al., Nucl. Instrum. Methods 157:47 (1978).
[23] A.N. Diddens et al., Nucl. Instrum. Methods 178:27 (1980).
[24] D. Bogert et al., IEEE Trans. Nucl. NS–29:336 (1982).
[25] J. Ranft, Particle Accelerators 3:129 (1972);
A. Baroncelli, Nucl. Instrum. Methods 118:445 (1974);
T.A. Gabriel et al., Nucl. Instrum. Methods 134:271 (1976).
[26] R. Bock et al., Nucl. Instrum. Methods 186:533 (1981).
[27] C.W. Fabjan and W.J. Willis, Proc. Calorimeter Workshop, Batavia, 1975, M. Atac ed., FNAL, Batavia, Ill. (1975), p. 1.
C.W. Fabjan et al., Phys. Lett. 60B:105 (1975).
[28] O. Botner, Phys. Scripta 23:555 (1981).
[29] A. Benvenuti et al., Nucl. Instrum. Methods 125:447 (1975).
[30] M.J. Corden et al., Phys. Scripta 25:5 (1982).
[31] A. Beer et al., Nucl. Instrum. Methods 224:360 (1984).
[32] T. Akesson et al., Properties of a fine sampling uranium–copper scintillator hadron calorimeter, submitted to Nucl. Instrum. Methods (1985).
[33] T. Gabriel and W. Selove, private communication.
[34] U. Amaldi and G. Matthiae, private communications.
[35] H. Abramowicz et al., Nucl. Instrum. Methods 180:429 (1981). See also, for earlier work, J.P. Rishan, SLAC 216 (1979).
[36] Results similar to those given in [35] were recently obtained by the WA78 Collaboration at the CERN SPS (P. Pistilli, private communication).

[37] C.W. Fabjan et al., Nucl. Instrum. Methods 141:61 (1977).
[38] W.J. Willis, Invited talk given at the Discussion Meeting on HERA Experiments, Genoa (1984).
[39] T. Akesson et al., Proc. ECFA-CERN Workshop on a Large Hadron Collider in the LEP Tunnel, Lausanne and Geneva, M. Jacob, ed., CERN 84-10 (1984).
[40] F. Binon et al., Nucl. Instrum. Methods 188:507 (1981).
[41] A.V. Barns et al., Phys. Rev. Lett. 37:76 (1970).
See also T. Ferbel, Understanding the fundamental constituents of matter, A. Zichichi, ed., Plenum Press, New York, NY (1978).
[42] J.A. Appel et al., Nucl. Instrum. Methods 127:495 (1975).
D. Hitlin et al., Nucl. Instrum. Methods 137:225 (1976).
R. Engelmann et al., Nucl. Instrum. Methods 216:45 (1983)
U. Micke et al., Nucl. Instrum. Methods 221: 495 (1984).
[43] M. Basile et al., A limited-streamer tube electron detector with high rejection power against pions, to be published in Nucl. Instrum. Methods (1985).
[44] J. Cobb et al., Nucl. Instrum. Methods 158:93 (1979).
[45] J. Ledermann et al., Nucl. Instrum. Methods 129:65 (1975).
[46] L. Baum et al., Proc. Calorimeter Workshop, Batavia, 1975, M. Atac, ed., FNAL, Batavia, Ill. (1975), p. 295.
A. Grant, Nucl. Instrum. Methods 131:167 (1975).
M. Holder et al., Nucl. Instrum. Methods 151:69 (1978).
[47] A. Bodek et al., Phys. Lett. 113B:77 (1982).
[48] K. Eggert et al., Nucl. Instrum. Methods 176 (1980) 217.
[49] Technical Proposal of the L3 Collaboration, CERN/LEPC/83-5 (1983).
[50] H. Gordon et al. (HELIOS Collaboration), Lepton production, CERN/SPSC 83-51 (1983).
[51] Design Report: An experiment at D0 to study antiproton-proton collisions at 2 TeV, December 1983.
[52] G. Arnison et al., Phys. Lett. 139B:115 (1984).
P. Bagnaia et al., Z. Phys. C. 24:1 (1984).
[53] W.J. Willis and K. Winter, *in* Physics with very high energy e^+e^- colliding beams, CERN 76-18 (1976).
[54] B.L. Beron et al., Proc. 5th Int. Conf. on Instrumentation for High-Energy Physics, Frascati, 1973. Laboratori Nazionali del CNEN, Frascati (1973), p. 362.
[55] Y. Chan et al., IEEE Trans. Nucl. Sci. NS-25:333 (1978).
[56] R. Batley et al., Performance of NaI array with photodiode readout at the CERN ISR, to be submitted to Nucl. Instrum. Methods
[57] M. Miyajima et al., Number of photo-electrons from photomultiplier cathode coupled with NaI (Tl) scintillator, KEK (Japan) 83-36 (1983).

[58] G.J. Bobbink et al., Nucl. Instrum. Methods 227:470 (1985).
[59] G. Blanar et al., Nucl. Instrum. Methods 203:213 (1982).
E. Lorenz, Nucl. Instrum. Methods 225:500 (1984).
[60] J. Ahme et al., Nucl. Instrum. Methods 221:543 (1984).
[61] J.A. Bakken et al., Nucl. Instrum. Methods 228:294 (1985).
[62] C. Laviron and P. Lecoq, Radiation damage of bismuth germanate crystals, CERN-EF/84-5 (1984).
[63] Ch. Bieler et al., Nucl. Instrum. Methods 234:435 (1985).
[64] M. Kobayashi et al., Proc. Int. Symp. on Nuclear Radiation Detectors, INS Tokyo, (1981). Inst. for Nuclear Study, Tokyo (1981), p. 465.
[65] D.E. Wagoner et al., A measurement of the energy resolution and related properties of an SCG1-C scintillation glass shower counter array for 1-25 GeV positrons, to be published in Nucl. Instrum. Methods (1985).
[66] B. Powell et al., Nucl. Instrum. Methods 198:217 (1982).
[67] W. Bartel et al., Phys. Lett. 88B:171 (1979).
[68] P.D. Grannis et al., Nucl. Instrum. Methods 188:239 (1981).
[69] K. Ogawa et al., A test of dense lead glass counters, to be published in Nucl. Instrum. Methods.
[70] R.M. Brown et al., An electromagnetic calorimeter for use in a strong magnetic field at LEP based on CEREN 25 lead glass and vacuum photo-triodes, presented at the IEEE Meeting on Nuclear Science, Orlando, Fla., 1984.
[71] C.A. Heusch, The use of Cherenkov techniques for total absorption measurements, preprint CERN-EP/84-98 (1984): invited talk given at the Seminar on Cherenkov Detectors and their Application in Science and Technology, Moscow, 1984.
[72] H. Grassmann, Untersuchung der Energieauflösung eines CsI(Tl) Testkalorimeters für Elektronen zwischen 1 GeV und 20 GeV, Universität Erlangen (1984).
H. Grassmann et al., Nucl. Instrum. Methods 228:323 (1985).
[73] M. Laval et al., Nucl. Instrum. Methods 208:169 (1983).
[74] D.F. Anderson et al., Nucl. Instrum. Methods 228:33 (1985)
[75] A. Kusumegi et al., Nucl. Instrum. Methods 185:83 (1981).
[76] H.H. Chen et al., Nucl. Instrum. Methods 150:579 (1984).
[77] E. Gatti et al., Considerations for the design of a time projection liquid argon ionization chamber, BNL 23988 (1978).
[78] K. L. Giboni, Nucl. Instrum. Methods 225:579 (1984).
[79] C. Cerri et al., Nucl. Instrum. Methods 227:227 (1984).
[80] J. Engler et al., Phys. Lett. 29B: 321 (1969).
[81] W.A. Shurcliff, J. Opt. Soc. Am. 41:209 (1951).
[82] R.C. Garwin, Rev. Sci. Instrum. 31:1010 (1960).
[83] G. Keil, Nucl. Instrum. Methods 89:111 (1970).

[84] W.B. Atwood et al., SLAC-TN-76-7 (1976).
[85] A. Barish et al., IEEE Trans. Nucl. Sci. NS-25:532 (1978).
[86] W. Selove et al., Nucl. Instrum. Methods 161:233 (1979).
[87] V. Eckardt et al., Nucl. Instrum. Methods 155:353 (1978).
[88] O. Botner et al., IEEE Trans. Nucl. Sci. NS-28:510 (1981).
[89] W. Hofmann et al., Nucl. Instrum. Methods 195:475 (1982).
[90] W. Kienzle, Scintillator development at CERN, CERN-NP Int. Report 75-12 (1975).
[91] W. Viehmann and R.L. Frost, Nucl. Instrum. Methods 167:405 (1979).
[92] J. Fent et al., Nucl. Instrum. Methods 225:509 (1984).
[93] H. Spitzer, Contribution to the Discussion Meeting on HERA Experiments, Genoa (1984);
J. Ahme et al., Novel readout schemes for scintillator sandwich shower counters, to be published.
[94] H.A. Gordon et al., Phys. Scripta 23:564 (1981).
[95] UA1 Collaboration, Technical report on the design of a new combined electromagnetic/hadronic calorimeter for UA1, CERN/SPSC/84-72 (1984).
[96] W. Kononnenko et al., Nucl. Instrum. Methods 214:237 (1983).
[97] L. Bachman et al., Nucl. Instrum. Methods 206:85 (1983).
[98] J. Borenstein et al., Phys. Scripta 23:549 (1981).
[99] H. Blumenfeld et al., Nucl. Instrum. Methods 225:518 (1984).
H. Burmeister et al., Nucl. Instrum. Methods 225:530 (1984).
[100] H. Schönbacher and W. Witzeling, Nucl. Instrum. Methods 165:517 (1979).
Y. Sirois and R. Wigmans, Radiation damage in plastic scintillators, submitted to Nucl. Instrum. Methods.
[101] G. Marini et al., Radiation damage of organic scintillation materials, CERN 'Yellow' Report, in preparation (1985).
[102] Usually, 'accelerated' tests are performed with levels of irradiation 10 to 10^6 times higher than those encountered in an experiment. Because radiation damage is frequently a function of both dose rate and integral dose, such tests are likely to indicate a higher dose tolerance than in the actual lower dose-rate experimental environment. R. Wigmans, private communication and Ref. [100].
[103] R.J. Madaras et al., Nucl. Instrum. Methods 160:263 (1979).
A.E. Baumbaugh et al., Nucl. Instrum. Methods 197:297 (1982).
A.M. Breakstone et al., Nucl. Instrum. Methods 211:73 (1982).
[104] Design report for the Fermilab Collider Detector Facility (CDF), FNAL (1981).
[105] V. Brisson et al., Phys. Scripta 23:688 (1981).
[106] W.J. Willis and V. Radeka, Nucl. Instrum. Methods 120:221 (1974).
[107] L.W. Alvarez, LRL Physics Note 672 (1968) unpublished.
[108] S.E. Derenzo et al., Nucl. Instrum. Methods 122:319 (1974).

[109] K. Masuda et al., Nucl. Instrum. Methods 188:629 (1981).
[110] T. Doke et al., Nucl. Instrum. Methods 134:353 (1976).
T. Doke, Portugal Phys. 12,1:9 (1981).
[111] G.R. Gruhn, private communication (1973).
[112] M. Conversi, Nature 241:160 (1973);
M. Conversi and L. Frederici, Nucl. Instrum. Methods 151:193 (1978).
[113] G.S. Abrams et al., IEEE Trans. Nucl. Sci. NS-27:59 (1980).
[114] V. Kadansky et al., Phys. Scripta 23:680 (1981).
[115] H.J. Behrend et al., Phys. Scripta 23:610 (1981).
[116] C. Nelson et al., Nucl. Instrum. Methods 216:381 (1983).
[117] J.A. Appel, Summary Session of the Gas Sampling Calorimeter Workshop, Fermilab FN-380 (1982);
J. Engler, Nucl. Instrum. Methods 217:9 (1983).
[118] M. Jonker et al., Nucl. Instrum. Methods 215:361 (1983).
[119] G. Battistoni et al., Nucl. Instrum. Methods 202:459 (1982).
[120] SLD Design Report SLAC-273 (1984).
[121] Presentations at the Discussion Meeting on HERA Experiments, Genoa, 1984.
[122] W.F. Schmidt and A.O. Allen, J. Chem. Phys. 52:4788 (1970).
[123] J. Engler and H. Keim, Nucl. Instrum. Methods 223:47 (1984).
[124] L. Onsager, Phys. Rev. 54:554 (1938).
[125] R.C. Munoz et al., Ionization of tetramethylsilane by alpha particles, Brookhaven Nat. Lab. C-2911 (1984), submitted to Chemical Physics Letters.
[126] G.G. Harigel, Nucl. Instrum. Methods 225: 641 (1984).
[127] P.G. Rancoita and A. Seidman, Nucl. Instrum. Methods 226:369 (1984).
G. Barbiellini et al., Nucl. Instrum. Methods 235:55 (1985).
[128] E. Gatti and V. Radeka, IEEE Trans. Nucl. Sci. NS-25:676 (1978).
[129] G. Charpak and F. Sauli, Ann. Rev. Nucl. Part. Sci. 34:285 (1984).
[130] H. Videau, Nucl. Instrum. Methods 225:481 (1984).
[131] G. Battistoni et al., Nucl. Instrum. Methods 176:297 (1980).
[132] H.G. Fischer and O. Ullaland, IEEE Trans. Nucl. Sci. NS-27:38 (1980);
M. Berggren et al., Nucl. Instrum. Methods 225:477 (1984).
[133] E. Iarocci, Nucl. Instrum. Methods 217:30 (1983).
[134] P. Campana, Nucl. Instrum. Methods 225:505 (1984).
[135] H. Aihara et al., Nucl. Instrum. Methods 217:259 (1983).
[136] B. Pope, Proc. DPF Workshop, Berkeley (1983), LBL-15973, p. 49.
[137] O. Botner and C.W. Fabjan, Measurements with the AFS calorimeter, unpublished note (1982).
[138] R. Diebold and R. Wagner, Physics at 10^{34} cm^{-2} s^{-1}, ANL-HEP-CP-84-87 and Proc. of the 1984 Summer Study of the Design and Utilization of the Superconducting Super Collider, Snowmass, Colo. (1984).
[139] P. Curie and A. Laborde, C.R. Acad. Sci. 136:673 (1903).

[140] S. Simon, Nature 135:763 (1935).

[141] T.O. Niinikoski and F. Udo, Cryogenic detection of neutrinos, CERN/NP Internal Report 74-6 (1974).
E. Fiorini and T. Niinikoski, Nucl. Instrum. Methods 224:83 (1984).

[142] B. Cabrera et al., Bolometric detection of neutrinos, Harvard preprint HUTP-84/A077 (1984).

[143] N. Coron et al., A composite bolometer as a charged-particle spectrometer, preprint CERN-EP/85-15 (1985).

[144] A.V. Vishnevskii et al., Moscow preprint ITEP-53 (1979).

[145] R. Bouclier et al., CERN-EP Internal Report 80-07 (1980).

[146] W.J. Willis, private communication.

[147] D.H. Perkins, Ann. Rev. Nucl. Part. Sci. 34:1 (1984).

[148] See, for example, Proc. 1980 International DUMAND Symposium, ed. V.J. Stenger. Honolulu, Hawaii (1981).

[149] R. Cady et al., Proc. 1982 DPF Summer Study on Elementary Particle Physics and Future Facilities, eds. R. Donaldson, R. Gustavson and F. Paige, p. 630.
R.M. Baltrusaitas et al., Phys. Rev. Letters 52:380 (1984).

[150] UA2 Collaboration, Proposal to improve the performance of the UA2 detector, CERN/SPSC 84-30 (1984).

[151] W.J. Willis, BNL 17522:207 (1972).

[152] See for example, Proc. DPF Workshop, Berkeley (1983), LBL-15973.

[153] E.D. Bloom and C.W. Peck, Ann. Rev. Nucl. Part. Sci. 33:143 (1983).

[154] Physics with very high energy e^+e^- colliding beams, CERN 76-18 (1976).

[155] E. Picasso, General Meeting on LEP, Villars-sur-Ollon, 1981, ed. M. Bourquin. ECFA 81/54, CERN, Geneva (1981), p. 32.

[156] The technical proposals for the four LEP experiments have the following LEP Committee numbers:
ALEPH, CERN/LEPC 83-2(1983);
DELPHI, CERN/LEPC/83-3 (1983);
OPAL CERN/LEPC/83-4 (1983);
L3, CERN/LEPC/83-5 (1983).

[157] W.K.H. Panofsky, Proc. Int. Symp. on Lepton and Photon Interactions at High Energies, Bonn, 1981, ed. W. Pfeil, Phys. Inst., Bonn (1981), p. 957.

[158] S.L.Wu, Phys. Reports 107:61 (1984).

[159] C. Rubbia, Physics results of the UA1 Collaboration at the CERN proton-antiproton Collider preprint CERN-EP/84-135 (1984): invited talk given at the Int. Conf. on Neutrino Physics and Astrophysics, Nordkirchen near Dortmund,1984.

[160] Experimentation at HERA, Proceedings of a Workshop jointly organized by DESY, ECFA, and NIKHEF, Amsterdam (1983).

[161] W. Selove, CERN/NP Internal Report 72-25 (1972).

[162] S. Almehed et al., CERN/ISRC/76-36 (1976).
[163] M.A. Dris, Nucl. Instrum. Methods 161:311 (1979).
[164] M. Block, UA1 Collaboration (CERN), unpublished note UA1-6, (1977).
[165] L. Rosselet, Proc. Topical Conf. on the Applications of Microprocessors to High-Energy Physics Experiments, Geneva, 1981, CERN 81-07 (1981), p. 316.
[166] R.L. Ford and W.P. Nelson, Stanford preprint SLAC-210 EGS, Version IV.
[167] T. Kondo et al., Talk given at the 1984 Summer Study on the Design and Utilization of the Superconducting Super-Collider, Snowmass, Colo., 1984.
DELPHI Progress Report, CERN/LEPC 84-16 (1984).
[168] A. Grant, Nucl. Instrum. Methods 131:167 (1975).
[169] H. Fesefeldt, Proc. Workshop on Shower Simulation for LEP Experiments, eds. A. Grant et al. CERN report in preparation.
[170] Proc. Workshop on Shower Simulation for LEP Experiments, eds. A. Grant et al. CERN report in preparation.

Reprinted, with permission, from *Physica Scripta*, Vol. 23, pp. 409-424 (1981).

FLUCTUATIONS IN CALORIMETRY MEASUREMENTS

by U. Amaldi

Abstract

Calorimeters are used in high energy physics to measure energy, position, direction and sometimes nature of a primary particle. Their properties are reviewed here with particular emphasis on the qualitative understanding of the relations between the physics of the cascade processes and the performances of actual instruments. The accent is put on energy measurements of photons and hadrons, but the limits achieved in determining the spatial position of electromagnetic and hadronic showers are also presented and discussed.

1. Introduction

In high energy physics calorimeters are blocks of matter in which the energy of a particle is degraded to the level of detectable atomic ionizations and excitations. They are used to measure not only the energy, but also the spatial position, the direction and, in some cases, the nature of the primary particle. Their performances, which improve with increasing energy, are limited both by the unavoidable fluctuations of the elementary phenomena through which the energy is degraded and by the technique chosen to measure the final products of the cascade processes. This paper sketches the status of the present knowledge of these two classes of limiting factors by treating electromagnetic showers (Section 2) and hadronic showers (Section 3). In Sections 2 and 3 energy and position measurements are discussed one after the other, but a larger emphasis is given to the discussion of energy fluctuations. Section 4 is devoted to the description of some techniques recently introduced to reduce the effect of different types of fluctuations and aims at stimulating new ideas.

It has to be stressed that, although we have a clear understanding of the elementary degradation processes, in most cases a gap exists between general cascade theories and the explanation of the experimental findings. In many instances Monte Carlo calculations fill the gap, however, the presentation of the subject is by necessity phenomenological and the selection of the basic data among the large variety of available results becomes a matter of personal choice. Further references can be found in the Proceedings of the Fermilab Workshop [1] and in the recent review article by Seigi Iwata [2].

2. Electromagnetic showers

2.1. *General properties*

The longitudinal development of an electromagnetic shower in matter is determined by the radiation length X_0. The lateral spread of a shower is mainly due to the multiple scattering of the electrons that do not radiate but have a large enough energy to travel far away from the axis. The energy of an electron that loses as much energy in collisions as in radiation has the name of critical energy ϵ, so that the natural transverse unit of a shower is the lateral spread of an electron beam of energy ϵ after traversing a thickness X_0:

$$R_M = \frac{E_s}{\epsilon} X_0, \quad (E_s = 21\ \mathrm{MeV}) \tag{1}$$

E_s is the usual constant appearing in multiple scattering theory [3]. The formulae giving the radiation unit X_0 and the Molière unit R_M are rather complicated. For rapid estimates one can use the approximate expressions

$$X_0 \simeq 180 \frac{A}{Z^2} \frac{\mathrm{g}}{\mathrm{cm}^2} \left(\frac{\Delta X_0}{X_0} < \pm 20\% \text{ for } 13 \leqslant Z \leqslant 92 \right)$$

$$\epsilon \simeq \frac{550}{Z}\ \text{MeV} \qquad \left(\frac{\Delta\epsilon}{\epsilon} < \pm\ 10\%\ \text{for}\ 13 \leqslant Z \leqslant 92\right)$$

$$R_{\mathrm{M}} \simeq 7\frac{A}{Z}\frac{\text{g}}{\text{cm}^2} \qquad \left(\frac{\Delta R_{\mathrm{M}}}{R_{\mathrm{M}}} < \pm\ 10\%\ \text{for}\ 13 \leqslant Z \leqslant 92\right) \tag{2}$$

We remark that the ratio ϵ/X_0 equals the minimum collision loss, which in the above approximation is simply

$$\left(\frac{\Delta E}{\Delta x}\right)_{\min} = \frac{\epsilon}{X_0} \simeq 3\frac{Z}{A}\frac{\text{MeV cm}^2}{\text{g}} \tag{3}$$

The quantity which is usually introduced to represent the spatial development of the shower is the differential distribution of the "track length" T, defined as the sum of the tracks of all the charged particles of the shower. The quantity T depends upon the cut-off energy E_{c}, the minimum kinetic energy of an electron (positron) that can be detected in the calorimeter. For $E_{\mathrm{c}} = 0$ all the electrons and positrons are detected and the total track length T/X_0 expressed in X_0 units is equal to E/ϵ. For increasing values of the cut-off energy the fractional useful track length F decreases as shown in Fig. 1, where we have plotted vs. E_{c}/ϵ results of calculations performed in various materials. The dashed line represents the result of an analytic calculation performed by Tamm and Belenky by using what is called "Approximation B" by Rossi [3]. In this approximation (a) all electrons lose a constant amount of energy ϵ per radiation length X_0 and (b) radiation phenomena and pair production at all energies are described by the asymptotic formulae for large energies. In this approximation the results are identical for all substances, provided one measures the thickness in X_0 units and energies in ϵ units. The dashed line represents the result of such a calculation for $E \gg E_{\mathrm{c}}$ ($z = 2.29\,E_{\mathrm{c}}/\epsilon$)

$$F(z) = 1 + z\,\mathrm{e}^z Ei(-z) \simeq \mathrm{e}^z\left[1 + z\ln\left(\frac{z}{1.526}\right)\right] \tag{4}$$

The last equation is valid within 10% for $z \leqslant 0.3$. The dotted curves of Fig. 1 represent numerical computations in lead and air by Richards and Nordheim, who assumed a continuous energy loss for the electrons [3], and indicate the difference between light and heavy materials. As expected, Approximation B (dashed line) is valid for *light* materials. Finally, the

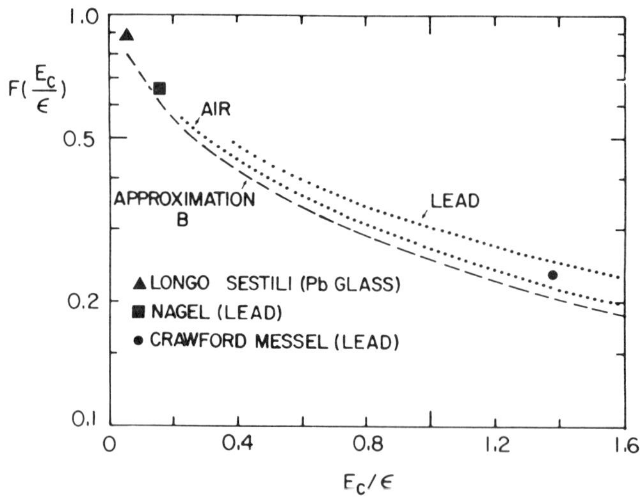

Fig. 1. Compilation of results concerning the fraction F of the total track length which is seen on average in a fully contained electromagnetic shower. The dotted lines represent the numerical calculations by Richards and Nordheim [3] and the points represent Monte Carlo results [4–6].

solid points give the results of various Monte Carlo calculations. We find that the data for different materials can be put on an almost universal curve by redefining the variable z in eq. (4):

$$z = 4.58 \frac{Z}{A} \frac{E_c}{\epsilon} \tag{5}$$

In summary the track length can be written with a good approximation as

$$T = F(z) \frac{E}{\epsilon} X_0 \tag{6}$$

and calorimetry is possible because, for any given value of E_c, the total track length is proportional to the energy E.

The distribution of the longitudinal track length, which is proportional to the longitudinal energy deposition, is well represented by the formula [6]:

$$\frac{dE}{dt} = E_0 \frac{b^{\alpha+1}}{\Gamma(\alpha+1)} t^{\alpha} e^{-bt}, \quad \left(t = \frac{x}{X_0}\right) \tag{7}$$

The parameters α and b are energy dependent and have been computed by Longo and Sestili for photon showers of less than 5 GeV in lead glass [6]. b is of the order of 0.5 and α is related to it through the position of the shower maximum $t_{max} = \alpha/b$ [7]. The median depth of the shower, i.e., the depth at which half of the incident energy is deposited, is given by the expression:

$$t_{med} = \left[\ln\left(\frac{E}{\epsilon}\right) + a\right] \quad (a = 0.4 \text{ or } 1.2 \text{ for } e\text{'s or } \gamma\text{'s}) \tag{8}$$

The median depth is related to the maximum of the shower $t_{max} \simeq t_{med} - 1.5$.

The knowledge of the longitudinal distribution of the shower allows us to compute the calorimeter length $L(98\%)$ needed to contain a fixed fraction of the incident energy. It turns out that

$$L(98\%) \simeq 3t_{med} \tag{9}$$

where t_{med} is given by eq. (8).

Equation (7) represents the *average* longitudinal distribution of a shower. For energy measurements even more important are the fluctuations in the shower development due to the statistical nature of the cascade processes. The r.m.s. values of the fluctuations for electromagnetic showers in liquid argon are given in Fig. 2(b) computed with a Monte Carlo program by Jensen, Amburgey and Gabriel [8]. Typically the fluctuations multiply or divide the average deposited energy by a factor 1.5. This factor does not contain enough information, because the energies deposited at different depths are strongly correlated. Figure 3 shows the correlations observed by Cerri and Sergiampetri in the energy deposited longitudinally in an argon calorimeter [9]. As expected, positive fluctuations in the region before the maximum of the shower correspond to negative ones on the tail and vice versa.

It has been stated above that the transverse development of a shower is governed by the Molière length R_M. This is not exactly true because, as shown in Fig. 4 (adapted from [10]), two different mechanisms contribute to the transverse distribution. The central part scales as R_M and is due to multiple scattering effects, while the peripheral part is mainly due to the propagation of the photons that are less attenuated in matter (10–

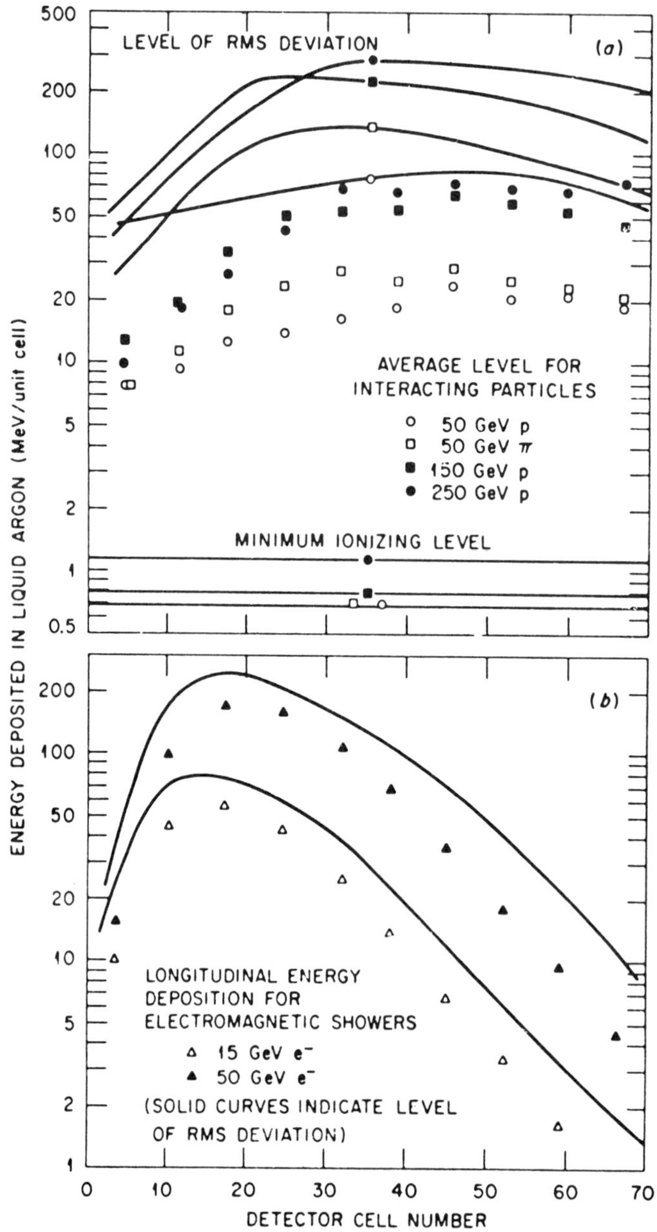

Fig. 2. Average longitudinal distribution of hadronic (a) and electromagnetic (b) showers in a lead–liquid argon calorimeter as computed by Jensen et al. [8]. The cell is made of 2 mm Pb and 1 mm of liquid argon for a total depth of 27 radiation lengths. The solid curves indicate the r.m.s. deviations from the mean values.

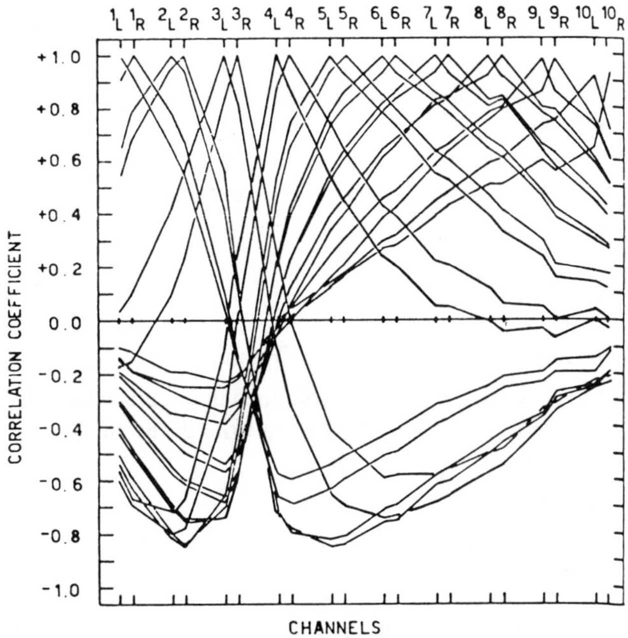

Fig. 3. Correlation coefficients between the energy deposited in the cells of an iron–liquid argon calorimeter. The measurements by Cerri and Sergiampietri [9] were made on a calorimeter with 400 layers each 1 mm iron + 1 mm argon thick. The layers were assembled in 20 cells of 1.3 radiation length. The measurement was performed with 25 GeV electrons.

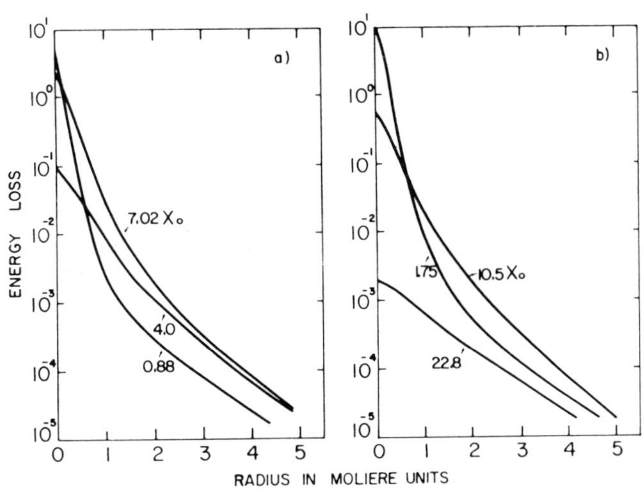

Fig. 4. Monte Carlo calculations by Yuda [10] on the lateral distribution of energy deposited by a 1 GeV shower in lead. Note the two different components, particularly visible at small depths.

20 MeV in lead). The spatial distribution of this component is determined by the minimum value of the photon attenuation coefficient, which has no simple dependence upon A and Z [11]. Still, 95% of the total energy is contained in a cylinder having the radius

$$R(95\%) \simeq 2R_{\mathrm{M}} \simeq 14 \frac{A}{Z} \frac{\mathrm{g}}{\mathrm{cm}^2} \qquad (10)$$

2.2. *Effect of containment on energy resolution*

The intrinsic energy resolution of an active calorimeter of infinite dimensions is limited by the statistics of the elementary processes and was computed to be [6]

$$\frac{\sigma(E)}{E} = \frac{\sigma(T)}{T} \simeq \frac{0.7\%}{\sqrt{E(\mathrm{GeV})}}; \quad (E_c/\epsilon = 0.5/11.8 = 0.04) \qquad (11)$$

This result applies to fully contained showers. Any lack of containment spoils the resolution. As a consequence not only the energy measurement is worsened but also the distribution of the output signals observed for monochromatic electrons acquires a tail extending towards zero. In the following this last effect will be neglected because we shall consider mainly small leakage losses.

The marble fine grained calorimeter of the CHARM Collaboration [12] has been used to determine the effect of longitudinal and lateral losses in a low-Z material ($\langle Z \rangle \simeq 13$). Figure 5 shows that the longitudinal losses are much more dangerous than the lateral ones. A 5% lateral loss spoils the energy resolution much less than a 2% longitudinal loss. Eqs. (9) and (10) give the rough calorimeter dimensions which correspond to such a containment. They are valid for energies of the order of 10 GeV. Prokoshkin [13] has extrapolated them up to 100 TeV and finds that at 100 GeV $L(98\%) \simeq 2.6 t_{\mathrm{max}}$ and at 1 TeV $L(98\%) \simeq 2.4 t_{\mathrm{max}}$. For showers in lead glasses the same paper gives for the degraded resolution the formula:

$$\text{lead glass:} \quad \frac{\sigma(E)}{E} \simeq \left[\frac{\sigma(E)}{E}\right]_{L=\infty} (1 + 4f + 50f^2) \quad (f < 0.1) \qquad (12)$$

where f is the fractional energy loss.

For a calorimeter of fixed dimensions, the fraction of the energy lost longitudinally increases with energy. Measurements

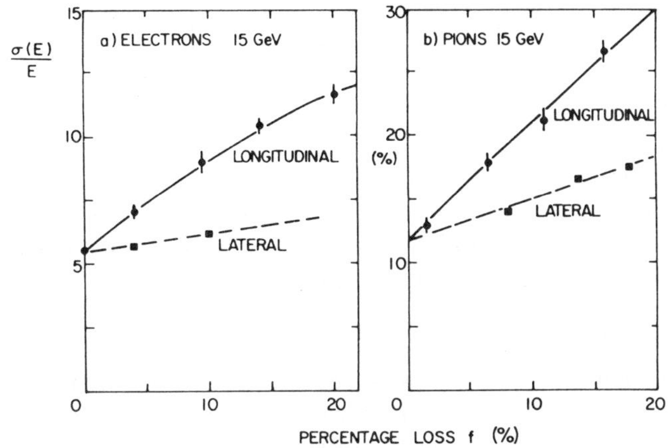

Fig. 5. Effect of the longitudinal and lateral losses on the energy resolution as measured for electrons (a) and for pions (b) by the CHARM Collaboration in a low-*Z* calorimeter.

in lead and iron of the resolution due to leakage fluctuations can be fit by an exponential form, $A \exp(-L/\zeta)$, where L is the calorimeter length and A and ζ depend weakly on incident energy [23]. For a fixed length the overall energy dependence of the resolution due to leakage is also weak, but does not follow any simple law.

2.3. *Homogeneous calorimeters*

We now pass to consider fully active calorimeters, typically homogeneous calorimeters in which the degraded energy is measured by collecting visible photons. The energy resolution of the best performing one, sodium iodide, is limited by leakage fluctuations. Indeed the resolution of a 24 radiation length long NaI crystal follows the law $\sigma(E)/E \simeq 0.9\%/E^{1/4}$ (E in GeV) [14]. The peculiar energy dependence and the fact that at 1 GeV this resolution is close to the intrinsic limit of eq. (8), while at 10 GeV it is larger by about a factor of two, indicate that the limiting factor is neither the cascade process nor the statistics of the photoelectrons (which would also give an $E^{-1/2}$ law). At variance with the case of NaI, the resolution obtainable with lead-glass counters is mainly limited by photon statistics. Typically $g = 1000$ photoelectrons are produced per GeV and, for fully contained showers, the resolution takes the form [13]

$$\left[\frac{\sigma(E)}{E}\right]_{L=\infty} = \left[\frac{\sigma_0^2(E)}{E^2} + \frac{1}{\xi g E}\right]^{1/2} \simeq 0.006 + \frac{0.03}{\xi^{1/2} E^{1/2}} \qquad (13)$$

where ξ is the ratio of the photocathode area to that of the radiator exit (typically $\xi \leqslant 0.5$) and $\sigma_0(E)$ is an energy dependent contribution which is due not only to shower fluctuations (eq. (8)) but also to the absorption of the Cerenkov light in the radiator itself. In Section 4 we shall come back to the use of light filters to reduce these types of fluctuations.

2.4. *Sampling fluctuations*

In a sampling calorimeter the degraded energy is measured in a number of sensitive layers interspersed by passive absorber. In this case the intrinsic resolution of eq. (11) is always negligible and the leakage contributions can be treated as done above in Section 2.2.

The dominating causes of fluctuations have to be discussed separately for two different classes of calorimeters: *digital* and *proportional* ones. In digital devices the fluctuations are dominated by "sampling fluctuations". These are the fluctuations in the total number of electron and positron tracks crossing the sensitive planes. In proportional calorimeters other effects have to be taken into account, as will be made clear in the following.

Let us first consider the fluctuations of the number of crossings N. For $E_c = 0$ the total track length T in a shower is given by eq. (6): $T = EX_0/\epsilon$. In Approximation B the shower has n lateral spread and in an infinitely long medium a set of planes equally spaced at distance x intercepts a number of tracks equal to the ratio T/x independently of the longitudinal distribution of the electrons and the positrons

$$N = \frac{E}{\epsilon}\frac{X_0}{x} = \frac{E}{\delta E} \qquad (14)$$

The last relation follows from the fact that ϵ/X_0 equals the minimum energy loss (eq. (3)); δE is then the energy lost by a minimum ionizing particle in traversing one layer of thickness x.

By counting the number of crossings N and by assuming that the crossings are independent and their number follows a normal distribution, one can measure the energy with an r.m.s. error which is due only to sampling fluctuations:

$$\left(\frac{\sigma(E)}{E}\right)_{\text{SAMPLING}} \simeq \frac{1}{\sqrt{N}} = \sqrt{\frac{\delta E}{F(z)E}} = 3.2\% \sqrt{\frac{\epsilon(\text{MeV})}{F(z)}} \times \sqrt{\frac{t}{E(\text{GeV})}} \qquad (15)$$

The factor $F(z)$, defined in eqs. (4) and (5), takes into account the shortening of the track length due to the cut-off energy E_c. As usual $t = x/X_0$ is the thickness of a calorimeter layer expressed in radiation lengths. To obtain a resolution as good as in a lead glass Cerenkov (eq. (13)) one must have $\epsilon t \simeq 1$, i.e., a glass Cerenkov counter is equivalent to a sampling calorimeter whose layers have thickness $t \simeq 1/\epsilon$ (MeV), i.e., $x \simeq 0.15X_0$ in lead. It has to be stressed that for very small values of t one does not expect that the low $t^{-1/2}$ continues to apply due to the correlations between the number of crossings detected in successive sensitive planes.

Equation (15) follows from Approximation B and applies to the counting of the total number of shower tracks traversing the sensitive plane immersed in a light material. In Fig. 6 we compare the spatial distributions of an electromagnetic shower in a low-Z and in a high-Z material: the transverse dimensions, measured in radiation lengths, are much larger in lead than in aluminium, because the Molière length scales as $R_M/X_0 = E_s/\epsilon$ (eq. (1)). Thus in a heavy material the electrons and the positrons form larger angles with respect to the shower axis than in a light one, and this effect is not considered in

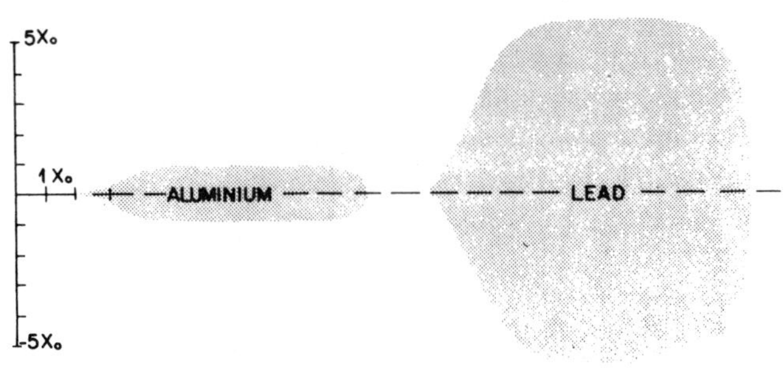

Fig. 6. Schematic representation of the cross-section of the volume that contains, on average, about 90% of the total energy of an electromagnetic shower. The units are radiation lengths on both axes.

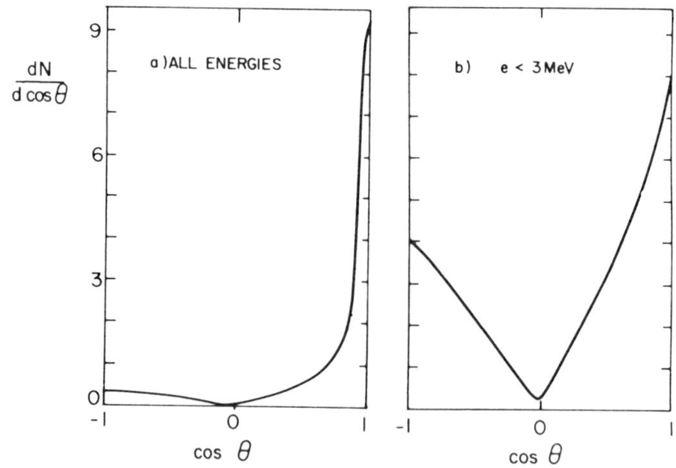

Fig. 7. Angular distributions of the electrons moving in the gaps of a lead quantameter with 48 samples for a total depth of $16X_0$ ($t = 1/3, E = 1\,\text{GeV}$) [15]. The electrons are cut at energies $e < 1\,\text{MeV}$. Numerical calculations indicate that the "isotropic" low energy component should have a distribution proportional to $\cos^2\theta$. The Monte Carlo by Fisher suggests a distribution $\sim |\cos\theta|$. According to this calculation $\sim 12\%$ of all the electrons move backwards in the gaps.

Approximation B. Figure 7 is taken from Fisher's Monte Carlo calculation [15] and shows that in lead a sizeable fraction of the low energy electrons form large angles θ with respect to the shower axis. To qualitatively understand the Z-dependence of this effect we have constructed a simplified model that divides all electrons which cross the active planes in two categories: (i) electrons that have $\theta = 0$ and energy $e_0 \leqslant e$ and (ii) electrons that are uniformly distributed and have energy $0 \leqslant e \leqslant e_0$. Figure 7 shows that in lead one should take $e_0 \simeq 4\,\text{MeV}$, i.e., $e_0/\epsilon \simeq 1/2$. The simple model, supplemented with this input from the Monte Carlo, allows the calculation of the fraction α of all the N crossings that belong to category (ii).

$$\alpha \simeq \frac{1}{2}\left(\frac{E_s}{\pi\epsilon}\right)^2; \quad (E_s = 21\,\text{MeV}; E_c = 1\,\text{MeV}) \qquad (16)$$

This equation shows that the number of large angle electrons decreases as ϵ^2, being large (of the order of 40%) in lead and small ($< 2\%$) in aluminium. Of course, α decreases when the cut-off energy E_c increases. Figure 8 shows how the fraction α depends upon E_c in our simple minded model. With active plates of thickness x_a, it is reasonable to assume that $E_c \simeq (x_a/2)(\Delta E/\Delta x)$, as indicated by the upper scale of Fig. 8. The curves then show in a *qualitative* way why dense active planes thicker than 1 g cm^{-2} are insensitive to the large angle electrons, while gas detectors with $d = 10^{-3}$–10^{-2} g cm^{-2} are expected to be severely affected by their presence. The figure also shows that in iron, and even more in aluminium, this phenomenon is much less important.

Let us now consider its effect on the sampling resolution, which for light material is given by eq. (15). To a track forming an angle θ with the shower axis, the sensitive layers appear at an effective distance $t/\cos\theta$, so that eq. (15) has to be corrected by an average factor $\langle\cos\theta\rangle^{-1/2}$. The angle θ must depend on R_M/X_0, i.e., on E_s/ϵ, but only a comparison with a Monte Carlo calculation can suggest a quantitative relation. As above, we use the calculation by Fisher [15] ($t = 1/3, E_c = 1$ MeV) to obtain, by comparing with eq. (15), the expression

$$\langle\cos\theta\rangle \simeq \cos\left(\frac{E_s}{\pi\epsilon}\right); \quad (E_s = 21\,\text{MeV}) \tag{17}$$

This expression is valid when the cut-off energy is not too large and from our simple model we infer that the condition is $E_c\,(\text{MeV}) \leqslant 10/\epsilon\,(\text{MeV})$. The sampling resolution then becomes (with $F(z)$ defined in eq. (4))

$$\begin{aligned}\left(\frac{\sigma(E)}{E}\right)_{\text{SAMPLING}} &= 3.2\% \sqrt{\frac{\epsilon\,(\text{MeV})}{F(z)\cos(E_s/\pi\epsilon)}}\sqrt{\frac{t}{E\,(\text{GeV})}} \\ &= 3.2\% \sqrt{\frac{\delta E\,(\text{MeV})}{F(z)\cos(E_s/\pi\epsilon)}}\sqrt{\frac{1}{E\,(\text{GeV})}} \end{aligned} \tag{18}$$

This equation gives for $t = 1/3$ and $z = 0$ in lead $\sigma(E)/E \simeq 6.5\%$, as computed by Fisher and, with a cut $E_c = 1.0$ MeV, corresponds to a number of tracks in lead $N \simeq 55E\,(\text{GeV})/t$, in good agreement with the Monte Carlo results by Nagel [4] who found $N \simeq 50E\,(\text{GeV})/t$.

Our formula for the sampling fluctuations deserves three remarks. First, to grasp the origin of the observed resolutions we have corrected the simple estimate of eq. (14) obtained in

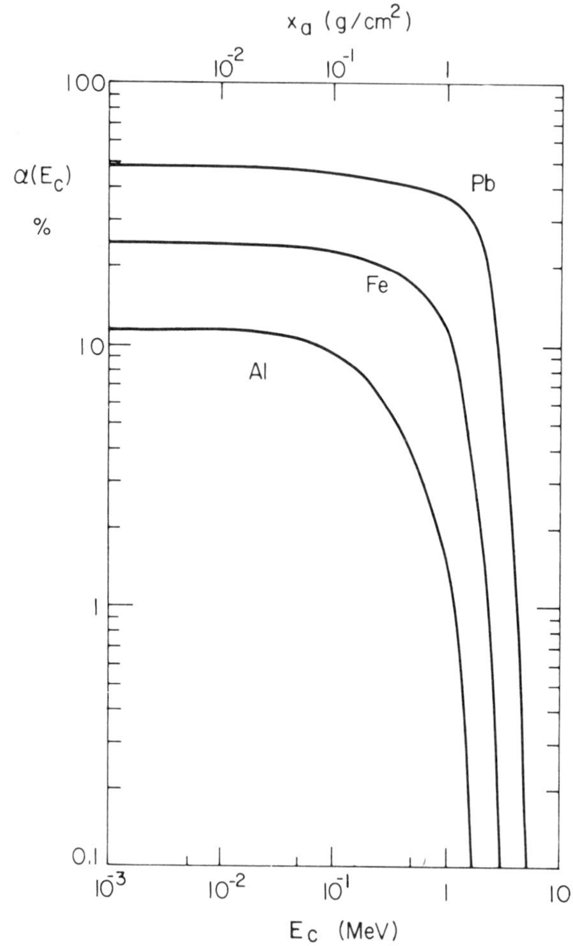

Fig. 8. By α we indicate the fraction of all the electrons and positrons traversing the gaps which belong to the "isotropic" component. The sampled fraction decreases with the cut-off energy E_c, i.e., with the thickness x_a of the active layers. The curves are the prediction of a very simple model and are only meant to show two features of the phenomenon: (i) the effect decreases drastically with Z; (ii) in low-Z quantameters one can have smaller cut-off energies and still achieve the same percentage reduction of the effect.

Approximation B by taking into account the cut-off energy E_c and the scattering of low energy electrons. In the factor $[F\langle\cos\theta\rangle]^{-1/2}$ we have lumped other less important effects neglected in Approximation B, in particular the fact that in a high-Z material the assumption of a constant absorption of gammas is incorrect even for energies larger than ϵ, so that there are more low energy photons and electrons than predicted by Approximation B. Second, it has to be remarked that the fact that eq. (14) overestimates the number of tracks in lead was noted a long time ago. Willis and Radeka [16] attributed the discrepancy to the fact that electrons and positrons are always created in pairs, so that the number of statistically independent crossings is $N/2$ and not N. We believe that this cannot be the main contribution to the widening of the resolution in lead because the argument would hold for light materials too and, for instance, it would predict a much too poor resolution for the marble calorimeter of the CHARM Collaboration: $\sigma(E)/E \simeq 28\%/E^{1/2}$ instead of the measured value $20\%/E^{1/2}$. The last remark has to do with the dependence upon the cut-off energy E_c of the two correcting factors $F(z)$ and $\cos(E_s/\pi\epsilon)$. With increasing E_c the first factor decreases while the second one increases, because the isotropic component of the shower is less sampled, as qualitatively shown in Fig. 8. We can thus expect a rough compensation between the two corrections and a wide range of validity for eq. (18).

Sampling fluctuations domine the resolution of *digital calorimeters*. The "flash calorimeter" based on the use of Conversi tubes [17] is the best known of them. In the most recent versions the sensitive planes are made of extruded plates of polypropylene forming tubular cells of $3.5 \times 5\ \text{mm}^2$ transverse dimensions [18]. The measured resolution for a calorimeter with lead plates of thickness $t = 0.9$ is $\sigma(E)/E \simeq 12\%/E^{1/2}$ (GeV), in the energy range 0.5–3 GeV, in good agreement with eq. 18 which gives $11\%/E^{1/2}$ (GeV) for $E_c \simeq 0$. The response of the calorimeter is non-linear above ~ 3 GeV, because the density of the tracks increases and there is more than one crossing per cell. For the chosen cell size this limits the achievable resolution at energies larger than 5 GeV to $\simeq 6\%$. Recently a multiwire detector consisting of an array of cubic cells operated in the limited streamer mode has been proposed as a new fast digital

Table I. *Comparison between measured and computed resolutions for scintillator calorimeters*

Ref.	Material	t	x_a ($g\,cm^{-2}$)	E (GeV)	R (%) Exp.	E_c (MeV)	z eq. (5)	$F(z)^{-1/2}$ eq. (4)	$\langle\cos\theta\rangle^{-1/2}$ eq. (17)	R (%) eq. (18)	R (%) eq. (23)
12	Al	1.0	3.0	10–50	20	3.0	0.168	1.16	1.00	23.0	61.4
23	Fe	0.3–1.5	0.65	0.2–2.5	16.9	0.65	0.068	1.09	1.03	16.1	33.7
23	Pb	0.3–1.5	1.3	0.2–2.5	12.6	1.3	0.328	1.21	1.29	13.2	13.5

calorimeter whose resolution would be limited by sampling fluctuations [19].

2.5. *Landau and path length fluctuations*

The sensitive material in a proportional calorimeter has either high density (solid or liquid) or low density (gas). In both cases other sources of fluctuations have to be considered. They dominate the behaviour of gas proportional calorimeters (multiwire quantameters) while contributing minor corrections to the sampling fluctuations in the case of solid or liquid detectors. One usually distinguishes two different contributions to the resolution: ***Landau fluctuations*** and ***path length fluctuations***.

Minimum ionizing particles traversing a thickness x of material give an asymmetric distribution of deposited energy e whose r.m.s. value has the form [20]

$$\frac{\sigma(e)}{e} \simeq \frac{2}{\ln (4W/E_m)} \tag{19}$$

where W is the energy above which on the average one δ ray is produced in the thickness x:

$$W\,(\mathrm{MeV}) = 0.15\frac{Z}{A}\,x(\mathrm{g\,cm^{-2}})] \tag{20}$$

and E_m is the minimum energy of a δ-ray, usually taken to be of the order of 30 eV. The tail of the Landau distribution is such that the contribution to the total energy resolution of N crossings is not quite proportional to $N^{-1/2}$, but for an order of magnitude estimate one can still combine eq. (19) and eq. (20) and write

$$\left(\frac{\sigma(E)}{E}\right)_{\mathrm{LANDAU}} \simeq \frac{1}{\sqrt{N}}\frac{2}{\ln\,[10^4\,x(\mathrm{g\,cm^{-2}})]} \tag{21}$$

For a detector thickness $x = 1\,\mathrm{g\,cm^{-2}}$ this increases the sampling fluctuation by less than 3%. In a gas quantameter $x \simeq 10^{-3}\,\mathrm{g\,cm^{-2}}$ and eq. (21) would predict a widening of the sampling resolution by a factor $\sim\sqrt{2}$. The actual situation is worse than this [20] since Landau formula underestimates the pulse height distribution in a gas by about a factor of two because it neglects electron binding [21, 22].

The wide spread of electron angles discussed above corresponds to large fluctuations in the path length that the electrons

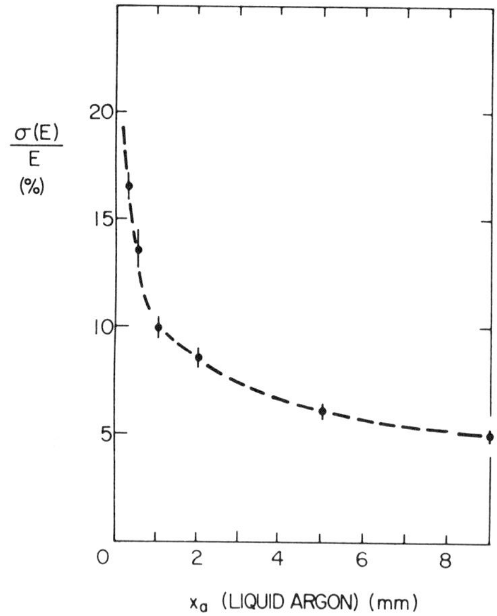

Fig. 9. Energy resolution vs. thickness of the active layers for 1 GeV electrons in an iron–argon calorimeter [15].

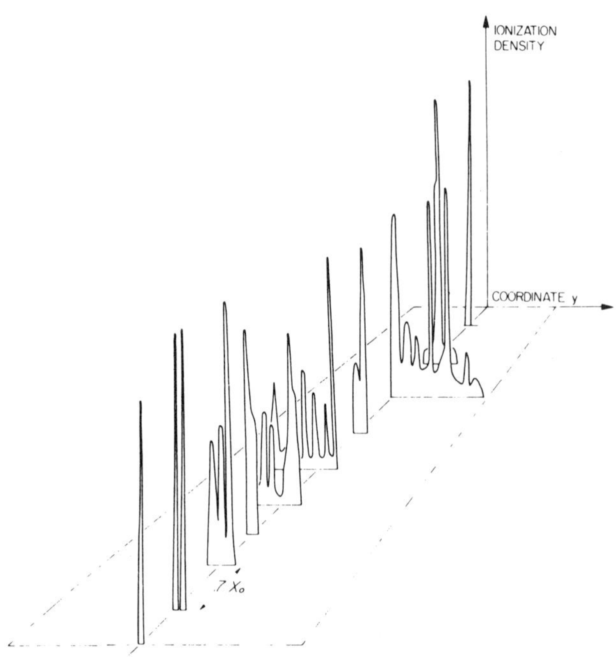

Fig. 10. Perspective view of the energy depositions in the gas gaps of a quantameter by a Monte Carlo generated 1 GeV electron shower. Note the large fluctuations happening, far from the shower axis, in the third and the fifth active layers.

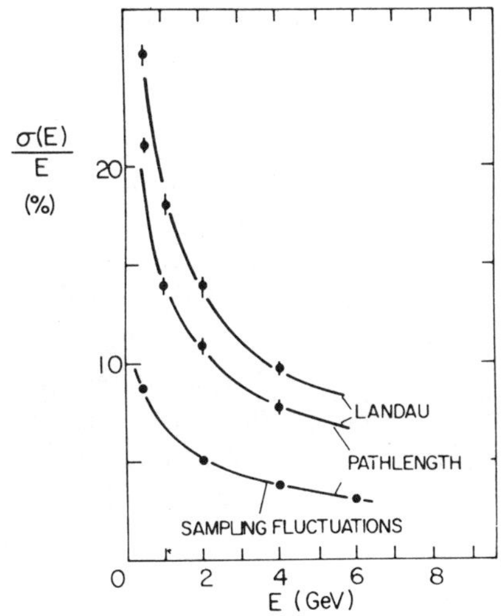

Fig. 11. Contributions of the sampling, path length and Landau fluctuations to the energy resolution of a lead–gas quantameter as computed by Fisher [15]. The last two effects give similar contributions (~ 12% at $E = 1$ GeV) which, combined quadratically with the sampling fluctuations (~ 7%), produce the overall resolution ~ $18\%/\sqrt{E}$.

themselves make in the active material of the calorimeter. These path length fluctuations are much larger in a gas than in a solid active material for two reasons: (i) the cut-off energy is much smaller in gas and low energy electrons that move along a sensitive layer leave much more energy than electrons which move perpendicularly to the plane; (ii) in a dense layer multiple scattering is much larger than in gas, so that the electrons tend to be scattered out of the dense layer with a corresponding reduction of the path length and of its fluctuations.

In summary Landau and path length fluctuations give small contributions in a calorimeter with dense proportional layers, and the sampling fluctuations are expected to dominate, as confirmed by the comparison reported in Table I, where we list computed and measured values for the quantity R defined through the relation:

$$\frac{\sigma(E)}{E} = R(\%) \sqrt{\frac{t}{E\,(\mathrm{GeV})}} \tag{22}$$

For iron and lead we have taken an experiment in which the data have been carefully corrected for leakage fluctuations and photoelectron statistics [23]. The "aluminium" point was obtained by the CHARM collaboration with a very large marble calorimeter equipped with thick scintillators (marble and aluminium have practically identical properties). By comparing the measured with the computed R-values we conclude that eq. (18) gives a good representation of the data [24]. Of course, it cannot substitute accurate measurements or detailed Monte Carlo calculations, but it has the advantage of showing in a transparent way the main corrections that have to be applied to the formula derived in Approximation B. It is worth remarking that our equation 18 for the sampling fluctuations differs greatly from the one proposed by Iwata, as shown in the last column of Table I, whose entries were computed with the formula [2]:

$$\left(\frac{\sigma(E)}{E}\right)_{\text{SAMPLING}} \simeq 2.0\sqrt{\epsilon\,(\text{MeV})}\,X_0(\text{g cm}^{-2})\sqrt{\frac{t}{E\,(\text{GeV})}}\,\% \quad (23)$$

2.6. *Energy resolution of multiwire proportional quantameters*

In liquid argon calorimeters with very thin sensitive layers the two reasons that make path length fluctuations not negligible in a scintillator sandwich calorimeter are no more valid. Figure 9 shows how the resolution deteriorates with decreasing argon thickness. It is clear that in multiwire proportional quantameters the resolution will be even worse than in an argon calorimeter. Very low energy electrons moving at large angles along a detector plane leave much more energy than $\theta = 0°$ electrons, also because of the larger ionization loss. This is shown in Fig. 10, which displays the results of a Monte Carlo calculation by G. Fisher and O. Ullaland. The energy deposited in some gas gaps is much larger than in others, and these high depositions are clearly due to electrons moving along the gap. Fisher has computed by Monte Carlo methods the effect of these path length fluctuations on the resolution of a gas device [15]. Some results for a lead–gas calorimeter are plotted in Fig. 11. They confirm the estimate based on eq. (21), from which we deduce that Landau fluctuations should increase the resolution by a factor ~ 1.4, and indicate that, in gas, path length fluctuations are as important as Landau fluctuations. Combined together, these two effects multiply the sampling resolution by a factor ~ 2. This is a well known problem in the use of gas quantameters [26] and measures to cure it have been proposed. Fisher has shown [15] that the resolution can be improved by eliminating the contributions to the ionization in the gas beyond a certain distance from the shower axis (Fig. 12). This can be done either by introducing walls to stop low energy electrons (as already partially happens if one uses proportional wire tubes instead of proportional wire chambers) or by an active cut on the measured distribution of the deposited energy in the gas. This last possibility is one of the main justifications for the development of the time projection quantameter by Fisher and Ullaland and of the drift collection calorimeter by Price [25].

2.7. *Transition effects*

Since in this paper we address the problem of fluctuations in calorimetry measurements with the aim of ascertaining the fundamental limitations, we do not treat other very important

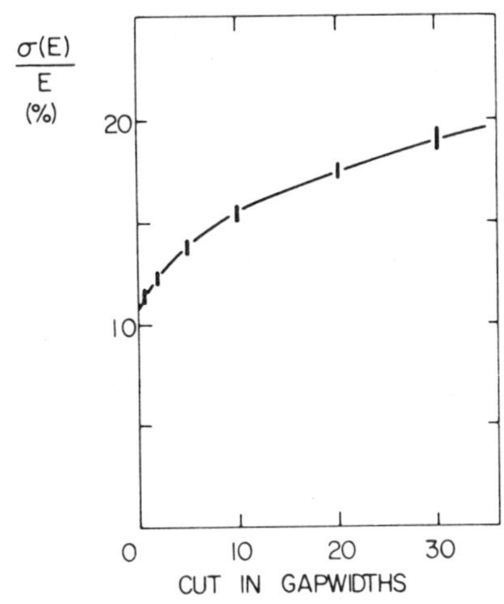

Fig. 12. Fisher's results on the improvement achievable by neglecting the energy deposited in the gas gaps far away from the shower axis. The distance from the axis is measured in gap widths. Multiple scattering is neglected in the Monte Carlo calculation. It introduces an effective cutoff that reduces the gain which is achievable with an external cut.

practical limitations such as photoelectron statistics, pedestal subtractions, calibration uniformity and so on. However, before closing this Section, we have to mention an important effect that occurs in sampling calorimeters and influences the absolute value of the observed deposited energy E_{vis}. This effect is apparent when a sampling calorimeter is calibrated by using particles of known momentum (for instance muons) moving perpendicularly through the active and passive layers of thickness x_a and x_p respectively. If P_a is the average pulse height seen in any one of the active layers, the "number of equivalent particles" in a fully contained shower producing a pulse height P_{sh} is

$$n_{eq} = \frac{P_{sh}}{P_a} \tag{24}$$

and the visible energy is

$$E_{vis} = \left[\left(\frac{\Delta E}{\Delta x}\right)_a x_a + \left(\frac{\Delta E}{\Delta x}\right)_p x_p\right] n_{eq} \tag{25}$$

where the average stopping powers have to be computed by taking into account the momentum of the particles used for the calibration. The experiments show that the ratio E_{vis}/E, where E is the energy of the incoming electron, is always smaller than 1. For instance, Stone et al. [23] found that for a lead-scintillator calorimeter this ratio is equal to 0.52 ($x_a = 0.63\,\mathrm{g\,cm^{-2}}$; $x_p = 24\,\mathrm{g\,cm^{-2}}$). Cobb et al. [27] report a ratio 0.7 for a liquid argon–lead calorimeter with $x_a = 0.28\,\mathrm{g\,cm^{-2}}$ and $x_p = 17\,\mathrm{g\,cm^{-2}}$. For the light marble calorimeter ($x_a = 3\,\mathrm{g\,cm^{-2}}$; $x_p = 23\,\mathrm{g\,cm^{-2}}$) the CHARM collaboration found $E_{vis}/E \simeq 0.85$. The results obtained by Cheshire et al. [28] in a tungsten scintillator calorimeter are plotted in Fig. 13 vs. the energy of the incoming particles. Note that in this case the first 5 radiation lengths are made of fully active CsI scintillators, so that the initial part of the cascade is much better sampled than in a normal calorimeter.

These very sizeable effects are often attributed to "transition effects" in the light active material, i.e., to the abrupt increase of the electron collision losses on crossing the boundary. Indeed, at such a boundary, while the materialization rate per radiation length of the photons remains unchanged, the collision losses per radiation length (i.e., the critical energy) increases by

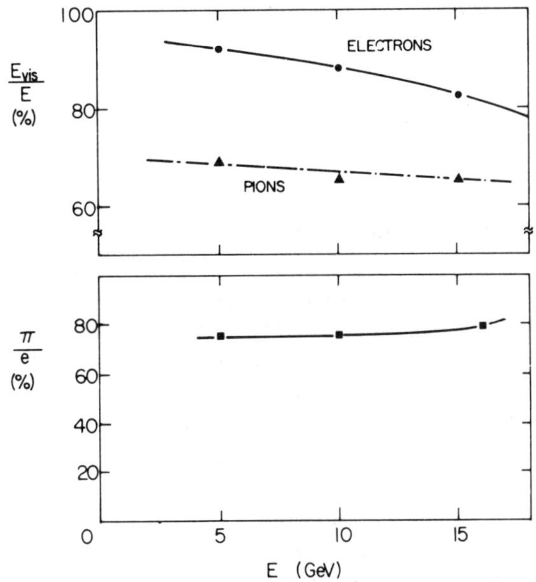

Fig. 13. Ratio of the visible energy to the incoming energy for electrons and pions illuminating the calorimeter of Cheshire et al. [28]. The lower part of the figure shows the energy dependence of the ratio between the pion and the electron visible energies. This ratio is ~ 0.75% The missing 25% is spent in undetected nuclear phenomena.

a large factor disturbing the photon–electron equilibrium with a consequent reduction of the electron flux. Some years ago the reduction expected in Approximation B was computed by Pinkau [29]. Direct measurements have also been performed [10, 30]. They show [10] that a 0.9 cm thick acrylic layer immersed in lead reduces the electron flux by ~ 20%. This is the difference, as measured with photosensitive films ($E_c \simeq 0$), between the fluxed upstream and downstream of the acrylic layer. Such a transition effect has not to be confused with two other phenomena which also reduce the number of equivalent particles n_{eq}. (A) when low-Z layers are interspersed in a block of heavy material the track length, which is proportional to E/ϵ, decreases because the average critical energy increases. (B) due to multiple scattering, the pathlength of electrons is relatively longer in the high-Z passive material than in the low-Z active layers. Equation (25) takes already into account effect (A), while effect (B) contributes, together with transition effects, to make the ratio E_{vis}/E smaller than 1. To our knowledge there is no detailed study of the relative contributions of these, and possibly other, effects to the response of sampling calorimeters to electromagnetic showers.

2.8. *Position measurements*

In this section we concentrate on the achievable accuracies in the measurement of the impact point of a gamma ray. The intrinsic resolution of such a measurement is very small, because the electron and the positron are created by the photon of energy E at angles of the order of $m_e c^2 \ln (E/m_e c^2)/E$ and their multiple scattering in less than one radiation length widens the spatial distribution to a negligible amount. The limitations are thus of practical nature and have to do with the longitudinal granularity of the calorimeter and with the transverse size of its cells.

Let us first consider the case of a matrix of total absorption counters whose lateral size d is comparable to the width of an electromagnetic shower. The IHEP group has built such a lead–glass detector with a size $d = 3.5$ cm and has developed an algorithm to deduce the coordinate of the impinging photon from the pulse heights observed in contiguous counters [31, 32]. The dependence of the r.m.s. value of transverse coordinate y on the size d of the cell is shown in Fig. 14(a) [13]. Figure 14(b) reproduces the results obtained by Amendolia et al. [33] with a similar detector: the spatial resolution improves close to the cell edges.

Spatial resolutions of 1 mm have been obtained in these detectors around 25 GeV. Since the number of shower particles grows with the energy E, the scaling law should be $\sigma(y) \simeq \sigma_0/E^{1/2}$ if the lateral correlations of the number of particles are negligible. This dependence was checked in the energy range 2–40 GeV [31]. Detectors of the tower type have the added advantage of very good two-photon separation ($\simeq 5$ cm in the above case) without suffering from the ambiguity problems typical of strip detectors. Sandwiches of lead and scintillators have also been used in the tower geometry to measure the position of a photon with a spatial resolution much smaller than the cell size [34].

In a different approach, the early part of the shower is

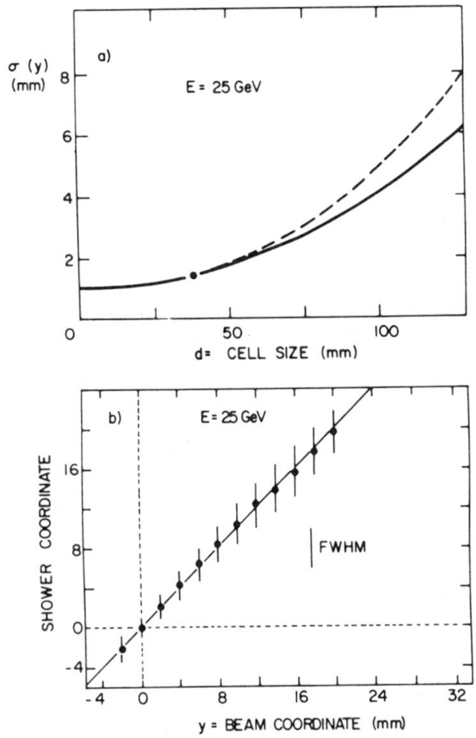

Fig. 14. (a) The GAMS detector of the IHEP group is made of lead glass blocks having a side $d = 3.5$ cm. The broken lines shows the d-dependence of the resolution for 25 GeV photons impinging at the centre of a block. The full line gives the average over block [13]. (b) Results of the Pisa group on the accuracy with which one can determine the photon position vs. the distance from the edge of the block which has $d = 3.5$ cm. The bars indicate the FWHM of the measurement [33].

sampled to determine its impact point. Multiwire proportional chambers offer the best spatial resolution and the results obtained compare well with expectations. The curve of Fig. 15(a) represents the predictions of a Monte Carlo simulation by Gabathuler et al. of the electromagnetic cascade with $E_c =$ 1 MeV [35]. It shows how the spatial resolution Δy (FWHM) obtained by using the proportional information varies with the width d of the detecting elements placed after 2.7 radiation lengths of lead glass. The measured points agree with the calculation. Figure 15(b) indicates that the distance between the

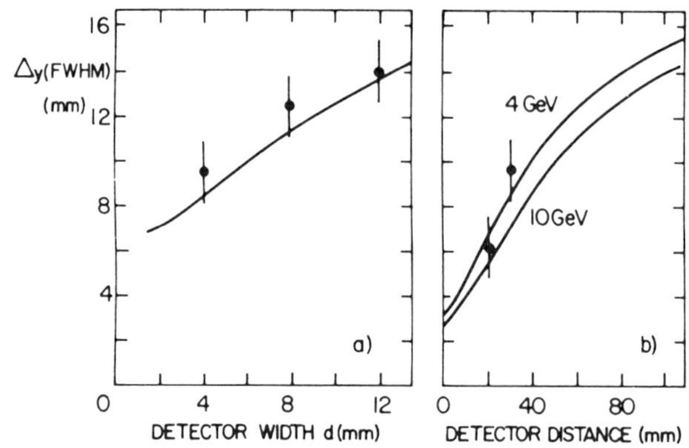

Fig. 15. (a) Monte Carlo predictions and data concerning the spatial resolution FWHM that can be achieved with a plane of proportional wires for 4 GeV photons [35]. (b) Effect of the distance between the showering material and the proportional wire plane for two different energies. The curves are Monte Carlo predictions.

showering block and the detector plane must be as small as possible because, due to the angular spread of the shower electrons, the resolution deteriorates quite rapidly with increasing distance. The Monte Carlo simulation shows that the achievable spatial resolution is practically unchanged when the thickness of the showering material is varied in the range 2–4 radiation lengths. In the energy range 10–50 GeV this method allows two-particle separations of about 3 cm with 5% confusion.

The above results can be scaled to different materials by expressing the transverse quantities in Molière units. Since in lead glass the Molière unit is $17\,\mathrm{g\,cm^{-2}} \simeq 40\,\mathrm{mm}$, to measure the impact point of a photon it is not worthwhile having proportional elements of width d smaller than $R_M/15$. In these conditions and for energies $E \simeq 5$ GeV, one single multiwire proportional plane can give a resolution Δy at FWHM of the order of $0.2 R_M \sqrt{\epsilon}/12$, where the lead glass data ($\epsilon = 12$ MeV) have been scaled according to the statistics of the number of tracks. For lead in bulk ($R_M = 16$ mm) this would correspond to $d \simeq 1$ mm and Δy (FWHM) = 2.5 mm, while for iron ($R_M = 18$ mm) with $d \simeq 1.5$ mm one would obtain Δy (FWHM) $\simeq 5$ mm. Any decrease in average density increases proportionally the achievable spatial resolution.

Better resolutions have been obtained with scintillator strips. This should not come as a surprise, since gas proportional wires have larger fluctuations than scintillators. The IHEP group measured the resolution achievable with scintillator strips of width $d = 5$, 10 and 15 mm immersed in lead, in lead scintillator and in iron scintillator sandwiches [36]. For large energies (20–50 GeV) the resolution does not deteriorate appreciably by going from $d = 5$ mm to $d = 15$ mm and in lead for $d = 15$ mm one has Δy (FWHM) $\simeq 3$ mm. Figure 16 shows the distributions of shower centres of gravity y_0 at various depths. By comparing with the proportional wire case discussed above, we conclude that scintillator strips can be much wider than proportional wire cells and still achieve the same spatial resolution. This concept has been used by the IHEP group in the construction of an iron scintillator sandwich which measures photons with $\sigma(y) \simeq$ 2 mm at energies of many tens of GeV, but has still $\sigma(y) \simeq$ 6 mm at 0.5 GeV [37]. Two photon separations of 5 cm have also been obtained with such a detector. Argon calorimeters have reached similar spatial resolutions and separations. For instance the CELLO group [38] has obtained $\sigma(y) = 3$ mm with 20 mm strips at $E = 1$ GeV.

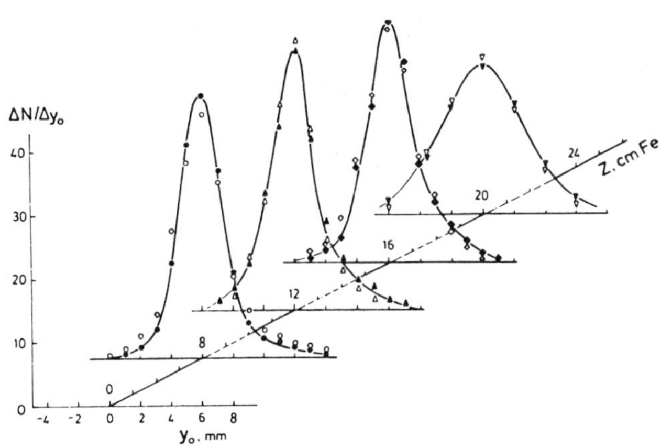

Fig. 16. Distribution of the centre of gravity y_0 of 40 GeV electromagnetic showers at different depths for two counter widths: $d = 5$ mm (solid points) and 15 mm (empty points) [36].

3. Hadronic showers

3.1. *General properties*

The development of hadronic showers in matter is so complicated that simplified analytical treatments are not available. However, the elementary physical processes are well understood and many Monte Carlo programs exist which simulate them. The most important properties of the interesting phenomena are summarized in Table II, which is mainly based on the presentations by Fabian, Willis and collaborators [39, 40] and by Sciulli [41].

Hadron production is insensitive to the energy and the type of the projectile and the multiplicity increases very slowly with the mass number of the target material. In pair production the photon inelasticity is 1, while the average inelasticity in a hadron interaction is one half, so that half of the energy is carried by leading particles. Neutral pions amount, on average, to 1/3 of the produced pions and their energy is dissipated in the form of electromagnetic showers. The fraction of the total energy which is dissipated in ionizations by electrons and charged hadrons fluctuates from event to event and is the main contribution to the energy resolution in hadron calorimetry because, while the electromagnetic energy and the energy of the charged hadrons are well sampled, a large fraction of the remaining energy is not seen. Indeed, the energy which goes either in breaking nuclei (binding energy) or in low energy neutrons is invisible; moreover many of the low energy particles (gammas and protons of few MeV) produced in the de-excitation of nuclei are badly sampled because of the short range and/or saturation effects in the active material. As pointed out by Sciulli [41] and reported in Table II, a sizeable fraction of the nuclear excitation energy goes into fast protons, i.e., protons having average energies of 150 MeV, which can be detected. Finally, muons and neutrinos emitted in the decay of pions escape detection in any reasonable size calorimeter. In a 40 GeV shower they give [39] a loss of $\sim 1\%$, which decreases with energy and will be neglected in the following.

The relative contributions of the above processes to the energy visible in a calorimeter have been determined by Monte Carlo calculations. The input formulae are by necessity approximate and the results obtained by various authors differ appreciably. This is apparent in Fig. 17, where we have summarized the

Table II. *Properties of the phenomena which determine the development of hadronic showers*

Phenomenon	Properties	Influences energy resolution through	Characteristic time	Characteristic length
Hadron production	Multiplicity $\simeq A^{0.1} \ln s$ Inelasticity $\simeq \frac{1}{2}$	π^0/π^+ ratio Binding energy loss	10^{-22} s	Abs. length $\lambda \simeq 35A^{1/3}$ g cm^{-2}
Nuclear deexcitation	Evaporation energy $\simeq 10\%$ Binding energy $\simeq 10\%$ Fast neutrons $\simeq 40\%$ Fast protons $\simeq 40\%$	Binding energy loss No sampling of n's Poor sampling of slow parts and γ's	10^{-18}–10^{-13} s	Fast neturons $\simeq 100$ g cm^{-2} Fast protons $\simeq 20$ g cm^{-2}
Pion and muon decays	Fractional energy of μ's and ν's $\simeq \frac{5\%}{\ln E\,(\mathrm{GeV})}$	Losses of ν's Losses of μ's	10^{-8}–10^{-6} s	$\gg \lambda$

results obtained by Ranft [42], Baroncelli [43] and Gabriel [44]. The first Monte Carlo was mainly devised to study shielding problems. Baroncelli's approximations are valid at high energies, while Gabriel accurately simulated the low energy and nuclear part of the cascade processes. Still, one would expect that around 10 GeV the three approaches should give consistent results, while Fig. 17 shows that the results are widely different. For instance, for 10 GeV protons the fraction of the total energy appearing in the form of electromagnetic showers is 30%, 45% and 25% respectively. (Since Baroncelli's calculation refers to positive pions, we have reduced the fraction reported in Fig. 17(b) by 10% to compare it with the other calculations.) The fractions going into nuclear binding energy plus "nuclear fragments", which are the particles produced in the phenomenon named nuclear de-excitation in Table II, appear to be 40%, 35% and 35% at 10 GeV. In this case the three calculations give similar results, all of them pointing to the well-known fact that about 1/3 of the total energy of a hadronic shower is practically invisible. The more detailed calculations by Gabriel show that 2/3 of this energy (i.e., $2/9 \simeq 20\%$ of the total) is totally lost because it is spent to compensate for the binding energy of the disrupted nuclei, while 1/3 of it (i.e., $\simeq 10\%$ of the total) goes into nuclear fragments. Table II shows that half of this energy ($\sim 5\%$ of the total) appears under the form of "fast" protons, i.e., protons of about 150 MeV. As pointed out by Sciulli [41], this energy can be sampled if the thickness of the calorimeter layer is smaller than the range of 150 MeV protons, i.e., less than ~ 3 cm of iron [41].

As already mentioned, shower fluctuations dominate the energy resolution of a hadron calorimeter. The average number of π's in a shower is small and its fluctuations large. This is illustrated in Fig. 18b, which shows the 1 r.m.s. band for the total number n_0 of π^0's in a shower initiated by π^+ of energy E in iron [45]. It is seen that this number increases as $\ln E$,

$$n_0 \simeq 5 \ln E\,(\text{GeV}) - 4.6 \qquad (E > 2.5\ \text{GeV}), \tag{26}$$

and that the r.m.s. value of the distribution, which is reasonably symmetric around the central value, is approximately equal to $\sqrt{n_0}$. Figure 18(a) shows the ± 1 r.m.s. band for the fraction of the total energy dissipated in the form of electromagnetic

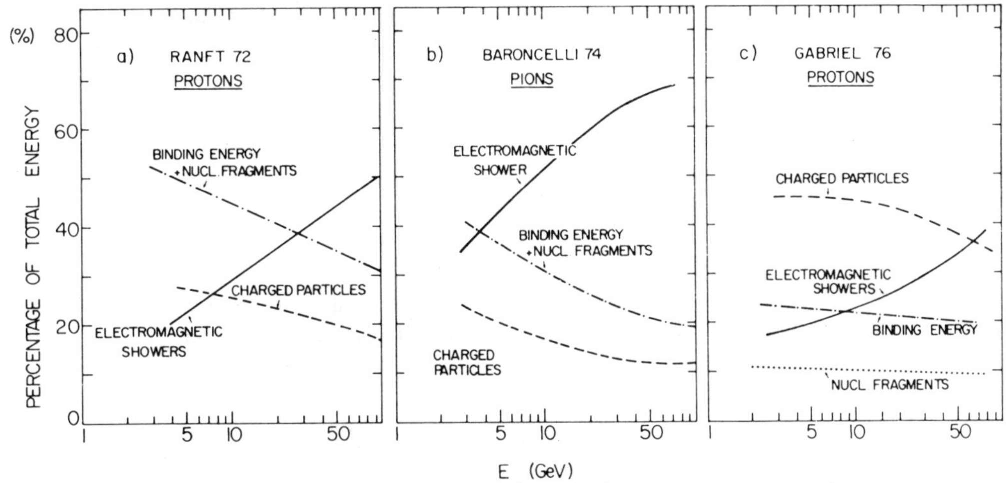

Fig. 17. Relative contributions of the most important processes to the energy dissipated by hadronic showers in iron. The Monte Carlo calculations are by Ranft [42], by Baroncelli [43, 45] and by Gabriel [44].

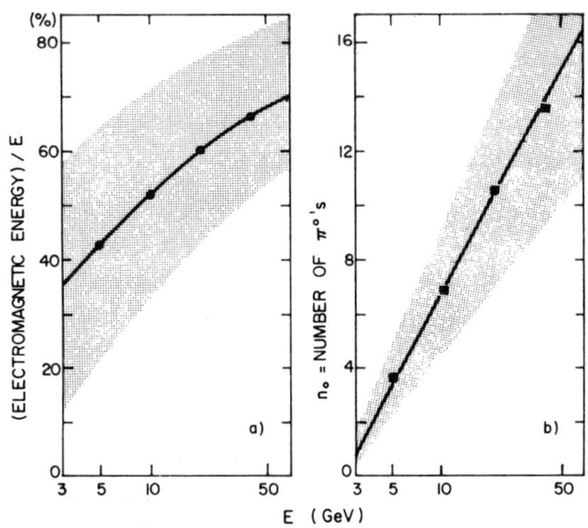

Fig. 18. The points are (a) the average ratio between the electromagnetic energy and the initial energy in iron and (b) the number n_0 of π^0's as computed by Baroncelli [45]. The bands represent ± 1 r.m.s. around the central value.

showers, as computed by Baroncelli's Monte Carlo. If all the other forms of energy were invisible, the width of the band would be directly related to the achievable energy resolution. Fortunately, as seen above, about half of the remaining energy is sampled and the electromagnetic fluctuations are partially compensated by the energy dissipated by relativistic charged particles and by fast protons.

3.2. *Effect of containment on energy resolution*

Figure 2(a) reproduces the Monte Carlo results of Jensen, Amburgey and Gabriel [8] and illustrates the fluctuations in the longitudinal development of hadronic showers induced by pions and protons in an iron–liquid argon calorimeter. By comparing the curves drawn in Fig. 2(a) and in Fig. 2(b) one gets a visual impression of the differences between electromagnetic and hadronic showers: fluctuations multiply the average energy deposited in one plane by a factor equal to ~ 5 in the case of hadrons, while the same factor is ~ 1.5 in the case of electrons and photons.

In a hadronic shower the energy deposited in a layer of matter initially rises as a function of depth (Fig. 2(a)). In iron this rapid rise is partly due to the electromagnetic component. Beyond the maximum the energy deposition is essentially due to the hadronic component and the decrease is slow. Iwata compiled many different empirical formulae that give the median depth of hadronic showers [2] and concluded that a good representation of the data is given by

$$x_{\text{med}} = [0.54 \ln [E(\text{GeV})] + 0.4]\, \lambda_0 \qquad (27)$$

where λ_0 is the nuclear absorption length. For rapid calculation one can use the approximate formula

$$\lambda_0 \simeq 35 A^{1/3}\, \text{g cm}^{-2} \qquad (28)$$

The longitudinal development of hadronic showers is almost energy- and particle-independent when the depth is measured in units of x_{med}. For this reason it is possible to give a general rule to contain a fixed fraction of the total energy longitudinally [2]:

$$L(95\%) \simeq 2.5 x_{\text{med}} \qquad (29)$$

Note that this formula parallels eq. (9) written down for electromagnetic showers. It implies that, for an iron calorimeter that has to detect 50 GeV hadrons, $L(95\%) \simeq 1.0$ m. This is an overestimate of the thickness really needed, since the formulae reported for iron by Holder et al. [46]

$$L(95\%) \simeq [6.3 \ln [E\,(\text{GeV})] \;+ \text{cm Fe}$$

and by Prokhoskin [13]

$$L(95\%) \simeq [9 \ln [E\,(\text{GeV})] + 40] \text{ cm Fe}$$

both give $L(95\%) \simeq 75$ cm.

The effect of an $f = 5\%$ loss on the hadron energy resolution can be read from the graph of Fig. 5(b), which shows that the longitudinal containment is more important than the lateral one, as for electromagnetic showers. Since 95% of the energy is contained in a cylinder having $R(95\%) \simeq \lambda_0$ [2], a good iron calorimeter must have a diameter of about 250 g cm^{-2}, or 35 cm for full density.

3.3. *Energy resolution in hadronic calorimetry*

The best known *homogeneous calorimeter* is the liquid scintillator neutrino detector described by Benvenuti et al. [47]. Its energy resolution for fully contained hadronic showers can be represented by the formula

$$\frac{\sigma(E)}{E} = 9\% + \frac{11\%}{\sqrt{E}\,(\text{GeV})} \tag{30}$$

and the ratio E_{vis}/E passes from 0.75 to 0.80 when the energy of the incident pions increases from 10 to 150 GeV. The large constant term of eq. (30) indicates that the resolution is not determined by the fluctuations of elementary processes proportional in number to the energy E, as the sampling fluctuations considered for electromagnetic showers. A large sodium iodide was also used to measure hadron energies [48], but in this case the containment corrections were too large to give useful information on the achievable energy resolution.

Sampling calorimeters are widely used in detecting hadron showers. No general formula can be given to describe the achievable resolution and its dependence upon energy, material and layer thickness and one has to resort either to Monte Carlo calculations or, preferably, to experimental data. In the present case sampling fluctuations give a relatively small contribution for small enough thickness of the passive layers. This has been experimentally proven by Fabjan et al. [40] with an iron–liquid argon calorimeter having 1.5 mm thick plates by comparing the signals in interleaved active layers. In this way each event is measured twice at a distance in iron of $\sim 1.2\,\mathrm{g\,cm^{-2}}$ and shower fluctuations can be separated from sampling fluctuations. It was found that without sampling fluctuations the energy resolution would be reduced by less than 10%.

Figure 19 is adapted from [2] and contains a selected sample of data on the dependence of the measured energy resolution upon the thickness of passive material x_p. The x_p-dependence is more complicated than in the case of electromagnetic showers. For $x_p \gtrsim 100\,\mathrm{g\,cm^{-2}} \simeq 13$ cm of Fe the electromagnetic showers are sampled every $\gtrsim 7$ radiation lengths and the resolution deteriorates linearly with x_p [49]. We find that, for $x_p \lesssim 100$ g $\mathrm{cm^{-2}}$, in iron the resolution behaves as

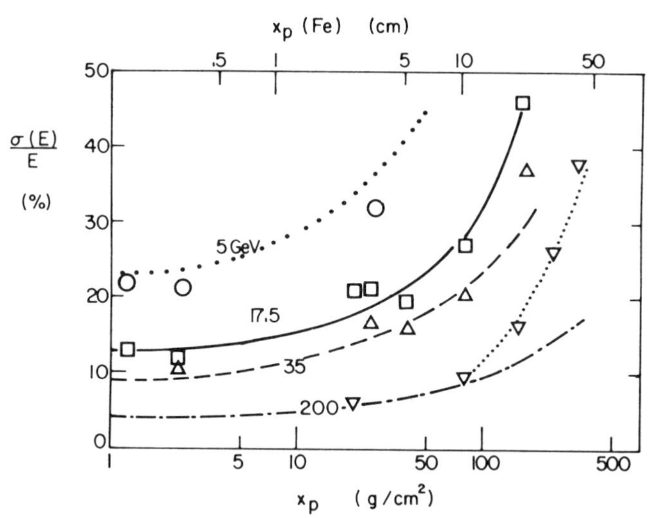

Fig. 19. The compilation by Iwata [2] on the thickness dependence of measured hadronic resolutions at 5, 17.5, 35 and 200 GeV is compared with the predictions of eq. (31) with $R' = 40\%$. The agreement is good below $\sim 100\,\mathrm{g\,cm^{-2}}$. The dotted curve above $100\,\mathrm{g\,cm^{-2}}$ is only meant to guide the eye through the 200 GeV points.

$$\left[\frac{\sigma(E)}{E}\right]^2 \simeq \left[\frac{50\%}{\sqrt{E\,(\mathrm{GeV})}}\right]^2 + \left[R'(\%)\sqrt{\frac{4t}{3E\,(\mathrm{GeV})}}\right]^2 \tag{31}$$

The first term is attributed to shower fluctuations. It is sizeable because, as discussed above, about one fourth of the hadron energy E is not sampled and the fluctuations on the missing 25% are very large. The second term has the typical dependence of the energy resolution for electromagnetic showers (eq. (22)) with the energy multiplied by a factor 3/4, which is approximately the fraction of the total energy dissipated in ionizations by electrons, positrons and charged hadrons. (For our present purposes there is no need to distinguish between a hadron and a lepton because, for equal energies dissipated in ionization and very low cut-off values ($E_c/\epsilon \simeq 0$), the total track length is the same.) The data collected by Iwata and presented in Fig. 19 are approximately reproduced by introducing the value $R' = 40\%$ in eq. (31). The results of the very recent measurement of the CDHS Collaboration [50] are presented in Fig. 20 together with the curves predicted by eq. (31) with $R' = 40\%$ and $R' = 30\%$. It appears that the second choice has to be preferred at low energies and for $x_p \lesssim 60\,\mathrm{g\,cm^{-2}} \simeq 8\,\mathrm{cm}$ while the first one gives an overall average fit at all energies. We shall see in Section 4.2

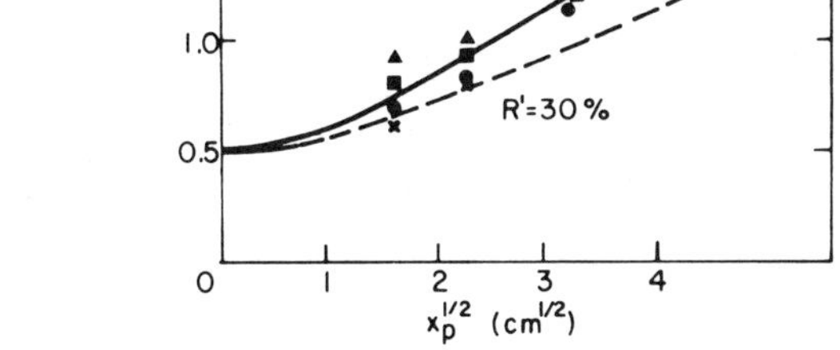

Fig. 20. Data of the CDHS Collaboration [50] are compared with eq. (31) for two values of the parameter R'.

that, for $x_p = 2.5$ cm, the high energy data points can be brought to coincide with the low energy ones with an appropriate algorithm, so that we conclude that $R' \simeq 30\%$ gives a good representation of optimized data. For the same calorimeter eq. (18) predicts a sampling electromagnetic resolution $\sigma(E)/E = 20\%/\sqrt{E}$ in agreement [51] with the measured value $23\%/\sqrt{E}$. The measured resolution corresponds to a value $R(\text{Fe}) \simeq 19\%$ for the parameter appearing in eq. (22), which is close [51] to the one measured by Stone et al. [23] and by Asano et al. [24]. By comparing hadronic and electromagnetic resolutions we conclude that in Fe

$$\frac{R'}{R} \simeq 1.5\text{–}2 \tag{32}$$

Why is this ratio not unity? We attribute at least part of the difference to the fact that charged and neutral pions form larger angles with the shower axis than the electrons of a pure electromagnetic shower. This would introduce in R a correcting factor $\langle \cos\theta \rangle^{-1/2}$ which is larger than the 1.03 given for iron by eq. (17) (see Table I). Moreover, the pion track length is insensitive to the cut-off E_c, at variance with the electron track, and shower fluctuations between the two types of energy depositions contribute to the t-dependent term in the resolution.

Equation (31) shows that it is not worthwhile reducing the plate thickness below 1 radiation length, i.e., ~ 2 cm of steel. In these conditions the energy resolution is dominated by shower fluctuations and is of the order of $50\%/\sqrt{E}$. One may expect that this limiting value decreases together with the invisible fraction of the total energy. Data of the CHARM Collaboration [12] seem to support such a view. In the marble calorimeter, which sees $\sim 85\%$ of the total energy, the hadronic resolution is $\sigma(E)\sqrt{E} = 53\%$, which agrees with eqs. (31) and (32) if the limiting resolution is $\sim 40\%/\sqrt{E}$. This is smaller than the $\sim 50\%/\sqrt{E}$ derived by fitting the data reported in Figs. 19 and 20 for various iron calorimeters, which typically see $\sim 75\%$ of the total energy around 10–20 GeV. However, the situation is more complicated than that, because by increasing the energy to 140 GeV, and thus increasing the fraction of visible energy to $\sim 85\%$, the CDHS Collaboration found that the resolution deteriorates with respect to the predictions of eq. (31) with a fixed value of R' (see Fig. 20). In our opinion the material and energy dependence of the limiting resolution is the most important open problem in hadron calorimetry.

3.4. *Position measurements*

Figure 21 shows the transverse distribution of a hadron shower as measured by the IHEP–IISN–LAPP Collaboration [52] with an iron-scintillator calorimeter having 5 cm wide and 1 cm thick hodoscope strips sandwiched between 2.5 cm thick steel plates. The shower can be parametrized as

$$\frac{\mathrm{d}E}{\mathrm{d}y} = a_1 \mathrm{e}^{-|y|/b_1} + a_2 \mathrm{e}^{-|y|/b_2} \tag{33}$$

with $a_1/a_2 \simeq 2$, $b_1 \simeq 2.2\,\mathrm{cm}$ and $b_2 \simeq 7\,\mathrm{cm}$. It follows that accurate measurements of the conversion point of a neutral hadron (neutron, K_0 etc.) can be made in iron if the cells are not wider than $2b_1 \simeq 4\,\mathrm{cm}$. The Bologna group [53] measured the r.m.s. value $\sigma(y)$ at different depths in iron with 2 cm wide strips. Figure 22 shows that $\sigma(y)$ is proportional to the distance from the shower vertex in the region from $\sim$ 15 to 55 cm, i.e., for distances larger than one interaction length. For smaller

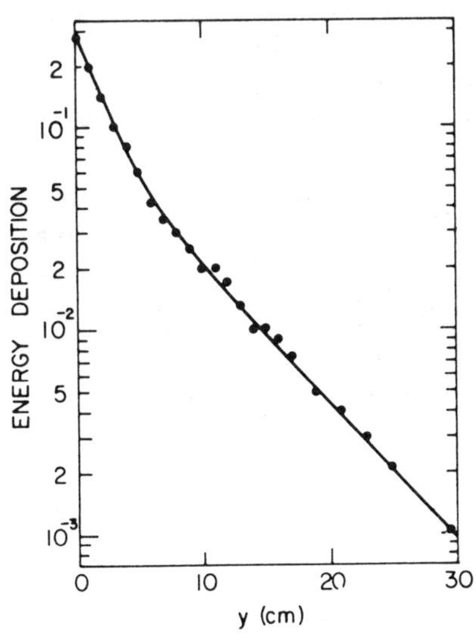

Fig. 21. Lateral distribution of a shower initiated by a 30 GeV anti-proton in the iron calorimeter of the IHEP–IISN–LAPP group [52]. Note the two components of different slope (eq. (33)).

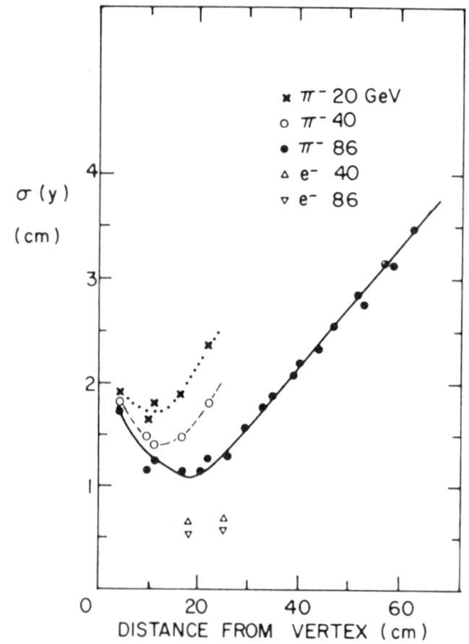

Fig. 22. r.m.s. value of the transverse distribution of a hadronic shower at different depths in iron as measured by the Bologna group [53].

distances $\sigma(y)$ increases, probably due to albedo effects, and at the vertex is ~ 18 mm, equal at all energies. The figure shows that with a single plane of scintillator strips 2 cm wide in iron one can reach spatial resolutions which are of the order of 15 mm.

With a multilayer structure better resolutions have been achieved. Binon et al. [52] have measured the resolution in an iron calorimeter with $x_p \simeq 20\,\mathrm{g\,cm^{-2}}$ as a function of the cell size d. The results are shown in Fig. 23 and indicate that a reduction of the counter width from $d = 5$ to $d = 2.5$ cm does not gain very much. The dependence is of the form

$$\sigma(y) = \sigma_0(y)\,\mathrm{e}^{d/d_0} \tag{34}$$

with $d_0 = 10$ cm. According to Prokoshkin [13] the energy dependence of $\sigma_0(y)$ is

$$\sigma_0(y) = \frac{15}{E}\left[\ln^4 E + \frac{E}{4}\right]^{1/2} \text{mm} \tag{35}$$

where E is, as usual, in GeV. Similar results have been obtained in the much less dense calorimeter of the CHARM Collaboration

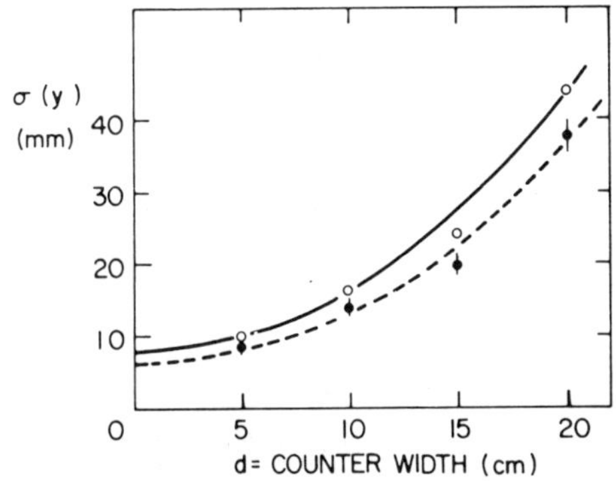

Fig. 23. Spatial resolution vs. strip width measured by the IHEP–IISN–LAPP Collaboration at 25 GeV (open points) and 40 GeV (solid points). The curves are fits based on eq. (34).

[12]. Figure 24 shows the energy dependence of the vertex determined by planes of 3 cm wide gas proportional tubes separated by marble layers 1 radiation length thick ($\simeq 25$ g cm^{-2}). The spatial resolution is $\sigma(y) \simeq 30$ mm at 40 GeV, worse than in an iron calorimeter with 3 cm strips because of both the lower density and the use of gas counters instead of scintillators.

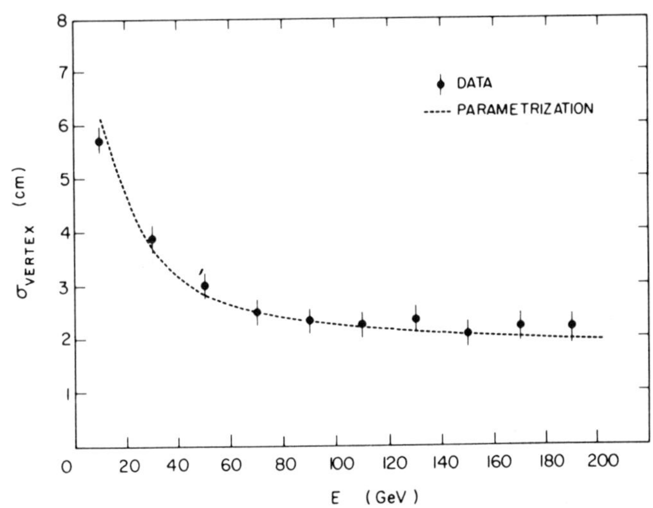

Fig. 24. Results of the CHARM Collaboration on the energy dependence of the spatial accuracy of the shower vertex in a light calorimeter equipped with $d = 3$ cm proportional wire tubes [12]. The parametrization is $\sigma_{\mathrm{VERTEX}} = [19.5/\sqrt{E} + 0.003E]$ cm with E in GeV.

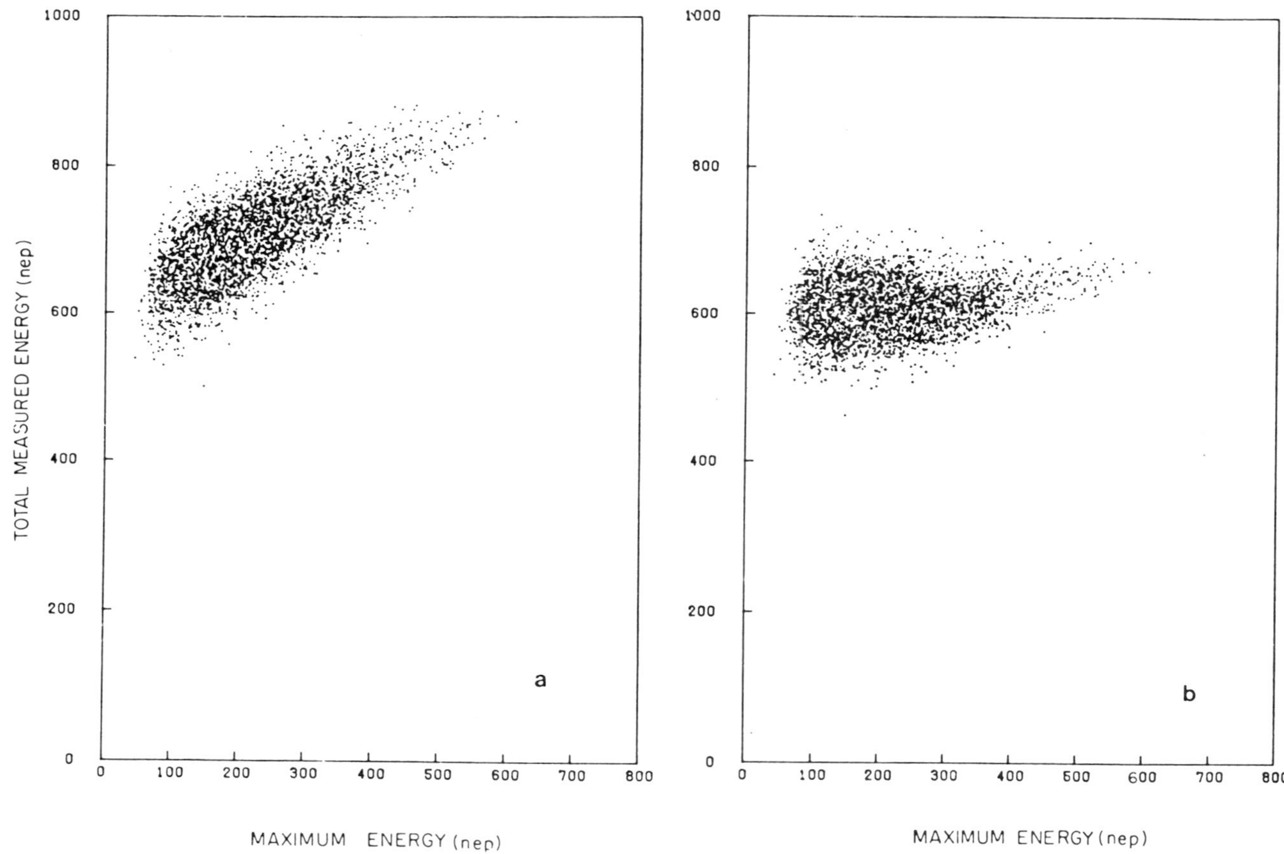

Fig. 25. (a) CDHS results on the correlation between the measured energy and the maximum energy deposited in any single counter for 140 GeV pions. The energy is measured in number of equivalent particles (eq. (24)). (b) Scatter plot of the same events after reducing the individual responses of the counters by a fraction proportional to the unweighted response (eq. (36)).

Finally, we remark that, for the iron calorimeter with $d = 5$ cm, Fig. 21 shows that two parallel showers overlap by $\sim 1\%$ when their vertices are separated by more than 25 cm and that the coordinates of two showers can be disentangled if they are separated by $\gtrsim 15$ cm [52].

4. Methods to reduce the effects of fluctuations

4.1. *By software*

As an example we consider the measurement of the hadronic energy in the iron-scintillator calorimeter of the CDHS Collaboration [50]. Figure 25(a) is a scatter plot of the total energy measured, when a large number of 140 GeV pions impinges on the calorimeter, vs. the maximum energy deposited in any single counter. The observed correlation agrees with the idea that larger total energies are observed for showers that contain a larger fraction of electromagnetic energy, which in turn produces high concentrations of ionization because in iron the radiation length ($14\,\mathrm{g\,cm^{-2}}$) is much shorter than the nuclear absorption length ($140\,\mathrm{g\,cm^{-2}}$). The hadronic energy resolution is plotted in Fig. 26 vs. the pion energy. The open

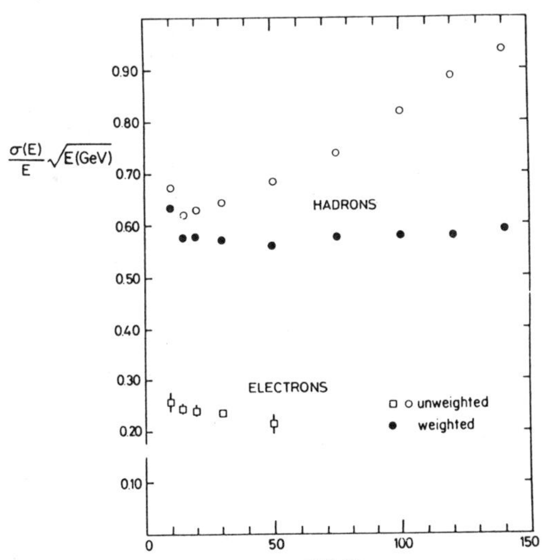

Fig. 26. Energy resolution of the 2.5 cm iron calorimeter of the CDHS Collaboration vs. the energy of the impinging particles. When the weighting procedure of eq. (36) is applied, the resolution improves, particularly at large energies.

points are the results of the standard analysis, while the full ones are obtained through a weighting procedure, an idea introduced by the Caltech–Fermilab group [54]. The response e_i of each counter was reduced according to the formula

$$e_i' = e_i \left(1 - \frac{C}{\sqrt{E}} e_i\right) \tag{36}$$

never allowing the correction factor to be smaller than 0.70. The optimum value for the constant was found to be $C = 0.03 n_{eq}^{-1/2}$. This procedure reduces the correlation between e_{max} and E_{vis} as shown in Fig. 25(b), and produces a resolution which is well fitted by the expression $\sigma(E)/E = 0.58/\sqrt{E}$ (E in GeV as usual).

This example illustrates the features of many software methods developed to reduce the effects of fluctuations: use the information content of the observed correlations and a qualitative understanding of the underlying phenomena to weight the data. Similar approaches are also adopted in spatial and angular measurements, when low weights are given to the energy deposited far away from the shower axis.

4.2. *By hardware*

As a first example we consider the measurement of electromagnetic energy with lead–glass counters. The low transparency of the glass for the blue part of the Cerenkov light spectrum makes the pulse height measured by a photomultiplier viewing the bottom of the block sensitive to the longitudinal shower distribution. It has been shown by the Monte Carlo calculation of Hanin and Stern [55] that, through this effect, the longitudinal fluctuations of the shower give a major contribution to the width of the pulse-height spectra. They also showed that, by excluding the wavelength shorter than 5000 Å with a light filter, the number of photoelectrons produced by a 10 GeV photon decreases by almost a factor 3, while the energy resolution passes from 7.7% (FWHM) to 5.0%. This example illustrates a general feature of the fight against fluctuations: when statistics do not dominate, even a qualitative understanding of the sources of fluctuations may suggest useful remedies.

The second example is too well known to be discussed in detail. It is the uranium hadronic calorimeter proposed by Fabjan et al. [40] to compensate the fluctuations due to the missing energy with the nuclear fissions produced by neutrons in uranium. With the argon–uranium calorimeter of [40] these authors showed that the hadron resolution is $\simeq 30\%/\sqrt{E}$, definitely superior to the resolution obtained with iron calorimeters (eq. (31)). The improve resolution goes together with an increase of the ratio between the visible energies for pions and electrons from $\pi/e \simeq 0.70$ in argon–iron to $\pi/e \simeq 1.05$ in argon–uranium. More recent experimental work [56] indicates that uranium plates 2 mm thick alternated with 2.5 mm scintillator layers give very good hadronic and electromagnetic resolutions: $32\%/\sqrt{E}$ and $14\%/\sqrt{E}$ respectively [57]. However, in this calorimeter a preliminary analysis gives a ratio $\pi/e \simeq 0.9$, still smaller than 1.

The discussions of Sections 2 and 3 suggest some possible lines of hardware development on the way to reach better energy resolutions. As far as electromagnetic showers are concerned, it seems that calorimeters with dense active layers thicker than $\sim 1\,\mathrm{g\,cm^{-2}}$ ($E_c \simeq 0.5$ MeV) are limited by sampling statistics and reach the resolution given by eq. (18). For thinner layers other sources of fluctuations play a role and the best resolutions are achieved when the detector directly counts the number of electrons traversing the gaps. This concept was tested by Conversi and collaborators by means of plastic flash tubes [18]. Since the main sources of fluctuations are the electrons moving along the gaps, Iarocci and collaborators suggested a detector based on a lattice of cubes, whose walls would stop the dangerous electrons [19]. A similar cube lattice working in Geiger mode would have some advantages and is at present under study [58]. A possibility, also mentioned by Sauli [59], is offered by the use of multistage chambers which detect the entrance point of a particle in the gap and not the total ionization deposited in the gas.

Hardware improvements in hadron calorimetry will follow from a better understanding of its limitation, and in particular of the relation between energy resolution due to shower fluctuations (which is of the order of $50\%/\sqrt{E}$ (GeV)) and ratio π/e between the visible energies produced by pions and electrons. To make this ratio closer to 1 one can think of reducing the response to electrons by constructing a calorimeter that has large transition effects, as probably already partially happens in a uranium-scintillator stack [56]. Another approach to improve the energy resolution can be based on an independent measurement of the energy going into nuclear processes. Some years ago a first attempt was made at measuring the energy going into nuclear fragments by detecting the "late" light produced by plexipop when traversed by heavily ionizing particles [60]. Other possibilities probably exist and it is hoped that this review may stimulate the reader to propose solutions to this challenging problem.

I would like to thank my colleagues of the CHARM Collaboration for many enlightening discussions on the subjects presented in this review.

References

1. Proc. of the Calorimeter Workshop, Batavia, May 1975, M. Atac (Ed.), Fermilab (1975).
2. Iwata, S., DPNU 13-80 (May 1980). This is a part of the report for the TRISTAN ep Working Group.
3. Rossi, B., High Energy Particles, Prentice-Hall, New York (1952).
4. Nagel, H. H., Z. Physik **186**, 319 (1965).
5. Crawford, D. F. and Messel, H., Phys. Rev. **128**, 2352 (1962).
6. Longo, E. and Sestili, I., Nucl. Instr. and Methods **128**, 283 (1975).
7. A useful numerical parametrization of the shower development is given by Abshire, G. et al., Nucl. Instr. and Methods **164**, 67 (1979).
8. Jensen, T., Amburgey, J. D. and Gabriel, T. A., Nucl. Instr. and Methods **143**, 429 (1977).
9. Cerri, C. and Sergiampietri, F., Nucl. Instr. and Methods **141**, 207 (1977).
10. Yuda, T., Nucl. Instr. and Methods **73**, 301 (1969).
11. Measurements of the transverse distributions in lead-scintillator shower detectors have been recently reported, together with two component fits to the data, in Ref. [7]. See also Nelson, W. R. et al., Phys. Rev. **149**, 201 (1966) and Ref. [10].
12. Diddens, A. N. et al., CHARM Collaboration, Nucl. Instr. and Methods, **178**, 27 (1980).
13. Prokoshkin, Yu. D., Proc. of the Second ICFA Workshop on Possibilities and Limitations of Accelerators and Detectors, Les Diablerets, 4–10 October 1979, U. Amaldi (Ed.), CERN, June 1980, p. 405.
14. Hughes, E. B. et al., Stanford University Report, No. 627 (1972).
15. Fisher, G., Nucl. Instr. and Methods **156**, 81 (1978).
16. Willis, W. J. and Radeka, V., Nucl. Instr. and Methods **120**, 221 (1974).
17. Conversi, M. and Gozzini, A., Nuovo Cimento **2**, 189 (1955).
18. Federici, L. et al., Nucl. Instr. and Methods **151**, 103 (1978).

19. Battistoni, G. et al., "A cube lattice multiwire detector", Report LNF-79/12(P), 12 Feb. 1979.
20. A more detailed discussion of this effect in a gas quantameter can be found in Katsura, T. et al., Nucl. Instr. and Methods **105**, 245 (1972).
21. West, D., Proc. Phys. Soc. **A66**, 306 (1953). For recent experiments see, for instance, Onuchin, A. P. and Telnov, V. I., Nucl. Instr. and Methods **120**, 365 (1974).
22. Blunck, O. and Leisegang, S., Z. Physik **128**, 500 (1950).
23. Stone, S. L. et al., Nucl. Instr. and Methods **151**, 387 (1978).
24. Very recent and accurate data have been obtained by Asano, Y. et al. (Nucl. Instr. and Methods **174**, 357 (1980)) with a liquid argon–iron and a liquid argon–lead calorimeter. They found R(Fe) = 17.1% and R(Pb) = 15.8% with 2 mm of argon as active material. Equation (18) gives R(Fe) = 16.5% and R(Pb) = 13%. While the iron data is well reproduced, the lead resolution is worse than predicted. We attribute the discrepancy to the path length fluctuations in the thin argon gap, which is more important when the angular spread of the electrons is larger (see Fig. 9).
25. Fisher, G. and Ullaland, O., Physica Scripta **23**, (1981); Price, L. E., "Drift collection calorimeters", Report ANL-HEP-CP-80-40, Physica Scripta **23**, 685 (1981).
26. For example, the lead–MWPC MAC detector has $R \simeq 24\%$ as described by Anderson, R. L. et al., IEEE Trans, Nucl. Sci., **NS-25**, 340 (1978). For further bibliography see [2].
27. Cobb, J. H. et al., Nucl. Instr. and Methods **158**, 93 (1979).
28. Cheshire, D. L. et al., Nucl. Instr. and Methods **126**, 253 (1975).
29. Pinkau, K., Phys. Rev. **139**, B1548 (1965).
30. Crannel, C. J. et al., Phys. Rev. **182**, 1435 (1969).
31. Akapdjanov, G. A. et al., Nucl. Instr. and Methods **140**, 441 (1977).
32. Binon, F. et al., Preprint IHEP 79-128, Serpukhov (1979).
33. Amendolia, S. R. et al., report Pisa 80-4 (29 May 1980), Contribution to the Conference on Experimentation at LEP, Uppsala, June 1980.
34. A resolution of ~ 12 mm with $10 \times 10\,cm^2$ cells was obtained by Hofmann, W. et al., Nucl. Instr. and Methods **163**, 77 (1979).
35. Gabathuler, E. et al., Nucl. Instr. and Methods **157**, 47 (1978).
36. Bushnin, Yu. B. et al., Nucl. Instr. and Methods **106**, 493 (1973).
37. Buchnin, Yu. B. et al., Nucl. Instr. and Methods **120**, 391 (1974).
38. CELLO Collaboration, Contribution to the Int. Conference on Experimentation at LEP, Uppsala, June 1980.
39. Fabjan, C. W. and Willis, W. J., Ref. [1].
40. Fabjan, C. W. et al., Nucl. Instr. and Methods **141**, 61 (1977).
41. Sciulli, F. J., Ref. [1].
42. Ranft, J., Part. Accelerators **3**, 129 (1972).
43. Baroncelli, A., Nucl. Instr. and Methods **118**, 445 (1974).
44. Gabriel, T. A. and Schmidt, W., Nucl. Instr. and Methods **134**, 271 (1976).
45. Baroncelli, A., Private communication.

46. Holder, M. et al., CDHS Collaboration, Nucl. Instr. and Methods **151**, 69 (1978).
47. Benvenuti, A. et al., HPWF Collaboration, Nucl. Instr. and Methods **125**, 447 (1975).
48. Hughes, E. B. et al., Nucl. Instr. and Methods **75**, 130 (1969).
49. Barish, B. C. et al., Nucl. Instr. and Methods **130**, 49 (1975).
50. Abramowicz, H. et al., CDHS Collaboration, CERN-EP/80-188, 14 October 1980, submitted to Nucl. Instr. and Methods.
51. Note that, the scintillator thickness being only $x_a = 5$ mm, one expects a deterioration due to path length fluctuations. Figure 9 gives, for the same amount of g cm^{-2}, a deterioration factor ~ 1.3.
52. Binon, F. et al., IHEP–IISN–LAPP Collaboration, A hodoscope calorimeter for high energy hadrons, CERN-EP/80-15, 7 February 1980, to be published in Nucl. Instr. and Methods.
53. Bollini, D. et al., Nucl. Instr. and Methods **171**, 237 (1980).
54. Dishaw, J. P., Thesis, SLAC Report-216, March 1979.
55. Hanin, V. A. and Stern, B. E., Nucl. Instr. and Methods **157** 455 (1978).
56. Botner, O., Physica Scripta **23**, (1981).
57. It can be noted that eq. (18) gives in this case for the sampling fluctuations $R \simeq 10\%$. We attribute the difference between 14% and 10% to photoelectron statistics which, in a similar calorimeter of the same group with copper plates, contributed $\simeq 8\%$. See: Botner, O. et al., Nucl. Instr. and Methods (to be published), and CERN-EP/80-126, 14 July 1980.
58. Gygi, E. and Schneider, F., EP Int. Report 80-08, 9 Dec. 1980.
59. Sauli, F., Physica Scripta **23**, 526 (1981).
60. Amaldi, U. and Matthiae, G., Unpublished.

Reprinted, with permission, from *Formulae and Methods in Experimental Data Evaluation, Vol. 2:* Articles on Physics and Detectors, ed. R. K. Bock, et al (The European Physical Society, CERN, 1984.

THE PHYSICS OF CHARGED PARTICLE IDENTIFICATION: dE/dx, CERENKOV AND TRANSITION RADIATION

W.W.M. Allison
Nuclear Physics Laboratory, Univ. of Oxford, United Kingdom

P.R.S. Wright
CERN, Geneva, Switzerland

I. INTRODUCTION

The analysis of most High Energy Physics experiments requires a knowledge of the 4-momenta, ($\underline{p}$,E) of the secondary particles. The 3-momenta, $\underline{p}$, are usually obtained by measuring the deflection of the trajectory of each of the particles in a magnetic field. A further measurement, be it of mass, energy or velocity, is needed to determine the fourth component of the 4-momentum, E, and fix a value for the mass M. Since M uniquely (for charged particles) identifies the internal quantum numbers, this measurement is generally referred to as "particle identification".

At low energies particle identification has traditionally been achieved by total absorption calorimetry, by time-of-flight measurement or by simple dE/dx measurements. Calorimetry becomes more difficult at higher energies, as the number of interaction lengths of absorber needed to contain the showers completely becomes large. More seriously, the energy resolution, $\Delta E/E$, required to distinguish different masses varies as E^{-2}, while the resolution achievable varies as $E^{-1/2}$. Time-of-flight measurements yield the velocity of the particles over a given distance. The time difference, Δt, for two particles of masses M_1 and M_2, with the same momentum, p, over a distance L, is given by:

$$\Delta t = 1/2\ (M_1^2 - M_2^2)\ Lc/p$$

Resolvable values of Δt are currently ~250ps and so, for a reasonable value of L of a few metres, Δt becomes unmeasurably small for $p > 1$ GeV/c. In the non-relativistic range energy loss (dE/dx) is proportional to $1/\beta^2$ (βc is the particle velocity). This gives a large difference in signal amplitude for different masses. As β approaches one however, this simple discrimination is no longer possible.

For these reasons other methods of particle identification have been developed for the relativistic range. They are based on the following physical effects:

a) Cerenkov Radiation: If a charged particle moves through a medium faster than the phase velocity of light in that medium, it will emit radiation at an angle determined by its velocity and the refractive index of the medium. Either the presence/absence of this radiation (in threshold Cerenkov counters), or a direct measurement of the Cerenkov angle (Differential or Ring-Image Cerenkov counters) can be used to give information on the particle velocity.

b) Energy loss (dE/dx): The rate of energy loss for a relativistic particle is a weak function of the $\beta\gamma$ of the particle $\left(\beta\gamma = p/Mc = \frac{v}{c}\left[1 - \frac{v^2}{c^2}\right]^{-1/2}\right)$. This dependence is shown in fig.1. In the non-relativistic region ($\beta\gamma<4$) the rate of energy loss falls to a minimum as $1/\beta^2$. Above $\beta\gamma \simeq 4$ the rate of energy loss rises again as $\log(\beta\gamma)$; this is the so-called relativistic rise. At $\beta\gamma$ of a few hundred the rate of energy loss saturates (the "Fermi plateau"). In solids and liquids the plateau is only a few percent above the minimum; in high-Z noble gases at atmospheric pressure it reaches 50-70%. An accurate measurement of the energy loss in the relativistic rise region provides a measurement of $\beta\gamma$, and hence of M.

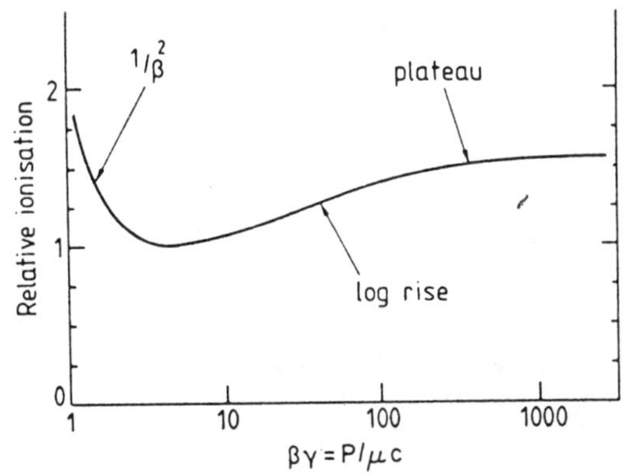

Fig.1. The dependence of ionisation in a gas on βγ.

c) Transition Radiation: When a highly relativistic particle ($\beta\gamma \gtrsim 500$) crosses the boundary between two dielectric media, x-ray photons are emitted. The energy of these photons is a function of βγ. Hence by measuring the transition radiation the mass of the particle may be obtained.

The above phenomena are intimately related, all depending on the dielectric properties of the medium and the velocity of the charged particle. Their treatment in standard texts is often cursory, disjointed, and without some of the details needed to design and predict the performance of a practical device. In Section II we discuss the theoretical ideas that link these three phenomena together. Section III demonstrates how these theoretical ideas may be applied to practical devices. Finally, in an appendix we discuss some particular practical problems and solutions for dE/dx devices.

An elementary introduction to Cerenkov radiation is given in the Nobel Prize lectures (Cerenkov 1958) and a full discussion is given by Jelley (1958) and Ter-Mikaelian (1972). The latter also treat the theory of Transition Radiation. Useful treatments are also to be found in Jackson (1962) and Landau & Lifshitz (1960). The theory of X-ray transition radiation is described by Garibyan (1957). The theory of dE/dx is described fully and with references in Allison & Cobb (1980).

Some elements of the development are to be found in Jackson (1975), Fano (1963) and Landau & Lifshitz (1960). The details of how it might be used for particle identification were first worked out by Alikhanov (1956).

II. THEORY

II.1 General Discussion

Any device that is to detect a particle must interact with it in some way. If the particle is to pass through essentially undeviated, this interaction must be a soft electromagnetic one. Let us consider a particle of mass M and velocity $\underline{v}$, which interacts with the medium of the detector via a photon of energy $\hbar\omega$ and momentum $\hbar\underline{k}$, as shown in fig.2. Conservation of energy and momentum require:

$$\hbar\omega\ (1 - \hbar\omega/2\gamma Mc^2) = \hbar\ \underline{k}\cdot\underline{v} - \hbar\ k^2/2\gamma M$$

With the restriction to soft collisions, $\hbar\omega \ll \gamma Mc^2$ and $\hbar k \ll v\gamma M$, this reduces to:

$$\omega = \underline{v} \cdot \underline{k} \tag{1}$$

The behaviour of a photon in a medium is described by the dispersion relation

$$\omega^2 - k^2c^2/\varepsilon = 0 \tag{2}$$

where ε is the relative electric permittivity or dielectric constant.

Eliminating ω and k between eqns.(1) and (2) gives:

$$\sqrt{\varepsilon}\ \frac{v}{c}\cos\theta_c = 1 \tag{3}$$

where θ_c is the angle between $\underline{v}$ and $\underline{k}$. This shows that, if $\frac{v}{c}\sqrt{\varepsilon}$ is greater than unity, a real angle θ_c exists for which free photons can be emitted (or absorbed). This is known as the Cerenkov angle and the free

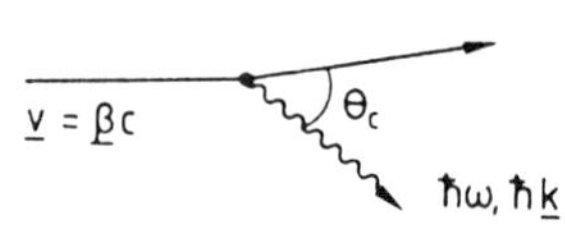

Fig.2. Photon emission in a medium.

photon flux as Cerenkov radiation. The velocity at which v is equal to $c/\sqrt{\varepsilon}$ is called the Cerenkov threshold. At lower velocities $\cos\theta_c$ is greater than unity, and free photons are not emitted in continuous media. In discontinuous media diffraction causes free photon emission even below the Cerenkov threshold (see II.3).

So far it has been assumed that ε is real. In practice this is only true below the ionisation threshold of the medium. Fig.3 shows ε as a function of frequency for argon gas as an example. There are three important frequency ranges:

a) The Optical region. For frequencies below the absorption region the medium is transparent, ε is real and greater than one, and Cerenkov radiation is emitted by particles with velocity above the threshold (which is always less than c). The emission of subthreshold Cerenkov radiation by discontinuities in the medium is called Optical Transition Radiation which is not important for

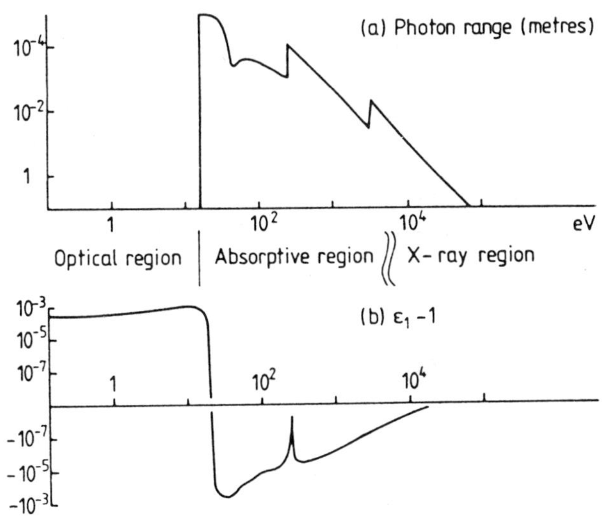

Fig.3. The dependence of ε for argon at normal density on photon energy,
a) imaginary part expressed as a range and
b) real part - 1 on a split log scale.

particle identification.

b) The absorption region. ε is complex and the range of photons is short. Absorption of the virtual photons constituting the field of a charged particle gives rise to the ionisation of the material that is measured in dE/dx detectors.

c) The X-ray region. The residual absorptive part of ε still causes a small contribution to be made to the tail of the dE/dx distribution but at frequencies some 30% above its K-edge the medium may be treated as being nearly transparent. There ε is less than one, however, and the Cerenkov threshold is greater than c. Nevertheless, the emission of sub-threshold Cerenkov radiation in the presence of discontinuities in the medium may still occur; this is known as X-ray Transition Radiation, and is exploited by Transition Radiation detectors.

In the following sections we derive formulae for the magnitude of the energy-loss signal, the flux of Cerenkov radiation, as well as the Transition Radiation effect. This is much simpler than is often supposed and in particular Quantum Electrodynamics is not needed at any stage. The formalism of Maxwell's Equations with semi-classical quantisation, as in Planck's description of black-body radiation, is all that is necessary, apart from the knowledge of ε for the media concerned.

II.2 Cerenkov radiation and dE/dx

II.2.1 Simple model

To discuss the relationship between Cerenkov radiation and dE/dx let us consider the field seen by an observer at a point (x,y) due to a particle moving along the x-axis. To simplify the picture we ignore the z-dimension - the effect of this is to give exponentials in place of Bessel functions which eases the manipulation without altering any principles of physics.

Such an observer sees an electromagnetic pulse as the particle goes by. This pulse is made up of a broad spectrum of frequency components and travels in the x direction with velocity v without dispersion. These conditions require that the phase velocity in the x direction for each Fourier component satisfies equation (1). This means that the components of the field are components of a static field as seen by an observer comoving with the particle. We therefore have $k_x = \frac{\omega}{v}$ and $k_x^2 + k_y^2 = \frac{\omega^2\varepsilon}{c^2}$ from equation (2), giving $k_y = \frac{\omega}{v}\left(\frac{v^2\varepsilon}{c^2} - 1\right)^{1/2}$. Let us call the phase velocity of light in the medium $u(\omega) = c/\sqrt{\varepsilon}$. Then this becomes:

$$k_y = \frac{\omega}{v}\left(\frac{v^2}{u^2} - 1\right)^{1/2}$$

There are two situations to which this formula can be applied:

a) v greater than u. Then k_y is real and the component of frequency ω represents a real travelling wave at an angle $\cos^{-1}(u/v)$. This is the case of Cerenkov Radiation.

b) v less than u. k_y is now purely imaginary and the component of frequency ω propagates as an evanescent wave in the transverse direction (just like the evanescent wave encountered in the phenomenon of total internal reflection):

$$\exp i(\underline{k}\cdot\underline{r} - \omega t) = \exp i\,\frac{\omega}{v}\,(x - vt)\,\exp - y/y_o$$

where the range y_o is given by

$$y_o = \frac{v}{\omega}\,(1 - v^2/u^2)^{-1/2}$$

Re-expressed in terms of the dimensionless variable, $\beta' = v/u$, $\gamma' = (1-\beta'^2)^{-1/2}$ and $\bar{\lambda} = u/\omega$, the free wavelength over 2π, the range becomes

$$y_o = \bar{\lambda}\,\beta'\,\gamma' \qquad (4)$$

This shows that the range of the field expands linearly with the dimensionless scaled variables $\beta'\gamma'$ as the Cerenkov threshold is approached. This expansion, often referred to as 'the relativistic expansion', depends only on the wave nature of the field and is responsible for the 'relativistic rise' of the ionisation cross section.

The velocity of light in vacuum, c, only enters the problem through the kinematics of the particle itself - which is strictly irrelevant to the soft electromagnetic field of the particle in a medium. However in practice we are interested in the dependence of the field on the dimensionless scaled variables appropriate to the vacuum, $\beta=v/c$ and $\gamma=(1-\beta^2)^{-1/2}$. Then the clear formula (4) appears more complicated:

$$y_o = \lambda\!\!\!^{-}\beta \left(\frac{1}{\gamma^2} + (1-\varepsilon)\,\beta^2\right)^{-1/2} \qquad (5)$$

The complexity however has only arisen from the change of variables.

In considering equation (5) we distinguish two cases of interest depending on whether ε is greater than unity or not:

a) $\varepsilon>1$. As shown in figure 3 this implies that the frequency is in the Optical region below the ionisation threshold. At increasing values of velocity the range increases until at $\beta'=1$ (and $\gamma'=\infty$), the range becomes infinite and we have a Cerenkov field.

b) $\varepsilon<1$. Here ω is above ionisation threshold (the absorption and x-ray regions). The range increases with increasing values of velocity as before but reaches an upper bound when $\beta'=c/u$. This plateau is given by equation (5) for ($\beta=1$, $\gamma=\infty$) by

$$y_o = \lambda\!\!\!^{-}\,(1-\varepsilon)^{-1/2}$$

The plateau, which gives a corresponding saturation in the relativistic rise of the ionisation cross section, starts to set in at velocities for which the two terms in the denominator of equation (5) are equal i.e.

$$\beta\gamma \sim (1-\varepsilon)^{-1/2}$$

For different frequencies the plateau becomes important for different values of $\beta\gamma$ depending on $\varepsilon(\omega)$. Nevertheless in any kind of sum over frequencies the same general effect is apparent. The dE/dx dependence shown in fig.1 is an example. At the highest values of $\beta\gamma$ the energy loss saturates ("the density effect"). This plateau sets an upper bound to the range of $\beta\gamma$ over which dE/dx may be used to measure velocity effectively. Since $(1-\varepsilon)$ scales with the density the onset of saturation varies as $\rho^{-1/2}$. For velocity discrimination at high values of $\beta\gamma$, ε must be close to unity and gases must be used.

There is a corresponding limit beyond which the Cerenkov field becomes asymptotic as a function of $\beta\gamma$. The Cerenkov angle is within a value $\delta\theta$ of its asymptotic value for values of $\beta\gamma$ given by

$$\beta\gamma \sim (\varepsilon-1)^{-1/4}\,(\delta\theta)^{-1/2}$$

and the flux is within 90% of its asymptotic value for $\beta\gamma$ greater than

$$\beta\gamma \sim 3(\varepsilon-1)^{-1/2}$$

Taking nitrogen at normal density as an example ($\varepsilon-1=594\times10^{-6}$) and $\delta\theta=10$ mrad the two $\beta\gamma$ values come out to be 64 and 123. The dE/dx signal also saturates around $\beta\gamma=100$. The need to work with low density gases is therefore evident in both Cerenkov and dE/dx techniques and the reason is essentially the same.

II.2.2 Careful calculation

Since it is a form of energy loss, Cerenkov Radiation should appear in a natural way from a rigorous calculation of dE/dx. Such a calculation may be made using the Photo Absorption Ionisation Model for any mixture of gases whose photo-absorption spectra are available. These calculations have been described in detail elsewhere (Allison & Cobb (1980)). A FORTRAN program and data files are available for further calculations. Here we describe the physics of the five main steps involved in the derivation of the energy-loss differential cross section.

Step 1

This is a conventional problem in Classical Electrodynamics. We solve Maxwell's Equations in a medium ($\underline{D}=\varepsilon\varepsilon_0\underline{E}$, $\mu=1$) due to a charge density, $\rho=e\delta^3(\underline{r}-\underline{\beta}ct)$, and current density, $\underline{j}=\underline{\beta}c\rho$, which together describe a charge moving with velocity, $\underline{\beta}c$. Working in SI units and the Coulomb Gauge one obtains

$$\phi(\underline{k},\omega) = \frac{e}{2\pi\varepsilon\varepsilon_o k^2}\,\delta(\omega - \underline{k}\cdot\underline{\beta}c)$$

$$\underline{A}(\underline{k},\omega) = \frac{e}{2\pi\varepsilon_o c^2}\,\frac{(\omega\underline{k}/k^2 - \underline{\beta}c)}{(\varepsilon\,\omega^2/c^2 - k^2)}\,\delta(\omega - \underline{k}\cdot\underline{\beta}c)$$

with

$$\underline{E}(\underline{r},t) = \frac{1}{(2\pi)^2}\iint [i\omega\,\underline{A}(\underline{k},\omega) - i\underline{k}\phi(\underline{k},\omega)]$$

$$\times \exp i(\underline{k}\cdot\underline{r} - \omega t)\, d^3k\, d\omega$$

Step 2

The energy loss is due to the component of this electric field in the direction $\underline{\beta}$ doing work on the particle itself at the point $\underline{r}=\underline{\beta}ct$:

$$\langle dE/dx\rangle = \frac{e}{\beta}\,\underline{E}(\underline{\beta}ct,t)\cdot\underline{\beta}$$

Step 3

This energy loss, which is expressed as an integral over Fourier components, is not a smooth rate of energy loss but needs to be re-interpreted as in semiclassical radiation theory as a probability of energy transfers, $E=\hbar\omega$. Thus

$$\langle dE/dx\rangle = -\int_o^\infty d\omega \int_{\omega/v}^\infty dk\; NE\,\frac{d^2\sigma}{dE\,dp}\,\hbar^2\,d\omega\,dk$$

where N is the electron density and $\frac{d^2\sigma}{dE\,dp}$ is the double differential cross section per electron. Doing the implied algebra we get

$$\frac{d^2\sigma}{dE\,dp} = \frac{e^2}{4\pi\varepsilon_o}\,\frac{2}{N\beta^2\pi\hbar^2}\left[p\left(\beta^2 - \frac{E^2}{p^2c^2}\right)\,\mathrm{m}\left(\frac{1}{\varepsilon E^2 - p^2c^2}\right) - \frac{1}{pc^2}\,\mathrm{m}\left(\frac{1}{\varepsilon}\right)\right]$$

This formula already shows the $1/\beta^2$ factor which dominates the rate of energy loss at non relativistic velocities.

Step 4

The only unknown parameter in this formula is ε. It is a complex function ($\varepsilon=\varepsilon_1+i\varepsilon_2$) of k and ω, and is essentially the structure function of the medium. A sufficiently reliable model (Allison & Cobb 1980) for ε may be derived from

- detailed photoabsorption spectra and their sum rules giving ε_2 for on mass-shell photons ($\omega=kc$), see Berkowitz (1979),
- the Kramers Kronig relation giving ε_1 in terms of ε_2,
- the dipole approximation which tells us that for small k $\varepsilon(k,\omega)$ is independent of k at fixed ω,
- constituent scattering from quasi-free electrons and the Bethe sum rule which together give values for ε in the large k off-mass-shell region.

Step 5

With this model it is possible to integrate the cross section over momentum transfer analytically. The differential cross section per electron per unit energy loss is:

$$\frac{d\sigma}{dE} = \frac{\alpha}{\beta^2\pi}\frac{\sigma_\gamma(E)}{E\,Z}\,\ell n\left[(1-\beta^2\varepsilon_1)^2+\beta^4\varepsilon_2^2\right]^{-1/2} + \frac{\alpha}{\beta^2\pi}\frac{1}{N\hbar c}\left(\beta^2-\frac{\varepsilon_1}{|\varepsilon|^2}\right)\Theta$$

$$+\frac{\alpha}{\beta^2\pi}\frac{\sigma_\gamma(E)}{E\,Z}\,\ell n\left(\frac{2mc^2\beta^2}{E}\right) + \frac{\alpha}{\beta^2\pi}\frac{1}{E^2}\int_0^E\frac{\sigma_\gamma(E')}{Z}\,dE' \qquad (6)$$

where $\alpha=e^2/4\pi\varepsilon_0\hbar c$ is the fine structure constant, ε_1 and ε_2 are the real and imaginary parts of the usual on-mass-shell dielectric constant and Θ is the phase of the complex expression $1-\varepsilon_1\beta^2+i\varepsilon_2\beta^2$. σ_γ is the atomic photo-absorption cross section of the gas.

In the first term, ignoring ε_2, we recognise a factor $\ell n(1-\beta^2\varepsilon_1)^{-1}$ or $\ell n\,\gamma'^2$ in the notation of our simplified discussion. This is responsible for the relativistic rise of the energy-loss cross section and its saturation as already discussed. In the Optical region σ_γ

vanishes and only the second term contributes. This describes energy loss by Cerenkov Radiation. In the absence of ε_2 the phase θ jumps from zero below threshold to π above. Multiplying by $N\hbar$ this term gives the flux of Cerenkov photons per unit path length

$$\frac{dn}{d\omega} = \frac{\alpha}{c}\left(1 - \frac{1}{\beta^2\varepsilon}\right) . \tag{7}$$

However when ε_2 and σ_γ do not vanish, the separate interpretation of this term in the cross section dissolves and it may even be negative. The last term in equation (6) comes from the constituent scattering from electrons. It is a Rutherford scattering term, shows no relativistic behaviour and, being the sole non-vanishing term for energy transfers E in the far X-ray region, describes δ-ray production. In general it is small and its nuisance value in dE/dx measurements is often over-stated.

II.3 Transition Radiation and Cerenkov Radiation

When a fast charged particle passes through a thin film of material, the wavefront of the Cerenkov Radiation, BC, shown in figure 4 is restricted in width. It follows that there is broadening of the Cerenkov angle due to diffraction. This Cerenkov radiation field may be represented in one of two ways:

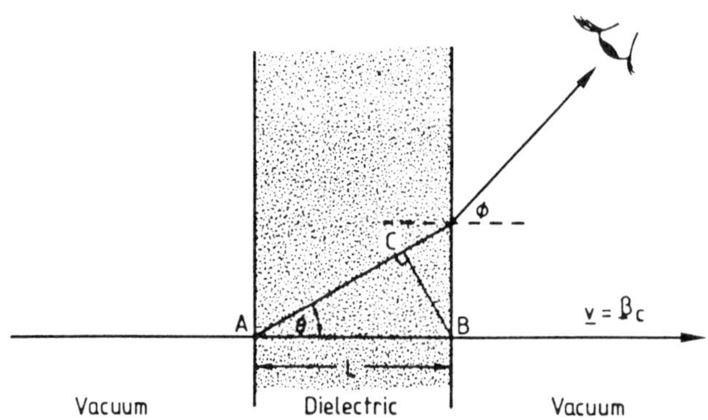

Fig.4. The emission of Cerenkov or Transition Radiation in a thin slab.

a) As the sum of many wavelets whose sources are distributed along the path AB. This is the conventional Cerenkov or "differential" representation of the field.

b) As the integral of this field due to sources at points A and B subtracted from one another. This field emitted from A and B with opposite sign is the Transition Radiation or "integral" representation of the field.

Using this integral transformation we now derive a formula for the Transition Radiation emitted from a foil starting from equation (7) for the Cerenkov flux. Since the radiation is emitted at an angle given by equation (3) we may rewrite the differential flux as

$$\frac{d^2n}{d\omega\, d\Omega} = \frac{\alpha}{2\pi c} \sin^2\theta \; \delta(\cos\theta - \frac{1}{\beta\sqrt{\varepsilon}})\; L$$

For L, the track length in the dielectric, small, the δ-function should be replaced by the single slit Fraunhofer diffraction function thus

$$\delta(\cos\theta - \frac{1}{\beta\sqrt{\varepsilon}}) \rightarrow \frac{L}{\lambda}\, \frac{\sin^2(\Phi/2)}{(\Phi/2)^2}$$

where λ is the wavelength in the film and $\Phi(\theta)$ is the phase difference between wavelets emitted from the two ends of the track in the foil, A and B:

$$\Phi(\theta) = \frac{2\pi L}{\lambda} (\cos\theta - \frac{1}{\beta\sqrt{\varepsilon}})$$

Since L is small, the broadening of the Cerenkov angle due to diffraction can be significant. Indeed it can be large enough that radiation is observed even when the Cerenkov angle itself is unphysical. For ω in the optical region this gives rise to Optical Transition Radiation at wide angles and at modest velocities, that is for values of β below Cerenkov threshold. This case is not technically important although historically it was the first to be described. The important case is the emission of Transition Radiation in the X-ray region for very large values of γ and at very small angles of order $1/\gamma$. Then ε is very close to unity and may be approximated as

$$\varepsilon = 1 - \frac{\omega_p^2}{\omega^2}$$

where ω_p is the plasma frequency, $(Ne^2/\varepsilon_o m)^{1/2}$, N being the electron density in the film†.

We now derive the formula for the flux of X-ray Transition Radiation from such a foil. Replacing the δ function and substituting for $\lambda = 2\pi c/\omega\sqrt{\varepsilon}$, we have

$$\frac{d^2n}{d\omega d\Omega} = \frac{\alpha\sqrt{\varepsilon}}{\pi^2\omega}\sin^2\theta\,\frac{\sin^2\left[\frac{\omega L}{2c}\left(\sqrt{\varepsilon}\cos\theta - 1/\beta\right)\right]}{\left(\sqrt{\varepsilon}\cos\theta - 1/\beta\right)^2}$$

However, referring to fig.4, we see that the radiation is observed, not in the medium, but in vacuum. Since $\sin\phi = \sqrt{\varepsilon}\sin\theta$, we have

$$\frac{d^2n}{d\omega d\Omega} = \frac{\alpha}{\pi^2\omega}\sin^2\phi\,\frac{\sin^2\left[\frac{\omega L}{2c}\left(\sqrt{\varepsilon - \sin^2\phi} - 1/\beta\right)\right]}{\left(\sqrt{\varepsilon - \sin^2\phi} - 1/\beta\right)^2}$$

where, since ε is near 1, we have ignored simple factors of ε, the effect of reflection and the change in solid angle $d\Omega$. However the above formula contains a serious error. It is not consistent with the application of the superposition principle to the film as a radiator. The net effect of the film as a radiator must be seen as the replacement of a slice of vacuum by a slice of dielectric. There is therefore a contribution describing the absence of the vacuum field from the slice. Its amplitude is of the same form (with $\varepsilon = 1$) but opposite phase. The $\sin^2$ factor in the numerator describing the dependence on the phase difference Φ is common. The correct formula is therefore

---

† A handy formula for the plasma frequency is given by $\hbar\omega_p = 29\left(\frac{\rho}{1000} Z/A\right)^{1/2}$eV, where ρ is the density in kg m^{-3}.

$$\frac{d^2n}{d\omega d\Omega} = \frac{\alpha}{\pi^2\omega}\sin^2\phi\,\sin^2\left[\frac{\omega L}{2c}\left(\sqrt{\varepsilon - \sin^2\phi} - 1/\beta\right)\right] \times$$

$$\left(\frac{1}{\sqrt{\varepsilon - \sin^2\phi} - \frac{1}{\beta}} - \frac{1}{\cos\phi - \frac{1}{\beta}}\right)^2$$

The pole of the first term, representing the Cerenkov field in the medium, is dominant in Optical Transition Radiation. The pole of the second term, representing the Cerenkov field in vacuum, is dominant in X-ray Transition Radiation at small angle ϕ and $\beta \sim 1$. This pole gives X-ray Transition Radiation its unique advantage that, provided the foil is held in vacuum, this term grows ever larger as the limit $\beta=1$ is approached. For small angles and $\beta \sim 1$ we get the simplified result

$$\frac{d^2n}{d\omega d\Omega} = \frac{\alpha}{\pi^2\omega}\phi^2 \cdot 4\sin^2\left[\frac{\omega L}{4c}\left(\omega_p^2/\omega^2 + \phi^2 + 1/\gamma^2\right)\right]$$
$$\times \left(\frac{1}{1/\gamma^2 + \omega_p^2/\omega^2 + \phi^2} - \frac{1}{1/\gamma^2 + \phi^2}\right)^2 \qquad (8)$$

This is the usual formula for the X-ray Transition Radiation from a thin film. It is the most convenient one to work with for practical calculations.

Equation (8) represents the interference between two equal and opposite amplitudes with relative phase given by the argument of the $\sin^2$ factor. In the absence of this interference effect the flux from each interface separately would be

$$\frac{d^2n}{d\omega d\Omega} = \frac{\alpha\phi^2}{\pi^2\omega}\left(\frac{1}{1/\gamma^2 + \omega_p^2/\omega^2 + \phi^2} - \frac{1}{1/\gamma^2 + \phi^2}\right)^2$$

This may be integrated over the solid angle, $d\Omega=2\pi\phi d\phi$, to give

$$\frac{dn}{d\omega} = \frac{2\alpha}{\pi\omega}\left[\left(\frac{1}{2} + \frac{\omega^2}{\gamma^2\omega_p^2}\right) \ell n \left(1 + \frac{\gamma^2\omega_p^2}{\omega^2}\right) - 1\right]$$

$$\simeq \frac{2\alpha}{\pi\omega} \ell n \left(\gamma\omega_p/\omega\right), \text{ if } \omega << \gamma\,\omega_p$$

which in turn may be integrated to give a total energy flux $\frac{\alpha}{3}\gamma\hbar\omega_p$, a typical photon energy of order $1/4\ \gamma\hbar\omega_p$ and about α photons per interface. The γ dependence of the yield is seen to come primarily from the hardening of the spectrum rather than an increase in the flux.

These latter results are somewhat academic, however, (though often quoted) since the interference term in equation (8) cannot be ignored in practice. Indeed because of this term the energy flux S, instead of increasing linearly with γ, saturates. This is called the Formation Zone Effect. We define a "formation zone" Z:

$$Z = \frac{\lambda}{\pi}\left(\frac{1}{\gamma^2} + \frac{\omega_p^2}{\omega^2} + \phi^2\right)^{-1}$$

Then the L dependent modulation factor in equation (8) becomes simply: $4\sin^2(L/Z)$. Negative interference is therefore general when $L \lesssim Z$. Since for a particular γ the significant values of ω are of order $\gamma\omega_p$ and ϕ is of order $1/\gamma$,

$$Z \sim \frac{\lambda\gamma^2}{3\pi}$$

Thus the Transition Radiation flux instead of increasing linearly with γ will saturate around $\gamma \sim (3\pi L/\lambda)^{-1/2}$. Alternatively this formula indicates the minimum foil thickness required to avoid saturation at a given value of γ.

III. CALCULATIONS FOR PRACTICAL DEVICES

III.1 Cerenkov counters

Cerenkov counters measure the particle velocity either by simply recording the flux of radiation as given by equation (7) (threshold counters) or by measuring its characteristic angle with respect to the line of flight of the particle, as given by equation (3) (Differential counters and Ring Imaging Cerenkov counters). They have been reviewed by Litt & Meunier (1973). Recent advances are described by Ypsilantis 1981, and examples given by Glass (1983) and Davenport (1983). The velocity range, the frequency range $\Delta\omega$ and the radiator length L are crucial to the number of photons emitted:

$$n = \frac{\alpha}{c} L \, \Delta\omega \, (1 + \beta_t\gamma_t)^{-1} \left[1 - \left(\frac{\beta_t\gamma_t}{\beta\gamma}\right)^2\right]$$

where $\beta_t\gamma_t$ is the $\beta\gamma$ value at threshold. The first bracket indicates that the length L required rises almost linearly with the desired value of $\beta_t\gamma_t$ - this follows directly from the need to get the appropriate value of ε. The range of $\beta\gamma$ in which the Cerenkov flux rises from threshold to 90% of its asymptotic value is given by:

$$\beta_t\gamma_t < \beta\gamma < 3\beta_t\gamma_t \quad \text{or} \quad (\varepsilon-1)^{-1/2} < \beta\gamma < 3(\varepsilon-1)^{1/2} .$$

The threshold is determined by the value of ε for the radiator. It is convenient, where possible, to choose a gas at atmospheric pressure as other pressures necessitate thick windows that introduce unwanted scattering and secondary interactions. Some useful values are given in the table for gases at 1 atm and 0°C:

	$(\varepsilon-1)*10^6$	$\beta_t\gamma_t$
Helium	72	118
Neon	134	86
Hydrogen	264	62
Argon	562	42
Nitrogen	594	41
Methane	888	33
Carbon dioxide	902	33
n-Pentane	3422	17

For values of ε closer to unity, low-density gas radiators employ sub-atmospheric pressure or high temperature. For instance a large multicell Cerenkov with high threshold employing helium at normal pressure and 300 degrees centigrade is operational as part of the European Hybrid Spectrometer at CERN (Ladron de Guevara 1983). For lower thresholds a special material, "silica aerogel", is available. This is a substance which may be prepared with refractive index in the range 1.01-1.05 (see Poelz and Riethmuller 1982 and refs quoted there).

The number of photons actually detected depends on the product of the number generated (according to equation (7)), the transmission of the optical components (including windows and the reflection coefficient of mirrors) and the quantum efficiency of the final detecting elements. The latter may be photomultipliers or a gas in a proportional counter with very low ionisation threshold such as TMAE (Ypsilantis 1981). The product of these factors determines an effective bandwidth since at long wavelengths the quantum efficiency of the detector falls to zero while at short wavelengths the transmission of the optical system cuts off. Although there is a considerable advantage in extending the sensitivity into the ultraviolet, the dispersion of the refractive index in this region smears the threshold, the Cerenkov angle and the velocity resolution. Noble gases have least dispersion and are usually chosen where the ultraviolet region is to be used. Care has to be taken to avoid radiators that scintillate. In experiments with a number of secondary particles the problem of identifying which photons belong to which particles has to be solved. On the other hand in Transition Radiation or dE/dx detectors the signals are usually superimposed on the particle trajectory.

III.2 Transition Radiation devices

Equation (8) gives the flux of X-ray Transition Radiation from a single foil in vacuum. For a foil in a gas the denominator of the last term needs an additional $+\omega_{pG}^2/\omega^2$, where ω_{pG} is the plasma frequency of the gas. The flux is so low that many such foils must be used to make an efficient radiator. However the observed flux is not N times the single foil flux, for a number of reasons:

First there is coherence between foils. For complete coherence we would have

$$\frac{dn_N}{d\omega} = \frac{\sin^2\left[N\left(\frac{\Phi+\Phi_G}{2}\right)\right]}{\left(\frac{\Phi+\Phi_G}{2}\right)^2} \cdot \frac{dn_1}{d\omega}$$

where $\Phi(\phi)$ is the phase difference across one foil introduced in II.3 and $\Phi_G(\phi)$ is the corresponding phase difference across the gap between foils. If each foil is typically a formation zone thick, so as not to suppress $\frac{dn_1}{d\omega}$, the radiation from one foil, $\Phi(\phi)$ is not small. It follows that Φ_G must not be small either and the foil separation must be greater than or equal to the formation zone in the gas otherwise there will be distructive interference between successive foils.

At low frequencies the effect of self absorption of the X-rays by the foils is also important. The general case is given by

$$\frac{dn_N}{d\omega} = \frac{1 + \exp(-N\sigma) - 2\exp(-N\sigma/2)\cos N(\Phi+\Phi_G)}{1 + \exp(-\sigma) - 2\exp(-\sigma/2)\cos(\Phi+\Phi_G)} \times \frac{dn_1}{d\omega}$$

where $\sigma = \mu d + \mu_G d_G$, the combined attenuation of a foil and a gap. These interference effects have been demonstrated experimentally for a small number of foils (Fabjan 1975) and have been analysed theoretically (Artru 1975). For a large number of foils such as is appropriate for an optimised detector the effect of multiple scattering can be important in reducing coherence over the whole stack. This happens when

$$\phi_{ms} = \frac{15}{\gamma M(\mathrm{MeV}/c^2)} \cdot \left(\frac{\ell}{\ell_{rad}}\right)^{1/2}$$

is of the same order as the peak angle $\phi \sim 1/\gamma$. The effect is therefore independent of γ, occurs at $\ell/\ell_{rad} = 10^{-3}$ for electrons and is negligible for other particles. There are no accepted handy formulae for including the effect of multiple scattering.

The most thorough detector optimisation has been described by the Willis group (Cobb 1977), which ignored multiple scattering. The flux of photons that they observed was some 30% less than expected. It is not known whether this was due to multiple scattering or mechanical tolerances. We note that their stack of 500 50 μm lithium foils represents $\ell/\ell_{rad}=16\times10^{-3}$. Many practical radiators consist not of films, but of random fibres or foam (Bauche 1982). The theory of such radiators (Garibyan 1975) employing a spread of radiator elements described by a gaussian distribution has been confirmed experimentally (Fabjan 1977). A radiator incorporating 20 carbon-fibre stacks forms part of the European Hybrid Spectrometer (EHS) at CERN and is in use for experiments (Commichau 1980, Struczinski 1983). To summarise although there are effects that reduce its importance, in practice the effect of interfoil coherence may not be ignored.

The optimum choice of radiator thickness is a compromise between attenuation and formation zone suppression. The best material has high electron density (according to section II.3 the energy flux scales with ω_p) and low Z (for low X ray attenuation). The table gives a list of materials that have been or might be used.

Transition Radiators

Material	Form	$\hbar\omega_p$ (eV)	Z	K-edge (eV)	Comment
Deuterium	Foam	7.8	1	14	
Lithium	Foil	14	3	55	Cobb 77
Lithium hydride		19	3	55	
Berylium	Foil	27	4	110	
Boron		31	5	190	
B_4C		32	6	280	
C	Fibres	28	6	280	Commichau 80/Bauche 82

Mylar ($C_5H_4O_2$)	Foil	24	8	530	Many tests
Polyethylene (CH_2)		19	6	280	

Interspersed with sections of radiator are the Transition Radiation photon detectors. To have a high quantum efficiency for a modest thickness they must contain an active absorber of high Z. To date Xenon-filled proportional chambers have been used for real devices, although many tests have used argon or krypton which are significantly less efficient. In a sequence of radiator stacks and gas filled chambers it is necessary to take into account absorption by materials in gas gaps and windows as well as the flux not absorbed by one detector which may reach the next. For this reason optimised stacks usually include a thicker radiator at the front.

The photons are emitted at an angle of order $1/\gamma$ with respect to the parent particle. It is not therefore usually considered practical to resolve the proportional chamber signal due to the Transition Radiation from the signal due to the dE/dx of the particle itself passing through the detector gas. It follows that signals from a Transition Radiation detector are generally the sum of dE/dx and Transition Radiation. There are two points to make. First, the relativistic rise of dE/dx, particularly pronounced in Xenon, often provides a helpful contribution in addition to the Transition Radiation. The effect may be assessed in tests by removing the radiators. Second, the broad spectra both of dE/dx and of Transition Radiation imply that a number of measurements is required. Typically between 2 (Cobb 1977) and 20 (Commichau 1980) measurements are used. Discrimination is achieved by taking the mean signal and comparing it with model distributions based on the analysis of known tracks. Fig.5 shows the γ dependence of the mean and FWHM of signals from the 20 stack EHS device (Commichau 1980, Struczinski 1983). In considering the probability of different masses at known momentum such phenomenological data are used rather than theoretical calculations.

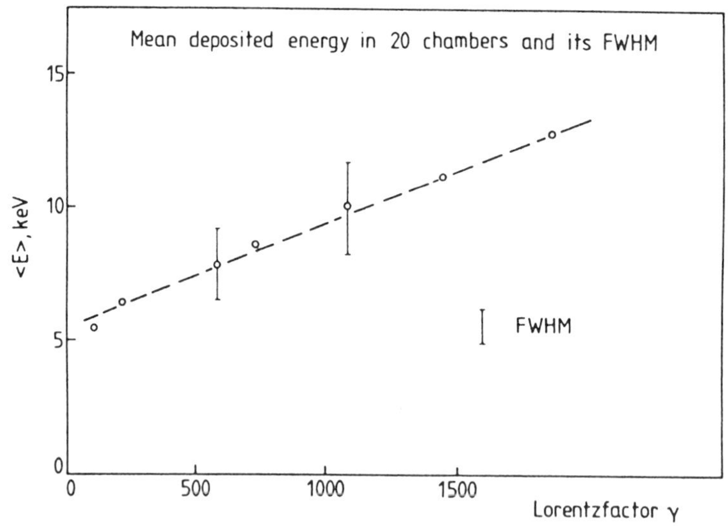

Fig.5. The observed dependence of a Transition Radiation detector signal on γ.

III.3 dE/dx detectors

III.3.1 Theoretical spectra and their velocity dependence

The cross sections described by equation (6) are large. For instance a relativistic particle passing through argon at atmospheric pressure makes a collision about every 300 μm. It is not possible to measure the differential cross-section directly on a collision-by-collision basis and deduce a value for the velocity. Instead, practical detectors measure the energy (or ionisation) deposited in a fixed path length ℓ in the gas more than a hundred times for each track Alikhanov (1956). The resulting energy-loss spectrum may be analysed to achieve the required resolution in $\beta\gamma$. In the following we show how to calculate the spectra and the resolution theoretically, how to analyse experimental spectra in practice and show examples of the velocity resolution that has been achieved in this way.

To calculate the spectrum of energy loss in a thickness ℓ of gas we sum over the collisions occurring by the convolution method as follows:

(1) Calculate the number of collisions, $n=N\ell\int\frac{d\sigma}{dE}\,dE$.

(2) Find ν, the number of times ℓ must be subdivided by 2 to reach a pathlength x in which the chance of a collision τ is less than 10% - and therefore the chance of two or more collisions much less than 1%.

(3) The energy-loss distribution in pathlength x is

$$F_x(E) = (1-\tau)\,\delta(E) + Nx\,\frac{d\sigma}{dE} + O(\tau^2)$$

(4) The energy-loss distribution in pathlength 2x is then

$$F_{2x}(E) = \int_0^E F_x(E-\Delta)\,F_x(\Delta)\,d\Delta$$

By ν applications of this formula, the desired spectrum, $F_\ell(E)$, may be derived. In comparing such spectra with ionisation measurements we assume that ionisation is directly proportional to energy loss. Although untenable in detail this assumption is consistent with available data and perfectly adequate for a discussion of dE/dx detectors.

Fig.6a shows dE/dx distributions for ℓ=3mms of pure argon at normal density calculated for different values of $\beta\gamma$. The changing shape arises from the vestigial effect of the shell structure. Fig.6b is a similar plot for ℓ=15mms of pure argon. Here the shape of the distribution has become smooth and may be matched for different values of $\beta\gamma$ by scale changes along the axes. The extent to which this is true is illustrated in fig.7 where the reduced width is plotted for different $\beta\gamma$. There is a small variation in width arising from the improved statistics when the ionisation is higher. We ignore this and assume that for ℓ fixed the distribution has a fixed universal shape F independent of $\beta\gamma$. A likelihood study shows that the resulting loss of information is small provided that the length ℓ is large enough (>1cm atm for argon).

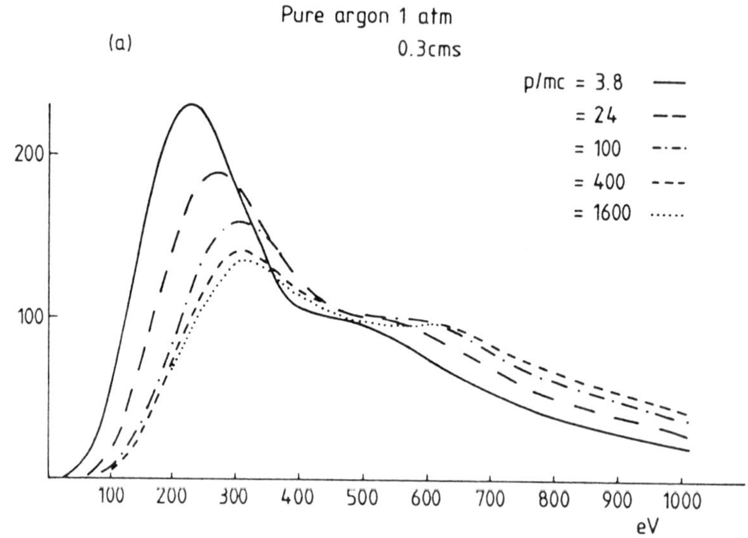

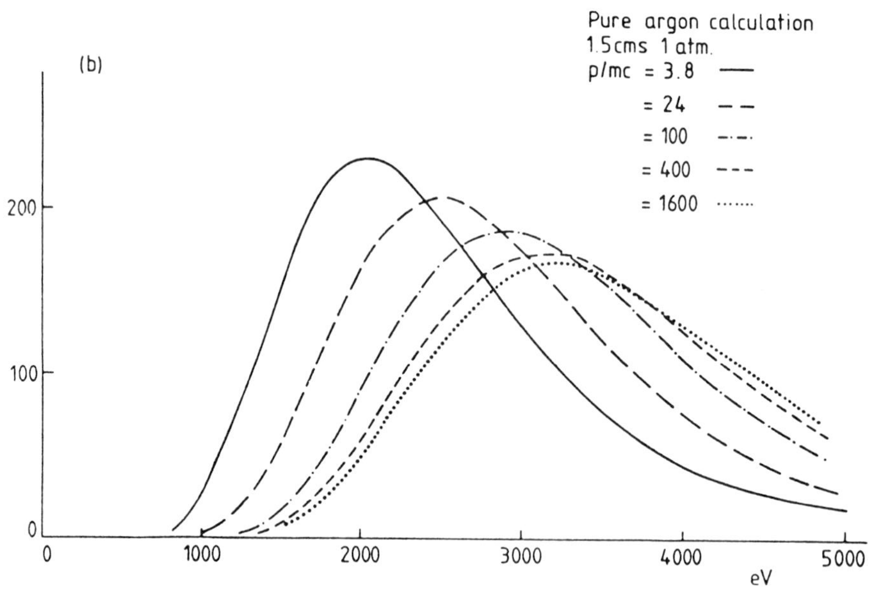

Fig.6. The calculated βγ-dependence of dE/dx spectra for a) 3 mms and b) 5 mms of argon.

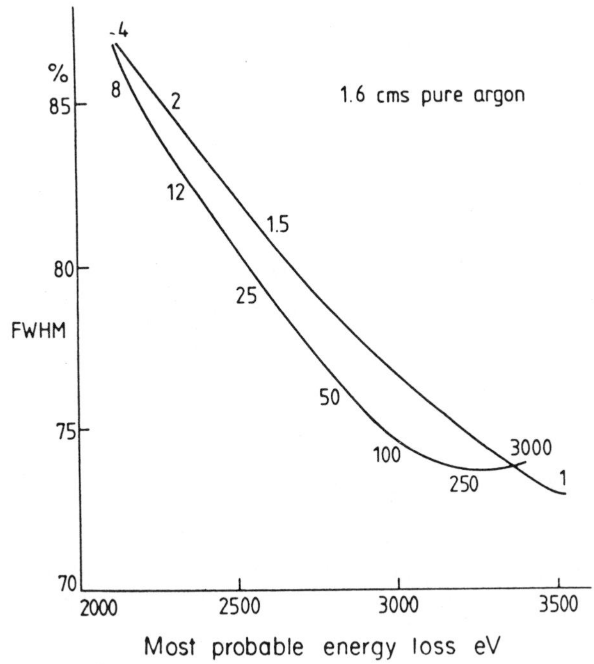

Fig.7. The calculated relative width of the energy-loss distribution as a function of the most probable energy-loss for 1.6 cms of pure argon at normal density. The numbers on the graph are the corresponding values of $\beta\gamma$.

III.3.2 Determining velocity from experimental measurements

The most efficient fitting method with these assumptions is a single parameter likelihood method which maximises

$$L(\lambda) = \prod_i F(E_i/\lambda)$$

where the measurements E_i are scaled by the parameter λ. λ is proportional to the ionisation of the track. Fig.8 shows a comparison of the theoretical distribution F for 1.6cms of argon/20% CO_2 at 1 atm with experimental data from ISIS2 on a large number of tracks (Allison 1974, 1982). The agreement is surprisingly good considering that instrumental effects such as resolution, crosstalk and diffusion have been ignored. With this level of agreement it is immaterial whether the

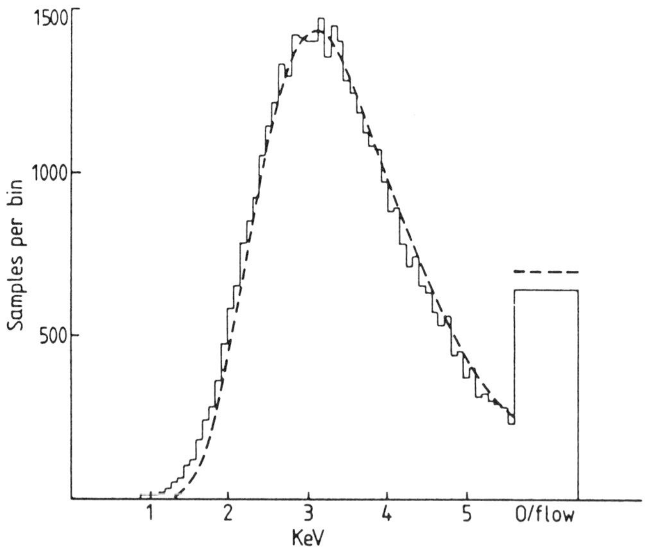

Fig.8. A comparison of theoretical and experimental dE/dx distributions. The experimental energy calibration is a free parameter.

theoretical or experimental distribution is used as likelihood function although in principle the latter is preferred. The advantages of the maximum likelihood method over the truncated mean method described below are threefold:

(1) It provides not only a value of λ but an internal error.

(2) It has a much superior dynamic range, providing unbiassed estimates of λ even when 95% of the spectrum is in the 'overflow bin'.

(3) It is unambiguous†.

Nevertheless in most cases it does not give a significantly superior ionisation resolution.

Fig.9 shows a theoretical analysis of the ionisation resolution, $\Delta\lambda/\lambda$

---

† Some other methods involve technical ambiguities. For example, how do you calculate the unbiassed mean of the lowest 40% of 99 measurements?

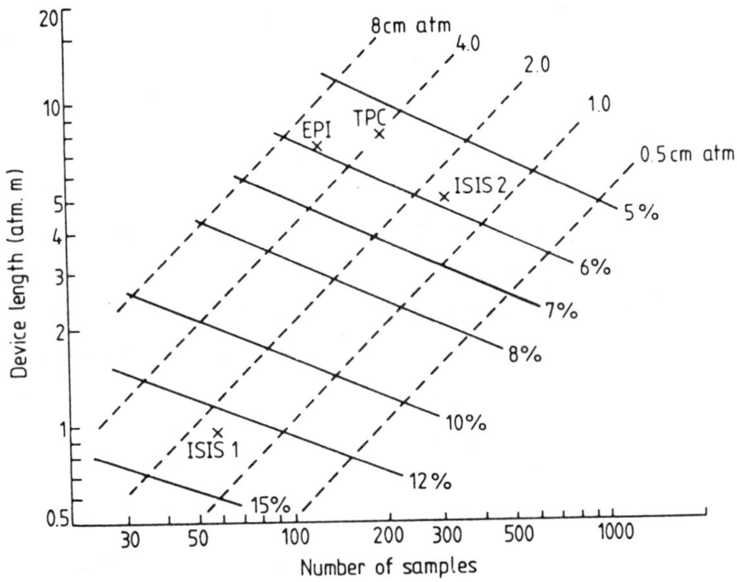

Fig.9. Calculated ionisation resolution (% FWHM) contours for argon with $\beta\gamma=100$. The crosses indicate the sizes and samplings of a number of devices. The dashed lines are loci of constant sample thickness.

(FWHM), that can be achieved with argon, as a function of the number of samples and the device length. The calculation uses the distribution for $\beta\gamma=100$. For thicknesses less than 1cm atm† the shape of the distribution changes significantly with $\beta\gamma$ as we have seen and the single parameter likelihood analysis described here is not optimal. Cruder methods of analysis such as the use of the mean of the lowest 50% of the signals (Lehraus 1978) yield similar values for the resolution in general but are significantly worse below 1cm atm because they ignore the information in the shape of the spectrum. However this is rather academic for practical detectors because there are significant effects

---

† To a first approximation the resolution depends only on the thickness in gm cm^{-2}. For gases this is conveniently measured as the product of path length times pressure where normal temperature is understood.

which make it very difficult to measure such thin samples independently on the scale needed (Lapique & Piuz 1980). For good separation of π/K/proton masses a FWHM resolution of order 6% is needed (2.5% RMS).

The calculated dependence of the ionisation on $\beta\gamma$ for argon (Allison & Cobb 1980) is compared with experiment (Lehraus 1978) in fig.10. The agreement is good. Calculations for a number of other gases, pressures and sample thicknesses have been published (Allison 1982). The dependence on choice of gas and sampling geometry favours fine sampling and the use of argon or possibly xenon. Fine sampling (down to 1cm atm) not only improves the resolution but even increases the relativistic rise a little.

III.3.3 The calibration problem

In practice the main problem is one of calibration. The likelihood fit yields a value of λ for each track from its many ionisation

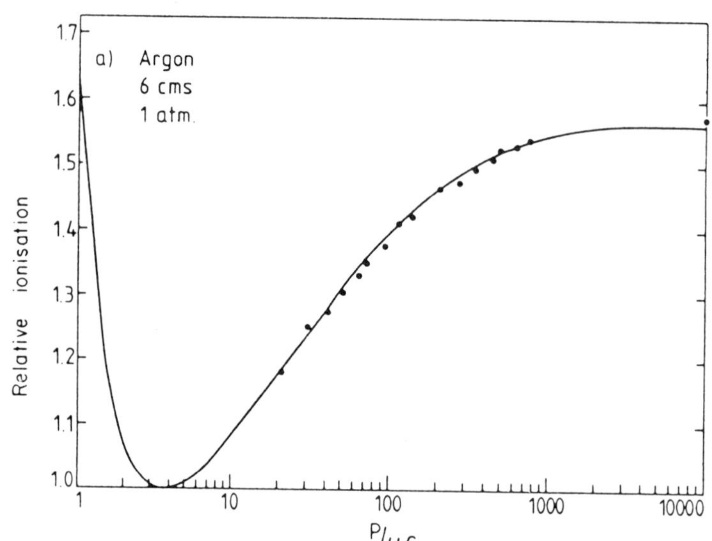

Fig.10. A comparison of the measurements of Lehraus 1978 with the calculated dependence of ionisation I/I_o on $\beta\gamma = P/\mu c$.

measurements. The relationship between λ and the ionisation relative to minimum, I/I_o, involves:

a) Track inclination (θ). For angles up to about 45° λ is proportional to secθ and may therefore be corrected by multiplying by cosθ.

b) Amplification variation - possibly depending on the position in the detector.

Calibration may be attempted using signals from X-ray or other sources. This is often not easy. Instead here we describe a method that uses the tracks themselves assuming that a majority are pions. It proceeds in two steps:

a) Assume that all tracks are pions ignoring tracks that may be non relativistic. Calculate for each track the calibration factor G equal to the ionisation expected for a pion of the known track momentum divided by the corresponding value of $\lambda\cos\theta$. The first estimate of the calibration factor uses the mean value of G. This calibration is wrong but only by a few percent because in general only a small fraction of tracks are not pions and their ionisation differs by not more than 10-15% from that of a pion.

b) Find those tracks whose momenta and ionisation errors are such that the expected ionisation for a pion is more than 4 standard deviations (say) from the electron or kaon hypothesis. Use the calibration from (a) on these to eliminate those tracks whose measured ionisation is more than 2.5 standard deviations from the pion expectation. Calculate a better value for the calibration factor by evaluating the mean G for the remaining tracks. This step may be iterated again although this brings little change.

Given the calibration and, for each track, the value of λ and its error, we may derive an experimental value of I/I_o. Knowing the track momentum this is compared with the expected value of I/I_o for each mass assignment using the known $\beta\gamma$ dependence for the gas, for example fig.10. Monte Carlo simulation studies show that it is the logarithm of I/I_o that is normally distributed rather than I/I_o itself. We therefore calculate a χ^2 with 1 degree of freedom for each of the masses e, π, K and p:

$$\chi^2_M = \frac{(Z_M^{th} - Z^{exp})^2}{(\Delta Z)^2}$$

where Z is ln (I/I_o). This is the ionisation χ^2 for mass M.

III.3.4 Example of performance that can be achieved

In the case of the data from ISIS2 (Allison 1974, 1982) in the NA27 EHS experiment at the CERN SPS this method of calibration was checked by looking at electron candidates identified by a lead-glass wall (Aguilar-Benitez 1983). From a sample of about 2×10^4 secondaries those that deposited between 0.8 and 1.3 times their energy (as measured by magnetic dispersion) in the lead-glass were chosen. Fig.11 shows for these tracks a plot of the χ^2 for the electron hypothesis against the χ^2 for the best hadron hypothesis. The tracks divide almost equally into three categories; those in the corner that are compatible at the 1% level with either an electron or hadron interpretation; a horizontal band of those that are consistent with a hadron interpretation but incompatible with an electron (the fraction of these at a few $\times10^{-3}$ is consistent with the efficiency of the lead-glass detector); and a vertical band of confirmed electrons inconsistent with hadrons. Fig.12 shows the χ^2 probability distribution for the electron mass for all tracks with hadron probability less than 1%. The flat distribution confirms that the error used, which was $106/\sqrt{N}$ % FWHM where N is the number of samples, is correct. This is in complete agreement with the prediction shown in fig.9. The mean value of N in practice was 240 out of the maximum of 320 samples per track. The loss of samples is due

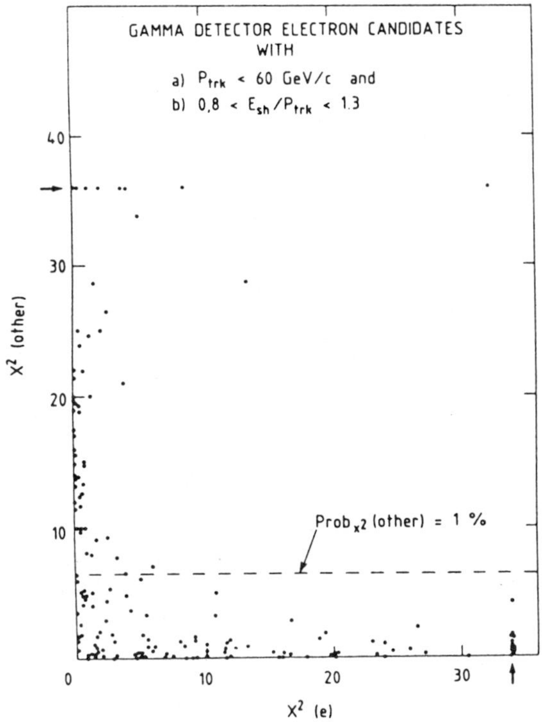

Fig.11. A scatter plot of ionisation chi-squares for electron candidates selected by the lead-glass wall. Values of χ^2 greater than 35 or so are shown superposed in line with the arrows.

mostly to track overlap (see appendix). The two tracks shown cross hatched in fig.12 have ionisations incompatible with any mass. They represent a measure of current residual effects in the analysis.

Fig.13 is a scatter plot of I/I_o versus log momentum for several thousand secondaries in the NA27 charm experiment. The ±3% bands expected for e, π, K and proton masses are shown. The plot shows separation according to the Rayleigh Criterion (1 FWHM or 2.36σ) for the following momentum ranges:

Mass pair	Momentum range (N=300)
e,π	Up to 25 GeV/c

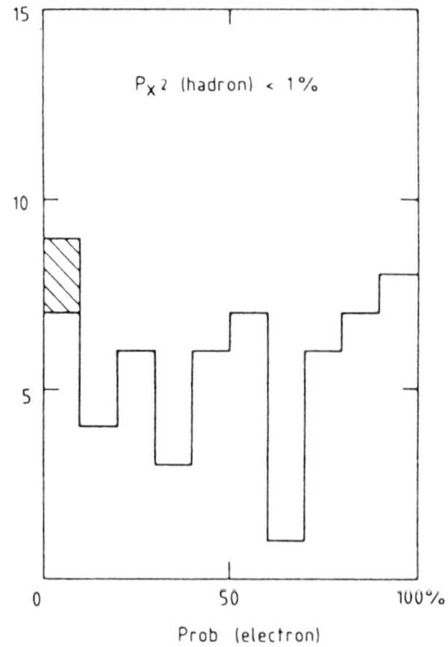

Fig.12. The ionisation probability distribution for those electron candidates whose ionisation is not compatible with any hadron.

π,k	2-60 GeV/c
k,p	5-40 GeV/c
π,p	1.5-150 GeV/c

Practical questions of separation depend on the confidence level sought and the relative fluxes concerned.

As a live example of dE/dx data we show in fig.14 the track hit data from ISIS2 for an event with a D^o decay uniquely identified by kinematics. Four charged tracks coming from the resolved D^o decay vertex in the bubble chamber are marked. The tracks shown are up to 5.12m long (horizontal axis) divided into 320 samples. The vertical axis corresponds to 4m (folded). For each individual hit shown there is a 7-bit pulse height. The pulse heights associated with each track are histogrammed and fitted as described earlier. Fig.15 shows the data and fits for three of the tracks from the D^o decay. The tracks are clearly distinguished as K or π and this identification is confirmed in the

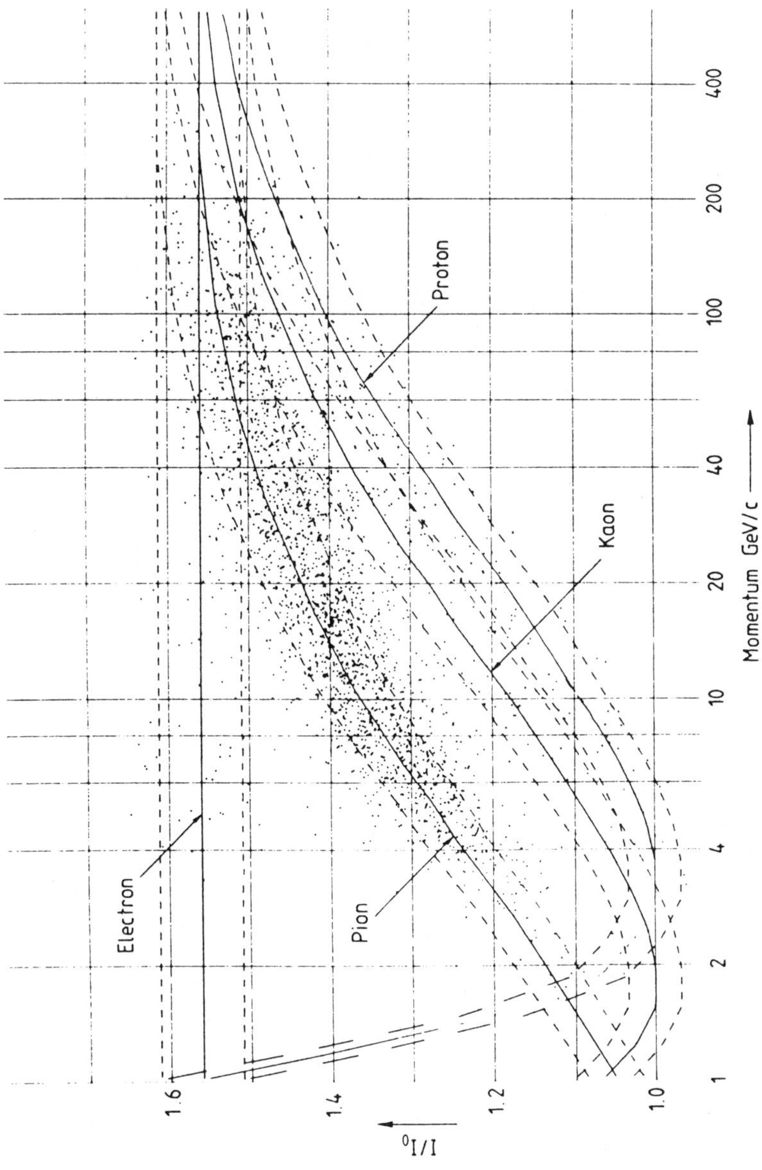

Fig.13. A scatter plot of measured values of ionisation I/I_0 versus log momentum. The bands shown for π, K and proton are discussed in the text.

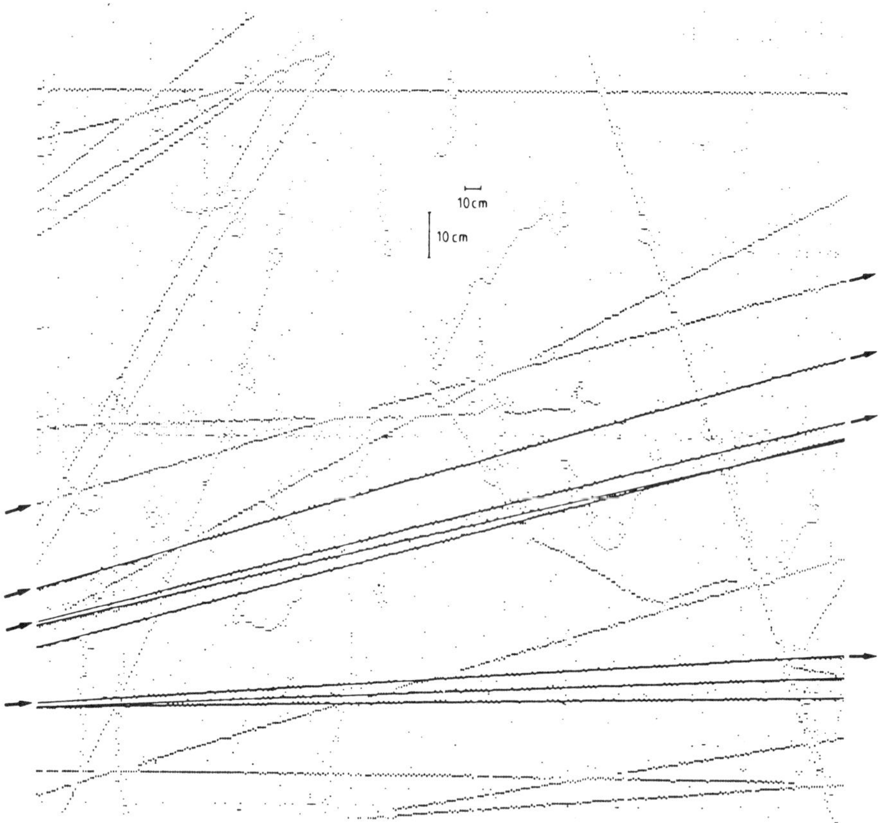

Fig. 14. The track-hit data in ISIS2 for a single event which includes a reconstructed D^{o}. The four charged tracks from its decay are marked by arrows.

unique kinematic fit based on momentum information alone. The situation is summarised in the table. The identification of the fourth track as a pion is poor but not ruled out.

Track	Momentum	# of samples	Ionisation ± RMS	Probability(%)			
	GeV/c	N	I/I_o	e	π	K	P
1	9.8π	237	1.434 ± 3.0%	.5	5.4	<0.1	<0.1

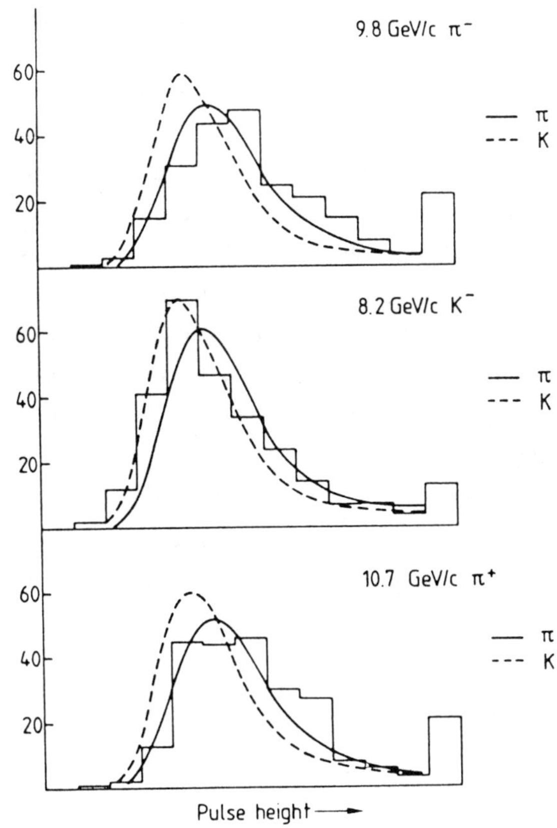

Fig.15. The dE/dx spectra of three of the charged tracks marked in fig.14. The superposed curves are the K and π predictions using the measured track momenta.

2	8.2K	275	1.139 ± 2.8%	<0.1	<0.1	70.6	1.9
3	10.7π	246	1.397 ± 2.9%	<0.1	42.5	<0.1	<0.1
4	5.1π	247	1.156 ± 2.7%	<0.1	0.2	4.2	<0.1

Reference to experience with other detectors may be found in Lehraus (1983).

Appendix: The solution of practical problems for dE/dx detectors

The description of relativistic particle identification by dE/dx given in III.3 presupposes the solution of a number of problems. These are discussed here with the solutions adopted for the ISIS detector (Allison 1974, Aguilar Benitez 1983, Allison 1982) but the points are common to any large drift chamber seeking the necessary control of systematic effects (Pleming 1977). ISIS is a pictorial drift chamber whose prime role is particle identification; tracking is a free but impressive by-product. Fig.16 shows a vertical beam-plane section of the upstream part of the chamber. Track signals drift to a single central wire plane perpendicular to the plane of the diagram. The multihit electronics can handle between 30 and 50 hits per event on each wire and have been described elsewhere (Brooks 1978). The data include the drift time and pulse height for each hit. The third coordinate is not measured and the up-down ambiguity is not resolved by the chamber alone.

	ISIS2
Acceptance	4 x 2m^2
Drift distance	2 x 2m
Voltage	100 KV
Maximum number of Samples	320
Volume	120m^3
Ionisation resolution, FWHM (with 320 samples)	6.0%

Fig.17 summarises the various relationships that determine the uniformity of track signal response which needs to be free of systematic effects at the 1% level. The diagram indicates the important role of the shape of the track signals. An individual electron arriving at an anode wire and generating an avalanche produces a pulse with a long tail of the form $1/t+t_0$ (Charpak 1970). This may be seen as the linear

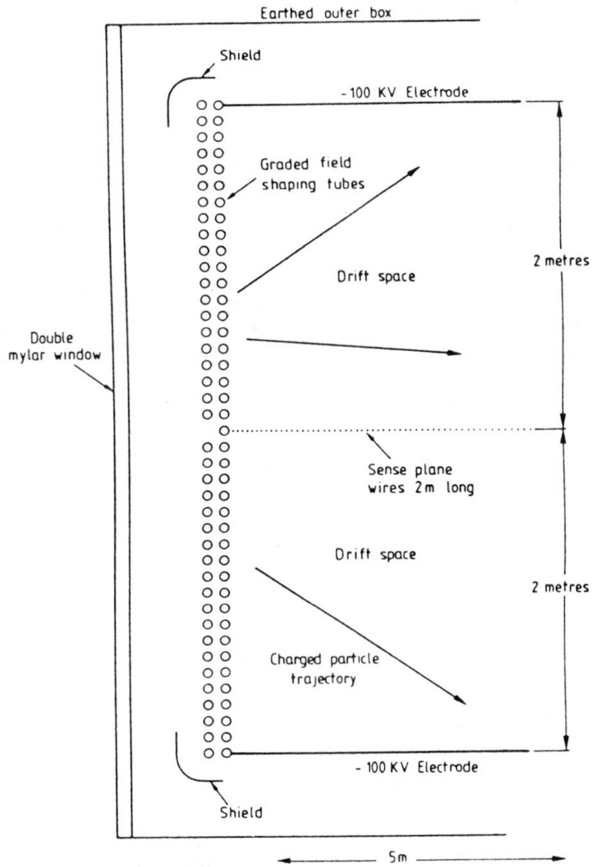

Fig.16. A diagram of ISIS2.

impulse response of the gas amplifier. Inverting to provide a short clipped pulse without undershoot is a linear problem. In ISIS it is done by a passive filter with three time constants - two is probably enough. Of course such correction need not reproduce a δ-function. It is sufficient to match the pulse to the bandwidth of the electronics which in turn must be chosen to integrate the signal sufficiently. A relativistic particle makes about 30 collisions and 100 secondary electrons in 1 cm atm of argon. Depending on the track angle, diffusion and collection geometry these 100 electrons will be spread up to 1 cm in

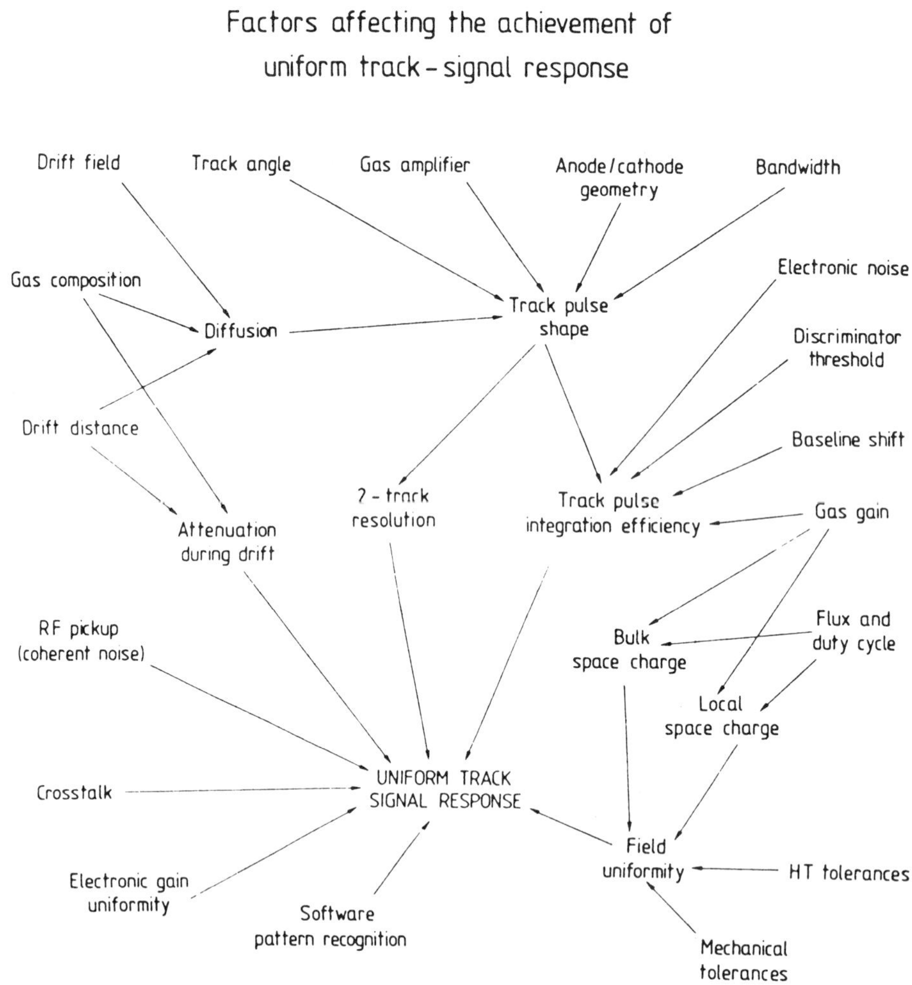

Fig.17

drift distance. With a drift velocity of 2cm/μs (or 20μm/ns) the mean time between electron arrival is 5ns. In ISIS the bandwidth of the electronics is 15 MHz. This implies a typical instantaneous signal charge of about 5-10$\times$G electrons referred to preamplifier input where G is the gas amplification factor (10^4 in ISIS). In terms of current this represents a typical peak 300-1000 namps with an rms noise of 18 namps.

The detailed effect of electronic noise and electron arrival time was studied in a Monte Carlo program to demonstrate that the signal-to-noise ratio is sufficient to trigger efficiently (>99%) on tracks and rarely on noise ($\lesssim$ 1 KHz). The collection geometry contribution to the track pulse shape comes from the variation of drift time for different drift paths near the anode wire. The effect of diffusion is important but may be minimised by using a suitable molecular "cooling" gas such as CO_2 which keeps the r.m.s. energy of the electrons near to $kT_e \sim 1/40$eV (Allison 1974). An electron that drifts a distance ℓ through a potential V has an r.m.s. displacement σ in each orthogonal dimension given by

$$\sigma/\ell = \sqrt{kT_e/eV}$$

where T_e is the electron temperature (assumed isotropic). Minimising σ has two other consequences. First, because the electrons are kept in thermal equilibrium with the gas as far as possible, it follows from kinetic theory that the drift velocity is not "saturated". This is in fact not seen as a disadvantage for the drift velocity is kept uniform by designing an electric field with the necessary uniformity (10^{-3}) which is needed anyway if the necessary gain uniformity is to be achieved (10^{-2}). The second consequence of the cooling gas is an increased sensitivity to residual oxygen concentration. The cooling gas catalyses the attachment of electrons to oxygen. Typically oxygen concentrations below 0.2 parts per million (by volume) must be achieved if loss by attachment is to be ignored (Allison 1974). Of course the exponential loss of pulse height with drift distance can be corrected if the loss is known. Calibrating this loss may mask other effects including imperfect track pulse integration efficiency.

The pulses have a variable shape as the effect of angle and diffusion varies from track to track. To measure the charge in a pulse it is therefore necessary to do a true gated integration of a slightly delayed track pulse with the front and back edge of the gate operated by a discriminator looking at the prompt track signal. The effects of electronic noise and modest gas amplification require the discriminator to have a finite threshold. Delays are incorporated in the logic to avoid missing part of the track signal due to the consequent slewing. A more insidious problem is the baseline shift at high rates. It is usual for the electronics to employ ac coupled signals to avoid the hazards of dc amplication. It follows that the mean signal amplitude is zero. In particular if the signal occupancy is f% within one ac coupling time constant, the baseline will shift by f% of the pulse height as shown in fig.18. Since the occupancy in ISIS for an event is 10% or more this is a serious effect. It cannot be computed in practice from the amplitude of earlier pulses partly because of uncertainties surrounding the frequent saturating signals. The problem is essentially nonlinear and cannot be corrected with passive filters. There are two solutions - either the background level between pulses can be digitised and used for off-line correction, or the background level signal can be fed back in the electronics itself. In ISIS the latter was chosen. Once this problem is solved excellent rejection of low frequency hum, or at least its linear consequences, can be achieved.

The need to reduce systematic biases to the 1% level leads to a very conservative two-track resolution. The values of the delays in the gated integration are designed to integrate 99% of the signal in the most unfavourable case as checked in the Monte Carlo simulation of the ISIS signal processing (Pleming 1977). The success of the optimisation may be judged from fig.19. The line superposed on the lower histogram is the expectation of Poisson statistics. All track hits separated by

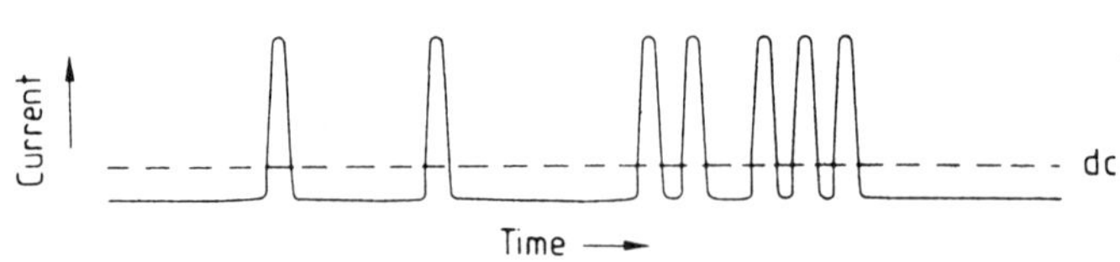

Fig.18. A pulse train illustrating the effect of ac coupling in creating a dc level shift.

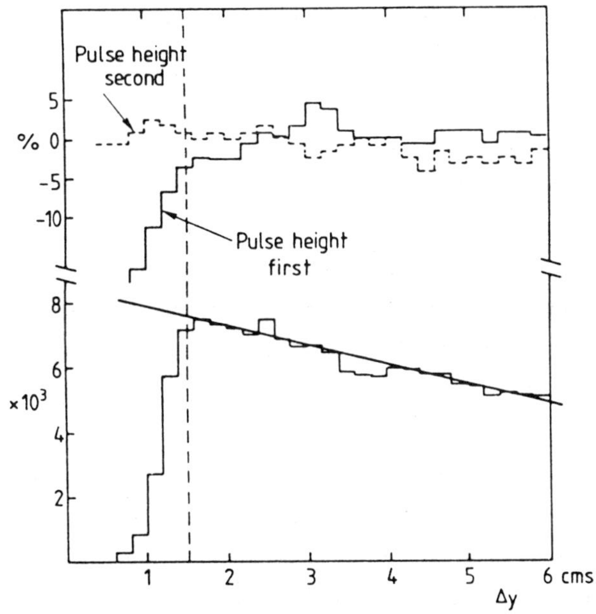

Fig.19. Lower histogram - the distribution of resolved track hits as a function of track separation. Upper plot - the dependence of mean pulse height on track separation.

more than 15mms are resolved. The mean pulse-heights for the first and second hits of such pairs show that resolved hits above 15mms are unaffected by the presence of the other at the level of 2-3%.

To achieve uniform track-signal response it is also necessary to have stable uniform gas amplification. This depends on the uniformity of the gas composition, the charge per unit length of the anode wire and its diameter. Since the gas amplification is an avalanche process involving 10 or more generations the uniformity required to achieve 1% gain stability is about 0.1%. (The measured dependence of gain on small variations of gas pressure, for example, shows a factor of 7). Since the voltage difference between the anode wire and all other conductors may be controlled relatively easily the problem is to control the dimensions such that the capacitance is sufficiently uniform. With careful design these problems can be solved. More difficult are the effects of space charge on the anode wire charge. These separate into two parts, bulk space charge and local space charge. Bulk (drift volume) space charge. As a result of cosmic rays, background and earlier events the drift volume always contains positive ions released during the gas amplification process. One can estimate roughly the density ρ knowing the gas amplification factor G, the mean ionisation rate (100e/cm) and the positive ion drift velocity (~1cm/s/volt/cm). Bulk space charge will be a problem for both spatial precision (at 10^{-3} level) and gain uniformity (at 10^{-2} level) if

$$\frac{\int \rho(z)\,dz}{\sigma} > 10^{-3}$$

The integral is the line integral of the space charge density through the drift volume and σ is drift electrode charge density per unit area. We have reported spectacular effects from such space charge. In ISIS we have cured the problem by gating the gas amplification off except during the bubble chamber sensitive time. The problem is a major constraint for any detector of this type. ISIS has no such effect with the following parameters:

Amplified flux (total)	1200 tracks/second
Drift field	5×10^{4} Volts/metre
Pressure	1 atmosphere
Gas gain	10^{4}
Drift distance	2 metres

The second space charge effect is local space charge in the immediate neighbourhood of the anode. Fig.20 shows a track crossing a series of anode wires at an angle θ. The avalanche for this track (ignoring diffusion) takes place over a length of wire $s\theta$ where s is the anode separation. This size of the avalanche is of order

$$G(\text{gain}) \times P(\text{bar}) \times 10(\text{electrons/mm}) \times s(\text{mm}) \times 1.6 \times 10^{-19}$$

giving a linear density of avalanche

$$GP\,\theta^{-1} \times 1.6 \times 10^{-18}\ \text{Coulombs/mm.}$$

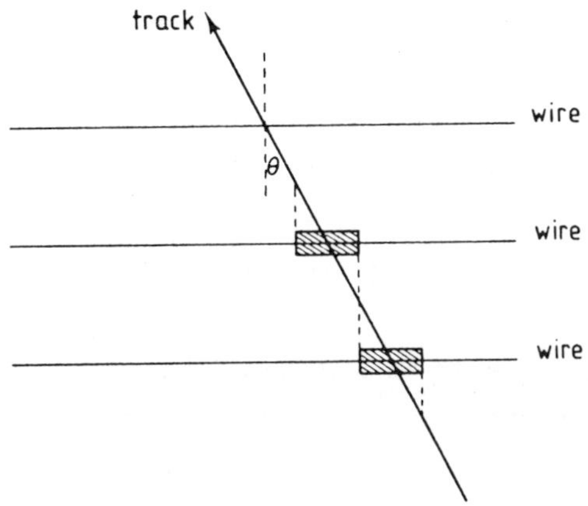

Fig.20. Where the gas avalanche occurs along the anode wires for a track at angle θ.

If this charge density is significant (at level 10^{-3}) compared with the anode charge density causing the avalance, the gain will be non linear. This is the so called "semiproportional" region of gas amplification but its singular dependence on angle is quite unsatisfactory. The charge density on a 25μm anode wire is typically 12pC/mm. In practice we observe no effect if GP θ^{-1} is less than 2×10^5. Tracks at angles less than 50 mrad in ISIS are avoided by rotating the chamber such that all tracks have significant values of θ. Detailed observations by Frehse (1978) confirm back-of-envelope estimates although diffusion during the drift helps to spread the avalanche out. However there is another effect that makes matters worse. Since the positive ions travel very slowly (5-10 m/s), any track that arrives within a msecond on the same section of wire as an earlier track will suffer from the influence of the earlier ions which are still only a few mms from the anode. In ISIS such an effect is observed (15% gain reduction) when the beam is finely focussed (instantaneous rate of order 2.5×10^4 tracks per sec per cm). There is no such effect with a defocussed beam.

Monitoring changes in the gas density and composition is a task that must be taken seriously. Temperature variations of a fraction of a degree or a millibar pressure change must be recorded. Thermal conductivity devices are good for monitoring binary gas mixtures, but measuring the oxygen contamination is harder. This is done with commercially available instruments based either on an aqueous alkaline electrolyte or a solid zirconia electrolyte. The trouble is that the former do not work with CO_2 and the latter run at high temperature. However available instruments have improved dramatically in the last few years.

Radio frequency pickup must be avoided. With very long anodes and sensitive low-noise electronics the problem demands careful design. A related problem is crosstalk of real signals between channels. This may have many causes of which capacitative crosstalk in the chamber or coupling through power supplies in the electronics are two. Another is diffusion. Because the dE/dx distribution is so highly skew the effect of diffusion crosstalk between nearest-neighbour samples causes the dE/dx peak not only to broaden but to shift to higher values with increasing drift distance. For a device that had either smaller samples or longer drift or was trying to get better resolution than ISIS2 this effect would become a problem. Diffusive crosstalk is positive while capacitative crosstalk is of negative sign. Generally Monte Carlo studies show that a few percent crosstalk between nearest neighbours is tolerable. Effects that couple all channels together are quite intolerable and lead to a dramatic loss in ionisation resolution.

To monitor the performance of the electronics in situ as well as keeping an eye on the software pattern recognition, two artificial "tracks" are included in every data readout. First there is a dummy track with a forced trigger whose pulse heights represent pedestal measurements (this is not plotted in the picture such as fig.14). Second there is a pulse generator signal sent to all channels in parallel and injected on the front of the preamplifier. This may be seen running across near the top in fig.14. Both the hardware and software treat this as any other track. Subsequently we find that this track has a pulse height resolution of <1% FWHM with individual measurements that have been successfully cleaned of overlapping signal regions. It is of course an important aspect of the design that in the main the same channels of electronics see every track so that problems of calibration between channels are of reduced importance. The detailed track filtering in the software is crucial. In fig.21 we show an expanded portion of fig.14. The hits are coded with different symbols depending on how they are used. The diamonds are the track hits used. The vertical bars are the track hits discarded as double on the grounds that they belong to two or more track vectors. The horizontal bars are

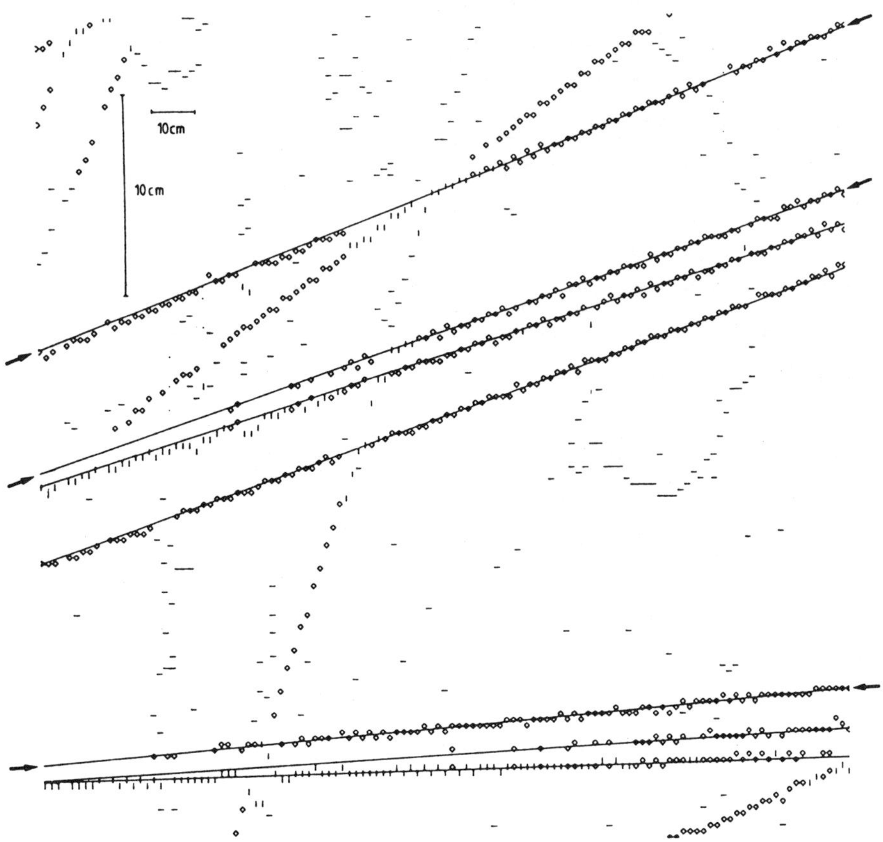

Fig.21. Detail of fig.14, 60 cms × 2 m long in space.

noise or δ-ray hits. The long sloping lines are the track vectors from the overall fit in the spectrometer. One can see by inspection that the contamination by double track hits is less than 1%.

References

Aguilar-Benitez (1983) Nucl.Inst.& Meth. 205, 79

Alikhanov (1956) Proc. CERN Symp.on High Energy Accelerators, Pion Phys.2, 87

Allison et al (1974) Nucl.Inst.& Meth. 119, 449

Allison and Cobb (1980) Ann.Rev.Nucl.i-5art.Sci. 30, 253

Allison (1982) Conf.on Instrumentation, SLAC-250, 61

Artru et al (1975) Phys.Rev. D12, 1289

Bauche et al (1982) Conf.on Instrumentation, SLAC-250, 122

Berkowitz (1979) "Photoabsorption, Photoionisation and Photoelectron Spectroscopy" Academic Press (New York)

Brooks et al (1978) Nucl.Inst.& Meth. 156, 297

Cerenkov (1958) in Nobel Lectures, Elsevier (1964), Amsterdam

Charpak (1970) Ann.Rev.Nucl.Sci. 20, 195

Cobb et al (1977) Nucl.Inst.& Meth. 140, 413

Commichau et al (1980) Nucl.Inst.& Meth. 176, 325

Davenport et al (1983) IEEE Trans.NS 30, 35

Fabjan (1975) Phys.Letters 57B, 484

Fabjan (1977) Nucl.Inst.& Meth. 146, 343

Fano (1963) Ann.Rev.Nucl.Sci. 13, 1

Frehse (1978) Nucl.Inst.& Meth. 156, 87

Garibyan (1957) JETP 33, 1403 [Soviet Physics JETP 6, 1079 (1958)]

Garibyan et al (1975) Nucl.Inst.& Meth. 125, 133

Glass et al (1983) IEEE Trans.NS 30, 30

Jackson (1962) "Classical Electrodynamics" (First edition) John Wiley, New York, §§14.9

Jackson (1975) "Classical Electrodynamics" (Second edition) John Wiley, New York, §§13

Jelley (1958) "Cerenkov Radiation" Pergamon (London)

Ladron de Guevara (1983) EHS Internal note.

Landau and Lifshitz (1960) "Electrodynamics of Continuous Media" Pergamon, Oxford §§85 & 86

Lapique and Piuz (1980) Nucl.Inst.& Meth. 175, 297

Lehraus et al (1978) Nucl.Inst.& Meth. 153 347

Lehraus (1983) Proc.of Vienna Wire Chamber Conf., Nucl.Inst.& Meth.

Litt and Meunier (1973) Ann.Rev.Nucl.Sci. 23, 1

Pleming (1977) D.Phil.thesis, Univ.of Oxford, Rutherford HEP/T/69

Poelz and Riethmuller (1982) Nucl.Inst.& Meth. 195, 491

Struczinski (1983) Internal EHS note.

Ter-Mikaelian (1972) "High Energy Electromagnetic Processes in Condensed Media" Wiley-Interscience (New York)

Ypsilantis (1981) Physica Scripta 23, 371

Reprinted, with permission, from *Nuclear Instruments and Methods in Physics Research*, Vol. 200, pp. 219-236 (1982).

A TWO-DIMENSIONAL, SINGLE-PHOTOELECTRON DRIFT DETECTOR FOR CHERENKOV RING IMAGING

by E. Barrelet, T. Ekelof, B. Lund-Jensen, J. Seguinot, J. Tocqueville, M. Urban, and T. Ypsilantis

1. Introduction

The Cherenkov light emitted by a high-energy particle traversing a dielectric medium may be focused to a ring image with the aid of a spherical mirror. The focus of the image is situated half-way between the mirror and its centre of curvature. The focal surface thus has the shape of a spherical shell with a radius half of that of the mirror, as illustrated in fig. 1.

The measurement of the radius of the Cherenkov ring allows a determination of the angle of emission of the Cherenkov light and thus of the particle velocity. If the particle momentum is also measured, this enables a determination of the mass of the particle and thereby its identity. A detailed analysis of the ring image optical quality and the concomitant resolution obtainable may be found elsewhere [1].

The basic technical problem in the utilization of this method is the detection of the position of single ultraviolet (UV) photons in the focal surface with good two-dimensional position resolution (~ 1 mm), high conversion ($\sim 50\%$) and counting efficiency ($\sim 90\%$).

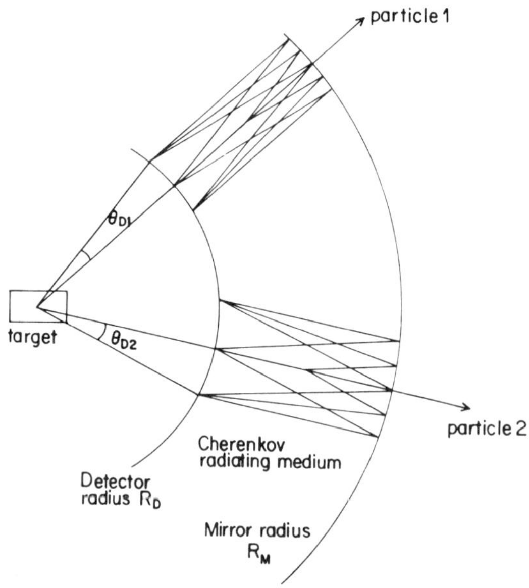

Fig. 1. Schematic diagram of ring-imaging geometry. The outer spherical surface is a mirror and the inner surface a detector. Photons emitted when two particles emerge from the interaction region, near the mirror centre of curvature, will be focused by the mirror into two separate rings with half-angles θ_{D1} and θ_{D2}. These ring radii are related to the particle velocity β_i by the relation $\theta_{Di} = \cos^{-1}(1/n\beta_i)$ where n is the refractive index of the gas (or liquid) radiating medium.

The conversion of photons to electrons is effected via the photoionization mechanism

$$h\nu + M \rightarrow M^+ + e^- , \tag{1}$$

where M is a suitable photosensitive molecule admixed with the detector gas. This process has been shown [2] to have a large quantum efficiency ($\sim 50\%$) for photons above the threshold energy E_t which is, for the molecules so far utilized, well into the UV region of the electromagnetic spectrum (i.e. $E_t \geqslant 5.4$ eV). High single-electron counting efficiency ($\geqslant 90\%$) has been achieved [3] by using pure methane (CH_4) as an amplifying gas in a multiwire proportional chamber (MWPC) geometry. As shown in fig. 2, we achieve unique two-dimensional determination of the photon conversion point by the use of a drift gap in which the photons are

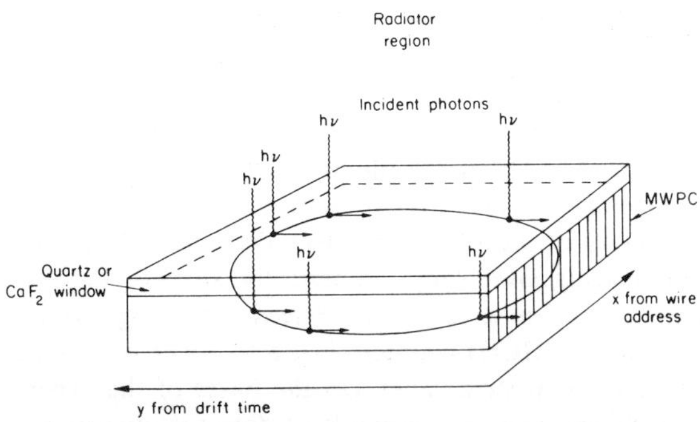

Fig. 2. Schematic of the photon drift detector. The UV photons produced by a charged particle in the radiator are reflected by a spherical mirror and form a ring image, in the focal plane of the mirror, just below the entrance window (quartz or CaF_2). The photons are converted to photoelectrons by photoionization of molecules in the drift volume. A uniform electric field parallel to the entrance window and perpendicular to the MWPC plane causes the electrons to drift to the MWPC plane where they are amplified, counted and timed. The carrier drift gas is methane plus TEA or TMAE.

converted and drift to a MWPC array where they are detected and timed [3]. In an experimental situation where several radiating particles are expected to emerge together in collimated jets, it is essential that the many near-by photon points in the focal surface can be reconstructed unambiguously in order to allow the pattern recognition of the circular images.

We have constructed a photon drift detector (type: time projection chamber), in accordance with the above criteria, which has a circular transmitting window of 127 mm diameter (open diameter 100 mm) sitting above a square drift gap of 140×140 mm^2 in area and 15 mm (or 45 mm) in depth. In this paper we describe the construction of the photon detector and the results of test measurements, performed in a 5–20 GeV/c π^- beam from the CERN Proton Synchrotron (PS), to detect the ionization electrons produced in the drift gap by through-going beam tracks. The parameters which affect the precision of spatial localization of the electrons, such as wire spacing, time binning, drift velocity, and diffusion, have been investigated. In addition, an unexpected non-uniformity in the drift field, due to incomplete polarization of the entrance window, has been observed and corrected. In a forthcoming paper we describe the ring images obtained with this detector, with triethylamine (TEA) and tetrakis (dimethylamine) ethylene (TMAE) as the photoionizing gases.

One of the aims of the present work is to explore the possibility of using this drift technique for UV photon ring-imaging over large, spherically shaped, detector areas (several m^2), as would be needed for ring-imaging Cherenkov (RICH) counters made to cover large solid angles (several steradians). On the bases of the experience gained with the photon drift detector discussed in this paper, we have designed and constructed a large modular counter unit having a large ratio of active to total area ($\sim 85\%$) and small matter density (~ 2 g/cm^2). Several such units can be mounted side by side at a slight angle to cover a large, approximately spherical surface.

2. Principle of the single-photoelectron drift detector

The basic idea of the drift detector is illustrated in fig. 2. Photons generated in the Cherenkov radiator enter the photosensitive drift region of the detector through a transparent window and converge to a ring image on the focal surface situated just behind this window. The drift region is filled with a suitable drift chamber gas such as methane, with a small admixture of a photoionizing vapour such as TEA or TMAE. These molecules may absorb a photon of energy E and produce a single photoelectron with probability $Q(E)$, i.e. the quantum efficiency for the process. The probability $Q(E)$ is null if $E < E_t$ with $E_t = 7.5$ eV for TEA and $E_t = 5.4$ eV for TMAE. The electron thus produced drifts in a uniform electric field, parallel to the window, into a MWPC plane where, after amplification in the intense electric field around a wire, the time of arrival (relative to an external trigger) of each electron is measured on the wire which it hits. The two coordinates (x, y) of the conversion point in the focal surface can thus be obtained for each detected photon, without ambiguity, from the measured drift time t ($y = v_D t$, where v_D is the electron drift velocity) and the corre-

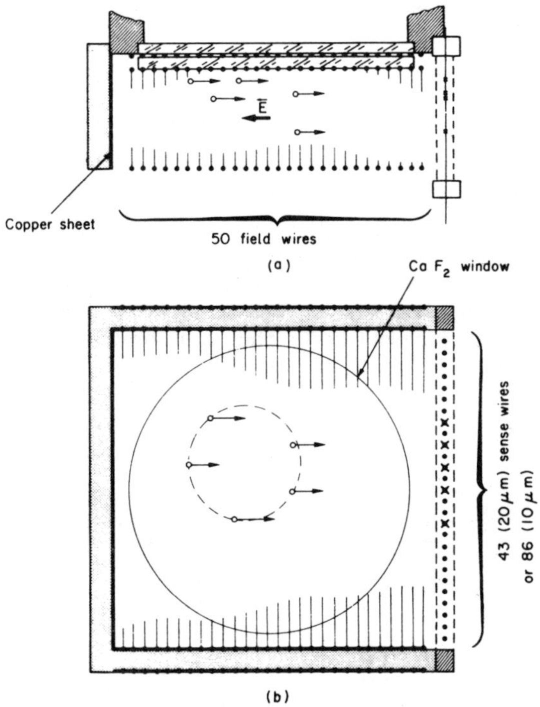

Fig. 3. A schematic side view (a) and top view (b) of the photon drift detector. A double entrance window is shown with field-shaping electrodes (full dots and lines) on both sides of the inner window as explained in section 3.3. Photons coming through the entrance windows are converted in the drift volume at points represented by open dots and drift in the uniform electric field $\boldsymbol{E}$ to the MWPC plane, where the impact points are shown by crosses. The cathodes of the MWPC are also shown (dotted lines). The large full circle in (b) shows the extent of the circular entrance window relative to the larger square drift volume. In a later design (Mark II) the entrance window is square (or rectangular) and covers the full drift volume.

sponding wire address (x). Fig. 3 shows the photon detector in two projections (side and top). The field-shaping electrodes are indicated by black dots and full lines, with the direction of the uniform electric field also indicated. A double entrance window is shown with field-shaping electrodes mounted on both sides of the inner window. The reason for this double window will be discussed in section 3. Also shown in fig. 3 are open dots corresponding to photon conversion points in the drift volume, and the direction of drift of these electrons towards the MWPC plane where the corresponding impact points are indicated by crosses. The depth of the conversion point in the detector depends on the photon mean free path for absorption in the detector gas. Typically, this absorption mean free path was 5 mm for TEA and 15 mm for TMAE; thus the drift gap depths were 15 mm and 45 mm, respectively, so as to ensure essentially complete (95%) photon absorption. Fig. 4

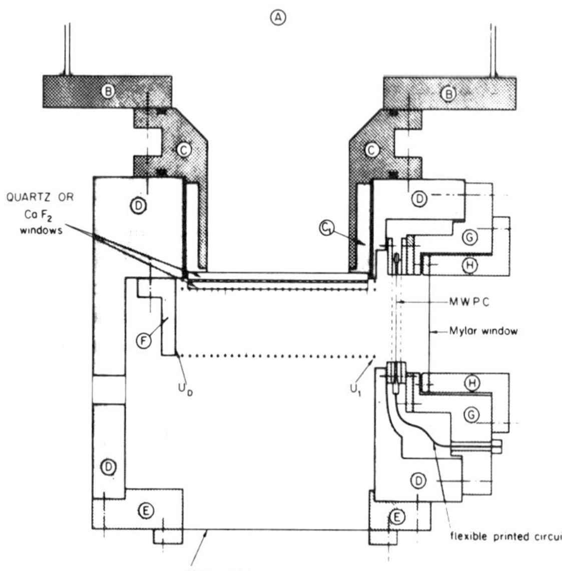

Fig. 4. Details of the photon drift detector: A – gas radiator volume; B – end flange of gas container (stainless steel); C – extension of gas container (stainless steel) in contact with radiator gas; C_1 – araldite extension (glued to C) onto which the outer entrance window is glued; D – support box (araldite) for drift gap and MWPC; E – rear closure for support box with Mylar window; F – drift volume with field-shaping electrodes and inner entrance window. Potential applied at the far side in U_D and the near side to the MWPC in U_1; G – MWPC array with wire mesh cathodes and wire anodes; H – closure of MWPC array with Mylar window.

shows a cross-section of the mechanical construction of the detector and shows its mounting relative to the Cherenkov radiator gas volume.

2.1. The photosensitive drift region

The parallel electric field in the drift region is defined by wires stretched between the two legs of a U-shaped Stesalite frame as illustrated in fig. 5. There are two layers of 50 equidistant wires, each of 140 mm length and 100 μm diameter, mounted with a pitch of 2.54 mm. The wires are interconnected through 1 MΩ resistors (tolerance 1%) in a voltage divider chain, as shown in fig. 6. The photoelectrons drift in the space between the legs and the wire planes towards the open side of the U-frame where the MWPC is placed. The thickness of the frame, which determines the depth of the photoabsorption region, was either 15 mm or 45 mm, depending on the absorption length of the photoionizing agent used (TEA or TMAE).

The connection between opposite equipotential wires, one in each wire plane, is made through parallel con-

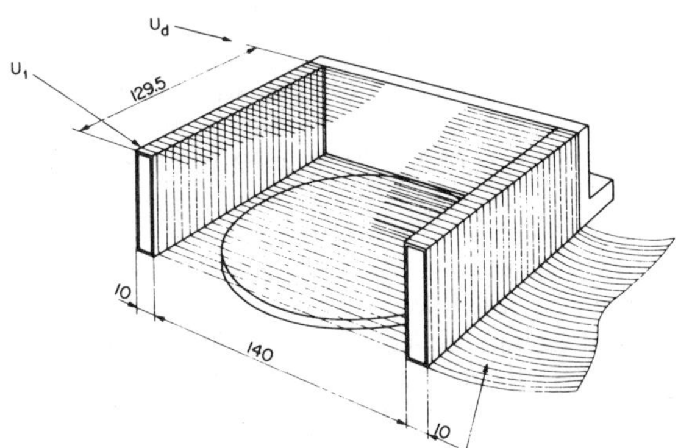

Fig. 5. View of the drift gap (F in fig. 4) with field-shaping electrode wires and Kapton printed-circuit channels covering the U arms. Also shown is the inner entrance window which abuts on the entrance field-shaping electrode wires. Not shown is a second plane of field-shaping wire electrodes which sandwich the inner entrance window.

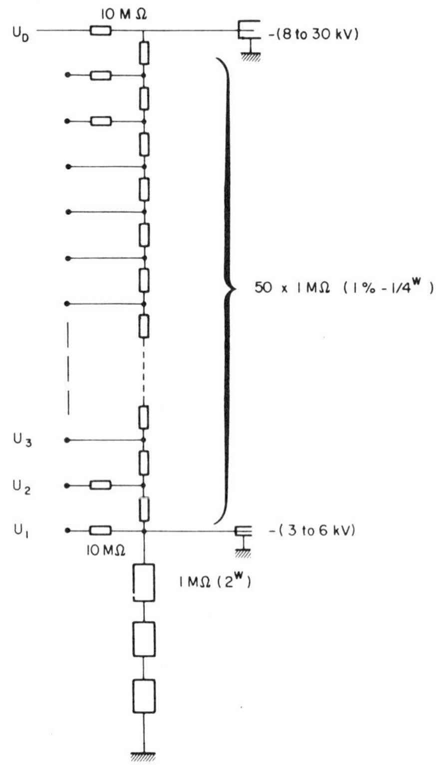

Fig. 6. Voltage resistor chain for field-shaping electrodes. The negative potentials U_D and U_1 are defined by external supplies. The black dots correspond to the wires of the field-shaping electrodes.

ductive lines deposited on a 100 μm thick Kapton foil using the printed-circuit technique. These foils are wrapped around the legs of the U-frame, thus making contact between the wire planes, on both the inside and the outside of each leg. This was done in order to avoid having a part of the surface of the U-frame where the potential would be undefined and where accumulation of electric charge leads to a transverse polarization of the frame material which gives rise to distortions of the electric field. This problem will be discussed in detail in a subsequent section (section 3).

The U-frame is mounted with one of its wire planes touching one of the entrance windows that are placed between the radiator gas volume and the detector gas volume. In order to have the same electric field on both sides of this window, a second field-shaping electrode wire plane is used on the outside of the window, as shown in fig. 3. As a high-voltage connection is needed between corresponding wires on either side of the window, it was technically a problem to have this window also providing the gas-tight separation needed between the detector gas and the radiator gas. Although a solution to this technical problem of using only one window exists (Mark II), we have first used the simple method of adding a second window outside the first one. This outer window is glued with Araldite onto an extension tube (C_1 in fig. 4) mounted on the radiator gas container so as to ensure gas tightness between detector and radiator gases. An obvious disadvantage of this solution is that, because of the limited transmission of the windows in the UV, the second window reduces somewhat the overall photon detection efficiency. In fig. 7 is shown the transmission of 5 mm thick fused quartz (Corning 7940) and calcium fluoride (Harshaw) entrance windows measured in the laboratory with a UV monochromator. We note that at the peak of TEA photoionization response (8.2 eV) the CaF_2 crystal has 68% transparency, whereas at the centre of TMAE response with fused quartz (6.7 eV) the quartz has 83%

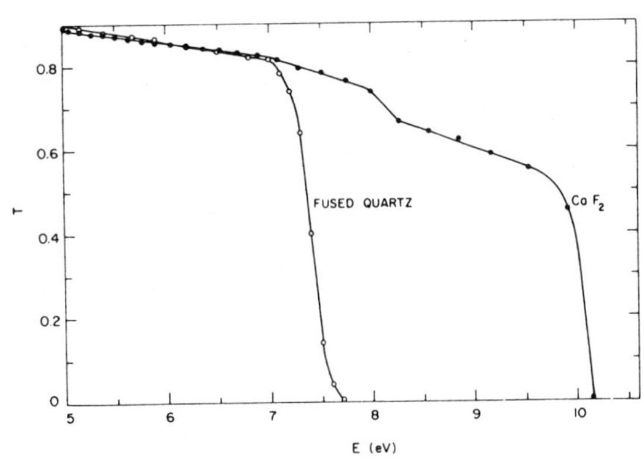

Fig. 7. The photon transmission vs photon energy of fused quartz (Corning 7940) and a CaF_2 (Harshaw) single crystal windows. Both windows are 5 mm thick and 127 mm in diameter.

transparency. An advantage of this solution, at this stage of experimentation, is that of flexibility. For example, we were able to vary the domain of spectral transmission of the detector by changing the inner window from CaF_2 to quartz when the photoionizing molecule was changed from TEA to TMAE. This provisional solution furthermore avoided the problem of field-shaping electrode wire planes in the argon radiator gas, which would cause a glow discharge when the electric field exceeds 1 kV/cm.

The detector volume is closed, on the side away from the entrance window, by a Mylar foil (see E in fig. 4). This foil is far enough away (7 cm) from the drift volume to cause a negligible influence on field uniformity in the gap. The third side of the U-frame is covered with a copper sheet, which is connected to the potential U_D defining the high-voltage end of the drift field. The U-frame is mounted inside the Araldite support box (see D in fig. 4). This support box contains the gas inlets and outlets, the high-voltage connections, the voltage division resistor chain, and the MWPC array (see G in fig. 4). The support box itself is mounted on an extension tube (see C in fig. 4) of the Cherenkov gas radiator.

2.2. The multiwire proportional chamber

The MWPC detects the lateral positions and the arrival times of the photoelectrons as they drift to the end of the electric field volume and into the MWPC array. Two different chambers were used, one having 43 anode wires of 20 μm diameter mounted with a pitch of 2.54 mm, the other having 86 anode wires of 10 μm diameter with a pitch of 1.27 mm. In both cases the wires are made of gold-plated tungsten, and the wire length is 6 cm. The cathode planes, placed 3.2 mm away on either side of the anode, consist of a woven stainless-steel mesh made of 50 μm wires with 500 μm spacing.

The MWPC chamber is mounted at position G as shown in fig. 4. In this position the chamber covers the end of the drift region, with the anode wires perpendicular to the entrance window plane and to the direction of the electric field.

The gap between the last wire of the field-shaping electrode (U_1) and the MWPC cathode was 14 mm, and it had been foreseen to insert two additional wire-mesh planes into this gap. This would have enabled the detector to be operated with gaseous preamplification [4]. The gap between the first and second additional meshes was to be used to cause a discharge avalanche, and the gap between the second additional mesh and the MWPC cathode as a transfer gap. However, the measurements reported in this paper were made without the preamplification and transfer stages, as sufficient amplification was obtained using methane as the amplifying gas coupled with very sensitive detection electronics. This left, however,a rather large gap (14 mm) between U_1 and the cathode of the MWPC, which has repercussions on the time resolution of the detector as will be discussed in section 5.

The MWPC is closed to the outside air by a Mylar foil mounted behind the outside cathode grid (see H in fig. 4). The electric potentials of the closest drift-gap wire (U_1) and the MWPC cathode have separate high-voltage leads and can thus be adjusted independently of each other. This allows us to maintain the same electric field in the drift region and in the gap between U_1 and the cathode when the MWPC voltage is varied.

2.3. The electronics and data-acquisition system

The basic layout of the electronics for the processing of the pulses from the anode wires of the MWPC is shown in fig. 8. The anode pulses are led through a 10 cm long flexible printed circuit-board conductor to the preamplifiers which are mounted outside on the support box. The detector electronics shown in fig. 9 are composed of fast MECL 10000 series amplifiers having a 100 MHz bandwidth. The preamplifiers (voltage gain of 50) are mounted on the detector and are connected to the anode (sense) wires of the MWPC by a flexible Kapton printed circuit, 10 cm in length. Every other channel on this circuit is connected to ground so as to minimize cross-talk between neighbouring wires. The resistance to ground on the sense wire is 1.5 kΩ, with a measured stray capacity of 16 pF and thus an entrance filter time constant of 24 ns. The preamplifiers are protected against sparking in the MWPC by a 270 Ω resistance in series with two back-biased diodes.

The preamplified analog signals are transferred to

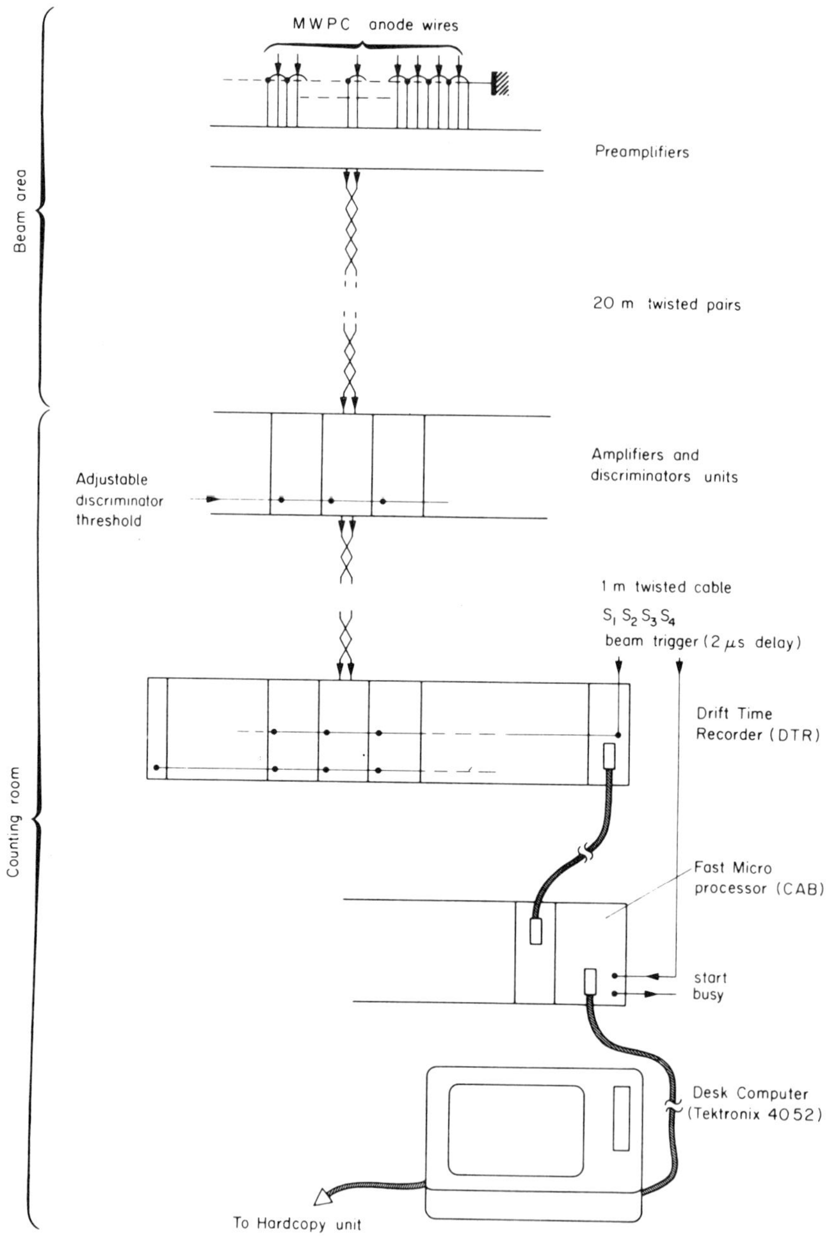

Fig. 8. Schematic diagram of drift detecor read-out.

the counting area via 20 m lengths of 120 Ω impedance twisted-pair cable, which reduces the bandwidth to about 80 MHz. In the counting area the analog signals are further amplified (voltage gain 60) by the first two stages of MC 10116 amplifiers (rated gain 12×12) with feedback, and discriminated in the third stage by a Schmitt trigger which gives 60 dB rejection of positive pulses induced on neighbouring wires. The discriminator has a hysteresis of 200 mV, which aids in noise rejection and allows for adjustment of the voltage threshold level by means of an external dc bias voltage. This threshold is adjustable for between 50 μV and 5 mV signals on the anode wire. The r.m.s. noise level at the entry to the preamplifier (when not connected to the MWPC wire) is 50 μV. When operating with the MWPC, this noise gives rise to a random counting rate of 10^2 Hz per wire when the discriminator level is set for a 150 μV signal on the wire (i.e. 0.1 μA on 1.5 kΩ), which for a capacity of 16 pF corresponds to a charge of 2.4 fC or 1.4×10^4 electrons.

The voltage response of the MWPC and preamplifier entry filter is, for one photoelectron, given by the relation

$$\begin{aligned} v(t) &= \frac{eAC}{C_i}\int_0^t \frac{e^{-\tau/R_iC_i}}{t+t_0-\tau}\,d\tau \\ &= \frac{eAC}{C_i} e^{-(t+t_0)/R_iC_i}\left[\ln\left(\frac{t+t_0}{t_0}\right) + \frac{t}{R_iC_i} \right. \\ &\quad \left. + \frac{(t+t_0)^2 - t_0^2}{4R_i^2C_i^2} + \frac{(t+t_0)^3 - t_0^3}{18R_i^3C_i^3} + \cdots \right] \end{aligned} \tag{2}$$

where A is the MWPC gain, $C = s/2\pi a$ is the capacity of a MWPC wire per unit distance (s is the wire spacing, a the cathode–anode distance), $R_i = 1.5$ kΩ and $C_i = 16$ pF, and t_0 is a constant:

$$\begin{aligned} t_0 &= \frac{s^2}{2CV\pi^2\alpha}\ln\left[1 + \frac{\pi^2 d^2}{8s^2}\right] \\ &= \frac{d^2}{16CV\alpha}\left[1 - \frac{1}{2}\left(\frac{\pi^2 d^2}{8s^2}\right) + \frac{1}{3}\left(\frac{\pi^2 d^2}{8s^2}\right)^2 - \cdots\right], \end{aligned} \tag{3}$$

where V is the cathode–anode potential, d the wire

diameter, and α the mobility of the positive ions ($cm^2/V \cdot s$). In the parameters of the detector ($R_i = 1.5$ kΩ, $C_i = 16$ pF, $s = 0.127$ cm, $d = 0.001$ cm, $V = 3.5$ kV, $a = 0.32$ cm, $\alpha = 2.26$) we find $v_{max}(\mu V) = 21 \times 10^{-4}A$ at $t = 7$ns, which decreases slowly to 10% of this value in about 150 ns. Taking 150 μV as the operating threshold level thus requires a gain $A = 7.0 \times 10^4$ so as to be sensitive to a single electron. Another operating condition of the detector was ($s = 0.254$ cm, $d = 0.002$ cm) which gives $v_{max}(\mu V) = 29 \times 10^{-4}A$ at $t = 10$ ns, hence single-electron sensitivity (150 μV threshold) when the gain $A = 5.2 \times 10^4$. The gain A of the methane-filled MWPC has been measured to be $\sim 1.8 \times 10^5$ by observing an extrapolated 5 V signal out of the preamplifier when the chamber was irradiated with 5.9 keV X-rays from a ^{55}Fe radioactive source. Since the voltage gain of the preamplifier is 50, this corresponds to a 100 mV signal on the wire. A converted X-ray in methane produces about 200 electrons. Thus the signal on the wire due to a single electron is 500 μV, which should be set equal to $28.6 \times 10^{-4}A$, hence the value $A = 1.8 \times 10^5$. This value may be a lower limit because of saturation effects due to such large signals. The chamber gain is thus 2 to 3 times bigger than necessary for single-electron sensitivity. This additional gain is important because of the large fluctuations in pulse height for single electrons (causing a loss in counting efficiency), and to minimize the slewing error in the arrival time measurement. This latter factor can be estimated to give rise to an r.m.s. error in the measured arrival time of ~ 8.5 ns$/\sqrt{12} = 2.5$ ns for a uniform pulse-height distribution.

The discriminated output (see figs. 8 and 9) for each wire is connected to a Drift Time Recorder (DTR Type 247 CERN) which digitizes the detection time of the input pulse, relative to a trigger pulse, at a frequency of 125 MHz. The DTR has 8 bits per channel and a memory which permits acceptance of up to 256 pulses on 16 wires (i.e. multi-hit capability). A time binning in 4 ns buckets is obtained by interpolation of the 8 ns period clock with a maximum time range of 2048 ns. The DTR modules are read via a standard CAMAC dataway by a fast microprocessor, called CAB (CAMAC Booster), developed at the Ecole Polytechnique. The microprocessor is based on the AM 2900 series (4-bit

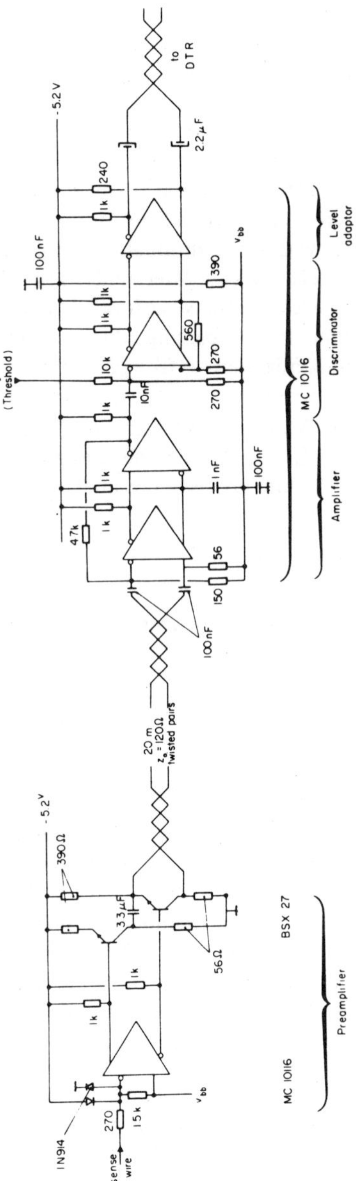

Fig. 9. Preamplifier, amplifier and discriminator circuits.

slices with 160 ns cycle time) and has a programming memory of 4K 24-bit words, a data memory of 4K 16-bit words controlled by an index register of 12 bits, a hard-cabled TRW multiplier/adder block of 16×16 bits, and an arithmetic and logic block ALU (AM 2901 series). The digitized data from the DTRs are transferred to CAB, where appropriate calculations are performed for each event. The resultant histograms are transferred to a Tektronix 4052 calculator for display on a cathode-ray screen and printed out on an associated hard-copy printer with the additional option of transfer of the data to 3M cassettes for later analysis off line. As an example of the calculations performed in CAB for straight-line beam tracks passing through the drift volume, the time digitizations t_i from the ith wire (wire position x_i) for N wires hit ($i = 1, \ldots, N$) are summed to give the moments $N\bar{x} = \Sigma x_i$, $N\bar{t} = \Sigma t_i$, $N\overline{x^2} = \Sigma x_i^2$, $N\overline{t^2} = \Sigma t_i^2$, and $N\overline{tx} = \Sigma t_i x_i$. We fit the straight-line hypothesis $t = ax + b$ using least-squares analysis to obtain

$$a = \frac{\overline{xt} - \bar{x}\bar{t}}{\overline{x^2} - \bar{x}^2} \quad \text{and} \quad b = \bar{t} - a\bar{x}. \tag{4}$$

We calculate also the variance σ of the straigth-line fit as

$$\sigma^2 = \frac{1}{N} \sum_{i=1}^{N} (t_i - ax_i - b)^2$$
$$= \overline{t^2} + a^2\overline{x^2} + b^2 - 2a\overline{tx} - 2b\bar{t} + 2ab\bar{x}, \tag{5}$$

which gives an estimate of r.m.s. error in a single point-time measurement. For each event (trigger) we calculate and histogram σ, N, $N(x_i)$, $N(t_i)$, where the latter two quantities are the number of times the ith wire is hit or fills the t_ith time bin. The probability that the ith wire is hit is then $N(x_i)/N$, which allows an estimate of the number of primary $\mathrm{d}E/\mathrm{d}x$ electrons produced in distance s (wire spacing), as will be discussed in section 3.1. We have tested this complete read-out chain by pulsing the cathodes of the MWPC, which induced an equal time signal on each wire (hence simulating a horizontal beam track $t =$ constant). A fit to these simulated tracks gave $\sigma = 3$ ns, which indicates the intrinsic time resolution of the electronics and binning. The σ expected with 4 ns time buckets is only 1.2

ns, the difference, in quadrature, of 2.7 ns being due to dispersion in the electronics.

3. Measurements of straight beam tracks

There are several potential difficulties connected with the operation of a drift chamber that has a long drift ($\geqslant 10$ cm), such as drift-field inhomogeneities causing image distortion and loss of drifting electrons, electron absorption by electronegative impurities in the drift gas, and electron diffusion, resulting in a decrease in the measurement precision. These problems were investigated by making measurements of the ionization created by relativistic charged beam particles traversing the drift region. The experimental layout is shown in fig. 10. The detector is oriented in the beam so that particles pass through the drift region between the two field-shaping electrode planes in a direction parallel to these planes and perpendicular to the MWPC wires. When oriented in position (a) of fig. 10, the charged particle produces ionization on all MWPC wires at equal times, whereas when tilted slightly to position (b) of fig. 10 the trajectory has a slight slope.

The measurements were made in the c_{13} test beam of the CERN PS. This beam is a secondary beam derived from an extracted 24 GeV/c proton beam. The beam momentum could be varied between 5 and 20 GeV/c,

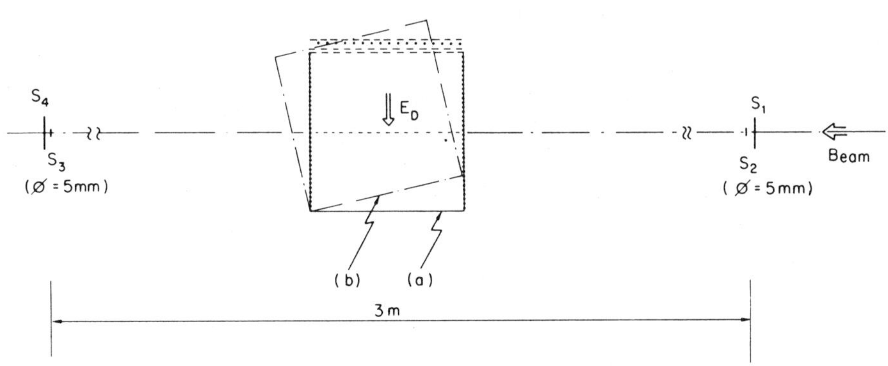

Fig. 10. Orientation of the drift detector to observe through-going beam tracks; (a) flat tracks at equal time; (b) sloping tracks at variable time.

with beam intensity between 5×10^3 and 5×10^4 negative particles (mostly pions) per 300 ns spill. The tests described here were done at 10 GeV/c with about 10^4 particles per burst. The trigger was defined by a fourfold coincidence between two pairs of scintillators, placed upstream and downstream of the detector; this corresponds to a beam of about 5 mm diameter, with a maximum divergence of 1.7 mrad (see fig. 10). The fourfold coincidence gave about 50 triggers/burst.

3.1. Measurements without the calcium fluoride window

Measurements were first made without the CaF_2 entrance window in the detector. A Mylar window, mounted at a distance of 7 cm from the closest field-shaping electrode plane, was used to close off the detector gas volume. The detector was aligned as described above, and the charge collection efficiency of different parts of the drift region was investigated by varying the position of the drift gap-width w relative to the beam, and the distance d of the beam from the MWPC wire plane. The measurements were performed using a 15 mm thick drift gap with an argon (70%) + methane (29%) + TEA (1%) gas mixture (volume flow ratios corrected by $\sqrt{M_A/M}$, where M_A and M are the molecular weights of air and the flowing molecule) and a 1 kV/cm drift field. The indicated argon purity was 99.995%, while the methane was 99.95%. Both gases were further purified by passage through oxisorb and dryer chemical purifiers. The TEA liquid purity was 99.5%. The predominantly argon filling was chosen in order to produce a large number (~ 5) of primary ionization electrons on each wire. The MWPC used had 43 wires at 2.54 mm pitch. Examples of data from two superimposed single tracks are shown in fig. 11. Note that the tracks have a slight slope and that every wire except the first two and the last one are hit. This is because the first and last wires of the MWPC are 100 μm in diameter (rather than 20 μm) and hence have negligible gain. The straight lines shown are from the least-squares fit (see subsection 2.3) and give a variance $\sigma = 6$ ns (using only wires 5 to 39 to minimize edge effects).

For N through-going beam tracks the number of digitizations N_i of the ith wire was measured to obtain a

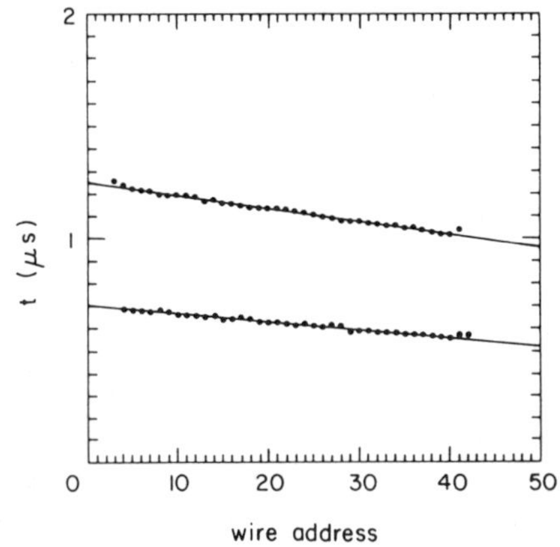

Fig. 11. Examples of two superimposed sloping tracks produced by 10 GeV/c pions passing through the drift detector.

counting efficiency $\epsilon_i = N_i/N$. Assuming that the number of primary ionization electrons that are detected by the ith wire is Poisson distributed with a mean value n_i, then the probability of not counting is $P(0) = e^{-n_i} = 1 - \epsilon_i$, hence

$$n_i = -\ln(1 - \epsilon_i). \tag{6}$$

If the efficiencies ϵ_i for 35 wires ($i = 5, \cdots, 39$) are reasonably uniform, we average the n_i values to obtain the average number of primary ionization electrons $n_{\text{pe}} = \bar{n} = (1/35)\Sigma n_i$ and its variance $\sigma_n = \sqrt{\overline{n^2} - \bar{n}^2}$. In fig. 12a we see the values of n_{pe} (error bars σ_n) versus the position w of the beam relative to the drift gap at a drift distance $d = 100$ mm. This scan shows $n_{\text{pe}} = 5.4$ in the centre of the gap, with a linear fall-off when approaching the field-shaping electrodes that is compatible with an increasing fraction of the 5 mm diameter beam passing outside the drift gap. The value $n_{\text{pe}} = 5.4$ corresponds to the number of primary electrons collected on one wire spacing (2.54 mm); hence we find a value 21.3/cm. In fig. 12b are shown the measured values of n_{pe} versus the drift distance d(when the beam is centred in the middle of the gap, i.e. $w = 7.5$ mm); this shows a plateau at $n_{\text{pe}} = 5.4$ with a similar fall-off when approaching the borders of the drift region as defined by the first (U_1) and last (U_{D}) field-shaping electrode wires. There is no indication of electron loss

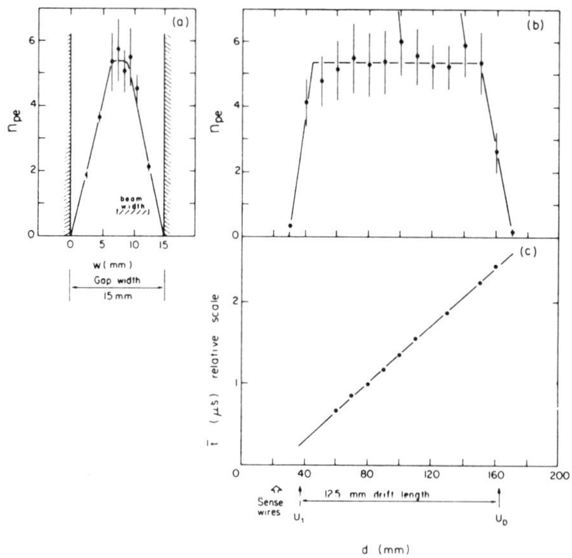

Fig. 12. (a) The average number of primary electrons n_{pe} collected on each wire when the beam is scanned across the drift gap w at a fixed drift distance $d = 100$ mm. (b) n_{pe} vs drift distance d when the beam is positioned in the gap centre $w = 7.5$ mm. (c) Relative drift time vs drift distance d at $w = 7.5$ mm.

after a drift of up to 12 cm. In fig. 12c is plotted the recorded drift time (for flat equal-time tracks) as a function of the drift distance d. The drift time increases linearly with d, indicating a drift velocity which is independent of d, with a slope corresponding to a drift velocity of 5.6 cm/μs.

The data shown in fig. 12 indicate that the drift field is homogeneous throughout the drift region when the field strength $E_D = 1$ kV/cm. Varying the drift field strength, the collection efficiency remains constant from 1 kV/cm down to 0.2 kV/cm, as shown in fig. 13a for a drift distance $d = 100$ and $w = 7.5$ mm (closed circles). In fig. 13b the measured variation of the drift time with E_D is plotted, showing that the maximum drift velocity occurs between 0.4 and 0.5 kV/cm in the argon (70%) + methane (29%) + TEA (1%) gas mixture used for these tests. Note also that the drift velocity does not plateau to a constant value. Increasing the voltage

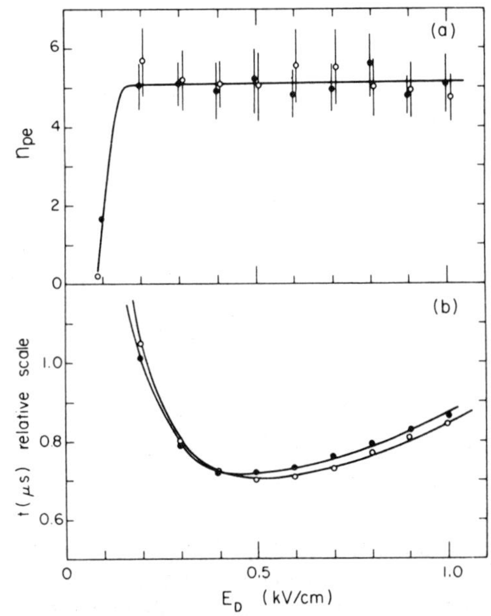

Fig. 13. (a) The average number of primary electrons n_{pe} collected on each wire vs the drift field E_D when the beam is centred in the gap $w = 7.5$ mm at a distance $d = 100$ mm. Closed circles are with no entrance window and open circles for a polarized CaF_2 entrance window. (b) Relative drift time vs drift time vs drift field E_D with conditions as in (a).

threshold of the discriminators, the number of collected primary electrons n_{pe} was found to be constant for a discriminator threshold up to twice the normal value (150 μV), as shown in fig. 14.

3.2. Measurements with an unpolarized calcium fluoride window

Performing the measurements described in the previous section with a calcium fluoride window installed outside the entrance field-shaping electrode plane (see fig. 5) gave results which strongly deviated from the earlier results and which changed with time. Fig. 15 shows the average number of primary electrons n_{pe} collected per wire, obtained with the beam in the median plane $w = 7.5$ at a distance $d = 100$ mm from the MWPC

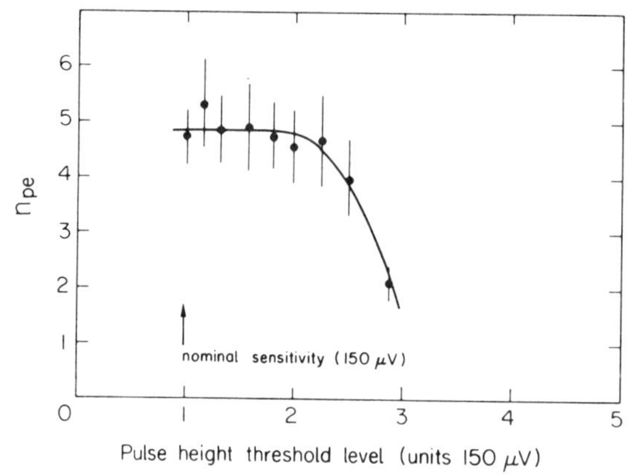

Fig. 14. The average number of primary electrons n_{pe} collected on each wire vs preamplifier threshold level.

anode plane as a function of elapsed time after the drift voltage was applied. During the first moments $n_{pe} = 5.4$ as before, but after half an hour n_{pe} falls to about 1, after which it slowly rises, during the following two hours, up to a value of 4. Further studies showed that n_{pe} continued to vary, with some tendency toward stability but never reached the original high wire-collection efficiency (i.e. $n_{pe} = 5.4$).

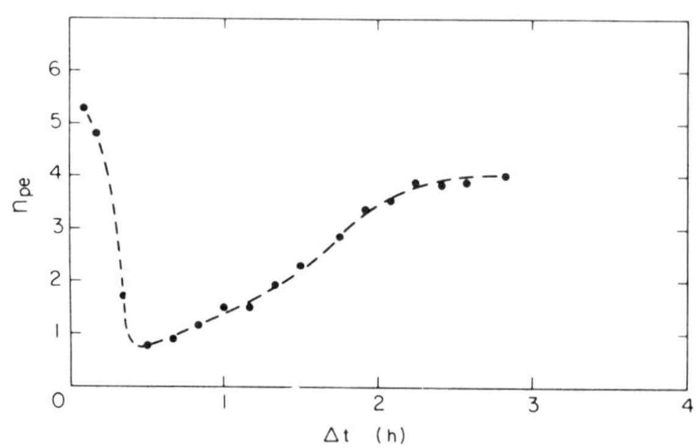

Fig. 15. Variation of average number of primary electrons collected per wire n_{pe} vs elapsed time Δt after turning on drift field with an unpolarized CaF_2 entrance window. Drift distance $d = 100$ mm, beam centred in gap $w = 7.5$ mm and $E_D = 1$ kV/cm.

Plotting the wire efficiency $\epsilon_i = N_i/N$ for each of the 43 anode wires for $N = 500$ events shows how the losses are distributed in the drift volume. In fig. 16, histograms collected with $d = 100$, $w = 7.5$ mm at different times after the drift voltage has been turned on, show how the deterioration of the collection efficiency occurs preferentially at the borders of the drift volume. Fig. 17 shows histograms collected at approximately the same time, but with the beam positioned at $w = 7.5$ mm and different distances d from the MWPC plane. The depletion is seen to be large at large drift distances ($d \geqslant 80$ mm).

From these data we concluded that the deterioration

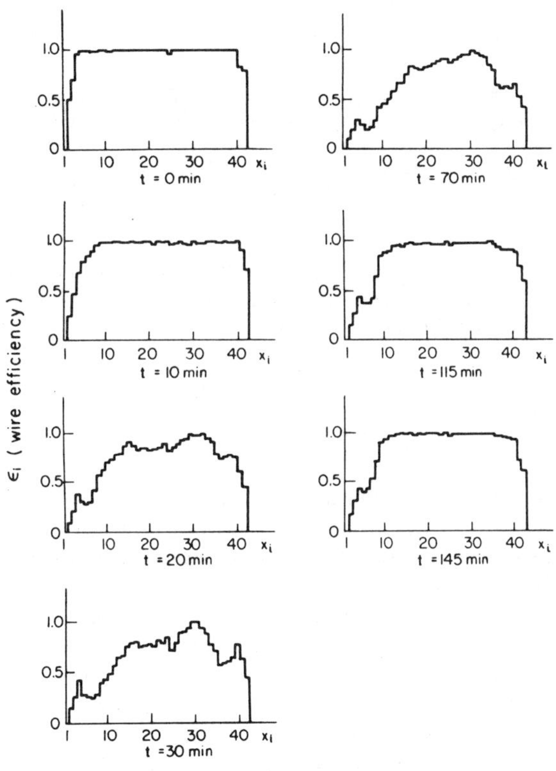

Fig. 16. Wire efficiency ϵ_i vs elapsed time after turning on the drift field when the CaF_2 entrance window is unpolarized. Drift distance $d = 100$ mm, $w = 7.5$ mm, $E_D = 1$ kV/cm.

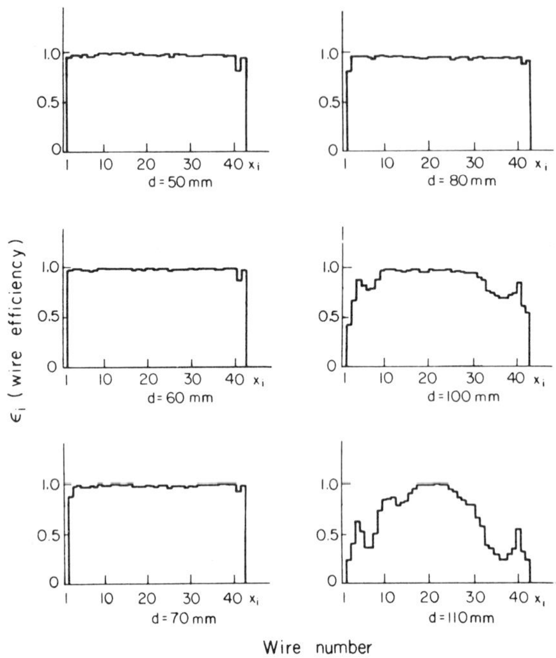

Fig. 17. Wire efficiency ϵ_i vs drift distance d when the CaF_2 entrance window is unpolarized. Drift field $E_D = 1$ kV/cm and beam centred in gap $w = 7.5$ mm.

in the charge collection efficiency is due to varying inhomogeneities in the electric drift field, which in turn are due to slow and irregular charging-up of the calcium fluoride entrance window.

In the measurements of figs. 15–17 the window was in direct contact with the field-shaping electrode plane, making charge collection from the window surface to the field wires possible. To check the influence of this collection, measurements were made with the window 0.2 mm away from the field-shaping electrode plane, again at $d = 100$ and $w = 7.5$ mm. Fig. 18 shows the time evolution in this situation. Again the collection efficiency deteriorates, but in this case depletion occurs in the central part of the drift region.

3.3. Measurements with the calcium fluoride window polarized

From the results described above, we tentatively concluded that in order to define a stable and homogeneous drift field inside the drift volume in the presence of the dielectric calcium fluoride window, the potential had to be defined also on the outside of this window. Having a field-shaping electrode plane on both sides eliminates uncontrolled charge accumulation on the outer window surface and polarizes the dielectric medium parallel to the window surface. Since the CaF_2 window was already glued in place, a second CaF_2 window was added, inside the drift volume, with field-shaping electrode planes on both sides of this inner window (see figs. 3 and 4). As already mentioned, this turned out to be convenient in later tests when this inner CaF_2 window was replaced by a fused quartz

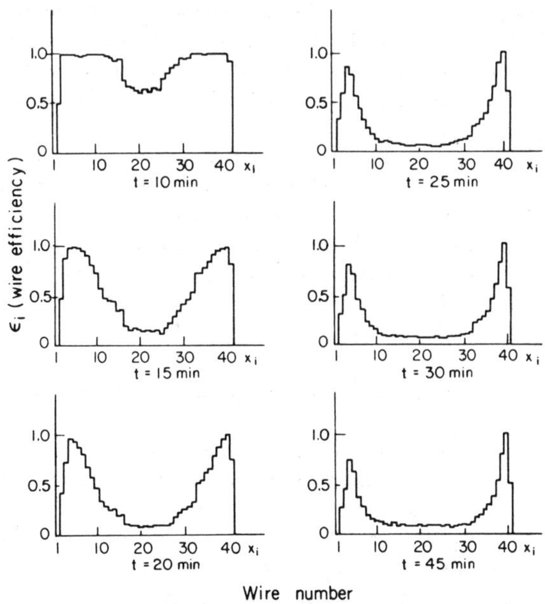

Fig. 18. Wire efficiency ϵ_i vs elapsed time after turning on drift field when the unpolarized CaF_2 entrance window is not in contact with the field-shaping electrode (separation 0.2 mm). Drift distance $d = 100$ mm, $w = 7.5$ and $E_D = 1$ kV/cm.

window for investigations with the photoionizing gas TMAE. In addition, to avoid charge accumulation in the legs of the U-frame, the potential-defining lines were made to encircle the U-frame legs.

In fig. 19a the results obtained with the window polarized are shown (open circles) and compared with the results obtained with the window unpolarized (closed circles). The collected charge (n_{pe}) remains high (≈ 5.4) up to a distance $d = 120$ mm, whereas with the window unpolarized it decreases significantly beyond $d = 80$ mm. Fig. 19b shows the scan over the gap width w (at $d = 100$ mm), again indicating more efficient collection with the window polarized (open circles) than for the window unpolarized (closed circles). The former is as good as the measurement without window (see fig. 12a). The collected charge (n_{pe}) as a function of drift field is plotted in fig. 20 (open circles) and is compared to the unpolarized window case (closed circles). This curve shows efficient collection down to a field of 0.2 kV/cm, as was observed when no window was present. These data (open circles) are directly compared with the no-window data in fig. 13 (closed circles). We note that for the window unpolarized the drift field required for efficient collection is very high ($E_D \geqslant 1$ kV/cm) when $d > 100$ mm.

The equality of the collected charge $n_{pe} = 5.4$ when the window is polarized and without window, and its independence with drift distance, indicate that electron loss through drift-field distortions and gas absorption is small and that electron collection is efficient throughout the drift volume.

4. Drift velocity measurements in various gas mixtures

To measure the drift velocity the detector was placed in the 10 GeV/c pion beam such that the beam was incident normally to the entrance window and parallel to the MWPC wires. In this geometry all the ionization electrons drift to only one or two wires of the MWPC where the drift time is measured at a given value of drift field strength E_D. The detector was then displaced a known distance, so as to increase (or decrease) the drift distance d, and the drift time was again measured. A

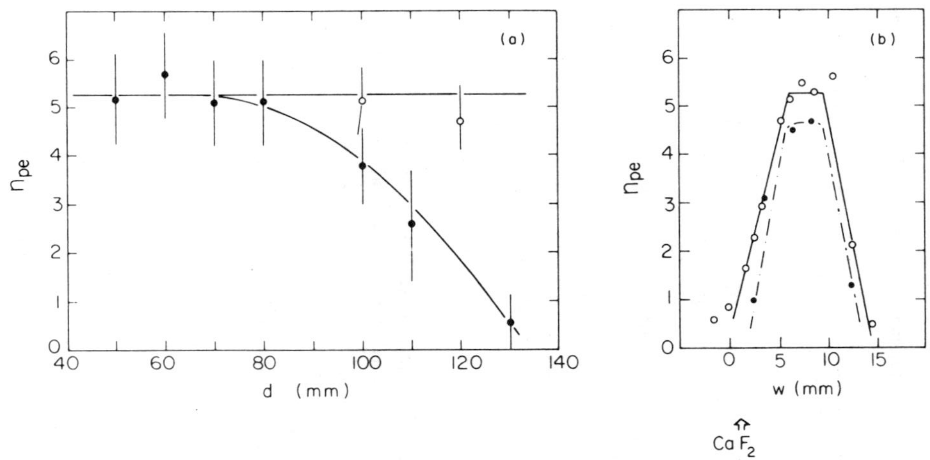

Fig. 19. The average number of primary electrons n_{pe} collected on each wire with the CaF_2 entrance window polarized (open circles) and unpolarized (closed circles) vs (a) drift distance d with $w = 7.5$ mm and $E_D = 1$ kV/cm, and (b) position of beam in the drift gap w with $d = 100$ mm and $E_D = 1$ kV/cm.

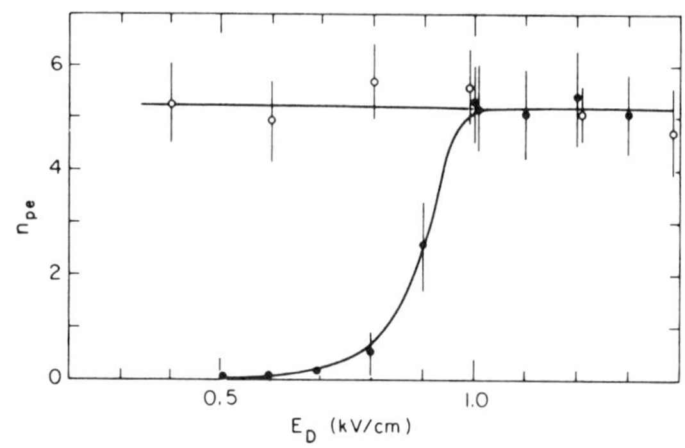

Fig. 20. The average number of primary electrons n_{pe} collected on each wire with the CaF_2 entrance window polarized (open circles) and unpolarized (closed circles) vs drift electric field E_D when the drift distance $d = 100$ mm and $w = 7.5$ mm.

plot of the measured drift time versus drift distance gave a straight line whose slope is the drift velocity v_D at the set value of electric field E_D.

The amplifying and drift carrier gas used in the photon drift detector should satisfy a number of requirements. It should have a large first Townsend coefficient so as to allow sufficient gas amplification; it should be opaque (or partially opaque) to the photons emitted in the avalanche amplification process so as to quench the discharge; it should have a small diffusion coefficient to allow accurate localization of the photon conversion point; and its drift velocity should not vary too strongly with drift field so as to allow stable operation. Furthermore, for use in Cherenkov ring imaging, it should be transparent over some range in the UV region where the photoionizing vapours (TEA or TMAE) are sensitive. As shown in ref. 3, methane satisfies all these requirements.

Various combinations of methane (CH_4), ethane (C_2H_6), isobutane (iso-C_4H_{10}), and carbon dioxide (CO_2) gases with small admixtures of TEA and TMAE were investigated. These gases are all transparent, to varying degrees, in the region up to 9 eV photon energy, as is shown in fig. 21. Also shown in this figure are our present best estimates of the quantum efficiency of TEA and TMAE. As can be seen from fig. 21 only methane can be used when the photoionizing vapour is TEA, whereas for TMAE, combinations of methane, ethane, and CO_2 are possible. The TEA and TMAE quantum efficiency curves shown have been measured with methane as a carrier gas so the high-energy cut-off reflects only the onset of methane absorption rather than an intrinsic decrease of quantum efficiency above 8.5 eV. Measurements using argon as the carrier gas (transparent to 11 eV) are planned so as to determine the quantum efficiency of TEA and TMAE in the 8.5–11 eV energy region.

The drift velocities measured are shown in fig. 22 versus the drift electric field. Note that the CH_4 + TMAE (curve a) and CH_4 + TEA (b) drift velocities plateau to a value of about 9.7 cm/μs above 0.6 kV/cm, whereas the argon + TEA (curve f) drift velocity has a maximum at 0.6 kV/cm and subsequently decreases. The other curves for CH_4, C_2H_6, iso-C_4H_{10}, and CO_2 show monotonically rising drift velocities with increasing electric field. The sharpest rise is shown by CH_4

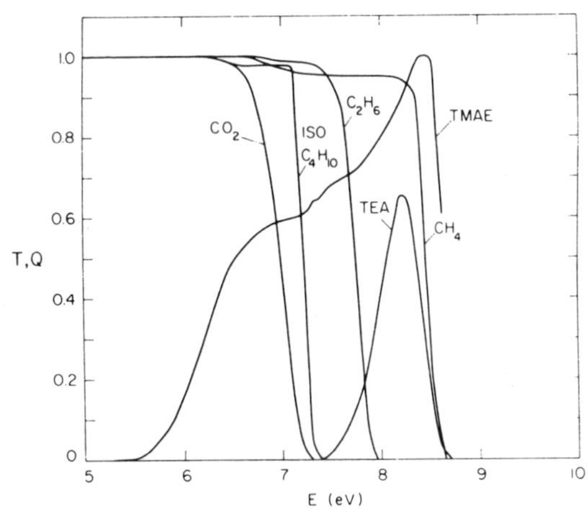

Fig. 21. Transmission T vs photon energy E of 10 cm long samples of methane (99.95%), ethane (99.97%), isobutane (99.95%), and carbon dioxide (99.9%) all treated by flowing through oxisorb and dryer chemical purifiers. Shown also are the quantum efficiencies Q of TEA and TMAE in methane. The decrease of Q at high energies (8.5 eV) is due to photon absorption in methane and does not reflect the true quantum efficiency of these molecules above 8.5 eV.

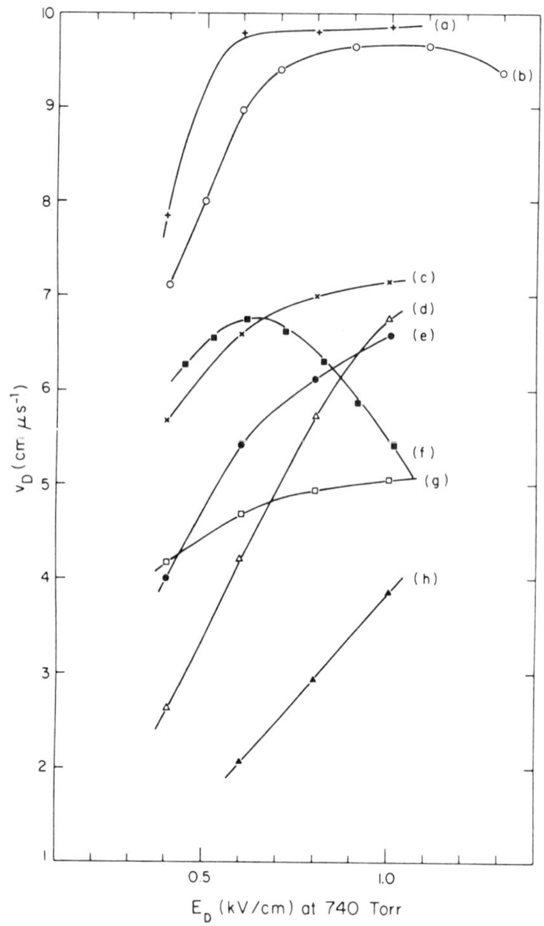

Fig. 22. Drift velocity v_D vs electric field E_D for (a) 100% CH_4 flowing through liquid TMAE (27°C), (b) 88.5% CH_4 with 11.5% CH_4 flowing through liguid TEA (3°C), (c) 85% CH_4 with 15% C_2H_6, (d) 97% CH_4 with 3% CO_2, (e) 89% CH_4 with 11% iso-C_4H_{10}, (f) 80% argon with 20% argon flowing through liquid TEA (3°C), (g) 100% C_2H_6 flowing through liquid TMAE (27°C), (h) 87% CH_4 with 13% CO_2. The argon gas purity was 99.995%, that of TEA was 99.5%; the TMAE purity was unknown and the other gases as listed in fig. 21.

with CO_2 mixtures (curves d and h). This strong dependence of v_D with E_D coupled with the low value of v_D at low electric fields ($E_D \leq 0.5$ kV/cm) argues against the use of CO_2. From the viewpoint of fastest drift the CH_4 + TMAE and CH_4 + TEA mixtures are clearly superior, i.e. a total drift time of 2 μs for 20 cm drift distance. Mixtures of CH_4 + iso-C_4H_{10}, even though slower, are interesting because of their increased ability to quench the avalanche discharge around a MWPC wire and the almost exact match with the quartz transmission cut-off. In fig. 23 are shown the measurements of drift velocity versus the percentage of isobutane in methane + isobutane mixtures for the given fixed value of the drift field E_D. Even though the drift velocity in an argon + TEA mixture was measured (curve f in fig. 22) it is not considered as a useful drift and amplifying gas mixture because of its large diffusion coefficient (see section 5) and because the single photoelectron counting efficiency is low without preamplification [3].

5. Accuracy of the measured drift coordinate

Straight lines were fitted to 10 GeV/c pion beam tracks as illustrated in fig. 11 and outlined in section 2.3. The variance σ [see eq. (5)] of the fit provides a measurement of the accuracy to which the drift time is determined. This quantity is limited by the amount of longitudinal diffusion of the drifting electron(s) in the gas, by the time binning and electronics dispersion, and by any inhomogeneities in the drift field. Fig. 24 (taken from ref. 5) shows the measured or calculated values of the transverse diffusion coefficient σ_t for a single electron versus the drift field E_D in pure argon, CH_4, iso-C_4H_{10}, and CO_2. The measured longitudinal diffusion coefficient σ_l (as well as σ_t) for an argon (87%) + CH_4 (10%) + iso-C_4H_{10} (3%) mixture used in the JADE detector [6] has been added in fig. 24. From this figure we note that the diffusion coefficient σ_t in argon is extermely large, so to obtain highest precision a predominantly argon gas mixture should be avoided. Methane, isobutane, and CO_2 all have small diffusion coefficients; however, as indicated earlier, the low drift velocities which characterize CO_2 mixtures mitigate against its use. Since efficient single-electron counting

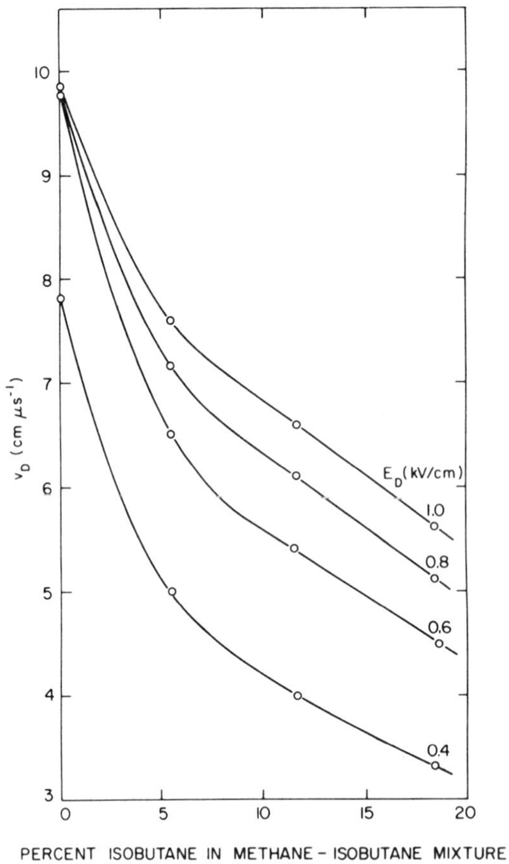

Fig. 23. Drift velocity v_D vs volume percent of isobutane in methane–isobutane mixtures for fixed values of the drift electric field E_D.

in methane has been proved possible, we have concentrated on measurements with methane and small admixtures of isobutane to enhance quenching. As seen from fig. 24, the longitudinal diffusion coefficient in the JADE gas mixture is about a factor of 2 smaller than the transverse coefficient. This inequality, $\sigma_l < \sigma_t$, has been shown [7] to be true for Ar, H_2, N_2, He, CO and C_2H_4 whereas for CH_4 and CO_2 these coefficients are equal. We plan to measure the longitudinal and transverse coefficients in methane + isobutane (or C_2H_4)

mixtures (with TEA or TMAE) using a pulsed (2 ns) UV light source to produce single electrons in a long drift (~60 cm) volume inside a solenoidal magnetic field.

The measurements reported in this section were obtained with a second drift gap of the same construction as described previously but with a 45 mm gap (instead of 15 mm) and a MWPC wire spacing of 1.27 mm instead of 2.54 mm with 10 μm diameter wires (instead of 20 μm). The gas mixture used was methane (89%) + isobutane (11%) at atmospheric pressure but at an

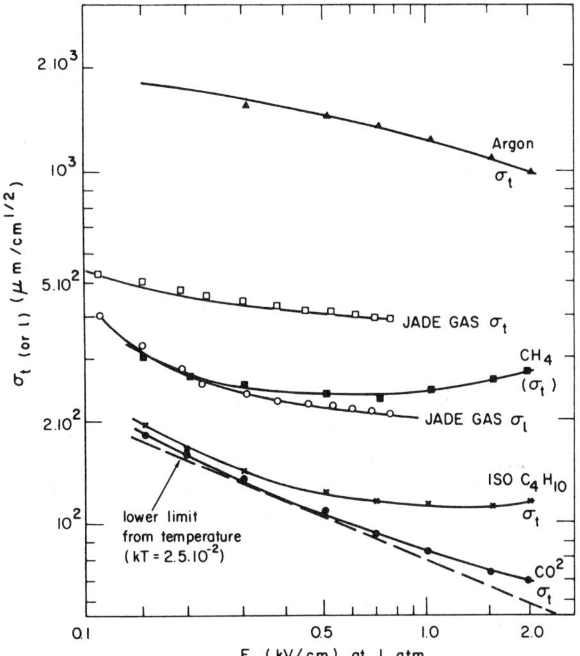

Fig. 24. Experimental values (from ref. 5) of the diffusion coefficents σ_t or σ_l vs electric drift field E_D at atmospheric pressure. The iso-C_4H_{10} curve is calculated. The values of σ_t and σ_l (from ref. 6) for the JADE gas (87% argon + 10% CH_4 + 3% iso-C_4H_{10}) have been added. The measured values of σ_l (from ref. 7) for C_2H_4 follow very closely the σ_t curve for iso-C_4H_{10} and the σ_l for CO_2 follows the σ_t curve shown for CO_2.

elevated temperature. The temperature of the outer support box was kept at 34°C, hence the gas temperature inside was probably higher than this value. Throughgoing beam tracks at a gap depth w were observed and the ionization electrons were collected on the MWPC wires after drifting a distance d in a drift electric field E_D. For each track a straight line fit was made with variance σ, and histograms of σ, N, $N(x_i)$, and $N(t_i)$ were obtained for 500 tracks as described in section 2.3. From the efficiency of each wire the number of primary electrons on each wire is calculated and this is averaged over the wires to obtain n_{pe}. The measurements are shown in fig. 25a, where n_{pe} is shown versus the position w of the beam in the gap. A constant value $n_{pe} = 1.5$ is observed throughout the 45 mm thick gap. Here, the value $n_{pe} = 1.5$ is smaller than the value 5.4 obtained previously (see fig. 12) because of a lower gas density, atomic charge, and wire spacing. Fig. 25b shows a scan in drift distance d, where a slight decrease in collection efficiency perhaps occurs at short drift distances. In fig. 25c the local drift velocity v_D (measured time difference Δt of two flat equal-time tracks with incremental drift distance Δd) is plotted versus drift distance d. We note a large increase in v_D between U_1 and U_{PC} indicating that field inhomogeneities remain in this 14 mm long region. Fig. 25d shows that n_{pe} and the variance σ are constant when the drift field E_D is varied from 0.4 kV/cm to 1.0 kV/cm. From the average constant value $n_{pe} = 1.5$ and wire spacing 0.127 cm we calculate that 11.8 primary electrons are produced in 1 cm of this CH_4 (89%) + iso-C_4H_{10} (11%) gas mixture at atmospheric pressure and temperature $\geqslant$ 34°C.

To measure the longitudinal diffusion coefficient, the variance σ^2(ns) of the straight line fit was plotted versus d as shown in fig. 26. A straight line $\sigma^2 = \sigma_0^2 + \sigma_1^2 d$ has been fit to the data giving $\sigma_0 = 10.2$ ns and $\sigma_1 = 2.81$ ns/cm$^{1/2}$. Multiplying these quantities by the drift velocity $v_D = 5.75$ cm/μs gives the space errors $\sigma_0 = 590$ μm and $\sigma_1 = 162$ μm/cm$^{1/2}$. The magnitude of the σ_0 error is due partially to electronics dispersion (2.7 ns) and to time binning (1.2 ns), but the remainder (9.7 ns) is attributed to the drift field inhomogeneity between U_1 and the MWPC cathode U_{PC}. This defect has been corrected in a later drift gap (Mark II) where $\sigma_0 = 5.6$ ns

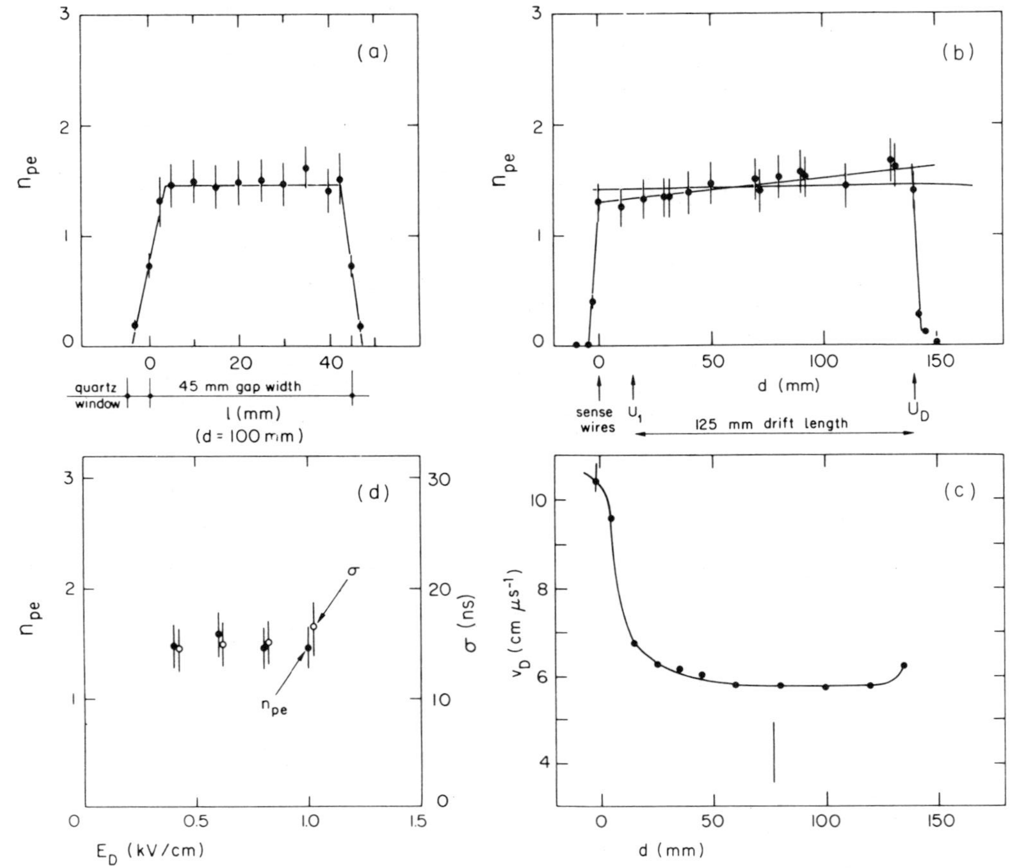

Fig. 25. The average number of primary electrons n_{pe} collected per wire vs (a) gap width w at drift distance $d = 100$ mm and $E_D = 0.6$ kV/cm, (b) drift distance d at gap width $w = 22.5$ mm and $E_D = 0.6$ kV/cm. (c) The local drift velocity v_D vs drift distance d at gap width $w = 22.5$ mm and $E_D = 0.6$ kV/cm. (d) The n_{pe} and the σ of the straight line fit vs drift electric field E_D at drift distance $d = 100$ mm and gap width $w = 22.5$ mm.

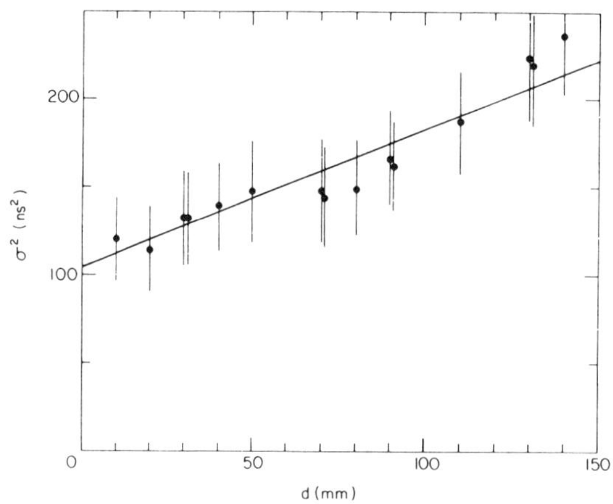

Fig. 26. The variance σ^2 of the straight-line fits plotted vs drift distance d for $w = 22.5$ mm and $E_D = 0.6$ kV/cm. The straight-line fit $\sigma^2 = \sigma_0^2 + \sigma_1^2 d$ has values $\sigma_0 = 10.2$ ns and $\sigma_1 = 2.81$ ns/cm$^{1/2}$.

has been realized. The single-electron longitudinal diffusion coefficient σ_ℓ is related [8] to σ_1 through the equation

$$\sigma_1 = \frac{\pi}{2\sqrt{3 \ln n}} \sigma_\ell, \qquad (6)$$

where n is the total number (primaries + secondaries) of electrons which contribute to the measurement of σ_1 when the electronics is sensitive to a single electron. The total number of electrons is $n = kn_{pe}$ with k between 3 and 5; however, the resulting value of σ_1 is comparatively insensitive to the exact value of k. Using $k = 4$ and $n_{pe} = 1.5$ we obtain $\sigma_\ell = 1.48\sigma_1$, hence $\sigma_\ell = 240$ μm/cm$^{1/2}$. This result may be compared with the data in fig. 24, which give the values $\sigma_t = 230$ μm/cm$^{1/2}$ for methane and $\sigma_t = 120$ μm/cm$^{1/2}$ for isobutane (calculated) at $E_D = 0.6$ kV/cm. Our result is $\sigma_\ell \approx \sigma_t$ for methane (since our gas mixture is predominantly methane), which agrees with the data of ref. 7 for σ_ℓ.

An improved drift cage (Mark II) has been constructed as shown in fig. 27. It has a 140 mm long drift distance between the extreme field-shaping electrodes U_D and U_1 with a transfer mesh at voltage U_T between U_1 and the MWPC cathode U_{PC}. The distance U_1 to U_T

is 4.8 mm, U_T to U_{PC} is 3.2 mm, and U_{PC} to the sense wire plane is 3.2 mm. The sense wires of 10 μm diameter and 50 mm length are spaced 1.27 mm apart. The transverse dimension of the drift volume is 185 mm and the detector plane contains 145 wires. As may be seen in fig. 27, this drift cage is self-contained (i.e. not mounted inside an outer support box), with three sides of the box, 40 mm deep, made of 1.6 mm thick Stesalite on both sides of which are glued Kapton printed-circuit sheets. These printed circuits have parallel conducting copper strips, 100 μm wide, every 2.54 mm. The fourth side of the box has a quartz rectangular entrance window (140 mm × 185 mm) on which are wound 100 μm diameter wire loops, also with 2.54 mm pitch. This window is glued to the Stesalite–Kapton drift cage, with the wire loops in electrical contact with the Kapton printed-circuit strips. A resistance chain mounted near the box provides the electric potentials to the strips which, because of the contact between the strips and the loops, are conducted to both sides of the quartz entrance window. The fifth side of the box is closed by a Stesalite plane, fully copper plated on the inside of the box and connected to the high potential U_D. The sixth and last side of the box is closed by the transfer mesh and MWPC array. The box is gas-tight and has gas

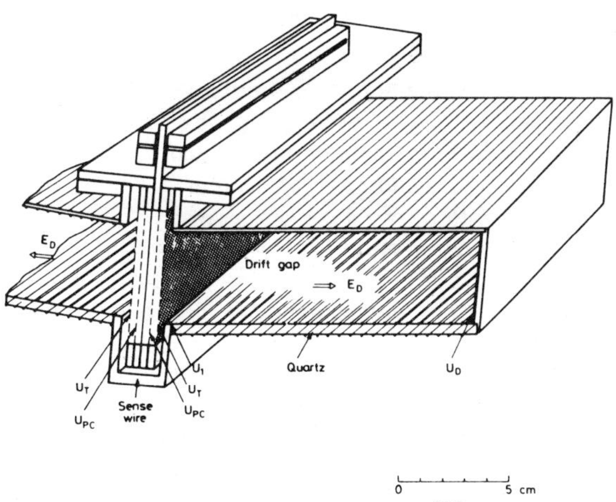

Fig. 27. Schematic view of the modular Mark II drift gap.

input and output leads located near the MWPC array. This unit is modular in the sense that it has been designed so that many such units can be mounted, for example, like the lines of longitude and latitude to cover a spherical surface.

The Mark II detector has been tested with through-going 10 GeV/c pion beam tracks, as described above, with two different gas mixtures (1) pure CH_4, and (2) CH_4 (89%) + iso-C_4H_{10} (11%) at atmospheric pressure and 20°C. The average number of primary electrons n_{pe} measured is shown in fig. 28a versus drift distance d for the two gas mixtures. We find $n_{pe} = 1.1$ and 1.8, respectively, where the latter number should be compared with the value $n_{pe} = 1.5$ found using the original (Mark I) field cage with the same CH_4 (89%) + iso-C_4H_{10} (11%) gas mixture (see fig. 25); however at an elevated temperature $\geqslant 34$°C. This difference in n_{pe} for the two measurements is attributed to a difference in gas densities caused by the disparity in temperatures. The local drift velocity v_D versus drift distance in Mark II is shown in fig. 28b, where we find $v_D = 9.9$ cm/μs and 5.4 cm/μs, respectively, with a rapid change of v_D only near the MWPC wire plane as expected. Thus a much better linearity of the electric field is observed compared to the Mark I drift cage (see fig. 25c).

In fig. 29 the variance of the straight line fits σ^2 (ns) is plotted versus d, and straight lines of the form $\sigma^2 = \sigma_1^2 + \sigma_2^2 d$ are fitted to these data. The parameters are:

for (1) $\sigma_0 = 5.8$ ns and $\sigma_1 = 1.74$ ns/cm$^{1/2}$;
and with $v_D = 9.9$ cm/μs, giving
$\sigma_0 = 574$ μm and $\sigma_1 = 172$ μm/cm$^{1/2}$;
for (2) $\sigma_0 = 5.4$ ns and $\sigma_1 = 2.80$ ns/cm$^{1/2}$;
and with $v_D = 5.4$ cm/μs, giving
$\sigma_0 = 292$ μm and $\sigma_1 = 151$ μm/cm$^{1/2}$.

Correcting for the total number of electrons $n = 4n_{pe}$ we obtain (1) $\sigma_\ell = 231$ μm/cm$^{1/2}$ and (2) $\sigma_\ell = 234$ μm/cm$^{1/2}$, respectively, for the two different gas mixtures. These numbers should be compared to $\sigma_\ell = 240$ μm/cm$^{1/2}$ obtained with the previous (Mark I) detector. It is not thought that these numbers are significantly different, so the average of $\sigma_\ell = 235$ μm/cm$^{1/2}$ represents the results of our longitudinal diffusion coef-

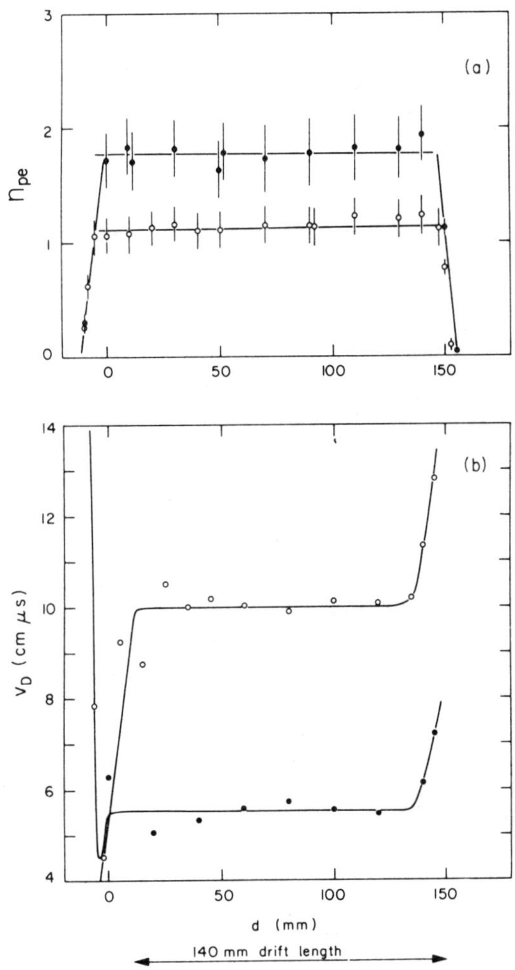

Fig. 28. (a) The average number of primary electrons n_{pe} collected per wire vs drift distance d in the Mark II drift gap with the beam centred in the gap and $E_D = 0.6$ kV/cm. The open circles are for pure methane and the closed circles for 89% $CH_4 + 11\%$ iso-C_4H_{10} gas mixture. (b) The local drift velocity vs drift distance d with conditions as in (a).

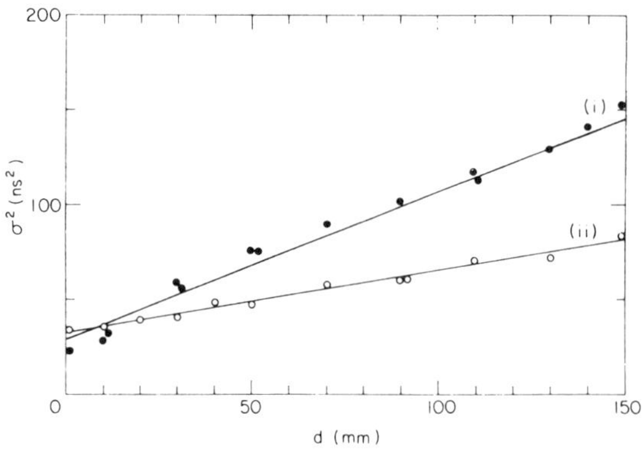

Fig. 29. The square of the variance σ of the straight-line fits for the Mark II drift gap plotted vs drift distance (1) for pure CH_4 and (2) for a mixture of 89% CH_4 + 11% iso-C_4H_{10}. The straight-line fits are of the form $\sigma^2 = \sigma_0^2 + \sigma_1^2 d$ with (1) $\sigma_0 = 5.8$ ns and $\sigma_1 = 1.74$ ns/cm$^{1/2}$, and (2) $\sigma_0 = 5.4$ ns and $\sigma_1 = 2.8$ ns/cm$^{1/2}$.

ficient measurements in pure CH_4 or CH_4 (89%) + iso-C_4H_{10} (11%) at a drift field of $E_D = 0.6$ kV/cm. There is, however, a significant difference in σ_0 for the Mark II detector. This value has been reduced to 5.6 ns, compared with 10.2 ns for the Mark I detector. This represents an approximately 4.7 ns contribution due to field inhomogeneities, compared to 9.7 ns for the Mark I detector, hence a significant improvement. In a slow gas mixture such as CH_4 (89%) + iso-C_4H_{10} (11%) 5.6 ns corresponds to an intrinsic spatial error $\sigma_0 \simeq 300$ μm, whereas for a fast gas such as CH_4 this error becomes $\sigma_0 = 500$ μm. The 2.7 ns contribution from electronics dispersion is not intrinsic and can largely be removed by calibrating each channel. However, this has not been done in treating the data reported in this paper.

6. Summary and conclusions

A drift detector, capable of localizing single photoelectrons, has been constructed and tested by measuring ionization tracks from relativistic charged particles. A first conclusion from these tests is that it is necessary to apply field-shaping conductive strips to both the inside and the outside of the dielectric walls surrounding the drift volume in order to have a stable and homogeneous electric field inside. It is furthermore necessary to have the drift volume surrounded by potential-defining strips all the way up to the MWPC cathode plane, so as to avoid field distortions which produce timing errors.

The spatial resolution of the detector is limited by several sources of error. In the drift (longitudinal) direction, one component of the error is proportional to the square root of the drift distance and the proportionality constant has been measured to be $\sigma_\ell = 235\ \mu\text{m/cm}^{1/2}$ for predominantly methane gas mixtures. A second component of the error which is independent of the drift distance is due to error in the drift time measurement. This component has been determined, in the best case, to be $\sigma_{o\ell} = 5.6$ ns corresponding to a 300 μm constant positional error. In the direction transverse to the electric field, there are also two error components, the first being again proportional to the square root of the drift distance with the proportionality constant given by the transverse diffusion coefficient σ_t. We have not here measured this quantity; however, classical Townsend electron diffusion experiments [9,10] give this quantity. The second component of the error in the transverse coordinate is due to the wire spacing of the MWPC that detects the drifting electrons. It has a value $\sigma_{0t} = s/\sqrt{12}$, where s is the wire spacing. For a drift over a maximum distance D, the average of the distance-dependent error is $\frac{2}{3}\sigma\sqrt{D}$. Adding the independent error components in the quadrature we obtain

$$\sigma_y^2 = \sigma_{o\ell}^2 + \tfrac{4}{9}\sigma_\ell^2 D, \quad \sigma_x^2 = \tfrac{1}{12}s^2 + \tfrac{4}{9}\sigma_t^2 D, \tag{7}$$

as the average square errors in the two directions (x, y) over the detector surface.

The intrinsic resolution of a Cherenkov ring image is given by $\Delta\theta_c$, the chromatic aberrations of the focused image [1]. This quantity depends on the optical dispersion of the Cherenkov radiating medium and is given approximately as $\Delta\theta_c = \Delta n/\sqrt{12n^2(n^2-1)}$, where n is the refractive index and Δn is the difference of refractive indices of the two extreme photon energies detected. The corresponding spatial resolution Δr is determined by the mirror focal length f as simply $\Delta r = f\Delta\theta_c$. An approximate optimization of error sources requires that the detector resolution $1/2(\sigma_x^2 + \sigma_y^2)$, averaged over the circular image, should be equal to $(\Delta r)^2$. Using eq. (7) to solve for the optimal drift distance D as

$$D = \frac{2(\Delta r)^2 - \sigma_{o\ell}^2 - s^2/12}{4(\sigma_\ell^2 + \sigma_t^2)/9}. \tag{8}$$

As an example, for $\Delta\theta_c = 1$ mrad, $f = 1000$ mm we find $\Delta r = 1$ mm. Taking $s = 1.27$ mm, $\sigma_{o\ell} = 0.3$ mm, $\sigma_\ell = \sigma_t = 0.235$ mm/cm$^{1/2}$ as measured, we find $D = 36.2$ cm, corresponding to a drift length more than double that of the present detector (Mark II). To cover e.g. 1 m^2 of detection surface three drift zones, each of 33.3 cm in length, are separated by three rows of MWPC wires, 1 m in lenght, each containing 787 wires. Such a detector, covering 1 m^2, has 5×10^5 resolution elements (pixels) at the cost of instrumenting $3 \times 787 = 2361$ proportional chamber wires. Such a detector allows reconstruction of the ring images without ambiguity, with sufficient precission, at minimal cost.

For shorter focal lenghts, Δr decreases; hence, from eq. (8), the diffusion coefficients must decrease if long (economical) drift distances are to be used. This can certainly be achieved with $CH_4 + CO_2$ drift gas mixtures where $\sigma_\ell = \sigma_t = 0.1$ mm/cm$^{1/2}$ have been measured [7,10]; however, as previously discussed, such mixtures have low drift velocities. Other gases such as C_2H_2, C_2H_4, C_3H_6, C_3H_8, and SiH_4 have been measured to have low values of $\sigma_\ell \approx \sigma_t = 0.14$ mm/cm$^{1/2}$, however with reasonable drift velocities (2–4 cm/μs) at 0.6 kV/cm drift field. The first four of the above gases are transparent below 7 eV (SiH_4 has not been measured) and so could be used for Cherenkov ring imaging with TMAE and quartz. We have not yet tried to count single photoelectrons with additions of these gases to methane but plan to do so in the near future.

We wish to acknowledge the aid and support of Dr. A. Wetherell and Dr. E. Gabathuler of CERN, Dr. P. Fleury of the Ecole Polytechnique, Dr. S. Kullander of the University of Uppsala and Dr. M. Froissart of Collège de France. The technical aid of D. Bernier, R. Bouhot, A. Brehm, C. Detraz, B. Goret, J.P. Hubert, C. Muratori, P. Quéru and C. Rivoiron, and the electronic design of Y. Kornelis were very important in this work, and to them we express our thanks.

References

[1] T. Ypsilantis, Physica Scripta 23 (1981) 370.

[2] J. Séguinot and T. Ypsilantis, Nucl. Instr. and Meth. 142 (1977) 377.

[3] T. Ekelöf, J. Séguinot, J. Tocqueville and T. Ypsilantis, Physica Scripta 23 (1981) 718.

[4] G. Charpak and F. Sauli, Phys. Lett. 78B (1978) 523. See also J. Séguinot, J. Tocqueville and T. Ypsilantis, Nucl. Instr. and Meth. 173 (1980) 283.

[5] V. Palladino and B. Sadoulet, Nucl. Instr. and Meth. 128 (1975) 323.

[6] H. Drum et al., Nucl. Instr. and Meth. 176 (1980) 333, from the data of B. Schmitt, Diplomarbeit, Phys. Inst. der Univ. Heidelberg (1980).

[7] E. B. Wagner, F.J. Davis and G.S. Hurst, J. Chem. Phys. 47 (1967) 3138.

[8] W. Farr, J. Heintze, K.H. Hellenbrand and A.H. Walenta, Nucl. Instr. and Meth. 154 (1978) 175.

[9] T.L. Cottrell and I.C. Walker, Trans. Faraday Soc. 63 (1976) 459.

[10] L.W. Cochran and D.W. Forester, Phys. Rev. 126 (1962) 1785.

DEVELOPMENT OF PROPORTIONAL COUNTERS USING PHOTOSENSITIVE GASES AND LIQUIDS

by D. F. Anderson

Abstract

An introduction to the history and to the principle of operation of wire chambers using photosensitive gases and liquids is presented. Their use as light sensors coupled to Gas Scintillation Proportional Counters and BaF_2, as well as their use in Cherenkov Ring Imaging, is discussed in some detail.

1. INTRODUCTION

In recent years there has been a great deal of effort spent on the development of photosensitive wire chambers. Although the field has a long history, the renewed interest can be traced back to the work of Seguinot and Ypsilantis.[1] They suggested adding a photosensitive gas to a proportional counter, fitted with a UV transparent window, as a means of detecting the UV photons produced as Cherenkov radiation in gases.

To demonstrate their idea, they constructed the instrument shown in Fig. 1. The light source consisted of a volume of pure argon excited by an alpha source. This produces light in the 9-11 eV range. The detector was a multiwire chamber filled with argon, CO_2, and benzene. The photosensitive material was benzene with a threshold for the

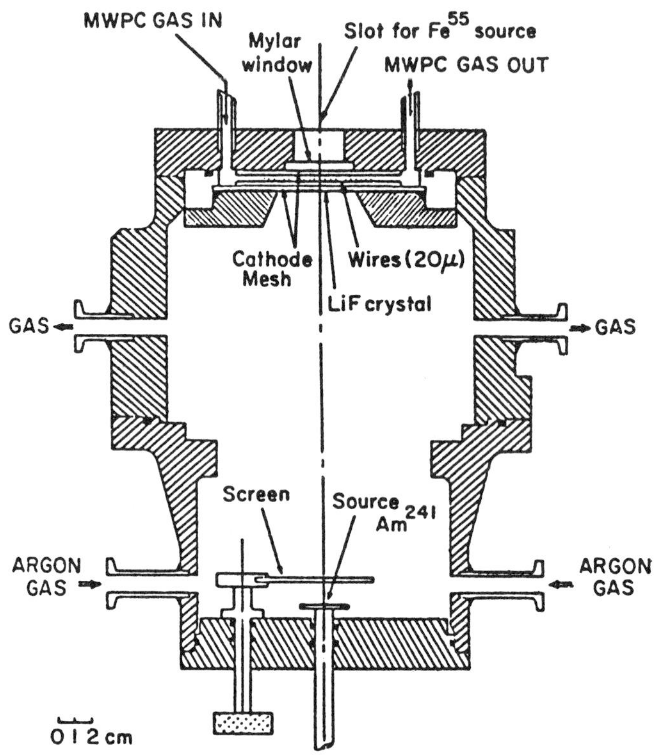

Figure 1. Photosensitive proportional counter and UV generator.

photoelectric effect of 9.15 eV (1355 Å). With this setup, they were able to show that single photon counting was possible.

At the same time as the above work, the use of photosensitive proportional counters was suggested independently by Bogomolov, Dubrovskii and Peskov for plasma studies.[2] They demonstrated their idea with a small multiwire chamber with a LiF window, filled with helium and toluene vapors. Toluene has an ionization threshold of 8.5 eV (1460 Å).

The development of wire chambers using photosensitive gases and, later, photosensitive liquids, has gone in three major directions. The first, is as a replacement for photomultiplier tubes, PMT, for gas scintillation proportional counters. The second, is for Ring Imaging Cherenkov detectors. The third, is as the photon sensor for the scintillator BaF_2. We will discuss all three fields here; although it is not possible to cover these topics completely in a work of this length. In particular, the field of Cherenkov detectors, which has been reviewed elsewhere,[3,4] will be the least complete. A fairly complete bibliography will be provided. We will also restrict ourselves to detectors that detect light, and are not just moderated by photons such as multistep avalanche detectors.[5]

2. GSPC

A gas scintillation proportional counter, GSPC, is a type of proportional counter with an energy resolution for X rays a factor of two better than conventional wire chambers. Although the designs of the GSPC vary dramatically, the

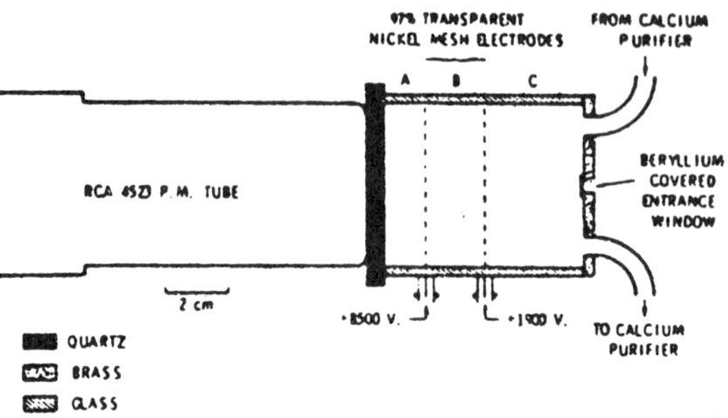

Figure 2. Parallel plate gas scintillation proportional counter[6].

principle is best exhibited in the parallel plate design in Fig. 2.

The GSPC consists of a volume of pure noble gas, usually at 1 atm. Region C of Fig. 2 is an absorption region, deep enough to stop entering X rays. The primary electrons liberated by the X ray are drifted out of this region by a small electric field and into region B. Here the electric field is adjusted such that the electrons gain enough energy between collisions with the gas, to excite the atoms but not ionize them. The excited atoms form a Rydberg molecule which de-excites giving a broad continuous spectrum. The emission spectra of argon, krypton and xenon are shown in Fig. 3. The UV light is then either converted in a wave-
on this GSPC.)

The noble gas of choice for GSPC's has been xenon. This is because of its high X-ray stopping power and its higher UV photon yield. Also, its softer spectrum makes light

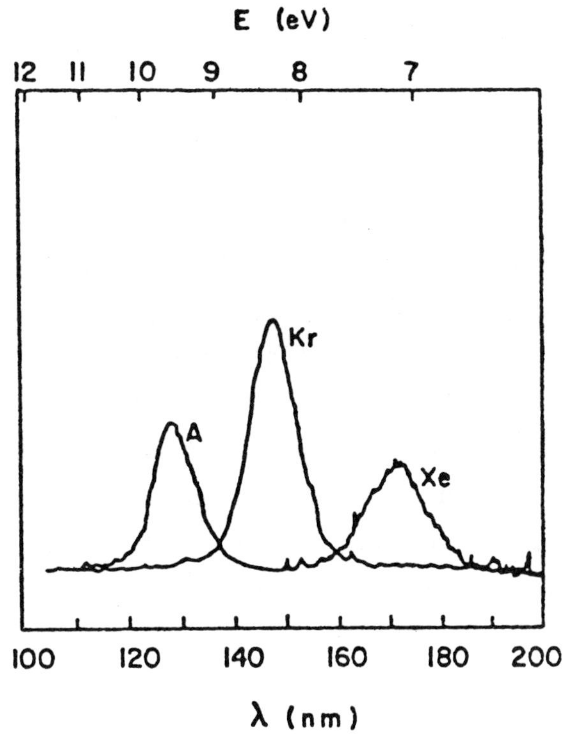

Figure 3. Argon, krypton and xenon emission spectra[7].

length shifter and detected by a glass PMT or detected directly by a quartz PMT. (See Refs. 8 and 9 for a review

detection easier. But, the quartz PMT's used are very expensive and greatly limit detector design.

It was first suggested by Policarpo[10] that a photo-sensitive wire chamber or photo-ionization detector, PID, could be used to replace the PMT for a GSPC. This would offer an instrument with the energy resolution of a GSPC, and the imaging capabilities of a wire chamber. Such an instrument should also be less expensive to construct and be much less sensitive to magnetic fields.

The first coupling of a GSPC to a photosensitive detector was achieved by Peskov.[11,12] The instrument, shown in Fig. 4, was a hybrid using two PMTs at the side of the GSPC, filled with argon plus a little xenon, for the energy determination. The position of an event was determined by an array of "end" type proportional counters looking from below. The photosensitive gas used was trimethylamine, sensitive in the 1050-1550 Å region.[11] An energy resolution of 12% FWHM was measured for 5.9 keV X rays, with a position resolution of about 0.3 mm. The energy resolution was

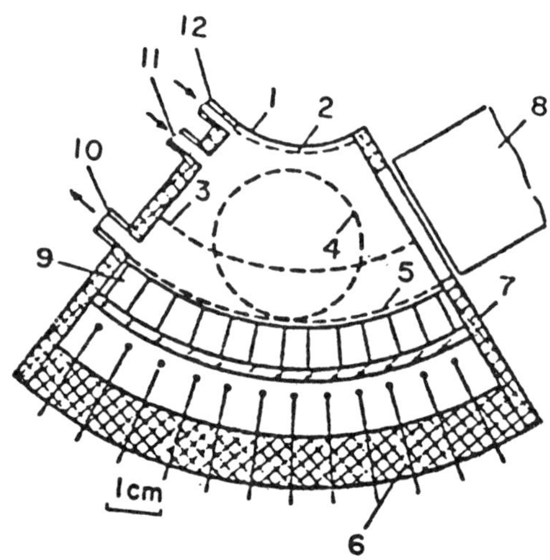

Figure 4. Gas scintillation proportional counter with PMT readout and photosensitive "end"-type proportional counter[11].

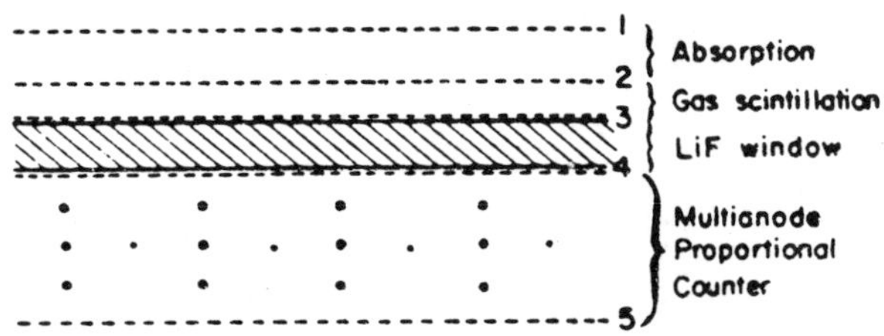

Figure 5. Gas scintillation proportional counter coupled to a photo-ionization detector[7].

improved to 9.7% FWHM with the use of a Penning gas mixture, differentially mixed.

The first demonstration of a true PID coupled to a GSPC was made by Charpak, Policarpo, and Sauli.[7] Their instrument is shown schematically in Fig. 5. The PID used triethylamine, TEA, with a threshold of 7.5 eV (1653 Å) as the photosensitive gas. The gas mixture was 83% argon, 3% TEA, and 14% methane. Because of the high threshold of TEA, the GSPC gas was Kr, and a LiF window separated the two sections. They measured an energy resolution of 10.8% FWHM for 5.9 keV X rays. This same group had first introduced TEA for Cherenkov Ring Imaging with Ypsilantis.[13]

The use of xenon as the GSPC gas was made possible by the introduction, by Anderson,[14] of tetrakis(dimethylamino)-ethylene, TMAE, with a threshold of only 5.36 eV (2315 Å). This lower threshold has the additional advantage of allowing quartz windows to be used which are considerably more rugged and less expensive than the flouride windows used with the other photosensitive gases.

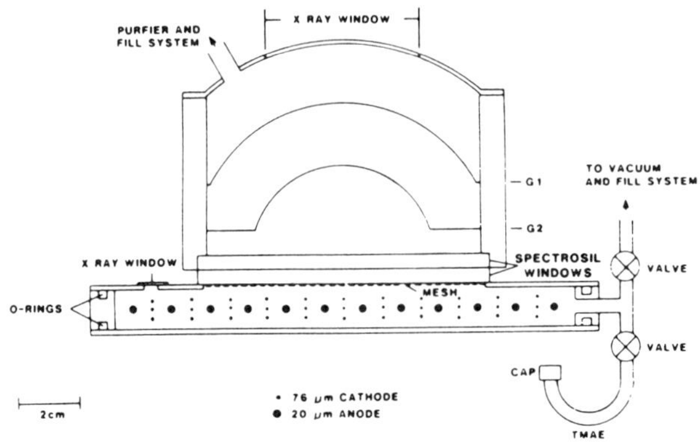

Figure 6. Curved grid gas scintillation proportional counter coupled to a photo-ionization detector[14].

The detector used in the above work is shown in Fig. 6. The PIPS was filled with 90% argon and 10% methane (P-10). Because of the low vapor pressure of TMAE (0.38 Torr at 21°C[15]) it made up only 0.05% of the gas. The GSPC was a curved grid type[16] with a 1 atm filling of pure xenon. An energy resolution of 9.5% FWHM was measured for 5.9 keV X rays.

The absorption length in TMAE of xenon light filtered by a quartz window was also measured. Although the average cross section was estimated at 38 Mb, its low vapor pressure yields an absorption lengths of 23 mm and 11 mm at 20°C and 30°C, respectively. This is much longer than the submillimeter absorption length typical for TEA.

The next logical step in the development of PID's coupled to GSPC was made by Ku and Hailey.[17] They built the first imaging PID. Their detector is shown in Fig. 7. The PID was a multiwire proportional counter with the event position determined from the induced signal on orthogonal cathode planes. The PID gas mixture was 0.3 Torr of TMAE and 580 Torr of P-20. The GSPC was filled with 1 atm of xenon. CaF_2 was used for the window between the two detectors. An energy resolution of 9% FWHM was measured for 5.9 keV X rays, and a position resolution of 0.9 mm was achieved. They also tested the same instrument with krypton in the GSPC, and TEA as the photosensor. Although the induced signal in the PID was narrower due to its shorter absorption depth, the position resolution measured was not better than with TMAE.

As an example of the kind of energy resolution possible with a GSPC coupled to a PID, Fig. 8 shows an X-ray pulse height spectrum from unpublished work by Anderson. The GSPC was filled with xenon and an imaging PID with TMAE was used. The X-ray source is ^{55}Fe at 5.9 keV, yielding an energy resolution of 7.9% FWHM. Note that the K_β line at 6.4 keV

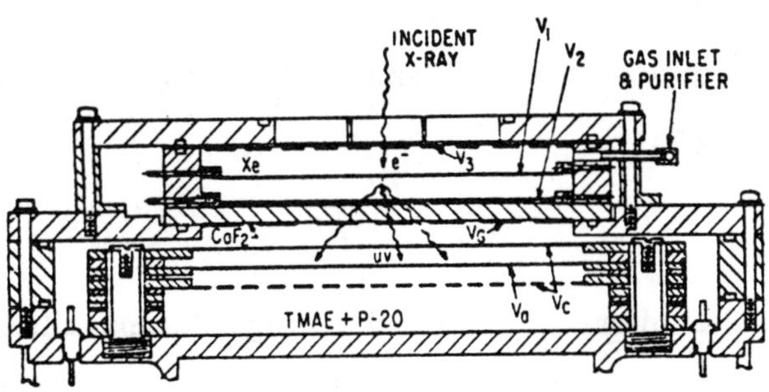

Figure 7. Imaging gas scintillation proportional counter[17].

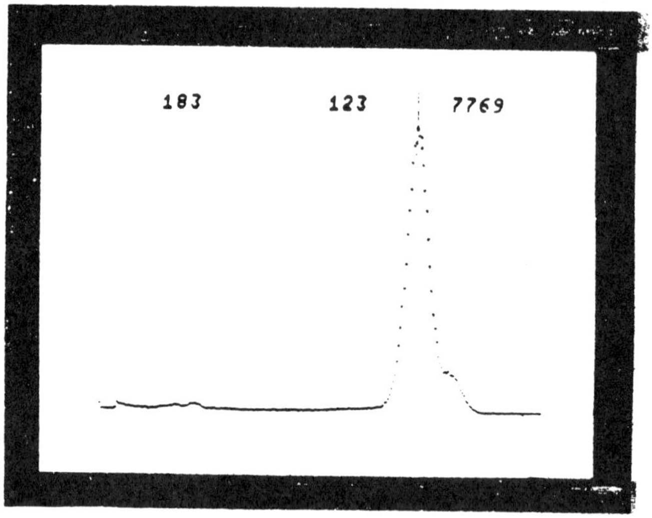

Figure 8. ^{55}Fe pulse height spectrum.

is starting to be resolved at the right. This is as good as the best measurements made with a PMT.

3. CHERENKOV RING IMAGING

The second major application of photosensitive wire chambers is particle identification using Ring Imaging Cherenkov RICH detectors. When a charged particle with velocity $\beta(\beta=v/c)$ passes through an optical medium, with index of refraction n, it emits UV photons at an angle θ_c, given by the Cherenkov relation:

$$\cos\theta_c = 1/\beta n \quad ,$$

for $\beta n > 1$. The photons emitted along the particle path are reflected by a spherical mirror of radius r onto a focal surface of radius r/2. (See Fig. 9.)[1] These photons fall on a circle of radius

$$R = \frac{r}{2} \tan \theta_c$$

As was first suggested by Seguinot and Ypsilantis,[1] a photosensitive proportional counter placed at the surface of radius r/2 could measure the value of R. Thus one knows the β of the particle. With a separate measurement of the particle momentum, the particle can be identified. It should also be noted that since the trajectory of the particle, and therefore the center of the Cherenkov ring, can often be determined by external measurements, a single detected photon is often enough to identify a particle.

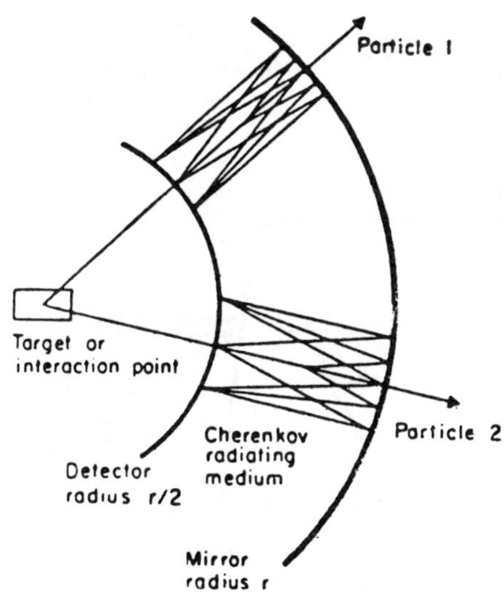

Figure 9. Schematic of ring-imaging Cherenkov experiment[1].

An important parameter for RICH detectors is the number of photons detected. This number can be calculated from

$$N = N_o \, L \sin^2 \theta_c \quad ,$$

where L is the length of the radiator. The figure of merit, N_o, gives the spectral response of the detector to the Cherenkov radiation. (Also see Ref. 18.)

As with the work done simultaneously with the GSPC, the introduction of the new photosensitive gases defined the UV transparent windows used as well as the radiators. For reference, Fig. 10 shows the absorption spectra of four photosensitive gases and the transmission of the UV windows available.[19]

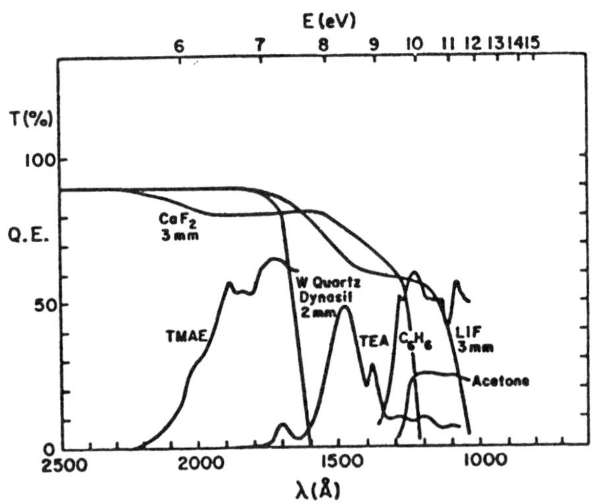

Figure 10. Absorption spectra of some photosensitive gases and transmissions of UV windows[19].

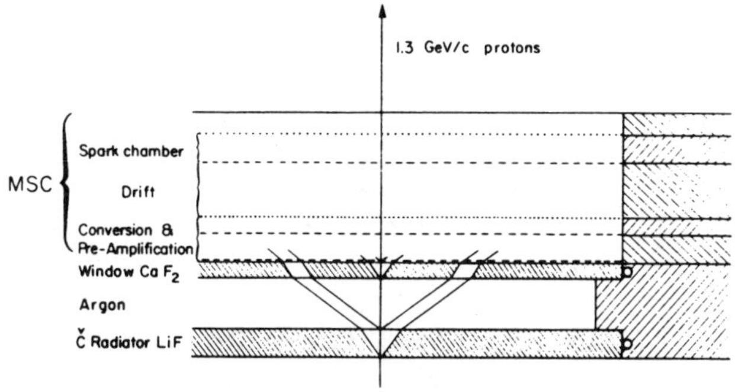

Figure 11. Schematic of ring-imaging Cherenkov detector experiment[1].

3.1 Spark Chamber Readout

The first work in RICH detectors was done with simple wire chambers with one-dimensional readouts.[20,21] The first two-dimensional Cherenkov ring was recorded by Charpak et al.,[13] with a wire chamber employing a spark chamber and video readout. Figure 11 shows a schematic of their detector. The instrument was a multistep avalanche chamber with a triggered spark chamber as a second step, separated by a drift region to prevent photon feedback and to allow for triggering time. The gas filling was helium with about 5% TEA. A CaF_2 window was used. Entering UV photons were converted to electrons by the TEA. These electrons were amplified in the first step of the detector which then caused a spark which was read out at the back by a TV camera.

Two Cherenkov radiators were used in the first test: a 5 mm thick LiF radiator, and then a 1 m long argon radiator

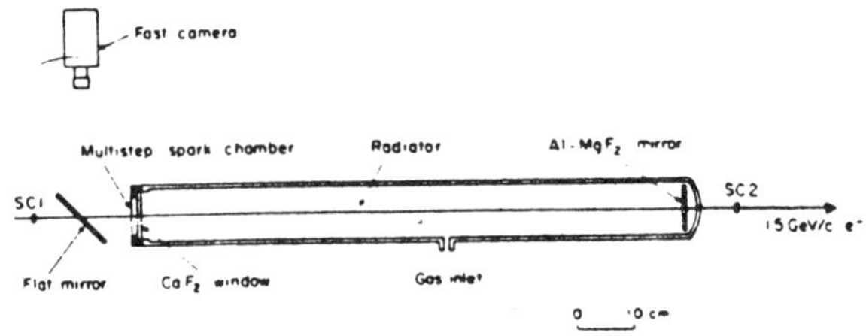

Figure 12. Experimental Cherenkov setup with argon radiator[13].

in the configuration of Fig. 12. For 1.5 GeV/c protons they detected 1.7 photons per event with the argon radiator. The integrated image of 30 events is shown in Fig. 13. The spot in the center is due to the direct ionization of the beam particles. The ring radius was R = 24.42 ± 0.22 mm with a standard deviation of 2.05 mm.

Figure 13. Cherenkov ring for 30 events with 1.5 GeV/c protons[13].

In later work the same group measured an average of 6.8 photons per event, for 10 GeV/c pions, with the radiator filled with 1.6 atm of argon. This yields an N_o of 80 cm^{-1}. In this work they also made the first demonstration of TMAE in a RICH detector.

3.2 Cathode Readout

The first RICH detector used in an actual high energy physics experiment is an experiment E-605 at Fermilab.[22,23] Two detectors are used with a single 15 m long helium radiator at 1 atm, used to identify hadrons between 50 and 200 GeV/c. The detectors consist of two multistep proportional counters each with an active area of 40 X 80 cm^2.

The photosensitive gas used is TEA (3%) with helium (97%). The UV windows are made of CaF_2, consisting of a mosaic of 10 X 10 cm^2 crystals. The detectors have absorption, preamplification and transfer regions, followed by a multiwire proportional counter. Cathode readouts give the x- and y-coordinate of the event, while the direct anode pulse provides the u-coordinate.

For 70 GeV/c pions, they measured an average of 2.5 detected photons per ring, or N_o = 24 cm^{-1}. The measured position resolution was 2 mm FWHM with an average ring radius of 67 mm. The predicted Cherenkov ring centers of all identified particles hitting one of their mirrors are super- imposed in Fig. 14.[23] There are about 1000 events with kaons visible inside a pion ring.

Another technique has been used by Coutrakon, et al.[24]

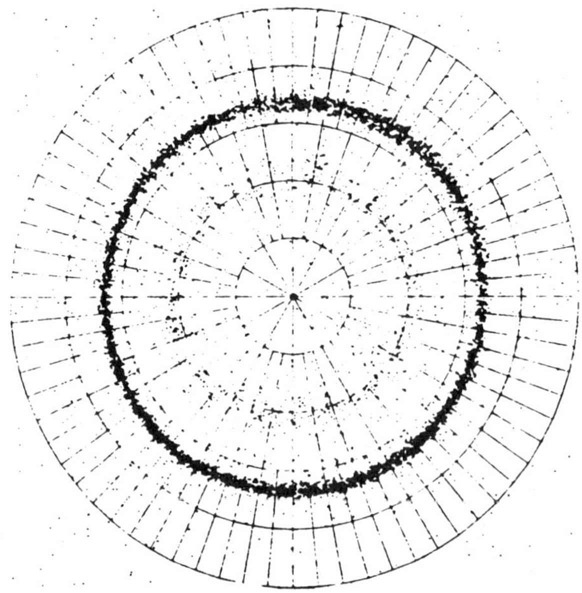

Figure 14. Cherenkov events from E605[23].

They built a test proportional counter with a combination of anode wire and cathode pad readout, using a 18 X 18 cm^2 CaF_2 window. The gas filling was CH_4 (97%) and TEA (3%). For their test they used a Cherenkov radiator consisting of 5 m of argon (20%) and helium (80%) at 1 atm. For 100 GeV/c muons they detected an average of about 4 photons per event, or $N_o = 40\ cm^{-1}$. Their position resolution was 4 mm FWHM with a ring size of R = 70 mm.

3.3 <u>TPC</u>

Another technique used for RICH detectors is the Time Projection Chamber, TPC, shown schematically in Fig. 15.[25] The electrons forming the Cherenkov ring are drifted side-

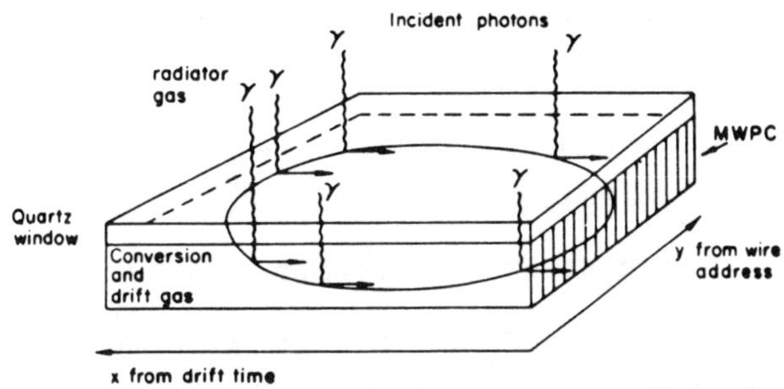

Figure 15. Schematic of a time projection chamber for Cherenkov ring imaging[15].

ways into a multiwire proportional counter. The y-coordinate of an electron is determined by the wire struck. The x-coordinate is given by the arrival time of the electron. The advantage of such a scheme is that a much larger number of electrons can be detected per ring with a small number of channels. The disadvantage is that with a typical drift velocity of 5-10 cm/μs for the electrons, a large detector will require a fairly long time between particles. This technique was first suggested for RICH applications and tested with a one-dimensional detector by Ekelof, et al.[26]

Cherenkov detectors using TPC readouts have been designed for a Delphi detector at LEP (CERN) and for the Omega spectrometer at CERN. A test of the Omega RICH detector has been carried out.[27] The photon detectors use TMAE with two quartz windows 80 X 20 cm^2. The electrons are drifted towards the center with a maximum drift of 20 cm. With a 2 m long argon radiator, they measured an average of 3.2 photons per 10 GeV/c pions with a position resolution of 1.7 mm

FWHM. Their final design with a 5 m Cherenkov radiator is now in operation.

A test detector for the Delphi experiment has also been tested.[28] It consisted of a TPC with a 20 X 20 cm^2 quartz window. The gas mixture was methane (80%), isobutane (20%), and TMAE. The Cherenkov radiator was 50 cm of isobutane. With 10 GeV/c negative particles, they detected an average of 10 photons per event yielding an $N_o = 72\ cm^{-1}$. The final design with a 170 cm drift is presently under test.

3.4 Needles

The work on needle detectors for Cherenkov experiments has been directed by G. Comby[29] at Saclay, although their use was first suggested for RICH by Seguinot and Ypsilantis. The photoelectrons are amplified in the strong field at the end of a needle (see Fig. 16), giving a large signal. The advantage is that a large number of photons can be detected

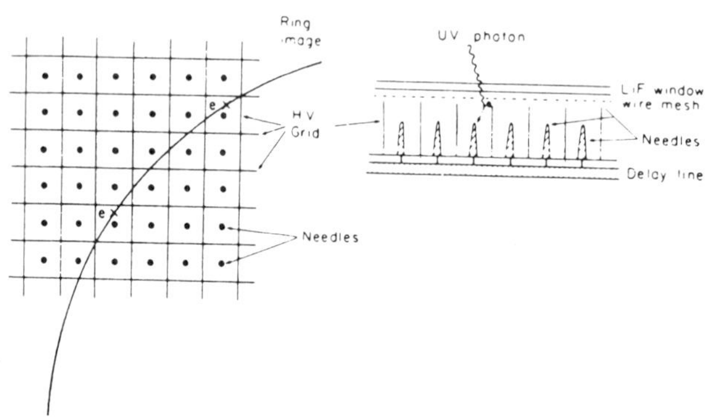

Figure 16. Schematic of a needle detector for Cherenkov ring imaging[1].

at once. The disadvantages are that they are difficult to construct, the spacial resolution is limited to the cell size, and a great deal of electronics is necessary.

There were constructed instruments with up to 2500 needles with different needle spacings. With an instrument of 1008 needles, a CaF_2 window, and using TEA as the photosensitive material, the light produced by 6 GeV/c pions in a 93 cm argon radiator was detected. They measured an average of 0.4 photoelectrons per event, or $N_o = 28\ cm^{-1}$.

3.5 Low Pressure RICH

One problem that has faced RICH detectors is that a high energy particle passing through the detector liberates on the order of 10^3 electrons. In a detector designed to detect one electron, this can be a serious problem, and often causes sparking or photon feedback. To avoid this, most detectors have either been operated out of the beam or with the region of the beam inactive in the detector. But, even these solutions are ineffective for the particles in the beam halo.

To address this problem the Charpak group at CERN[30] has constructed a small test low pressure chamber with an absorption, a preamplification and a transfer region. This was followed by a multiwire proportional counter in one configuration. In a second design, they used a parallel plate region and a 2 X 2 cm^2 "wedge and strip" printed circuit readout.[31] The detector had a CaF_2 window. The gas fillings were TMAE plus 5 to 20 Torr of pure isobutane. For single photons they measured a high detection efficiency and better than 1 mm FWHM position resolution.

The advantage of operating a RICH detector at such a low pressure is that it reduces the amount of charge liberated by a passing particle by up to two orders of magnitude. Also the use of "wedge and strip" cathode readouts allow high event multiplicity with much fewer channels for a non-TPC design.

Also see Ref. 32 for additional works on the subject of RICH.

4. BaF_2 SCINTILLATOR

With the introduction of TMAE, with its low photo-ionization threshold, it was natural to search for a solid scintillator that emitted light in a region detectable by the gas. The first, though marginal, success was reported by Anderson.[15] The scintillator was BaF_2, which to date, emits light of the shortest wavelength found. It is also the only solid scintillator yet to be coupled to a photosensitive gas, though the search goes on.

As a sintillator, BaF_2 has many desirable characteristics. As a comparison, Table I lists some properties of BaF_2, along with those of BGO and NaI(Tl).[33] Unlike the other two, BaF_2 has two emission maxima. The shorter component peaks at 225 nm and contains about 20% of the light. This component also has a decay constant 1000 times faster than those of long component and about 500 times

Table I

Properties of Three Scintillators

	BaF_2	BGO	NaI(Tl)
Density (g/cm^3)	4.9	7.1	3.7
Radiation length (cm)	2.1	1.1	2.6
dE/dx (min)(MeV/cm)	~6	8	4.8
Linear attenuation coefficient at 511 keV (cm^{-1})	0.47	0.92	0.34
Peak emission (nm)	225 310	480	410
Decay constant (ns)	0.6 620	300	250
Index of refraction	1.56	2.15	1.85
Light yield (photons/MeV)	2×10^3 6.5×10^3	2.8×10^3	4×10^4
Hygroscopic	No	No	Yes

Ref 33.

faster than those of BGO and NaI(Tl). Thus the short component is often referred to as the fast component.

In the first work, a small BaF_2 crystal was coupled to a single wire proportional counter filled with TMAE, argon

(90%) and methane (10%) at 600 Torr. For gamma rays in the 1 MeV range, detection efficiencies were measured at a few percent. This poor efficiency was due to the small overlap of the BaF_2 spectrum with the detection response of TMAE in the gas phase, and to the poor light collection efficiency of the first setup.

4.1 Liquid Photocathode

In an attempt to improve the photon detection efficiency for BaF_2, a "liquid" photocathode was conceived.[34] It is quite easy to calculate the ionization threshold, E_{TH}, for liquid TMAE (electrons detected in the liquid) and for a liquid TMAE photocathode, E_{LPC}.[35,36] These are given by

$$E_{TH} = IP + P_+ + V_o$$

and

$$E_{LPC} = IP + P_+ \qquad (\text{for } V_o < 0)$$

Here IP is the ionization potential in the gas phase and P_+ is the polarization energy of the positive ion. V_o is the ground-state energy, with respect to the vacuum, of a free electron in the liquid. Using the value $V_o = -0.28$ eV[37] and the estimate of $P_+ = -1.26$ eV, we have $E_{TH} = 3.82$ eV (3246 Å) and $E_{LPC} = 4.0$ eV (3100 Å). The reason the photocathode threshold is higher is because the electrons released in the liquid must overcome V_o to escape. Thus there are low energy electrons in the liquid that, in

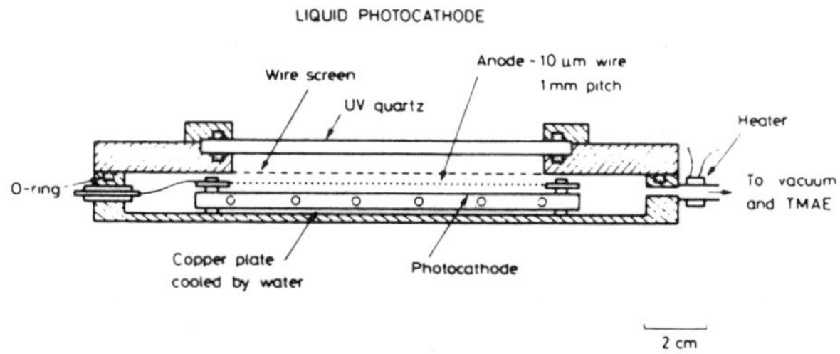

Figure 17. Liquid photocathode[34].

principle, could be drawn from the surface if the electric field were high enough.

The first liquid photocathode is shown in Fig. 17. It consisted of a water-cooled copper plate for condensing the TMAE and a low-pressure wire chamber. The TMAE is extremely thin on the surface and thus the instrument can be operated in any orientation. The optical window was made of quartz. The counter gas was only 3 Torr of isobutane to increase the photoelectron collection efficiency. In that work, the counter was operated warm with the photocathode kept at 18°C. Using filters, a threshold of about 4.3 eV was estimated, and a much higher efficiency for the light from BaF_2 was also measured. The estimated quantum efficiency for the short component of BaF_2 was about 4%.

A second instrument was made that incorporated a BaF_2 crystal, 2.5 cm thick and 12 cm in diameter, as shown in Fig. 18. Again a low pressure counter was used, filled with 3 to 9 Torr of isobutane. This low pressure makes the wire chamber insensitive to the passing particles, while

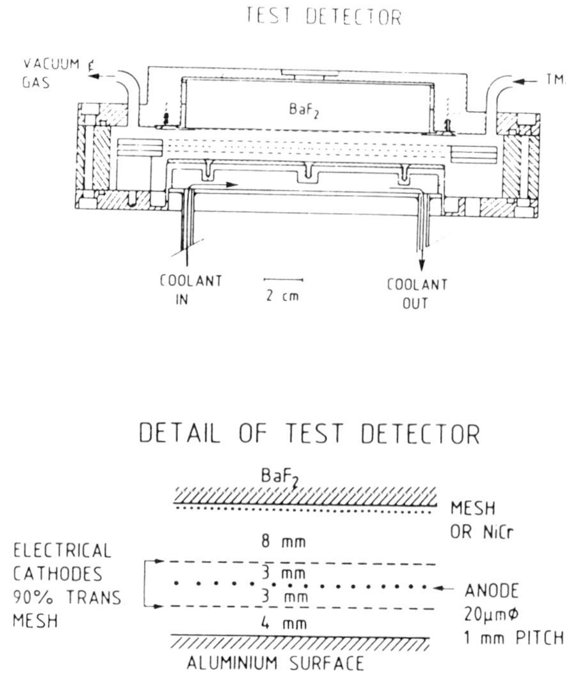

Figure 18. BaF_2 coupled to a liquid photocathode[38].

enhancing the sensitivity to UV photons. The TMAE was condensed on a cold surface at 2°C.

Using 350 MeV alpha particles, the electron component of the signal had a rise time of only 10 ns, and the ion drift time was only slightly over 1 μs. A timing resolution of 540 ps FWHM was achieved. An energy resolution of 28.5% FWHM was also measured for 970 MeV protons losing about 18 MeV in the crystal. Using the same instrument with a cathode pad readout, a position resolution of about 3 mm FWHM was measured.

The estimated number of photons detected in this final configuration was approximately 10^4 per GeV. It is expected that if the electric field at the surface of the liquid photocathode is increased, this number in principle could be increased.

4.2 Adsorption and Gas

It was noted[38] that when TMAE is added to a detector, all the metallic surfaces become photosensitive. To utilize this effect, a complete BaF_2 calorimeter tower was constructed,[39] that used TMAE and low pressure chambers, but without cooling. A schematic of the tower is shown in Fig. 19. The instrument consisted of 14 crystals varying in thickness from 1 cm to 5 cm. All were 12.6 cm in diameter. The total thickness of BaF_2 was 40.5 cm, making 19.3 radiation lengths. Each crystal was preceded by its own wire chamber. The surfaces of the BaF_2 crystals were coated with a 15 Å layer of NiCr, which is a UV transparent conductive coating. When TMAE is introduced into the

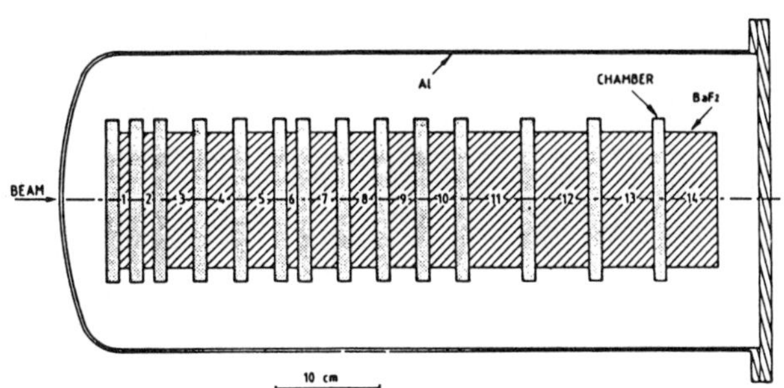

Figure 19. BaF_2 coupled to a liquid photocathode[38].

detector, the adsorbed gas acts as a photocathode. There is also some contribution from the gas volume. But the amplification in a low pressure chamber is almost parallel plate amplification. Thus the contribution from gas is probably fairly small.

The pulse-height spectra for 108 MeV and 200 MeV electrons are shown in Fig. 20. The resolutions are 30% and 20% FWHM, respectively. But when correcting for the energy leakage out the side of the tower, a resolution of

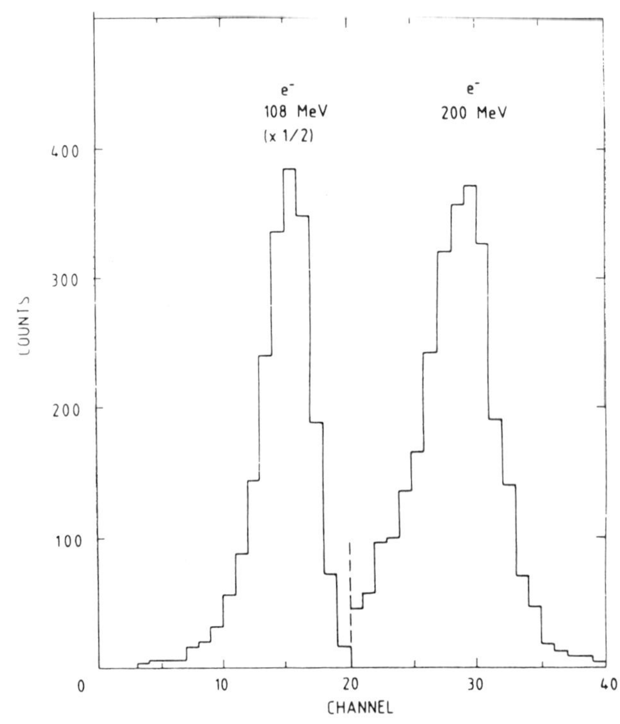

Figure 20. Pulse height spectra for 108 MeV and 200 MeV electrons[39].

$\sigma/E = 2.5\% \ E^{1/2}$ (GeV) was determined. This correction was made using the EGS Monte Carlo simulation program.

The energy deposit per crystal is shown in Fig. 21 for 108 MeV and 200 MeV electrons, as well as for 108 MeV/c pions and muons. The energy deposited in crystal 6 appears low because it is 1 cm thick and the surrounding crystals are 2.5 cm thick. This figure shows the advantage of using wire chambers over phototubes. One can achieve high segmentation and measure both the lateral and longitudinal development of the shower.

In a new work by the Charpak group,[40] the effect of increasing the temperature of the detector to increase the TMAE pressure was measured. They found that the energy resolution for 662 keV gamma rays increases up to around 70°C and then remained constant. The limiting resolution was 81% FWHM.

For a review on the subject of low pressure counters, see Refs. 41 and 42.

5. Conclusion

The modern field of proportional counters using photosensitive gases and liquids is a new one, with a growing number of applications and investigators. Chambers of a large fraction of 1 m^2 are now possible with quantum efficiencies approaching 50%, and with millimeter position resolution. Only the three major applications of the GSPC, RICH and BaF_2 have been addressed here. The techniques being developed are certainly applicable to other fields

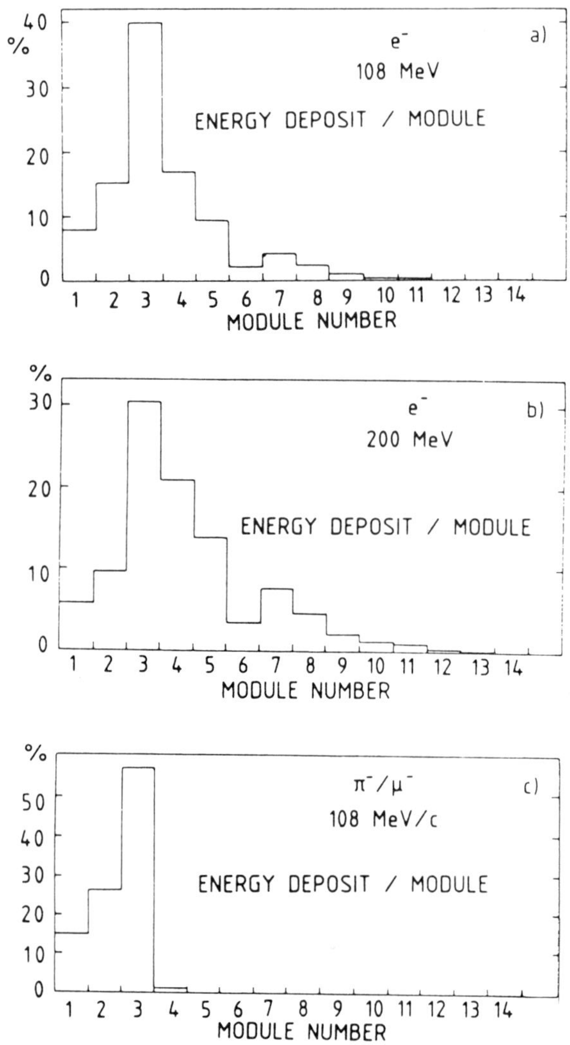

Figure 21. Energy deposit per crystal for 108 MeV electrons, 200 MeV electrons, and 108 MeV/c pions and muons. Crystal 6 is only 40% as thick as its neighbors.[39]

such as the counting of Lyman Alpha photons in atomic physics.

The study of photosensitive gases and liquids is in need of much good academic work. Much of the existing work has been done by pragmatic investigators with a specific problem to solve. And thus, they were often denied the luxury of careful academic studies. (The work of T. Ypsilantis and collaborators is a marked exception.) Academic institutions should be encouraged to become involved. An example of a subject that would be suitable for academic study is the liquid photocathode. Such questions as spectral response, the effect of temperature, and the role of the electric field are yet untouched.

References

1. J. Seguinot and T. Ypsilantis, Nucl. Instrum. Methods 142 (1977) 377.
2. G. D. Bogomalov, Yu. V. Dubrovskii and V. D. Peskov, Instrum. Exp. Tech. 21 (1978) 778.
3. G. B. Coutrakon, IEEE Trans. Nucl. Sci. NS-31 (1984) 27.
4. T. Ypsilantis, "Ring Imaging Cerenkov Counters," IEEE Trans. Nucl. Sci. NS-32 (1985).
5. G. Melchart, G. Charpak and F. Sauli, IEEE Trans. Nucl. Sci. NS-27 (1980) 124.
6. H. E. Palmer, IEEE Trans. Nucl. Sci. NS-22 (1975) 100.
7. G. Charpak, A. Policarpo and F. Sauli, IEEE Trans. Nucl. Sci. NS-27 (1980) 212; M. Alegria Feio et al., Nucl.

Instrum. Methods 176 (1980) 473; M. Saleta S.C.P. Leite et al., Nucl. Instrum. Methods 179 (1981) 295.

8. A. J. P. L. Policarpo, Space Sci. Instrum. 3 (1977) 77.
9. D. F. Anderson, "Gas Scintillation Proportional Counters," Proceedings of the 1981 INS International Symposium on Nuclear Radiation Detectors," Tokyo, 23-26 March, 1981.
10. A. J. P. L. Policarpo, Nucl. Instrum. Methods, 153 (1978) 389.
11. V. D. Peskov, Instrum. Exp. Tech. 23 (1980) 507.
12. G. F. Karabadzhak, V. D. Peskov and E. R. Podolyak, Nucl. Instrum. Methods, 217 (1983) 56.
13. G. Charpak et al., Nucl. Instrum. Methods 164 (1979) 419; G. Charpak et al., Nucl. Instrum. Methods 180 (1981) 387.
14. D. F. Anderson, Nucl. Instrum. Methods 178 (1980) 125.
15. D. F. Anderson, IEEE Trans. Nucl. Sci. NS-28 (1981) 842.
16. D. F. Anderson et al., Nucl. Instrum. Methods 163 (1979) 125.
17. William H.-M. Ku and Charles J. Haley, IEEE Trans. Nucl. Sci. NS-28 (1981) 830; C. J. Hailey, W. H.-M. Ku and M. H. Vartanian, "An Imaging Gas Scintillation Proportional Counter for the Detection of Subkiloelectron-volt X-rays", paper presented at "Workshop on X-ray Astronomy and Spectroscopy." Held at the NASA Goddard Space Flight Center, Greenbelt, MD, 5-7 October 1981. Columbia Astrophysics Lab Contribution No. 214; William H.-M. Ku,

Charles J. Hailey and Michael H. Vartanian, Nucl. Instrum. Methods 196 (1982) 63.

18. J. Seguinot, J. Tocqueville and T. Ypsilantis, Nucl. Instrum. Methods 173 (1980) 283.
19. A. Breskin et al., IEEE Trans. Nucl. Sci. NS-28 (1981) 429.
20. R. S. Gilmore et al., Nucl. Instrum. Methods 157 (1978) 507.
21. J. Chapman, D. Meyer and R. Thun, Nucl. Instrum. Methods 158 (1979) 187.
22. M. Adams et al., Nucl. Instrum. Methods 217 (1983) 237; Ph. Mangeot et al., Nucl. Instrum. Methods 216 (1983) 79; G. Coutrakon et al., IEEE Trans. Nucl. Sci. NS-29 (1982) 323; R. Bouclier et al., Nucl. Instrum. Methods 205 (1983) 403.
23. H. Glass et al., IEEE Trans. Nucl. Sci. NS-30 (1983) 30.
24. G. Coutrakon, S. Dhawan and M. Izycki, "Cherenkov Ring Imaging Using a Proportional Wire Chamber with Cathode Readout," submitted to Nucl. Instrum. Methods, 1984.
25. E. Barrelet et al., Nucl. Instrum. Methods 200 (1982) 219.
26. T. Ekelof et al., "The Cherenkov Ring-Imaging Detector: Recent Progress and Future Development," CERN-EP/80-115 (1980).
27. M. Davenport et al., IEEE Trans. Nucl. Sci. NS-30 (1983) 35.

28. L. O. Eek et al., IEEE Trans. Nucl. Sci. NS-31, No. 2 (1984) 949.

29. G. Comby and P. Mangeot, IEEE Trans. Nucl. Sci. NS-27 (1980) 106; G. Comby and P. Mangeot, IEEE Trans. Nucl. Sci. NS-27 (1980) 111; G. Comby et al., IEEE Trans. Nucl. Sci. NS-29 (1982) 328.

30. W. Dominik et al., "Possible Applications of Low-Pressure Multistep Chambers to Cherenkov Ring Imaging," presented at the Int. Conf. on Instrumentation for Colliding-Beam Physics, Novosibirsk, USSR, 15-21 March 1984, CERN-EP/84-29.

31. C. Martin et al., Rev. Sci. Instrum. 52 (1981) 1067; R. S. Gilmore et al., IEEE Trans. on Nucl. Sci. NS-28 (1981) 435; O. H. W. Siegmund et al., IEEE Trans. Nucl. Sci. NS-30 (1983) 503.

32. S. Durkin, A. Honma and D.W.G.S. Leith, "Preliminary Results on Tests of a Cherenkov Ring Imaging Device Employing a Photoionizing PWC," SLAC-Pub-2186 (1978); S. H. Williams et al., "An Evaluation of Detectors for a Cherenkov Ring-Imaging Chamber," SLAC-Pub-2412 (1979); G. Charpak and F. Sauli, "Use of TMAE in a Multistep Proportional Chamber for Cherenkov Ring Imaging and Other Applications," CERN-EP/83-128; Stephen H. Williams, "Cherenkov Ring Imaging Detector Development at SLAC," SLAC-Pub-3360 (1984).

33. M. Laval et al., Nucl. Instrum. Methods 208 (1983) 169; R. Allemand et al., "New Developments in Fast Timing

with BaF_2 Scintillator," Communication LETI/MCTE/82-245, Grenoble, France (1982); BGO-NaI(Tl) Comparison, paper distributed at the International Workshop on Bismuth Germanate, Princeton University, 1982; M. R. Farukhi and C. F. Swinehart, IEEE Trans. Sci. NS-18 (1971) 200.

34. D. F. Anderson, Phys. Lett. 118B (1983) 230.
35. Y. Nakato, M. Ozaki and H. Tsubomura, J. Phys. Chem. 76 (1972) 2105; R. A. Holroyd and R. L. Russell, J. Phys. Chem. 79 (1983) 483.
36. D. Anderson et al., Nucl. Instrum. Methods 225 (1984) 8.
37. R. A. Holroyd and D. Anderson, "The Physics and Chemistry of Room-Temperature Liquid Ionization Chambers," submitted to Nucl. Instrum. Methods (1984).
38. D. Anderson et al., Nucl. Instrum. Methods 217 (1983) 217.
39. D. Anderson et al., "Test Results of a BaF_2 Calorimeter Tower with a Wire Chamber Readout," CERN-EP/84-82; submitted to Nucl. Instrum. Methods.
40. M. Suffert, R. Bouclier and G. Charpak, "Influence of the Temperature on the Response of the Fast Component of BaF_2 Scintillators Coupled to a Photomultilier or a Photosensitive Wire Chamber," to be submitted to Nucl. Instrum. Methods.
41. A. Breskin, Nucl. Instrum. Methods 196 (1982) 11.
42. A. Breskin, G. Charpak and S. Majewski, Nucl. Instrum. Methods 202 (1984) 349.

Reprinted, with permission, from *Nuclear Instruments and Methods*, Vol. 120, pp. 221-236 (North-Holland Publishing Co., 1974).

LIQUID-ARGON IONIZATION CHAMBERS AS TOTAL ABSORPTION DETECTORS*

by W. J. Willis and V. Radeka

A new detector for the measurement of energy by total absorption, based on the use of multiple-plate ion chambers, is described. The use of liquid argon as the working medium and optimized readout results in an electronic noise contribution to the resolution of less than 0.1 GeV, in a large detector. The use of thin plates, 0.1 radiation length, ensures that sampling fluctuations are small The technique allows absolute calibration and very good gain stability. Tests on a detector large enough to absorb a high-energy electromagnetic shower are described, where the energy resolution is limited by the residual sampling fluctuations.

1. Principles and limitations of calorimetric detectors

If a high-energy particle enters a sufficiently large block of matter, all of its energy will be transformed into ionization and eventually into heat, with certain important exceptions. Thus, detectors relying on total absorption to measure particle energy are often called

* This work was performed under the auspices of the U.S. Atomic Energy Commission.

calorimeters. The ionization released is a good measure of the energy, and we shall be describing an ionization calorimeter which functions by collection of the charge due to the ionization, without charge multiplication within the device.

The design of calorimeters has been reviewed by Murzin[1]), but we give here a brief discussion of the limitations of these devices to explain why we have adopted our solution. We will not discuss here the complex mechanisms by which the particles develop cascade showers and gradually convert their energy into ionization. The unavoidable exceptions to the statement that all the energy appears in the calorimeter are the origin of basic limitations to the precision of energy measurements. These are:

i) particle leakage back out of the surface through which the high-energy particle enters;
ii) the energy carried off by neutrinos, and, in practice, some muons;
iii) the energy required to remove nucleons from complex nuclei.

Without exception, the effects which limit the energy resolution are more serious for incident hadrons than for incident electrons or photons. Thus, the first effect above can be appreciable for incident hadrons of low energy ($\lesssim 2$ GeV), but is always rather small for electrons. Effect (ii) comes mainly from positive pions which come to rest, which then convert about 135 MeV into undetected energy. This effect can be calculated[2]), and is most prominent for incident pions in the several-GeV region. Effect (iii) is very small for electrons but can average about 30% of the total for hadrons, falling quite slowly with energy[3]).

Other effects which may be present in practical detectors are:

iv) energy leakage from the sides and the back of the absorber;
v) sampling fluctuations;
vi) saturation of response on densely ionizing particles;
vii) non-uniform response;
viii) noise.

Effect (iv) is determined by the size of the absorber as measured by radiation lengths for the electromagnetic component, or by interaction lengths for the hadronic component. Roughly speaking, this effect requires that the length of the absorber be greater than 10 to 20 radiation lengths, and greater than 5 to 10 interaction lengths, depending on the particle energy and the resolution desired. In most materials (aluminum and heavier) the radiation length is much shorter than the interaction length, and a detector designed for electrons and photons is smaller than one designed for hadrons.

The sampling fluctuation effect (v) occurs when there are portions of the absorber in which ionization is not collected. Many calorimeters introduce plates of a dense material and sample the ionization in an active material between the plates, for example.

The saturation effect (vi) occurs particularly when the ionization is measured indirectly by means of scintillation light. In organic scintillators especially, the light output per unit track length approaches a limit as the rate of ionization increases. Electromagnetic showers deliver almost all of their ionization near the minimum rate, but hadron cascades often contain many heavily ionizing particles. The effect amounts to about 20% of the total on the average[4]). A device which measures ionization directly may be similarly affected by columnar recombination of the ions, though this effect vanishes in strong electric fields.

A potential limitation (vii) is due to the non-uniform response over different portions of the volume, an important design consideration for scintillation detectors where the light is collected with a low efficiency.

Noise (viii) can be of several sorts: photon statistics in a scintillation detector; amplifier noise in the case of ionization detectors; random pileup of background particles; and, in the case of multiparticle detectors, energy leakage from one particle of a given event entering the spatial domain of another particle.

The energy resolution of the calorimeter will be determined by the fluctuations in the above effects. Unfortunately, many of these effects lead to distributions which are not at all Gaussian in shape, and have long tails. In most applications, tails which give too large a response are particularly to be avoided. Fortunately, most of the above effects tend to give tails on the low side of the response curve, with the exception of the non-uniform response effect in scintillators and in some circumstances, the sampling fluctuation effect.

The net effect is to allow the possibility of very good resolution ($\lesssim 1\%$) for electrons and photons in the few-GeV region[5]), while for hadrons the fundamental limit is nearer 10% [1,2]). The resolution improves at higher energies ($\propto E^{-\frac{1}{2}}$). The results so far attained with hadrons are somewhat poorer than the potential limit[6]).

2. The ion-chamber approach

In searching for a technique which would reach a satisfactory compromise among the effects discussed in section 1, while attaining a short average interaction length and radiation length, we arrived at the concept of an ion chamber with a large number of thin plates. If the number is large enough, the sampling fluctuations can be held down while keeping the average density high. The uniformity of charge collection can be essentially perfect.

For simplicity and to obtain accurate calibration properties, we prefer to measure the ionization in the ion-chamber mode without any electron multiplication. The amount of charge collected in the active medium must be some reasonable fraction of the total if the ratio of signal to amplifier noise is to be satisfactory. This implies that the active medium must be a condensed material which does not attach electrons and has a high electron mobility. We believe that liquid argon satisfies the requirements better than any other material:

i) it is dense (1.4 g/cm^3);
ii) it does not attach electrons;
iii) it has a high electron mobility (~ 5 mm/μs at 1 kV/mm);
iv) the cost is low (\$0.14→0.50/kg, depending on source and quantity);
v) it is inert, in contrast to flammable scintillators;

vi) it is easy to obtain in a pure form and easy to purify;

vii) many electronegative impurities are frozen out in liquid argon.

The disadvantage is that the container must be insulated for liquid-argon temperature (86 K).

Some of the properties[7]) of such a device are illustrated in table 1 for the configuration we have used in the tests described in this paper, with 1.5 mm steel plates immersed in liquid argon (LA).

One sees that the sampling is indeed very fine in that a single cell represents less than a tenth of a radiation length and the energy loss across a cell is less than a tenth of the critical energy. The average interaction length is quite short, however, and the average radiation length is short compared to the interaction length. (Reasonable configurations do not allow much shorter interaction lengths, but the use of high-Z plates would provide radiation lengths as low as 15 mm.) Transition effects are also small, since the properties of steel and argon are similar.

The observable amount of charge is calculated in the following. The most important fact to be noted is that the ionization chamber with liquid argon is a single-carrier device as far as charge collection is concerned. Positive ions due to their very low mobility contribute

TABLE 1

One cell: 1.5 mm steel + 2.0 mm LA.

	Steel	LA	Average in chamber
ρ, g/cm^3	7.9	1.4	4.2
dE/dX, MeV/cm	11.6	2.2	6.2
Radiation length, cm	1.77	13.5	3.5
E_{crit}, MeV	21	30	23
Radiation lengths per cell	0.084	0.015	0.099
Interaction length, cm	12.9	65	24
Dielectric constant		1.6	

$dE(LA)/dE(LA + steel) = 0.20$

little to the signal charge in the short time electrons take to drift across the gap. The basic relations for the current and charge waveforms for planar electrode geometry are illustrated in fig. 1. Fig. 1(a) is for one ion pair. The current, due to one carrier, is determined as e/t_d, by the drift time t_d across the gap. The charge measured in the external circuit is determined by the ratio of the distance traversed and the electrode spacing, $Q_s(x)/e = (d-x)/d$. Due to this, electrons uniformly distributed across the interelectrode gap produce an induced signal equal to one half of their charge on the average. The resulting current and charge waveforms are different for localized ionization and for uniform ionization across the gap. It is interesting to note that for uniform ionization three quarters of the observable charge is "collected" in one half the drift time across the gap.

We compute, now, the amount of charge for the calorimeter configuration described above. The energy loss per ion pair in argon is 26.4 eV [8]). The energy loss per observed electron charge is then 52.8 eV. Taking into account the fact that 20% of the energy is deposited in the LA in the configuration we have chosen, we see that the amount of energy deposit necessary to give a signal corresponding to one electron is 264 eV. In other units one GeV gives a signal of 0.61 pC (picocoulombs).

The mobility of electrons in liquid argon and its use in small ion chambers have been studied a number of times in the past, and all the necessary data are published[9,10]). In order to verify that we could reproduce these results in apparatus constructed on a larger scale in a way which could be extrapolated to very large structures, we built the test chamber shown in fig. 2. Two interchangeable electrode structures were provided. The one shown is for high-energy particles. The other one consists simply of two plates, on one of which an α source was electroplated.

The charge amplifier used on this chamber has greater bandwidth and much less noise than those used in previous studies, so that it is possible to examine

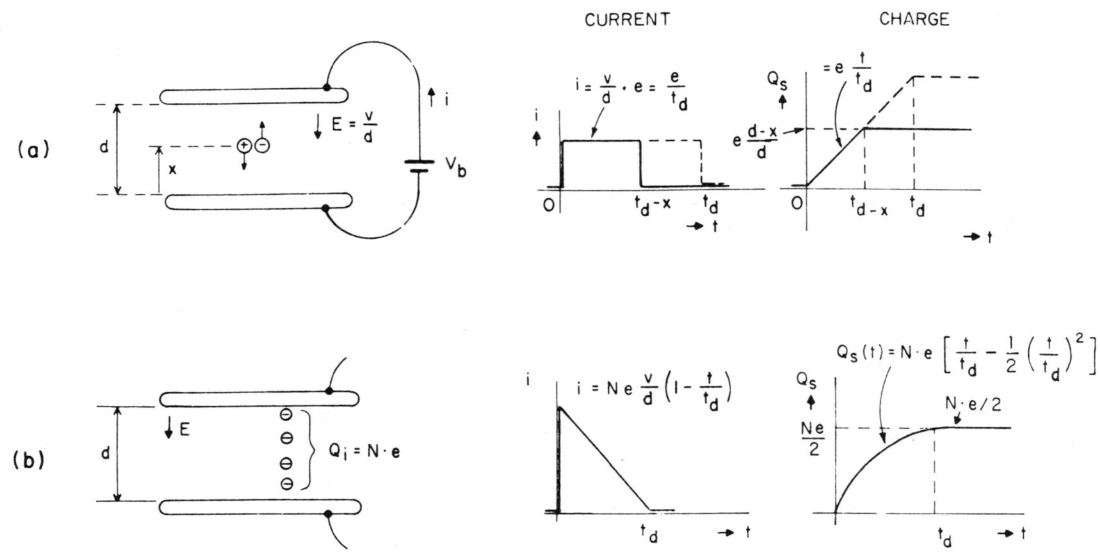

Fig. 1. Charge collection in liquid-argon ionization chamber, (a) single ion pair (solid lines), and for ions formed at the negatively biased plate (dashed lines); (b) uniform ionization.

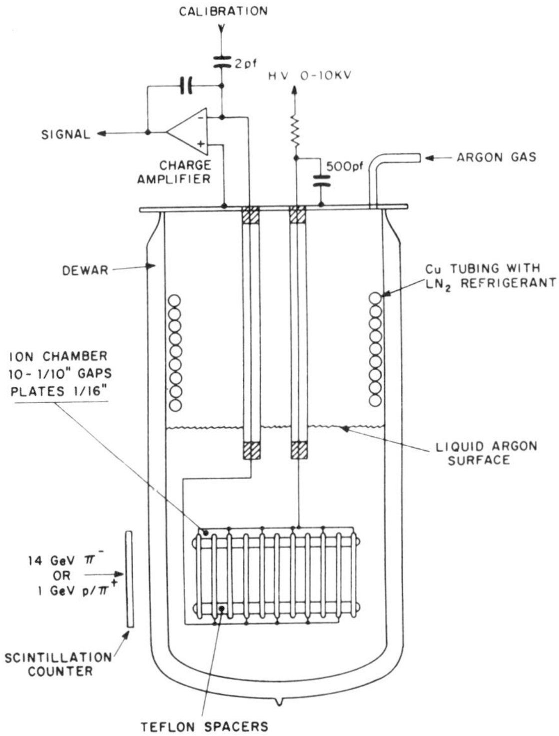

Fig. 2. Small ionization chamber for testing of charge-collection properties.

the pulse shape due to individual particles in some detail. The pulses have the expected shapes as shown in fig. 3 for an alpha-particle source (localized ionization at one electrode), and for high-energy particles (uniform ionization across the gap). These show no evidence for the loss of electrons as they move through the liquid argon (this is apparent from the linear dependence of charge vs time for α particles). The corresponding charge distributions are shown in fig. 4. The dependence of the amount of charge collected versus the electric field in the liquid argon is shown in fig. 5. The value on the plateau ($\approx 4 \times 10^3$ electrons/mm) is approximately the value expected on the basis of the measurements of the average energy required to create an ion pair in gaseous argon. Since not much of the charge is lost in transit, the decrease in charge collected at low fields must be due to columnar recombination near the point of primary ionization. The availability of alpha particles as well as minimum-ionizing particles (muons, or electrons from a beta source) has allowed the convenient measurement of charge collection in argon of different purity, as shown by the different curves on fig. 5. The charge collected from minimum ionizing particles is not very sensitive to the purity, but the charge from the very densely ionizing alphas is quite sensitive to purity, especially when the electric

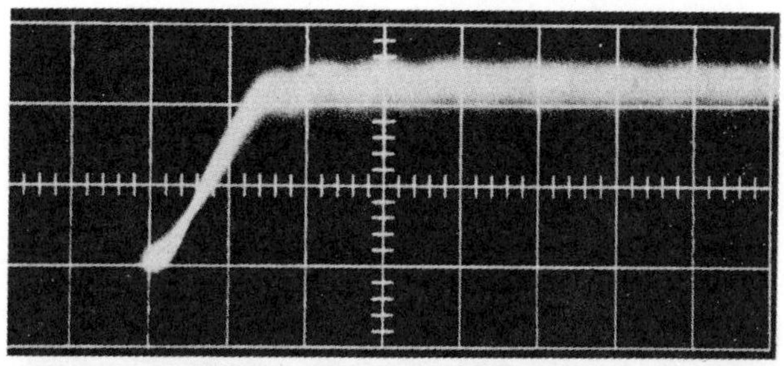

(a)

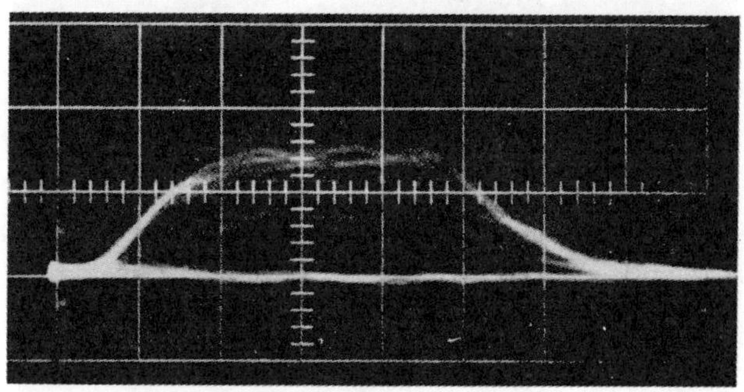

(b)

Fig. 3. Charge as a function of time for 5.5 MeV α particles (a), (horizontal scale 0.1 ós/div., electrode spacing 1 mm); and for minimum-ionizing particles (b) (horizontal scale 0.2 ós/div., electrode spacing 2.5 mm). In (b), the trailing edge is due to the 1 μs delay line clipping in the amplifier. (Note agreement with theoretical waveforms in fig. 1.)

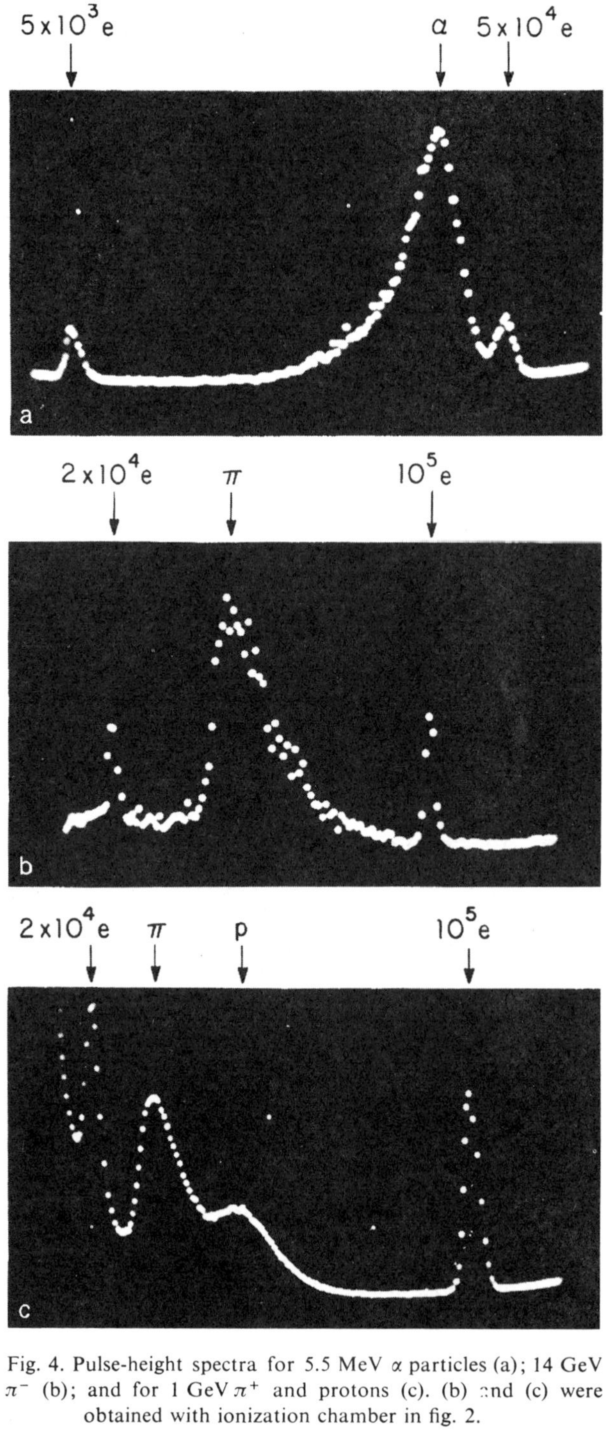

Fig. 4. Pulse-height spectra for 5.5 MeV α particles (a); 14 GeV π^- (b); and for 1 GeV π^+ and protons (c). (b) and (c) were obtained with ionization chamber in fig. 2.

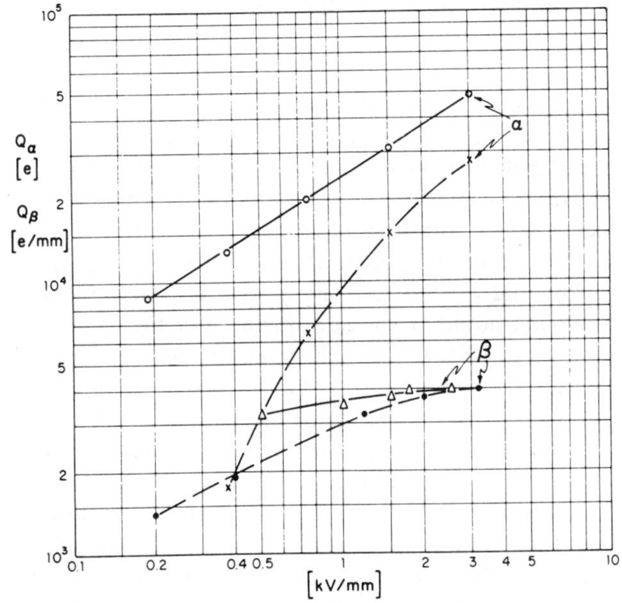

Fig. 5. Measured charge as a function of electric field for 5.5 MeV α particles and for minimum-ionizing particles with different purity of argon.

o, Δ – argon distilled from welding-grade argon supplied in liquid form; the same results are obtained from bottled high-purity argon gas ("gold label");

× – welding-grade argon gas with higher concentration of oxygen (≈ 50 ppm);

● – another bottle of welding-grade argon.

field is low. The best charge collection was observed in the argon recondensed from gas evaporated from welding-grade liquid, or from high-purity gas cylinders.

The drift time, t_d, is about 180 ns/mm. It does not vary with fairly high concentrations of impurities discussed above.

The conclusion of this small-scale test was that relatively fast operation with almost complete electron collection was easy to obtain, and that a large-scale test was justified.

3. Electronic noise and energy resolution

In order to measure the charge from large ionization chambers with negligible fluctuations due to the electronic noise an optimized signal-processing system is essential. The basic elements of an optimum charge-measurement system for this case, and the relation of electronic noise to energy resolution as a function of detector size, are discussed here.

In the preceding section we have determined that a substantial quantity of charge can be produced in a sampling liquid-argon calorimeter. For the configuration given, this is 0.61 picocoulombs per GeV of the energy converted into showers. Large calorimeters involve a large number of plates, and therefore a large capacitance. The large test detector described in this paper has a capacitancc of about 10^5 pF. Charge measurement is performed by observing a fraction of the total charge produced by ionization which is diverted onto the charge-measuring amplifier. The best low-noise amplifying devices presently available are junction field-effect transistors which have an input capacitance of the order of 10 pF. Connected directly to the ionization chamber, such an amplifier "sees" only one part in 10^4 of the total charge, which is not large enough compared to the amplifier noise. This is an obvious case of impedance mismatch between the detector and the amplifier. A much better charge sharing (and a higher signal-to-noise ratio) can be achieved if the detector and amplifier are matched by using a transformer.

The detector–amplifier circuit configuration we have developed, with provisions for detector biasing and for charge calibration, is shown in fig. 6. An equivalent circuit showing the essential elements for noise analysis is given in fig. 7. We present here the noise results required to calculate the contribution of the electronic noise to the fluctuations determining the energy resolution of the calorimeter. A discussion of various noise sources in charge amplifiers is presented in ref. 11, and the effects of pulse shaping on noise are described in ref. 12. A more detailed analysis of the case presented here will be given in a separate report[13]).

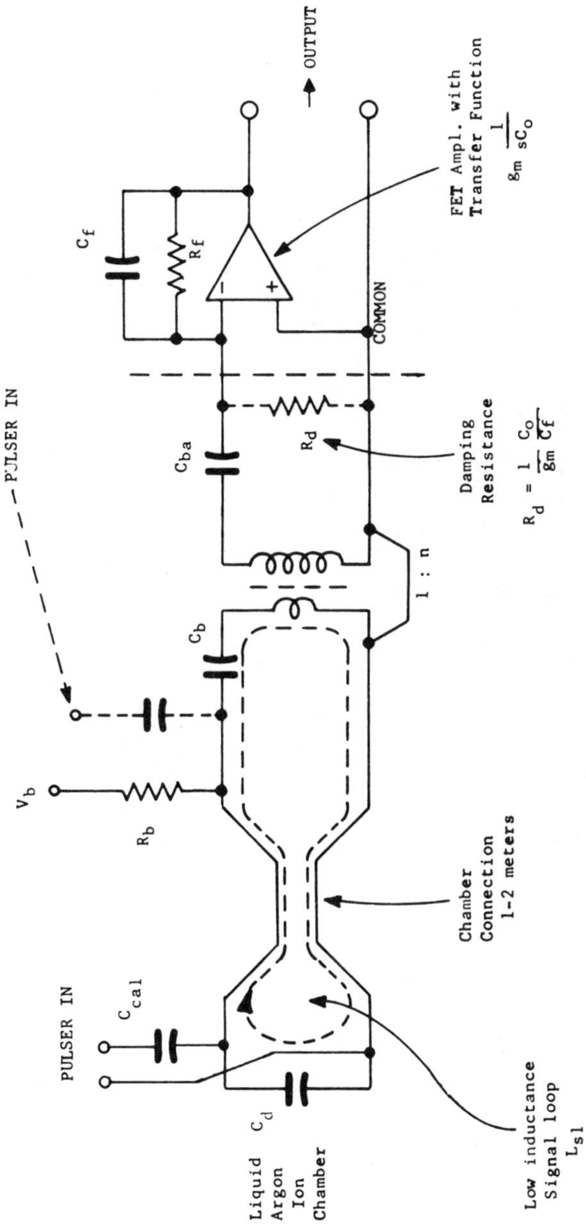

Fig. 6. Detector–preamplifier circuit configuration. $C_{cal} = 10$ pF, $C_{ba} = 0.1\ \mu$F.

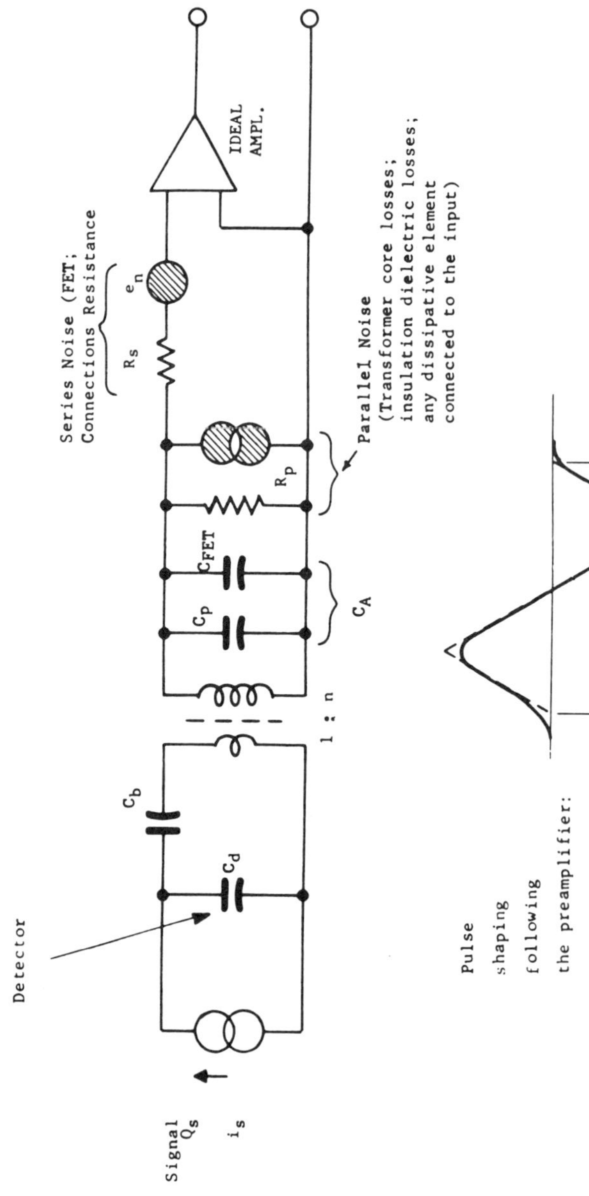

Fig. 7. Equivalent circuit of detector and preamplifier for noise analysis.

The principal component of electronic noise which cannot be avoided is due to the fluctuations related to the gain mechanism of the amplifying element. In the case of the field-effect transistors, this is the thermal noise of the conducting channel. Other noise source can be present, as series resistance of the detector–amplifier connections and dissipative elements connected to the input of the amplifier (transformer ferrite core losses, insulation dielectric losses and biasing resistors). Noise due to these sources can be made negligible by proper design. We express noise in terms of a signal producing the same output as noise, that is as an "equivalent noise charge" (*ENC*). We calculate the noise for the circuit in fig. 7 by the methods described in refs. 11 and 12, and we obtain,

$$\overline{ENC^2} = 8\, e_n^2 \frac{1}{\lambda}\left[\frac{C_d}{n} + nC_A\left(1 + \frac{C_d}{C_b}\right)\right]^2 \quad \text{series noise,}$$

$$+ \frac{2}{3} kT \frac{\lambda}{R_p} n^2 \left(1 + \frac{C_d}{C_b}\right)^2 \quad \text{parallel noise.} \quad (1)$$

Detector, amplifier, and circuit parameters are defined as follows:

C_d = detector capacitance;
C_b = detector decoupling capacitance:
C_A = amplifier input capacitance = $C_{FET} + C_p$ = field-effect transistor input capacitance plus stray capacitances;
n = transformation ratio;
e_n = rms spot noise [$V/Hz^{\frac{1}{2}}$] measured for input field-effect transistor.

Expressed differently, the mean-square series noise voltage per unit bandwidth is given by the Nyquist formula,

$$e_n^2 = 4\, kTR_s\,, \quad (2)$$

where R_s is equivalent series noise resistance of the field-effect transistor, k is Boltzman constant and T is temperature. The relation (1) is valid for triangular bipolar shaping, or approximately for bipolar Gaussian shaping, where the resolving time λ is defined as in fig. 7. Only the numerical coefficients of the two terms change with pulse shaping[12]).

R_p = total loss resistance in parallel with the input of the amplifier at frequency $1/\lambda$.

There is an optimum transformer ratio which for the general case with parallel noise is given by

$$n_{\text{opt}} = \left[\frac{C_d^2}{C_A^2 + \dfrac{1}{48} \dfrac{\lambda^2}{R_p R_s}} \frac{1}{\left(1 + \dfrac{C_d}{C_b}\right)^2} \right]^{\frac{1}{4}}.$$

Parallel noise is not inherent to the amplification mechanism and it should be made negligible (by reduction of all losses, i.e., $R_p \to \infty$). The optimum transformer ratio in the case of negligible parallel noise is then:

$$n_{\text{opt}} = \left[\frac{C_d}{C_A[1 + (C_d/C_b)]} \right]^{\frac{1}{2}}. \tag{3}$$

The condition $(\overline{ENC^2})_p \leq \frac{1}{3}(\overline{ENC^2})_s$ (that is for less than 5% increase in ENC due to parallel noise) requires,

$$R_p \geq \frac{1}{16} \frac{\lambda^2}{R_s C_A^2} \left[1 + \left(\frac{n_{\text{opt}}}{n} \right)^2 \right]^{-2}. \tag{4}$$

The minimum equivalent noise charge due to the series noise is then,

$$ENC_{s\,\text{opt}} = 4\sqrt{2}\, e_n \left[C_d C_A \left(1 + \frac{C_d}{C_b} \right) \right]^{\frac{1}{2}} \frac{1}{\lambda^{\frac{1}{2}}}. \tag{5}$$

The noise for non-optimal transformer ratios is

$$\frac{ENC_s}{ENC_{\text{opt}}} = \frac{1}{2} \left(\frac{n_{\text{opt}}}{n} + \frac{n}{n_{\text{opt}}} \right). \tag{6}$$

In order to achieve the equivalent noise charge given by the relation (5), the charge [$1/(2n)$ of the total charge in the optimum case!] must be transferred from the detector to the amplifier within a time short compared to the resolving time as determined by the pulse shaping.

This time is limited by the inductance L_{sl} of the whole connection loop shown in fig. 6. This loop includes the chamber-plate connections, the transmission line, the decoupling capacitance and the leakage inductance of the transformer. It can be shown that in order to realize a resolving time λ the inductance of the connection loop must satisfy the following condition:

$$L_{sl} \leq \frac{\lambda^2}{64\,\pi^2\,C_A}\,\frac{1+(n/n_{opt})^2}{n^2}$$

$$= \frac{\lambda^2}{64\,\pi^2}\,\frac{1+(C_d/C_b)}{C_d}\left[\left(\frac{n_{opt}}{n}\right)^2+1\right]. \tag{7}$$

We give an example with typical parameter values,

Ion-chamber capacitance:	C_d	$= 5\times 10^4$ pF
Decoupling capacitance:	C_b	$= 10^5$ pF
FET input cap. + ampl. and	C_{FET}	$= 18$ pF
transformer stray cap.:	C_A	$= 30$ pF
FET noise (for Texas Instr.	R_s	$= 60\ \Omega$
devices SFB 8558, BF 817):	e_n	$= 1$ nV/Hz$^{1/2}$
Resolving time:	λ	$= 0.6\ \mu$s

From eq. (3), the optimum transformer ratio $n_{opt} = 33.3$. From eq. (5), the equivalent noise charge is $ENC_{s\,opt} = 6.9\times 10^4$ rms electrons. For the calorimeter configuration described in table 1, the energy deposit necessary to give a signal corresponding to one electron is 264 eV for pure argon. The rms noise in terms of energy corresponds to 18.2 MeV. The noise contribution to the linewidth, or the "electronic resolution", is thus fwhm$\approx$43 MeV. If no transformer were used ($n = 1$), the noise would be increased according to eq. (6) by 16.7 times (fwhm would be $\approx$0.7 GeV)! Noise measurements confirm these calculations. (Contamination of LA increases the value of energy required to give a signal corresponding to one electron, and of course, the equivalent noise expressed in terms of energy.)

We calculate now the conditions for the parallel loss resistance and for the inductance of connections. From eq. (4), $R_p \geq 10^5\ \Omega$, and from eq. (7), $L_{sl} \leq 34$ nH.

According to eq. (6), the noise increases little (by $\approx 4\%$) if n/n_{opt} is decreased to 0.75, that is $n = 25$. A lower value of n offers several advantages. The conditions for R_p and L_{sl} are more easily satisfied; with $n/n_{opt} = 0.75$, $R_p \geq 5.4 \times 10^4\ \Omega$ and $L_{sl} \leq 67$ nH. In addition, damping of the two resonances of the transformer circuit is increased.

Dominant losses in the input circuit are due to the ferrite core of the transformer. The above criterion for negligible noise can be satisfied easily with Ferroxcube 3D3 or 4C4 cores. Many other types of ferrites are not satisfactory and may result in higher noise. In the example given above, the transformer was wound with 4 times 2 turns in parallel (to reduce leakage inductance) in the primary and with 50 turns in the secondary (giving an inductance of about 1 mH).

The charge preamplifier used in all tests with both small and large test chambers is shown in fig. 8. The capacitance in feedback in conjunction with the 90° phase shift in the amplifier is used for damping of the transformer circuit. (This is a new technique described in some detail in ref. 12.) An input of 500–1000 Ω is required for damping. Pulse shaping is shown in fig. 9. Bipolar shaping was chosen as the most suitable for the signal conditions in the experiments envisaged for large calorimeters, that is, measurements of rare large signals in the presence of very high rates of low-energy background events.

The fundamental lower limit of noise is given by eq. (5). To achieve this limit, the capacitance matching of the detector and the amplifier is essential. An alternative to the transformer matching is, at least in principle, the connection of a number of amplifiers in parallel. With respect to the field-effect transistor noise, equal equivalent noise charge is obtained by n^2 field-effect transistors connected in parallel as with a transformer with a ratio n (for $n = 25$, 625 transistors in parallel would be required!). Equivalent to paralleling amplifiers is subdivision of the ionization chambers into n^2 sections with one amplifier for each section. Carrying through the analysis of transformer matching

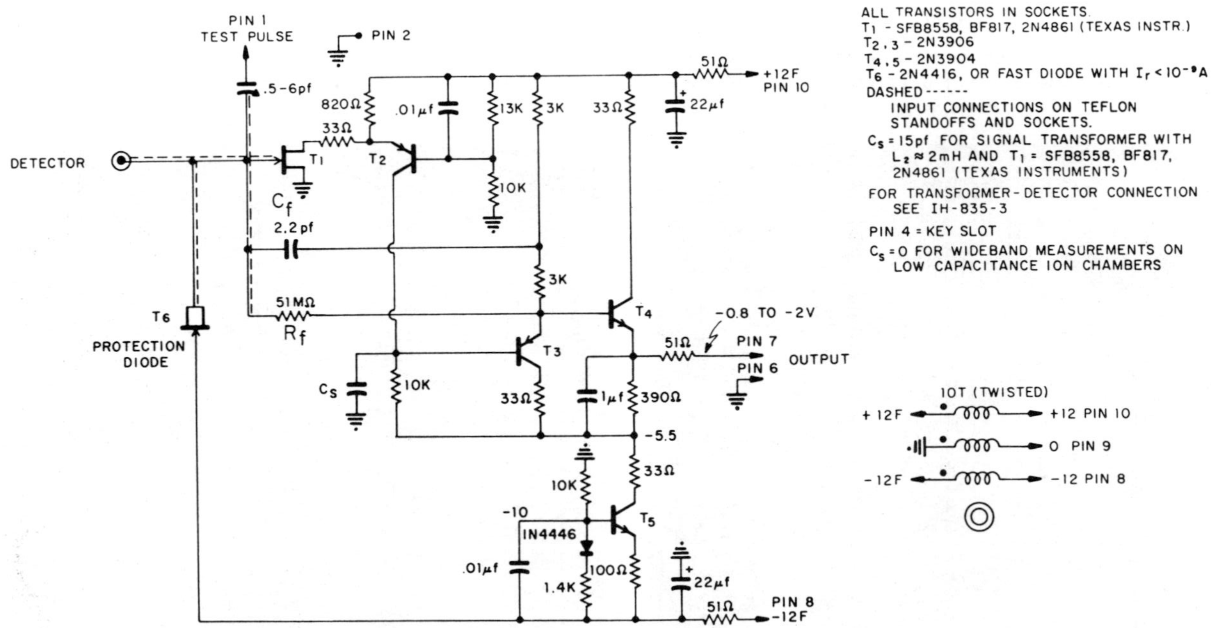

Fig. 8. Charge preamplifier used with both small (fig. 2) and large (fig. 10) liquid-argon ionization chambers. For use with low-capacitance ionization chambers as in fig. 2, R_f should be 1000 MΩ.

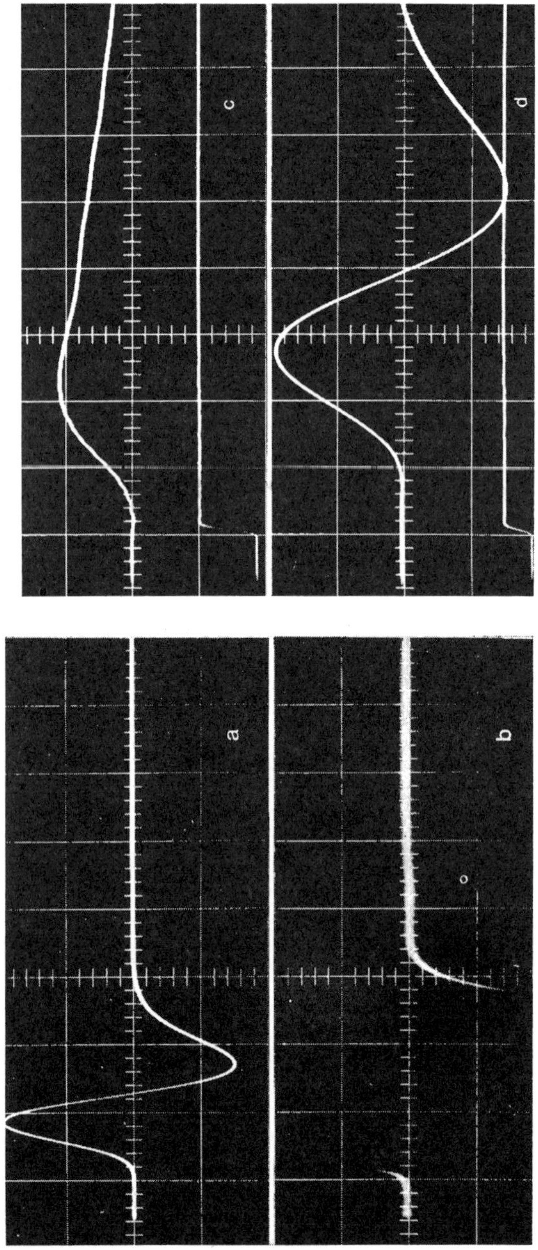

Fig. 9. System waveforms with bipolar shaping: (a) shaping amplifier output, horizontal scale 0.5 μs/div.; (b) same as (a), magnified so that electronic noise is visible; (c) preamplifier response to step calibration pulse, horizontal scale 0.1 μs/div.; (d) response of a system with a fast shaping amplifier, horizontal scale 0.1 μs/div.

and detector subdivision, the following conclusion is reached:

The equivalent noise charge of the detector is proportional to the square root of its total capacitance and it is independent of the number of separate sections, provided that each section is matched to its amplifier, and that the signals from all sections are linearly summed.

The amplifier determines the noise by its parameter $e_n^2 C_A$. Stray capacitances should be kept low so that $C_A \rightarrow C_{FET}$. The product $e_n^2 C_{FET} \propto R_s C_{FET}$ is basically determined by the geometry (carrier transit time) of the field-effect transistor. The values for R_s and C_{FET} given in the example represent state-of-the-art. The ratio C_d/C_b should ideally be $\ll 1$. The cost of decoupling capacitors determines this ratio in practice as 1/3 to 1/2. The equivalent noise charge is inversely proportional to the square root of the resolving time. The upper limit for the resolving time is determined by the event rates and pileup considerations. Assuming the shortest resolving time allowed by the charge collection time ($\lambda \approx 0.6$ μs), the equivalent noise charge can be expressed as a function of detector size (capacitance) for the purposes of approximate noise estimation as

$$ENC \approx 3 \times 10^5 \, C_d^{\frac{1}{2}} \, [\mu\text{F}] \quad \text{rms electrons.} \qquad (8)$$

The electronic resolution for the ratio of liquid argon to steel given in table 1 (264 eV/electron) expressed in terms of deposited energy is

$$\text{fwhm} \approx 0.18 \, C_d^{\frac{1}{2}} \, [\mu\text{F}] \quad \text{GeV}.$$

4. Construction and argon handling

In order to test these ideas at full scale, we have constructed a chamber which is large enough to contain fully an electromagnetic shower, shown in fig. 10. It has 200 one-tenth radiation length cells of the sort described in table 1. The plates are in the form of hexagons with a diameter of 230 mm. The plates are supported on four fiberglass rods, and the gaps are maintained by fiberglass spacers at the two corners of the hexagon without rods.

Fig. 10. Large test chamber with 200 steel plates, 1.5 mm thick, with 2.0 mm gaps.

The plates are connected to a readout line made of Teflon impregnated printed-circuit board by a short wire soldered to each plate. The inductance of these connections is acceptable because there is a large number of them in parallel, but the leads for the whole section of plates must be treated carefully to keep the inductance satisfactorily low. The cable carrying the signal to the amplifier is made of low impedance strip cable normally intended for use in spark-chamber pulse distribution. A multilead two-sided printed-circuit connector is soldered to the cable and connected to the readout bus.

The readout is arranged in groups: 25 cells; 25 cells; 100 cells; and 50 cells, or: $2\frac{1}{2}$, $2\frac{1}{2}$, 10, and 5 radiation lengths. The two small sections use 8 Ω cable, while the large sections use 4 Ω cable. There are, in fact, two readouts for each group in the manner shown in fig. 11. We have effectively two independent inter-

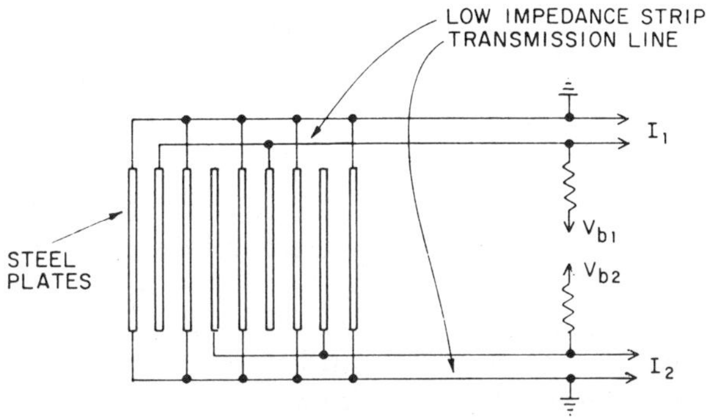

Fig. 11. Electrode connections providing for two interleaved chambers (e.g., for measurement of sampling fluctuations).

leaved ion chambers which may be used to obtain two measurements of the energy of a single particle, and to perform other differential measurements to be described in the next section.

The cables, about 1.5 m long, passed out of the vessel by means of a feedthrough which consisted of an epoxy resin forced around and between the strip cables. It was found that the cables could be placed in immediate contact without cross talk problems.

The transformers and preamplifiers were placed in an aluminum box mounted just above the cryostat. This system forms a closed electrical shield for the chambers and low-level electronics. Dry nitrogen was introduced into the box to avoid corona in air. The connections at the ion-chamber end are all submersed in liquid argon.

The bias voltage was tested up to 5 kV, though most runs were made at around 2 kV. We had found in the test chamber that steel electrodes cut in a shear would break down at about 7 kV. If the edges are rounded the breakdown voltage is above 10 kV for a 2 mm gap.

The chamber was suspended in an available 300 l cryostat. Argon was introduced as a gas and liquified by means of a heat exchanger made of about 20 m of 12 mm copper tubing fed with liquid nitrogen. After filling with liquid argon to a level just below the cooling coils, the argon system was sealed and the argon vapor pressure was maintained at (0.7 ± 0.04) atm gauge by regulating the nitrogen flow with a simple on–off pressure switch.

The argon was obtained from a local welding supply house in 130 l containers, in liquid form, though we used only the vented gas. This was found to be quite pure, as judged by measurements of the electron collection in the test chamber described earlier. We avoided using the last few percent in each container, fearing that it might be enriched in higher boiling point impurities, a belief which was supported by some later measurements.

Other measurements have been made using welding argon in standard gas cylinders. A variation in quality is observed, with the oxygen content varying from 5 to (occasionally) 50 ppm. This is probably a function of the past history of the cylinders, since the supplier does not take great care to flush out the cylinders when refilling them with this grade of argon. This oxygen can be removed by a purifier we made, and is then satisfactory for use. The purifier consists of about 15 kg of a dispersed copper catalyst (BASF) which has been reduced with hydrogen at 120 °C. This material will reduce the oxygen content below 1 ppm while operating at room temperature. Another possibility is to add a trace of hydrogen and use a DEOXO catalyst to remove oxygen.

A reasonable tolerance on the amount of oxygen is 2–5 ppm, which may easily be measured with a Hirsch electrolytic oxygen analyzer, available from a number of manufacturers. In general, we have found that impurities other than oxygen are harmless because they are not appreciably electronegative (nitrogen) or have negligible vapor pressures in liquid argon (water). We have found one other impurity to be harmful; some variety of freon contained in a foamed plastic.

The particles enter this assembly through walls of about 12 mm of aluminum, and pass through about 100 mm of liquid argon before reaching the first plate. Although electrons may radiate some of their energy before reaching the chamber, by adding additional radiator at different distances upstream, we have verified that essentially all of the energy is captured.

5. Results

When a 7 GeV/c negative beam was introduced into the calorimeter, the charge spectrum seen in fig. 12(a)

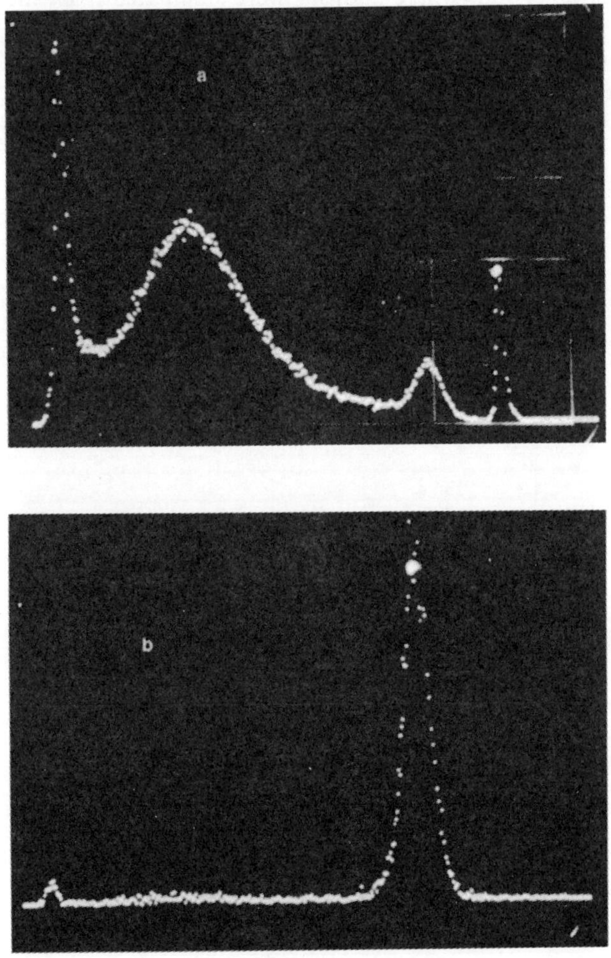

Fig. 12. Charge spectrum with large test chamber for 7 GeV/c negative beam. (a) peaks from left to right: muons, π's, electrons, calibration pulser; (b) electrons enhanced with a Cherenkov detector.

was observed on a pulse-height analyzer. One sees peaks corresponding to muons (and some hadrons) passing straight through the ion chamber, pions depositing about half their energy, electrons depositing all their energy, and the last narrow peak which is a calibration of 2 pC.

The width of the muon peak was measured on another run (where hadrons were eliminated from the beam) and found to be 27% (fwhm). The width of a calibration peak at the same average charge was 19%. Thus we can observe the natural Landau distribution of the energy loss of a minimum-ionizing particle without too much distortion by the amplifier noise. The muon peak was used to measure the charge collection as a function of high voltage, with the results shown in fig. 13. The plateau is seen to occur at a rather low voltage, but at a value lower than the ideal value. This was verified by measurements in the small chamber on the samples of argon removed from the

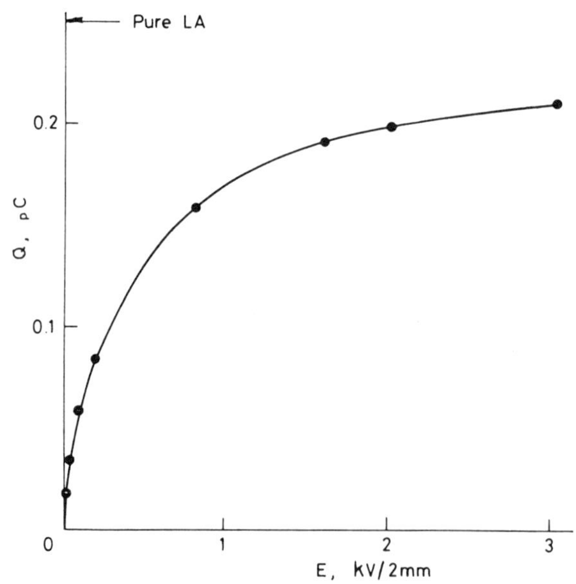

Fig. 13. Charge as a function of voltage in an experimental run with large test chamber.

large detector. It was found to be due to the freon impurity mentioned earlier.

The distribution of energy deposited by pions (and the small fraction of other negative hadrons) is characteristic of that expected in the present absorbing volume, as determined experimentally by Hofstadter and Hughes[5]), for example. This distribution in itself is not very sensitive for answering questions of interest, but we shall show below that the correlations with differential measurements from the interleaved ion chambers can give most valuable results.

The response for electrons can be studied more carefully with a trigger using a threshold Cherenkov counter, as shown in fig. 12(b). (Some muon and pion Cherenkov accidental triggers are still visible.) The electron peak width is about 7.6% fwhm, but it depends on the amount of material in the beam line[18]). The calorimeter was placed at the end of a long test beam, and there was a considerable amount of equipment in the beam: seven spark chambers, a number of scintillation counters, and 30 m of air, an estimated total of about 0.15 radiation lengths, most of it so distant that the radiation can miss the calorimeter. This radiative straggling is apparent in spectra taken with normal conditions [fig. 14(a)], and with one extra set of upstream chambers in the beam [fig. 14(b)].

The broadening of the electron energy distribution due to straggling was calculated as follows. From measurements with added radiators at different positions we concluded that about one third of the radiated energy was captured in the detector, thus not representing an apparent energy loss. Accordingly, we use an effective radiator thickness of 0.10 radiation lengths. The electron energy distribution was calculated from the theory of Eyges[14]). To correspond to our analysis procedure used to determine the peak width, we consider the distribution for energy losses only less than 7%. Average electron energy is then 98% of the initial energy, with a rms variance of 1.66%. We assign an error of 25% to the last number. The energy spread in the beam itself should be less than one percent and can be neglected. A quadratic unfolding of this

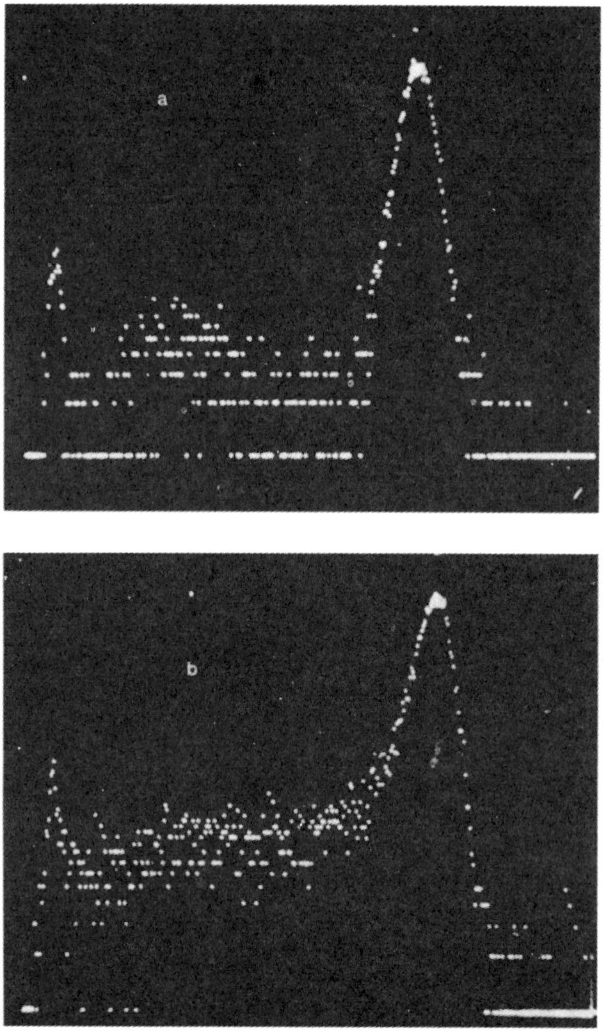

Fig. 14. Electron spectra displayed on a logarithmic scale covering three decades. The effect of radiative straggling can be seen by comparing (a), taken with less material in the beam, with (b), where another piece of equipment was in the beam some distance in front of the detector. Signals from muons and pions are also visible.

straggling from the observed width results in a value for the intrinsic resolution of $(2.8 \pm 0.3)\%$ rms, or a fwhm of $(6.5 \pm 0.7)\%$. A reliable lower limit obtained by another method is described below. Data for electrons of 2.5 $[\sigma_{\text{sum}} = (5.2 \pm 0.2)\%]$ and 11 $[\sigma_{\text{sum}} = (2.6 \pm 0.3)\%]$ GeV displayed behavior consistent with a resolution proportional to $E^{-\frac{1}{2}}$.

Another approach to the resolution was a direct measurement of the sampling fluctuation. To accomplish this we make use of the division of the ion chamber into independent interleaved sections. These were used with equal high voltage so that in the absence of fluctuations the output signals should be equal. The signals from the interleaved chambers, summed over longitudinal subdivisions, were input to a difference circuit. The gain and timing were accurately checked with calibration pulses. A display of the output of the difference circuit is shown in fig. 15. This is the superposition of many oscilloscope traces. One sees two features: a continuous band representing the electronic noise, and a larger range of signals in time with the incident particle (a 7 GeV electron). The gain is such that the summed output would be many times off scale. The shape of the distribution due to the particle is that expected from the difference of two

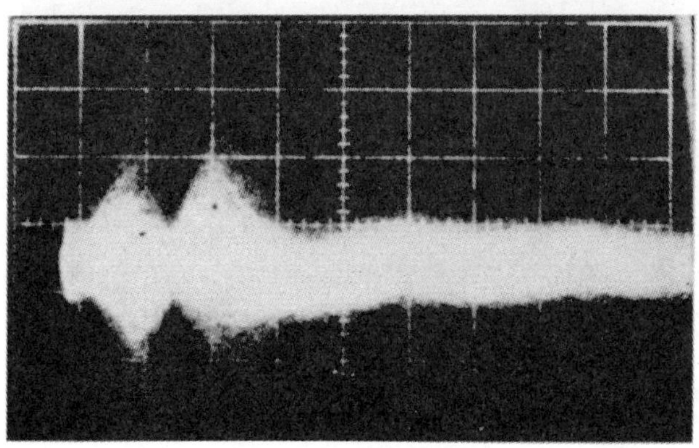

Fig. 15. Difference signals from two interleaved chambers for 7 GeV electrons (see text).

bipolar signals with different amplitudes. The first point to note is that the *average* of the signals due to the particles is very near to zero. This shows that the average of the signals in the two channels was the same (to $\leq 1/2\%$). One should appreciate that no adjustments were performed during the experiment to make these gains equal. Therefore one could expect to have the same reproducibility between different elements of a large multielement array of detectors. This, we believe, is one of the most valuable properties of these detectors.

The width of this distribution is a direct measurement of the sampling fluctuations. This is displayed most clearly in fig. 16, showing a three-decade logarithmic display of the distribution of difference signals from the interleaved chambers with 7 GeV electrons. The parabolic shape characteristic of a Gaussian distribution in this display is evident over the whole range of fluctuations. The rms width of the difference distribution is normalized to the (nearly constant) value for the sum, with the result:

$$\sigma_{\text{diff}} = \left(\frac{I_2 - I_1}{I_2 + I_1}\right)_{\text{rms}} = (2.8 \pm 0.1)\%.$$

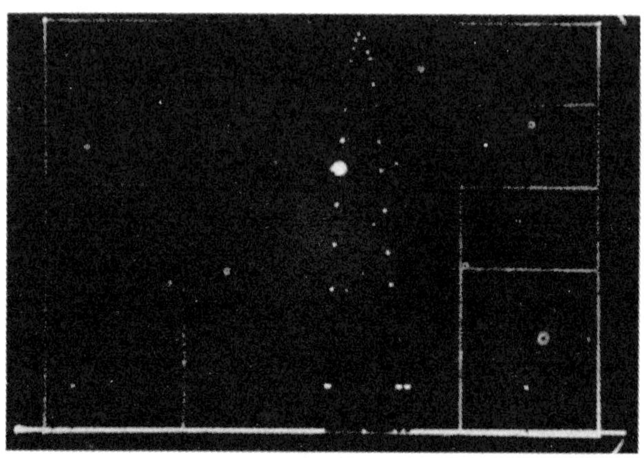

Fig. 16. Distribution of difference signals from two interleaved chambers with 7 GeV electrons. The three-decade logarithmic display shows a Gaussian distribution.

Subtracting the electronic noise quadratically would lead to a value of 2.6% for the part of σ_D due to shower fluctuations, 93% of the total.

$$\sigma_{\mathrm{sum}} = \frac{(I_1+I_2-\langle I_1+I_2\rangle)_{\mathrm{rms}}}{I_1+I_2} = \sigma_{\mathrm{diff}} = (2.8\pm 0.1)\%.$$

This is, in fact, compatible with the value for the resolution measured directly. We conclude that the sampling fluctuation dominates the resolution, even for thin plates ($\sim$0.1 radiation length).

An estimate of the sampling fluctuations can be obtained in a very simple way, which shows that the resolution observed is approximately that to be expected.

The total number of cells traversed is equal to the electron energy divided by the energy loss per cell, which in our case is one tenth of the critical energy[15]). To obtain an upper limit on the expected fluctuation from this source, we take half of this number of samples, since electrons are created in pairs, and neglect the correlation between samples on the same track, and assume that all electrons stop in the plates, so that each electron crossing the liquid argon deposits the same energy. Then the number of samples is:

$$n_e = 10\,\frac{E_e}{E_{\mathrm{crit}}}\,\frac{1}{2},$$

and the resolution due to sampling fluctuations is:

$$\sigma_{E_e}/E_e = 1/\sqrt{n_e}\,.$$

or for $E_e = 7$ GeV,

$$\sigma_{E_e}/E_e = 2.35\%,$$

just a bit smaller than that observed.

It is interesting to use this technique to study the sampling fluctuations in hadron showers, a subject on which there is little firm information. To be sure, we see only half the energy for a typical hadron due to the relatively small size of our detector. Thus our results will be characteristic of the central part of the shower rather than the extremes, a part which contains a disproportionate fraction of pi-zero energy compared to low-energy neutron stars. This might be expected to give somewhat smaller sampling fluctuations.

Our approach was to eliminate electrons from the beam by introducing a distant lead absorber, and to place a single-channel analyzer on the summed signals. The difference distribution was then recorded as correlated with different parts of the hadron spectrum. The results are summarized in fig. 17, expressed as the σ_D, i.e., the equivalent sampling-fluctuation contribution to the resolution. The electron point is also shown. We have subtracted electronic noise, a small correction. We note that hadrons which manage to deposit most

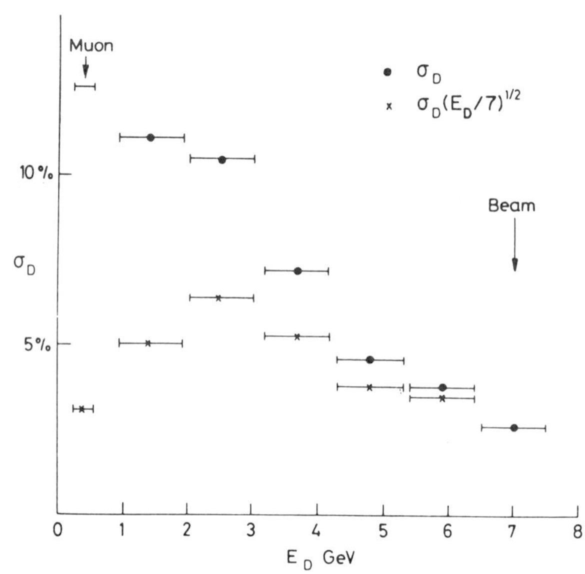

Fig. 17. Sampling fluctuations determined from the difference signal from interleaved chambers, as a function of total energy deposited. The point at the beam energy is due to electrons, the lowest-energy point is due to muons, and the intermediate points are due to hadrons. Also shown are the points scaled by the square root of the energy. σ_D is in percent of E_D.

of their energy in this absorber display fluctuations similar to the electrons. This is not surprising since they must be mainly electromagnetic in character. On the other hand, the typical hadrons display much larger sampling fluctuations, larger than that expected simply by an inverse-square-root dependence on the sum signal, as shown by the set of points where the width has been scaled by a factor proportional to the square root of the energy deposited. This is probably due to the presence of stars in these events, with a number of tracks of short range, difficult to sample. For energy releases near that of a straight-through particle, the average of the difference signal drops abruptly to that given by the Landau distribution for ionization.

We have also studied the difference distribution for non-equal high voltages on the two interleaved chambers:

$$\text{HV on } \mathrm{I}_1 = 3.0\ \text{kV}$$

$$\text{HV on } \mathrm{I}_2 = 0.2\ \text{kV}.$$

This leads to unequal signals with incident electrons, which would obscure the effects which interest us. Accordingly, we make an analog computation of

$$D = kI_1 - I_2,$$

and adjust k so that the average of D is zero for incident electrons: $k = 0.29$. It is then found to average zero for straight-through particles also.

The interest in this procedure lies in the possibility that it may be used to measure the relative contributions of electromagnetic and hadronic showers in an individual interaction, thus allowing a possible correction for saturation and binding-energy effects. The expected effect is due to saturation of the low-voltage chamber on densely ionizing hadrons. The measurements were made with a single-channel analyzer selection of a given region of the hadron spectrum. As expected, for energy releases near the maximum possible, the average difference was still near zero, as for electrons. For events near the peak of the hadron spectrum, the distribution of the quantity D develops an asymmetry in the expected direction, as shown in fig. 18. Most events

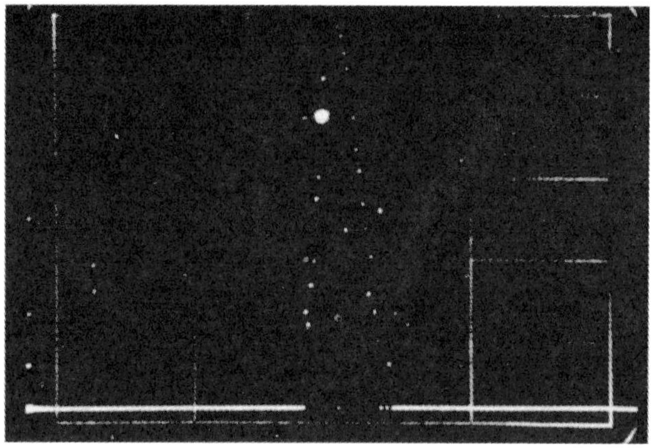

Fig. 18. Distribution of difference signals for hadrons with different voltages on the two interleaved chambers.

show D/S equal to zero within a few per cent, but there is an appreciable fraction with a shift of $D/S \sim 0.1$. Because of somewhat smaller collected charge at the time of this test, due to argon contamination and the effect of the factor k, electronic noise makes a substantial contribution to the width of the D distribution. These results seem to encourage hope that at higher energies particularly, this technique can be used to make event-by-event corrections for the saturation effects themselves, and the nuclear binding loss which is closely correlated.

A measurement of the timing jitter inherent in the detector was made with 4 GeV incident electrons. A zero-crossing discriminator was used to stop a time-to-pulseheight converter which was started by a coincidence of several small scintillation counters in the beam. The observed time distribution has a fwhm of 14 ns. It is interesting to note that this value was exactly the time spread one would predict from the electronic noise level observed at that time, together with the slope of the signal as it passes through the zero-crossing. Thus, we would expect the time resolution to vary inversely with energy. Better timing could be achieved by a shorter shaping time and, of course, a larger signal obtained with uncontaminated LA.

The energy distribution of electrons and hadrons in the different longitudinal subdivisions of the detector was studied to determine the usefulness for hadron–electron discrimination. If one supposes that the momentum of the particle is known from a measurement in a magnetic field, the particle is identified as an electron if the observed energy is consistent with this value, within an error which we took as 14% full width in our measurement, so that electrons were accepted with high efficiency. To examine the hadron spectrum, electrons were eliminated by a 50 mm lead radiator, inserted before the last bending magnet, 60 m upstream of the detector. Then the fraction of hadrons satisfying the electron condition was 0.0005.

One might expect that the twenty radiation-length detector would be longer than optimum for this purpose. When we examined the distribution of energy deposited by electrons in the first fifteen radiation lengths, we found that the width is almost the same as in the full detector (8 % wider), but the peak is displaced downward by 3 %. Thus, although the hadron energy deposited in the smaller detector is indeed less, the hadron rejection does not turn out to be better.

If the detector is to be used in circumstances where particle momentum is not separately measured, the hadron rejection arises from two factors. First, criteria may be set up on the longitudinal distribution of energy within the detector, whence electrons are identified by requiring a large fraction of the observed energy to appear in the front part of the detector. Second, if the experimental situation is such that the number of hadrons falls rapidly with increasing energy, the fact that the average hadron energy must be considerably higher to produce a given signal in the detector gives rise to a substantial rejection factor. This factor can be evaluated, given a hadron spectrum, from the distribution of signals arising from hadrons, which has been shown in fig. 12(a).

Here we give some properties of the electron identification by the longitudinal distribution of energy. In order to obtain a good efficiency for electrons, the first subdivision of the detector must be chosen deep enough to produce a signal distribution which peaks at a sufficiently high value so that the fraction of cases with small signals is low. Our measurements are in accord with shower theory[16]). We found that the first 25 plates gave a distribution with too many small signals, but that the first 50 plates, or five radiation lengths, were satisfactory. Then we may inspect the distribution with hadrons incident to evaluate the hadron rejection. In doing this, it is interesting to search for a correlation between the two hadron rejecting factors. That is, the cases which give large signals from hadrons in the first five radiation lengths which may be interpreted as electrons are just those in which the hadron signal in the whole detector was anomalously large.

Indeed, such a correlation was found. When a hadron spectrum is taken for the first 50 plates with no condition on the total energy loss, the result shows about 10% of the events which overlap the electron distribution. If a requirement is set that the total energy release in 200 plates is more than 75% of that expected for electrons, about 25% of the hadrons overlap the electron distribution in the first 50 plates.

6. Future developments and applications

The reader might be interested in the applications we have foreseen for these detectors. The most important open question is the possible energy resolution which can be obtained with incident hadrons. To provide an answer, we are constructing a calorimeter of the required size, one meter diameter and two meters long. Another extension of the technique is the use of high-Z plates in the first part of the calorimeter. This has two beneficial effects. With an average critical energy about half that of steel, the hadron rejection should be a factor of two better. Also, the electron track length is twice as great, leading to a reduction in the sampling fluctuations by the square root of two.

Another development underway is to introduce subdivision into narrower electrodes (strips) near the shower maximum, to provide space resolution with an estimated precision of a few mm.

One of the desirable properties of a detector which measures energy directly is that one has the possibility of measuring large signals in the presence of a high rate of smaller signals. In this way, one may overcome the problem of very high counting rates in an experiment where the interaction rate must be very high if one is to obtain a reasonable rate of extremely rare events. If the rare events are characterized by unusually large amounts of energy, one has to worry about small signals in so far as they broaden the resolution by pileup. We have chosen to use bipolar shaping so that the pileup does not produce a net shift in the measurement of a large signal. To illustrate these points, suppose we have a high rate of "background" events with energy E_b, at a rate of n per second. We desire to measure $E_m \gg E_b$. The shaping $w(t)$ of the signals is assumed to be a triangular bipolar form of total length λ. Then the average value measured for E_m remains E_m for all n, but with a pileup variance of

$$\overline{\Delta E^2} = nE_b^2 \int_{-\infty}^{\infty} w(t)^2 \, dt,$$

or

$$\Delta E_{rms} = E_b(n\lambda/3)^{\frac{1}{2}} .$$

Thus, for cases of plausible interest, one might allow $n\lambda \simeq 10$–100, illustrating the possibility of working at quite high rates.

An illustration of a large-scale application which relies on a number of the properties of liquid-argon calorimeters is given in a design study of an Impactometer, a detector designed to cover the full solid angle around the interaction point of high-energy beams, performing analog computations of quantities of physical interest using the direct measurements of energies and angles in the calorimeter[17]).

We wish to thank Drs A. M. Thorndike and R. R. Rau for their encouragement. We are indebted to Dr D. Berley and A. P. Schlafke, A. R. Blummert, H. Thorwarth, R. J. Gibbs, and the target operation group for help with the cryogenics and mechanical support. We also wish to thank Dr J. Fischer and F. C. Merritt,

D. Stephani, C. Z. Nawrocki, and R. W. Dillingham of the BNL Instrumentation Division for their valuable help in all phases of this work. Drs S. Ozaki and E. D. Platner kindly provided the strip cables, and Dr F. Turkot lent the oxygen analyzer.

References

1) V. S. Murzin, Principles and application of the ionization chamber calorimeter, in: *Progress in elementary particle and cosmic ray physics* (J. G. Wilson and S. A. Wouthuysen, ed., North-Holland Publishing Co., Amsterdam, 1967). Also, see W. V. Jones Phys, Rev. **D 1** (1970) 2201.

2) Ibid., p. 262, see also ref. 4.

3) W. V. Jones, Proc. 11th Int. Conf. on *Cosmic rays* (Budapest, 1969), also following reference.

4) T. A. Gabriel and K. C. Chandler, Particle Accelerators **5** (1973) 161.

5) R. Hofstadter, E. B. Hughes, W. L. Lakin and I. Sick, Nature **221** (1969) 228; and E. B. Hughes et al., Nucl. Instr. and Meth. **75** (1969) 130.

6) J. Engles, W. Flauger, B. Gibbard, F. Mönnig, K. Runge and H. Schopper, Nucl. Instr. and Meth. **106** (1973) 189.

7) The values are from the table in the Review of Particle Properties, Rev. Mod. Phys. **45** No. 2, part II (1973).

8) W. P. Jesse and J. Sadaukis, Phys. Rev. **107** (1957) 766.

9) N. Davidson and A. E. Larsh, Phys. Rev. **77** (1950) 706.

10) L. S. Miller, S. Howe and W. E. Spear, Phys. Rev. **166** (1968) 871.

11) V. Radeka, Proc. Intern. Symp. on *Nuclear electronics*, Vol. 1 (Versailles, France, 1968) p. 46-1.

12) V. Radeka, Signal, noise and resolution in position-sensitive detectors, BNL 18377, IEEE Trans. Nucl. Sci. NS-**21**, no. 1 (Feb. 1974) 51.

13) V. Radeka, Signal processing for liquid argon ionization chambers, (to be published).

14) L. Eyges, Phys. Rev. **76** (1949) 264.

15) B. Rossi, *High energy particles* (Prentice Hall, New York, 1952) p. 224.

16) H. Messel and D. Crawford, *Electron-photon shower distribution function* (Pergamon Press, Oxford, 1970).

17) W. J. Willis, An instrument for the measurement of large total transverse momentum; the impactometer. Appeared in ISABELLE–Physics Prospects, BNL 17522, 1972, p. 207. CRISP #72-15.

[18]) One may ask for evidence of shower transition effects between the iron and argon layers. These would give a ratio of charge deposited by electrons and by muons different from that predicted on the basis of energy loss. Indeed, we find that this ratio is $(80 \pm 5)\%$ of that predicted from energy loss, though this measurement may be affected by systematic errors, such as the value of the beam energy. We are investigating different plate materials which should allow us to arrive at a better understanding of this point.

Reprinted, with permission, from *Portugal Phys.*, Vol. 12, fasc. 1-2, p. 9 (1981).

FUNDAMENTAL PROPERTIES OF LIQUID ARGON, KRYPTON AND XENON AS RADIATION DETECTOR MEDIA

by T. Doke

ABSTRACT — The fundamental properties of liquid argon, krypton and xenon required from a point of view of radiation detector media, such as W-values, Fano factors, electron drift velocities, etc., have been measured during the past decade. These results are summarized and the author's considerations for its physical understanding are presented. Also, the possibilities of application of these liquids to nuclear radiation detectors are discussed.

1 — INTRODUCTION

Since Alvarez suggested the possibility of the use of liquefied rare gases, such as liquid argon or liquid xenon, as detector media of counters to be used in experiments of elementary particle physics [1], some trials for developing liquid argon or liquid xenon detectors were undertaken [2-9]. In the middle of the 1970's, liquid argon was sucessfully used as detector medium of calorimeters for high energy gamma-rays or electrons [10-15]. After a few years, a proposal of a «liquid argon time projection chamber» (LATPC), which is a new type three dimensional position sensitive detector with large sensitive volume and is used for neutrino detection, has been made by high energy physicists [16] and at present, the fundamental experiments for LATPC are in progress [17, 18].

Since the end of the 1960's, we began the studies on the fundamental properties of liquid rare gases [19-26] required for applications as radiation detector media, independently of Alvarez's suggestion, and recently have developed some new type liquid xenon detectors [27-30].

On the basis of the results obtained so far [19-30], in this article the fundamental properties of liquid rare gases as detector media are summarized and considerations are made regarding the physical understanding of their properties and the possibilities of application of these liquids to nuclear radiation detectors.

2 — IONIZATION

In the design of a liquid rare gas detector which is operated in the ionization mode it is necessary to know the mean number of ion pairs produced in the liquid by the ionizing radiation and its fluctuation around the mean value. The former can easily be estimated by knowing the W-value in the liquid, defined as the average energy required to produce an ion pair and the latter by knowing its Fano factor, which expresses the degree of fluctuation of the ionization.

Here, we will discuss the W-values and the Fano-factors of liquid argon, liquid krypton and liquid xenon recently measured or estimated by us and briefly describe the reasons why W-values in liquid rare gases are near the values in the gaseous state, rather than those in semiconductors, although Fano-factors are close to the values in semicondutors rather than those in the gaseous state.

2.1 — *W-values in liquid argon, krypton and xenon*

For measurement of the W-value in liquid rare gases, the steady current method by irradiation with X-rays or alpha-rays has often been used by several investigators until recently [19, 20, 31-34]. However, this method is not suitable for precise measurement of W-values, because in X-ray irradiation, it is difficult to accurately determine the absorbed energy in the liquid medium and in alpha-particle irradiation, it is also difficult to completely collect the charge produced by alpha-particles. To overcome these difficulties of the measurement technique, we tried to use the electron pulse method and energetic conversion electrons as ionizing radiation, for measurement of W-values in liquid rare gases [21, 22]. This method has the two following advantages: 1) the energy of individual pulses is known without any uncertainty, 2) the saturation of ionization pulses can easily be achieved because the specific ionization of energetic electrons is considerably low compared with that of alpha-particles. The W-values in liquid argon and liquid xenon obtained by this method are shown in Table 1, as well as those previously obtained by the steady current method [20]. From this table, it is clear that the accuracy of the determination of W-values by this method is remarkably improved, but the W-values obtained by both methods are in good agreement within the experimental errors.

It is a well known fact that the ratios of W-values to ionization potentials in rare gases are nearly 1.7 and this was semi--quantitatively explained by Platzman on the basis of the result

TABLE 1 — The W values in the gas and liquid phases of argon, krypton and xenon. The number in parenthesis shows the ratio, W_{gas}/I or W_{liq}/E_g.

Liquid	I (eV)	W_{gas} (eV) (W_{gas}/I)	E_g (eV)	Experimental W_{liq} (eV)		Calculated W_{liq} (eV) (W_{liq}/E_g)
				Steady current method	Electron pulse method (W_{liq}/E_g)	
Ar	15.76	26.4 (1.68)	14.3	23.7 ± 0.7	23.6 ± 0.3 (1.65)	24.4 (1.70)
Kr	14.00	24.1 (1.72)	11.7	20.5 ± 1.5		20.2 (1.72)
Xe	12.13	21.9 (1.81)	9.28	16.4 ± 1.4	15.6 ± 0.3 (1.68)	15.7 (1.69)

calculated for helium gas [35]. The existence of the electron band structure in solid rare gases, such as solid argon or solid xenon, has been already confirmed, but in the liquid state is confirmed only for liquid xenon. Let us assume that the same electron band structure as that in solid rare gases also exists in the liquid state. In this case, it is considered that the band gap energy E_g in the liquid state corresponds to the ionization potential I in the gaseous state. From such a view-point, in liquid rare gases, the ratio W/E_g of the W-value to the band gap energy will be used in place of the ratio W/I in the gaseous state. As seen from Table 1, however, the values of W/E_g in liquid rare gases are almost equal to the values of W/I in the gas state and differ from those in semiconductors such as silicon or germanium (~ 3). This fact can be explained from the phenomenological theory given by Shockley [36] and developed by Klein [37].

Namely, Shockley proposed the following relation for the balance of the energy dissipated in semiconductors in order to phenomenologically understand the ionization mechanism [35],

$$W = \overset{+}{E_f} + \overset{-}{E_f} + E_i + r\,E_r\,, \tag{1}$$

where $\overset{+}{E_f}$ and $\overset{-}{E_f}$ are the mean energies of a subionization hole and electron respectively which are finally transferred to the lattice, E_i is the energy absorbed in production of an electron-hole pair, and rE_r is the energy transferred to the lattice while a free electron or a free hole, with energies higher than E_i, crosses the mean distance for electron-ion pair production. Assuming that the widths of the conduction band and the valence band are wider than the band gap and E_j is equal to E_g, Shockley derived the following formula for the W-value in a semiconductor (*):

(*) 0.6 E_g is obtained assuming that electrons or holes, with energies lower than E_j, distribute proportionally to their level density which is proportional to the square root of the kinetic energy.

$$\begin{aligned} W &= 2\,E_f + E_g + r\,E_r \\ &= 2 \times 0.6\,E_g + E_g + r\,E_r \\ &= 2.2\,E_g + r\,E_r \end{aligned} \tag{1'}$$

After that, Klein obtained the following improved formula assuming that $E_i = 3\,E_g/2$,

$$W = (14/5)\,E_g + r\,E_r \tag{1''}$$

A good agreement between both values of dW/dE_g from the above formulas and the experimental results is obtained. By fitting the formula (1″) to the experimental results, rE_r was estimated to be between 0.5 and 1.0 eV.

Now, let us apply such considerations to liquid rare gases, assuming that they have the same electron band structures as in solid state. In solid rare gases such as solid argon and xenon, there exists conduction band but no valence band. Therefore, E_f^+ should be neglected. Since E_g is larger than ten times rE_r we can also neglect the last terms in the formulas (1), (1′) and (1″). Thus, we can obtain the following relation between the W-value and Eg in liquid rare gases,

$$W = 1.6\,E_g \qquad \text{(from Shockley's formula)}$$
$$W = 1.9\,E_g \qquad \text{(from Klein's formula).}$$

These ratios (1.9 or 1.6) roughly agree with the experimental values of 1.65 for liquid argon [21] and 1.68 for liquid xenon [22]. More accurate estimation of W-values in liquid rare gases on the basis of the solid model will be made in the next section.

By admixing a small amount of a gas whose ionization potential is lower than the first excited state of the main gas, enhanced ionization is often observed in rare gases. Such a phenomenon is called Jesse effect. From the analogy with the Jesse effect in rare gases, it is expected that the same effect will occur in liquid rare gases. For example, the apparent ionization potential of a xenon atom doped in solid argon is

10.5 eV [38], which is lower than the energy of excitons in solid argon ($\gtrsim$ 12.0 eV) [38]. This clearly shows that enhanced ionization is expected in xenon doped liquid argon. Actually, we observed such an enhanced ionization in the experiment of xenon doped liquid argon [24]. Figure 1 shows the variation of the ionization yield as a function of the doped xenon concentration. The enhanced ionization is also expected when some molecular gases are doped into liquid rare gases, but the mixing of molecular gases often leads to reduction of the pulse height of electron induced signals [23], because of the loss of electrons due to electron attachment to the molecular gases. Therefore, the observable enhancement of ionization may be limited to the case of mixing between rare gases.

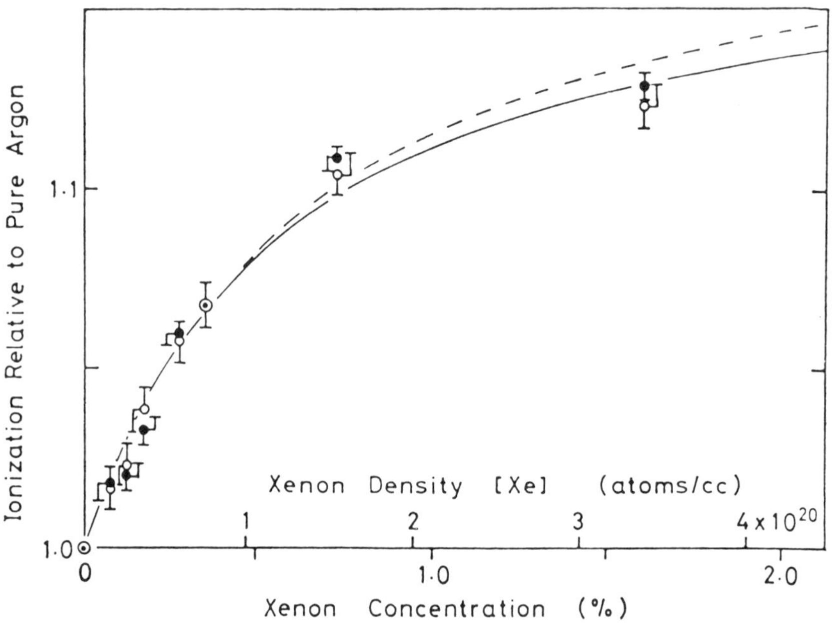

Fig. 1 — Relative ionization yield as a function of the Xe concentration, for the ionization measured at 17 kV/cm(o) and for the saturation value estimated from the 1/I versus 1/E plot(•), where I is the ionization yield and E is the electric field.

2.2 — *Estimate of W-values on the basis of the solid model*

In this section, we describe more accurate estimations of W-values in liquid rare gases on the basis of the solid model. Namely, it is possible to accurately estimate the W-value in liquid rare gases, if its band structure and its oscillator strength are given. At present, we know the band structures in solid state obtained by theoretical calculation [39] and the oscillator strengths in solid state derived from the photo absorption spectra of these solid [40]. Assuming that these data are also applicable to liquid state, we try to estimate each term in the following energy balance equation previously applied to gases by Platzman [35]

$$W_1 / E_g = (E_i / E_g) + (E_{ex} / E_g)(N_{ex} / N_i) + (\varepsilon / E_g), \qquad (2)$$

where W_1 is the W-value in liquid rare gas, N_{ex} is the number of excited atoms at an average expenditure E_{ex}, N_i is the number of ions produced at an average energy expenditure E_i and ε is the average kinetic energy of subionization electrons. Here, E_i is estimated as a mean value of the gap energy in the electron momentum space. The ratios E_{ex}/E_g and N_{ex}/N_i are also estimated with the optical approximation using the oscillator strength spectra of solid rare gases. In these estimations, it is assumed that all the excitations which lie in the continum above E_g dissociate to electron-hole pairs immediately. The estimation of the energy ε is made under the assumptions that the subionization electrons have energy less than E_i and distribute proportionally to the state density dn/dE for the energy levels. The results obtained in this way are shown in the last column of Table 1. The optical approximation used in this calculation is not valid for collisions due to low energy secondary electrons, which are a main part of collisions in the slowing down process of the primary particle. As seen from the table, nevertheless, the agreement between the estimated values and the experimental ones is very good.

From extrapolation of the enhanced ionization in xenon doped liquid argon to high concentration of xenon, the value of N_{ex}/N_i can be also estimated under the assumption that the excitons with energy higher than the ionization potential of the doped xenon atom contribute to the enhanced ionization. Then the estimated value of N_{ex}/N_i was 0.19 ± 0.02 [24], which is in good agreement with that (0.21) used in the estimation of the W-value. This fact shows that the assumptions made in the estimation of W-values are reasonable. In the next section, therefore, we will try to estimate the Fano-factors in liquid rare gases by the use of the results obtained for the W-values.

2.3 — *Fano-factors in liquid rare gases*

The fundamental formula for the fluctuation of the number of ions produced by an ionizing particle when all its energy is absorbed in a stopping material was given by Fano [41]. For convenience of calculation, the Fano's formula is transformed as follows [42, 43]:

$$\begin{aligned} F &= F_1 + F_2 + F_3 \\ &= (N_{ex}/N_i)\,[1 + (N_{ex}/N_i)]\,(\overline{W_{ex}^2}/W^2) + [\overline{(\varepsilon_i - W_i)^2}/W^2] \\ &\qquad + (N_{ex}/N_i)\,[\overline{(\varepsilon_{ex} - W_{ex})^2}/W^2] \qquad (3) \end{aligned}$$

where $W_{ex} = E_{ex}$, $W_i = E_i + \varepsilon$, and ε_i or ε_{ex} are the energy absorbed per ionization collision or per excitation collision in a large number of collisions in the slowing down process of an ionizing particle in matter, respectively. In this formula, the first term F_1 is due to redistributions of the numbers of excited and ionized atoms, the second term F_2 and the third term F_3 are due to the energy loss fluctuations in ionization and excitation, respectively. In the calculation of Fano-factors, the values of N_{ex}/N_i and ε at the end of the ionization process should be used, because the fluctuation of the number of ion pairs produced by the ionizing particle is then determined and is not affected by the process of the excitation collisions after that. The values of N_{ex}/N_i, E_i, E_{ex} and ε at the end of the ionization process for liquid argon, krypton and

TABLE 2 — Quantities appearing in the energy balance equation for liquid argon, krypton and xenon.

Liquid	E_i (eV)	E_{ex} (eV)	N_{ex}/N_i	ε (eV)
Ar	15.4	12.7	0.21	6.3
Kr	13.0	10.3	0.10	6.13
Xe	10.5	8.4	0.06	4.65

xenon, obtained by the method described in the previous section, are given in Table 2. Table 3 shows the Fano-factors in liquid rare gases calculated from formula (3) by using these values and they are clearly small compared to those in the gaseous state. This is mainly attributed to the small values of N_{ex}/N_i in liquid state. From the view-point of detector application, in particular, it should be noted that the Fano-factors in liquid krypton and liquid xenon are comparable to those in semiconductors such as silicon and germanium.

We can also estimate the Fano-factor in xenon doped liquid argon by using the formula for gas mixtures, derived by Alkhazov et al. [42]. The result is given by the following formula,

$$F_m = 0.107 - 0.067\ \sigma ,$$

TABLE 3 — F_1, F_2, F_3 and F (Fano-factor) in liquid argon, krypton and xenon for the solid model.

Liquid	F_1	F_2	F_3	F
Ar	0.076	0.027	0.004	0.107
Kr	0.032	0.024	0.001	0.057
Xe	0.019	0.021	0.0006	0.041

where σ is the probability of deexcitation followed by an additional ionization. Considering the practical use as detector medium, therefore, we estimate the Fano-factor for xenon doped (1.6 %) liquid argon, whose ionization relative to that in pure liquid argon is 1.13 ($\sigma \sim 0.68$),

$$F_m = 0.064 .$$

2.4 — *Energy resolution in liquid rare gas chambers*

Let us estimate the energy resolutions when these liquid rare gases are used as detector media in an ionization pulse chamber. In order to get the energy resolution of the chamber, first, the electronic noise level in the pulse amplification system must be given. We assume $N_{n\ eq} = 65e$ as the r.m.s. value of the noise equivalent charge in the electronic system, which can easily be achieved by using FETs (commercially available) kept at low temperature. If the W-value and the Fano factor for the detector media are known, the ultimate energy resolution ΔE_T, which is expressed by full width at half maximum (fwhm), is calculated from the following formula

$$\Delta E_T = (\Delta E_n^2 + \Delta E_j^2)^{1/2}$$

where $\Delta E_n = 2{,}36\ W\ (\mathrm{eV})\ N_{n\ eq} \times 10^{-3}$ keV, and $\Delta E_j = 2{,}36 \times \sqrt{E_\gamma\ (\mathrm{MeV})\ FW\ (\mathrm{eV})}$ keV and E_γ is the energy of ionizing radiation. The ultimate energy resolutions in liquid argon, liquid krypton and liquid xenon were calculated for $E_\gamma = 1$ MeV, using the above formula. The results are shown in Table 4 [25]. From this table, it is clear that the ultimate energy resolutions for these liquid chambers are from 3 keV to 5 keV. In the conventional Ge(Li) detector with large volume, the actual energy resolution (fwhm) for gamma-rays of 1 MeV is about 1.5 keV. Therefore, the fwhm obtained in the liquid xenon ionization chamber is expected to be near that of the Ge(Li) detector. This encourages the development of a liquid xenon gamma-ray spectrometer.

TABLE 4 — Ultimate energy resolutions (fwhm) in liquid rare gas ionization chambers.

Liquid	ΔE_n	ΔE_i(kev) for $E\gamma$ = 1MeV	ΔE_T keV) for $E\gamma$ = 1MeV
Ar	3.57	3.77	5.19
Kr	2.99	2.49	3.89
Xe	2.36	1.88	3.02

In order to check the theoretical estimation as mentioned above, we tried to measure the energy resolution for 0.569 MeV gamma-rays emitted from ^{207}Bi source using a small size liquid xenon gridded ionization chamber [22]. Figure 2 shows the relative energy resolution expressed by fwhm versus the electric field strength as well as the electronic noise level, corresponding

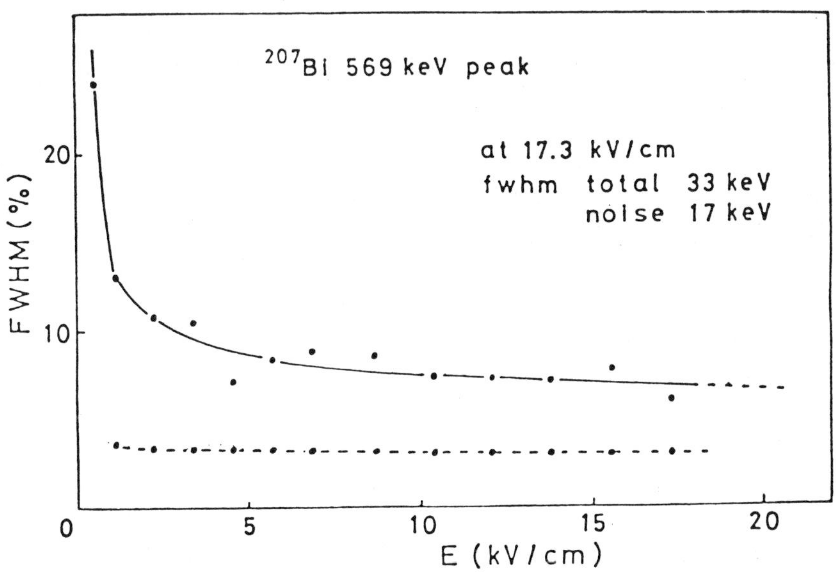

Fig. 2 — Variation of the energy resolution (fwhm) for gamma-rays of 569 keV with the electric field. Dashed line shows the level of electronic noise.

to 17 keV, which is not so good. This curve is in fairly good agreement with that recently obtained by using a large volume liquid xenon chamber [29] and that of a Russian group using a small size chamber [44]. From this figure, it is clear that the energy resolution (about 6 % fwhm) is about twice the electronic noise level even for the electrice field of 17 kV/cm, although the fwhm value still decreases with increase of the electric field. Such a large value of fwhm can not be attributed to attachment of electrons to electro-negative impurities in the liquid, because it had been experimentally shown that the pulse height of the ionization signals scarcely depends on the drift distance of electrons. Contributions to the energy resolution other than the electronic noise, such as the positive ion effect caused by the shielding inefficiency of the grid and the rise time effect of the ionization pulse, are considerably smaller than that of the electronic noise. Also, it may be difficult to explain such a large discrepancy between the theoretical estimation and the experimental results by the non-saturation effect of the collected charge for the applied electric field, because the difference between the collected charge at 17 kV/cm and its saturated value obtained by Onsager's theory is estimated to be only 3 % [22]. However, it is necessary to check whether the energy resolution is improved in much higher electric field or not, because at present the applicability of Onsager's theory to the ionization due to fast electrons in liquid rare gases is not sufficiently tested.

2.5 — *Electron multiplication in liquid and solid rare gases*

Several years ago, we tried to observe the occurrence of electron multiplication in liquid argon, and xenon doped or organic molecule doped liquid argon, by using a simple cylindrical counter with a center wire of about 5 μm in diameter, but could not find it before electrical breaking occurs. To confirm the electron multiplication in solid argon or solid xenon as observed by Pisarev's group [45], furthermore, we also tried to test it using a parallel plate solid argon filament chamber [46]. In this test, a maximum multiplication factor of about ten was observed for a tungsten filament wire of 5 μm in diameter. With increase of the anode voltage, however, the rise time of the output pulses became long and its decay time rapidly increased. At last, the pulses became unobservable because of pile-up. At present, these phenomena are interpreted assuming that the electron multiplication in solid argon occurs in a thin layer of gaseous argon near the wire surface and the slow component of the pulse is caused by space charge effect of electrons trapped in the imperfections of the interface between gas and solid. After these efforts, it was concluded that the electron multiplication to be useful in nuclear radiation detectors occurs only in liquid xenon as has already been confirmed by Derenzo et al [47].

The typical curves of charge gain versus applied voltage for a liquid xenon cylindrical counter with a center wire 5 μm in diameter, obtained with internal conversion electrons from ^{207}Bi and collimated ^{137}Cs gamma rays, are shown in Fig 3 [48]. Here, the unit gain is equal to the saturated pulse height which is obtained in a small gridded ionization chamber filled with liquid xenon. The maximum gain obtained so far is about 200.

According to our experiences, the resolution of liquid xenon proportional counters becomes poor with the increase of the applied voltage. This degradation of resolution seems due to local irregularities of the surface of the center wire. In practical applications of such a counter, this fact should be taken into consideration.

Derenzo et al. also tried to estimate the first Townsend coefficient in liquid xenon from their experimental data [3]. Namely they firstly derived the analytical formula for the variation of the charge gain with the applied voltage and fitted the curves obtained from the formula to the experimental data for three center wires of different diameter, by adjusting the values of W, recombination constant, attachment probability and first Townsend coefficient as unknown parameters. Then it was found that the first Townsend coefficient was 27 times larger than that in xenon gas with the same density as that of liquid xenon. However, this figure seems to be wrong, because the W-value and the

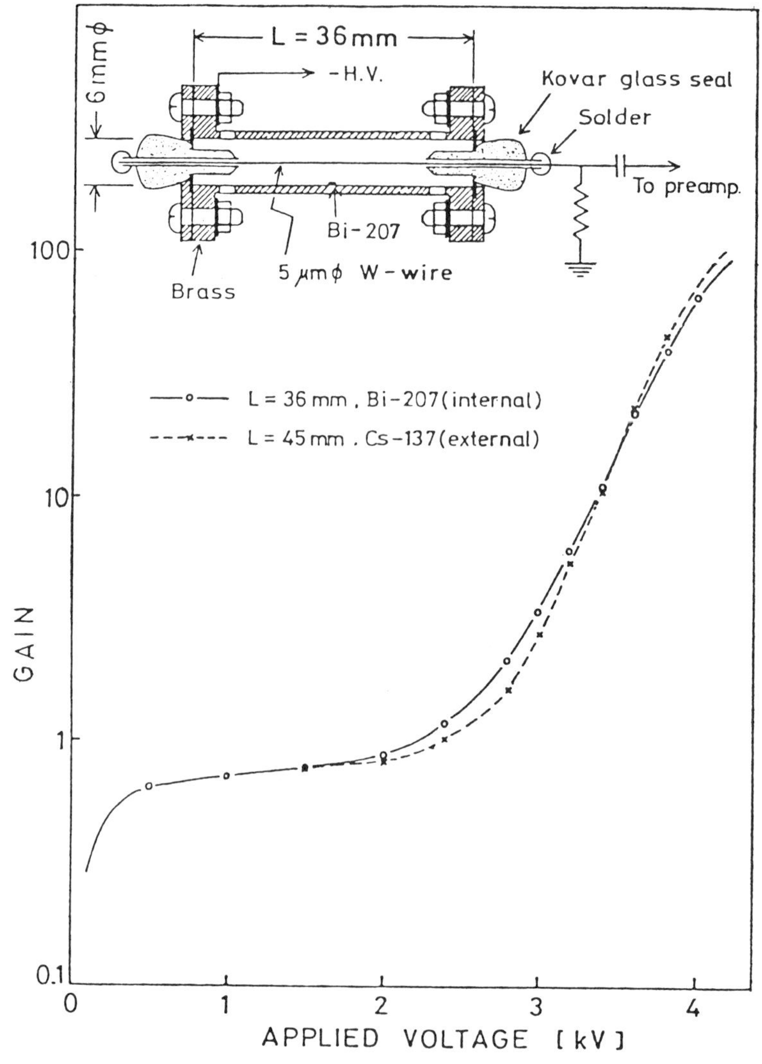

Fig. 3 — Charge gain versus applied voltage for a proportional counter having a center wire of 5 μm in diameter, as shown in the inserted figure. The solid line represents the gain for an internal ^{207}Bi source and the dashed line for external irradiation with collimated ^{137}Cs gamma-rays.

recombination constant obtained as well as the first Townsend coefficient are inconsistent with those accurately measured by us [22, 23]. Therefore, we are trying to estimate the real value of the first Townsend coefficient by treating the W-value and the recombination constant as known parameters.

3 — RECOMBINATION AND ATTACHMENT

When an ionization chamber is used as an energy spectrometer for nuclear radiation, the charge produced by the ionizing radiation must be collected completely in the collector electrode in order to determine its intrinsic energy resolution. In the electron pulse chamber, the characteristics of electron charge collection are firstly determined by recombination between electrons and ions. In addition, the electrons attaching to electro-negative gas during the drift to the collector can not effectively contribute to the induced voltage in the collector within the short measuring time, of the order of μsec. Therefore, the existence of electronegative gases in a liquid rare gas deteriorates the characteristics of the charge collection in an electron pulse chamber.

Here, we describe the present status in our understanding about the recombination and the attachment processes in liquid rare gases.

3.1 — *Recombination between electrons and ions*

If the electron-ion pairs produced by a minimum ionizing particle with low specific ionization are independent of each other, we can understand the recombination process between electrons and ions by Onsager's theory [49]. When an electron originated by ionization is slowed down to thermal energy, according to the theory, if the electron is within a certain distance from its parent ion where the Coulomb energy is equal to thermal energy, it can not escape from the influence of the parent ion and the electron-ion pair recombines. Converseley, if the thermalizing point of the electron is outside that critical distance, it is free from the influence of the parent ion even if the external electric field is zero. Actually, it seems that there exists a large number of free electrons, without recombination with the parent ion or other ions, during a considerably long time (> msec) in absence of electric field. However, it is difficult to completely understand the recombination process between electrons and ions in liquid rare gases only by Onsager's theory, because the mean interval of electron-ion pairs produced by ionizing radiation in liquids is comparable to the critical distance even for minimum ionizing particle and, under such a condition, the assumption of the existence of ion pairs which are independent of each other becomes unreal. Nevertheless, we consider that a rough explanation of the characteristics of charge collection in liquid rare gases is possible by Onsager's theory.

Figure 4 shows the characteristics of charge collection in liquid argon and liquid xenon observed by using internal con-

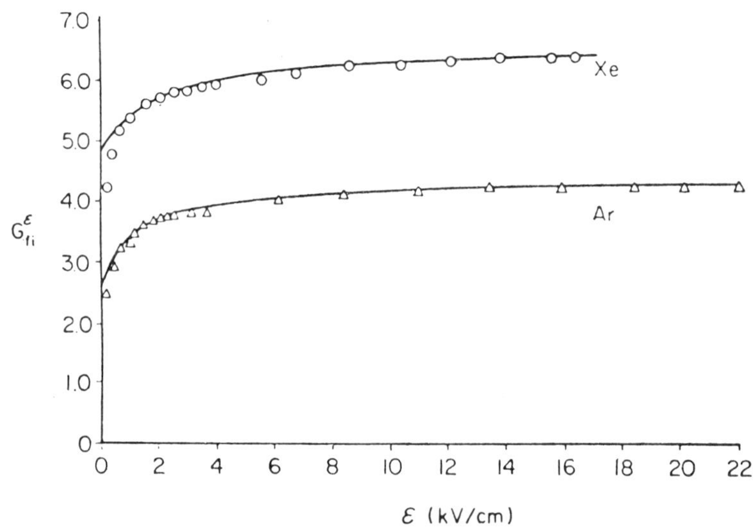

Fig. 4 — Saturation characteristics of collected charge versus electric field for liquid argon and xenon. G_{fi} is the free ion yield per 100 eV of absorbed energy at the field strength ε.

version electrons of ^{207}Bi [50]. In the figure, the solid curves are the theoretical ones obtained by fitting Onsager's theory to the experimental data and the agreement is good. According to Onsager's theory also, the initial slope of the charge collection curve in liquid xenon is smaller than in liquid argon.

However, the rise of the charge collection curve in liquid xenon is steeper that in liquid argon for electric fields lower than 1 kV/cm. This is different from the above prediction of Onsager's theory. Such a difference is caused by the effect of columnar recombination which occurs between electrons and ions other than the parent ion and the attachment of electrons to electro-negative impurities in the rare gas liquid. By extrapolating the theoretical charge collection curve, obtained from the fitting to the experimental data for electric fields larger than 1 kV/cm, to the low electric field region, we can estimate the fraction of recombination free electrons to the total number of electrons produced by the radiation for zero electric field. The values obtained in this way are 0.53 ± 0.04 for liquid argon and 0.73 ± 0.04 for liquid xenon, respectively [50]. These values seem to be an overestimate, because of the ambiguity in the low electric field region. This problem will be again discussed in the section on direct scintillation.

3.2 — *Electron Attachment*

In an ionization chamber with a large sensitive volume, such as total absorption chambers or time projection chambers for neutrino detection, the electrons produced by ionizing radiation are required to drift a long distance without losses due to electron attachment to electro-negative impurities in the liquid rare gases. From the point of view of the liquid argon time projection chamber, Chen et al. made an experiment to estimate how long electron drift lengths can be achieved in liquid argon supplied through a purifier, using a drift distance of a few centimeters, and showed that attenuation lengths longer than 35 cm are achievable at the electric field of 2 kV/cm [17]. Recently, they constructed a 50 liter liquid argon test chamber with a maximum drift distance of 30 cm and showed that an attenuation length of 55 cm is achievable at a drift field of 1.6 kV/cm [18]. Now, let us consider oxygen molecules as a typical electro-negative gas in liquid argon. Its cross section for electron attachment (or attenuation coefficient of drifting charge) in liquid argon at the electric field of 10^2 to 10^4 V/cm has already been measured by Zaklad [51], Hofmann et al [52] and Bakale et al [53].

Their results are shown in Figure 5 as well as the results in liquid xenon obtained by Bakale et al [53]. The values of the electron attachment cross section in liquid argon are in good agreement. Using these values, the upper limit of the concentration of oxygen molecule in liquid argon used in Chen's experiment is estimated to be 2.6 ppb. Such a purity was obtained

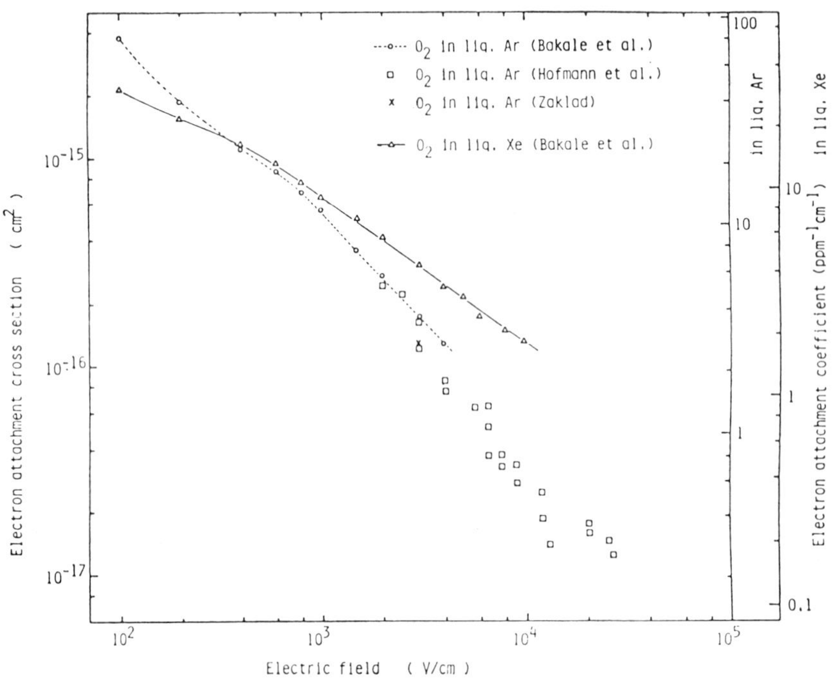

Fig. 5 — Variation of the electron attachment cross section with the electric field in liquid argon and xenon.

by passing a sequence of a Hydrox purifier and molecular sieves maintained at 196 K.

Complete collection of the charge produced by ionizing radiation in liquid xenon is more difficult than that in liquid argon, because a larger amount of electro-negative impurities are dissolved in liquid xenon due to a temperature higher than that of liquid Ar [51]. In addition, the cross section for electron attachement by oxygen molecules is about one and a half times larger than that in liquid argon as seen in Fig. 5. Recently, we observed the reduction of the ionization pulse height for the drift distance of 5 cm in a liquid xenon drift chamber using collimated gamma-rays [30]. The result is shown in Figure 6.

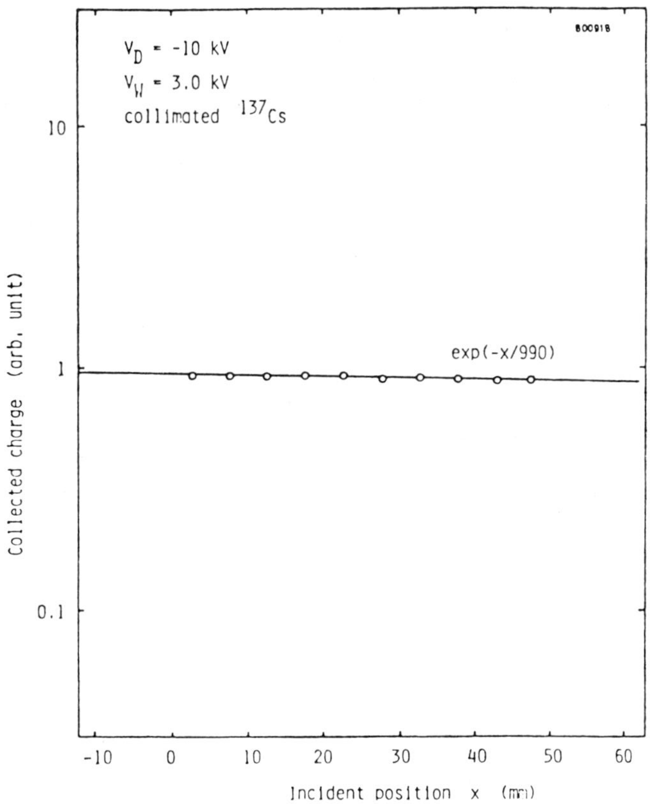

Fig. 6 — Attenuation of collected charge versus drifting distance in liquid xenon.

In this case, the attenuation length of electrons in liquid xenon was one meter. Assuming the cross section of electron attachment shown in Fig. 5, we can estimate the concentration of oxygen molecules in the liquid xenon used in our experiment to be 1.8 ppb, which is better than that obtained in liquid argon by Chen et al., in spite of the difficulty in purification of xenon. This shows that our test chamber and gas purification systems, in which a titanium-barium getter [48] is used, are superior to the systems of Chen et al.

4 — ELECTRON DRIFT VELOCITY AND DIFFUSION

The drift velocity of electrons determines the time response of radiation detectors and the diffusion of electrons in the detector medium gives the limit of the accuracy of position determination. From the analogy with the mixing effect on the electron drift velocity in gaseous state, several years ago, we studied electron drift velocities in liquid argon mixed with small amounts of molecular gases and found a remarkable increase of the drift velocity for mixtures with methane or ethylene [23]. If the diffusion coefficient and the drift velocity of electrons in liquids are experimentally obtained as a function of the electric field strength, we can estimate the momentum transfer cross sections and the agitation (random) energies of electrons in the liquid state. This information gives us not only the spread of electrons drifting under an electric field in liquids, but also the understanding of the mixing effect as mentioned above. Therefore, we also tried to measure the diffusion coefficients of electrons in liquid argon [26, 57] and liquid xenon [57].

In this section, these results are shown and some relevant considerations are made.

4.1 — *Electron drift velocities*

As mentioned above, we measured the variations of electron drift velocities by admixing various kinds of molecular gases (200 to 5000 ppm) into liquid argon [23]. These results are

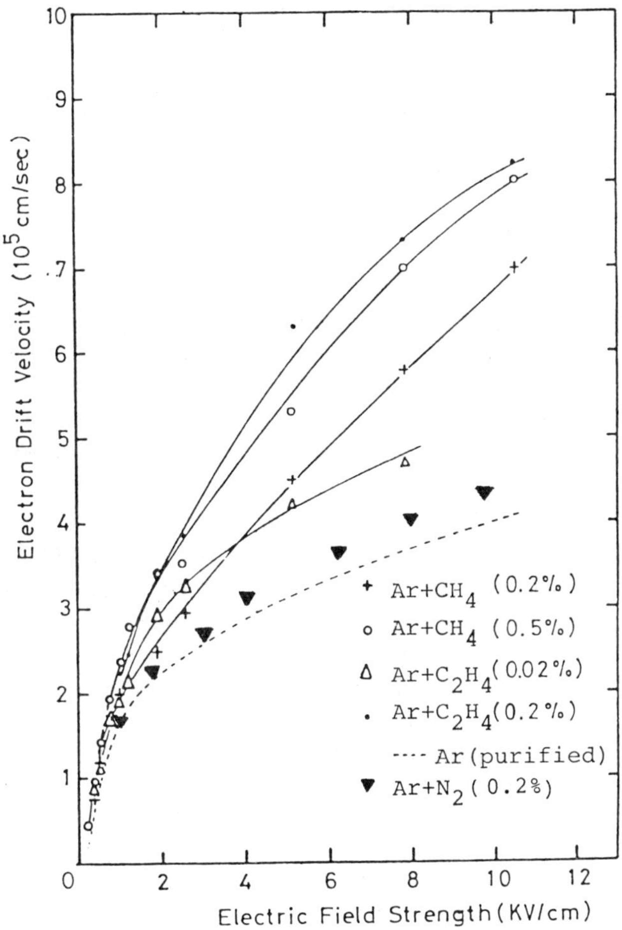

Fig. 7 — Variation of the drift velocity of electrons in liquid argon, liquid argon-nitrogen, - methane and - ethylene mixtures as a function of the electric field.

shown in Figure 7 as well as the variation of the electron drift velocity for pure liquid argon as a function of electric field. As seen from the figure, in the liquid argon-nitrogen mixture, no significant change in the electron drift velocity was observed, while in liquid argon-methane and -ethylene mixtures, a considerable increase was observed. The degree of the increase in the electron drift velocity, however, is not so large as expected directly from the analogy with the mixing effect in the gaseous state. This fact is due to the lack of the Ramsauer--Townsend effect in liquid argon as shown in the next section. Here, it should be noted that admixing of molecular gases into liquid rare gases results in a considerable reduction of pulse height. This is attributed to attachment of drifting electrons to electronegative gases included in the mixing molecular gas or in the rare gas itself. In applications of mixtures of liquid rare gases and molecular gases to detector media, great care must be taken regarding this problem.

Figure 8 shows the variations of the electron drift velocities in liquid xenon and liquid argon with the electric field for a wide range, as well as in gaseous argon and xenon with densities corresponding to the liquid state. These curves are drawn on the basis of the data recently obtained by several investigators [54, 55, 56]. From the figure, it is clear that the electron drift velocity in liquid is larger than that in gas over the whole region. In particular, the difference is remarkable in xenon. From the view-point of radiation detectors it should also be noted that liquid xenon is suitable as detector medium of drift chambers, because the electron drift velocity in liquid xenon is almost constant for electric fields higher than 3 kV/cm.

4.2 — *Diffusion coefficient of electrons*

The group of electrons produced by the ionizing radiation in liquid rare gas gradually spreads during drifting along the lines of electric force by the diffusion process. The process is determined by the agitation velocity of electrons V_{ag} ($\propto <\varepsilon>^{1/2}$, $<\varepsilon>$ being the agitation energy) and the momentum transfer cross section of electrons σ in the liquid. If the spread of the electron group ($\propto (Dt)^{1/2} = (Dd/\mu E)^{1/2} \propto (D/\mu)^{1/2}$, where D is the diffusion coeffiecient, d the drift distance, μ the mobility and E the electric field strength) in the liquid is measured as a function of the electric field, we can get the agitation energy of electrons from Einstein's relation $eD/\mu = kT = 2<\varepsilon>/3$. If we know the electron drift velocity or the electron mobility

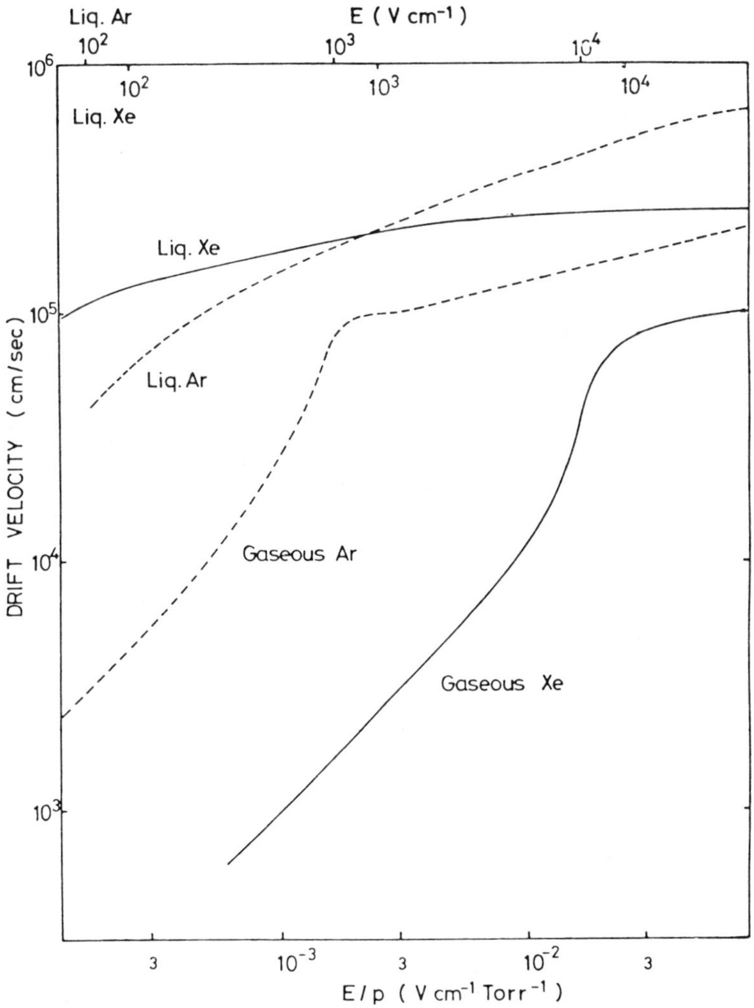

Fig. 8 — Variation of the drift velocity of electrons in liquid argon and xenon, and in gaseous argon and xenon with the same atomic density as in liquid state, with the reduced electric field.

in the liquid, we can also get the diffusion coefficient of electrons by using that relation. Figure 9(a) and 9(b) show the variation of $< \varepsilon >$ in liquid argon and liquid xenon with the electric field strength obtained from measurements of the spread of electrons

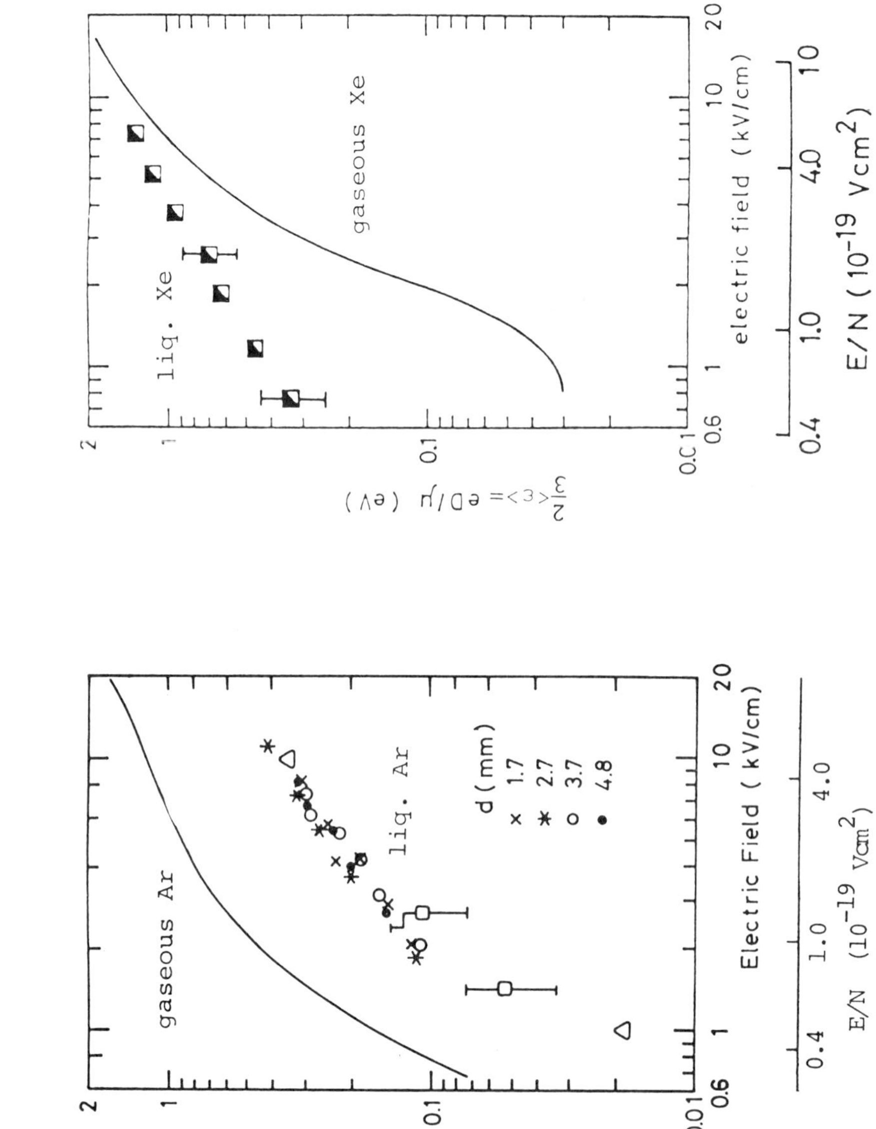

Fig. 9 — Field dependence of $2/3 \langle \varepsilon \rangle = eD/\mu$ (eV) in liquid argon (a), and liquid xenon (b). Solid curves show the results for the gaseous states, respectively. N, in the lower horizontal scale, is the atomic density in the liquid state.

as mentioned above [26, 57]. For comparison with the agitation energy of electrons in gas, the curve for argon or xenon gas with the same density as in the liquid state is also shown in each figure. As seen from the figures, the values of $<\varepsilon>$ of electrons in liquid argon is several times lower than in gaseous argon, while those in liquid xenon is a few to ten times larger than in gaseous xenon.

Figure 10 shows the variation of diffusion coefficients of electrons in liquid argon and liquid xenon with the electric field strength, which was obtained from the data of Fig. 9 (a) and (b). The diffusion coefficients in liquid argon are 10 to 20 cm^2/sec for the electric field of 2 to 11 kV/cm and clearly smaller than

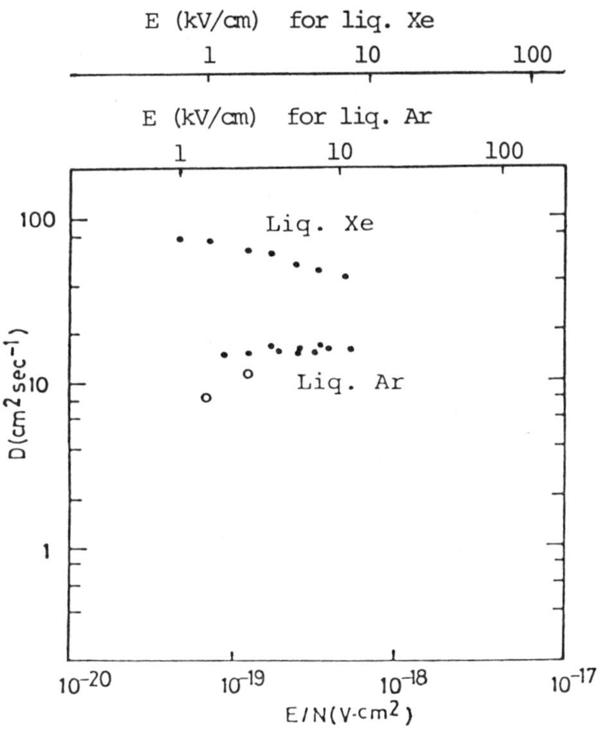

Fig. 10 — Field dependence of the diffusion coefficients of electrons.

those (17 to 37 cm^2/sec) in gaseous argon with the same density as in the liquid state. This shows that liquid argon is superior to high pressurized gaseous argon as a detector medium for position sensitive detectors. On the other hand, the diffusion coefficients in liquid xenon are several times larger than those in liquid argon. This means that the diffusion coefficient in liquid xenon is larger than that in gaseous xenon with the same density as in the liquid state, because the diffusion coefficient in gaseous xenon is smaller than in gaseous argon. Nevertheless, the diffusion coefficient in liquid xenon is still fifty times smaller than that in xenon gas at one atmosphere and small enough to make negligible the fluctuation of the center of gravity of the electron cloud distribution, less than 1 μm for the drift distance of 2 mm.

From the agitation energy $<\varepsilon>$ of electrons and the electron drift velocity w or electron mobility μ, we can estimate the momentum transfer cross section by using the simple formula $\sigma_t = eE/Nw\,(2m < \varepsilon >)^{1/2}$, derived from the relation $w = eE\lambda/m < v >$, where N is the atomic density, e and m are the charge and mass of the electron, respectively, λ is the mean free path and $< v >$ is the average agitation velocity of the electrons. Figure 11 shows the variation of the momentum transfer cross sections of electrons in liquid argon and liquid xenon with the electric field strength, which was obtained from the above formula using the data shown in Fig. 9 (a) and (b). For comparison, the variation of the cross section with the electric field in gaseous argon with the same density as that of liquid argon is also shown by the solid line in the figure. This figure clearly shows the lack of the Ramsauer-Townsend minimum in liquid argon and liquid xenon as predicted by Lekner [58], compared with the curve for gaseous argon.

5 — SCINTILLATION

The scintillations in liquid rare gases due to ionizing radiations were studied by Northrop and Gursky [59], about twenty years ago. After that, however, such scintillations have been

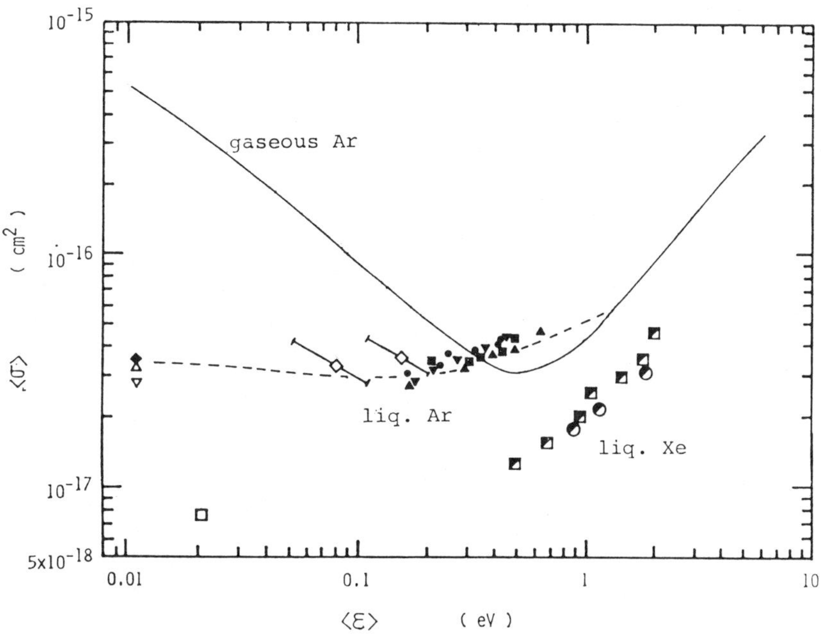

Fig. 11 — Variation of the momentum transfer cross section as a function of $<\varepsilon>$ in liquid argon and xenon. The solid line shows the result for gaseous argon and the dashed line the result obtained by Lekner's theory.

scarcely used in the field of nuclear experiments. Although it is sure that its application is not so easy as implied by Northrop and Gursky, the decay time of the scintillations from liquid rare gases such as liquid argon or xenon is very fast [64-68], and so they are useful as triggering pulses in the case of fast counting. Recently, furthermore, we found the so called proportional scintillation in liquid xenon [27, 28], which is the same phenomenon as that in rare gases. This finding will open the way to applications wider than that at the present. In this section, we describe some results on the direct scintillation and the proportional scintillation obtained in our experiments.

5.1 — *Direct scintillation*

Excited atoms R^* produced by ionizing radiation form excited molecules R_2^* through collision with other atoms on the ground state and ultraviolet photons are emmited in transitions from the lowest excited molecular state of R_2^* to the dissociative ground state. On the other hand, ionized atoms R^+ produced by ionizing radiation also form excited molecules through the following processes: i) $R^+ + R \rightarrow R_2^+$, ii) $R_2^+ + e \rightarrow R^{**} + R$, iii) $R^{**} \rightarrow R^*$ and iv) $R^* + R \rightarrow R_2^*$ and then the excited molecules also give rise to ultraviolet photons. We call this type of scintillation «recombination scintillation». The mean wave lengths of these photons are 1300 Å for liquid argon, 1500 Å for liquid krypton and 1750 Å for liquid xenon, respectively.

The intensity ratio of scintillation from excited atoms to recombination scintillation will be given as N_{ex}/N_i if there is no radiationless transition process of the excited molecular state to the dissociative ground state, which does not seem to be theoretically probable for low temperatures as in liquid argon or liquid xenon. To justify such a theoretical consideration, we measured the variations of the scintillation intensities in liquid argon and liquid xenon as a function of electric field using internal conversion electrons from ^{207}Bi. The results are shown in Figure 12 [60]. From this figure, it is clear that the scintillation intensity decreases with increase of the applied electric field, but even under electric fields higher than 10 kV/cm the scintillation intensities remain 32 % of that without the electric field for liquid argon and 26 % for liquid xenon. These values are larger than the theoretical values of 17 % for liquid argon and 5.7 % for liquid xenon, which are estimated from $N_{ex}/N_i + N_{ex}$. This discrepancy may be explained by taking into consideration the existence of recombination free electrons as mentioned in section 3. Namely, the recombination rate between recombination free electrons and ions is very slow and as the result, we can not observe the scintillation produced from such a recombination by the fast pulse techniques, which are widely used in the field of nuclear experiments. So, assuming that the recombination free electrons do not contribute to the scintillation, we can estimate the reduction factor of scintillation light to be 31 % for liquid argon and

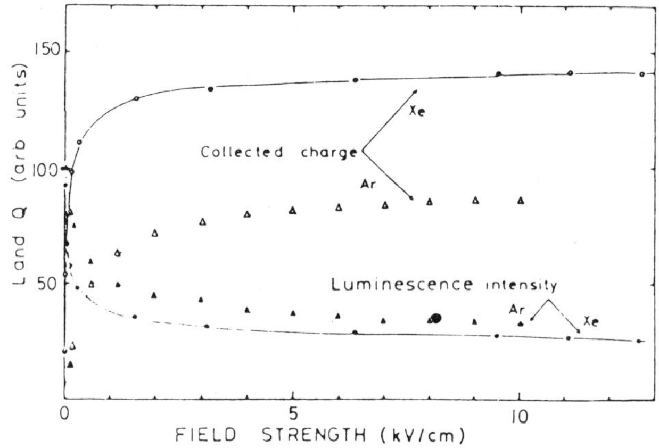

Fig. 12 — Variation of the relative luminescence intensity L and collected charge Q in liquid argon and liquid xenon with the applied electric field strength for 1 MeV electrons.

18 % for liquid xenon, using the fractions of recombination free electrons obtained by the analysis on the basis of Onsager's theory [50]. These values roughly agree with the experimental ones.

If the effect of recombination free electrons on the scintillation intensity in liquid rare gases is clear for fast electrons, it is considered that the scintillation yield per unit absorbed energy due to fast electrons may be smaller than that due to alpha-particles, which produce comparatively higher specific ionizations along the track where recombination efficiently occurs. To make sure of the matter, we tried to measure the dE/dx dependence of the scintillation yield in liquid argon using several kinds of ionizing particles [61, 62]. Figure 13 shows the results obtained. The data in the figure are normalized to the scintillation yield due to alpha-particles and show that the scintillation yield due to fast electrons is about 15 % lower than that due to alpha-particles. However, we could not observe the reduction of one-third in the scintillation yield due to alpha-particles. By using fission fragments from ^{252}Cf, on the other hand, we found a great reduction of the scintilation yield in the high specific ionization region [61]. To compare with NaI (Tl) crystals, the

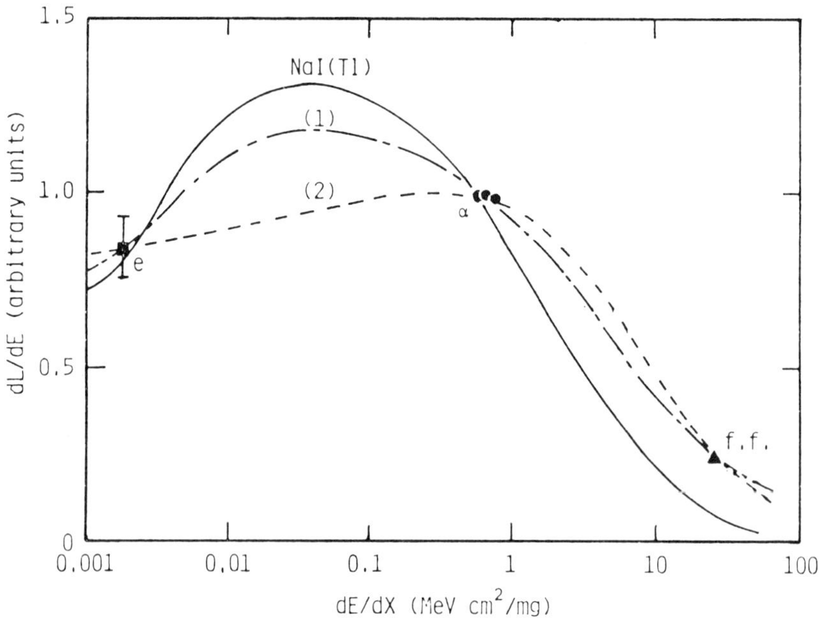

Fig. 13 — Scintillation yield per unit absorbed energy dL/dE as a function of the specific ionization dE/dx in liquid argon. Solid curve shows the dE/dx-dependence of scintillation yield in NaI(Tl) crystals. Scintillation yields in the figure are normalized to that due to alpha-particles. For dashed curves (1) and (2), see the text.

variation of the scintillation yield with the energy loss rate of the incident particle for those scintillators [63] is also shown by a solid curve in the same figure. As seen from this figure, it seems that the experimental data naturally fit to the curve (1), like that for NaI (Tl) crystals, rather than to the curve (2), which is comparatively flat compared with curve (1). The peak seen in curve (1) may be explained by considering the effect of recombination free electrons in the low specific ionization region and the quenching effect in the high specific ionization region; and by the same considerations, the question why the scintillation yield due to fast electrons is only 15 % lower than that due to alpha-particles may be solved. To justify such a consideration, at present, we are planning to measure the scintillation yields due to protons of several tens MeV and other heavy ions.

Apart from the theoretical view point as mentioned above, let us compare the dE/dx-dependence of the scintillation yield in liquid argon with that in NaI (Tl) crystals. As a whole, the curve of scintillation yield versus dE/dx for liquid argon is comparatively flat compared with the curve for NaI (Tl) crystals. In particular, the reduction of scintillation yield due to fission fragments in NaI (Tl) crystals is remarkable compared with that in liquid argon. This means that the quenching effect in the high specific ionization region in liquid argon is smaller than that in NaI (Tl) crystals.

For comparison of the scintillation yield in liquid argon with that in NaI (Tl) crystals, let us try to estimate the relative light yields from the widths of the pulse height distributions of scintillations from liquid argon, liquid xenon and NaI (Tl) crystals. The apparatus used for observation of scintillation from liquid rare gases is shown in Figure 14 as well as one

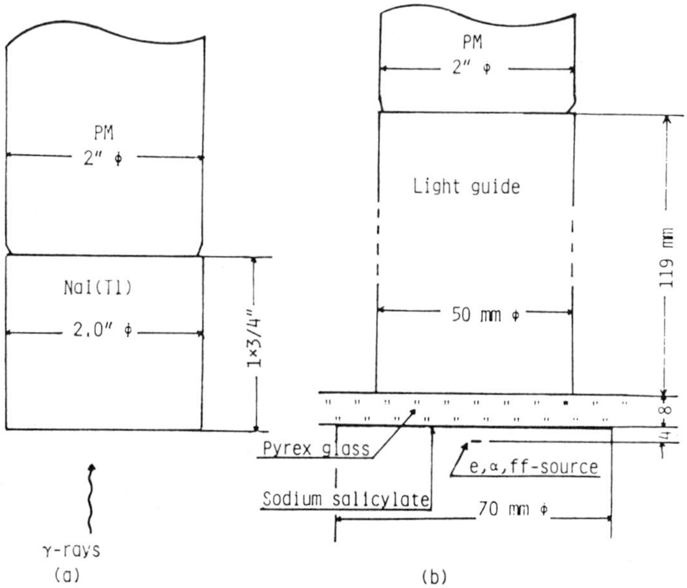

Fig. 14 — Geometrical arrangements of the apparatus for liquid argon and xenon (a), and for NaI(Tl) crystal detector (b) used in the measurements for Table 5.

used for the NaI (Tl) detector. As seen from the figure, we used sodium salicylate film, whose conversion efficiency to visible photons is nearly constant over a wide range of wave lengths, coated on the surface of a Pyrex glass window as wave lenght shifter. The resolution (fwhm) of scintillation light due to alpha-particles expressed in % for liquid argon and liquid xenon are shown in Table 5 as well as that of 1 MeV gamma--rays in the NaI (Tl) detector. From these values, the relative photon intensity ratio for argon and xenon, S_{Xe}/S_{Ar} is estimated to be 1.27, assuming that the resolution is proportional to the square root of the number of emitted photons. This value is in fairly good agreement with the theoretical one estimated from N_{ex}/N_i and W_{liq} (*). Also, the last column in the table shows the relative scintillation yields of liquid argon and liquid xenon, when the scintillation yield of NaI (Tl) crystal is assumed to be unity. In this estimation, also, we assumed that the fraction of photons incident upon the surface of sodium salicylate to the

TABLE 5 — Comparison of the energy resolution and the light yield for liquid argon or liquid xenon scintillation counters and NaI (Tl) detectors.

Scintillator	Energy resolution (fwhm) for 6 MeV alpha-particles	Energy resolution (fwhm) for 1 MeV electrons	Relative light yield
Liquid Ar	10.3 ± 0.5 %	(25.2 %) (*)	$1.08 \begin{smallmatrix}+1.92\\-0.52\end{smallmatrix}$ (**)
Liquid Xe	9.1 ± 0.5 %	(22.3 %) (*)	$1.37 \begin{smallmatrix}+2.44\\-0.67\end{smallmatrix}$ (**)
NaI (Tl)		6.0 %	1.00

(*) These values were estimated from the energy resolutions for 6 MeV alpha-particles.
(**) These errors arise from the uncertainties of light reduction factors in the light guide and conversion efficiencies of sodium salicylate used in the measurement.

(*) Namely, $$\frac{S_{Xe}}{S_{Ar}} = \frac{W_{liqAr}\,(1 + N_{ex}/N_i)_{Xe}}{W_{liqXe}\,(1 + N_{ex}/N_i)_{Ar}} = 1.32$$

total number of photons is 0.42, the conversion efficiency of sodium salicylate to visible photons is 0.5 ± 0.2 and the reduction factor of light in the light guide is 0.5 ± 0.2. These results show that the relative scintillation yields per unit absorbed energy for liquid argon and liquid xenon are comparable to that of NaI (Tl) crytals.

The counting rate capability of scintillation counters is limited by the decay time of the scintillation. The scintillation from liquid rare gases has two decay components of a few nano--seconds and a few microseconds, which correspond to the life times of the singlet state and the triplet state of the excited molecule, respectively. Table 6 shows both decay time constants τ_1 and τ_2 for scintillations from liquid argon, liquid krypton and liquid xenon, excited by fast electrons or alpha-particles, and the ratio between both amplitudes when the decay is expressed by $A_1 \exp(-t/\tau_1) + A_2 \exp(-t/\tau_2)$ [64-68]. The time constants quoted in the table show small variations for the various measurements. However, the difference in τ_2 in liquid xenon excited by fast electrons and alpha-particles is caused

TABLE 6 — τ_1, τ_2 and A_1/A_2 for scintillations from liquid argon, krypton and xenon, excited by fast electrons and alpha particles.

Liquid (Temperature)	Excitation (Ref. n.º)	Electric field (kV/cm)	τ_1 (ns)	τ_2 (ns)	A_1/A_2
Liquid Ar (94K)	e(Ref. 64)	6	5.0	860	7.8
»	e(»)	0	6.3	1020	13.5
»	e(Ref. 67)	0	2.4	1100	14.6
(90-100K)	e(Ref. 68)	0	4.6	1540	
» »	α(»)	0	4.4	1100	
Liquid Kr (120K)	e(Ref. 64)	4	2.1	80	0.9
»	e(»)	0	2.0	91	0.4
»	e(Ref. 67)	0	2.0	85	0.49
Liquid Xe (179K)	e(Ref. 64)	4	2.2	27	0.6
»	e(»)	0		34	
»	α(Ref. 67)	0	3.0	22	25
»	α(Ref. 66)	0	4.0	27	

by recombination between electrons and ions. In the case of electron excitation, the time required for recombination τ_r is longer than τ_1 and comparable to τ_2 in liquid xenon. So, if a high electric field is applied, the component due to recombination does not appear. For the alpha-particle excitation, such a slow component does not appear because the recombination rapidly occurs due to the high density of electron-ion pairs produced by this particle. On the other hand, in liquid argon, this phenomenon is not seen because τ_r is shorter than τ_1 even for the electron excitation.

Recently, Kubota et al. [67] observed a decrease in the decay times for the slow component with an addition of Xe (>lppm), N_2 (~2 %) or CO_2 (~1 %) in liquid argon excited by fast electrons. Such a mixing effect is promissing for application to scintillation counters with fast time response.

The scintillation yields in liquid argon and liquid xenon are comparable to those in NaI (Tl) crystals and the dE/dx-dependence of the scintillation yield is smaller. Also, these liquid scintillators have a faster time response than NaI (Tl) detectors. However, a considerable percentage of the photons emitted in the scintillators is absorbed by surrounding materials, because there are no good reflectors for ultraviolet photons. At present, therefore, we can not achieve a good energy resolution by liquid rare gas scintillators. If a wave length shifter that can be doped into liquid rare gases without any quenching effect is found the above difficulty may be overcome.

5.2 — *Proportional scintillation*

In the gas scintillation proportional counter «photon-multiplication» occurs along the path of a drifting electron in the region of high electric field around the center wire. This is called «proportional scintillation», because the intensity is proportional to the number of electrons initially produced by ionizing radiation. In liquid rare gases, we found that such a proportional scintillation occurs only in liquid xenon. The photons in the proportional scintillation are emitted from deexcitation of excited molecules to the dissociative ground state, as in the direct scintillation. The proportional scintillation is produced only in the region very close to the surface of the center wire. Therefore, the rise time of the scintillation is comparable to or slightly longer than that of the direct scintillation, if the range of an incident particle is negligibly small. Accordingly, a counting rate of 10^5 counts/sec will be achievable in proportional scintillation as in direct scintillation. These properties show that the proportional scintillation counter is suitable as a position sensitive detector with fast time response as well as a proportional counter. If the gain of photon-multiplication is sufficient, furthermore, it is expected that in principle a good energy resolution for gamma-rays, as determined only by the Fano factor, is achievable as in gas proportional scintillation counters for X-ray detection [69]. This means that by the use of the proportional scintillation, a liquid xenon gamma-ray spectrometer with a high energy resolution, which is not affected by the electronic noise of the preamplifier, can be developed.

From the same point of view, we measured the light gain versus the voltage applied to the center wire for liquid xenon proportional scintillation counters with different center wires of 4, 6, 8.5, 10, 11 and 20 μm in diameter [28]. The results are shown in Figure 15 as well as the curves of charge gain. From the analysis of these results, we found that the increase in the photon gain with the applied voltage in liquid xenon is approximately explained by assuming the linear relation between the photon gain and the electric field strength as in gaseous xenon; and the threshold field strength for production of photons in liquid xenon ($4 - 7 \times 10^5$ V/cm) is nearly equal to the value calculated by considering liquid xenon as gaseous xenon of 520 atm. Also, the number of photons emitted by one electron in the proportional scintillation process was estimated to be about five for a 20 μm wire at the applied voltage of 5 kV. This means that if a center wire of 50 μm in diameter is used in a proportional counter and 12 kV is applied the photon gain per one electron is expected to be about 30 [70]. This value is large enough to make the electronic noise effect negligible.

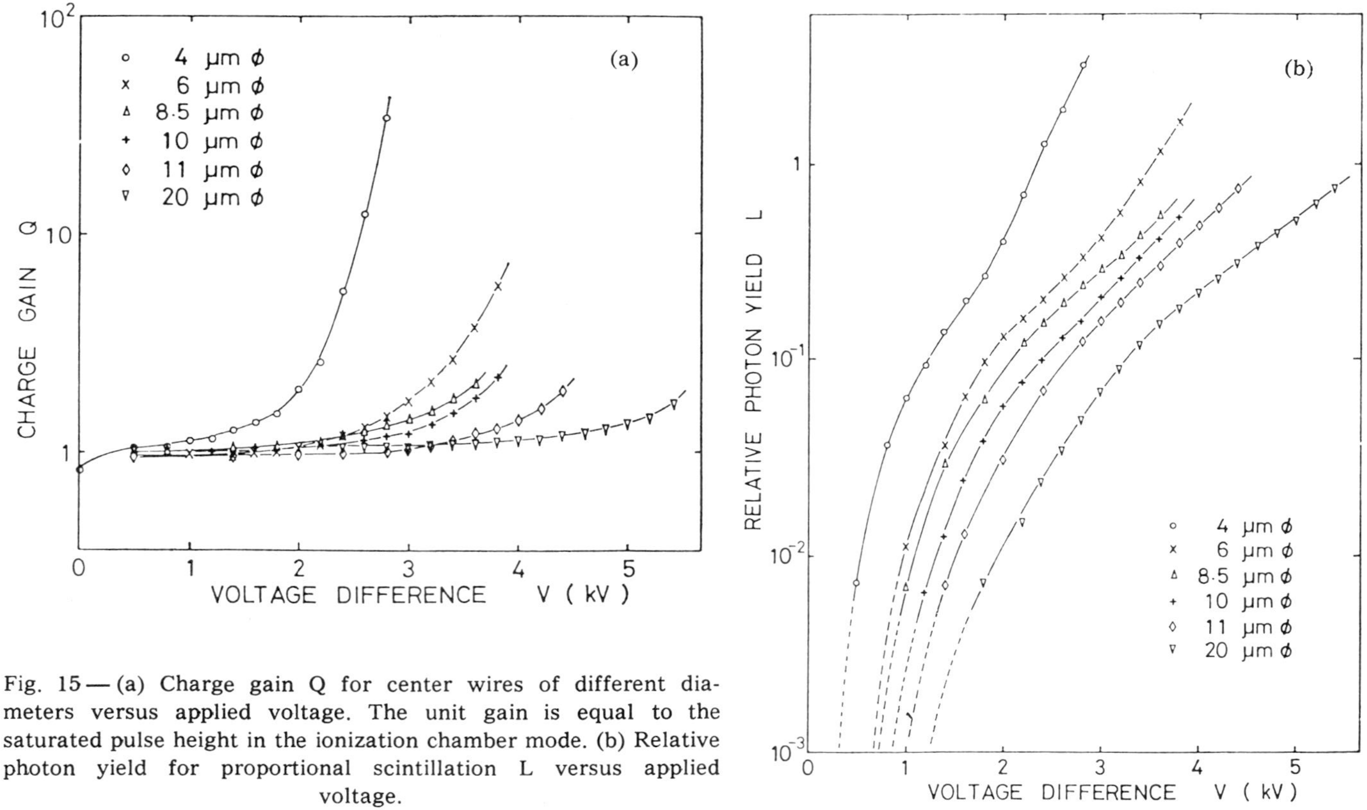

Fig. 15 — (a) Charge gain Q for center wires of different diameters versus applied voltage. The unit gain is equal to the saturated pulse height in the ionization chamber mode. (b) Relative photon yield for proportional scintillation L versus applied voltage.

6 — POSSIBILITIES OF APPLICATION TO NUCLEAR RADIATION DETECTORS

Some possibilities of application of liquid rare gases to nuclear radiation detectors have already been described in each section. Here, they are summarized as a whole and some ideas for developing liquid rare gases detectors are presented. Liquid rare gases can be used as detector media in the following four detector modes as in gaseous detectors; i) ionization mode, ii) scintillation mode, iii) proportional scintillation mode and iv) proportional ionization mode. In the modes, i) and iii) they will be used as energy spectrometers with good energy resolution or as position sensitive detectors with good position resolution for minimum ionizing particles or gama-rays. Mode ii) will correspond to fast counters or detectors for heavy ions. In the mode iv) they will be used only as position sensitive detectors, because its signals are useful only for timing.

Now, let us consider in detail the possibilities of application to nuclear radiation detectors of liquid argon and liquid xenon.

In liquid argon, the photon- and electron-multiplications do not occur as mentioned before, and so, liquid argon can be used only as detector medium in the ionization and scintillation modes. Nevertheless, it is expected to be a good detector medium in large size detectors, such as calorimeters for energy measurement of high energy gamma-rays or high energy electrons and time projection chambers for neutrino detection, because of its cheapness and the easiness of its treatment. Recently, the possibility of «photo-ionization detectors» in liquid phase was suggested by Policarpo [70]. The most practical medium for this type of detector is liquid argon doped by a small amount (~ several ten ppm) of organic compounds with low ionization potential. Such a liquid photo-ionization detector (LPID) is expected to have a large detection efficiency over a wide wavelength range of ultraviolet photons and a better position resolution than gaseous PID.

On the other hand, it is not so easy to construct a large size liquid xenon detector as compared with liquid argon, because of its high cost and its high sensitivity to impurities. However, liquid xenon is suitable as detector medium for gamma-ray detectors, because of its high atomic number and its high atomic density. As a gamma-ray spectrometer with good energy resolution, a gridded ionization chamber or a proportional scintillation chamber filled with liquid xenon may be used. Also, liquid xenon is suitable as detector medium in the drift chamber, because the drift velocity of electrons in liquid xenon is almost constant for electric fields higher than 3 kV/cm. If a liquid xenon proportional scintillation chamber is used with a liquid xenon PID, we can use the detector system as a position sensitive detector as well as an energy spectrometer for gamma--rays. Furthermore, it is possible to construct a kind of time projection chamber for gamma-rays by combining drift chambers and gridded ionization techniques.

Finally, we would like to suggest a new type liquid rare gas detector, which uses both signals of ionization and scintillation. Considerablc part of thc scintillation from liquid rarc gases arises form recombination between electrons and ions. In particular, this is remarkable for heavy ions. On the other hand, only a small amount of charge produced by a heavy ion is observable. Thus, the scintillation signal and charge signal are complementary for heavy charged particles and it is considered that some linear combination of scintillation signal and charge signal will be proportional to the energy of the particle. Such a detector will be useful as a total absorption detector for high energy heavy ions. Also, the ratio of the scintillation signal to the charge signal depends on the kind of particle. Therefore, such a technique may be used for particle identification of heavy ions.

ACKNOWLEDGMENT

The author would like to express his thanks to Professor A. J. P. L. Policarpo for having made possible the realization of this review article.

REFERENCES

[1] L. W. ALVAREZ, *Lawrence Radiation Laboratory Physics Note*, No. 672 (1968).

[2] R. A. MULLER, S. E. DERENZO, G. SMADJA, B. SMITH, R. G. SMITS, H. ZAKLAD and L W. ALVAREZ, *Phys. Rev. Letters*, **27**, 532 (1971).

[3] S. E. DERENZO, T. S. MAST, H. ZAKLAD and R. A. MULLER, *Phys, Rev.*, **A9**, 2582 (1974).
[4] S. E. DERENZO, A. R. KIRSCHBAUM, P. H. EBERHARD, R. R. ROSS and F. T. SOLMITZ, *Nucl. Instr. Meth.*, **122**, 319 (1974).
[5] M. C. GADENNE, A. LANSIART and A. SEIGNEUR, *Nucl. Instr. Meth.*, **124**, 521 (1975).
[6] J. PRUNIER, R. ALLEMAND, M. LAVAL and G. THOMAS, *Nucl. Instr. Meth.*, **109**, 257 (1973).
[7] A. LANSIART, A. SEIGNEUR, J. MORETTI and J. MORUCCI, *Nucl. Instr. Meth.*, **135**, 47 (1976).
[8] B. A. DOLGOSHEIN, V .N. LEBEDENKO and B. U. RODIONOV, *JETP Letters*, **11**, 351 (1970).
[9] B. A. DOLGOSHEIN, A. A. KRUGLOV, V. N. LEBEDENKO, V. P. MIROSHICHENKO and B. U. RODIONOV, *Sov. J. Particles Nucl.*, **4**, 70 (1973).
[10] G. KNIES and D. NEUFFER, *Nucl. Instr. Meth.*, **120**, 1 (1974).
[11] J. ENGLER, B. FRIEND, W. HOFMANN, H. KLEIN, R. NICKSON, W. SCHMIDT-PARZEFALL, A. SEGAR, M. TYRRELL, D. WEGENER, T. WILLARD and K. WINTER, *Nucl. Instr. Meth.*, **120**, 157 (1974).
[12] W. J. WILLIS and V. RADEKA, *Nucl. Instr. Meth.*, **120**, 221 (1974).
[13] D. HITLIN, J. F. MARTIN, C. C. MOREHOUSE, G. S. ABRAMS, D. BRIGGS, W. CARITHERS, S. COOPER, R. DEVOE, C. FRIEDBERG, D. MARSH, S. SHANNON, E. VELLA and J. S. WHITAKER, *Nucl. Instr. Meth.*, **137**, 225 (1976).
[14] C. J. FABJAN, W. STRUCZINSKI, W. J. WILLIS, C. KOURKOUMELIES, A. J. LANKFORD and P. REHAK, *Nucl. Instr. Meth.*, **141**, 61 (1977).
[15] C. CERRI and F. SERGIAMPRIETRI, *Nucl. Instr. Meth.*, **141**, 207 (1977).
[16] C. RUBBIA, *CERN EP Internal Report*, 77-8 (1977).
[17] H. H. CHEN and J. F. LATHROP, *Nucl. Instr. Meth.*, **150**, 585 (1978).
[18] H. H. CHEN and P. J. DOE, *IEEE Transaction on Nucl. Sci.*, **NS-28** 454 (1980).
[19] S. KONNO and S. KOBAYASHI, *Sci. Pap. Inst. Phys. Chem. Res.*, **67**, 57 (1973).
[20] T. TAKAHASHI, S. KONNO and T. DOKE, *J. Phys.* **C7**, 230 (1974).
[21] M. MIYAJIMA, T. TAKAHASHI, S. KONNO, T. HAMADA, S. KUBOTA, E. SHIBAMURA and T. DOKE, *Phys. Rev.*, **A9**, 1438 (1974).
[22] T. TAKAHASHI, S. KONNO, T. HAMADA, S. KUBOTA, A. NAKAMOTO, A. HITACHI, E. SHIBAMURA and T. DOKE, *Phys. Rev.*, **A12**, 1771 (1975).
[23] E. SHIBAMURA, A. HITACHI, T. DOKE, T. TAKAHASHI, S. KUBOTA and M. MIYAJIMA, *Nucl. Instr. Meth.*, **131**, 249 (1975).
[24] S. KUBOTA, A. NAKAMOTO, T. TAKAHASHI, S. KONNO, T. HAMADA, M. MIYAJIMA, E. SHIBAMURA, A. HITACHI and T. DOKE, *Phys. Rev.*, **B13**, 1649 (1976).
[25] T. DOKE, A. HITACHI, S. KUBOTA, A. NAKAMOTO and T. TAKAHASHI, *Nucl. Instr. Meth.*, **134**, 353 (1976).
[26] E. SHIBAMURA, S. KUBOTA, T. TAKAHASHI and T. DOKE, *Phys. Rev.*, **A20**, 2547 (1979).

[27] M. Miyajima, K. Masuda, Y. Hoshi, T. Doke, T. Takahashi, T. Hamadä, S. Kubota, A. Nakamoto and E. Shibamura, *Nucl. Instr. Meth.*, **160**, 239 (1979).

[28] K. Masuda, S. Takasu, T. Doke, T. Takahashi, A. Nakamoto, S. Kubota and E. Shibamura, *Nucl. Instr. Meth.*, **160**, 247 (1979).

[29] K. Masuda, A. Hitachi, Y. Hoshi, T. Doke, A. Nakamoto, E. Shibamura and T. Takahashi, *Nucl. Instr. Meth.*, **174**, 439 (1980).

[30] K. Masuda, T. Takahashi and T. Doke, *Nucl. Instr. Meth.* (to be published).

[31] H. A. Ullmaier, *Phys. Med. Biol.*, **11**, 95 (1966).

[32] N. V. Klassen and W. F. Schmidt, *Can. J. Chem.*, **47**, 4286 (1969).

[33] M. G. Robinson and G. R. Freeman., *Can. J. Chem.*, **51**, 641 (1973),

[34] S. Huang and G. R. Freeman, *Can. J. Chem.*, **55**, 1838 (1977).

[35] R. L. Platzman, *Int. J. Appl. Rad. Isotopes*, **10**, 116 (1961).

[36] W. Shockley, *Czech. J. Phys.*, **B11**, 81 (1961).

[37] C. A. Klein, *J. Appl. Phys.*, **39**, 2029 (1968).

[38] G. Baldini and R. S. Knox, *Phys. Rev. Letters*, **11**, 127 (1963).
G. Baldini, *Phys. Rev.*, **137**, A508 (1965).

[39] U. Rossler, *Phys. Status Solidi* **B42**, 345 (1970), **B45**, 483 (1971).

[40] G. Keitel, DESY F41-70/7 (1970).
P. Schreber, DESY F1-70/5 (1970).

[41] U. Fano, *Phys. Rev.*, **72**, 26 (1947).

[42] G. D. Alkhazov, A. P. Komar and A. A. Vorobev, *Nucl. Instr. Meth.*, **48**, 1 (1967).

[43] H. A. Bethe and J. Ashkin, *Experimental Nuclear Physics* (ed. E. Segre, Wiley, New York, 1973) Vol. 1, Part 2.

[44] I. M. Obodovsky and S. G. Pokachalov, *Low Temp. Phys.*, **5**, 829 (1979).

[45] A. P. Pisarev, *Sov. Phys.* JETP, **36**, 823 (1973).

[46] E. Shibamura, A. Hitachi, M. Miyajima, S. Kubota, A. Nakamoto, T. Takahashi, S. Konno and T. Doke, *Bull. Sci. Eng. Res. Lab. Waseda Univ.*, **69**, 104 (1975).

[47] S. E. Derenzo, D. B. Smith, R. G. Smits, H. Zaklad, L. W. Alvarez and R. A. Muller, UCRL-20118 (1970).

[48] M. Miyajima, K. Masuda, A. Hitachi, T. Doke, T. Takahashi, S. Konno, T. Hamada, S. Kubota, A. Nakamoto and E. Shibamura, *Nucl. Instr. Meth.*, **134**, 403 (1976).

[49] L. Onsager, *Phys. Rev.*, **54**, 554 (1938).

[50] T. Takahashi, S. Konno, A. Hitachi, T. Hamada, A. Nakamoto, M. Miyajima, E. Shibamura, Y. Hoshi, K. Masuda and T. Doke, *Sci. Pap. Inst. Phys. Chem. Res.*, **74**, 65 (1980).

[51] H. Zaklad, UCRL-20690 (1971).

[52] W. Hofmann, U. Klein, M. Schulz, J. Spengler and D. Wegener, *Nucl. Instr. Meth.*, **135**, 151 (1976).

[53] G. BAKALE, U. SOWADA and W. F. SCHMIDT, *J. Phys. Chem.*, **80**, 2556 (1976).
[54] J. L. PACK, R. E. VOSHALL and A. V. PHELPS, *Phys. Rev.*, **127**, 2084 (1962).
[55] L. S. MILLER, S. HOWE and W. E. SPEAR, *Phys. Rev.*, **166**, 871 (1968).
[56] Y. YOSHINO, U. SOWADA and W. F. SCHMIDT, *Phys. Rev.*, **A14** 438 (1976).
[57] E. SHIBAMURA, S. KUBOTA, T. TAKAHASHI and T. DOKE, *Proc. Int. Seminar on Swarm Experiments in Atomic Collision Research*, Tokyo, 47 (1979).
[58] J. LEKNER, *Phys. Rev.*, **158**, 130 (1967).
[59] J. A. NORTHROP and J. C. GURSKY, *Nucl. Instr. Meth.*, **3**, 207 (1958).
[60] S. KUBOTA, A. NAKAMOTO, T. TAKAHASHI, T. HAMADA, E. SHIBAMURA, M. MIYAJIMA, K. MASUDA and T. DOKE, *Phys. Rev.* **B17**, 2762 (1978).
[61] A. HITACHI, T. TAKAHASHI, T. HAMADA, E. SHIBAMURA, A. NAKAMOTO, N. FUNAYAMA, K. MASUDA and T. DOKE, *Phys. Rev.*, **B23**, (1981) in press.
[62] A. HITACHI, T. TAKAHASHI, T. HAMADA, E. SHIBAMURA, N. FUNAYAMA, K. MASUDA, J. KIKUCHI and T. DOKE, *Proc. INS Intern. Symposium on Nuclear Radiation Detectors*, Tokyo, (1981) in press.
[63] R. B. MURRAY and A. MEYER, *Phys. Rev.*, **122**, 815 (1961).
[64] S. KUBOTA, M. HISHIDA and J. RUAN, *J. Phys.*, **C11**, 2645 (1978).
[65] S. KUBOTA, M. HISHIDA, M. SUZUKI and J. RUAN, *Phys. Rev.*, **B20**, 3486 (1979).
[66] S. KUBOTA, M. SUZUKI and J. RUAN, *Phys Rev.* **B21**, 2632 (1980).
[67] S. KUBOTA, M. HISHIDA, M. SUZUKI and J. RUAN, *Proc. INS Intern. Symposium on Nuclear Radiation Detectors*, Tokyo, (1981) in press.
[68] M. J. CARVALHO and G. KLEIN, *J. Luminescence*, **18/19**, 487 (1979).
[69] A. J. P. L. POLICARPO, M. A. F. ALVES, M. C. M. DOS SANTOS and M. J. T. CARVALHO, *Nucl. Instr. Meth.*, **102**, 337 (1972).
[70] A. J. P. L. POLICARPO, *Proc. INS Intern. Symposium on Nuclear Radiation Detectors*, Tokyo, (1981) in press.

Vol. 21, No. 1, pp. 51-64, February 1974.

SIGNAL, NOISE AND RESOLUTION IN POSITION-SENSITIVE DETECTORS*

by V. Radeka

Abstract

An analysis is presented of signal, noise and position resolution relations for some of the most interesting position-sensing methods. "Electronic cooling" of delay line terminations is introduced in order to reduce noise in the position-sensing with delay lines. A new method for terminating transmission lines and for "noiseless" damping which employs a capacitance in feedback is presented. It is shown that the position resolution for the charge division method with resistive electrodes is determined only by the electrode capacitance and not by the electrode resistance, if optimum filtering is used.

*This work was performed under the auspices of the U.S. Atomic Energy Commission

1. Introduction

There has been a great proliferation of various readout methods for position sensitive detectors. One would imagine that in each particular case the best solution is arrived at by 1) the detector design to maximize the significant signal, 2) reduction of noise at its physical source, and 3) optimum filtering of signal and noise. Many imaginative contributions have been made to the detector design. It can be stated, however, that the noise reduction and filtering have been neglected to the point where detector design and operation become critical in order to obtain a large enough signal to overcome noise.

In interpolating readouts the position resolution is determined by the noise which originates either in the position sensing medium or in the amplifier and other components external to it. In readouts for multiwire proportional chambers based on one amplifier per wire it is usually assumed that the signal is large and cost and size are dominant in design considerations. Recently applications have become of interest where the signal magnitude is limited by the detector design (smaller anode spacing), or where linear signal information is needed. In all cases involving semiconductor detectors, and gas or liquid ionization detectors with little or no gain, noise becomes important.

This paper concentrates on point 2), that is, reduction of noise at its physical source. Optimum filtering is determined by both signal and noise and by the information (amplitude or time) required so that a great variety of cases is possible. A simple approach based on filter weighting function analysis is given here, and the treatment of this subject is limited to several examples.

There are two prevalent interpolating readout methods in use and under development at present. One is based on a delay line as a position sensing medium.[1,2,3] The other is based on distributed R-C line.[4-9] The two are entirely different. In the first one the position-sensing medium is non-dissipative in principle, and the noise is generated external to it in the terminations and amplifiers. In this paper "electronic cooling" of delay line terminations is introduced in order to reduce noise. A new method for terminating transmission lines and for "noiseless" damping by a capacitance in feedback is presented.

An R-C line is a dissipative position-sensing medium. With proper choice of line and amplifier parameters the noise in signal amplifiers is negligible, and all the noise is generated in the position-sensing medium. Thus the noise and the position resolution are determined by the R-C line and by the filtering. It is shown here that for detectors with short charge collection times the position resolution is independent of the time required to process one event. Detectors with smaller resistance, better timing and energy resolution can be made with the same position resolution.

The case of single amplifier per wire (detector segment)

is discussed with respect to noise limits, location of amplifiers, and sensitivity to spurious signals in interconnections.

2. Characterization of Signal and Noise

2.1 Signal

The main features of signals of the two principal kinds of ionization detectors are briefly described here. The elementary current and charge waveforms for gas multiplication proportional detectors and for planar semiconductor detectors are illustrated in Fig. 1. The current results from the motion of charge carriers in the electric field. Therefore, distinctly different waveforms result for the cylindrical geometry of the proportional counter than for the planar geometry of the semiconductor detector.

For point ionization in a proportional detector the electrons reach the multiplication region simultaneously. If the detector is operated in the proportional mode the charge multiplication takes place in the high field region near the center wire (anode). The electrons resulting from the multiplication are collected on the center wire after traversing a very small potential difference. Their contribution to the total observed charge is thus very small. In this case most of the signal is due to the sheath of positive ions moving toward the cathode. The current as a function of time is of the form:[10]

$$i(t) = \frac{Q_m}{2t_o \ell n(b/a)} (1 + t/t_o)^{-1} \quad , \qquad (1)$$

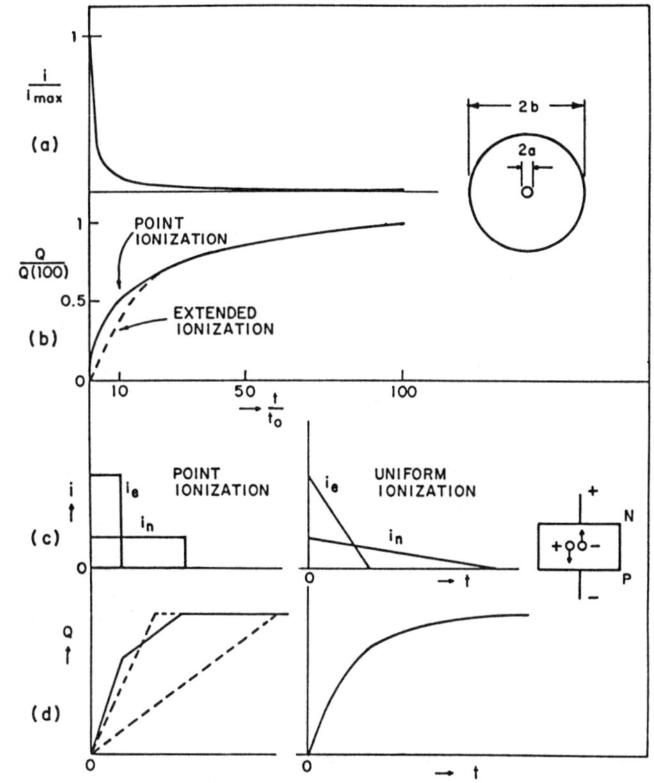

Figure 1. Signal waveforms.

Proportional detector:

(a) current for point ionization

(b) charge for point ionization (solid) and for extended ionization (dashed)

Planar semiconductor detector (for point ionization in the middle and for uniform ionization between the electrodes):

(c) current

(d) charge.

and the charge

$$Q(t) = \frac{Q_m}{2\ln(b/a)} \ln(1 + t/t_o) \text{ for } \frac{t}{t_o} \le (\frac{b}{a})^2 \quad . \qquad (2)$$

$Q_m = N_o A_g q_e$, N_o = number of electrons produced by ionization, A_g = gas gain, b and a are cathode and anode radius. t_o is determined by the mobility of positive ions μ_p, b and a, and by the voltage between anode and cathode; $t_o = [a^2 \ln(b/a)]/2\mu_p V_o$.

For extended ionization tracks, the total signal is obtained by superposition of signals due to individual primary electrons arriving at different times to the multiplication region. This affects the early portion of current and charge waveforms, Fig. 1(b), and it is of great significance for timing. The logarithmic nature of the charge as a function of time has consequences in many applications. Where event timing resolution is required, only the early portion of the signal is useful, and thus very little charge is utilized. For measurements based on the quantity of charge (energy) longer filtering times are required. The whole function is rarely observed in practice, and the measurements of gas gain are usually biased due to the low frequency cutoff of filters used.

Multiwire proportional chambers are frequently operated in the saturated gain mode. It has been suggested[2,11] that the gas multiplication takes place far enough from the anode so that the avalanche electrons contribute significantly to the signal by an amount of charge which is collected in a much shorter time (10-20 nsec) than the charge due to positive ions. This mode cannot be used where (linear) information about the charge produced by an ionizing event is required.

Charge collection times in semiconductor position-sensitive detectors are usually very short ($<$ 100 nsec), so that the signal current can be considered as an impulse in most cases.

2.2 Noise

Some simple noise relations are presented here in order to be able to determine the noise contributions by various circuit components. Detectors based on ionization represent a capacitive source of charge. The basic equivalent circuit of detector and amplifier is shown in Fig. 2. The signal is represented as a current source in parallel with the total input capacitance. The series noise is related to the amplification mechanism. The parallel noise is due to imperfections of the amplifier and the detector (leakage currents) and due to dissipative elements (represented by R_p) connected to the input. The noise is expressed in terms of equivalent noise charge. The noise at the input is important only if it contributes to the output of the filter, and the knowledge of the transfer function of the whole system is essential to determine the equivalent noise charge. The whole system from the detector to the output of the filter is described by its weighting function (which is the mirror image in time of the impulse response for time-

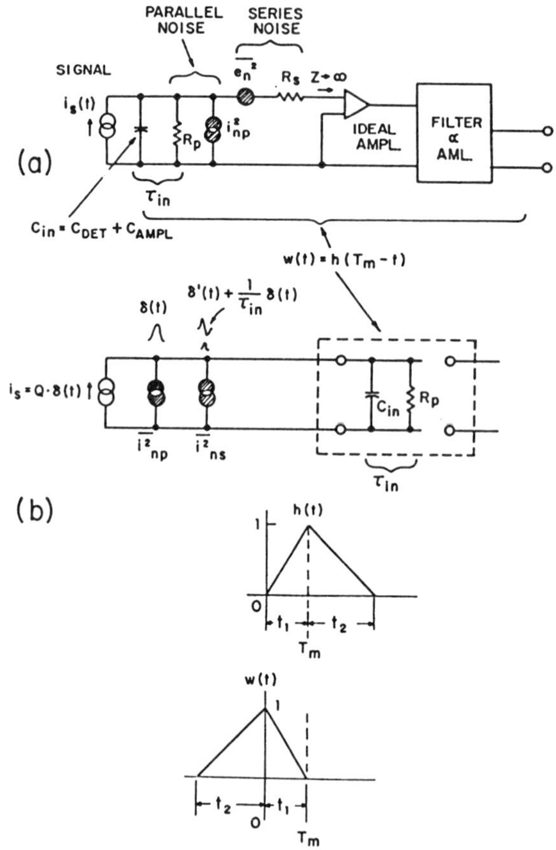

Figure 2. Equivalent circuit of detector and amplifier for noise analysis.

(a) location of noise sources in the circuit

(b) eq. circuit with noise sources transformed to the input C_{in} = detector capacitance + amplifier input capacitance.

invariant filters). A more detailed discussion of sources of noise is given in Ref. 12, and signal processing and noise calculations are described in Ref. 13.

We can think of the parallel noise as a random sequence of current impulses (delta functions). Similarly, we can think of the series noise as a random sequence of voltage impulses. The series noise can also be represented by a current generator in parallel with the input C-R network, as shown in Fig. 2(b). The current noise spectrum is obtained upon multiplication by $C_{in}\,(j\omega + 1/\tau_{in})$, where $\tau_{in} = C_{in} R_p$. Consequently each voltage impulse $v_o\delta(t)$ is converted into $v_o C_{in}\left[\delta'(t) + \frac{1}{\tau_{in}}\,\delta(t)\right]$, the sum of a doublet and an impulse. The mean square equivalent noise charge is obtained by adding up independent contributions to the filter output by all impulses and doublets, according to the Campbell's theorem. (It can be shown that the mean square contributions to the filter output by the pairs of impulses and doublets with the same time origin are uncorrelated, and thus the impulses and doublets due to the series noise can be treated as independent in the summation of their mean square contributions.) The equivalent noise charge for these two noise generators is given by the following relations,

Series noise:

$$\overline{ENC_s^2} = \frac{1}{2}\,\overline{e_n^2}\,C_{in}^2 \int_{-\infty}^{\infty} [w'(t)]^2 + \frac{1}{\tau_{in}^2}\,[w(t)]^2 dt \qquad (3)$$

Parallel noise:

$$\overline{ENC_p^2} = \frac{1}{2}\,\overline{i_n^2} \int_{-\infty}^{\infty} [w(t)]^2 dt \quad , \tag{4}$$

where: e_n = rms eq. series noise voltage per $Hz^{1/2}$

i_n = rms noise current per $Hz^{1/2}$ from parallel sources

$\tau_{in} = C_{in} R_p$ = input circuit time constant

e_n expressed in terms of series eq. noise resistance:

$$\overline{e_n^2} = 4\ kTR_s \quad . \tag{5}$$

i_n can be generated as thermal noise in the resistors in parallel with the input,

$$\overline{i_n^2} = 4\ kT \frac{1}{R_p} \quad , \quad \text{or} \tag{6}$$

as shot noise in the current I_o into the input (leakage currents, transistor base current, etc.),

$$\overline{i_n^2} = 2\ q_e\ I_o \quad . \tag{7}$$

The noise spectrum of the charge amplifier is characterized by the noise-corner time constant, which can be expressed in terms of various noise parameters as,

$$\tau_c = C_{in} \frac{e_n}{i_n} = C_{in}\ (R_s\ R_p)^{1/2} \quad , \tag{8}$$

$$\text{or,} \quad \frac{\tau_c}{\tau_{in}} = \left(\frac{R_s}{R_p}\right)^{1/2} \quad . \tag{9}$$

The weighting function is normalized to its maximum value so that the impulse signal (charge) from the detector is recorded with a weight of unity. The limits of integration, $-\infty$, ∞, imply that the integration is carried out for all non-zero values of $w(t)$ and $w'(t)$. The noise contribution of a specific part of the weighting function is determined by integration over that part only. The weighting function determines the magnitude and the relative importance of various noise contributions. For inspection of the filtering properties of any weighting function, it is necessary to evaluate the integrals:

$$I_1 = \int_{-\infty}^{\infty} [w'(t)]^2 dt = \frac{a_{F1}}{\tau_F} \tag{10}$$

$$I_2 = \int_{-\infty}^{\infty} [w(t)]^2 dt = a_{F2}\tau_F \tag{11}$$

τ_F is the "width parameter" of the weighting function in time, and a_{F1} and a_{F2} are nondimensional "form factors".

a_{F1} represents the effect of steep parts of the weighting function (the high frequency limit), and it determines the contribution of the series (or "doublet") noise. It is easy to compute these integrals for a given weighting function. For purposes of quick evaluation of noise and of the weighting function, a piece-wise linear approximation of the weighting function can be used. A straight line should be fitted to the steepest parts of the function in order to obtain a conservative estimate of the series noise. The values of integrals I_1 and I_2 for various line segments are

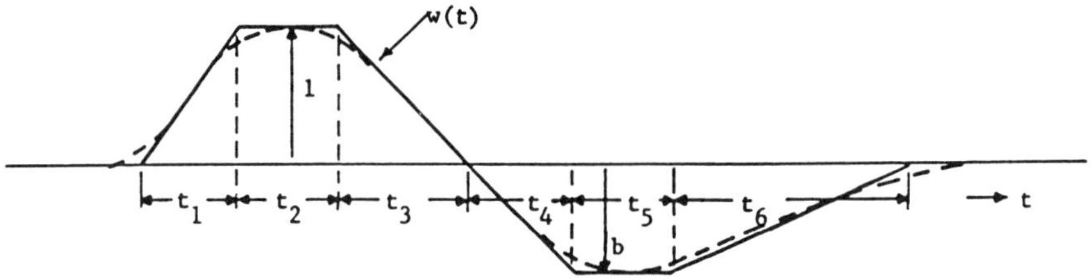

$$I_1 = \frac{1}{t_1} + 0 + \frac{1}{t_3} + b^2 \frac{1}{t_4} + 0 + b^2 \frac{1}{t_6}$$

$$I_2 = \frac{t_1}{3} + t_2 + \frac{t_3}{3} + b^2 \frac{t_4}{3} \; b^2 t_5 + b^2 \frac{t_6}{3}$$

Figure 3. Evaluation of weighting function integrals, Eqs. (10) and (11) by piecewise linear approximation.

given in Fig. 3. Weighting functions can be so approximated with an accuracy of noise estimates of better than 10%. For the purpose of inspection of Eq. (3) we assume a unipolar triangular weighting function as shown in Fig. 2, and with $t_1 = t_2 = t_m$. Then the two integrals are

$$I_1 = \frac{2}{t_m} \tag{12}$$

$$I_2 = \frac{2}{3} t_m \tag{13}$$

Eqs. (3) and (4) can be written in the form,

Series noise: $$\overline{ENC_s^2} = \frac{\overline{e_n^2} \, C_{in}^2}{t_m} \left[1 + \frac{1}{3} \left(\frac{t_m}{\tau_{in}}\right)^2\right] \tag{14}$$

Parallel noise: $$\overline{ENC_p^2} = \frac{1}{3} \overline{i_n^2} \, t_m \tag{15}$$

If $t_m/\tau_{in} \ll 1$, Eq. (14) reduces to the classical case of the charge amplifier (R_p is very large),

$$\overline{ENC_s^2} = \frac{\overline{e_n^2}\, C_{in}^2}{t_m} \qquad (16)$$

In a number of instances with position-sensitive detectors $t_m/\tau_{in} >> 1$. Let us assume, as an example, that $t_m/\tau_{in} = 2$. Then the two terms in Eq. (14) are:

$$\overline{ENC_s^2} = \frac{\overline{e_n^2}\, C_{in}^2}{t_m} (1 + 4/3) = 2.33\ (\overline{ENC_s^2})_{min} \quad . \qquad (17)$$

In addition R_p generates parallel noise:

$$\overline{ENC_p^2} = \frac{1}{3} \left(\frac{t_m}{\tau_c}\right)^2 \frac{\overline{e_n^2}\, C_{in}^2}{t_m} \qquad (18)$$

Thus, if τ_{in} is smaller than the width of the weighting function, there is a significant increase in the contribution by the series noise, simply because the signal is attenuated before it even arrives at the amplifier.

The lowest noise can be achieved with no dissipative elements connected to the input, Eq. (16). The filtering or "pulse shaping" should be performed after amplification, so that no noise is added by dissipative filter components. The current from the detector can be measured by shaping (differentiating) the signal after amplification.

3. Transmission Line Termination with Electronically Cooled Resistance

The need for damping arises in three typical cases: 1) there is a resonant circuit in the input, or a transformer in conjunction with a capacitive signal source, and an aperiodic response is required; 2) waveform measurement from a capacitive signal source by "differentiation" of the charge signal at the input; and 3) transmission line termination. These cases are illustrated in Fig. 4(a). It is a common practice to use a resistor to achieve the required damping. In all these cases the required damping or terminating resistor is the only dissipative element in the input circuit, and its noise exceeds the amplifier noise.

,It has long been recognized that damping can be achieved with resistance in feedback, for example to damp a galvanometer,[14] where the noise added by the "active resistance" is less than it would be by the physical resistor. Referring to Fig. 4(b), an inverting amplifier with a <u>real</u> gain G_o and a resistance R_f in feedback results in an apparent input resistance,

$$R_{in} = \frac{R_f}{|G_o|+1} \approx \frac{R_f}{|G_o|} \tag{19}$$

R_f represents a noise source in parallel with the input, and its noise current is much smaller than it would be from a resistor of value R_{in}, Eqs. (4) and (6).

This technique is limited to low frequencies because the amplifier gain should be real. It also poses stability problems at higher frequencies, due to the phase shift in the amplifier and in the combination of the feedback resistance and the capacitance at the input.

The new approach is based on exploiting the natural pro-

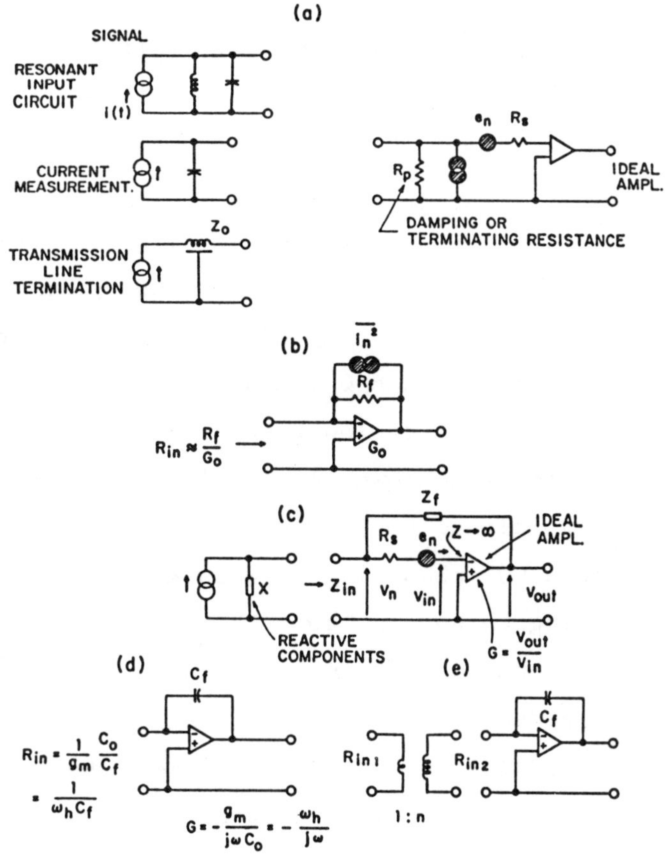

Figure 4. Electronically "cooled" terminations and damping

(a) classical damping with resistor.

(b) classical "cooled" damping with resistance in feedback.

(c) eq. circuit for analysis of "cooled" damping.

(d) "cooled" damping with capacitance in feedback.

(e) use of transformation to realize low values of "cooled" resistances.

perties of an elementary operational amplifier. For such an amplifier the gain expressed as a function of frequency is given by

$$G(j\omega) = -\frac{\omega_h}{j\omega} , \tag{20}$$

where ω_h is angular frequency at which the gain becomes unity. It is assumed that any additional poles correspond to frequencies higher than ω_h, and that the gain reaches its d.c. value at a frequency lower than the lowest frequency of interest. As is well known for amplifiers where the gain is inversely proportional to frequency ("6 dB per octave"), there must be a 90° phase lag associated with this dependence, according to Bode's law. This phase lag must be offset by a 90° phase lead in feedback in order to achieve a real input impedance (resistance). Thus a capacitance is required in feedback in order to realize a damping resistance. Necessary relations can be derived from the basic feedback circuit in Fig. 4(c). It is simple to show that the input impedance is given by

$$Z_{in} = \frac{Z_f}{(-G+1)} \approx \frac{Z_f}{-G} , \tag{21}$$

(where Z_{in}, Z_f and G are functions of either $j\omega$ or the Laplace transform variable s). We assume a capacitance in feedback, as shown in Fig. 4(d),

$$Z_f = \frac{1}{j\omega C_f} . \tag{22}$$

The input impedance is then,

$$Z_{in} = \frac{1}{\omega_h C_f} = R_{in} \quad , \tag{23}$$

i.e., the input resistance is equal to the absolute value of the reactance in feedback at the unity-gain frequency of the amplifier. Thus, damping can be realized without connecting any dissipative elements to the input. The damping cannot be noiseless since it is realized by a physical amplifier. It is interesting to find the effective noise temperature of the damping resistor realized by feedback. Due to the feedback the equivalent series noise voltage appears at the input terminal,

$$v_n = - \frac{e_n}{1 - \frac{1}{G}} \approx - e_n \quad .$$

Thus the open circuit mean square noise voltage per $H_z^{1/2}$ (the physical spectral density) at the input terminal is

$$\overline{v_n^2} = 4\ kTR_s \quad . \tag{24}$$

Since it appears to be generated by the source with an internal resistance R_{in}, it can expressed in the form

$$\overline{v_n^2} = 4k \left(T \frac{R_s}{R_{in}}\right) R_{in} \tag{25}$$

or in the form of a current source,

$$\overline{i_n^2} = \overline{v_n^2} / \ R_{in}^{\ 2} = 4k \left(T \frac{R_s}{R_{in}}\right) \frac{1}{R_{in}} \quad . \tag{25a}$$

Thus the effective temperature of the damping resistance

R_{in} is

$$T_f = \frac{R_s}{R_{in}} T \quad . \tag{26}$$

The improvement in noise over the physical ("warm") damping resistor depends on its value. The equivalent series noise resistance for field-effect transistors and bipolar transistors is in the range 50 - 200 Ω. If a low value of R_{in} is required, a transformer can be used. A high value of R_{in} is realized by feedback, and then transformed to the required low value R_{in1}, Fig. 4(e). The effective temperature is then,

$$T_f = \frac{R_s}{n^2 R_{in1}} T \quad . \tag{27}$$

The upper limit to the cooling factor $n^2 R_{in1}/R_s$ is determined by the practical considerations in the transformer design (low parasitic inductance and capacitance, low noise).

A basic circuit configuration for cooled damping with capacitance in feedback is shown in Fig. 5. This is a complementary cascode used in charge amplifiers.[12] The open loop gain is

$$G = g_m \frac{1}{j\omega C_o} \quad , \tag{28}$$

and the input resistance becomes

$$R_{in} = \frac{1}{g_m} \frac{C_o}{C_f} \quad . \tag{29}$$

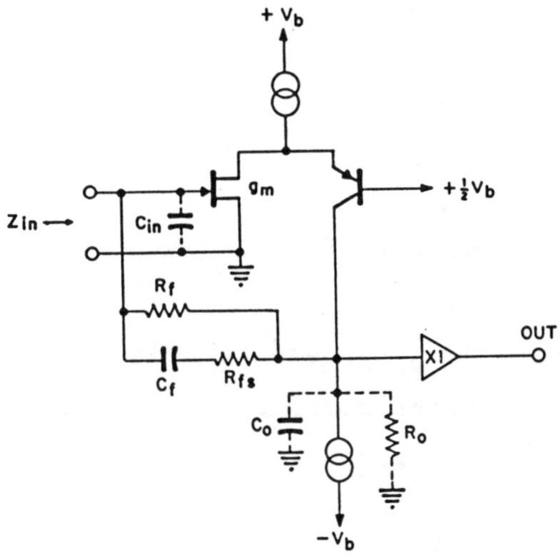

Figure 5. Basic circuit configuration for "cooled" damping with capacitance.

In this discussion it was assumed that the gain is given by Eq. (20) and that in the region of interest $|G(j\omega)|>>1$. We now consider the complete expression for all frequencies. Assuming negligible loading of the amplifier output by the feedback network, the open loop gain is given by

$$G(j\omega) = - g_m \frac{1}{j\omega C_o + 1/R_o} . \tag{30}$$

($1/R_o$ is the output conductance of the current source and the input conductance of the unity gain amplifier in Fig. 5). The input impedance for a capacitance in feedback ($R_f \to \infty$, $R_{fs} = 0$) is then,

$$Z_{in} = \frac{1}{g_m} \frac{C_o}{C_f} \left[(1 + 1/j\omega C_o R_o)^{-1} + j\omega/\omega_h\right]^{-1} \tag{31}$$

At frequencies where $\omega C_oR_o >> 1$,

$$Z_{in} \approx \frac{1}{g_m}\frac{C_o}{C_f}(1 + j\omega/\omega_h)^{-1} \quad , \tag{32}$$

or $Z_{in} \approx \left(\frac{1}{R_{in}} + j\omega C_f\right)$, where R_{in} is given by Eq. 29, and

$\omega_h = g_m/C_o$. The term $j\omega C_f$ represents only the feedback capacitance. The important component of input impedance is the one due to the feedback,

$$Z_{in} = \frac{1}{g_m}\frac{C_o}{C_f} + \frac{1}{j\omega C_f(g_mR_o)} \quad . \tag{33}$$

This is a series combination of the resistance R_{in} and the capacitance $C_fg_mR_o$. In the frequency region of interest for damping the second term should be negligible, i.e. $\omega C_oR_o >> 1$.

We note parenthetically that Eq. (33) represents input impedance of the "operational integrator" or charge amplifier. It has not been generally recognized that in most cases the first term dominates.

It is assumed here that the biasing resistor R_f is large, so that its noise is negligible. It is interesting to note that R_f appears at the input as an inductance. By using Eq. (21), it follows:

$$L_{in} = \frac{C_o}{g_m}R_f = \frac{R_f}{\omega_h} \quad . \tag{34}$$

This is another way of interpreting the periodic impulse response of an operational amplifier with resistance in feedback and capacitance at the input. Damping resistance can be realized at all frequencies from dc to beyond ω_h by making $R_f C_f = R_o C_o$. This eliminates the imaginary term in Eq. (33).

The feedback network can be used to compensate for the effect of parasitic input capacitances. With R_{fs} in series with C_f in feedback,

$$Z_{in} = \frac{1}{C_f \omega_h} + j\omega \frac{R_{fs}}{\omega_h} \qquad (35)$$

The second term represents an inductance in series with the resistance R_{in}, and it can be used for "high frequency peaking" in transmission line terminations.

It was assumed throughout that additional poles correspond to frequencies higher than ω_h ($\omega_1 \geq 4\ \omega_h$), which can be achieved with high frequency transistors. An aperiodic response can be achieved more easily with fewer additional poles in the loop, if a short rise time is required. In such a case feedback is taken from the output of the cascode, as shown in Fig. 5, rather than from the output of the unity gain amplifier.

The output rise time for impulse signals is determined by the time constant,

$$\tau_r = R_{in}\ C_{in} = \frac{1}{g_m} \frac{C_o}{C_f} C_{in} \qquad (36)$$

as is well known for charge amplifiers.[12]

Damping or resistive termination by capacitance in feedback results in an amplifier with a real input impedance which integrates the input current. Further pulse shaping (filtering) should take into account this integration. The current waveform can be obtained by differentiation.

4. Position Sensing with Delay Lines

Position sensing with delay lines is based on conversion of position information into a time delay. In this method position sensing electrodes of the detector are connected at uniform spacing to a delay line. The signal can be coupled either capacitively, or directly.[2,3,15] Direct coupling can be achieved also by making the detector electrodes a part of the helical coil of the delay line.[2,3] If cathodes of a multiwire proportional chamber, or strips on either side of a semiconductor detector, are connected to the delay line, the well known interpolating property results. The image of the signal charge is seen by several electrodes, and by determining the centroid of the composite signal propagating along the delay line, the position can be determined with an error smaller than the wire spacing or the projected length of an inclined particle track. This method, where the position appears as the difference between the arrival times of the signals at the ends of the delay line, is illustrated in Fig. 6(a).

The delay line is a non-dissipative position sensing medium in principle. The noise in this system is generated in the terminations and in the amplifiers. The position resolution is ultimately limited by the statistics of the spatial charge distribution in the detector. The electronic

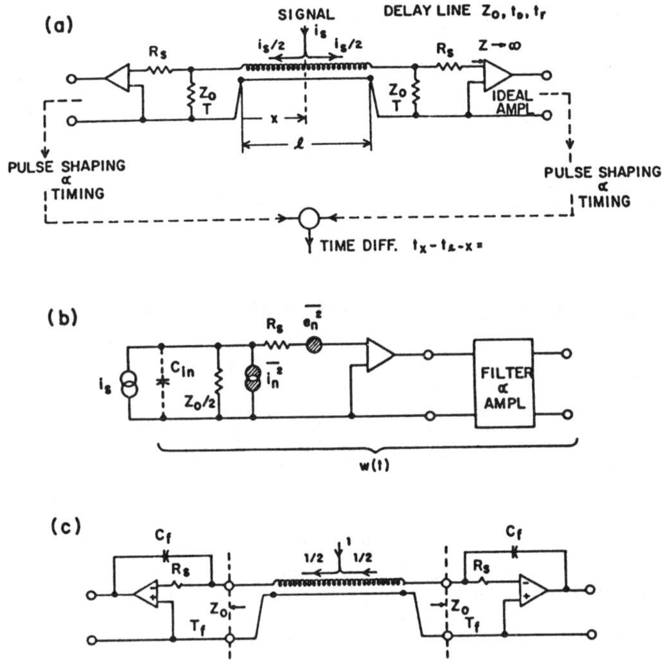

Figure 6. Position sensing with delay lines.

(a) basic circuit with "warm" resistance terminations Z_o, T.

(b) eq. circuit of the "warm" terminations case for noise analysis.

(c) delay line position sensing with "cooled" terminations with capacitance in feedback.

noise determines the magnitude of the signal required to achieve this resolution.

We derive first the expression for noise in the conventional system with "warm" resistance terminations. The signal and noise sources for the circuit in Fig. 6(a) are shown in the equivalent circuit in Fig. 6(b). We use the results of the analysis in Section 2. With a terminated delay line the impedance of the input circuit is real, and clearly the time constant $\tau_{in} = C_{in} Z_o/2$ will be much smaller than the width of the weighting function. In this case the first term in the expression (3) for series noise becomes negligible, and we have for the noise from the amplifier,

$$\overline{ENC_s^2} = \frac{1}{2} \overline{e_n^2} \frac{1}{(Z_o/2)^2} I_2 = 8 \ kT \frac{R_s}{Z_o^2} I_2 \quad . \tag{37}$$

The noise from both terminations is in parallel with the input. Using Eqs. (4) and (6) with the values from Fig. 6(b), the noise from the terminations is:

$$\overline{ENC_p^2} = 4 \ kT \frac{1}{Z_o} I_2 \quad . \tag{38}$$

The total noise for "warm terminations":

$$\overline{ENC^2} = 4 \ kT \frac{1}{Z_o} \left(1 + 2 \frac{R_s}{Z_o}\right) \int_{-\infty}^{\infty} [w(t)]^2 dt \tag{39}$$

We now consider "cooled" terminations introduced in the preceding section. The delay line and the preamplifier circuit is then as in Fig. 6(c). Referring to Eqs. (3), (10) and the equivalent circuit in Fig. 2, we have again $\tau_{in} \ll \tau_F$, and we substitute Z_o for R_p. The remaining second term in Eq. (3) represents the equivalent noise charge due to the series noise of one amplifier,

$$\overline{ENC^2_{s1}} = \frac{1}{2}\,\overline{e^2_n}\,\frac{1}{Z_o^2}\,I_2 = 2\,kT\,\frac{R_s}{Z_o^2}\,I_2 \quad . \tag{40}$$

The active termination at the other end adds an equal amount of noise due to the series noise of its amplifier. As shown in the preceding section, a transformer can be used if a low value of input resistance is required. This is equivalent to dividing $\overline{e^2_n}$ by n^2. The total equivalent noise charge is then, for electronically "cooled" terminations:

$$\overline{ENC^2_s} = 4k\left(T\,\frac{R_s}{n^2 Z_o}\right)\frac{1}{Z_o}\int_{-\infty}^{\infty}[w(t)]^2 dt \quad . \tag{41}$$

The ratio of noise temperatures for the two cases gives the improvement factor due to cooling,

$$\eta_f^2 = \frac{T_w}{T_f} = \frac{n^2 Z_o}{R_s}\left(1 + 2\,\frac{R_s}{Z_o}\right) \tag{42}$$

Z_o = characteristic impedance of the delay line, R_s = eq. series noise resistance of the amplifier, n = transformation ratio between the delay line and the amplifier.

η_f represents the reduction in rms noise. The signal amplitude can be decreased by η_f while giving the same position resolution as with "warm" terminations. Let us assume, as an example, $Z_o = 1000\ \Omega$, $n = 1$, $R_s = 50\ \Omega$. Then

$$\eta_f \geq \sqrt{22} = 4.7$$

R_s is determined by the input amplifying device. A higher ratio can be achieved for delay lines with higher characteristic impedance Z_o. For low impedance delay lines a transformer is necessary. (Practical improvement factor for existing systems with warm terminations may be higher since in many cases the amplifiers used have not been designed for low noise.)

What is the relative position resolution with the delay line position sensing? The resolution is determined by several factors: the detector design, particle type, energy, angle of incidence, etc. As the length of the delay line is increased, the dispersion in the delay line limits the timing resolution, and the resolution becomes determined largely by the signal-to-noise ratio. The best resolution in this case is achieved with an optimum filter for timing. This filter should be matched to both the signal and noise. The noise in this case is white. The optimum filter for timing is the derivative of the optimum filter for amplitude measurement. The optimum filter for amplitude measurement is in this case (with white noise) a mirror image in time of the signal waveform from the delay line. This is illustrated in Fig. 7 on a piecewise linear approximation of an odd current function. The optimum filter for timing results from minimization of the ratio of noise to the slope of the signal at the output of the filter. The filter weighting function is generally bipolar and the maximum slope is at zero crossing, which results in well-known antiwalk properties. As is known from the filter theory, the minimum ratio for a given signal and noise is given by,

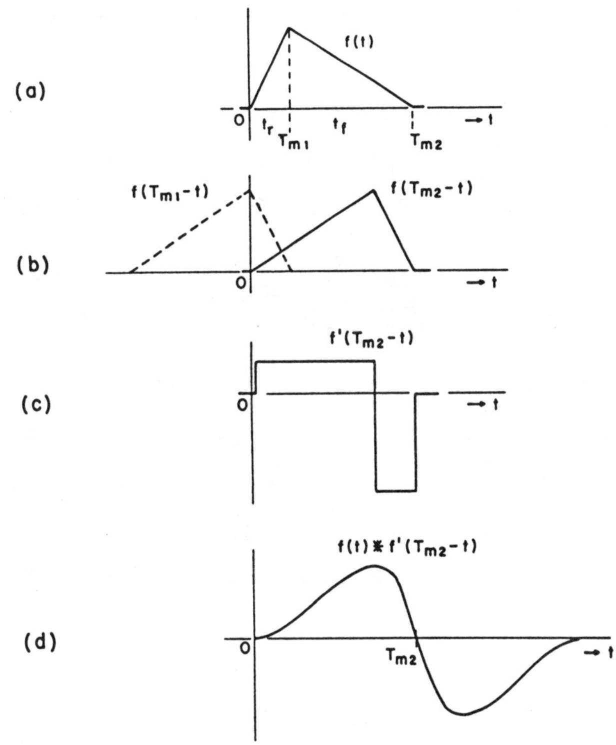

Figure 7. Optimum filter for timing in presence of white noise (method of derivation).

(a) signal waveform.

(b) optimum filter for amplitude measurements.

(c) optimum filtering for timing - derivative of (b).

(d) output waveform.

$$\frac{\overline{v_n^2}}{(d\, v_{out}/dt)^2} = (\delta t)^2 = \frac{W_o}{A^2 \int_{-\infty}^{\infty} [f'(t)]^2 dt} , \qquad (43)$$

where W_o = white noise spectral density, A = signal amplitude, and f(t) is the signal waveform.

Let us assume that the current signal at the output of the delay line is approximated by a gaussian, described by the total charge and by the full width at half maximum,

$$i(t) = \frac{2.35}{\sqrt{2\pi}} \frac{Q_s}{t_{FW}} \exp\left[-2.76 \left(\frac{t}{t_{FW}}\right)^2\right] \quad . \tag{44}$$

The optimum weighting function is a derivative of the gaussian. The noise current spectral density in the case of cooled terminations is determined from the relations (40) and (41) assuming no filtering, i.e. $I_2 = 1$. Then we can calculate the variance in timing from Eqs. (43) and (44) as

$$\overline{(\delta t)^2} = 0.55 \frac{\overline{e_n^2}\, t_{FW}^3}{Z_o^2\, Q_s^2} \quad . \tag{45}$$

We can define the quality factor of the delay line as the ratio of the total delay to the time dispersion t_{FW},

$$\theta_D = \frac{t_D}{t_{FW}} \quad . \tag{46}$$

The relative position resolution is determined by the timing error from the two outputs. The noise from the two outputs is uncorrelated for timing, since the noise from one amplifier is added with a delay t_D to the other one. Thus,

$$\frac{\delta \ell}{\ell} = \sqrt{2}\, \frac{\delta_t}{t_D} \quad . \tag{47}$$

With (45) and (46) it follows:

$$\frac{\delta \ell}{\ell} = 2.46 \, \frac{1}{\theta_D} \, \frac{e_n}{Z_o Q_s} \, t_{FW}^{1/2} \quad [\text{FWHM}] \quad . \tag{48}$$

Let us assume, as an example, $\theta_D = 15$, $Z_o = 500\ \Omega$, $e_n = 10^{-9}\ \text{V/Hz}^{1/2}$, $t_{FW} = 10^{-7}$ sec.

To achieve a resolution $\delta\ell/\ell = 10^{-3}$, a signal is required, $Q_s = 1.04 \times 10^{-13}$ C $= 6.5 \times 10^5$ e. Without electronic cooling a signal of about 3×10^6 e would be required.

Cooled terminations are simple to realize. They involve only a judicious use of a few amplifier components which had to be used also in the case of warm terminations. An amplifier used as a cooled termination with a multiwire proportional detector[15] is shown in Fig. 8. The amplifier is intended for low impedance delay lines, $Z_o = 200$ and $500\ \Omega$. Two bipolar transistors are used in parallel to reduce the effect of base and emitter bulk resistances and achieve $R_s \approx 30\ \Omega$. Above $Z_o = 500\ \Omega$ field-effect transistors should be used, since the base current noise becomes comparable to the (series) emitter current noise. An amplifier for damping in the range above $500\ \Omega$ is shown in Fig. 9. The value of R_{in} is adjusted by trimming C_o. g_m is maintained constant by the dc feedback control of the drain (collector) current. In this analysis we have assumed that the noise due to the losses in the delay line is negligible. We found this to be the case for the lumped parameter lines we measured. The noise measurement can easily be performed in the system according to Fig. 6(c). Delay line is replaced

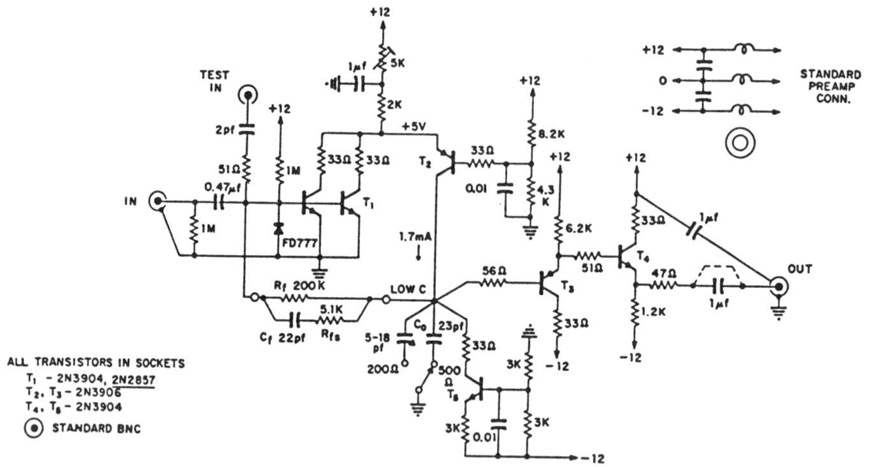

Figure 8. An amplifier with bipolar transistor input for cooled damping or termination with R_{in} = 200 and 500 Ω. $R_{in} = \frac{1}{g_m}\frac{C_o}{C_f}$. Peaking at high frequencies may be used for delay lines with Z_o = 500 Ω. R_{fs} = 0 for cases where peaking is not necessary. Series eq. noise resistance $R_s \approx 30$ Ω.

by a short connection between the amplifiers, which leaves the same relation of impedances in the circuit. There should be no difference in noise if the delay line noise is negligible.

5. Position Sensing with Resistive Electrodes

Resistive electrodes have been used in various forms for position sensing in gas proportional detectors and in semiconductor detectors. There are two basically different

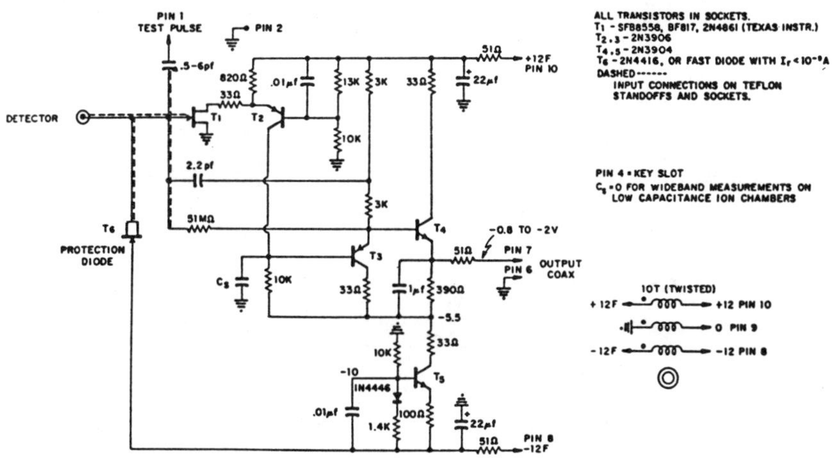

Figure 9. An amplifier with field-effect transistor input for cooled damping in the range $R_{in} \geq 500\ \Omega$.

$$R_{in} = \frac{1}{g_m} \frac{C_o}{C_f/2} .$$ Series eq. noise resistance $R_s \approx 60\ \Omega$. Peaking at high frequencies may be achieved by adding a resistor in series with C_f.

methods for position sensing with resistive electrodes. Due to the electrode capacitance the resistive electrode represents in both cases a diffusive R-C line. The two methods are illustrated in Fig. 10. In the charge division method[4-7,17,18] the position is determined from the ratio of charge flowing out of the ends of the resistive electrode terminated into low impedances. In the so-called rise time method[7,8,9] the resistive electrode represents a part of an

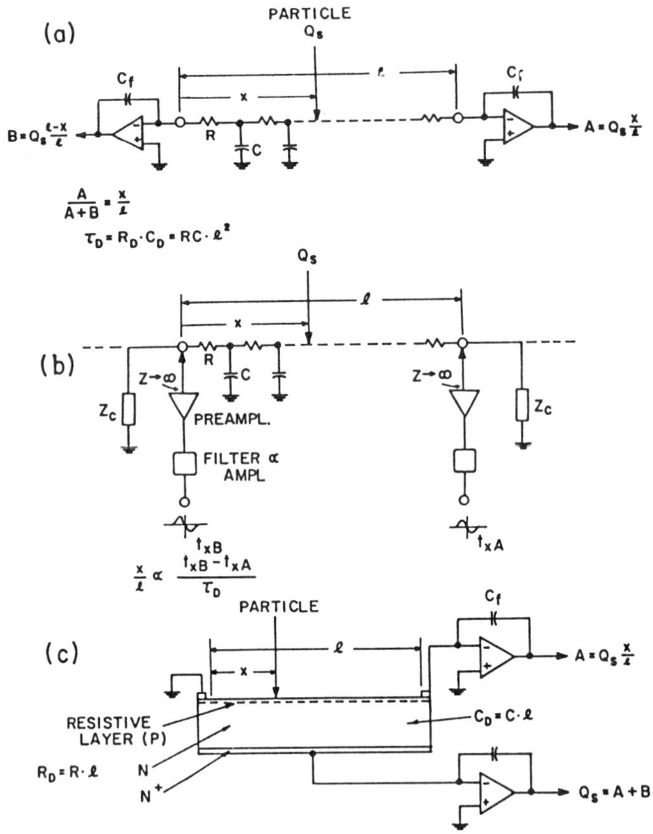

Figure 10. Resistive electrode (R-C line) as position-sensing medium.

(a) charge division method.

(b) diffusion time method.

(c) alternative configuration for charge division illustrated for a semiconductor detector with a resistive layer.

infinite diffusive line (if it is terminated into its characteristic impedance). The position is determined from the difference in the signal diffusion time from the point where the charge is collected. In the charge division method the electrode capacitance is a nuisance - it limits the resolution for all variables of interest: energy, position and event arrival time as well as the resolving time. In the diffusion time method the electrode capacitance is an essential element. With the electrode resistance it determines the relation of the diffusion time to the position. In both methods signal processing in relation to R-C line parameters is critical in order to achieve the best position resolution, linearity of the output signal with the particle position, and maximum event rates. We outline here an analysis from which the optimum signal processing and the limits of performance can be determined.

All the performance parameters can be expressed in terms of the R-C line parameters,

$$R_D = R\,\ell \ ,$$

$$C_D = C\,\ell \ , \qquad (49)$$

$$\tau_D = R_D\ C_D = RC\,\ell^2 \ ,$$

where R and C are resistance and capacitance per unit length, ℓ is the length of the line, and τ_D = the product of the total line resistance and capacitance is the line "time constant".

We first consider the requirement for linearity of the output signal with particle position. Due to the diffusive nature of the line, it takes time after a charge impulse is delivered until the ratio $A/(A+B) \approx x/\ell$ is established.

The charge at one end of the electrode (connected to zero impedance at both ends) as a function of time and position x/ℓ is given by,[7]

$$\frac{Q(t,x)}{Q_s} = 1 - \frac{x}{\ell} - \sum_{m=1}^{\infty} \frac{2}{m\pi} \, \mathrm{Si}\left(\frac{m\pi x}{\ell}\right) \exp\left(- \frac{m^2\pi^2}{\tau_D} t\right) \tag{50}$$

The dominant time constant of the series for transient phenomena in the line is τ_D/π^2. For linear relation between $Q(t,x)$ and the position, the sum in Eq. (50) should be negligible. The time required for the position nonlinearity to be less than 0.2% is,

$$t \geq \frac{1}{2} \tau_D \tag{51}$$

Since the signal waveform $Q(t,x)$ is position-dependent, we have to consider the problem of variable ballistic deficit in the selection of optimum filtering. In such a case a trapezoidal weighting function[16] gives the best signal-to-noise ratio with minimum resolving time. The form of this function is determined by the relation of the noise generated in the resistive electrode and the amplifier (series) noise.

For $t \geq 1/2\ \tau_D$ the admittance of the resistive electrode shorted at one end can be approximated by two components,[7]

$$Y = \frac{1}{R_D} + j\omega \frac{C_D}{3} \quad . \tag{52}$$

Then the equivalent circuit in Fig. 2 can be used for an approximate noise analysis, where R_D generates the "parallel noise" and CD/3 determines the contribution by the amplifier series noise. The noise corner time constant is for this case,

$$\tau_C = \left(\frac{C_D}{3} + C_A\right)\left(R_D \; R_s\right)^{1/2} \tag{53}$$

Since the diffusive line represents a dissipative position sensing medium, and its noise is inherently present with the signal, we should at least not add any other noise. To make the amplifier series noise negligible, one can conclude from relations (3), (4), (10) and (11) that the following conditions should be satisfied:

for noise sources: $$\frac{R_s}{R_D} \ll 1 \tag{54}$$

for the filter weighting function $$\left(\frac{I_2}{I_1}\right)^{1/2} \gg \tau_C \tag{55}$$

The condition (55) emphasizes that the weighting function must not have infinitely steep parts. On the basis of this condition we can determine the ratio of the sloped parts to the flat part for the filter weighting function for a given set of values R_s, C_A, R_D, C_D. Conversely, we can assume a trapezoidal weighting function with a minimum easily realizable ratio of the sloped parts to the flat part, and determine the condition for the detector and amplifier parameters. Referring to Fig. 3, we assume for an even, unipolar, trapezoidal function, $t_1 = t_3 = 0.2\ \tau_F$, $t_2 = 0.6\ \tau_F$, $b = 0$, where τ_F is the base width of the trapezoid. Then $I_1 = 10/\tau_F$, $I_2 = 0.733\ \tau_F$. The condition (55) with (53) becomes then a stricter condition for R_s/R_D,

$$\frac{R_s}{R_D} \ll 0.42\ (1 + 3\ C_A/C_D)^{-2} \quad . \tag{56}$$

The resistive electrode generates noise in parallel with the input, Eqs. (4), (6) and (11),

$$\overline{ENC_p^2} \approx 2\ kT \frac{1}{R_D} a_{F2} \tau_F \quad \text{for} \quad \tau_F \geq \frac{1}{2} \tau_D \quad . \tag{57}$$

For the trapezoidal function assumed, $a_{F2} = 0.733$, and

$$\overline{ENC_p^2} \approx 1.47\ kT \frac{1}{R_D} \tau_F \quad \text{for} \quad \tau_F \geq \frac{1}{2} \tau_D \quad . \tag{57a}$$

If we set $\tau_F = 0.8\ \tau_D$ so that the flat top of the trapezoidal function is equal to $1/2\ \tau_D$, we get for the minimum noise from the resistive electrode,

$$\overline{ENC_p^2} \approx 1.17\ kT \frac{\tau_D}{R_D} = 1.17\ kT\ C_D \quad . \tag{58}$$

This noise adds to the position signal Q_A (the noise in $Q_A + Q_B$ is much smaller since for $\tau_F > \tau_D/2$ it is anti-correlated at the two ends of the electrode). The optimum position resolution (for position nonlinearity $< 0.2\%$) is then,

$$\frac{\Delta \ell}{\ell} \approx 2.35 \frac{ENC_p}{Q_s} = 2.54 \frac{(kTC)^{1/2}}{Q_s} \quad (FWHM) \tag{59}$$

The position resolution for the charge division method with optimum filtering is determined only by the electrode capacitance and not by the electrode resistance. The resolving time is determined by both the electrode capacitance and the resistance, and it is equal to the width at the base of the weighting function, $\tau_F = 0.8\ \tau_D$.

In addition to the position resolution two other parameters are frequently of interest. These are the noise in the energy signal (the sum $Q_A + Q_B$) and the position dependent timing walk in the sum signal. The equivalent circuit for the common electrode in Fig. 10(c) and for the sum signals in Fig. 10(a) can be approximated by the detector capacitance C_D in series with a resistance $R_D/12$ (Ref. 7). The noise contribution to the sum signal by the resistive electrode can be determined from the equivalent circuit in Fig. 2 and from Eqs. (3) and (5), substituting $R_D/12$ for R_s, C_D for C_{in} and $R_p \to \infty$. This noise contribution increases with both R_D and C_D, and it decreases with the width of the weighting function. The optimum weighting function for the energy measurement has a narrower flat part and longer sloped parts than the optimum function for the position measurements. The position-dependent timing walk for the timing signal derived from the sum signal is about $\tau_C/10$.

The lower limit for τ_D is determined by the charge

collection time in the detector. The charge division method is insensitive to charge collection time variations, and as can be shown, the charge collection time can be longer than τ_D and/or τ_F, while preserving the position signal linearity. The only consequence is a smaller signal in relation to the noise, and thus poorer position resolution. For the case where τ_D is shorter than the charge collection time, τ_F can be increased from the optimum value for an impulse signal, to optimize the position resolution.

For a given (smallest possible) capacitance C_D, R_D should be reduced until $\tau_D/2$ is equal to the detector charge collection time. This is in order to achieve the shortest resolving time (higher event rates), better energy resolution, and smaller timing walk.

We take as an example a semiconductor detector with a resistive electrode. With $\ell = 5$ cm, $C = 50$ pF, the charge required for a position resolution $\Delta\ell/\ell = 10^{-2}$ is $Q_s \geq 6.5 \times 10^5$ e, and $ENC_p = 3 \times 10^3$ rms e. With a charge collection time of less than 100 nsec, τ_D should be reduced to about 200 nsec; i.e., $R_D = 4 \times 10^3\ \Omega$. The filter, as discussed previously, is a trapezoidal filter with base width $\tau_F = 0.8\ \tau_D = 160$ nsec. Such a filter can be approximated by single delay line clipping and RC integrating networks.

As another example we take a proportional detector with single wire with capacitance $C = 10$ pF. In order to achieve a resolution $\Delta\ell/\ell = 10^{-3}$, a signal charge $Q_s \geq 2.8 \times 10^6$ is required. A careful inspection of the charge collection waveform is required to determine if the signal-to-noise ratio improves enough to warrant an increase in τ_F. This is due to the logarithmic nature of the charge collection with time, Fig. 1. If R_D is given (by the construction limitations of the detector) then for point ionization, an increase of τ_F above 0.8 τ_D would result in poorer resolution. In some cases of extended ionization (high pressure neutron detectors) the signal increase may exceed the noise increase. There is presently a gap in resistance values between metal alloy wires and carbon coated quartz filaments. A 1m long metal alloy wire may have a resistance of 2 to 4 X 10^3 Ω, if it is to be of adequate diameter for the detector operation and for the strength. With C $\approx$ 10 pF, this would give τ_D = 20 to 40 nsec. Such a detector would be capable of handling event rates of 10^6 sec^{-1} with 2 to 4% pileup probability.

Some properties of the diffusion time method have been described previously.[8,9] An exact comparison of all the properties of the two methods has not been made as yet. There is some experimental evidence that the position resolution as limited by noise is the same. The resolving time for the charge division method, as shown in this section, is about 0.8 τ_D for optimum position resolution and linearity. The resolving time for the diffusion time method, as achieved in practice,[8,9,19] is several times τ_D. There are some practical advantages in each of the methods. The charge division requires a ratio digitizer, while the diffusion time method requires only a time digitizer. The former requires a simple preamplifier with low input impedance. These preamplifiers can be some distance away from the detector, and the connecting transmission lines can be terminated by the input resistance of the preamplifiers (as shown in Sections 3 and 4). The diffusion time method requires R-C line terminations into its characteristic impedance, and preamplifiers with high impedance and short connections to the detector. The diffusion time method requires uniformity of both the resistance and capacitance in the design of the detector.

The difficulties of performing division have been somewhat exaggerated in the literature. A fast and accurate (logarithmic) divider has been constructed and operated.[17] Another approach, first proposed by Miller, et al.,[18] is based on a dividing analog-to-digital converter. With this approach the circuits are likely to be simpler than with a separate divider.

6. Position Sensing for Multielectrode Detectors with One Amplifier per Electrode

In readouts for multiwire proportional chambers based on one amplifier per wire, and where only position information is required, it is usually assumed that the signal is large and that electronic noise in the amplifier is not a problem. Due to cost and size considerations in the design of these amplifiers, their charge sensitivity is low. Since noise is not of concern in such applications, the circuit techniques used result usually in higher noise than justified. The problem of inadequate sensitivity arises in cases where the signal magnitude is limited by the detector design (smaller anode spacing). A number of applications have become of interest where linear signal information is needed (detection of transition radiation, identification of particles by ionization loss, x-ray detectors in astrophysics). In all cases involving semiconductor detectors and gas or liquid ionization detectors with little or no gain, charge sensitivity and noise become important.

What is the lowest noise one can achieve? Assuming detectors with a low electrode capacitance (10 - 100 pF) no transformers can be used in practice for detector amplifier matching. The limit is determined by the series noise of the amplifying device, assuming that the noise from parallel sources is negligible. The equivalent noise charge is given by Eq. (16),

$$\overline{ENC_s^2} = \frac{\overline{e_n^2}\, C_{in}^2}{t_m} \qquad (60)$$

An optimum weighting function was assumed. In this case this is a triangle with base width 2 t_m. In practice, a gaussian function can be achieved, with a small increase in noise for equal width. We assume C_{in} = 25 pF for the electrode (wire) + amplifier capacitance and a short resolving time t_m = 25 nsec. For better bipolar transistors and field-effect transistors $e_n \approx 10^{-9}$ V/Hz$^{1/2}$. Then $ENC_s \approx 10^3$ rms e.

The equivalent noise charge can be reduced by increasing the resolving time. For t_m = 0.5 μsec, we obtain $ENC_s \approx 220$ rms e.

The next question is, can one use a bipolar transistor or must one use a field-effect transistor. This is determined by the relation of the resolving time to the noise corner time constant. To make the parallel noise negligible, the following condition must be satisfied.

$$t_m < \frac{1}{2} \tau_c = C_{in} (R_s R_p)^{1/2} \tag{61}$$

For the above example with $t_m = 25$ nsec, this requires that $R_p \geq 5 \times 10^4\ \Omega$. For the example with $t_m = 0.5\ \mu$sec, the noise resistance in parallel must be much higher, $R_p \geq 5 \times 10^6\ \Omega$. To be able to compare the noise contribution by the base current shot noise to that of a biasing resistor, we can use the "50 millivolt rule". It can be derived from Eqs. (6) and (7). If the shot noise due to a current with mean value I_o is equal to the thermal noise from resistance R_p at T = 295°K, then the "50 mV rule" for equal noise from R_p and I_o:

$$R_p I_o \approx 50 \text{ mV} \quad . \tag{62}$$

For a base current $I_{Bo} = 10^{-5}$ A, $R = 5 \times 10^4\ \Omega$, $R_p I_o \approx 0.5$ V, which means that the base current noise is higher than the noise from R_p and that it does not satisfy the condition (61).

Thus even at very short resolving times field-effect transistors give better noise performance. If ultimate noise performance is not required; bipolar transistors can be used. The noise can be determined from the relations (3) and (4). In general, with bipolar transistors the width of the weighting function should not be larger than necessary for processing of the signal otherwise an increased noise due to the parallel sources would result.

How simple can a low noise amplifier be? Figure 9 illustrates a design of a low noise amplifier for high resolution spectrometry. However, if the requirements on the linearity, dynamic range and open loop gain are somewhat relaxed, simple low noise circuits result. The simplest feedback charge amplifier is based on the configuration in Fig. 5, and it employs three transistors, as shown in Fig. 11(a).

The next question of current interest with multi-wire proportional chambers is whether the preamplifiers can be remote to the detector. The equivalent noise charge increases linearly with capacitance at the input, so that much higher noise would result. The amplifier noise may still be sufficiently small if large signals are available. The main problem in this case is pickup noise from the surrounding electrical equipment. Ground loop decoupling and common mode rejection are ineffective with high impedance sources, since these are difficult to balance, and the noise signal induced is not shorted in the source. A circuit configuration for "minimum electronics at the detector" and with common mode rejection is shown in Fig. 11(b). The eq. series noise resistance of the input device is increased by only about 100 Ω. The circuit also illustrates transmission line termination with a transistor in the common base configuration.

Remote amplifiers usually require transmission line

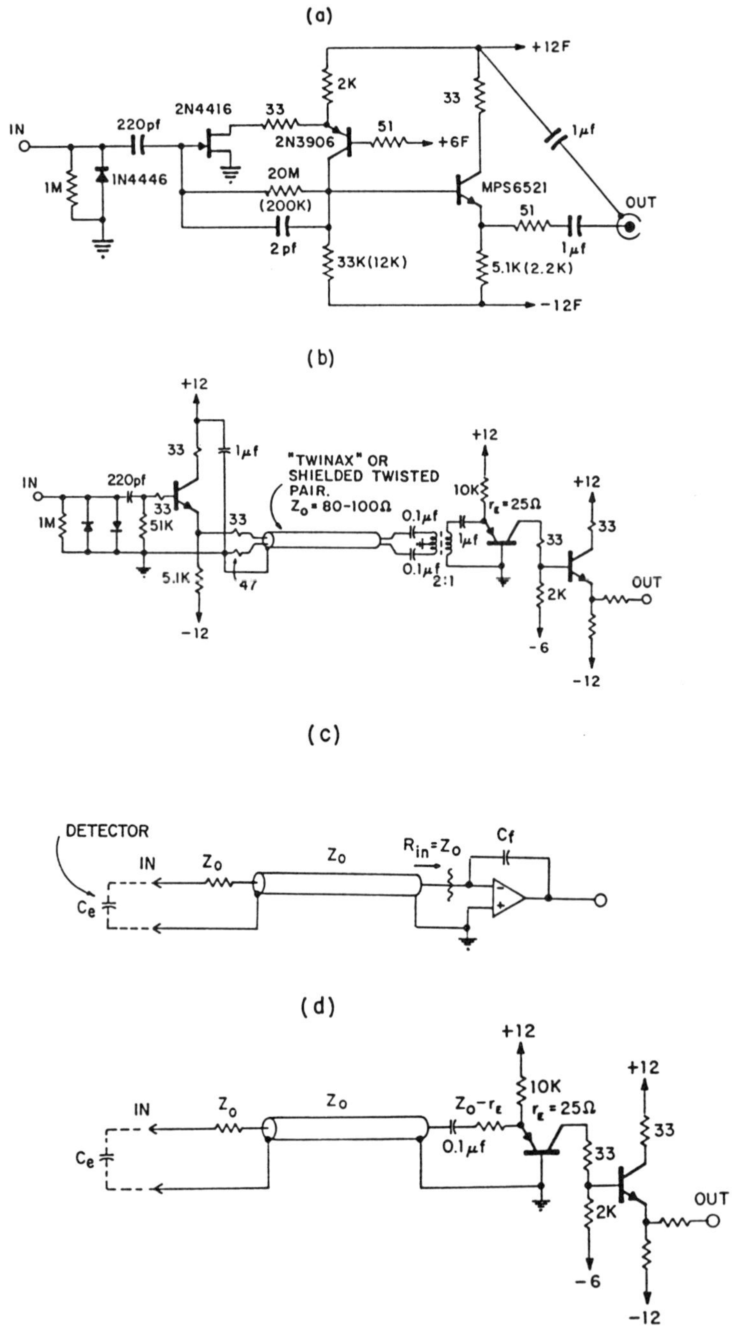

Figure 11. Linear amplifier configurations for multielectrode detectors.

termination. Instead of wasting the signal in the termination resistance, the input impedance of the amplifier can be used. One method is based on the feedback amplifier with resistive input impedance ("cooled" termination), Fig. 1(c), as described in Section 3. The other is based on a common base configuration, Fig. 11(d). In both cases a series sending end terminating resistor reduces reflections at high frequencies.

7. Discussion

The most interesting results are the resolution limitations due to noise in the position-sensing methods with delay lines and with resistive electrodes, Eqs. (48) and (59). It

---

(a) a simple low-noise amplifier. Bipolar transistors can be used in place of the field-effect transistors when longer filter weighting function and minimum noise are not required (resistor values are indicated in parenthesis).

(b) the "minimum electronics at the detector" amplifier. Common mode rejection by transformer. Voltage gain from the input of the emitter follower to the output ≈ 20.

(c) remote amplifier. Termination by "cooled" resistance.

(d) remote amplifier. Termination by the emitter dynamic resistance.

appears that the position sensing with delay lines requires a smaller signal for equal relative position resolution. An important condition for this is that "cooled terminations" should be used to achieve the improvement given by Eq. (42). This conclusion applies to the direct coupling to the delay line. With capacitive coupling to a delay line, a substantial attenuation of the signal may occur. With "warm" resistance terminations and amplifiers not optimized with respect to noise, the advantage of delay line position sensing may disappear. An improvement, as with "cooled" terminations, is important since it extends the useful operating range of proportional detectors, and it may make possible simultaneous position and ionization loss measurements.

It should be emphasized that these two methods are different and not applicable in all cases, and that each poses different requirements on the electrode configuration and construction of the detector. The same applies to the position sensing methods as to the detectors - no single solution satisfies all different applications. The analysis presented is concerned with noise limitations. In some cases a more detailed consideration of the charge collection statistics may be required. For example, with delay line position sensing, optimum filtering may be dependent on the particle track inclination. In general, methods based on timing are more dependent on the charge collection statistics than is the charge division method.

One subject this paper avoids carefully is any comparison of readout methods for particular applications. The choice of the detector and the readout is primarily a question of optimum experiment design and then of the system design. To mention a few variables entering in such considerations: single event or multiple event detection, event rate, particle type and density of ionization, track inclination, detection efficiency, timing and energy resolution requirements, and last but not least, cost distribution among the detector, readout electronics and data processing software.

Acknowledgements

Many useful discussions with J. Fischer, J. L. Alberi, E. Gatti, S. Iwata, E. Beardsworth and M. J. LeVine are gratefully acknowledged.

References

1. R. Grove, I. Ko, B. Leskovar and V. Perez-Mendez, Nucl. Instr. & Meth. 99 (1970) 381.
2. D. M. Lee, S. E. Sobottka and H. A. Thiessen, Nucl. Instr. & Meth. 104 (1972) 179.
3. J. R. Gilland and J. G. Emming, Nucl. Instr. & Meth. 104 (1972) 241.
4. K. Louterjung, et al., Nucl. Instr. & Meth. 22 (1963) 117.
5. W. R. Kuhlmann, et al., Nucl. Instr. & Meth. 22 (1966) 40.
6. G. Kalbitzer and W. Melzer, Nucl. Instr. & Meth. 56 (1967) 301.

7. R. B. Owen and M. L. Awcock, IEEE Trans. Nucl. Sci. NS-15 (1968) 290.

8. C. J. Borkowski and M. K. Kopp, Rev. Sci. Instr. 39 (1968) 1515.

9. C. J. Borkowski and M. K. Kopp, IEEE Trans. Nucl. Sci. NS-17 (1970) 340.

10. G. R. Richer, Jr. and J. J. Gomez, Rev. Sci. Instr. 49 (1969) 227.

11. S. Dhawan, IEEE Trans. Nucl. Sci. NS-20 (1973) 166.

12. V. Radeka, Proc. Intern. Symp. Nuclear Electronics, Vol. 1 (Versailles, France, 1968) p. 46-1.

13. V. Radeka, IEEE Trans. Nucl. Sci. NS-15 (1968) 455.

14. C. Kittel, Elementary Statistical Physics, John Wiley & Sons, Inc., New York, 1967, p. 152.

15. J. Fischer, S. Iwata, V. Radeka, E. Beardsworth, M. LeVine, to be published.

16. V. Radeka, Nucl. Instr. & Meth. 99 (1972) 525.

17. J. Alberi, J. Fischer, V. Radeka, L. Rogers, to be published.

18. G. L. Miller, N. Williams, A. Senator, R. Stensgaard and J. Fischer, Nucl. Instr. & Meth. 91 (1971) 389.

19. C. J. Borkowski and M. K. Kopp, "Recent Developments and Applications of Position Sensitive Proportional Counters Using RC-Line Signal Encoding", IEEE Trans. on Nucl. Sci., same issue.

Reprinted, with permission, from *Reports on Progress in Physics*, Vol. 43, pp. 1145-1189 (1980).

Monte Carlo theory and practice

F James

1. Introduction and definitions

1.1. Definition

A Monte Carlo technique is any technique making use of random numbers to solve a problem. (We assume for the moment that the reader understands what a random number is, although this is by no means a trivial point and will be treated later in some detail.)

The above definition should be supplemented by a somewhat narrower but more enlightening definition as given by Halton (1970): the Monte Carlo method is defined as representing the solution of a problem as a parameter of a hypothetical population, and using a random sequence of numbers to construct a sample of the population, from which statistical estimates of the parameter can be obtained.

Let us express the solution of the problem as a result F, which may be a real number, a set of numbers, a yes/no decision, etc. The Monte Carlo estimate of F will be a function of, among other things, the random numbers used in the calculation. The introduction of randomness into an otherwise well-defined problem produces solutions with rather special properties which, as we shall see, are sometimes surprisingly good.

1.2. Simulation

Historically, the first large-scale calculations to make use of the Monte Carlo method were studies of neutron scattering and absorption, random processes for which it is quite natural to employ random numbers. Such calculations, a subset of Monte Carlo calculations, are known as direct simulation, since the 'hypothetical population' of the narrower definition above corresponds directly to the real population being studied. However, as those involved were well aware, the numerical results obtained were perfectly 'deterministic' and, in principle, obtainable by classical computational techniques (in fact, integration). Whether or not the Monte Carlo method can be applied to a given problem does not depend on the stochastic nature of the system being studied, but only on our ability to formulate the problem in such a way that random numbers may be used to obtain the solution. This can be seen by inverting the neutron scattering problem and considering first the classical solution in terms of a complicated multidimensional integral. The value of this integral is quite non-random, but happens also to be the solution of a problem involving random processes. The Monte Carlo method may be applied wherever it is possible to establish equivalence between the desired result and the expected behaviour of a stochastic system.

The problem to be solved may already be of a probabilistic or statistical nature, in which case its Monte Carlo formulation will usually be a straightforward simulation, or it may be of a deterministic or analytic nature, in which case an appropriate Monte Carlo formulation may require some imagination and may appear contrived or artificial. In any case, the suitability of the method chosen will depend on its mathematical properties and not on its superficial resemblance to the problem to be solved. We shall see how Monte Carlo techniques may be compared with other methods of solution of the same physical problem.

1.3. Integration

At least in a formal sense, all Monte Carlo calculations are equivalent to integrations. This follows from the definition of a Monte Carlo calculation as producing a result F which is a function of random numbers r_i. Let us assume for simplicity the usual case that the r_i are uniformly distributed between zero and one. Then the Monte Carlo result $F = F(r_1, r_2, \ldots, r_n)$ is an unbiased estimator of the multidimensional integral

$$I = \int_0^1 \ldots \int_0^1 F(x_1, x_2, \ldots, x_n)\, \mathrm{d}x_1\, \mathrm{d}x_2 \ldots \mathrm{d}x_n$$

or, stated another way, the expectation of F is the integral I. (When the problem to be solved is explicitly the problem of integrating a function f, the F above is not to be identified with f but rather the Monte Carlo estimate of its integral.) This formal equivalence will allow us to lay a firm theoretical justification for Monte Carlo techniques and will also lead us to many results of practical importance.

2. Mathematical foundation for Monte Carlo integration

In this section we will define some basic statistical terms and invoke some of the important results of mathematical statistics to lay a formal foundation for the validity of Monte Carlo calculations. The results of this section will be important to the later sections, so we will try to make it complete, but since many readers will already be familiar with this material, no attempt is made to be mathematically rigorous. Those who wish a more detailed treatment are urged to consult an independent text, such as Eadie *et al* (1971). Those who still remember their elementary statistics are advised to skip directly to §2.6.

2.1. Random variables and distributions

A *random variable* is a variable that can take on more than one value (generally a continuous range of values), and for which any particular value that will be taken cannot be predicted in advance. Even though the value of the variable is unpredictable, the *distribution* of the variable may well be known. The distribution of a random variable gives the probability of a given value (or infinitesimal range of values). Since we will usually be working with continuous variables, we define

$$g(u)\,\mathrm{d}u = P[u < u' < u + \mathrm{d}u].$$

The function $g(u)$ is the ***probability density function*** of u and gives the probability of finding the random variable u' within $\mathrm{d}u$ of a given value u. This is the most usual way for physicists to express the way u' is distributed, although it is sometimes more convenient mathematically to use the ***integrated distribution function*** defined as the definite integral of g from minus infinity to u:

$$G(u) = \int_{-\infty}^{u} g(x)\,\mathrm{d}x$$

$$g(u) = \mathrm{d}G(u)/\mathrm{d}u.$$

Note that $G(u)$ is a monotonically non-decreasing function taking on values from zero to one, and that g is always normalised so that its integral over all u is one.

A function of a random variable is, of course, itself a random variable, although it will in general be distributed differently from its argument. The functions $G(u)$ and $g(u)$ defined above are, however, not to be considered as random variables since they are functions of the variable u rather than the random variable u'.

2.2. *Independence of random variables*

Let us consider two random variables u' and v'. In order to specify completely the distribution of u' and v', we now require a function of two variables, say $h(u, v)$, and the ensuing mathematics becomes considerably more complicated. However, an important special case is when the function $h(u, v)$ can be factored exactly into a product of two functions, each of which depends only on one variable, $h(u, v) = p(u)q(v)$. In this case we say that u' and v' are stochastically ***independent*** since the distribution of u' does not depend on the value of v' and vice versa.

When more than two variables are considered, the concept of independence becomes more complicated, and it is no longer sufficient to consider only the dependence of pairs of variables. Indeed, it is possible to have all pairs of variables independent and still have dependence among triplets or higher combinations of variables. For example, let r and s be two independent random variables, each uniformly distributed between zero and one, and consider the three new variables:

$$x = r$$

$$y = s$$

$$z = (r + s) \bmod 1.$$

Now each of the three random variables x, y, z is also uniformly distributed between zero and one, and all pairs (x, y), (y, z) and (x, z) are independent (knowledge about the value of one member of a pair gives no information about the value of the other member). However, the three are clearly dependent, since knowledge of any two determines the third completely.

2.3. *Expectation, variance, covariance*

The mathematical *expectation of a function* $f(u')$ is defined as the average or mean value of the function

$$E(f) = \int f(u)\, dG(u) = \int f(u)\, g(u)\, du$$

where $G(u)$ is a distribution function giving the distribution of the independent variable u'. Usually the u' will be uniformly distributed between a and b: $dG = du/(b-a)$, so that the expectation becomes

$$E(f) = \frac{1}{b-a} \int_a^b f(u)\, du.$$

Similarly the *expectation of a variable* u' is the average value of u:

$$E(u') = \int u\, dG(u) = \int u\, g(u)\, du.$$

The *variance* of a function or variable is the average of the squared deviation from its expectation and is most conveniently defined in terms of the expectation:

$$V(f)=E[f-E(f)]^2=\int [f-E(f)]^2\, dG.$$

Note that calculating the expectation requires one integration and the variance involves one more integration.

The square root of the variance is called the *standard deviation*. It is more physically meaningful than the variance since it has the same dimensions as its argument but the square root makes it more clumsy to manipulate mathematically. The standard deviation can most easily be interpreted as the root-mean-square deviation from the mean.

Considering expectation and variance as operators, we may verify some simple rules for applying these operators to linear combinations of variables. Let x and y be random variables and c be a constant. Then

$$E(cx+y)=cE(x)+E(y) \tag{2.1}$$

$$V(cx+y)=c^2V(x)+V(y)+2cE[(y-E(y))(x-E(x))]. \tag{2.2}$$

Expectation is therefore a linear operator, whereas variance is not linear. The last term in the above expression for the variance is called the *covariance between x and y* and is zero if x and y are *independent*. If this term is positive, x and y are said to be positively correlated, and if negative, x and y are negatively correlated. Note that x and y may be uncorrelated (i.e. their covariance may be zero) even if they are not independent, but if they are independent they must also be uncorrelated. Note also that even though the variance operator is not linear, the following relationship holds if x and y are independent variables:

$$V(x+y)=V(x)+V(y) \qquad x, y \text{ uncorrelated.}$$

2.4. The law of large numbers

The law of large numbers concerns the behaviour of sums of large numbers of random variables. Let us choose n numbers u_i randomly with probability density uniform on the interval from a to b, and for each u_i evaluate the function $f(u_i)$. This law says that the sum of these function values, divided by n, will converge to the expectation of the function f. That is, as n becomes very large,

$$\frac{1}{n}\sum_{i=1}^{n} f(u_i) \to \frac{1}{b-a}\int_a^b f(u)\, du. \tag{2.3}$$

In statistical language, the left-hand side of (2.3) is a *consistent estimator* of the integral on the right-hand side, since (under certain conditions) it converges to the exact value of the integral as n approaches infinity. The 'certain conditions' involve the behaviour of the function f, since it must of course be integrable, and we will generally require that it be everywhere finite and at least piecewise continuous (it may have a finite number of discontinuities in the interval under consideration).

Since the left-hand side of (2.3) is just the Monte Carlo estimate of the integral on the right-hand side, the law of large numbers can be interpreted as a statement that the Monte Carlo estimate of an integral is, under 'certain conditions', a consistent estimate, i.e. it converges to the correct answer as the random sample size becomes very large.

2.5. Convergence

It is worthwhile discussing at this point the meaning of convergence in the statistical context, since it is considerably more complex than the more familiar convergence of calculus. We recall that in calculus, the sequence $\{A\}$ is said to converge to B if for any arbitrarily small positive quantity δ, an element of $\{A\}$ can be found such that all the succeeding elements of $\{A\}$ are guaranteed to be within δ of B.

In the statistical context, the 'guarantee' must be replaced by a statement of probability, so that the corresponding definition becomes: $A(n)$ is said to converge to B as n goes to infinity if for any probability $P[0<P<1]$, and any positive quantity δ, a k can be found such that for all $n>k$ the probability that $A(n)$ will be within δ of B is greater than P. Note that this is quite weak, in that no matter how big n is, $A(n)$ can never be guaranteed to be within a given distance of B.

This risk, that convergence is only given with a certain probability, is inherent in Monte Carlo calculations and is the reason why this technique was named after the world's most famous gambling casino. Indeed, the name is doubly appropriate because the style of gambling in the Monte Carlo casino, not to be confused with the noisy and tasteless gambling houses of Las Vegas and Reno, is serious and sophisticated. The apparent contradiction between the unpredictability of the gambling process and the seriousness of the results is one of the fascinating aspects of the Monte Carlo method which has been responsible for a great deal of the interest shown in the method but has also resulted in considerable confusion and misunderstanding. This point will come up again, especially in our discussion of random numbers.

2.6. The central limit theorem

Whereas the law of large numbers tells us that the Monte Carlo estimate of an integral is correct for 'infinite' n, the central limit theorem tells us approximately how that estimate is distributed for large but finite n. This very important theorem says essentially that the sum of a large number of independent random variables is always normally distributed (i.e. a Gaussian distribution), no matter how the individual random variables are distributed, provided they have finite expectations and variances and provided n is 'large enough'. How large n has to be depends, of course, on the individual distributions, but in practice the convergence to the Gaussian distribution is surprisingly fast, even when the underlying distributions are, for example, uniform, as we shall see in an example in the following section.

The Gaussian distribution is completely specified by giving its expectation a and variance s^2. We denote by $N(a, s^2)$ the distribution whose density is Gaussian with mean a and variance s^2:

$$f(x)=\frac{1}{s\sqrt{2\pi}}\exp\left[-(x-a)^2/2s^2\right]$$

we can complete the statement of the central limit theorem by giving the expectation and variance of the (Gaussian) distribution resulting from summing a (large) number of independent random variables. This expectation and variance will, of course, depend on the expectations and variances of the individual distributions and can be calculated immediately using (2.1) and (2.2). Let the n independent random variables x_i have distributions with finite expectations e_i and variances v_i. Then $S=\Sigma x_i$ will have expectation $E(S)=\Sigma e_i$ and variance $V(S)=\Sigma v_i$. This is an exact result even for finite n, which follows from (2.1) and (2.2). The fact that the distribution of S is asymptotically Gaussian is the important part of the theorem which enables us to turn our knowledge of $E(S)$ and $V(S)$ into statements of probability about the value of S for a given trial.

2.6.1. Example: Gaussian random number generator. The central limit theorem allows us to construct a Gaussian random number generator, given any other kind of random number generator, simply by taking sums of random numbers. Let us see how this works in practice, using a uniform random number generator which we assume for the moment to be given. We will denote the sum of n uniform random numbers as R_n, so that R_1 will be a random number distributed uniformly (between zero and one). Then R_2 will be distributed as in figure 1(b), i.e. with a density function which is a triangle. This kind of distribution is familiar to gamblers using dice, where the outcome is the sum of two numbers uniformly distributed between one and six. The extreme values of the sum (2 and 12) are the most unlikely, and the middle value (7) is the most probable. R_3 is distributed as shown in figure 1(c), i.e. a parabolic spline function with knots at 1 and 2 (i.e. three different parabolas joined at the points $x=1$ and $x=2$, with the first derivative continuous at these points), which is beginning to look like the well-known bell-shaped Gaussian curve. R_4 is a cubic spline function, and higher sums are higher-order spline functions which approximate more and more closely the Gaussian distribution. After R_5 or R_6 the

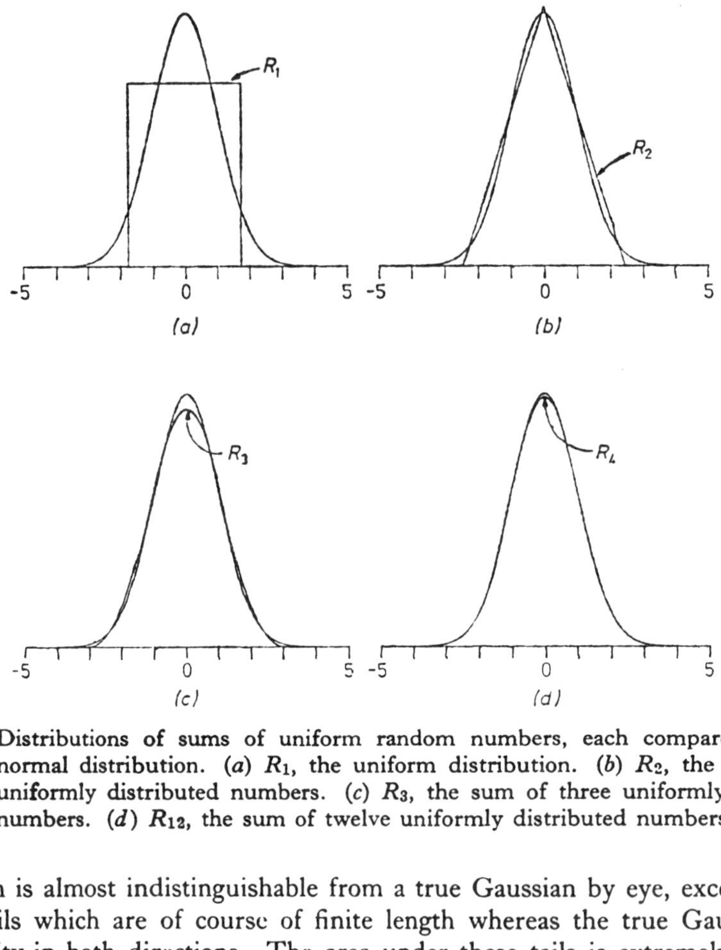

Figure 1. Distributions of sums of uniform random numbers, each compared with the normal distribution. (*a*) R_1, the uniform distribution. (*b*) R_2, the sum of two uniformly distributed numbers. (*c*) R_3, the sum of three uniformly distributed numbers. (*d*) R_{12}, the sum of twelve uniformly distributed numbers.

distribution is almost indistinguishable from a true Gaussian by eye, except for the extreme tails which are of course of finite length whereas the true Gaussian tails go to infinity in both directions. The area under these tails is extremely small, so the discrepancy in probability content is negligible for many applications, but care must be taken since the tails may be the most important feature.

Since the expectation and variance of the uniform distribution are, respectively, $\frac{1}{2}$ and $\frac{1}{12}$ (by straightforward calculation from the definitions of expectation and variance), we have

$$E(R_n) = n/2 \qquad V(R_n) = n/12.$$

Usually we want a standard Gaussian distribution, i.e. with mean zero and variance one. We therefore take

$$\frac{R_n - n/2}{n/12} \to N(0, 1).$$

A convenient choice for a practical Gaussian random number generator is $n=12$, which reduces simply to $R_{12}-6$. The properties of this generator will be discussed below in §7.

2.7. *Résumé: mathematical properties of the Monte Carlo method*

Let us consider again (2.3), where the left-hand side is the n-point Monte Carlo estimate of the integral on the right-hand side, the u_i being truly random numbers uniformly distributed between the integration limits a and b. The mathematical properties of this estimate are rather general properties of numerical results of Monte Carlo calculations, which we outline here.

(i) If the variance of f is finite, the Monte Carlo estimate is *consistent*, i.e. it converges to the true value of the integral for very large n.

(ii) The Monte Carlo estimate is *unbiased* for all n, i.e. the expectation of the Monte Carlo estimate is the true value of the integral. This follows directly from the linearity of the expectation operator.

(iii) The Monte Carlo estimate is asymptotically *normally distributed* (approaches a Gaussian density).

(iv) The *standard deviation* of the Monte Carlo estimate is given by $\sigma=\sqrt{V(f)}/\sqrt{n}$. This result is true for all n but is only useful insofar as the estimate is Gaussian-distributed (true only for 'large' n).

3. From Buffon's needle to variance-reducing techniques

In this section we present one of the earliest real Monte Carlo calculations, that of Buffon's needle, and examine some of its properties. We will see that its most important and worst property is its slow convergence (low efficiency). We then present a series of techniques known collectively as 'variance-reduction', designed to improve this efficiency.

3.1. *Buffon's needle: hit-or-miss Monte Carlo*

Although it is hard to imagine nowadays doing Monte Carlo calculations without a high-speed computer, the technique was first investigated and used long before the existence of electronics. One such early calculation, known as Buffon's needle (Buffon 1777), was used to calculate the value of π. It is a good example of the use of the Monte Carlo method to solve a problem which has no immediate statistical interpretation and which we are accustomed to attacking with more traditional mathematical tools.

The 'calculation' proceeds as follows. Lay out on the floor a pattern of parallel lines separated by a distance d (the stripes of an American flag will do). Repeatedly throw 'randomly' a needle of length d onto this striped pattern. Each time the needle lands in such a way as to cross the boundary between two stripes, count a 'hit'. When the needle does not cross a boundary, count a 'miss'. After a given (large) number of tries, estimate π by twice the number of tries (hits + misses) divided by the number of hits.

The above recipe is based on the fact that the probability of a hit is $2/\pi$. This can be calculated very easily as follows. Let the angle between the needle and the perpendicular to the stripes be equal to a, then the projection of the needle onto this perpendicular is of length $d|\cos(a)|$ and the distance between stripes is d. For a given angle a, the probability of a hit is clearly the ratio of these two lengths, $d|\cos(a)|/d = |\cos(a)|$. Since all angles are equally likely, the average value of $|\cos(a)|$ can be calculated by integrating $|\cos(a)|$ over its range and dividing by the range. By symmetry it is sufficient to integrate over one quadrant, say from 0 to $\pi/2$, where the integral is just one, and the probability is therefore $2/\pi$.

Estimating this probability by the actual ratio of hits to random tries is called *hit-or-miss Monte Carlo* and is in general the least efficient Monte Carlo method. Let us calculate the expected accuracy after n tries. The number of hits follows a binomial distribution with expectation np (where p is the probability of a hit, $2/\pi$) and variance $np(1-p)$ (Eadie *et al* 1971, p44). The variance of $2/\pi$ is therefore $p(1-p)/n$ and the standard deviation is the square root of this. Converting this to the standard deviation on π gives $2{\cdot}37/\sqrt{n}$. (We have to know π to calculate this result, but it could also be estimated from the data.) This means that the uncertainty on the value of π is

after	100 tries:	0·2374
after	10 000 tries:	0·0237
after	1 000 000 tries:	0·0024.

These uncertainties are intolerably high compared with those of almost any other method of calculating π. In addition, physical biases are difficult to eliminate, as will be discussed below in connection with the generation of truly random numbers. We can therefore conclude that Buffon's needle represents an amusing exercise and a good example of the application of the Monte Carlo method in an unexpected domain unrelated to stochastic phenomena, but that it should not be used in practice to calculate π. Now let us see how to improve upon it, still within the general framework of the Monte Carlo method.

3.2. Integration: crude Monte Carlo

Consider doing the Buffon needle calculation on a computer. We would choose a random angle a and a random distance x from the edge of the stripe pattern along the direction perpendicular to the stripes (the outcome is clearly independent of

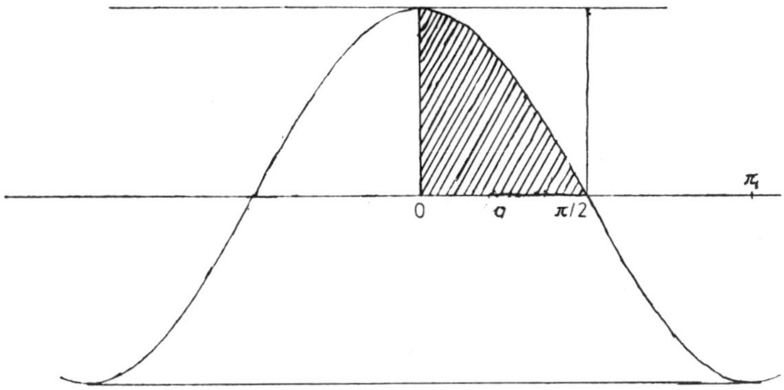

Figure 2. Buffon's needle as an integration problem.

translations along the direction of the stripes). On these (a, x) axes, figure 2 shows the region corresponding to a hit, namely the area between the a axis and the curve cos (a). The calculation is equivalent to the integration of cos (a).

Let us therefore perform this integration using *crude Monte Carlo* instead of the hit-or-miss variety, by straightforward application of the method of §2, choosing randomly values of a and averaging the values of $|\cos(a)|$. It is easily verified that this results in a standard deviation smaller by a factor of 0·82. This is a general result: crude Monte Carlo is always more efficient than hit-or-miss Monte Carlo, since hit-or-miss can be considered as crude Monte Carlo on a step function taking on only values zero or one, and of all functions bounded between zero and one with a given expectation, the step function has the largest variance.

Another way of looking at the comparison between crude and hit-or-miss is the following. For a given angle a, the probability of a hit is $|\cos(a)|$. Instead of finding the expectation of this value by direct averaging (crude Monte Carlo), we take it as the probability of actually generating a hit. In order to make a hit with probability $|\cos(a)|$, generate another random number x, $0<x<1$, and call it a hit whenever $x<|\cos(a)|$. This is less efficient, but it does mean that all the values entering into the average are equal to one (or zero), which may be advantageous in some situations. In many practical calculations it may correspond to using 'unweighted' rather than 'weighted' events, by taking the weight as the probability of accepting the event. In terms of pure Monte Carlo efficiency, this unweighting procedure is always disadvantageous, but it may improve the efficiency of other parts of the calculation, as we shall see later.

3.3. Classical variance-reducing techniques

From the results of §2, the square of the uncertainty on a Monte Carlo integral is

$$s^2 = V(f)/n.$$

This uncertainty can be decreased by increasing n, but this improves (converges) very slowly. Another way is to try to decrease the effective variance $V(f)$. We have already seen one example of changing the variance in comparing crude with hit-or-miss Monte Carlo. In this subsection we introduce the most important techniques for variance-reduction.

3.3.1. Stratified sampling. We may feel intuitively that the reason why Monte Carlo integration has such a large uncertainty is that the points are chosen unevenly, and that if the points were more uniformly distributed the fluctuations would be smaller. Intuition is not always right as we shall see in §4, but there is at least one way to make the point distribution more uniform which we can show will produce in general an improvement in the variance. Since it is a special case of a more general technique of controlling the distribution of points, let us first present the general technique.

Mathematically, stratified sampling is based on the fundamental property of the Riemann integral:

$$\begin{aligned} I &= \textstyle\int_0^1 f(u)\,\mathrm{d}u \\ &= \textstyle\int_0^a f(u)\,\mathrm{d}u + \int_a^1 f(u)\,\mathrm{d}u \qquad 0<a<1. \end{aligned}$$

The splitting up of the integral into pieces is a common technique in adaptive numerical quadrature, but the properties of this technique in the framework of Monte Carlo integration are somewhat different. The technique consists, in the general case, of dividing the full integration interval (or space) into sub-intervals (sub-spaces), and choosing n_j points in the jth sub-interval, whose length (volume) we will denote by $\{j\}$. Then, instead of adding the contributions from all points directly, partial sums are formed over each interval, and the partial sums are added, weighted proportionally to $\{j\}$ and inversely to n_j. This yields a result with the variance

$$s^2 = \sum_j \frac{\{j\}}{n_j} \int_{\{j\}} f^2(x)\,\mathrm{d}x - \sum_j \frac{1}{n_j} \left| \int_{\{j\}} f(x)\,\mathrm{d}x \right|^2$$

which is of course just the sum of the variances of the individual pieces. If the intervals $\{j\}$ and the numbers of points n_j are chosen carefully, this can lead to a dramatic reduction in the variance compared with crude Monte Carlo, but it can also lead to a *larger* variance, so something must be known about the function in order to use this technique most advantageously.

Suppose we do not know anything about the function and simply divide the space into equal volumes $\{j\}$, choosing in each volume equal numbers of points n_j (uniform stratification). It is easily verified from the above formula, using the triangle inequality, that uniform stratification cannot increase the variance and will in general decrease it if the expectation of the function is different in the different sub-regions. In particular, if the stratification is into just two equal regions $\{1\}$ and $\{2\}$, the improvement in variance is

$$D(s^2)=\frac{1}{n}\left|\int_{\{1\}} f(x)\,\mathrm{d}x-\int_{\{2\}} f(x)\,\mathrm{d}x\right|^2.$$

Since this cannot be negative, uniform stratification can be seen to be a safe method but the improvement in variance may be arbitrarily small.

In real calculations, additional complications may arise. In many-dimensional integration, for example, it may not be at all straightforward to divide the integration region into sub-regions of known volume. Computational overheads in time and memory space may also be prohibitive.

3.3.2. Importance sampling. We have seen that a large variation in the value of the function f leads to a large uncertainty in the Monte Carlo estimate of its integral. Conversely, Monte Carlo calculations will be most efficient when each point (event) has nearly the same function value (weight). This can be arranged by choosing a large number of points in regions of the sampling space where the function value is largest and compensating for this overpopulation by reducing the function values in these regions. In this way the reweighted function values become more nearly constant and the effective variance is reduced.

Mathematically, importance sampling corresponds to a change of integration variable:

$$f(x)\,\mathrm{d}x \to f(x)\,\mathrm{d}G(x)/g(x).$$

Points are chosen according to $G(x)$ instead of uniformly, and f is weighted inversely by $g(x)=\mathrm{d}G(x)/\mathrm{d}x$. The relevant variance is now $V(f/g)$, which will be small if g has been chosen to be close to f in shape.

To apply importance sampling to a function f, a function g must be found such that:

(i) $g(x)$ is a probability density function, i.e. it is everywhere non-negative and is normalised so that its integral over the sampling space is unity.

(ii) $G(x)$, the integral of g, is known analytically. This is an integrated distribution function, which increases monotonically as a function of x, from zero to one.

(iii) Either the function $G(x)$ can be inverted (solved for x) analytically or, alternatively, a g-distributed random number generator is available.

(iv) The ratio $f(x)/g(x)$ is as nearly constant as possible, so that the variance $V(f/g)$ is small compared with $V(f)$.

Importance sampling then proceeds as follows. Choose values of G randomly and uniformly between zero and one: for each G, solve for x, and evaluate $f(x)/g(x)$, taking the sum of these ratios as the result.

Although importance sampling is undoubtedly one of the most basic and useful Monte Carlo techniques, it suffers in practice from a number of drawbacks:

(i) The class of functions g which are integrable and of which the integral can be inverted analytically, is small: essentially the trigonometric functions, exponentials, and polynomials of very low degree, and some combinations of these. Of course the inversion can be done numerically, but this is usually slow and somewhat clumsy or else inaccurate.

(ii) True multidimensional importance sampling is extremely clumsy for all but the simplest functions, so that it is usually used one-dimension-at-a-time in multidimensional problems.

(iii) It is *unstable* in the sense that if the function g becomes very small, f/g becomes very large and in general its variance also. In particular, if g goes to zero somewhere where f is not zero, $V(f/g)$ may be infinite and the usual technique of estimating the variance from the sample points may not detect this fact if the region where $g=0$ is small. It is therefore dangerous to choose functions g which go through zero, or which approach zero quickly (such as Gaussian functions).

On the positive side, importance sampling is the only general method for removing infinite singularities in the integrand f, by using a sampling function g with a similar singularity in the same place.

3.3.3. Control variates. The control variate method is similar to importance sampling in that one again seeks an integrable function g which approximates the function to be integrated f, but this time the two functions are subtracted rather than divided. Mathematically, this technique is based on the linearity of the integral operator:

$$\int f(x)\,\mathrm{d}x = \int \{f(x) - g(x)\}\,\mathrm{d}x + \int g(x)\,\mathrm{d}x.$$

Now, if the definite integral of g over the entire interval is known, the only uncertainty comes from the integral of $(f-g)$, which will have a smaller variance than f if g has been chosen carefully.

The method of control variates is more stable than importance sampling, since zeros in g cannot induce singularities in $(f-g)$. Another advantage over importance sampling is that the integral of the 'approximating function' g need not be inverted analytically.

3.3.4. Antithetic variates. Usually Monte Carlo calculations make use of random numbers (points) which are *independent* of each other, at least in principle. The method of antithetic variates deliberately makes use of correlated points, taking advantage of the fact that such correlation may be negative as well as positive. We recall from (2.2) that the variance of the sum of two function values f' and f'' is just the sum of the individual variances when the random points where the function is evaluated are chosen independently, but that in the general case an additional term is present:

$$V(f'+f'')=V(f')+V(f'')+2\,\mathrm{cov}\,(f',f'').$$

If we can arrange to choose points such that f' and f'' are negatively correlated, a substantial reduction in variance may be realised. This requires knowledge of the function f, and it is not easy to give general methods for accomplishing this negative correlation. Hammersley and Handscomb (1964, pp60–5) discuss this in some detail and give further references. For our purposes it will suffice to give a simple example to see how the technique works in general.

Suppose that it is known that $f(x)$ is a monotonically increasing function of x. Then choose x_i randomly and independently as usual, uniformly distributed between the integration limits (say, 0 to 1), but instead of forming the sum of $f(x_i)$ we take one-half of the sum of $\{f(x_i)+f(1-x_i)\}$. Then each time x_i is small, resulting in a small value of $f(x_i)$, $1-x_i$ and thus $f(1-x_i)$ will be large, and vice versa. The partial sums $\{f(x_i)+f(1-x_i)\}$ will therefore be more constant than the individual function values and have a lower variance. Looked at in another way, we are taking the average of the estimate of the integral of $f(x)$ and the estimate of the integral of $f(1-x)$ using the same points x, and since these two functions are highly (negatively) correlated, the variance of the sum is less than the sum of the variances.

3.4. Adaptive variance-reducing techniques

With the possible exception of uniform stratification, all the variance-reduction methods described above require some advance knowledge of the behaviour of the function, and if misapplied may easily lead to a degradation of the Monte Carlo efficiency rather than an improvement, not to mention the additional labour factor involved in the application of the variance-reduction. A natural extension is toward *adaptive* techniques which learn about the function as they proceed, preferably requiring no *a priori* knowledge about the function. Similarly inspired techniques abound in numerical quadrature where it is probably safe to say that most automatic function integration is done using adaptive methods. Truly adaptive methods for Monte Carlo integration are less common, perhaps because they are rather difficult to realise (and easy to misinterpret). We shall consider three examples which should serve to illustrate the problems involved and ideas that have proved to be useful. The programs I shall describe here are all designed for multidimensional integration of general functions, especially badly behaved functions with spikes and large variances.

3.4.1. Sheppey and Lautrup's RIWIAD. The program RIWIAD of Sheppey and Lautrup is one of the earliest to be used with success on difficult multivariate functions on the hypercube. It first divides the full hypercube evenly into a number of sub-hypercubes and estimates the integral and its variance in each hypercube by crude Monte Carlo (uniform stratification). Based on the values found in each sub-volume, it then adjusts the boundaries to form new hyper-rectangles such that sub-volumes are smaller where the function is larger, and the process is continued. At each step, an estimate of the integral and its uncertainty is made in each sub-volume, and the interval boundaries are modified to improve the next stratification. A running weighted average of the integral estimates and uncertainty estimates is maintained, and the procedure stops when the desired uncertainty is achieved.

RIWIAD has several drawbacks. The stratification boundaries are always parallel to the original parameter axes and always run along the whole length of the hypercube, dividing all the volumes through which they go, even if the previous results indicated that some of these sub-volumes did not have to be divided. Worst of all, the weighted average of partial results produces a bias due to the correlation between the estimate of the expectation and the estimate of the variance. Suppose, for example, that the function has a narrow spike, and that on the first step no point falls in the spike. Both the integral and its variance will be estimated too low. Then on the next step, a point hits the spike; this time the estimates are both about right, but since the variance is large the value gets a low weight and the overall estimate remains too low. The program never recovers from such an incident since it never forgets an early value even if later experience shows it to be a bad estimate.

3.4.2. Friedman's adaptive importance sampling. A more recent program of J Friedman (unpublished, superseded by his more recent effort described immediately below) uses a quite different approach. The program is divided into an exploratory phase and an evaluation phase, and none of the function values found in the exploratory phase are used explicitly in the evaluation. This avoids the bias due to the way the exploratory points are chosen, at a modest cost in efficiency. The exploratory phase is used to establish a control function which will be used for the importance sampling of the evaluation phase. The control function is a sum of Cauchy (Breit–Wigner) peaks, whose positions and shapes correspond to those of the function to be integrated, as determined respectively by a peak search using a function-minimising routine, and an eigenvector analysis of the covariance of the function around each peak. Cauchy-shaped peaks are used because they tend to zero more slowly than Gaussian peaks, helping to avoid the instability problem mentioned above.

Although this program is an improvement over RIWIAD for most functions, it also has several drawbacks in practice and is unsuitable for functions which cannot be approximated by a small number of peaks.

3.4.3. Friedman's DIVONNE2 *with recursive partitioning.* A more recent offering of Friedman, called DIVONNE2 (Friedman 1977a, b), represents a synthesis of the ideas seen to be most valuable in the above programs, together with some more modern ideas in multidimensional data structures. It consists of two separate programs, the first of which performs a recursive multidimensional partitioning (stratification) of the function parameter space, and the second does a stratified-sampling Monte Carlo integration based on this partitioning.

The goal of the partitioning is to produce sub-volumes in which the *range* of function values, as determined by function-minimisation techniques, is as small as possible. The partitioning program retains the drawback of RIWIAD that partition boundaries must be parallel to the parameter axes, but since the partitioning is recursive (only one sub-volume is split in two at each step, not a whole row), the algorithm eventually tends to liberate itself from the orientation of the axes.

The partitioning algorithm has other applications than integration and can be used, for example, in conjunction with a specially designed random number generator to generate points in the parameter space distributed according to the function f (see the subsection below on generating random numbers according to empirical distributions).

The actual integration need not be performed using Monte Carlo. Other methods are offered as options in the program, but in practice this choice does not seem to make much difference in the accuracy obtained, and Monte Carlo is usually used because it gives a reasonably accurate uncertainty estimate.

4. Comparison with numerical quadrature

In order to decide whether a Monte Carlo method should be applied to a given problem, it is reasonable to see how it compares with other available methods. In the case of integration, alternative numerical techniques have been the subject of extensive studies for centuries, and the widespread use of computers has led to considerable practical experience in this field. The current section is a brief review of the properties of numerical quadrature as it is commonly practised today, for the purposes of comparison with Monte Carlo. This is not intended to be a complete or detailed account of any quadrature techniques but is intended only to give the properties of most use in deciding whether to use quadrature at all.

4.1. One-dimensional quadrature

Unless otherwise stated, numerical quadrature is always done in one dimension. Some of the reasons for this will appear later, but certainly a prime motive for sticking with one dimension is the beauty and elegance of the methods that have been developed for one dimension.

All quadrature formulae approximate the value of the integral by a linear combination of function values:

$$I_q = \sum_{i=1}^{n} w_i f(x_i).$$

Different formulae correspond to different choices of the points x and the weights w. Crude Monte Carlo could be considered a quadrature formula with unit weights and points chosen uniformly but randomly.

4.1.1. Trapezoidal rule. This simplest of all rules consists of dividing the total interval into n sub-intervals and approximating the integral over each sub-interval by the area of the trapezoid inscribed under (or over) the curve to be integrated. The sum of these approximations reduces to the average of the $n+1$ function values multiplied by the length of the interval (in fact, the end points must be added with a factor one-half, but this important detail can be considered as a boundary correction and is not relevant to our arguments here). For large n, we can think of the function expressed as a Taylor series expansion about each of the n points: then the constant terms and the first derivative (linear) terms will be integrated exactly by the trapezoidal rule, and to the extent that higher-order terms are of decreasing importance, the largest contribution to the error will come from the second derivative (constant curvature) terms. This error is proportional to the sagittas of the curve segments over each band, and these sagittas will each be proportional to the square of the distance between successive points. Therefore if the function is evaluated at n equally spaced points, the uncertainty on the integral should be proportional to $1/n^2$ for large n.

Recall that for Monte Carlo integration, the convergence was only like the square root of n, so that where increasing n by a factor of 100 only buys you one more decimal digit with Monte Carlo, you get *four* digits with the trapezoidal rule. This is especially interesting because the two methods are so similar. Indeed, the methods are identical except that points are chosen equally spaced in one case and randomly in the other, and the randomness apparently causes us to lose a factor of four in convergence rate (decimal digits per factor of 100 increase in n). Before seeing what randomness gives us in return for this disastrous convergence rate, let us consider still more impressive convergence rates of other quadrature methods.

4.1.2. Higher-order quadrature. By choosing the points and weights appropriately, it is possible to integrate exactly polynomials of higher degree and therefore achieve higher convergence rates. The next step after the trapezoidal rule is Simpson's rule which requires three points on a given interval and integrates exactly all polynomials of degree three. The highest possible degree for a given number of points is achieved with Gauss quadrature formulae which integrate exactly all polynomials of degree $2n-1$ (or less) with n carefully chosen points and n corresponding weights. The numerical values of these points and weights, as well as the basic properties of Gaussian quadrature, are given by Stroud and Secrest (1966).

The theoretical convergence rate for Gauss quadrature is enormously higher than for Monte Carlo, but some of its other properties are not so nice. The uncertainty is not easy to estimate, error-bound formulae being given in terms of the values of higher derivatives of the function over the interval, which are much harder to calculate than the integral itself, so are essentially useless in practice. In addition, the validity of the error-bound formulae depends on continuity properties of the function and its derivatives, which may not be known. In practice, one is forced to use 'overkill', aiming at a precision much higher than that required, and uncertainties, if estimated at all, are usually estimated by comparing the results of more than one different Gauss rule on the same interval. Unfortunately, the nature of these rules is such that the best way to combine the results of two different Gauss rules over the same interval is to throw away the lower-order result and keep only the higher. Practical experience indicates also that there is no advantage in going to extremely high orders, and that beyond about 12 or 15 points it is usually better to split the interval and apply a lower-order rule several times. This indication of the breakdown of the polynomial philosophy is discussed below.

4.1.3. Adaptive quadrature. The quadrature rules described above are all fixed-point rules, i.e. the points and weights are fixed in advance. Adaptive quadrature, on the other hand, is an attempt to attain a prescribed accuracy by adapting the quadrature method to the function. The most common class of adaptive methods consists in using a fixed-point rule and an error-bound estimate, then dividing the interval into two or more pieces, usually of equal length, if the error-bound estimate exceeds the required value. The same procedure is then applied recursively to each sub-interval until all sub-intervals satisfy the error bounds, or until the sum of all estimated uncertainties reaches an acceptable level. The most common strategies are compared by Malcolm and Simpson (1975).

Most computer centres offer one or more 'automatic integration' programs based on adaptive quadrature of the above type. These programs differ mainly in the fixed-point rule used and in the method of obtaining an estimate of uncertainty which, as we have seen, is not always straightforward. Because of problems in obtaining reliable estimates of uncertainty, the better programs aim for a certain amount of overkill, but may be unreliable nonetheless. For example, a spline function which appears smooth to the eye has discontinuous higher-order derivatives which tend to produce poor results with high-order Gauss rules and consequently adaptive quadrature based on them. Other problems with adaptive quadrature are discussed by Lyness and Kaganove (1976).

4.2. Multidimensional quadrature

Numerical quadrature formulae are based on the study of orthogonal polynomials, which are well understood in one dimension. For higher dimensionalities the mathematical basis is not as well understood, and practical studies are much more recent and less extensive. We outline here briefly the current situation.

4.2.1. Multidimensional region boundaries. In one dimension, only three 'different' regions of integration need to be considered: finite, semi-infinite and infinite. Choosing one particular interval in each class, all other intervals can be mapped onto one of the three by a linear mapping, which conserves all the convergence properties of any integration method. In general in this review, we consider only the finite interval. Simple non-linear transformations are available to transform semi-infinite and infinite intervals into the unit interval, and this is a standard way to perform integration over infinite intervals, but these transformations do modify the properties of quadrature rules. For Monte Carlo integration, these transformations do not affect the n dependence of the convergence, but the function whose variance determines the uncertainty of the estimate is, of course, the transformed function.

In more than one dimension, the situation is quite different. Already in two dimensions, and restricting ourselves to finite regions, there are an infinite number of 'different' regions which cannot be transformed into each other by linear transformations. For example, a circle is fundamentally different from a square, in the sense that a quadrature formula for a square will not have the same properties when applied to a circle.

The standard Monte Carlo technique for dealing with odd-shaped regions is to embed the region in the smallest hyper-rectangle that will surround it and integrate over the hyper-rectangle, throwing away the points that fall outside the inner region. This leads to some inefficiency of course, due to the rejected points, but is capable of dealing in a straightforward way with essentially any finite region. Such a general technique does not work for numerical quadrature methods, since it introduces discontinuities on the boundary of the inner region, thus destroying some of the nice convergence properties.

The ability of Monte Carlo to integrate over complicated multidimensional regions (albeit not always very efficiently) is one of its most valuable properties, since it is often the only known technique capable of handling such problems. Purists may be right in saying that this only expresses our ignorance of better methods, but for people with real problems to be solved, it does represent a way out.

4.2.2. Extension of one-dimensional rules. For rectangular regions, which are after all the most common, multidimensional quadrature rules can be formed by straightforward extension of one-dimensional rules. Such rules, known as product rules, generally preserve the properties of the one-dimensional rules of which they are extensions, but only at the cost of increasing the number of points exponentially with the dimensionality. Thus a product rule requiring n function evaluations in one dimension will require n^2 evaluations in two dimensions, n^3 in three dimensions, and so on. This slows down the effective convergence rate in d dimensions by a factor $1/d$ in the exponent as shown in the table below.

Uncertainty as a function of number of points n	In one dimension	In d dimensions
Monte Carlo	$n^{-1/2}$	$n^{-1/2}$
Trapezoidal rule	n^{-2}	$n^{-2/d}$
Simpson's rule	n^{-4}	$n^{-4/d}$
Gauss rule	n^{-2m+1}	$n^{-(2m-1)/d}$

Since the convergence of Monte Carlo is independent of dimensionality, there is always some d above which Monte Carlo converges faster than any fixed quadrature rule. Thus Simpson's rule in more than eight dimensions converges more slowly than Monte Carlo and a ten-point Gauss rule converges more slowly than Monte Carlo in more than 38 dimensions, even assuming that the function has the nice continuity properties required by these higher-order rules.

But suppose we actually try to apply a ten-point Gauss rule in 38 dimensions. This requires at least 10^{38} function evaluations, which is clearly unfeasible. This brings up two new points.

(i) *The feasibility limit* is the largest number of function evaluations we can afford to make. Depending on the computer resources available, the feasibility limit will usually be between 10^5 and 10^{10} points for functions which can be evaluated reasonably fast. This limits the use of a ten-point Gauss rule to five dimensions for someone with moderate computer resources, or ten dimensions for someone with 'unlimited' computer resources. Figure 3 shows that, except for very-low-order rules, the feasibility limit is reached long before the crossover point where Monte Carlo converges faster than quadrature, so that the theoretical convergence rates for high-order rules in high dimensionalities will remain purely theoretical.

(ii) *The growth rate* is the smallest number of *additional* function evaluations needed to improve the current estimate. Monte Carlo estimates can be improved by adding a single point, but at the other extreme Gauss rule estimates can only be improved by going to a higher-order rule, requiring $(m+1)^d$ additional points

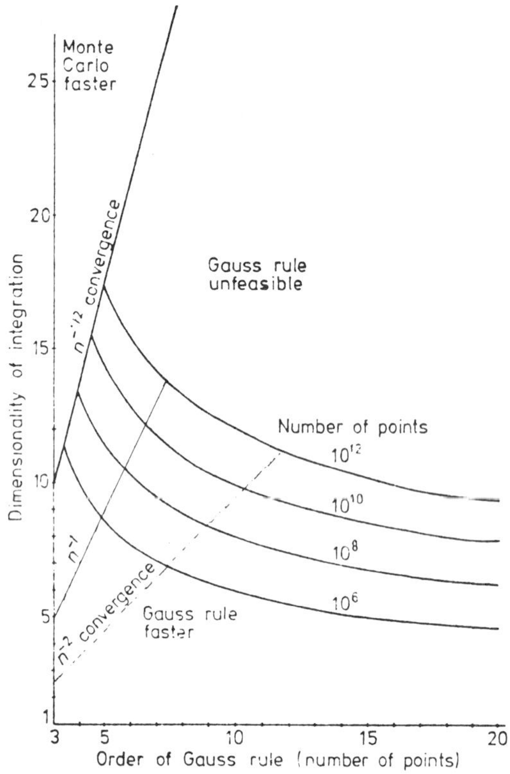

Figure 3. Comparison of Monte Carlo integration and numerical quadrature in many dimensions.

or by sub-dividing the space, which requires at least $2m^d$ additional points even for the simplest partitioning. In both cases all the previous Gauss points must be thrown away.

One way to get around the problems of feasibility limit and growth rate has been suggested by Tsuda (1973). He uses a rule with far too many points actually to evaluate, and then applies the standard Monte Carlo technique of sampling the resulting terms randomly. He reports good results for this combination of quadrature and Monte Carlo, but the reasons behind this success are not clear to me. It could be that the use of points of a quadrature rule guards against any two points being too close, and therefore ensures a certain uniformity of distribution even if only a random subset of these points is actually used (this explanation was suggested to me by J Friedman).

4.2.3. Multidimensional rules. The situation is greatly improved if truly multi-dimensional quadrature rules are used instead of product rules. Unfortunately, good quadrature rules are not known for many regions, dimensionalities and orders. The situation is well described in the article of Haber (1970) and in the book of Stroud (1971), of which we summarise some of the more important results here.

As in one dimension, multidimensional formulae can be found which will integrate exactly any polynomial of degree less than or equal to some degree r. In addition we may require the formulae to satisfy two important criteria.

(i) That all the weights be positive. This is important in order to avoid numerical instabilities arising from cancellation of large terms of opposite sign, and seems also to make the formulae more robust with respect to the validity of the polynomial assumption.

(ii) That all the points used lie within the region of integration. This seems such an elementary requirement that one is surprised to discover that many formulae in d dimensions do not possess it, even for convex regions.

If we restrict ourselves to formulae satisfying the above requirements, very few generally applicable formulae have been found. Even for the simplest region, the hypercube, the only known formulae valid for all dimensionalities and even reasonably close to the theoretical efficiency limit are of degree 2 and 3, as summarised in the table below. We see that there exist low-order multidimensional formulae con-

Degree r	Best known n	n (Gauss)
2	$2d+1$	
3	$2d$	2^d
5	$O(d^5)$, not found	3^d
>5	?	$\left(\frac{r+1}{2}\right)^d$, r odd

siderably better than the Gauss rule, and in fact some higher-order formulae of comparable theoretical efficiency are known, but they do not have all positive weights for all r. Stroud has shown that a formula of degree $r=5$ exists, with n of the order of d^5 and positive weights, but to my knowledge no one has as yet found it.

4.2.4. Adaptive multidimensional quadrature. Like non-adaptive quadrature, adaptive quadrature is much better developed in one dimension than in many dimensions, since the problems mentioned above for multidimensional quadrature, in general, clearly make adaptivity difficult too. Nevertheless, several attempts have been made, of which we will mention a few that have been published. They appear to be reasonably successful, at least for small dimensionalities (up to 6).

(i) Van Dooren and de Ridder (1976) have published an algorithm not too different from Friedman's DIVONNE2 (1977a, b), except that the former use multidimensional extensions of one-dimensional Gauss rules instead of Monte Carlo for the basic integration technique and their sub-division of regions is always into two equal parts.

(ii) Genz (1972) presents an algorithm especially interesting for its use of extrapolation methods, but the multidimensional adaptivity does not seem to result in a great improvement in efficiency.

(iii) Kahaner and Wells (1979) use an interesting technique based on simplices rather than hypercubes. Their basic thesis is that the lack of good adaptive quadrature procedures in many dimensions is mainly due to problems in organisation of multidimensional data structures. Their work is as much an exercise in programming as numerical analysis and it presents many interesting ideas in both areas. It probably points in the direction where we can expect the most significant advances. From a practical point of view, their program is not of much interest since it is written in a language (MADCAP) not generally available.

4.3. The Monte Carlo paradox

Some of the conclusions to be drawn from the comparison of Monte Carlo integration with numerical quadrature are somewhat surprising and call for deeper consideration.

(i) In one dimension, the perfectly 'regular' trapezoidal rule converges much faster than the identical rule with randomly distributed points, but in many dimensions a random distribution leads to faster convergence than the perfectly regular grid.

(ii) Just to confuse matters further, the random distribution which is superior to the regular distribution in many dimensions can nevertheless be improved by making it more uniform, either by stratified sampling as we have already seen, or through quasi-Monte Carlo which is discussed below.

The explanation of this paradox is that our intuitive feeling for what constitutes 'uniformity' in distribution, based on one-dimensional knowledge, is not quite right for higher dimensions. For example, consider the projection of the point distribution onto one axis, for the hyper-rectangular grid of points. Great spikes appear in this projection whenever we come to a 'hyper-row' of points, which no longer looks very uniform; the projections of a random distribution are more uniform in this sense. (See Sobol (1979) for a simple and convincing example of this.) In the section on quasi-Monte Carlo, we will define and discuss a more precise measure of uniformity (or non-uniformity) called discrepancy, which will explain this paradox.

Furthermore, the volume of multidimensional space is always very big, so that points are always far apart, which negates the very basis of quadrature rules.

4.3.1. The 'polynomial hangup'. Let us look more carefully at the theoretically fast convergence rate of high-order quadrature rules. This is related to the 'polynomial hypothesis' dear to the hearts of quadrature experts. For low orders, it is hard to find fault with the polynomial hypothesis; the zero-degree polynomial is certainly the simplest function and it is reasonable to expect a good integration method to be able to integrate it exactly. (Even Monte Carlo does that, by the way!) Similarly, a first-degree (straight-line) polynomial naturally comes next in the scale of complicatedness as perceived by the human eye, but who is to say that a parabola is simpler or smoother than, for example, a sine function or an exponential? Is there a justification for seeking methods that integrate exactly polynomials of degree r, when the function to be integrated is not a polynomial?

We may seek such a justification in Taylor's theorem. This theorem states that under certain conditions any function can be expressed as a polynomial of degree r, plus a remainder term. The conditions are that the function and all its derivatives should be continuous; the coefficients of the polynomial are given in terms of these derivatives evaluated at the point where the independent variable is equal to zero. The usefulness of the theorem comes from the cases where the remainder term becomes very small as r increases, but the theorem says nothing about when this can be expected to be true.

Indeed, there is nothing special about the polynomial in this respect. Other theorems give conditions under which general functions can be expressed as other infinite series (e.g. trigonometric series, Fourier series) and conditions under which the series can be truncated with a given remainder. The property that makes the Taylor series special is that under very general conditions the higher-order terms can indeed be neglected *in the neighbourhood of zero* (the point about which the expansion is performed). Unfortunately, this has very little to do with the macroscopic properties of the function which are important for integration over a large region, especially a multidimensional region which is always large.

In practice, polynomials are notoriously bad at approximating functions over large intervals, so we should not be surprised that related quadrature rules sometimes give unsatisfactory results. Experience shows spline functions to be good at approximating a wider class of functions, and although spline functions are piecewise polynomials they are not polynomials, and indeed have discontinuous derivatives of some degree at the knots. Integrating spline functions is, of course, easy if you know where the knots are, although Gauss rules generally fail without this knowledge.

5. Random and pseudo-random numbers

In principle, a random number is simply a particular value taken on by a random variable (which was defined above). However, in Monte Carlo studies, one often uses the word 'random' with various other, quite different, meanings. Here it is usually applied to sequences of numbers which, once they have been determined, are not at all random in the statistical sense but may have some properties which are similar to the properties of a truly random sequence. To be precise one must distinguish three different types of sequences: *truly random*, *pseudo-random* and *quasi-random*. (The first two of these are described in this section, and the third in the section on quasi-Monte Carlo.)

Unfortunately, it is common to confuse the *randomness* properties of a sequence with its *distribution*. This is unnecessary, since the two are quite independent. A perfectly random sequence may have any distribution (uniform, Gaussian, etc), whereas a perfectly uniformly distributed sequence may be not at all random.

5.1. Truly random numbers

A sequence of truly random numbers is unpredictable and therefore unreproducible. Such a sequence can only be generated by a random physical process, for example radioactive decay, thermal noise in electronic devices, cosmic ray arrival times, etc. If such a physical process is used (properly) to generate the random numbers for a Monte Carlo calculation, there is no theoretical problem, since the theory outlined above is sufficient justification, provided there is no physical defect in the apparatus.

In practice, however, it turns out to be very difficult to construct physical generators which are fast enough (one needs typically hundreds of floating-point numbers per second) and at the same time accurate and unbiased. Faced with these practical difficulties, very few large-scale calculations have been made using such generators.

One important exception is the work of Frigerio and Clark (1975) and Frigerio *et al* (1978). They used a radioactive alpha-particle source and a high-resolution counter turned on for periods of 20 ms, during which time they counted, on average, 24·315 decays. Whenever the count was odd, they recorded a zero-bit, and when even, a one-bit, all written to magnetic tape. A careful correction was made to eliminate the bias due to the fact that the probability of an odd count is not exactly one-half (the bias could have been removed without even knowing this probability, using the method given in the next subsection). Their apparatus yielded about 6000 31-bit truly random numbers per hour. These numbers have been stored on magnetic tape, subjected to a number of tests for 'randomness' and used in Monte Carlo calculations. (Copies of the tape, containing 2·5 million truly random numbers, are available from the Argonne National Laboratory Code Center, Argonne, Illinois 60439, USA.)

To illustrate the practical problems of physical bias in truly random generators, let us again consider the Buffon needle experiment. First of all, the width of the stripes must be constant and equal to the length of the needle to within the accuracy ultimately desired for the final result, which is not so hard if we only want one or two figure accuracy, but will clearly prevent us from going much further. Also, an unbiased decision procedure must be found for the cases when the needle almost crosses a boundary. Thirdly, we must ensure that the actual distribution of angle and position of the needle is uniform. The angular distribution may be made uniform by spinning the needle very fast as it is thrown, provided the surface is very flat and of homogeneous friction properties. The distribution of needle position will not be uniform but may be expected to follow some Gaussian distribution about the point where the thrower aims. In practice, one would determine the width of this distribution experimentally and carry out a rather complicated correction of the type performed by Frigerio *et al* as described above.

5.1.1. Bias removal technique. It often happens when generating truly random numbers, as in the example just above, that the major problem is in determining the exact distribution (i.e. the bias of the apparatus), whereas the 'truly random' property is guaranteed by the nature of the physical process used. In these circumstances, a very useful trick to eliminate the bias is the following.

Suppose we are given a truly random sequence of zeros and ones, but where the probabilities $P(0)$ and $P(1)$ may not be exactly one-half. Using this original sequence, we produce a second sequence in the following way. Consider *pairs* of bits in the sequence, and if the two bits in the pair are the same, reject both bits; if the two bits are different, accept the second bit (always rejecting the first of each pair). The new sequence thus formed is guaranteed to have zeros and ones with equal probability as long as there was no correlation between the bits of the original sequence. This can be seen easily by calculating $P'(0)$ and $P'(1)$, the probabilities of zero and one in the new sequence, in terms of $P(0)$ and $P(1)$, the original probabilities. Since a zero can only come from a one followed by a zero, $P'(0)=P(1)P(0)$ and similarly $P'(1)=P(0)P(1)$. These probabilities must therefore be equal no matter what $P(0)$ and $P(1)$ are. Unfortunately $P'(0)$ and $P'(1)$ do not add up to one because the probability of rejecting a pair entirely is $P^2(0)+P^2(1)$, which must be greater than or equal to one-half. In addition, half the bits are lost because a pair yields at most one bit, so the efficiency of the procedure is at most 25% but it allows the use of a basic generator which is of unknown bias, as long as this bias is nearly constant in time. (Any method using an explicit correction for bias must also know the exact time dependence of this bias.)

The efficiency of this method is easily seen to be $P(0)P(1)$, which is equal to $P(1-P)$, where P is either $P(0)$ or $P(1)$. This means that, for heavily biased original sequences, the efficiency is approximately equal to the probability of the less probable bit.

5.2. Pseudo-random numbers

The random numbers most often used in real calculations are those known as pesudo-random, which are generated according to a strict mathematical formula and therefore reproducible and not at all random in the mathematical sense but are supposed to be indistinguishable from a sequence generated truly randomly. That is, someone who does not know the formula is not supposed to be able to tell that a formula was used rather than a physical process. The theory outlined in §2 is generally assumed to hold for Monte Carlo results calculated with pseudo-random numbers as well as with truly random numbers.

Unfortunately, there is no way to generate such numbers, which are both truly random and not truly random. This has not prevented people from using pseudo-random sequences (often with considerable success), closing one eye to the theoretical impossibility of it all. In this subsection we discuss how this is done in practice.

5.2.1. From mid-squares to multiplicative generators. Perhaps the earliest pseudo-random number generator was that of Von Neumann known as 'mid-squares'. Given a starting number of r digits, the first 'random' number is the middle $r/2$ digits of this number. Then the first 'random' number is squared (forming another number of r digits) and the middle $r/2$ digits of this square are the second 'random' number, etc. The digits may be decimal, octal, binary or in any other base. If the original number is chosen carefully, this method can yield a reasonably long string of numbers which appear random, but the properties of this generator, to the extent that they are known at all, are not very good, and it is not used any more. First of all, this generator is characterised by a *period*, since if any number reappears the entire sequence from the first appearance to the second will reappear. This is a rather general property of pseudo-random generators, including those commonly used today. Also, certain numbers reproduce themselves immediately (for example, zero), which means that those numbers can never appear unless the period is one.

It may appear that the mid-squares method cannot possibly be very good because it is not complicated enough. The naive approach then consists in 'improving' the unacceptable method by making it more complicated. An excellent example of how one might do this is given by Knuth (1969, pp4–6). His 'super-random' generator is so complicated that one could never hope to understand its properties, and turns out nevertheless to be very bad. The lesson to be learned is that a simple generator, whose properties (and weaknesses) are known, is always to be preferred to a complicated generator of unknown properties. A corollary of this lesson is that it is not easy to 'improve' a bad pseudo-random generator by making it more complicated. Such an exercise cannot add any true randomness and usually serves only to shorten the period by using up several numbers to produce one. Exceptions to this are the shuffling technique discussed below in connection with quasi-random numbers and the Dieter–Ahrens generator also discussed below.

Indeed the pseudo-random generator most widely used is even somewhat simpler than mid-squares; it is the method attributed to D H Lehmer, known as *multiplicative congruential* or *linear congruential*. Given a modulus m, a multiplier a and a starting value r_0, the method generates successive pseudo-random numbers by the formula

$$r_i = ar_{i-1} \pmod{m}.$$

A variation known as the *mixed congruential* generator requires, in addition, an additive constant b:

$$r_i = ar_{i-1} + b \pmod{m}.$$

The two generators have very similar properties and will be considered together. For both generators, m is invariably chosen as 2^t, where t is the number of bits in the representation of an integer on the computer being used, so that in practice the algorithm consists of multiplying two numbers of t bits each, yielding a number of $2t$ bits, of which the *lower* (least significant) t bits are retained as the next 'random' number. These integers are then converted to floating-point numbers in the range zero to one by dividing by m.

5.2.2. The early approach: maximum period. It turns out to be a relatively easy problem in number theory to give the conditions for a congruential generator to attain the maximum period, which is generally of length $m/4$. Early theoretical results therefore concerned primarily this aspect with very little progress on other properties. This gave rise to a large number of generators with long periods, which were then subjected to 'tests for randomness' and the ones for which no 'non-random' behaviour could be discovered were used. Often these generators were later found to be unacceptable but continued to be used by those who had not yet stumbled upon the unfortunate properties†.

The 1960s may be termed the 'dark ages' of pseudo-random generators, characterised by an enormous number of articles (mostly unpublished) purporting to show, on the basis of 'tests' as described below, that one pseudo-random generator was better or worse than another.

5.2.3. Testing pseudo-random generators. Since there was in the early days no good theory about the behaviour of pseudo-random number generators, it was necessary to resort to 'tests of randomness' in order to certify a given generator as 'good'. These tests usually consist of forming some function of a given string of pseudo-random numbers and comparing the value of this function with the expected value

† The best example of this is RANDU which was distributed by IBM with their 360 series and was found almost immediately to be very poor. One can still find articles being published today by people just getting around to making this painful discovery.

of the same function of truly random numbers. For example, the simplest test would be to take the average of the first n numbers from a pseudo-random generator, which should be close to 0·5, the expectation of the average of truly random numbers uniformly distributed between zero and one. The variance of the average for truly random numbers being $n/12$, the square root of this quantity is the expected standard deviation, so we expect that 95% of the strings of n numbers will have an average within two such standard deviations of 0·5. If our pseudo-random generator yields an average which falls outside this range, we say that it fails that test at the 5% level. Of course, even a truly random sequence would fail such a test 5% of the time, but that is just too bad.

In practice, one uses somewhat more complicated tests, based on more complicated functions. These tests have names such as the runs test, poker test, etc. Some tests are felt to be more sensitive than others, but since one does not in principle know what kind of 'non-randomness' to look for, it is not possible to measure the power of a test in any precise way. The most common tests are described abundantly in the literature (e.g. Ahrens *et al* 1970) and summarised in Knuth (1969).

Since there is an infinite number of possible functions that could be applied to each of the possible sequences coming from a pseudo-random generator, no generator can be 'tested' thoroughly. The most interesting such function is just the calculation for which the pseudo-random numbers are needed, and the (unknown) correct answer to this problem provides yet another test of the generator—indeed, the only test we really care about. The philosophy of pseudo-Monte Carlo could therefore be stated in these words: if a pseudo-random number generator has passed a certain number of tests, then it will pass the next one, where the next one is the answer to our problem. It is, of course, not known in general why it should pass this next test, except for the fact that it is not known why it should not.

A somewhat different kind of test was used by J Lach (1962, unpublished) who was suspicious because results using the IBM 709 pseudo-random generator produced fluctuations greater than expected. He simply plotted the random number distribution on a cathode ray display and observed the 'non-randomness' by eye. Taking pairs of numbers as (x, y) coordinates of points, no obvious correlations were seen, but when triplets (x, y, z) were considered and (x, y) were plotted only for $z < 0{\cdot}1$, the resulting point distribution showed a structure of slanting bands, with all the space between the bands completely empty of points. The pseudo-random generator was later corrected by changing the multiplier so that the particular effect observed by Lach disappeared, but what Lach had observed was later showed by Marsaglia to be a defect inherent in all generators of this type (see next subsection).

My personal feeling about testing is that it is best to avoid it through a deeper theoretical understanding of the generator. (In the case of the multiplicative congruential generator, the important properties are now known exactly; see below.) If testing must be done, I prefer visual tests of the type used by Lach, since these tests are not only rather sensitive to the kinds of 'non-randomness' we are interested in, but may also give some insight into the properties of the generator.

5.2.4. The Marsaglia effect. In his classic paper *Random numbers fall mainly in the planes* Marsaglia (1968) finally brought some genuine understanding into the occult art of pseudo-random number generation. He showed that if successive d-tuples from a multiplicative congruential generator are taken as coordinates of points in d-dimensional space, all the points will lie on a certain finite number of parallel hyperplanes, this number always being not greater than a certain function of d and the bit length of integer arithmetic on the machine. We give some values of this function in the table below.

Maximum number of hyperplanes $= (d!\,2^t)^{1/d}$				
Number of bits(t)	$d=3$	$d=4$	$d=6$	$d=10$
16	73	35	19	13
32	2953	566	120	41
36	7442	1133	191	54
48	119086	9065	766	126
60	1905376	72520	3064	290

Furthermore, it is usually the case that the points lie on more than one such set of hyperplanes, making an extremely regular pattern rather than the 'random' distribution desired. (Of course, it is true that any points must lie on some set of hyperplanes, but truly random points would lie on a much larger number of such planes.) We can use the table above to decide the maximum dimensionality for which we care to use such random numbers to perform, for example, numerical integration, based on the word length of our machine. For machines with long words, the limit is probably beyond anything we would be likely to need, but with integers of 36 bits and less, care must be taken.

Note that Marsaglia completely explained the effect observed earlier by Lach, and which was 'corrected' by changing the multiplier of the generator. Lach was observing the hyperplanes in three-dimensional space and taking a slice in one of the dimensions produced the bands when projected onto the other two dimensions. Changing the multiplier may have increased the number of planes, and certainly changed their orientation, so that the effect then appeared to go away. Lach was using a computer with 36-bit integers, so that it should have been possible to get a good distribution in only three dimensions.

5.2.5. The Dieter–Ahrens solution. About the same time as Marsaglia was dicovering the hyperplanes, he and others were investigating multiplicative generators in more detail (Marsaglia 1972) and found ways to determine, for example, the exact distribution of pairs of numbers (Dieter 1971), and the autocorrelation function (Dieter and Ahrens 1971). The result of all this work is a good understanding of both the good and bad properties of such generators, as well as how to find good multipliers. Dieter and Ahrens (1979)† show that the way around the Marsaglia hyperplane problem is to use compound multiplicative congruential generators of the form

$$r_i = (ar_{i-1} + br_{i-2}) \pmod{m}$$

which will increase the number of hyperplanes by a factor $2^{(t/d)}$ provided the constants a and b are chosen carefully. The hyperplanes do not go away but their number may be increased arbitrarily by adding more terms as above.

5.2.6. Good pseudo-random generators. On a computer with integer length t bits, the best simple multiplicative generator is probably that proposed by Ahrens *et al* (1970), where the multiplier is

$$a = 2^{t-2}\,\tfrac{1}{2}(\sqrt{5} - 1).$$

(You may recognise the famous 'golden section' constant here.) In practice, the constant a is determined for a given value of the integer length t by multiplying 2^{t-2} into a very precise value of the golden section constant ($=0{\cdot}618\,033\,988\,749\,894\,848\,204\,5868$) and rounding to the nearest integer congruent to 5 (mod 8). This will yield a generator with period 2^{t-2} and good distribution properties.

On CDC 6000, 7000 and Cyber machines, it is unfortunately not easy to take advantage of the full 60-bit words, since integer multiplication is performed only on 48 bits (for compatibility with floating-point numbers which have 48-bit mantissas). For such computers, the value $t = 48$ is therefore appropriate, and the constant a is

$$a = (1170\,673\,633\,457\,725)_8 = (43\,490\,275\,647\,445)_{10}$$

which has a period of $2^{46} = 70\,368\,744\,177\,664$.

On IBM 370 and IBM-compatible computers, the 32-bit integer arithmetic makes simple generators somewhat risky for large calculations. With only 31 significant bits available, the maximum period is 2^{29} or about 500 million. Since it is dangerous to come anywhere close to exhausting the period (exhausting the period would give a perfectly uniform distribution since all numbers would be generated) it is not too difficult to imagine calculations where a better generator is needed. In this case I recommend using the McGill University package 'Super-duper'

† We are grateful to the authors of this book for providing a pre-publication version of the first seven chapters.

(available from Professor G Marsaglia, School of Computer Science, McGill University, PO Box 6070, Montreal, Canada). The basic generator of this package combines two methods to give a period as long as one would expect from a 64-bit machine.

5.2.7. Machine-independent pseudo-generators. It is sometimes convenient to have a random number generator which produces exactly the same numbers on any computer. Assuming that we want floating-point numbers between zero and one, we therefore choose the precision of the lowest-precision machine we are likely to use and simulate that precision on other computers. (On computers with longer words, the lower bits will be zero.) Such a generator will, in general, not be optimal on any machine, either in terms of period or of speed, but we will show here that it can be implemented, in FORTRAN, on most larger computers. It can then be used to test programs and compare and continue calculations across changes of computer.

If we choose IBM 32-bit words as our minimum precision, such a generator, called RN32 (CERN Program Library†), has been implemented as follows. As default starting integer use the value 65 539. Multiply the previous (or starting) integer ('seed') by 69 069. Keep only the lower 31 bits of the result. This 31-bit integer becomes the seed for the next number. We get a floating-point pseudo-random number from the seed by masking off the lower 8 bits to assure exact floating-point representation of the integer, floating it, and multiplying the result by the exact floating representation of 2^{-31}.

Differences in FORTRAN and floating-point representations require slightly different implementation on different machines. We show on p1174 as examples the CDC and IBM versions.

With the default seed shown, the first two numbers produced by these generators are approximately‡:

$$R1 = 0{\cdot}107\,915\,04\ldots$$

$$R2 = 0{\cdot}587\,475\,06\ldots.$$

5.2.8. Practical computing considerations. The usage of random number generators from FORTRAN programs requires some special considerations of a practical nature. Perhaps the most important of these stems from the fact that most pseudo-random generators, like the one above, are coded as FORTRAN functions rather than sub-

† Programs in this library are made generally available. Further information may be obtained from: Program Library, Division DD, CERN, 1211 Geneva 23, Switzerland.

‡ The numbers produced by different computers are exactly the same if represented as binary fractions, but the exact decimal representation requires many more digits than we reproduce here and more than your computer is likely to give in a printout.

```
      FUNCTION RN32(IDUMMY)
C                                   CDC VERSION,  F.JAMES, 1978
C              IY IS THE SEED, CONS=2**-31
      DATA IY/65539/
      DATA CONS /16611000000000000000B/
      DATA MASK31/17777777777B/
      IY = IY * 69069
C         KEEP ONLY LOWER 31 BITS
      IY = IY .AND. MASK31
C         SET LOWER 8 BITS TO ZERO TO ASSURE EXACT FLOAT
      JY = IY .AND. 07777777777777777400B
      YFL = JY
      RN32 = YFL*CONS
      RETURN
C         ENTRY TO INPUT SEED
      ENTRY RN32IN
      IY = IDUMMY
      RETURN
C         ENTRY TO OUTPUT SEED
      ENTRY RN32OT
      IDUMMY = IY
      RETURN
      END

      FUNCTION RN32(DUMMY)
C                                  IBM VERSION,  F.JAMES, 1978
C              IY IS THE SEED, CONS=2**-31
      DATA IY/65539/
      DATA CONS/Z39200000/
      IY = IY * 69069
C         ASSURE LEFTMOST BIT ZERO (POSITIVE INTEGER)
      IF (IY .GT. 0)  GO TO 6
      IY = IY + 2147483647 + 1
    6 CONTINUE
C         SET LOWER 8 BITS TO ZERO TO ASSURE EXACT FLOAT
      JY = (IY/256)*256
      YFL = JY
      RN32 = YFL*CONS
      RETURN
C         ENTRY TO INPUT SEED
      ENTRY RN32IN(IX)
      IY = IX
      RETURN
C         ENTRY TO OUTPUT SEED
      ENTRY RN32OT(IX)
      IX = IY
      RETURN
      END
```

routines. Strictly speaking, this is not in accordance with the rules of FORTRAN, since random number generators are not functions of their arguments only, they have 'side effects', namely they set up the next number. Since they are functions, the FORTRAN compilers reserve the right to optimise them out of existence by replacing each function evaluation by the constant value of the function. For example

$$X = \mathrm{RANDOM}(1) + \mathrm{RANDOM}(1)$$

could be compiled as if it were

$$X = 2{\cdot}0 * \mathrm{RANDOM}(1)$$

which is of course not the same thing at all. The well-known way around this is to do something like

X = RANDOM(I) + RANDOM(I + 1)

in order to fool the compiler into thinking the two calls have different arguments and must therefore be called twice. Similar problems may arise when calls to random number generators appear in DO loops. Of course, the proper way around this is to use random number generators coded as subroutines rather than as functions. This may be somewhat clumsier to use, but is much safer.

In many applications, the actual time taken to generate the random numbers may be important. In earlier days this was usually the case, and it is still a point of great pride among programmers to chop half a microsecond off the generation time, even though it may be quite negligible compared with the rest of the calculation. In cases where generation time is important, several tricks may be used.

One is, of course, to code the generator in assembler, which is often done anyway since the operations needed may be easier to code in assembler. Even better is to code the generator 'in-line' in the calling program to avoid the overhead of a subroutine call, which is usually the greater part of the time spent in getting a random number. The standard CDC FORTRAN function RANF causes the compiler to produce fast in-line code. Although the multiplier used by RANF is not the best, the generous effective word-length of 48 bits still produces random numbers good enough for most applications, so I would advise CDC users to call RANF whenever speed of generation is an important consideration.

Often a calculation requires n-tuples of random numbers, in which case it is much more efficient to use a subroutine that returns n random numbers at a time rather than calling a single generator n times, because of the overhead in the call. As an example, the CERN Program Library subroutine NRAN (V105) generates n random numbers in one call on the CDC 7600 about seven times as fast as n calls to RNDM (V104), for large n, even though the two routines use exactly the same method of generation (with different multipliers for 'independence').

Sometimes it is desirable to have exactly the same sequence of random numbers in one calculation as you had in the previous calculation, and sometimes it is equally important that the sequence be different. Many generators therefore offer different ways of initiating the sequence. Most generators use a default starting value (like RN32 above) and therefore always produce the same sequence unless requested otherwise. Such generators often allow inputting and outputting the seed value, so that at the end of a run the current seed value can be output and read back in at the beginning of the next run to continue the sequence (this is the case with RN32). In this way, different sequences can be forced by inputting different starting seeds. Still other generators use 'random' starting seeds obtained by using the time of day and date from the system clock and transforming that into an appropriate integer. This removes all control from the user and even adds some element of truly random unpredictability.

6. Quasi-Monte Carlo

The theoretical difficulties and practical success of pseudo-random numbers have given rise to another type of sequence known as quasi-random. (In English usage, 'pseudo-' means false, and 'quasi-' means almost, but in the technical context of random numbers their meanings are somewhat different and much more precise.) Quasi-random sequences are not even intended to appear random but only to give the right answer to the problem at hand. Thus they are more satisfactory since they are not based on an illusion, but on the other hand they must in principle be tailored to the problem at hand. Since this problem can often be reduced to multiple integration, the tailoring becomes ready-to-wear in practice and the theory applicable to most cases.

6.1. The quasi-random philosophy

The concept of quasi-random numbers arises from the realisation that the mathematical randomness of pseudo-random numbers is neither attainable in theory nor necessary in practice, and it is more meaningful to assure that the 'random' sequence has the necessary properties to produce the desired result. For example, in multiple integration and in most simulation studies, each multidimensional point or simulated event is considered independently of the others and the order in which they appear is immaterial. That is, correlations between successive points (events) is usually of no importance—this aspect of randomness can safely be abandoned for most calculations. Another aspect which can be abandoned is the degree of fluctuation about uniformity for certain distributions—in many cases a ***super-uniform*** distribution is, in fact, more desirable than a truly random distribution with uniform probability density.

Since we have now dropped all pretense of randomness, the reader may object at this point to retaining the name Monte Carlo. Strictly speaking he is right, but it is probably more justified to enlarge the concept of Monte Carlo to include the use of quasi-random sequences. Quasi-Monte Carlo is indeed rather a downward (in dimensionality) extension of Monte Carlo than an upward extension of one-dimensional quadrature, since it retains some fundamental properties of Monte Carlo such as applicability to spaces of very high dimensionality, performance nearly independent of dimensionality, very small growth rate, even for high dimensionalities, and robustness with respect to the continuity properties of the function. In addition, the theory of quasi-Monte Carlo outlined below is much closer to that of true Monte Carlo than to that of quadrature.

6.2. *The theoretical basis of quasi-Monte Carlo*

6.2.1. The discrepancy of a point set. Let us here introduce a measure of non-uniformity valid for any dimensionality, called *discrepancy* (see Weyl (1916), or secondary references Zaremba (1968, 1972) or Stroud (1971)). Consider the unit hypercube in d dimensions, with each coordinate of $\boldsymbol{x}$ varying from zero to one, and we are given a set of n points, the ith point having coordinates $\boldsymbol{x}_i$. The function $v(\boldsymbol{x})$ gives the integrated number of points, from the origin to the point $\boldsymbol{x}$ (the empirical distribution function). The corresponding volume from the origin to the point $\boldsymbol{x}$ is just given by the product of the coordinates of the point $\boldsymbol{x}$, and the local discrepancy g at $\boldsymbol{x}$ is defined as the difference between the number of points in this volume and the expected number based on the volume:

$$g(\boldsymbol{x})=\frac{v(\boldsymbol{x})}{n}-x_1x_2\ldots x_d.$$

One can then define various measures of global discrepancy by taking different norms of the function g. The most common are the *extreme discrepancy* given by the maximum of the absolute value of g for all $\boldsymbol{x}$, and the *mean square discrepancy* given by the integral of the square of g over all $\boldsymbol{x}$. The general term 'discrepancy' is sometimes loosely applied also to the global measures.

Since we will use discrepancy to *test the hypothesis* of uniformity of a point distribution, it is not surprising that this measure is already well-known to statisticians, who will recognise extreme discrepancy as the Kolmogorov statistic and mean-square discrepancy as the Smirnov–Cramer–Von Mises statistic for testing compatibility of distributions (see Eadie *et al* 1971, pp268–70).

If the extreme discrepancy of a point set approaches zero as the number of points approaches infinity, the (infinite) set of points is said to be *uniform*. We refer to this as uniformity in the sense of Weyl, to distinguish it from more common meanings of the word. A truly random point set in a finite-dimensional space can be shown to be uniform in this sense. In quasi-Monte Carlo we will use non-random points which are also uniform (for infinite sets) or which have low discrepancy (for finite sets).

6.2.2. The convergence of quasi-Monte Carlo integration. The theorems given in this subsection concern the approximation of a multidimensional definite integral by an unweighted sum of function values over a set of points. The function to be integrated will be assumed to be of finite *variation*. A precise definition of variation is not very enlightening and is beyond the scope of this article (see Zaremba 1968, Stroud 1971); we give here only a rough idea sufficient for an understanding of the results presented below. The variation in quasi-Monte Carlo theory plays the role of variance in true Monte Carlo, being also a measure of the non-constancy of the function. For a differentiable function of d variables, the variation can be thought of as an average of the absolute values of the dth mixed partial derivatives. Integrable functions of interest to physicists (with at most a finite number of discontinuities) have a finite variation.

The following theorems form the mathematical basis for integration by quasi-Monte Carlo.

(i) (Weyl 1916). If a definite integral is estimated by an unweighted sum of function values over a set of points, the estimate will converge to the true value of the integral as the number of points approaches infinity if and only if the point set is uniform in the sense of Weyl. This theorem is the equivalent of the law of large numbers for true Monte Carlo, and gives the conditions under which the quasi-Monte Carlo estimate is consistent.

(ii) (Hlawka, see Zaremba 1968). If a definite integral is approximated by an unweighted sum of function values over a finite set of points, the resulting error will be bounded by the product of the discrepancy of the point set and the variation of the function.

(iii) (Roth and others, see Kuipers and Niederreiter 1974, Zaremba 1968). The discrepancy of a point set cannot be made smaller than a certain value, which depends on the number of points n and the dimensionality d. Attempts to find point sets which achieve this fundamental lower limit have been successful only in a small number of cases.

(iv) (Korobov, see Stroud 1971). The discrepancy of the first n points of an infinite point set cannot decrease as a function of n any faster than $1/n$ for large n. The second theorem above implies that the estimate of the integral will converge to the correct answer as fast as the discrepancy of the point set converges to zero, and the fourth theorem gives us hope that this could be as fast as $1/n$, compared with the much slower square root of n for true Monte Carlo. Unfortunately, it is not generally known how to generate points which attain the lower discrepancy bound, but one can at least generate points with considerably lower discrepancy than the expectation of a truly random set.

After reading the above theorems, we should not be surprised to learn that the expected value of the extreme discrepancy of a set of n truly random points decreases with n like $1/\sqrt{n}$ for large n in any number of dimensions.

6.3. *Quasi-random number generators*

Because theorem (iv) of the last subsection applies only to infinite sequences, we must distinguish here between finite quasi-random sequences of n numbers where n is fixed in advance, and the first n numbers of an infinite sequence. The latter will clearly be more convenient to use since it can be extended if necessary, but the above theorems indicate that we might be able to get a better discrepancy if we fix n.

6.3.1. Good lattice points. Optimal points for function integration are generated by fixing n (and the dimensionality d) and actually minimising the extreme discrepancy of the n points with respect to their positions. The computational complexity of such a calculation being overwhelming, only an approximate minimum-discrepancy solution can be found for anything but a very small point set. A considerable amount of theoretical work has been done on d-dimensional lattices (Kuipers and Niederreiter 1974, Zaremba 1972) but this approach has not yet produced techniques of great interest for large calculations, except for the Korobov sequences described below.

6.3.2. Finite Korobov sequences. Korobov considered sets of points restricted to belong to certain families characterised by different expressions for the coordinates, with each expression containing some free parameters. The values of these parameters were then optimised by requiring a minimum extreme discrepancy. Probably the most successful Korobov family is the parallelepiped lattice, where successive points $\boldsymbol{x}$ are given by:

$$\boldsymbol{x}_k = \left.\frac{ak}{N}\right|_{\text{mod }1}, \left.\frac{bk}{N}\right|_{\text{mod }1}, \ldots, \left.\frac{\mathrm{d}k}{N}\right|_{\text{mod }1}, \qquad k=1, N$$

where $a, b, \ldots, d$ are coefficients to be determined in order to optimise the discrepancy for the given value of the number of points N and the dimensionality. Discussion of Korobov sequences and references to the original Russian articles can be found in Stroud (1971) and Zakrzewska *et al* (1978). The latter article describes a program for multiple integration using Korobov sequences. These sequences can also be used as an option in DIVONNE2 (Friedman 1977a), and extensive tables of optimal coefficients for generating Korobov sequences are given in Keast (1972).

6.3.3. The Richtmyer generator. This generator is the equivalent of the Korobov parallelepiped family described just above, but for infinite N and 'any' d. Since one can no longer optimise the coefficients, it is apparently sufficient to use 'irrational' numbers in order to avoid a short period. Since truly irrational numbers cannot be represented in computers, it has been suggested to use the square roots of the first few prime numbers. Thus one gets the simple formula for the jth coordinate of the ith quasi-random point:

$$x_{ij} = iS_j, \text{ mod } 1$$

where S_j is the square root of the jth prime number.

In theory this generator is supposed to have very good properties for an infinite number of points, and its discrepancy should decrease like $1/n$ for very large n. The problem is then to make it behave well for small n (which may still be very large in practice) without destroying the asymptotic behaviour. This is done, first of all, by observing the two-dimensional distributions of the first few thousand numbers of two of the coordinates. When a pair is seen to be badly distributed, one of the corresponding S values is dropped from the table and replaced by a higher root prime. Of course, this observed distribution would in principle improve with larger n, but one does not know how large, so it is better in practice to be careful.

The second method for improvement of short-term behaviour of such quasi-random generators is the *shuffling* technique, which assures that all the numbers from the generator will be used, but not quite in the order in which they are generated. Usually another (pseudo-)random generator is used for the shuffling, which is performed using a buffer (usually 10 or 20 words per dimension), and selecting the next quasi-random number pseudo-randomly from the buffer of the appropriate coordinate, filling the used location in the buffer with the next quasi-random number in the corresponding sequence. This yields points different from those of the unshuffled generator but preserves the super-uniform distribution of each of the coordinate values.

6.3.4. The van der Corput generator. The formula of van der Corput corresponds to expressing the integers in a system of base P, reversing the digits, putting a point in front, and interpreting the resulting sequence as fractions in the base P. P is any prime, so the ith coordinate is generated using this formula with P being the ith prime number. For example, for $P=2$ this gives the results shown in the table below.

Decimal integer	Binary integer	Binary fraction	Decimal fraction
$j=1$	1	0·1	0·5
2	10	0·01	0·25
3	11	0·11	0·75
4	100	0·001	0·125
5	101	0·101	0·625
6	110	0·011	0·375
7	111	0·111	0·875
8	1000	0·0001	0·0625

This generator has properties similar to those of the Richtmyer generator, except that it seems to behave much better for smaller n. In spite of the apparent computational complexity, it can be made fast, thanks to a relatively simple algorithm for implementing it, due to Halton (1960).

As with the Richtmyer generator, this method can be improved by shuffling. A particularly effective scrambling technique, based on explicit minimisation of the discrepancy for this generator, is given by Braaten and Weller (1979).

7. Non-uniform random numbers

Up to now we have been almost exclusively concerned with uniformly distributed random numbers, either with uniformly distributed probability of occurrence or for quasi-random sets, a distribution as uniform as possible (sometimes called 'super-uniform', since it is more uniform than a truly random set with uniform probability density). In this section we discuss the problem of generating random numbers such that the probability of obtaining a number in a given range is not uniform but follows some other distribution.

Generating non-uniform distributions is very important in many applications, where the physical phenomena being simulated are known to follow certain other distributions. The most important of these are the Gaussian (or normal) and exponential distributions for continuous variables, and the Poisson and binomial distributions for discrete variables. Many other distributions may be required for special applications, and many different techniques are known for generating them. We present here only a brief review of the most important methods with some indication of where to look for more. It is assumed throughout this section that an appropriate generator of uniformly distributed random numbers is available for use in generating the non-uniform distributions.

7.1. Gaussian generators

The Gaussian distribution is one of the most important in statistical and physical calculations and also one of the richest in terms of different methods proposed for generating random numbers.

7.1.1. Using the central limit theorem. This method has already been described above in §2.6.1. It is not exact, although it may be good enough for many purposes, and the absence of points in the extreme tails may even be desirable in some cases. It is also not especially fast, but may be faster than some other methods when a good generator of arrays of uniform numbers is available. (Note that this method, like most of those given in this section, must not make use of a quasi-random uniform number generator, since serial correlations in the uniform generator lead to distortions in the distribution of the output random numbers.)

As a word of warning, I should point out an interesting mistake sometimes made in connection with this generator. It arises from the realisation that the central limit theorem of course works for differences as well as sums, so that taking the sum of six uniform numbers minus the sum of six other uniform numbers would be as good as taking the sum of twelve uniform numbers and subtracting six. Some clever people decide therefore to use twelve uniform numbers to generate *two* random Gaussian deviates, once using a sum and once with differences. It is certainly true that this gives two (approximately) Gaussian numbers, but they are unfortunately highly correlated. Correlation has also been the source of some concern about the simple generator of §2.6.1, since any correlations in the uniform generator would produce deviations from the Gaussian distribution of the sum.

7.1.2. The transformation method. Since the Gaussian probability function cannot be integrated in terms of the usually available functions, it is not straightforward to apply a transformation from uniform to Gaussian-distributed variables. There is, however, a clever method of transforming two independent uniform variables u and v into two independent Gaussian variables x and y:

$$x=(-2\ln u)^{1/2}\cos(2\pi v)$$

$$y=(-2\ln u)^{1/2}\sin(2\pi v).$$

This method is exact and easy to program but is not quite as fast as it may appear, since it requires calculation of a logarithm, square root, sine and cosine, all of which are reasonably time-consuming operations.

An improvement on the above method is the polar method of Marsaglia:

(i) Generate uniform random numbers u and v.

(ii) Calculate $w=(2u-1)^2+(2v-1)^2$.

(iii) If $w>1$, go back to (i).

(iv) Return $x=uz$ and $y=vz$, where $z=(-2\ln w/w)^{1/2}$.

This variation eliminates the sine and cosine at the slight expense of $\simeq 21\%$ rejection in step (iii) and a few more arithmetic operations.

7.1.3. The Forsythe–Von Neumann method. This is an ingenious method for generating random numbers in any distribution of the form:

$$f(x)=c\exp[-G(x)] \qquad 0<G(x)<1,\ a<x<b$$

based on the fact that if you:

(i) choose u_0 uniformly between a and b;

(ii) calculate $t=G(u_0)$;

(iii) generate uniformly $u_1, u_2, \ldots, u_k$, $0<u_i<1$ where k is determined by the condition:

$$t>u_1>u_2> \ldots u_{k-1}<u_k$$

then the probability that k is odd is $P(t)=e^{-t}$.

Therefore, whenever k is even, reject that value of u_0 and go back to (i). When k is odd, accept that u_0 as a member of a sample from f. Unfortunately, the fact that the range of G must be from zero to one requires some fiddling to use this technique for generating from the Gaussian distribution, but some good methods are based on it (see Ahrens and Dieter 1973).

7.1.4. Compound methods. Many other techniques have been proposed for generating Gaussian random numbers, and the best (fastest exact) methods are composed by combining several of these techniques. The general idea is to use a fast approximate method most of the time, and then with a carefully calculated (small) probability, one draws from a 'corrective' distribution which just makes up for the approximation in the first technique. In addition, different regions under the Gaussian curve are attacked using different techniques, with the region first being chosen using an auxiliary random number. Such methods are often somewhat complicated to program and require a table of constants used to choose regions, methods, corrections, etc. A detailed account of a good compound method is given in Dieter and Ahrens (1973), and a summary of many good methods, both simple and compound for the Gaussian distribution, is given in Ahrens and Dieter (1972).

7.1.5. Generating correlated Gaussians. The above subsections deal only with the generation of one-dimensional Gaussians, which can be used directly for multi-dimensional Gaussian distributions only when the different variables (dimensions) are uncorrelated (i.e. when the covariance matrix is diagonal). For the general case of multidimensional Gaussian variables with a general covariance matrix V, uncorrelated standard Gaussian variables may be used when transformed as indicated here. Let $\boldsymbol{z}$ be a standard normal random vector (i.e. independent Gaussian-distributed components with zero mean and unit variance), then a unique lower-triangular matrix C exists such that

$$\boldsymbol{x} = C\boldsymbol{z} + \boldsymbol{m}$$

and $(\boldsymbol{x} - \boldsymbol{m})$ has the covariance matrix

$$V = CC'$$

where C' is the transpose of C.

Given V, the matrix C can be calculated by using the following recursive formulae (the 'square root' method):

$$c_{i1} = v_{i1}/\sqrt{v_{11}} \qquad 1 \leqslant i \leqslant m$$

$$c_{ii} = \left| v_{ii} - \sum_{k=1}^{i-1} c_{ik}{}^2 \right|^{1/2} \qquad 1 < i \leqslant m$$

$$c_{ij} = \frac{v_{ij} - \sum_{k=1}^{i-1} c_{ik} c_{jk}}{c_{jj}} \qquad 1 < j < i \leqslant m.$$

In practice, one usually wants a large set of random vectors all generated with the same covariance matrix V, so the matrix C is computed once at the beginning of the program and then used each time a random Gaussian vector is wanted.

7.2. *All other known distributions*

A vast number of transformations, tricks and formulae are known for generating random numbers according to different distributions. For example, given two uniform numbers, their sum is distributed according to a triangular distribution, and the largest of the two is distributed like $\sqrt{u}$. An extraordinarily complete and very dense collection of such techniques is given in Everett and Cashwell (1972).

7.3. *Empirical distributions*

It often happens that one wants to generate random numbers distributed according to some probability density f which is not any of the usual distributions but may, for example, have been determined empirically from measurements on a particular complex system.

7.3.1. The rejection (hit-or-miss) method. One can, of course, always use the hit-or-miss method if the probability density f is bounded and its upper bound is known. In this method, one simply chooses points randomly and uniformly in the space, using the function value at each point (divided by the maximum function value) as the probability of accepting the point. A point is then accepted if and only if $f/f_{\max}$ is greater than a uniform random number chosen between zero and one. This well-known technique becomes very inefficient when the variance of f is large, in which case nearly all the points are rejected. For this reason it is usually better to use one of the methods described below.

7.3.2. Distribution given as histogram. A distribution in the form of a histogram is usually represented as a vector of frequencies, where the first value is the relative frequency of points desired in the first bin, etc. These frequencies must first be normalised so that their sum is unity, then it is usually convenient to form the cumulative distribution, where the ith number in the cumulative distribution vector is the sum from one to i of the numbers in the corresponding density vector. (The last number in the cumulative vector is therefore always equal to one.) To generate a random number according to the histogram, one first generates a uniform number u_0 and then looks for the first position in the cumulative distribution vector where the value is greater than u_0. This is the bin in which that random number should be generated. It may of course be very inefficient to do this search sequentially (at least for long vectors), and a better method would be to do it by a binary search technique. (The CERN library program HISRAN uses this method.)

A still faster method, although much more complicated, is that of the Marsaglia tables, described in Ahrens and Dieter (1972).

7.3.3. Distribution given as function. To randomly sample according to a one-dimensional distribution given as a smooth function, the usual technique is first to determine the percentiles of this distribution, i.e. the points on the independent variable axis where the integral of the function takes on given values (called percentiles because they are chosen so that the integral over each interval is a given percentage, often 1%, of the total). This is the inversion of the cumulative distribution function. The result of this relatively time-consuming operation is a set of x values which can then be used to generate random numbers very rapidly by direct interpolation in the table of x now considered as a function of F. The CERN library program FUNRAN uses this method with four-point polynomial interpolation in a table of 100 values.

7.3.4. Multidimensional distributions. Multidimensional distributions given as histograms may, of course, be treated exactly as for one dimension. However, when the desired distribution is given as a smooth function, the method outlined above cannot be extended in a straightforward manner, and would anyway require multidimensional tables and multidimensional interpolation, which either consume considerable time and space or are quite inaccurate, especially when the function involved has a large variance.

The problem of randomly sampling a space of high-dimensionality is closely related to that of multidimensional integration, so it is reasonable to look at integration methods for indications on how to proceed. Indeed the recursive partitioning method of Friedman (1977b) is directly applicable and DIVONNE2 (Friedman 1977a) has as an option the generation of points according to the function. This is because the aim of the partitioning algorithm is to delimit regions in which the function variance is small, after which one can efficiently apply hit-or-miss generation or simply produce weighted points.

8. Applications

In Monte Carlo calculation, the step from theoretical understanding to correct results is often far from trivial. Unlike analytical calculations where gross errors usually produce results which are obviously absurd, subtle bugs in Monte Carlo 'reasoning' easily give rise to answers which are completely wrong but still appear sufficiently reasonable to go unnoticed. If only for this reason, it is indispensable to consider a few examples, particularly those which illustrate the most notorious traps for the unwary.

8.1. The uncertainty of a weighted average

The results given here can be derived easily from the definitions of mean and variance but are included here because they are of such central importance in real calculations. We suppose that (as is the usual case) the result of our calculation is an average over a set of terms which we will call weights w_i. We further assume that this average is Gaussian-distributed in accordance with the central limit theorem and wish to determine the standard deviation of this distribution. In order to estimate the average and its standard distribution it is necessary to accumulate:

(i) the sum of the weights, W;
(ii) the sum of the squares of the weights, Q;
(iii) the total number of entries, N.

Then it follows from §2 that the best estimate of the average is just W/N and that the standard deviation of this is $D=(1/N)(Q-W^2/N)^{1/2}$.

For the important case when most of the weights are zero (for example, for one bin of a histogram when most of the events go into other bins), the second term under the square root in the expression for D is negligible compared with the first term and the result is simplified considerably.

In the other limit, when all weights are equal (and non-zero), the two terms under the square root cancel and the standard deviation is of course zero. In practice it may not appear to be zero because of rounding error in the computer, which is especially serious for this particular calculation. For this reason, the sums Q and W should be accumulated in double precision, and it is necessary to test that rounding has not caused the argument of the square root to become negative.

8.2. Integration over a triangle

One of the fundamental advantages of the Monte Carlo method is the ability to easily handle problems with awkward integration regions (inter-dependent integration limits). However, as this example shows, there are a variety of different ways to handle these problems and not all of them are correct.

Consider the integration of the function g over the two-dimensional region specified as

$$I=\int_{x=0}^{1}\int_{y=0}^{x} g(x, y)\,\mathrm{d}y\,\mathrm{d}x.$$

We give four ways of estimating this integral by Monte Carlo.

8.2.1. The obvious way.

(*a*) Choose a random number x_i between zero and one.
(*b*) Choose another random number y_i between zero and x_i.
(*c*) Take the sum of $g(x_i, y_i)$ repeating steps (*a*) and (*b*).

A simple graphical representation of this method shows that it gives the wrong answer. While it is true that this procedure would yield points only in the allowed region (the lower triangle in figure 4), it would give the same expected number of points along each vertical line in the figure, producing a much higher density of points on the left-hand side than on the right.

8.2.2. The rejection method.

(*a*) Choose a random number x_i between zero and one.
(*b*) Choose another random number y_i also between zero and one.
(*c*) If $y_i > x_i$, reject the point and return to (*a*).
(*d*) Accumulate the sum of $g(x_i, y_i)$ for the remaining points.

This method, although correct, has the disadvantage of using only half the points generated, i.e. it is equivalent to integrating over the whole square, but considering the function to be zero on the upper triangle.

8.2.3. The folding method (a trick).

(*a*) Choose *two* independent random numbers r_1 and r_2, each between zero and one.
(*b*) Set x_i = larger of (r_1, r_2).

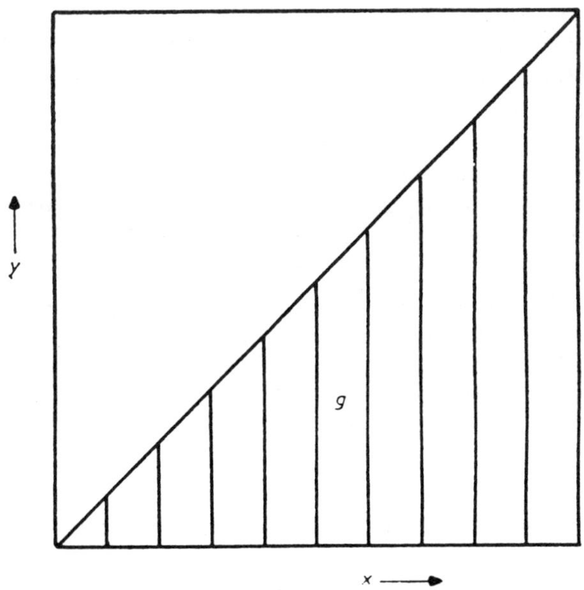

Figure 4. A triangular integration region.

(*c*) Set y_i = smaller of (r_1, r_2).
(*d*) Sum up $g(x_i, y_i)$ as before.

This method is equivalent to choosing points r over the whole square, then folding the square about the diagonal so that all points x, y fall in the lower triangle. It is clear that this gives a constant point density without any rejection, and is therefore correct and more efficient than the rejection method.

8.2.4. The weighting method.

(*a*) Choose a random number x_i between zero and one.
(*b*) Choose another random number y_i between zero and x_i.
(*c*) Take the sum of $2x_i g\,(x_i, y_i)$, repeating the steps above.

In this method, the points are chosen 'incorrectly' as in the obvious method, but the bias is corrected by applying the weighting function which happens to be just $2x$ in this case. This method may or may not be more efficient than folding, depending on the function g. In particular it will be more efficient whenever the variance of xg is smaller than the variance of g. If nothing is known *a priori* about g, it is usual to avoid weighting if possible.

8.3. Programs for real-life calculations

At this point the reader should already be convinced that the possibilities for undetected gross errors in Monte Carlo calculations are numerous. Of course, there is nothing special about Monte Carlo in this respect; complex systems lead to complex calculations and errors can be made on many levels, from the logical understanding of the problem and method all the way down to typing errors in the programs and data. In fact, the Monte Carlo method offers unique opportunities to verify the results of complicated calculations, especially in the case of simulations.

The basic principle is to output not only the number you are interested in but also as many other intermediate and accessory results as possible, especially those for which you know in advance what answer to expect. Even if you are only interested in the global average of some quantity, print out a histogram of the quantity as a function of some other interesting quantity. This generally costs little or nothing extra in a big calculation, and may give considerable insight into the system being studied (if the expected distribution is not known in advance) or allow a powerful check of the correctness of the computation (if the expected distribution is known). I find it convenient to use a general histogramming package such as the generally available HBOOK (CERN Program Library) which allows one to look at an entire one- or two-dimensional empirical distribution in very readable format with only two or three simple lines of FORTRAN. The quantities which you should look at will of course depend on the problem, but a general rule is to examine the quantity of interest in one more dimension than is required, if possible.

8.4. Splitting and killing in sequential simulations

In this subsection we consider simulation calculations in which each 'event' (member of the hypothetical population) consists of a sequence of elementary interactions. Examples of such calculations would be:

(i) Simulation of the traffic flow in a city, where elementary interactions would be car turning left, turning right, parking, breaking down, having an accident, etc.

(ii) Simulation of neutrons or charged particles traversing matter, where elementary interactions would be scattering, decay, absorption, etc.

In these calculations it may be necessary to assign to each elementary interaction a weight proportional to the probability of that interaction. The weight of an entire event is then the product of the weights of its component interactions and the final results of the simulation will be averages over these total weights. As we have seen, the uncertainties of these averages are minimised when the weights are equal. The efficiency of the calculation can therefore be improved by using the following techniques for reducing the variance of the weight distribution.

(i) *Splitting.* After each elementary interaction, compare the accumulated product of weights with the average product at that point for the other events. If it is significantly greater than the average, split the event into two (or more) events from that point on, each one having half (or less) of the above-mentioned accumulated product. In practice, this may be complicated to implement using programming languages which do not explicitly support recursiveness.

(ii) *Killing.* Compare the accumulated product as above, and if it is significantly *less* than the average, either kill (reject) the whole event before finishing it, or continue with the weight increased to the average. The probability of killing the event should be $1-r$, where r is the ratio of the accumulated product weight to the average accumulated product at that point.

It should be clear that it is of no use to apply the killing technique after the entire event has been generated, but only during intermediate steps to avoid the rest of the calculation. Splitting may be performed after the entire event has been generated if this is more convenient, but the decision to split should be made on the basis of the accumulated product weight at the point at which the event is to be split.

8.5. Multiparticle phase space

One of the richest areas of Monte Carlo calculations has been the integration of the relativistic phase space of multiparticle reactions in high energy and nuclear physics. For a reaction with k outgoing particles, the phase-space volume element is basically the $3k$-dimensional momentum space element, but the true dimensionality is reduced to $3k-4$ by a four-dimensional delta function expressing the conservation of energy and momentum. Whenever k is greater than four or five, the complexity of these integrals becomes overwhelming and they can only be performed by numerical techniques, usually only by Monte Carlo. Unfortunately, this interesting problem is much too vast to be treated here and we will merely point to the most important references on the subject.

(i) The classic work on the subject is the monograph of Hagedorn (1964).

(ii) A more recent and extensive treatment, also much more oriented toward practical Monte Carlo calculations, is the book of Byckling and Kajantie (1973).

(iii) The most recent techniques for enriching the region of low momentum transfer are summarised in the review article of Carey and Drijard (1978), which could be considered as an update to the book of Byckling and Kajantie. The techniques reviewed in this paper are very important since one finds, in practice, that in high-energy collisions only a small part of phase space is actually populated, namely that corresponding to peripheral or low-momentum-transfer events.

8.6. Sampling from a finite population

In many fields, particularly in astronomy, plasma physics, fluid dynamics, etc, it is a common problem to simulate the behaviour of a large but finite number of objects (stars, electrons, molecules, etc) which interact with one another. A typical step in such a simulation is the calculation of the force or potential at one object by summing the contributions due to all the other objects. Although the number of objects is finite it may be so large that it is not possible to perform the entire sum, and some approximation must then be made using a smaller sample of objects. Three possible approaches are:

(i) *A fixed-point rule.* Based on some additional knowledge of the physics or the geometry of the problem, it may be possible to average over some fixed set of points. Such a formula would be highly problem-dependent, and the uncertainty of the result would depend on the distributions involved, perhaps in a very complicated way.

(ii) *Random sampling with replacement.* In this method, objects are chosen randomly and one does not 'remember' which objects were already chosen, so that some may be taken more than once. The population thus becomes infinite, and the theory developed earlier applies just as if it were any other Monte Carlo calculation: the uncertainty on the potential is the standard deviation of the individual contributions, divided by the square root of the sample size.

(iii) *Sampling without replacement.* This method resembles (ii), except that one explicitly avoids taking the contribution from any one object more than once. The final convergence must be better than (ii), since one eventually reaches zero error when all contributions have been taken, but since by definition we cannot consider all contributions, it is the convergence rate in the early part of the sequence that matters. This convergence rate starts out equal to that of (ii), only improving slowly as the number of contributions taken becomes a significant fraction of the total. The price paid for this small improvement is having to remember which contributions were already chosen. Also the improvement may not be usable if it is too hard to calculate.

References

Ahrens J H and Dieter U 1972 *Commun. ACM* **15** 873–82

—— 1973 *Math. Comp.* **27** 927–37

Ahrens J H, Dieter U and Grube A 1970 *Computing* **6** 121–38

Braaten E and Weller G 1979 *J. Comp. Phys.* **33** 249–58

Buffon 1777 *Essai d'arithmetique morale*

Byckling E and Kajantie K 1973 *Particle Kinematics* (New York: Wiley)

Carey D and Drijard D 1978 *J. Comp. Phys.* **28** 327–56

Dieter U 1971 *Math. Comp.* **25** 855–83

Dieter U and Ahrens J H 1971 *Numer. Math.* **17** 101–23

—— 1973 *Computing* **11** 137–46

—— 1979 *Pseudo-Random Numbers* (New York: Wiley)

Eadie W T, Drijard D, James F E, Roos M and Sadoulet B 1971 *Statistical Methods in Experimental Physics* (Amsterdam: North-Holland)

Everett C J and Cashwell E D 1972 *A Monte Carlo Sampler. Los Alamos Scientific Laboratory Informal Rep.* LA-5061-MS

Friedman J 1977a *DIVONNE2, a program for multiple integration and adaptive importance sampling. SLAC Computation Research Group Tech. Memo* CGTM No 188

—— 1977b *Trans. Math. Software* submitted

Frigerio N A and Clark N 1975 *Trans. Am. Nucl. Soc.* **22** 283–4

Frigerio N A, Clark N and Tyler S 1978 *Toward Truly Random Numbers. Argonne National Laboratory Rep.* ANL/ES-26 Part 4

Genz A 1972 *Comp. Phys. Commun.* **4** 11–5

Haber S 1970 *Siam Rev.* **12** 4

Hagedorn R 1964 *Relativistic Kinematics* (New York: Benjamin)

Halton J H 1960 *Numer. Math.* **2** 84–90

—— 1970 *Siam Rev.* **12** 1–63

Hammersley J M and Handscomb D C 1964 *Monte Carlo Methods* (London: Methuen)

Kahaner D K and Wells B 1979 *ACM Trans. Math. Software* **5** 86–96

Keast P 1972 *Department of Computer Science, University of Toronto, Toronto, Canada. Tech. Rep.* 40

Knuth D 1969 *The Art of Computer Programming* vol 2 (Reading, Mass.: Addison-Wesley) pp1–160

Kuipers L and Niederreiter H 1974 *Uniform Distribution of Sequences* (New York: Wiley)

Lyness J N and Kaganove J J 1976 *ACM Trans. Math. Software* **2** 65–81

Malcolm M A and Simpson R B 1975 *ACM Trans. Math. Software* **1** 141–6

Marsaglia G 1968 *Proc. Nat. Acad. Sci.* **61** 25–8

—— 1972 *Applications of Number Theory to Numerical Analysis* (New York: Academic) pp249–85

Sobol I M 1979 *Siam J. Numer. Anal.* **16** 790–3 (Erratum **16** 1080)
Stroud A H 1971 *Approximate Calculation of Multiple Integrals* (Englewood Cliffs, NJ: Prentice-Hall)
Stroud A H and Secrest D 1966 *Gaussian Quadrature Formulas* (Englewood Cliffs, NJ: Prentice-Hall)
Tsuda T 1973 *Numer. Math.* **20** 377–91
van Dooren P and de Ridder L1976 *J. Comp. Appl. Math.* **2** 207–10
Weyl H 1916 *Math. Ann.* **77** 313–52
Zakrzewska K, Dudek J and Nazarewicz N 1978 *Comp. Phys. Commun.* **14** 299–309
Zaremba S K 1968 *Siam. Rev.* **10** 303–14
Zaremba S K (ed) 1972 *Applications of Number Theory to Numerical Analysis* (*Proc. Symp. Centre for Research in Mathematics, University of Montreal, 9–14 September 1971*) (New York: Academic)

Reproduced, with permission, from *Nucl. Instrum. Methods*
A265 pp. 273–290 (1988)

HIGH RESOLUTION HADRON CALORIMETRY

Richard Wigmans

NIKHEF-H, Amsterdam, The Netherlands

Abstract

The components that contribute to the signal of a hadron calorimeter and the factors that affect its performance are discussed, concentrating on two aspects: energy resolution and signal linearity. Both are decisively dependent on the relative response to the electromagnetic and the non-electromagnetic shower components, the e/h signal ratio, which should be equal to 1.0 for optimal performance. The factors that determine the value of this ratio are examined. The calorimeter performance is crucially determined by its response to the abundantly present soft neutrons in the shower. The presence of a considerable fraction of hydrogen atoms in the active medium is essential for achieving the best possible results. Firstly, this allows one to tune e/h to the desired value by choosing the appropriate sampling fraction. And secondly, the efficient neutron detection via recoil protons in the readout medium itself reduces considerably the effect of fluctuations in binding energy losses at the nuclear level, which dominate the intrinsic energy resolution. Signal equalization, or compensation ($e/h = 1.0$) does not seem to be a property unique to ^{238}U, but can also be achieved with lead and probably even iron absorbers.

1. Introduction

Large hadron calorimeters will be crucial components in any experiment at future high energy colliding beam machines. Calorimeters offer unique possibilities to recognize rare events at the trigger level, and to measure the characteristics of these selected events under severe background conditions, with a compact instrument. Achieving the best possible energy resolution is closely linked to the capability of detecting new phenomena in high energy physics. The energy resolution determines, for example, the mass resolution for detecting new particles that decay into jets (Higgs), the limits on missing (transverse) energy that can be handled (SUSY particle production), and the kinematic range accessible for data analysis (ep scattering). In order to be able to design calorimeters optimally it is, therefore, of vital interest to understand the factors that limit the energy resolution in as much detail as possible.

Calorimeters for the detection of electromagnetically interacting particles (e, γ) are well understood and can be reliably simulated in great detail by Monte Carlo techniques (e.g. EGS4). This is by no means the case for hadron calorimeters. To date there exists no computer program that is capable of simulating the complete development of a hadronic cascade and predicting reliably the response of a given calorimeter to it. As a consequence, progress in this area of detector technology had to proceed on a trial-and-error basis for a long time. Many results obtained in this process were considered extremely confusing.

The crucial role of the so-called e/h signal ratio for the energy resolution was already recognized in the early days of hadron calorimetry [1]. The R807 group at CERN demonstrated that energy resolutions of $35\%/\sqrt{E}$ could be achieved for hadron detection with their uranium/plastic-scintillator sandwich calorimeter, which had an e/h ratio close to the desired value 1.0 [2]. Measurements performed by HELIOS recently showed that the energy resolution of this detector scales with $E^{-1/2}$ up to the TeV regime, and that deviations from signal linearity are negligible over 3 orders of magnitude in energy [3].

Measurements performed with noncompensating ($e/h \neq 1$) devices, however, showed considerable deviations from linearity, and energy resolutions σ/E that do not improve as $E^{-1/2}$ with increasing energy [4]. The constant term that has to be introduced in order to describe the energy resolution is typically 5% and, therefore, dominates the result at energies beyond 100 GeV, where the new physics has to be found.

However, the use of ^{238}U as absorber material is by no means a guarantee for superior performance. Prototype tests by SLD and D0, using liquid argon readout, gave disappointing results, to the extent that SLD even abandoned the idea of using uranium and switched to a lead absorber that gave results of similar quality. Also the combination uranium/plastic-scintillator turned out not to be magic: WA78 found overcompensation ($e/h \approx 0.8$) and rather poor energy resolutions [5]. On the other hand Brückmann et al. from the ZEUS Collaboration, who basically copied the HELIOS design for their prototype studies, found equally good performance [6].

Recently, understanding of hadron calorimetry has considerably improved. We systematically investigated in detail the response of a sampling calorimeter to the various components of a hadron shower [7]. It turned out that the calorimeter performance in terms of energy resolution and signal linearity is crucially determined by its response to the abundantly present soft neutrons in the shower. The presence of a considerable fraction of hydrogen atoms in the readout medium is essential for the best possible performance. Firstly, this allows one to tune e/h to the desired value (1.0) by choosing the appropriate sampling fraction. And secondly, the efficient neutron detection via recoil protons in the readout medium itself reduces considerably the effect of fluctuations in binding energy losses at the nuclear level, which dominate the intrinsic energy resolution. The confusing experimental results mentioned before were quite nicely reproduced in this study.

In this paper, the theoretical framework that lead to this improved understanding is briefly described. In section 2, we will discuss how the e/h value affects the calorimeter performance. In section 3, we describe how the properties of the passive and active calorimeter materials determine its e/h value. In section 4, the factors that determine the energy resolution of practical calorimeters are examined. An outlook is given in section 5. Since this paper has to be necessarily brief, only a rather global description can be given. For details the reader is referred to the original work [7].

2. Effects of the e/h ratio on the calorimeter performance

Since calorimetry is based on statistical processes, the measurement accuracy increases with energy, contrary to tracking detectors based on momentum analysis in a magnetic field. This is a very attractive feature. When a particle develops a shower in a block of matter that we call a sampling calorimeter, some fraction of its energy is transformed into a measurable signal, usually a pulse of electrical charge. This pulse is the result of all the charged particles generated in the shower development, that traverse and ionise the active layers of the detector. The energy resolution for detecting the original particle is determined by the fluctuations occurring in this process, which affect the size of the resulting signal. Amongst these we mention the following:

(a) Sampling fluctuations, due to the fact that only a (small) fraction of the energy is deposited in the active layers and, therefore, contributes to the calorimeter signal. It has been shown [8] that if this fraction is larger than $\sim 1\%$, the effect is dominated by fluctuations in the total number of gap crossings by charged particles generated in the shower. Therefore, the contribution of this source of fluctuations is, to first order, for a given absorber material proportional to the square root of the thickness of the absorber plates, and insensitive to the type and thickness of the active layers. The latter part of this statement is not correct if the active layers are so thin (expressed in g/cm^2), that the statistical fluctuations in the number of ionization processes per gap crossing become important. This is for example the case for gaseous media, where Landau and path-length fluctuations dominate those mentioned before. In *electromagnetic* sampling calorimeters, sampling fluctuations are the major contribution to the energy resolution.

(b) Fluctuations in the fraction of the energy deposited in the form of ionizing particles. These effects usually dominate the energy resolution of *hadron* calorimeters. An obvious component comes from incomplete shower containment, i.e. from secondary particles like μ, ν, n, K_L^0, etc. that (partly) escape detection. A somewhat less obvious, but usually far more important source of fluctuations comes from the energy losses at the nuclear level. The energy needed to release the nucleons and nucleon aggregates like α-particles that are bound in atomic nuclei is lost for detection. The fraction of the initial hadron energy spent in this way is considerable, up to $\sim 40\%$ on average for the nonelectromagnetic part of hadron showers in high-Z elements. Due to the enormous variety of possible nuclear interactions, the fluctuations about this number are a major contribution to the energy resolution of hadron calorimeters.

The latter class of effects has no equivalent in electromagnetic showers. This has consequences for calorimetric hadron detection. Firstly, the signal distribution for hadrons of energy E will be broader than for electromagnetic showers at the same energy, because of the larger fluctuations. Secondly, the average response (signal per unit of energy) will have a different value for electromagnetic and hadronic showers ($e/h \neq 1$). Because of the losses mentioned before, the latter will in general be smaller ($e/h > 1$).

In the development of a shower generated by a high-energy hadron some fraction (f_{em}) of the energy is usually spend on the production of π^0s and ηs. A hadron shower, therefore, has in general an electromagnetic (em) and a nonelectromagnetic component. Because of the reasons just given, the calorimeter response to these components will be different (see fig. 1). The fluctuations in f_{em} are large and non-Gaussian. Moreover, $\langle f_{em} \rangle$ increases (logarithmically) with energy. If $e/h \neq 1$, this has the following consequences for the calorimeter performance:

(1) The fluctuations in f_{em} give an additional contribution to the energy resolution.

(2) Because of the non-Gaussian nature of the fluctuations in f_{em} the energy resolution σ/E will not improve as $E^{-1/2}$ with increasing energy.
(3) The calorimeter signal is not proportional to the hadron energy (alinearity).
(4) The measured e/h signal ratio is energy dependent.

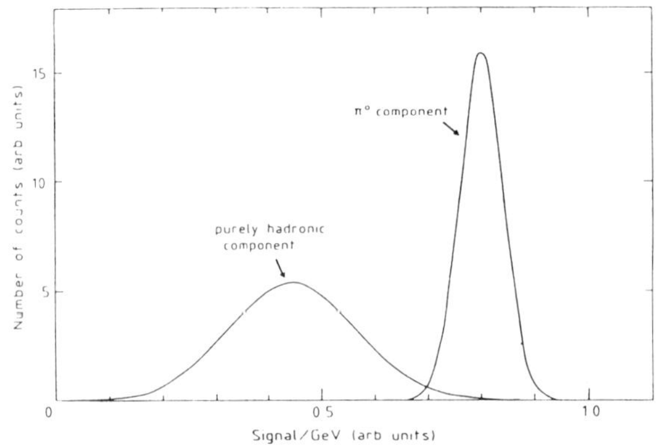

Fig. 1. The calorimeter response to the em and non-em components of a hadron shower.

All these effects have been experimentally observed. This is illustrated by fig. 2, which shows experimental data on calorimetric hadron detection with an iron/plastic-scintillator ($e/h \approx 1.4$) calorimeter [4], and with uranium/plastic-scintillator [3] ($e/h = 1$). The energy resolution obtained with iron is much worse than for uranium, and it does not scale as $E^{-1/2}$. The calorimeter response for iron rises with energy, and consequently the e/h signal ratio decreases as the energy is increased. Such effects are absent for uranium.

To avoid the confusion caused by the latter effect, we prefer to use the energy-independent quantity e/h^{intr}, given by the ratio of the mean values of the two curves in fig. 1. The relation between e/h^{intr} and e/h is given by

$$e/h(E) = \frac{e/h^{\text{intr}}}{1 - \langle f_{\text{em}}(E)\rangle\left[1 - e/h^{\text{intr}}\right]},$$

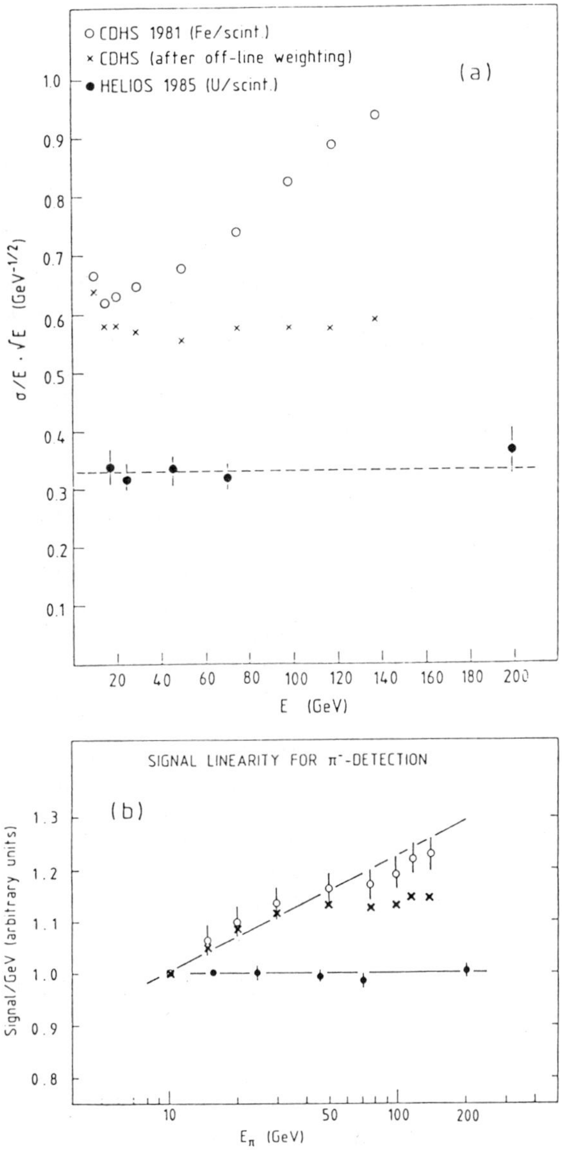

Fig. 2. Experimental data obtained with an iron/plastic-scintillator calorimeter (the open circles), and with uranium/plastic-scintillator (full circles). The crosses represent the off-line weighted iron results. Data are given on energy resolution (a) and linearity (b) as a function of energy.

where

$$\langle f_{\rm em}(E)\rangle = 0.1 \ln E(\text{GeV})$$

is a reasonable approximation in the energy range 10–100 GeV [9]. At 10 GeV, where e/h ratios are frequently given, the difference between the measured e/h value and $e/h^{\rm intr}$ is usually small, less than 10%.

We investigated the effects of $e/h \neq 1$ on the calorimeter performance with a very simple Monte Carlo simulation, making some reasonable assumptions on the widths of the distributions in fig. 1, n $\langle f_{\rm em}(E)\rangle$, and on the fluctuations in $f_{\rm em}(E)$. The results are given in fig. 3, where curves for the energy resolution and the calorimeter response are plotted for various values of e/h (10 GeV). The data on the energy resolution are plotted using an abscissa linear in $E^{-1/2}$. This shows that the deviation from $E^{-1/2}$ scaling can be very reasonably described as a constant term, the value of which is determined by e/h. This figure makes it very clear that, in particular at high energies, the value of $e/h^{\rm intr}$ is crucial for the energy resolution that can be obtained. For the CDHS calorimeter ($e/h = 1.36$ at 10 GeV), the resolution σ/E does not become better than $\sim 7\%$, even at the highest energies. For calorimeters with $e/h^{\rm intr} = 1$, the energy resolution will continue to improve with increasing energy, until limitations due to instrumental effects become important. HELIOS have experimentally shown that energy resolutions close to 1% are feasible for such devices [3].

If $e/h = 1$, the contribution of fluctuations in $f_{\rm em}$ to the energy resolution are eliminated. It has been advocated that there are other ways to (partly) achieve this goal. The em and non-em shower components have very different characteristic dimensions, particularly in high-Z absorber materials. If the calorimeter is sufficiently finely segmented one might, therefore, determine $f_{\rm em}$ on an event-by-event basis and hence eliminate, at least partly, the effects of fluctuations in this number. The CDHS Collaboration developed an algorithm for this purpose, with which they were able to restore $E^{-1/2}$ scaling for the energy resolution of single pion detection in their very fine-grained detector (read out every 7 radiation lengths in depth) [4]. Unfortunately, alinearities of $\sim 10\%$ remained after this procedure, which

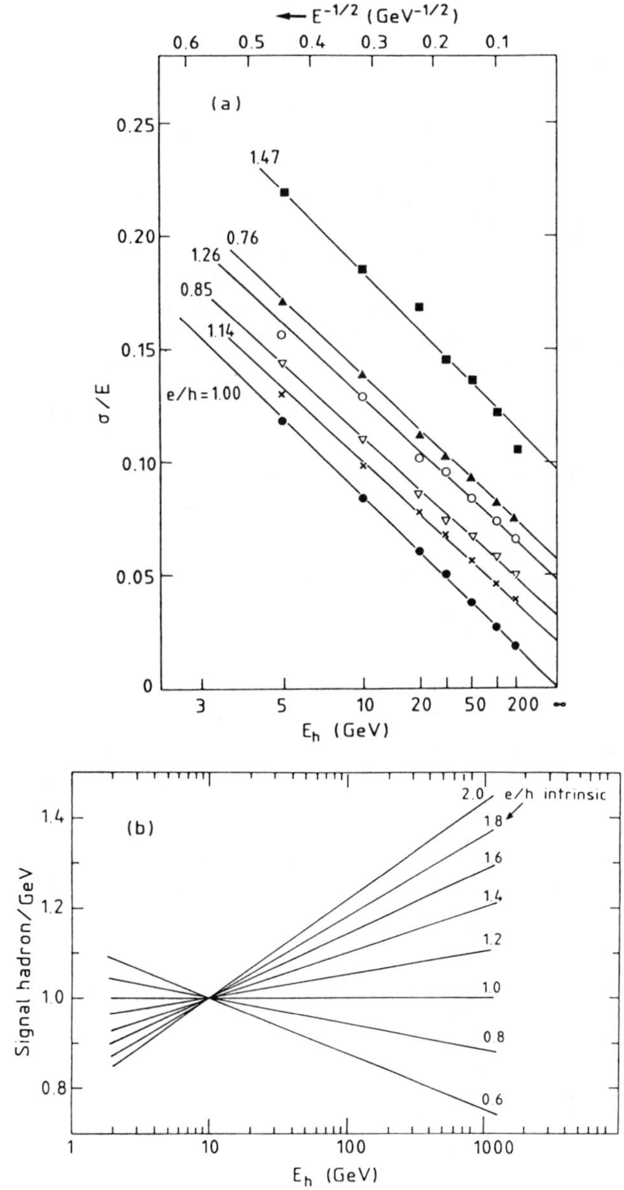

Fig. 3. Monte Carlo simulation of the effects of $e/h \neq 1$ on the energy resolution and signal linearity of hadron calorimeters.

involved energy-dependent weighting factors (see fig. 2). This means that the procedure will probably not work (as well) for the detection of jets, where the showers generated by individual hadrons overlap, and where the number of jet particles and their energy is not a priori known. In any case, these methods have never been experimentally demonstrated to improve the resolution under such conditions.

Because of these reasons, we think that a calorimeter with $e/h^{\mathrm{intr}} = 1$ has essential advantages, which cannot be compensated for by a very fine-grained readout in case $e/h \neq 1$. We will now examine which factors determine the e/h signal ratio, and how this ratio is affected by the properties of the calorimeter materials.

3. The calorimeter response to the shower components

3.1. The procedure of calculating e/h

In this section we will investigate the calorimeter response to the various components of a hadron shower. For a particular type of secondary particles, the signal will be determined by two factors:

(1) The fraction of the energy used to ionise the active layers.

(2) The ionization density in these layers. Saturation effects may reduce the signal from densely ionizing particles, e.g. soft protons, considerably.

We distinguish the following classes of secondary particles:

(a) High-energy photons and electrons from π^0 and η decay, that form the em shower component.

(b) Ionizing hadrons (charged pions, kaons, protons, etc.).

(c) Soft neutrons.

(d) Soft γs from nuclear deexcitation.

As a scale for the calorimeter signal, the response to a minimum-ionizing particle (mip) will be used. The fraction of the energy deposited by such a particle in the active layers of a given sampling calorimeter is called the sampling fraction f_{samp}, and can be calculated immediately from the mean $\mathrm{d}E/\mathrm{d}x$ values for mip's in the calorimeter materials:

$$f_{samp} = \frac{dE/dx[1 \text{ readout layer}]}{dE/dx[1 \text{ readout} + 1 \text{ absorber layer}]} .$$

In practical calorimeters f_{samp} amounts to about 5–10% for plastic scintillator or liquid argon (LAr) readout, ~ 1% for Si, and ~ 10^{-4} for wire chambers.

Relating the responses of the 4 mentioned types of particles to mip's, one can schematically write the e/h signal ratio as follows

$$e/h^{\text{intr}} = \frac{e/\text{mip}}{f_{\text{ion}}\ \text{ion}/\text{mip} + f_{\text{n}} n/\text{mip} + f_{\gamma}\gamma/\text{mip}} ,$$

where f_{ion}, f_{n} and f_{γ} are the average fractions of the energy in the non-em component that are deposited in the form of ionizing hadrons, neutrons and nuclear γs. We will now discuss the various terms of this expression.

3.2. The response to high-energy electron and photon showers (e / mip)

The signal from an em shower is the result of the ionization of the active layers by all the electrons and positrons generated in the shower development which traverse these layers. Naively, one might therefore expect that this signal is equal to the one from muons traversing the detector, which lose an amount of energy $\Delta E = E_{\text{em}}$. This is in practice usually not the case. Experimentally, one observes that in calorimeters with high-Z absorber material, the em signal is considerably suppressed with respect to mip's ($e/\text{mip} < 1$). If the Z-values of the active and passive media are about equal, as in Fe/LAr, e/mip is close to 1.

For historical reasons these phenomena are usually called transition effects, since it was assumed that they should be attributed to mechanisms affecting the shower composition at the boundary between layers of materials with different Z. This is not true.

The explanation of the transition effects is closely linked to the way in which soft γs ($E_{\gamma} < 1$ MeV) that are generated in large numbers in an em shower are absorbed in a mixture of materials. At these low energies, the photo-electric effect is an important mechanism. The cross section for this process is proportional to Z^5, while the ionization cross section for charged particles is to first order proportional to Z. In a calorimeter consisting of high-Z absorber material and low-Z active layers, the soft γs will, therefore, almost exclusively interact with absorber atoms. Only if the interaction occurs sufficiently close to a boundary can the photo-electron produced in this process contribute to the measured signal. To illustrate this point, it can be mentioned that the mean range of a 1 (0.1) MeV electron in uranium is only 0.4 mm (20 μm). As a result, the calorimeter response to an em shower is suppressed with respect to mip's.

Analysis of shower simulation data from EGS4 clearly supports this conclusion. Needless to say that, for obtaining accurate results, the program should be run with cutoff values that are sufficiently low to correctly describe the peculiarities of the energy deposition by the soft component. Although we are dealing with very soft particles, the total amount of energy carried by them is considerable. The evolution of the computer time needed for the simulation when lowering the cutoff values for E_γ and E_e, is already an indication for this. Fig. 4 shows that in lead or uranium calorimeters ~ 40% of the total energy of a 10 GeV em shower is deposited through ionization by particles softer than 1 MeV, in which energy region the photo-electric effect is the dominating process.

For what concerns the response of high-Z sampling calorimeters, em showers consist of two rather well-separable components: The hard part (> 1 MeV), that can be reasonably described as a bunch of mip's, and the soft part (< 1 MeV) for which the detector is less efficient. This has some quite interesting consequences for (hadron) calorimetry, all of which are confirmed by EGS4 calculations:

(a) The e/mip ratio changes with the depth of the shower, since the relative contribution of both components is a function of depth: The shower gets softer as the depth increases. In a longitudinally segmented calorimeter, calibrated with muons, the energy deposited by an electromagnetic shower in individual segments will be wrongly measured if the shower exceeds a segment boundary, although the *total* energy may be correct [3].

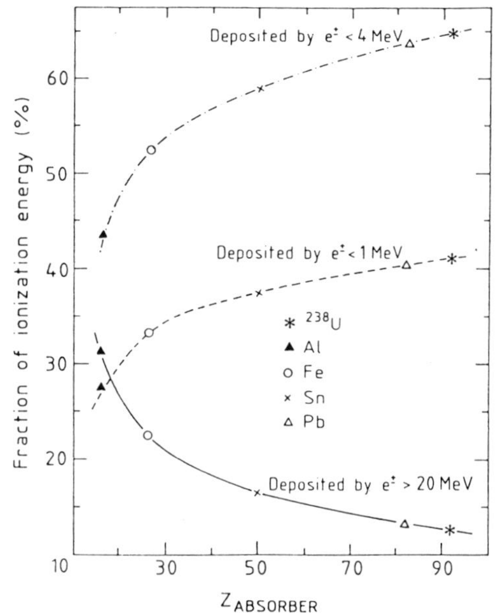

Fig. 4. The fraction of the energy of a 10 GeV electron shower that is deposited through ionization by electrons and positrons slower than 1 or 4 MeV, or faster than 20 MeV, as a function of the Z-value of the calorimeter absorber. Results from EGS4 Monte Carlo simulations.

(b) One can tune the e/mip and hence the e/h value to a certain extent, since the relative contribution of the soft component to the calorimeter signal depends on the thickness of the absorber layers and also on the thickness of the active layers (see fig. 5). This can be understood as follows. Only photo-electrons produced close to a boundary contribute to the signal. If they manage to reach the active layers they are usually so soft that they do not penetrate very far. If either the absorber or the readout thickness is decreased, the "effective" boundary region becomes a larger fraction of the calorimeter as a whole and, therefore, the calorimeter response is increased. These effects are mainly important for very fine sampling.

(c) In a high-Z calorimeter one may further reduce e/mip by inserting a low-Z foil in between the

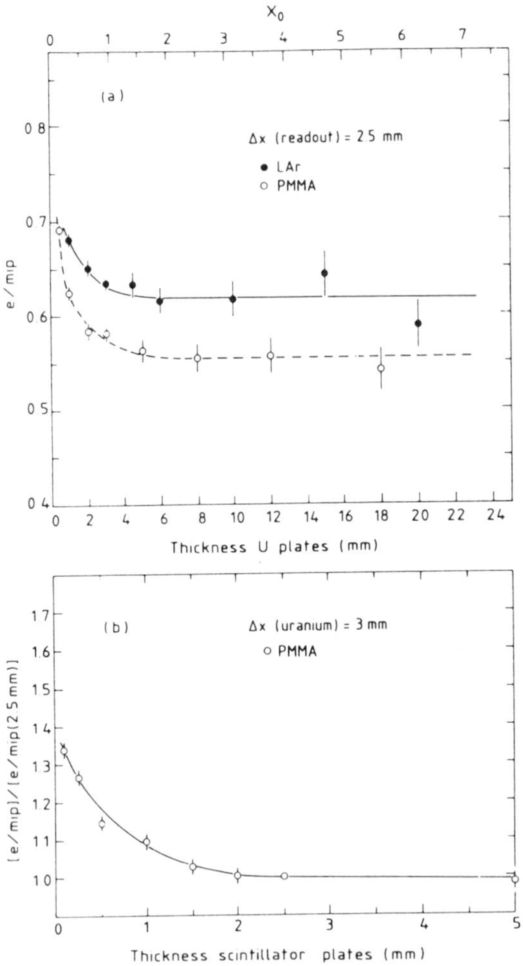

Fig. 5. The e/mip signal ratio as a function of the thickness of the active (a) and passive (b) layers for uranium calorimeters. Results from EGS4 calculations.

active and passive layers. This prevents photo-electrons produced in the high-Z material from reaching the active layers. This is illustrated in fig. 6, where the results of EGS4 calculations on a uranium/plastic-scintillator calorimeter are shown. The uranium plates are wrapped in stainless steel, the thickness of which is varied. This wrapping decreases the "intrinsic" e/mip ratio by at maximum ~ 8% at 0.5 mm Fe. If the wrapping material is made thicker than that, interactions by the shower particles in iron will become more and more important, and tend to increase e/mip because $Z_{Fe} \ll Z_U$.

The detector inefficiency for the soft component of em showers in high-Z materials is an important ingredient for compensation, since it brings the em response closer to the non-em one. As we shall see later it is, however, not enough.

3.3. The response to nuclear gammas (γ / mip)

In the non-em part of hadronic shower development large numbers of γs are produced from nuclear deexcitation. Their energy spectrum depends on the characteristic nuclear level structure of the nuclides involved. In general, the vast majority of these γs have an energy below 2 MeV.

In the previous subsection we saw that the transition effects observed for high-energy em showers are caused by the reduced detection efficiency for the soft γ shower component. The calorimeter signal suppression for nuclear γs will, therefore, even be stronger than for these showers. This is confirmed by EGS4 calculations.

Since the mean photo-electron range is extremely energy-dependent, the detection efficiency decreases sharply below 1 MeV (fig. 7). At 0.511 MeV, the energy of the very abundant e^+ annihilation γs, the signal is already reduced by a factor 7 compared to mip's, for the calorimeter in this example (HELIOS). The γ/mip signal ratio needed for the computation of e/h is sensitively dependent on the γ energy spectrum. For the measured spectrum of ^{238}U fission γs that was used throughout the calculations, we found γ/mip to be ~ 0.4 in uranium calorimeters versus 0.6 for e/mip. The EGS4 calculations showed also that the other effects found for em showers (figs. 5, 6) were enhanced for a nuclear γ-ray spectrum.

Nuclear fission is an important source of nuclear γs in uranium calorimeters. We measured that in the non-

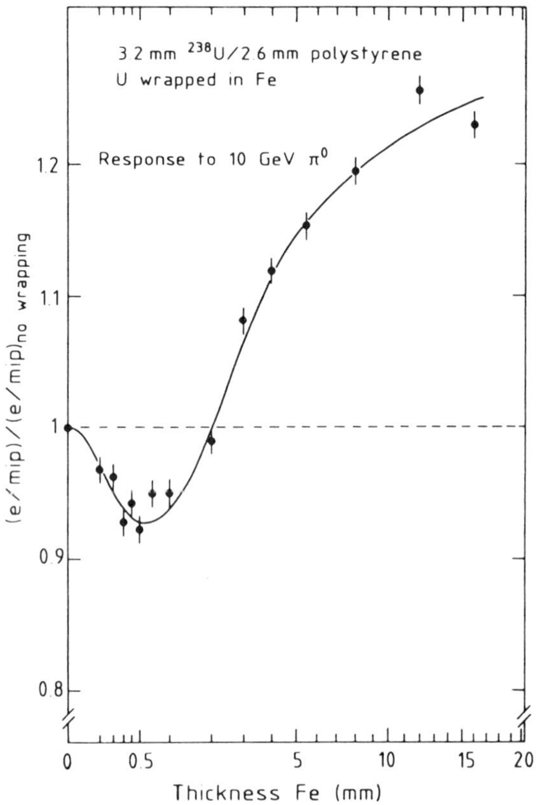

Fig. 6. The effect of stainless steel wrapping of the absorber plates on the e/mip value of a uranium/plastic-scintillator calorimeter. The vertical scale is normalized to the case where no steel is applied. Results from EGS4 calculations.

em part of hadron showers on average ~ 10 fissions per GeV are generated [10]. In each fission, on average 7.4 MeV is released in the form of prompt γs [11]. Other processes like inelastic neutron scattering and thermal neutron capture further increase the fission-related nuclear γ component to a considerable fraction of the initial particle energy. The use of uranium absorber in hadron calorimeters was originally advocated because of this extra energy, which was assumed to make up for the inevitable losses in the non-em part of the shower

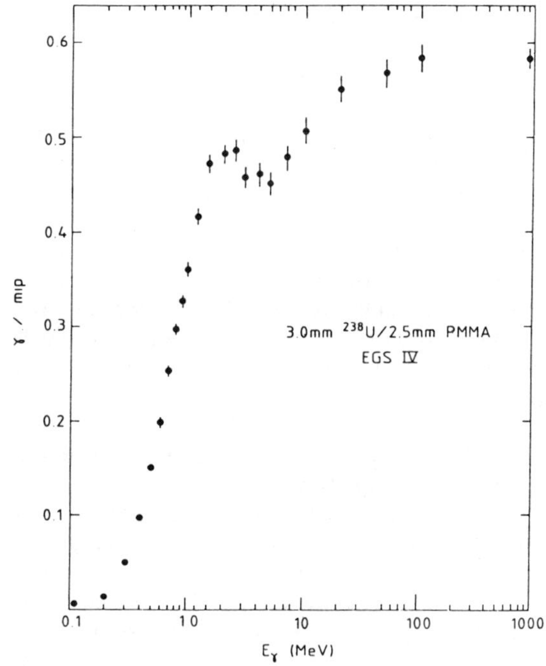

Fig. 7. The γ/mip signal ratio as a function of E_γ for a uranium/plastic-scintillator calorimeter. Results from EGS4 calculations.

(compensation) [12]. The reduced efficiency for detecting these γs limits the effect of this, however, to the extent that the condition $e/h^{\text{intr}} = 1$ cannot be achieved in U/LAr calorimeters.

3.4. The non-em shower part (f_{ion}, f_n, f_γ)

The analysis of the calorimeter signal from em showers has taught us an important lesson that is also relevant for understanding the details of the response to the non-em part. The lesson is that the signal is decisively determined by the peculiarities of the processes occurring at the very last stages of the shower development. Although the energy is very low, the number of particles is so large that a considerable fraction of the total energy is deposited through these processes.

In the last stages of the development of the non-em shower part the production of pions, kaons, etc. is energetically impossible. It turns out that even at very high energies, the contribution of such particles to the calorimeter signal is small: Typically, only 10–15% of the non-em energy is deposited through ionization by charged pions. An easy way to see this is by realizing that charged pions lose on average only ~ 0.2 GeV per nuclear interaction length through ionization of the medium that they traverse. A 100 GeV pion that produces in its first interaction 10 other charged pions will, therefore, have lost only $\sim 2\%$ of its energy when the secondary pions undergo strong interactions. The average pion multiplicity in these secondary interactions is probably something like 2–3 (adding another 4–6% to the $\mathrm{d}E/\mathrm{d}x$ loss). In further generations new pion production becomes rapidly negligible.

The remaining 85–90% of the shower energy is spent on nuclear processes. It is used to release nucleons and nucleon aggregates from atomic nuclei (binding energy losses), and to give kinetic energy to the reaction products. Usually the nuclei also emit γs in these processes (see section 3.3). Apart from these γs, protons and neutrons may contribute to the calorimeter signal. The range of the nucleon aggregates and of the recoiling target nuclei is so small that their kinetic energy is virtually lost for detection.

The way in which the protons and neutrons lose energy and contribute to the calorimeter signal is very different. The vast majority of the protons will lose their complete kinetic energy by ionization. Only the fast component ($E_{\mathrm{kin}} > 100$ MeV) may cause new nuclear reactions to occur at a non-negligible level. The neutrons exclusively lose kinetic energy through strong interactions. In these reactions new neutrons, protons and γs may be produced, and additional binding energy lost. In particular the fast neutrons produced in the initial phase of the shower may develop quite complicated showers themselves.

The numerical details of the nuclear reaction mechanisms which are important for our purpose (e.g. nuclear spallation) are extensively discussed in ref. [7]. Here we only give the main conclusions. The neutron multiplicity in hadron showers developing in a material ($Z \gg 1$)

is much larger than the proton multiplicity, particularly if Z is large. This has two reasons:

(a) More neutrons than protons are available in the target nuclei.

(b) At low energies the nuclear Coulomb barrier will prevent protons from escaping.

In the reactions that start the "nuclear" phase of the shower development, more kinetic energy will therefore be carried away by neutrons than by protons. Most of this kinetic neutron energy will, however, be lost as nuclear binding energy in a series of subsequent reactions, in which the neutron multiplicity increases considerably, and in which also some new protons may be produced. In each nuclear reaction on average 2–3 MeV is released in the form of nuclear γs.

At some stage in the shower development the neutrons have degraded so much in energy that they cannot release any more new protons from target nuclei. For high-Z absorbers this is the case below $E_n^{kin} \approx 20$ MeV. What we need to know at this point in order to be able to compute e/h^{intr} (see sect. 3.1) are the fraction of the non-em energy deposited through ionization by protons (the major component of f_{ion}), the remaining kinetic neutron energy (f_n), and the total number of nuclear reactions (needed for f_γ). Nuclear γs have only eaten up 2–3% of the energy at this stage. The other components depend obviously on the Z of the absorber material. Roughly one may say that $f_{ion} \approx Z/A$, while f_n varies from 8% for Fe to 15% for ^{238}U. The rest of the energy is invisible (binding energy, target recoil, nucleon aggregates). The contribution of nuclear fission in uranium becomes mainly important beyond this point in the shower development, i.e. for $E_n < 20$ MeV, because that is where the neutron multiplication and the energy gain from this process count most.

3.5. The response to ionizing hadrons (ion / mip)

As was shown in the previous subsection, this shower component is dominated by protons (70–75%). The remaining contribution comes from pions, kaons, etc. These were considered mip's. The calorimeter signal

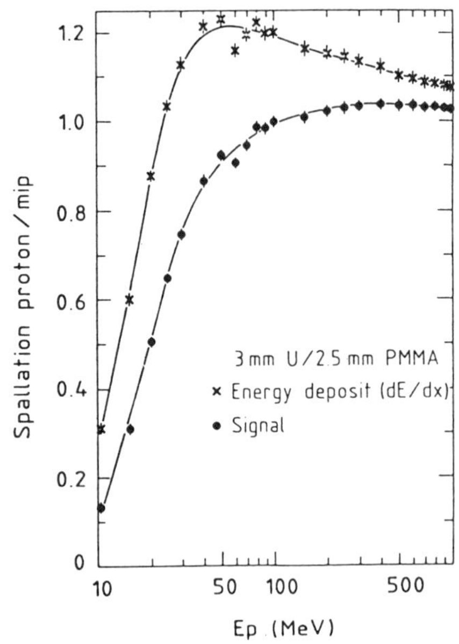

Fig. 8. The energy deposit and the calorimeter signal for stopping protons relative to mip's as a function of the proton energy, for a 3 mm U/2.5 mm plastic scintillator calorimeter.

ratio p/mip may deviate from 1 because of the following reasons:

(1) The range of the spallation protons is limited (3 mm in U at 40 MeV). Sampling inefficiencies for the soft part of the spectrum *decrease* the p/mip value.

(2) The sampling fraction (see sect. 3.1) for nonrelativistic particles may be very different from the mip value. The magnitude of this effect is very energy- and Z-dependent. Especially for high-Z absorbers p/mip may considerably *increase*.

(3) Saturation or recombination effects in the readout material *reduce* the response for densely ionizing particles, and hence the p/mip value.

(4) Multiple scattering of very soft protons *decreases* p/mip if $Z_{\text{abs}} \gg Z_{\text{readout}}$.

The combined effect of all these phenomena was investigated with a Monte-Carlo program. The ion/mip signal ratio turned out to depend on the material combination, on the proton energy spectrum, on the saturation properties of the readout material, and on the thickness of the active layers. The first three effects are obvious from 1–4; the last effect is presumably due to the fact that if the thickness of the active layers is increased, a larger fraction of the protons will stop inside them and hence produce a relatively small signal compared to the amount of energy deposited (saturation).

An example of the results is given in fig. 8. This figure illustrates the energy dependence and the saturation effects. For practical calorimeters with saturating readout materials (plastic scintillator, LAr), ion/mip signal ratios were found in the range 0.85–1.0. For a nonsaturating medium like Si the response would be ~ 15% larger.

3.6. The response to soft neutrons (n / mip)

Soft neutrons are the most interesting shower component, and the most crucial one for the calorimeter response. In the previous subsections we saw that the calorimeter response to the other shower components depends on rather global material properties, like Z_{abs}. In order to be able to properly evaluate the neutron contribution, we have to go rather deeply into nuclear physics details. Nuclear level structures and cross sections for individual reactions are important for the result.

For example, the fact that the first excited state in the "double magic" nucleus ^{208}Pb lies as high as 2.6 MeV above the ground state level (which prevents inelastic neutron scattering below this energy), means that the e/h ratio for a Pb/plastic-scintillator calorimeter would change considerably if the isotopes ^{204}Pb, ^{206}Pb, and ^{207}Pb, which can be excited by neutrons down to 0.6 MeV, were not present. Small impurities with anomalously large cross sections for certain neutron-induced reactions, like Cd, may also have a large impact.

Fortunately, neutron scattering is one of the most systematically studied areas in physics. The wealth of existing experimental information allows us to calculate rather precisely how the neutrons will lose their kinetic energy in a given calorimeter. Fig. 9 shows cross sec-

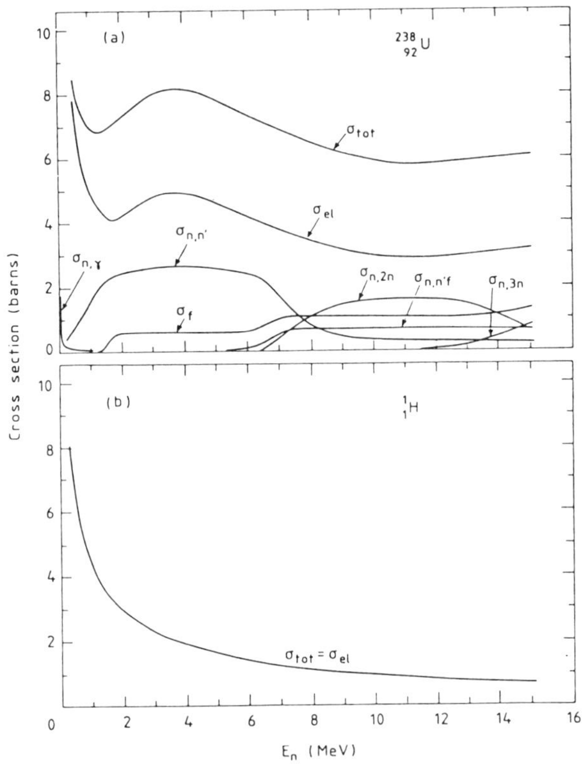

Fig. 9. The cross sections for neutron-induced reactions on ^{238}U (a) and hydrogen (b) nuclei.

tions that are relevant for the calculations of neutron transport in a uranium/plastic-scintillator calorimeter. Proton production off uranium nuclei no longer occurs at these low neutron energies. However, the neutrons can still multiply themselves, through $(n, 2n)$ or $(n, 3n)$ reactions where additional binding energy is lost, or through nuclear fission. The latter process may occur down to $E_n^{kin} \approx 1.5$ MeV. In inelastic scattering kinetic neutron energy is converted into γs, but also the probability for this process sharply decreases below 2 MeV.

The last step in the neutron's lifetime as a free particle is capture. In this process, the energy that was needed to liberate the neutron from the nuclear field is gained back and emitted by the capturing nucleus in the form of γs. The cross section for this process is only considerable when the neutron is practically thermalized. The dominating process to get rid of the last 1–2 MeV is elastic scattering. The cross section for this process is very large, more than 4 b (mean free neutron path less than 5 cm). However, the energy loss per collision is very small, 1/239 on average. The reaction products, recoiling target nuclei, are undetectable.

The situation changes completely if the calorimeter also contains hydrogen in the readout material. Hydrogen is well-known to be very efficient in thermalizing neutrons. In each elastic collision, on average half of the kinetic neutron energy is transferred. At 1 MeV the cross section for this process is 4 b, which corresponds to a mean free neutron path of ~ 5 cm in polystyrene, and ~ 4 cm in TMP. The reaction products of this process are recoil protons. These are produced inside the active layers, and contribute *directly* to the calorimeter signal, i.e. they are not sampled like mip's. The range of 1 MeV protons in polystyrene is 22 μm and, therefore, in practice the kinetic energy of protons produced in this way will almost *completely* be deposited in the active calorimeter layers.

In calorimeters that contain hydrogen atoms, neutron–proton scattering is a very important source of energy loss, because at low energies there is almost no competition from other processes. This may be concluded from an experimental study on the distribution of radioactive nuclides produced by the neutrons from hadron showers [10]. Fig. 10 shows the longitudinal distributions of ^{239}Np, produced from thermal neutron capture by ^{238}U, in different uranium calorimeter configurations. It turned out that the distribution in uranium/plastic-scintillator is very different from the ones in uranium calorimeters not containing hydrogen. The neutrons were much more rapidly thermalized in the scintillator case and, therefore, must have lost a considerable fraction of their kinetic energy in elastic collisions with hydrogen nuclei. The reduced number of fissions measured in the scintillator calorimeter supports this conclusion as well.

We studied the transport of the soft neutron shower component in a given calorimeter structure with a

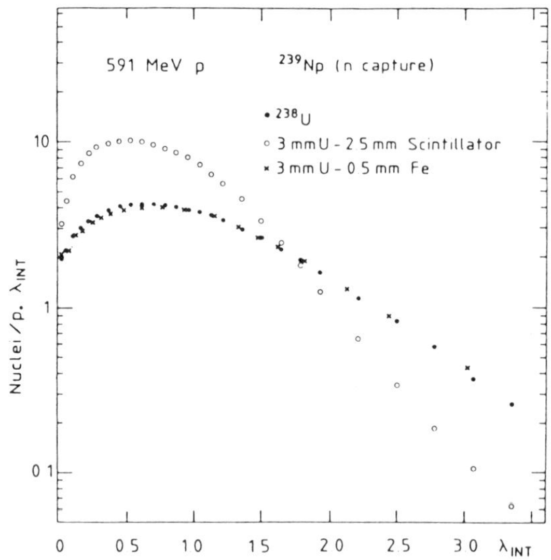

Fig. 10. The longitudinal distributions of ^{239}Np nuclei (from neutron capture in ^{238}U) produced by hadron showers in various calorimeter configurations.

Monte-Carlo program that contained all the relevant information on the nuclear physics: The cross sections for all possible reactions in the absorber and readout materials, the energy transfer in these reactions, the physics of fission neutrons and γs, etc. Since the recoil protons are very densely ionizing, the saturation characteristics of the readout materials are of crucial importance. We used Birk's law to convert from energy to signal (L):

$$L/E = \left[\int_0^{x_{max}} \left(1 + kB\frac{dE}{dx}\right) dx\right]^{-1}.$$

The values of the saturation constant kB are such that this meant a reduction of a factor 4–5 for materials like LAr and plastic scintillator. Essential for the results was also the input neutron spectrum. We used a linear combination of two Maxwell distributions to account for the soft evaporation and fission neutrons on the one hand, and for the degraded spallation neutrons on the other hand. The validity of these assumptions was checked and found in excellent agreement with all the experimental data from the neutron-induced activation analysis mentioned before [10].

Fig. 11 shows an example of the results for a 3 mm U/3 mm plastic-scintillator calorimeter. Below 4 MeV, more than half of the kinetic neutron energy is converted into recoil protons, and below 1 MeV even more than 70%. The remaining part of the energy is converted into γs (inelastic scattering, fission) and invisible forms (binding energy, target recoil). Note that the sampling fraction, i.e. the part of the energy that mip's deposit in the active layers, amounts to only ~10% for this calorimeter. The recoil protons will therefore substantially contribute to the calorimeter signal.

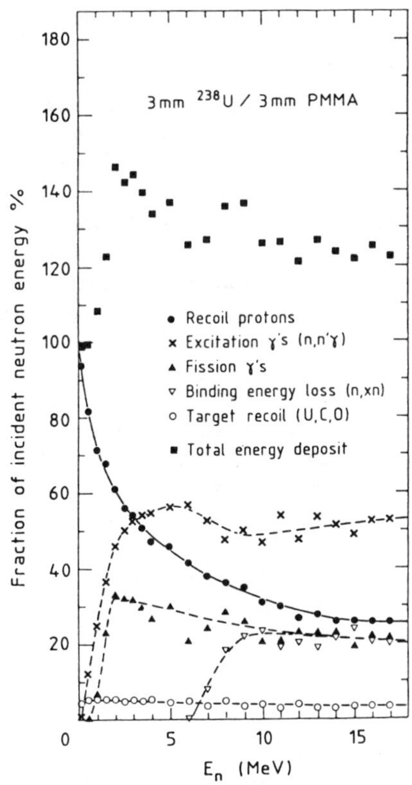

Fig. 11. The various forms in which the kinetic neutron energy is deposited in a 3 mm U/3 mm plastic-scintillator calorimeter. Because of fission contributions, the total energy can be larger than 100%.

A nice example to illustrate this comes from the L3 Collaboration, who tested uranium calorimeters with proportional wire chamber readout [13]. When using isobutane (containing hydrogen) the hadronic response was found to be about twice as high as for Ar/CO_2 (no hydrogen), while the electron response was almost gas-independent (see fig. 12). The difference is due to n–p scattering. Similar differences occur between plastic-scintillator (containing hydrogen) and LAr (no hydrogen) readout. These are smaller than for gaseous media because of saturation effects, but yet considerable.

The Monte-Carlo studies revealed another very interesting effect. It turned out that the n/mip signal ratio, or the relative contribution of recoil protons to the calorimeter signal, is sensitively dependent on the sampling fraction for mip's (f_{samp}, see sect. 3.1). If f_{samp} gets smaller, n/mip increases; if f_{samp} gets larger, n/mip decreases. This can be understood as follows.

If the thickness of the absorber plates is increased, the signal from mip's will decrease proportionally. The energy transferred to recoil protons in the active layers by the neutrons is *not* sampled. We saw before that at low neutron energies, there is almost no alternative for this process and, therefore, the fraction of the neutron energy deposited in this form decreases much slower than the mip signal. In other words, if the sampling fraction is decreased, the n/mip signal ratio and hence the total non-em calorimeter response increases, and the e/h signal ratio decreases. The sampling fraction is, therefore, a crucial parameter for the performance of hadron calorimeters with hydrogenous readout.

Fig. 13 shows some results of the calculations for uranium/plastic-scintillator calorimeters. Instead of the sampling fraction, we used the more practical variable R_{d}, defined as the ratio of the thicknesses (in mm) of the passive and active calorimeter layers; R_{d} is approximately inversely proportional to f_{samp}. In fig. 13a the n/mip signal ratio is given as a function of R_{d}. The neutron signal is split up in its various contributing components. The dominating recoil proton contribution strongly increases with R_{d}. The consequences of this for the calorimeter performance are of decisive importance, because of the relatively large contribution of the

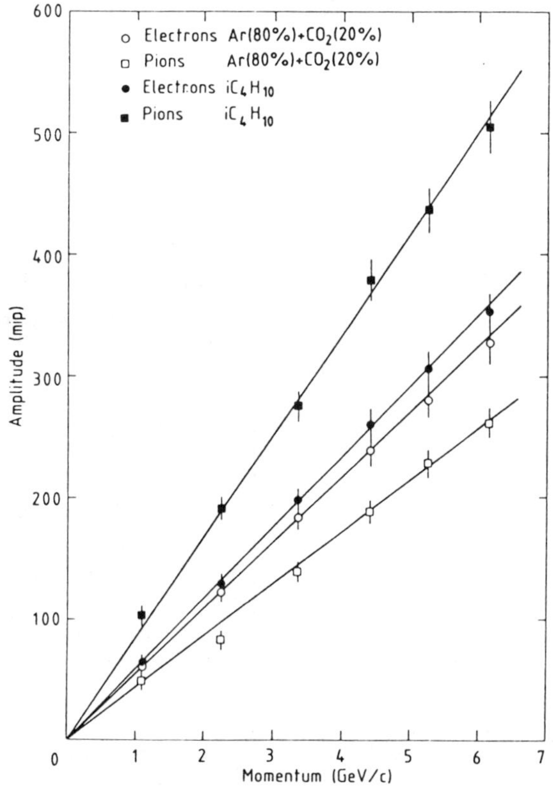

Fig. 12. The response of a uranium gas-sampling calorimeter to pions and electrons for different gases. Data from ref. [13].

soft-neutron shower component to the calorimeter signal. In the case of the compensating HELIOS calorimeter [2,3] ($R_d = 1.2$), these neutrons contribute almost 40% to the total non-em part of the signal, and for the overcompensating WA78 detector [5] ($R_d = 2$–3) even 50%.

Fig. 13a also shows that the neutrons not only contribute to the calorimeter signal through recoil protons but also through the γs that they produce by fission, inelastic scattering and capture. This contribution is limited because of the small γ/mip signal ratio (sect. 3.3). Nevertheless, a considerable fraction of the energy is deposited in this form, in particular through the

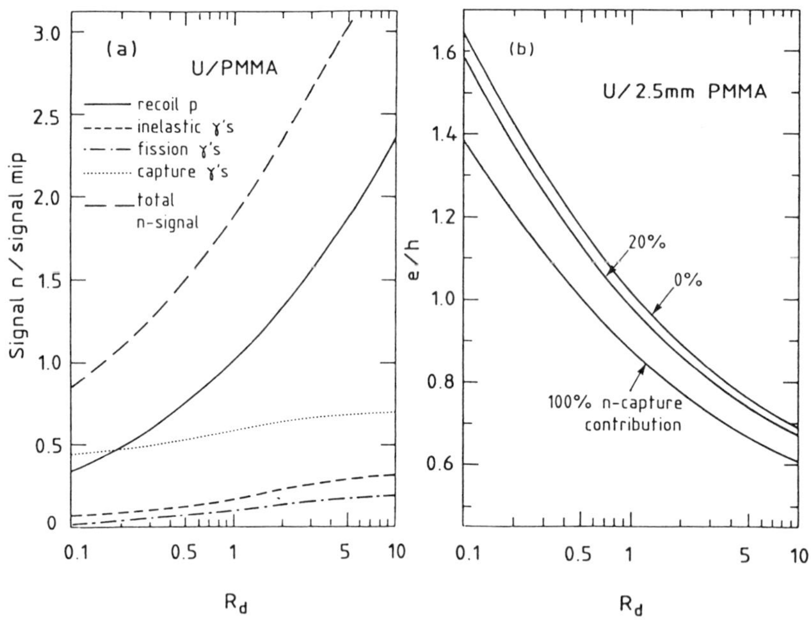

Fig. 13. The signal ratio n/mip split up in its components, for U/plastic-scintillator calorimeters, as a function of the ratio of the thicknesses of passive and active layers R_d (a). The signal ratio e/h as a function of R_d, assuming that 0, 20 or 100% of the γs released in thermal neutron capture contribute to the calorimeter signal (b).

neutron capture process. We measured that in uranium ~ 50 neutrons/GeV are produced in the non-em shower part [10]. If these are all captured, 240 MeV/GeV is released as capture γs. Their contribution to the calorimeter signal depends on the capture cross section, on the calorimeter size and, since thermalization is a slow process, on the signal integration time.

Fig. 13b shows the sensitivity of the e/h signal ratio to the fraction of the neutrons that contribute to the signal through capture (f_{capt}). Calculations by Brückmann [14] indicate that in 100 ns, a typical gate time for scintillator readout, $f_{capt} \approx 20\%$. Measurements on the evolution of the e/h signal ratio as a function of the gate time clearly confirmed the predicted effect [3,6]. For LAr readout, the signal integration time is usually longer (~ 1 μs) and, therefore, f_{capt} may be up to ~ 80% for sufficiently large detectors.

3.7. The e/h signal ratio

The e/h signal ratio was calculated in the way described in sect. 3.1. It was shown in sections 3.2–3.6, that the results depend on a large number of factors, of which we recall:

(a) The Z of the active and passive media (e.g. for e/mip, γ/mip, f_{ion}, f_n).
(b) The hydrogen content of the active medium (n/mip).
(c) The saturation properties (kB value) of the active medium (ion/mip, n/mip).
(d) The nuclear level structure of the absorber (n/mip).
(e) The size of the detector ($f_{capt} \to f_\gamma$).
(f) The cross section for thermal neutron capture in the absorber (f_γ).
(g) The signal integration time (f_γ).
(h) The thickness of the passive layers (e/mip, γ/mip).
(i) The thickness of the active layers (e/mip, γ/mip, ion/mip).
(j) The sampling fraction (n/mip).

Because of the interesting consequences of the latter effect, e/h curves are presented as a function of the sampling fraction, or rather R_d, for a fixed choice of the readout thickness (figs. 13b, 14). These figures show that the e/h signal ratio for uranium calorimeters with hydrogenous readout decreases rather sharply with increasing R_d, i.e. with increasing absorber plate thickness. The e/h ratio can therefore be tuned through the sampling fraction, and for the hydrogenous materials shown (various plastic scintillators, TMP) an ideal sampling fraction ($e/h = 1$) exists.

The sensitivity of the e/h ratio to the sampling fraction depends on two factors:

(1) The fraction of hydrogen atoms in the active medium. Since TMP contains more hydrogen atoms than plastic scintillator, the e/h signal ratio of TMP calorimeters is more sensitive to changes in the sampling fraction.

(2) Saturation of the active medium. Less saturation causes a larger sensitivity, and consequently leads to a larger optimal sampling fraction. This is illustrated in fig. 14, where the predictions for TMP readout are given for two different kB values. Contrary to plastic scintillators, the kB value has not been measured for soft protons in TMP, so far. The figure shows that this is a crucial parameter for the calorimeter performance.

For calorimeters with nonhydrogenous active layers (e.g. LAr, Si), the effect of the sampling fraction on the relative contribution of neutrons to the calorimeter signal (fig. 13a) is absent. The fact that e/h rises for small R_d values is caused by the effect of the absorber layer thickness on the electromagnetic response (fig. 5). It turned out that the e/h ratio depends quite sensitively on the contribution of captured neutrons to the hadronic calorimeter signal (sect. 3.6), which can be tuned through the signal integration time. Nevertheless, it looks as if the compensation condition cannot be reached for LAr. For Si readout it might just be possible. The difference between LAr and Si comes from the fact that Si does not saturate in detecting the spallation protons (sect. 3.5) and, therefore, gives a larger response to the non-em shower component.

The calculations were performed for infinitely large detectors. We investigated how sensitive the predictions are with respect to the various assumptions that went into the calculations. In fact fig. 13b is an example of this. Other uncertainties concern f_{ion}, f_n and the spallation proton spectrum (sections 3.4, 3.5). It turned out

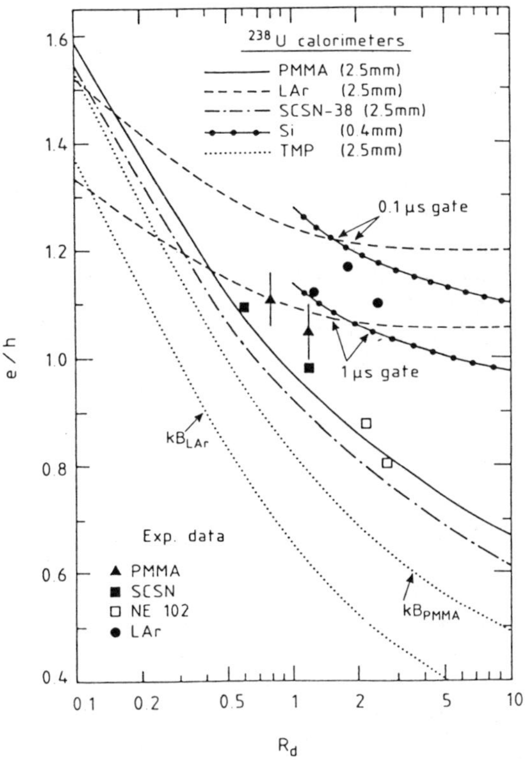

Fig. 14. The signal ratio e/h for uranium calorimeters employing different readout materials, as a function of the ratio of the thicknesses of absorber and readout layers. Results of experimental measurements are included.

that these uncertainties concerning details of the hadronic shower development may considerably affect the absolute value of the vertical scale, but *not* the variation of e/h with the sampling fraction, nor the difference between the various active materials. Therefore, the absolute e/h signal ratio for a given absorber/readout combination cannot be predicted from scratch more accurately than to within 5–10%, but one reliable experimental measurement will make possible a reliable prediction of the optimal configuration.

The experimental results obtained with uranium calorimeters so far seem to support this conclusion. In particular, the predicted trend of a decreasing e/h ratio for increasing R_d in the case of scintillator readout is experimentally very well confirmed. The same is true for the differences between scintillator and LAr. Experimental results for TMP and Si readout are unfortunately still lacking.

We also did extensive calculations for calorimeters with Pb or Fe absorber. The lead results are most interesting. The differences with uranium mainly concern the amount of energy carried by the soft-neutron component, because of the absence of fission and because of the fact that the Pb nucleus is less neutron-rich than uranium. Fig. 13a shows that this can be compensated for by choosing a smaller sampling fraction (larger R_d). This increases n/mip and hence brings the contribution of neutrons to the calorimeter signal at the appropriate level. Again, this trick works only for hydrogenous readout materials. For scintillator, e/h was predicted to become 1.0 for $R_d \approx 4$, i.e. if the lead plates are made 4 times as thick as the scintillator plates.

In sect. 3.6 we saw that γs from thermal neutron capture represent a considerable fraction of the total hadronic energy, since it is by emitting these γs that the nuclei render the binding energy lost when the neutrons were released. In plain lead calorimeters the contribution of this process to the signal will be negligible. The cross section for (n, γ) reactions is 0.17 b, against 2.7 b for uranium. This corresponds to a mean free path of ~ 180 cm for a thermal neutron in lead (8 cm in U). Cadmium has a capture cross section of 2450 b. If one would add ~ 1% Cd to the lead, thermal neutrons would be efficiently captured. This would in particular be interesting for LAr calorimeters with a long signal integration time. Also in U/LAr one might profit from adding a little Cd, since the energy released in neutron capture by Cd (9.0 MeV) is almost twice as large as in ^{238}U (4.8 MeV).

The calculations showed that even Fe calorimeters could be made compensating, at very small sampling fractions. Experimental data from the WA1 Collaboration seem to support this conclusion [4,7].

A final word should be said about homogeneous hadron calorimeters. The mechanisms that are described in this section and that make compensating calorimeters possible, are based on the fact that we are dealing with *sampling* calorimeters: Only a (small) fraction of the shower energy is deposited in the active layers, and by carefully tuning variables one may equalize the em and non-em *response*. This does not work for homogeneous devices, like the fully sensitive liquid argon tank proposed by Ypsilantis [15]. In the non-em shower part inevitably losses will occur that cannot be compensated for, e.g. the binding energy required to release protons or nucleon aggregates from their nuclear environment. The e/h^{intr} signal ratio will, therefore, always have to be larger than 1.0 for such devices.

4. The energy resolution of hadron calorimeters

After having discussed the e/h signal ratio and the factors that determine its value, we now return to the aim of this paper, i.e. trying to understand the energy resolution of hadron calorimeters, and the factors the limit their performance in this respect. The energy resolution of a hadron calorimeter with separate passive and active layers is determined by four factors:
(1) Sampling fluctuations.
(2) Detector imperfections.
(3) Deviations from $e/h^{\text{intr}} = 1$.
(4) The intrinsic energy resolution.

The *sampling fluctuations* were discussed in sect. 2. Fabjan et al. [12] have measured their contribution to the energy resolution of Fe/LAr and U/LAr calorimeters. The results are usually quoted as [9]

$$\sigma_{\text{samp}}/E = 0.09\sqrt{\Delta E(\text{MeV})/E(\text{GeV})}\ ,$$

where ΔE is the energy deposited by a mip in one sampling layer.

Detector imperfections will cause deviations from $1/\sqrt{E}$ scaling, and therefore limit the high-energy performance. It is believed that for well-designed calorimeters their contribution can be limited to $\sigma/E = 1\%$. Recently, HELIOS measured $\sigma/E = 1.9\%$ for the total energy distribution of reaction products from $^{16}O + W$ collisions at 3.2 TeV, at running conditions [3]. This

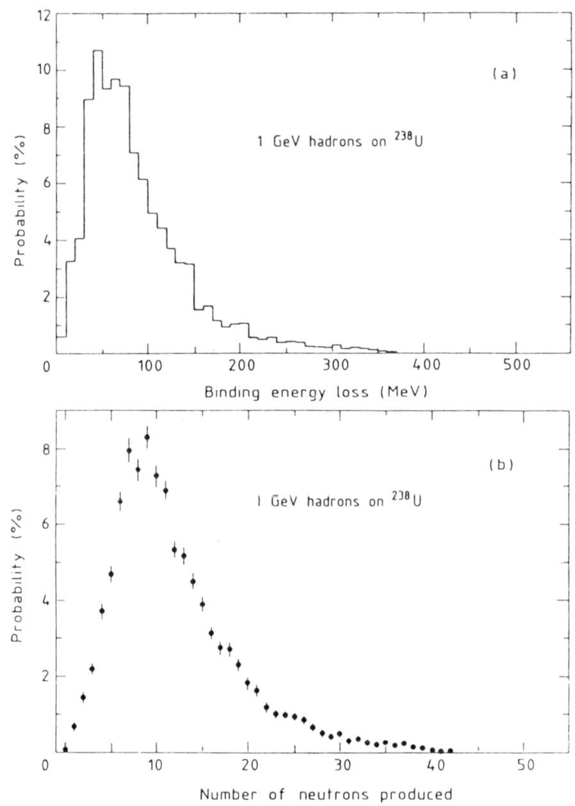

Fig. 15. The binding energy lost in spallation reactions induced by 1 GeV hadrons on ^{238}U nuclei (a), and the number of neutrons produced in such reactions (b). Results of calculations described in ref. [7].

impressive result supports the previous statement. In our calculations we assumed the detectors to be perfect.

As was shown in sect. 2, *deviations from* $e/h^{\mathrm{intr}} = 1$ contribute a constant term to the energy resolution σ/E. The value of this constant may be obtained from fig. 3a. In this section we will concentrate on the part that remains, i.e. on the *intrinsic energy resolution* for compensating calorimeters.

In hadron showers, a considerable fraction of the non-em energy is spent on nuclear binding energy losses, on average up to $\sim 40\%$ in high-Z materials. The

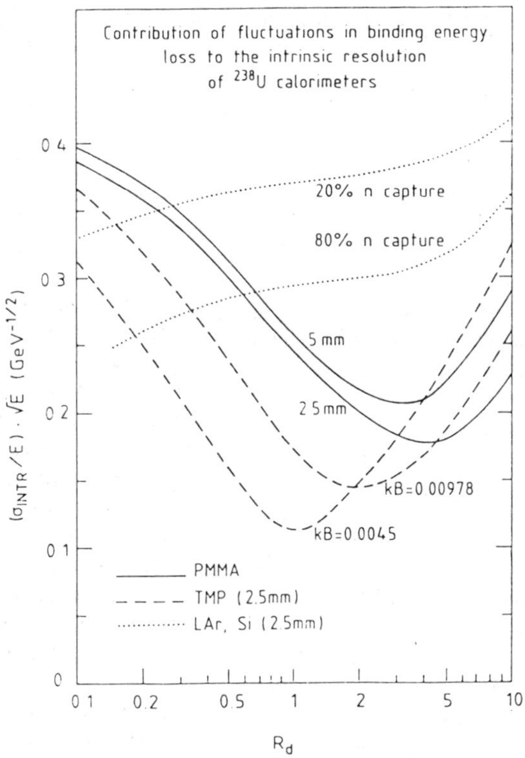

Fig. 16. The contribution of fluctuations in nuclear binding energy loss to the intrinsic resolution of various types of uranium calorimeters, as a function of the ratio of the thicknesses of the passive and active layers.

fluctuations about this average are huge. This is illustrated in fig. 15a, where the binding energy lost in the various reactions that may occur when 1 GeV protons hit ^{238}U nuclei is given. The total binding energy loss in a hadron shower (ΔB) is a convolution of many distributions of this type. We estimated that the contribution of event-to-event fluctuations in ΔB to the energy resolution amounts to $(30–35\%)/\sqrt{E}$, if the effects of neutron- and γ-detection are neglected. However, these effects are very important.

The total kinetic energy carried by neutrons (E_n^{tot}) is evidently correlated with ΔB (see fig. 15). A relatively large binding energy loss means that a relatively large fraction of the hadron energy is transferred to neutrons as well; detection of these particles corrects, at least partially, for the relatively small signal from ionizing particles in the event concerned, and vice versa. If the neutrons are detected with a good energy resolution themselves, i.e. if there is a strong correlation between E_n^{tot} and the neutron contribution to the calorimeter signal, the intrinsic resolution due to fluctuations in ΔB may be considerably reduced.

The energy resolution for neutron detection was studied with the same Monte Carlo package that was used to calculate the n/mip signal ratio (sect. 3.6). By transporting many neutrons through a given calorimeter structure, the signal distribution for a fixed E_n^{tot} was obtained. It turned out that the correlation between E_n^{tot} and the neutron contribution to the calorimeter signal may considerably improve if elastic n–p scattering in the active calorimeter layers occurs. The degree of correlation depends on the same factors that determined the n/mip signal ratio, i.e. on the hydrogen content, the kB value, and the sampling fraction. In addition, the thickness of the layers is important for the detection of the γ component.

With these results we calculated the total intrinsic energy resolution, assuming perfect correlation between ΔB and E_n^{tot}. The validity of this assumption breaks down at low E and/or low Z_{abs}. Results for uranium calorimeters are given in fig. 16. This figure shows that the intrinsic energy resolutions for practical detectors with hydrogenous readout material are considerably better than for LAr on Si, and may be as low as $20\%/\sqrt{E}$ for compensating calorimeters (note that the minima in the curves do not coincide with $e/h = 1$). The same conclusion holds for calorimeters with other absorbers.

This yields a new and independent argument in favour of hydrogenous readout material: The energy resolution of a U/Si calorimeter, even if this combination can be made compensating, will be inferior to plastic-scintillator or TMP calorimeters with $e/h = 1$.

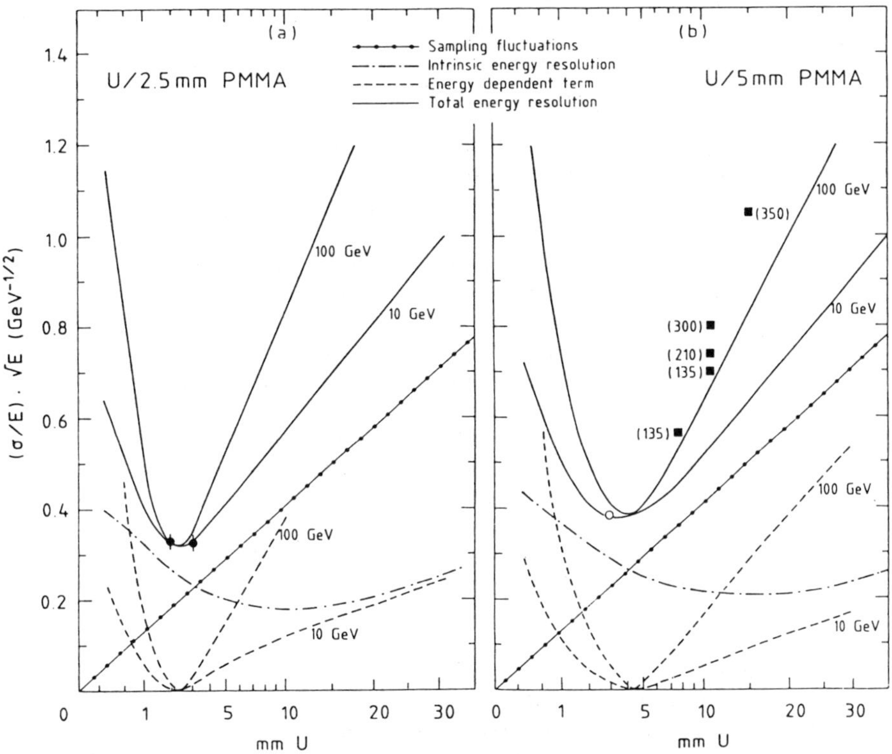

Fig. 17. The total energy resolution and the various factors contributing to it, for detection of 10 and 100 GeV hadrons in U/PMMA calorimeters, as a function of the uranium plate thickness. The thickness of the scintillator plates amounts to 2.5 mm (a) or 5 mm (b). The numbers in brackets denote other energies at which experimental results were reported.

The total hadronic energy resolution was computed as

$$\frac{\sigma}{E} = \sqrt{\frac{a_0^2 + a_1^2 t}{E}} + b .$$

The intrinsic resolution (a_0) and the sampling fluctuations (a_i, proportional to the square root of the absorber layer thickness t), are added in quadrature. The constant term b vanishes for $e/h = 1$, and causes $\sigma/\sqrt{E}$ to be energy dependent in other cases. Figs. 17–19 show some results of the calculations. Many more results are given in ref. [7]. The figures show $\sigma/\sqrt{E}$ for 10 and 100 GeV, and the various contributions to it, for calorimeters with a given thickness of the active layers, as a function of the thickness of the absorber layers. The abcissa is plotted linear in the square root of the latter, which yields a straight line for the contribution of sampling fluctuations. Experimental results are included in the figures.

Fig. 19 concerns uranium calorimeters with PMMA (polymethyl metacrylate, a plastic scintillator) readout. The total energy resolution reaches a minimum value for an absorber thickness close to the one for which $e/h = 1$. This is certainly true at high energies. At low energies, effects of sampling fluctuations and intrinsic contributions may slightly shift the optimal plate thickness. The curves for the total energy resolution at 10 and 100 GeV have a common point if $e/h = 1$: σ/E scales as $c/\sqrt{E}$, with $c = 0.32$ and 0.38 for 2.5 and 5 mm scintillator, respectively. These numbers are in good agreement with experimental values reported [2,3,16]. The results obtained by the WA78 Collaboration for nonoptimal configurations [5] confirm the predicted violation of $c/\sqrt{E}$ scaling.

In fig. 18 results for U/LAr calorimeters are given. As for the e/h signal ratio, two cases were considered, $f_{capt} = 0.20$ and 0.80, respectively (see sect. 3.7). The predictions are rather sensitive to this parameter. If f_{capt} is increased, the energy resolution improves and $\sigma/\sqrt{E}$ becomes less energy dependent. Since $e/h \neq 1$ for all configurations, $\sigma/\sqrt{E}$ will always increase with E. The experimental results are in better agreement with the low f_{capt} value. The small size of the test calorimeters with which they were obtained is likely to be (partly) responsible for this.

Fig. 19 shows results for lead/plastic-scintillator calorimeters. As was discussed in sect. 3.7, this combination can be made compensating if the absorber layers are ~ 4 times as thick as the readout layers. Inspired by this prediction, Kötz et al. from the ZEUS Collaboration tested a 10 mm Pb/2.5 mm plastic scintillator calorimeter [17]. They measured an energy resolution σ/E of $43\%/\sqrt{E}$ for pion detection, scaling with $1/\sqrt{E}$ over the whole energy range they scanned (3–75 GeV), while $41\%/\sqrt{E}$ was predicted. This is a remarkable result. Earlier, the same collaboration measured a considerably worse resolution ($60\%/\sqrt{E}$, fig. 19b) for a lead/plastic-scintillator calorimeter with a 4–5 times larger sampling fraction [16] (4–5 mm Pb/5 mm scintillator). According to common wisdom, thicker absorber and thinner scintillator plates should have considerably deteriorated this result. Instead, an enormous improvement was observed, in agreement with our predictions.

5. Outlook

The quality of experiments at a future supercollider will primarily be determined by the quality of the calorimetry. The calorimeters will have to deliver the information needed for reducing the event rates by ~ 7 orders of magnitude, while retaining those containing signatures of possible new physics. One will want to have the best possible calorimetry for that purpose. Energy resolution is an important issue in this respect, although by far not the only one. Three recent developments mark the considerable progress in this field:

(1) The excellent energy resolutions measured by the HELIOS Collaboration at very high energies ($\sigma/E < 2\%$ at 3.2 TeV).

(2) The experimental demonstration of a compensating lead calorimeter, with $\sigma/E = 43\%/\sqrt{E}$.

(3) The fact that (2) confirmed a theoretical prediction.

The first result implies that one may successfully aim for hadron calorimetry at the 1% precision level, for experiments at a multi-TeV machine. The second result shows that this goal can be achieved without uranium.

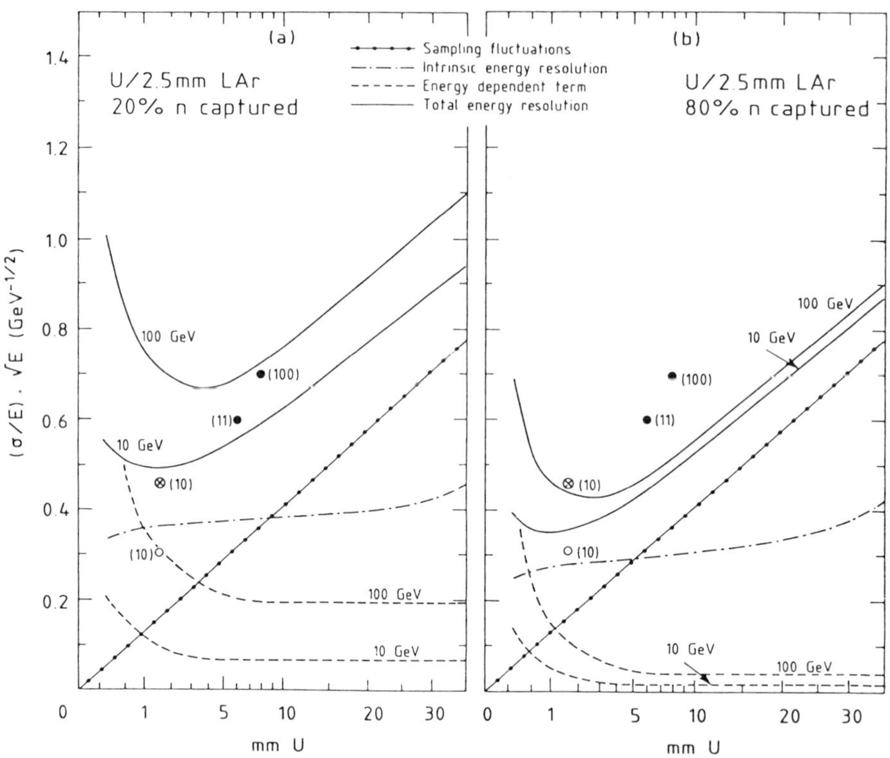

Fig. 18. The total energy resolution, and the various factors contributing to it, for detection of 10 and 100 GeV hadrons in U/LAr calorimeters, as a function of the uranium plate thickness. The liquid argon gap is 2.5 mm wide. The calculations assume that 20% (a) or 80% (b) of the thermalized neutrons are captured in the detector and contribute to the calorimeter signal.

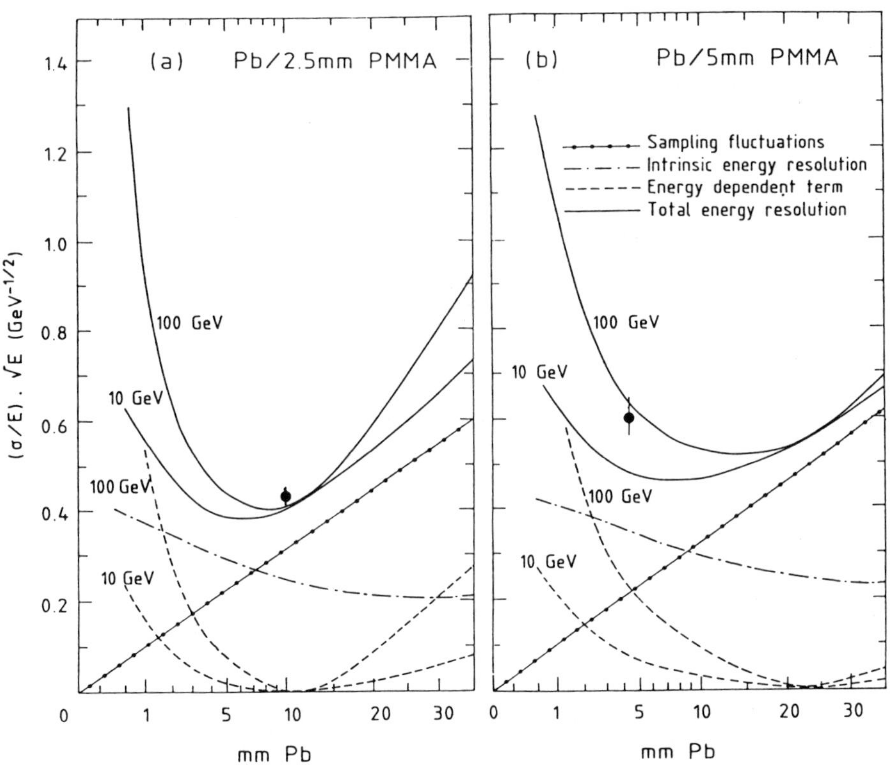

Fig. 19. The total energy resolution and the various factors contributing to it, for detection of 10 and 100 GeV hadrons in Pb/PMMA calorimeters, as a function of the thickness of the lead plates. The thickness of the scintillator plates is 2.5 mm (a) or 5 mm (b).

The third result means that we understand *how* to achieve this goal.

In order to reach this level of precision, the calorimeter should have $e/h^{\rm intr} = 1$ to within $\sim 5\%$. This is definitely possible for four combinations: U/plastic-scintillator, Pb/plastic-scintillator, U/warm-liquid and Pb/warm-liquid. In all these cases energy resolutions $\sigma/E \approx 30\%/\sqrt{E}$ seem possible for practical detectors. Compensation can probably also be reached for U/LAr if a substantial fraction (more than 10%) of methane can be added, and for U/Si if the silicon is sandwiched between thin foils of hydrogenous material, e.g. polyethylene [18]. The energy resolution is expected to be considerably worse for these combinations, $\sim 50\%/\sqrt{E}$. In the first case this is due to the small sampling fraction (thick U-plates) needed, in the second case the absence of the effects that compensate for fluctuations in the nuclear binding energy loss limit the energy resolution [19].

An essential condition for achieving good energy resolution is also that the calorimeter structure be the same throughout the complete sensitive volume. The energy resolution of the U/TMP calorimeter being built by the UA1 Collaboration [20] will, expectedly, not nearly be as good as the HELIOS and ZEUS ones, if only because of the fact that it is longitudinally subdivided into sections with very different sampling fractions (i.e. $e/h^{\rm intr}$ values).

Another important criterion for choosing a particular calorimeter will be compactness. In this respect it is interesting to remark that the effective nuclear interaction length ($\lambda_{\rm int}^{\rm eff}$) of *compensating* uranium and lead calorimeters using the same readout are practically the same, because of the quite different sampling fractions needed for compensation. In spite of its much larger density, uranium does not give any advantage in this respect. If only because of the well-known disadvantages of uranium, the lead alternatives deserve to be thoroughly explored [21]. For plastic scintillator, $\lambda_{\rm int}^{\rm eff}$ amounts to ~ 20 cm. For TMP, $\lambda_{\rm int}^{\rm eff}$ depends sensitively on the kB saturation constant, but is likely to be considerably larger than 20 cm. Silicon readout yields the most compact devices ($\lambda_{\rm int}^{\rm eff} \approx 15$ cm for U/Si).

Other important selection criteria concern the hermiticity, granularity, charge collection time, radiation resistance and calibration of the candidate calorimeter. The latter may well turn out to be the major limiting factor in the era of high resolution calorimetry.

Hadron calorimetry has become a mature scientific activity. The R&D activities needed for designing future detectors will get a solid basis.

References

[1] C.W. Fabjan and W.J. Willis, in: Proc. Calorimeter Workshop, ed., M. Atac, Batavia Illinois (1975) p. 1.

[2] T. Akesson et al., Nucl. Instr. and Meth. A241 (1985) 17.

[3] T. Akesson et al., Performance of the Uranium/Plastic Scintilator Calorimeter for the HELIOS Experiment at CERN, CERN-EP/87-111.

[4] See for example H. Abramowicz et al., Nucl. Instr. and Meth. 180 (1981) 429.

[5] M. de Vincenzi et al., Nucl. Instr. and Meth. A243 (1986) 348.

[6] H. Brückmann and H. Kowalski, The Dependence of Calorimeter Responses on the Gate Width, DESY, Hamburg, ZEUS-Internal Note 86/026 (1986).

[7] R. Wigmans, Nucl. Instr. and Meth. A259 (1987) 389.

[8] U. Amaldi, Phys. Scripta 23 (1981) 409.

[9] C.W. Fabjan and T. Ludlam, Ann. Rev. Nucl. Part. Sci. 32 (1982) 335.

[10] C. Leroy, Y. Sirois and R. Wigmans, Nucl. Instr. and Meth. A252 (1986) 4.

[11] V.M. Gorbachev, Y.S. Zamyathin and A.A. Lbov, Nuclear Reactions in Heavy Elements, a Data Handbook (Pergamon, London, 1980).

[12] C.W. Fabjan et al., Nucl. Instr. and Meth. 141 (1977) 61.

[13] Y. Galaktionov et al., Nucl. Instr. and Meth. A251 (1986) 258.

[14] H. Brückmann, Hadron Sampling Calorimetry, preprint DESY/86-155 (1986).

[15] T. Ypsilantis, these Proceedings (Advances in Experimental Methods for Colliding Beam Physics) Nucl. Instr. and Meth. A265 (1988).

[16] J. Engelen and H. Tiecke, Performance of a Hadron Test Calorimeter for the ZEUS Experiments, preprint NIKHEF-H/86-18 (1986).

[17] R. Klanner, these Proceedings (Advances in Experimental Methods for Colliding Beam Physics) Nucl. Instr. and Meth. A265 (1988) 200.

[18] T. Akesson et al., in: Proc. of the Workshop on Physics at Future Accelerators, La Thuile and CERN (1987); CERN-EP/87-88.

[19] R. Wigmans, Signal Equalization and Energy Resolution for Uranium/Silicon Hadron Calorimeters, to be published.

[20] See M. Albrow et al., Performance of a Uranium/Tetramethylpentane Electromagnetic Calorimeter, CERN-EP/87-55 (1987).

[21] P. Jenni et al., The High Resolution Spaghetti Hadron Calorimeter, NIKHEF-H/87-07 (1987).